THEORY
OF
PLASTICITY

J. Chakrabarty

Professor of Mechanical Engineering
Texas A & M University System
Formerly at the University of Birmingham, England

D0164572

McGraw-Hill Book Company

New York   St. Louis   San Francisco   Auckland   Bogotá   Hamburg
Johannesburg   London   Madrid   Mexico   Montreal   New Delhi   Panama
Paris   São Paulo   Singapore   Sydney   Tokyo   Toronto

To my wife
Swati

This book was set in Times Roman by Eta Services Ltd.
The editor was Anne Duffy, the production supervisor was Marietta Breitwieser;
the cover was designed by Mark Wiebaldt.
Project supervision was done by Albert Harrison, Harley Editorial Services.
R. R. Donnelley & Sons Company was printer and binder.

THEORY OF PLASTICITY

1234567890DOCDOC89876

ISBN 0-07-010392-5

Library of Congress Cataloging-in-Publication Data
Chakrabarty, J.
    Theory of plasticity.
    Includes indexes.
        1. Plasticity. I. Title.
TA418.14.C48 1987   620.1'1233   85-23647
ISBN 0-07-010392-5

CONTENTS

ABOUT THE AUTHOR

Doctor J. Chakrabarty received his Ph.D. in Mechanical Engineering in 1966 from the Imperial College of Science and Technology, London, under the supervision of Professor J. M. Alexander. After spending a couple of years at the Middle East Technical University, Ankara, as an Assistant Professor, he accepted a faculty position at the University of Birmingham, England, where he remained till 1980. He subsequently moved to the United States as a Professor of Mechanical Engineering at the University of Utah. Professor Chakrabarty has made important contributions in the area of plasticity through the publication of numerous papers in professional journals of international reputation.

PREFACE

During recent years, there has been considerable interest in the application of the macroscopic theory of plasticity to engineering problems associated with structural designs and the technological forming of metals. The need for a comprehensive text book on plasticity, incorporating the most recent developments of the subject, has been strongly felt for some years. This book has been written primarily to meet the needs of graduate and research students of Mechanical, Civil, and Metallurgical Engineering, although some of the material in the book is also suitable for undergraduate students and practicing engineers. In order to discuss the various topics as fully as possible, it has been found necessary to treat the subject matter in two volumes, of which the first one is now presented to the reader.

The first chapter of the book deals with the analysis of stress and strain rate, and introduces the definition of the stress rate. The second chapter discusses the yield criteria, stress-strain relations, uniqueness theorems, and extremum principles. A series of physical problems where elastic and plastic strains are simultaneously important are discussed in Chaps. 3 and 5. A detailed account of the limit analysis of framed structures is given in Chap. 4 as a logical continuation of the treatment of the bending of beams. The remaining chapters of the book deal with the theory and application of slipline fields, an area that has received the greatest attention in the literature. The basic theory is explained in Chap. 6, which includes the recent analytical and numerical methods of solution of the plane strain problem. A variety of practical problems involving steady, pseudosteady, and nonsteady states of plastic flow are thoroughly discussed in Chaps. 7 and 8. Several numerical tables are presented in the Appendix to facilitate the computation of slipline field solutions.

Tensor or suffix notation is introduced in the first chapter, where the summation convention and the associated algebraic operations have been explained for the benefit of those readers who are unfamiliar with them. The suffix notation is a convenient shorthand for writing the general equations, and is practically indispensable in the derivation of general theorems. Bessel functions are extensively used in the latter half of Chap. 6 for the analytical solution of boundary-value problems involving slipline fields. These sections may be omitted during the first reading, provided the results used in the subsequent chapters for the solution of special problems are taken for granted. I have made an earnest endeavor to make the treatment of each problem as complete as is warranted by the present state of knowledge. A large number of exercise problems are provided at the end of each chapter to enable the student to test his or her mastery of the subject. There is more material in this book than can be covered in a one-semester course on plasticity, so that the instructor has sufficient flexibility in the selection of topics.

References to original papers and books relating to plasticity and its applications have been given in numerous footnotes throughout the book. The literature in the field of plasticity is so extensive that I have been compelled to restrict myself mainly to publications that appeared in English. The reader would be able to form a list of publications in other languages from some of the references cited in this book. Although an exhaustive bibliography has not been attempted, I wish to express my sincere regrets for any inadvertent omission of important publications.

I would like to thank the following professors for reviewing the manuscript: David H. Allen, Texas A & M University; Nicholas J. Altiero, Michigan State University; James M. Gere, Stanford University; Kerry S. Havner, North Carolina State University; Philip G. Hodge, University of Minnesota; Francis T. C. Loo, Clarkson University; Huseyin Sehitoglu, University of Illinois; and David J. Unger, Ohio State University.

I take this opportunity to record my profound appreciation of the cooperation offered by the officers of McGraw-Hill Book Company during the planning and production of the book. I am indebted to Albert Harrison, Harley Editorial Services for his ready cooperation while dealing with the proofs.

J. Chakrabarty

ONE

STRESSES AND STRAINS

1.1 Introduction

The theory of plasticity is the branch of mechanics that deals with the calculation of stresses and strains in a body, made of ductile material, permanently deformed by a set of applied forces. The theory is based on certain experimental observations on the macroscopic behavior of metals in uniform states of combined stresses. The observed results are then idealized into a mathematical formulation to describe the behavior of metals under complex stresses. Unlike elastic solids, in which the state of strain depends only on the final state of stress, the deformation that occurs in a plastic solid is determined by the complete history of the loading. The plasticity problem is, therefore, essentially incremental in nature, the final distortion of the solid being obtained as the sum total of the incremental distortions following the strain path.

A metal may be regarded as macroscopically homogeneous and isotropic when the small crystal grains forming the aggregate are distributed with random orientations. As a result of plastic deformation, the crystallographic directions gradually rotate toward a common axis, producing a preferred orientation. An initially isotropic material thereby becomes anisotropic, and its mechanical properties vary with direction. The development of anisotropy with progressive cold work and the resulting strain-hardening are too complex to be successfully incorporated in the theoretical framework. In the mathematical theory of plasticity, it is generally assumed that the material remains isotropic throughout the deformation irrespective of the degree of cold work. Since the strain-hardening characteristic of a metal in a complex state of stress can be related to that in uniaxial tension or compression, it is necessary to examine the uniaxial stress-strain behavior in some detail before considering the general theory of plasticity.

The plastic deformation in a single crystal is generally produced by slip, which is the sliding of adjacent blocks of the crystal along definite crystallographic planes, called slip planes. The boundary line separating the slipped region of a crystal from the neighboring unslipped region is called a *dislocation*. The movement of the dislocation, which is responsible for the slip, is initiated by a line defect causing a local concentration of stress. Slip usually occurs on those planes which are most densely packed with atoms. The magnitude and direction of the relative movement in a slip is specified by a vector known as the *Burgers vector*. A dislocation is said to be one of unit strength when the magnitude of the Burgers vector is equal to one atomic spacing. The terms *edge dislocation* and *screw dislocation* are used to describe the situations where the Burgers vector is normal and parallel respectively to the dislocation line. In general, a dislocation is partly edge and partly screw in character, and the dislocation line forms a curve or a closed loop.†

In a polycrystalline metal, the crystallographic orientation changes from one grain to the next through a narrow transition zone, or grain boundary, which acts as an effective barrier to slip. Dislocations pile up along the active slip planes at the grain boundaries, the effect of which is to oppose the generations of new dislocations. When the applied stress is increased to a critical value, the shear stress developed at the head of the dislocation pile-up becomes large enough to cause dislocation movement across the boundary. The dislocation pile-up is mainly responsible for strain-hardening of the metal in the early stages of plastic deformation. The rate of hardening of the polycrystalline metal is always higher than that of the single crystal, where the increase in yield stress is caused by dislocations interacting with one another and with foreign atoms serving as barriers . The dislocation interactions control the yield strength of a polycrystalline metal only in the later stages of the deformation.

If the temperature of the strain-hardened metal is progressively increased, the cold-worked state becomes more and more unstable, and the material eventually reverts to the unstrained state. The overall process of heat treatment that restores the ductility to the cold-worked metal is known as *annealing*. The temperature at which there is a marked decrease in hardness of the metal is known as the recrystallization temperature. The dislocation density decreases considerably on recrystallization, and the cold-worked structure is replaced by a set of new strain-free grains. The greater the degree of cold-work, the lower the temperature necessary for recrystallization, and smaller the resulting grain size.‡

In ductile metals, under favorable conditions, plastic deformation can continue to a very large extent without failure by fracture. Large plastic strains do occur in

---

† For a complete discussion, see A. H. Cottrell, *Dislocations and Plastic Flow in Crystals*, Clarendon Press, Oxford (1953); W. T. Read, *Dislocations in Crystals*, McGraw-Hill Book Company, New York (1953); J. Friedel, *Dislocations*, Addison-Wesley Publishing Company, Reading, Mass. (1964); F. R. N. Nabarro, *Theory of Crystal Dislocations*, Clarendon Press, Oxford (1967); D. Hull, *Introduction to Dislocations*, 2d ed., Pergamon Press, Oxford (1975).

‡ See, for example, G. E. Dieter, *Mechanical Metallurgy*, chap. 5, 2d ed., McGraw-Hill Book Company, New York (1976). See also R. W. K. Honeycombe, *The Plastic Deformation of Metals*, 2d ed., Edward Arnold, London (1984).

many metal-working processes, which constitute an important area of application of the theory of plasticity. While elastic strains may be neglected in such problems, the continued change in geometry of the workpiece must be allowed for in the theoretical treatment. Severe plastic strains are produced locally in certain mechanical tests such as the hardness test and the notch tensile test. The significance of these tests cannot be fully appreciated without a knowledge of the extent of the plastic zone and the associated state of stress. Situations in which elastic and plastic strains are comparable in magnitude arise in a number of important structural problems when the loading is continued beyond the elastic limit. Structural designs based on the estimation of collapse loads are more economical than elastic designs, since the plastic method takes full advantage of the available ductility of the material.

1.2 The Stress-Strain Behavior

(i) *The true stress-strain curve* The stress-strain curve of an annealed material in simple tension is found to coincide with that in simple compression when the true stress σ is plotted against the true or natural strain ε. The *true stress*, defined as the load divided by the current cross-sectional area of the specimen, can be significantly different from the *nominal stress*, which is the load per unit original area of cross-section. Let l denote the current length of a tensile specimen and dl the increase in length produced by a small increment of the stress. Then the true strain increases by the amount $d\varepsilon = dl/l$. If the initial length is l_0, the total strain is $\varepsilon = \ln(l_0/l)$, called the true or *natural strain*.† For a specimen uniformly compressed from an initial height h_0 to a final height h, the magnitude of the natural strain is $\varepsilon = \ln(h_0/h)$. The conventional or *engineering strain* e, on the other hand, is the amount of extension or contraction per unit original length or height. It follows that $\varepsilon = \ln(l + e)$ in the case of tension, and $\varepsilon = -\ln(l - e)$ in the case of compression. Thus ε becomes progressively lower than e in tension, and higher than e in compression, as the deformation is continued in the plastic range.

Figure 1.1 shows the true stress-strain curve of a typical annealed material in simple tension. So long as the stress is sufficiently small, the material behaves elastically, and the original size of the specimen is regained on removal of the applied load. The initial part of the stress-strain curve is a straight line of slope E, which is known as Young's modulus. The point A represents the proportional limit at which the linear relationship between the stress and the strain ceases to hold. The elastic range generally extends slightly beyond the proportional limit. For most metals, the transition from elastic to plastic behavior is gradual, owing to successive yielding of the individual crystal grains. The location of the yield point B is, therefore, largely a matter of convention. The corresponding stress Y, known as

---

† The concept of natural strain has been introduced by P. Ludwik, *Elemente der Technologischen Mechanik*, Springer Verlag, Berlin (1909). The natural strains associated with successive deformations are additive, but the engineering strains are not.

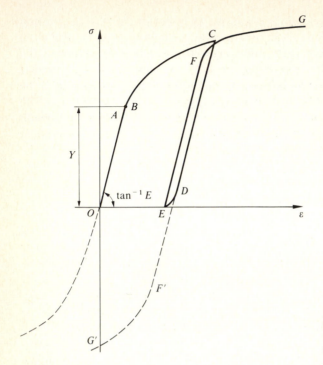

Figure 1.1 True stress-strain curve of metals with effects of unloading and reversed loading.

the yield stress, is generally defined as that for which a specified small amount of permanent deformation is observed. For theoretical purposes, it is often convenient to assume a sharp yield point defined by the intersection of a pair of straight lines, one of which is a continuation of OA and the other a tangent to the stress-strain curve at a point slightly above B.

Beyond the yield point, the stress continually increases with further plastic strain, while the slope of the stress-strain curve, representing the rate of strain-hardening, steadily decreases with increasing stress. If the specimen is stressed to some point C in the plastic range and the load is subsequently released, there is an elastic recovery following the path CD which is very nearly a straight line† of slope E. The permanent strain that remains on complete unloading is equal to OE. On reapplication of the load, the specimen deforms elasticity until a new yield point F is reached. Neglecting the hysteresis loop of narrow width formed during the loading and unloading, F may be taken as coincident with C. On further loading, the stress-strain curve proceeds along FG, virtually as a continuation of the curve BC. The curve EFG may be regarded as the stress-strain curve of the metal when prestrained by the amount OE. The greater the degree of prestrain, the higher the new yield point and the flatter the strain-hardening curve. For a heavily prestrained metal, the rate of strain-hardening is so small that the material may be regarded as approximately nonhardening or ideally plastic.

† L. Prandtl, *Z. angew. Math. Mech.*, **8**: 85 (1928).

A generic point on the stress-strain curve in the plastic range corresponds to a recoverable elastic strain equal to σ/E, and an irrecoverable plastic strain equal to $\varepsilon - \sigma/E$. If the stress is plotted against the plastic strain only, and the material is assumed to have a sharp yield point, the resulting curve will begin at $\sigma = Y$. Let H be the slope of the true stress-strain curve excluding the elastic strain, and T the slope of the curve including the elastic strain, for a given value of the stress σ. The quantities H and T are known as the *plastic modulus* and the *tangent modulus* respectively. A stress increment $d\sigma$ produces an elastic strain increment $d\sigma/E$ and a plastic strain increment $d\sigma/H$, while the total strain increment is $d\sigma/T$. Hence the relationship between H and T is

$$\frac{1}{T} = \frac{1}{E} + \frac{1}{H} \tag{1}$$

In an annealed material, H is considerably greater than T at the initial yielding, but these two moduli rapidly approach one another as the strain is increased. The difference between H and T becomes insignificant when the slope is only a few times the yield stress. At this stage, the elastic strain increment becomes negligible in comparison with the plastic strain increment. When the total strain is sufficiently large, the elastic strain itself is negligible. The stress-strain behavior at sufficiently large strains is identical to that of a hypothetical material in which E is infinitely large. Such a material is regarded as rigid/plastic, since it remains undeformed so long as the stress is below the yield point, while the subsequent deformation is entirely plastic.

Suppose that a specimen that has been completely unloaded from a tensile plastic state, represented by the point C, is reloaded in simple compression (Fig. 1.1). The stress-strain curve will then follow the path DF', where the new yield point F' corresponds to a stress that is appreciably smaller in magnitude than that at C. This phenomenon is known as the *Bauschinger effect*,† which occurs in real metals whenever there is a reversal of the stress. The subsequent strain-hardening follows the path $F'G'$, and approaches the stress-strain curve in compression as the loading is continued. The lowering of the yield stress in reversed loading is mainly caused by residual stresses that are left in the specimen on a microscopic scale due to the different stress states in the individual crystals. The Bauschinger effect can, therefore, be largely removed by a mild annealing. In the theory of plasticity, it is generally necessary to neglect the Bauschinger effect, the material being assumed to have identical yield stresses in tension and compression irrespective of the previous cold-work.

Some metals, such as annealed mild steel, exhibit a sharp yield point followed by a sudden drop in the stress, which remains approximately constant during a small amount of further straining. The sharp peak is known as the upper yield point, which is usually 10 to 20 percent higher than the lower yield point represented by the constant stress. At the upper yield point, a lamellar plastic zone, known as *Lüder's band*, inclined at approximately 45° to the tensile axis, appears at

---

† J. Bauschinger, *Zivilingenieur*, **27**: 289 (1881).

a local stress concentration. During the subsequent elongation under constant stress, several Lüder's bands appear and gradually spread over the entire specimen. After a total yield point elongation of about 10 percent, the stress begins to rise again due to strain-hardening, and the stress-strain curve then continues as before. The yield point drop is suppressed by a light cold-work, but the phenomenon reappears after the metal has been rested for several days at room temperature, or several hours at a relatively high temperature.†

(ii) *Some consequences of work-hardening* A longitudinal extension in the tensile test is accompanied by a contraction in the lateral direction. The ratio of the magnitude of the lateral strain increment to that of the longitudinal strain increment is known as the *contraction ratio*, denoted by η. In the elastic range of deformation, the contraction ratio has a constant value equal to Poisson's ratio v. When the yield point is exceeded, the plastic part of the lateral strain increment for an isotropic material is numerically equal to one-half of the longitudinal plastic strain increment. Since the ratio of the elastic parts of the lateral and longitudinal strain increments is equal to $-v$, the total lateral strain increment in uniaxial tension is

$$d\varepsilon' = -\tfrac{1}{2}\,d\varepsilon + (\tfrac{1}{2} - v)\,d\varepsilon^e$$

where $d\varepsilon^e$ is the elastic part of the longitudinal strain increment $d\varepsilon$. In view of the relationship $d\varepsilon^e = d\sigma/E = (T/E)\,d\varepsilon$, the contraction ratio becomes

$$\eta = -\frac{d\varepsilon'}{d\varepsilon} = \tfrac{1}{2} - (\tfrac{1}{2} - v)\frac{T}{E} \qquad (2)$$

Since the slope of the stress-strain curve decreases fairly rapidly in the early stages of strain-hardening, the contraction ratio rapidly approaches the asymptotic value of 0.5 as the strain is increased in the plastic range.‡ For a material having a sharp yield point, the contraction ratio changes discontinuously at this point to a value that depends on the initial rate of strain-hardening. When the tangent modulus becomes of the same order as that of the current yield stress, $\eta \simeq 0.5$, and the incremental change in volume becomes negligible.

The standard tensile test is unsuitable for obtaining the stress-strain curve of metals up to large values of the strain, since the specimen begins to neck when the rate of hardening decreases to a critical value. At this stage, the increase in load due to strain-hardening is exactly balanced by the decrease in load caused by the diminution of the area of cross section. Consequently, the load attains a maximum at the onset of necking. The longitudinal load at any stage is $P = \sigma A$, where A is the current cross-sectional area and σ the current stress, and the corresponding volume of the specimen is lA, where l is the current length. Using the constancy of volume,

---

† In addition to low-carbon steel, yield point phenomenon has been observed in aluminum, molybdenum, and titanium alloys.

‡ For an experimental investigation on the variation of the contraction ratio, see A. Shelton, *J. Mech. Eng. Sci.*, **3**: 89 (1961).

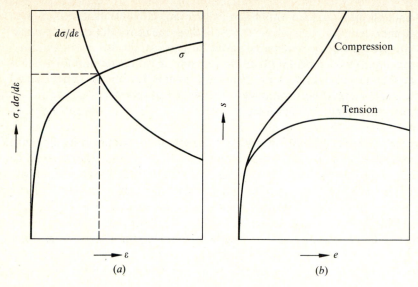

Figure 1.2 Peculiarities in tension and compression. (a) Location of point of tensile necking; (b) nominal stress versus engineering strain.

the maximum load condition $dP = 0$ may be written as

$$\frac{d\sigma}{\sigma} = -\frac{dA}{A} = \frac{dl}{l}$$

Since dl/l is equal to $d\varepsilon$, the condition for the onset of necking becomes

$$\frac{d\sigma}{d\varepsilon} = \sigma \qquad (3)$$

When the true stress-strain curve is given, the point on the curve that corresponds to the tensile necking can be located graphically from the fact that the slope at this point is equal to the current stress (Fig. 1.2a). A heavily prestrained metal will obviously neck as soon as the yield point is exceeded. Since $d\varepsilon = de/(1 + e)$, the condition for necking can be expressed in the alternative form

$$\frac{d\sigma}{de} = \frac{\sigma}{1 + e}$$

It follows that the maximum load corresponds to the point of contact of the tangent to the (σ, e) curve from the point $(-1, 0)$ on the negative strain axis.† The tensile test becomes unstable when the load reaches its maximum. The deformation is confined locally in the neck, while the remainder of the specimen recovers

---

† A Considere, *Ann. ponts et chausses*, **6**: 574 (1885). An interesting discussion has been given by C. R. Calladine, *Engineering Plasticity*, Chap. 2, Pergamon Press, Oxford (1970).

elastically under decreasing load until fracture intervenes. The stress distribution in the neck assumes a triaxial state which varies through the cross section of the neck. The test no longer provides a direct measure of the stress-strain behavior. Although the stress-strain curve may be continued by introducing a correction factor that requires careful measurements of the geometry of the neck,† the experimental difficulties render the method unsuitable for practical purposes.‡

The strain-hardening characteristic of metals at large strains is most conveniently obtained by compressing a solid cylindrical specimen between a pair of parallel platens. In the absence of efficient lubrication, the compression test is complicated by the fact that the friction at the platens restricts the metal flow at the ends of the specimen, causing barreling as the compression proceeds. Since homogeneous compression is thus prevented by friction, the stress-strain curve cannot be derived by the direct measurement of the load and the change in height of the specimen. In actual practice, the difficulty is overcome by using several cylinders with different initial diameter/height ratios, subjecting them to the same load each time on an incremental basis, and then extrapolating the results at each stage to obtain the strain corresponding to zero diameter/height ratio.§ Since the barreling would theoretically disappear for a specimen of infinite height, the extrapolation method eliminates the frictional effect.

Homogeneous deformation in the simple compression test can be achieved by inserting PTFE (polytetra fluoroethylene) films of suitable thickness between the specimen and the compression platens. As well as producing effective lubrication, the PTFE films are themselves compressed so as to exert radial pressure to the material near the periphery. This inhibits the barreling tendency, except when the film thickness is too small. An excessive film thickness, on the other hand, produces bollarding in which the diameter of the specimen becomes bigger at the ends than at the middle. For a given specimen, there is an optimum film thickness for which neither barreling nor bollarding would occur. The compression should be carried out incrementally, renewing the PTFE films after each load application. Using the constancy of volume, the load required during the homogeneous compression may be written as

$$P = \sigma A = \frac{\sigma A_0}{h_0/h} = \frac{\sigma A_0}{1 - e}$$

where A_0 is the original area of cross section of the specimen. The graph for P against e shows an upward inflection and rises continuously without limit (Fig. 1.2b). Setting $d^2P/de^2 = 0$, and using the fact that $d/d\varepsilon = (1 - e)d/de$, the condition

---

† P. W. Bridgman, *Trans. A.S.M.E.*, **32**: 553 (1944); N. N. Davidenkov and N. I. Spiridonova, *Proc. Am. Soc. Test. Mat.*, **46**: 1147 (1946). See also E. R. Marshall and M. C. Shaw, *Trans. A.S.M.E.*, **44**: 716 (1952); J. D. Lubahn and R. P. Felgar, *Plasticity and Creep of Metals*, p. 114, Wiley and Sons, New York (1961).

‡ A dynamic analysis for the development of the neck has been given by N. K. Gupta and B. Karunes, *Int. J. Mech. Sci.*, **21**: 387 (1979).

§ The extrapolation method has been developed by G. Sachs, *Zeit. Metallkunde*, **16**: 55 (1924), M. Cook and E. C. Larke, *J. Inst. Metals*, **71**: 371 (1945), A. B. Watts and H. Ford, *Proc. Inst. Mech. Eng.*, **169**: 1141 (1955).

for inflection is found as

$$\left(\frac{d}{d\varepsilon} + 2\right)\left(\frac{d\sigma}{d\varepsilon} + \sigma\right) = 0 \tag{4}$$

which defines the corresponding point on the true stress-strain curve. This point is most conveniently located if the stress-strain curve is represented by an empirical equation. In view of the incompressibility of the material, the nominal stress is $s = \sigma \exp(\varepsilon)$ in compression and $s = \sigma \exp(-\varepsilon)$ in tension.

The work done in changing the height of a specimen from h to $h + dh$ in simple compression is $-P\,dh$, where P is the current axial load. The incremental work done per unit volume of the specimen is therefore equal to $-P\,dh/Ah$ or $\sigma\,d\varepsilon$. It follows that during the homogeneous compression of a specimen from an initial height h_0 to a current height h, the work done per unit volume is given by the area under the true stress-strain curve up to a total strain of $\ln(h_0/h)$.

(iii) *Empirical stress-strain equations* For theoretical computations, it is often necessary to represent an experimentally determined stress-strain curve by an empirical equation of suitable form. When the material is rigid/plastic, it is frequently convenient to employ the Ludwik power law†

$$\sigma = C\varepsilon^n \tag{5}$$

where C is a constant stress, and n is a strain-hardening exponent usually lying between zero and 0.5. The equation predicts a zero initial stress and an infinite initial slope, except for $n = 0$ which represents a nonhardening rigid/plastic material. The higher the value of n, the more pronounced is the strain-hardening characteristic of the material (Fig. 1.3a). Since $d\sigma/d\varepsilon = n\sigma/\varepsilon$ in view of (5), it follows from (3) that the magnitude of the true strain at the onset of necking in simple tension is equal to n. The work done per unit volume during a homogeneous extension or contraction is easily shown to be $\sigma\varepsilon/(1 + n)$, where σ and ε are the final values of stress and strain.

The simple power law (5) may be readily modified by including a constant term Y representing the initial yield stress. The stress-strain equation then becomes

$$\sigma = Y(1 + m\varepsilon^n) \tag{6}$$

where m and n are dimensionless constants. Although this formula represents the strict rigid/plastic behavior of metals, it does not give a better fit for an actual stress-strain curve over a wide range of strains. When $n = 1$, the above equation represents a linear strain-hardening, which is a reasonable approximation for heavily prestrained metals. A more successful formula, due to Swift,‡ is the generalized power law

$$\sigma = C(m + \varepsilon)^n \tag{7}$$

---

† P. Ludwik, *Elem. Technol. Mech.*, Springer Verlag, Berlin (1909).
‡ H. W. Swift, *J. Mech. Phys. Solids*, **1**: 1 (1952).

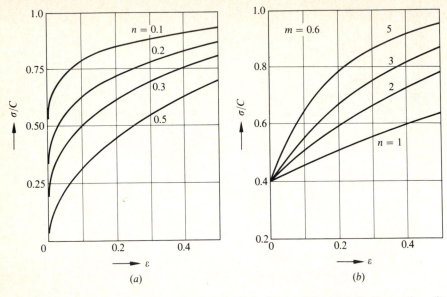

Figure 1.3 Empirical stress-strain curves for rigid/plastic materials. (*a*) Ludwik equation; (*b*) Voce equation.

where C, m, and n are empirical constants. The stress-strain curve represented by (7) can be obtained from that given by (5) if the stress axis is move along the positive strain axis through a distance m. Hence m may be regarded as the amount of prestrain in a material whose stress-strain curve in the annealed state corresponds to $m = 0$, the value of n remaining the same. If a given prestrained metal is represented by both (5) and (7), the value of n in the two cases will of course be different. The instability strain in simple tension according to the Swift equation is $n - m$ for $m \leqslant n$ and zero for $m \geqslant n$.

For certain applications involving rigid/plastic materials, it is convenient to use an equation suggested by Voce.† In its simplest form, the Voce equation may be written as

$$\sigma = C(1 - me^{-n\varepsilon}) \tag{8}$$

where e is the exponential constant. The curves corresponding to varying m and n approach the asymptote $\sigma = C$ (Fig. 1.3*b*). However, C is unlikely to be the saturation stress of a given metal as the rate of hardening becomes vanishingly small. The rapidity with which the asymptotic value is approached is represented by n. The coefficient m defines the initial state of hardening, the fully hardened material corresponding to $m = 0$. The slope of the stress-strain curve given by (8) is equal to $n(C - \sigma)$, which varies linearly with the stress.

† E. Voce, *J. Inst. Metals*, **74**: 537 (1948). See also J. H. Palm, *Appl. Sci. Res.*, **A-2**: 198 (1948).

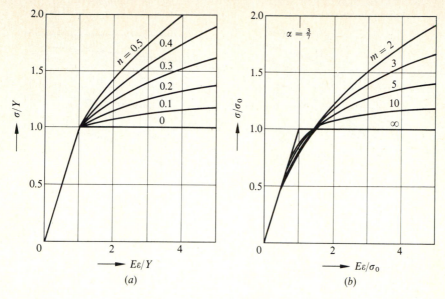

Figure 1.4 Empirical stress-strain curves for elastic/plastic materials. (*a*) Modified Ludwik equation; (*b*) Ramburg-Osgood equation.

When the elastic and plastic strains are of comparable magnitudes, it is necessary to replace ε in the preceding equations by the plastic strain ε^p. Considering the power law (5), the plastic part of the strain may be assumed to vary as σ^m, where $m = 1/n$, Since the elastic part of the strain is equal to σ/E, the total strain may be expressed by the Ramburg-Osgood equation†

$$\varepsilon = \frac{\sigma}{E}\left\{1 + \alpha\left(\frac{\sigma}{\sigma_0}\right)^{m-1}\right\} \tag{9}$$

where σ_0 is a nominal yield stress and α a dimensionless constant. The slope of the stress-strain curve given by the above equation continuously decreases from the value E at the origin (Fig. 1.4*b*). At the nominal yield point $\sigma = \sigma_0$, the plastic strain is α times the elastic strain, and the secant modulus is $E/(1 + \alpha)$. The tangent modulus at any point of the curve is given by

$$\frac{E}{T} = 1 + \alpha m\left(\frac{\sigma}{\sigma_0}\right)^{m-1} \tag{10}$$

The second term on the right-hand side is equal to E/H in view of (1). The stress-strain curve for a range of materials can be reasonably fitted by equation (9) with $\alpha = 3/7$. For a nonhardening material ($m = \infty$), the equation degenerates into a pair of straight lines meeting at the yield point $\sigma = \sigma_0$.

† W. Ramberg and W. R. Osgood, *NACA Tech. Note*, 902 (1943).

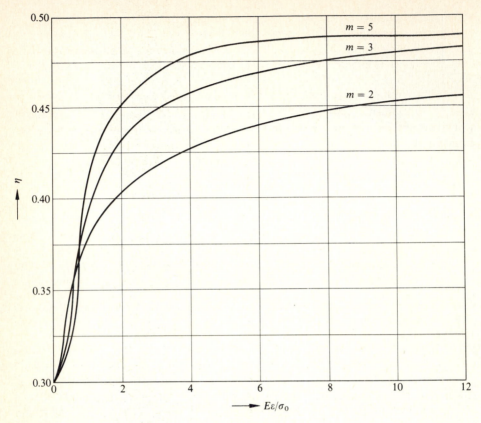

Figure 1.5 Variation of the contraction ratio with longitudinal strain in uniaxial tension according to the Ramberg-Osgood stress-strain equation ($v = 0.3$).

The contraction ratio η determined from (2) and (10) is plotted against $E\varepsilon/\sigma_0$ in Fig. 1.5, assuming $\alpha = 3/7$. Due to the nature of the Ramberg-Osgood equation, a variation of η is predicted even in the elastic range of straining. The contraction ratio increases very rapidly in the neighborhood of the yield point, following which η approaches the value 0.5 in an asymptotic manner. The actual value of η is seen to be reasonably close to 0.5 while the total strain is still of the elastic order of magnitude.

It is sometimes more convenient to employ a stress-strain equation where the curve in the plastic range is expressed by a simple power law, the material being assumed to have a definite yield point at $\sigma = Y$. The empirical representation then becomes

$$\sigma = \begin{cases} E\varepsilon & \varepsilon \leqslant \dfrac{Y}{E} \\[2ex] Y\left(\dfrac{E\varepsilon}{Y}\right)^n & \varepsilon \geqslant \dfrac{Y}{E} \end{cases} \tag{11}$$

where n is generally less than 0.5. The slope of the stress-strain curve given by (11) changes discontinuously from E to nE at the yield point (Fig. 1.4a). The tangent modulus at any point in the plastic range is n times the secant modulus. The empirical curve is effectively the Ludwik curve whose initial part is replaced by a chord of slope E.

The Ramburg-Osgood curve represents a continuous transition from the elastic to the plastic behavior expressed by a single equation when the material work-hardens. A similar curve for the ideally plastic material is given by the equation

$$\sigma = Y \tanh \left(\frac{E\varepsilon}{Y} \right)$$

which is due to Prager.† The curve having an initial slope E gradually bends over to approach the yield stress Y in an asymptotic manner. The approach is so rapid that σ is within 1 percent of Y when ε is only $4Y/E$. The tangent modulus at any point on the curve is equal to $E(1 - \sigma^2/Y^2)$, and the corresponding plastic modulus is $E(Y^2/\sigma^2 - 1)$. These moduli soon become negligible while the strain is still quite small.‡

(iv) *Influence of pressure, strain rate, and temperature* The tensile test of ductile materials under superimposed hydrostatic pressure has revealed that the yield point and the uniform elongation are unaffected by the applied pressure, but the strain to fracture increases with the intensity of the pressure. The increased ductility of the material is caused by the lateral compressive stresses which inhibit the formation of microcracks that lead to fracture. Test results for both tension and compression of brittle materials under fluid pressure indicate that there is a certain critical pressure above which the material behaves in a ductile manner.§ The stress-strain curves for axially compressed limestone cylinders under uniform fluid pressures acting on the curved surface are shown in Fig. 1.6, where σ denotes the axial compressive stress in excess of the confining pressure p. Each curve corresponds to a particular confining pressure expressed in atmospheres.¶ Some materials are found to suffer a certain amount of permanent volume change when

---

† W. Prager, *Rev. Fac. Sci., Univ. Istanbul*, **5**: 215 (1941); *Duke Math. J.*, **9**: 228 (1942).

‡ Other forms of stress-strain equation are sometimes used for the derivation of special solutions. See, for example, R. Hill, *Phil. Mag.*, **41**: 1133 (1950), and J. Chakrabarty, *Int. J. Mech. Sci.*, **12**: 315 (1970).

§ The pressure can be accurately measured from the change in resistance of a manganin wire immersed in the pressurized fluid. A detailed account of the experimental investigations regarding the effect of hydrostatic pressure on metals has been presented by P. W. Bridgman, *Studies in Large Plastic Flow and Fracture*, McGraw-Hill Book Company, New York (1952), and by H. Ll. D. Pugh (ed.), *Mechanical Behavior of Materials under Pressure*, Elsevier, Amsterdam (1970).

¶ Experimental results on the compression of marble and limestone cylinders under fluid pressure have been reported by Th. von Karman, *Z. Ver. deut. Ing.*, **55**: 1749 (1911), and by D. T. Griggs, *J. Geol.*, **44**: 541 (1936).

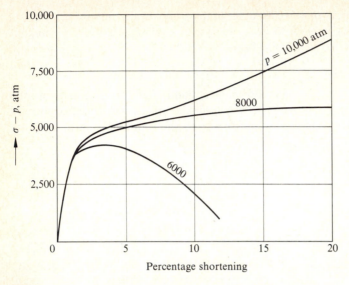

Figure 1.6 Behavior of limestone cylinders under axial thrust and lateral pressure (*after Griggs*).

subjected to hydrostatic pressures of exceedingly high magnitude, although the change is negligible in ordinary situations.†

Plastic instability is found to occur in cylindrical bars when subjected to lateral fluid pressures of sufficient magnitude.‡ The phenomenon is caused by a slight non-uniformity in distortion of the unconstrained surface which is exposed to fluid pressure. When the material is ductile, the longitudinal strain at the onset of necking is exactly the same as that in uniaxial tension, but the cross section of the neck is greatly reduced before fracture. Brittle materials, which normally fracture with no significant plastic strain under simple tension, are found to deform beyond the point of necking when tested under lateral fluid pressure. Moreover, the uniform strain at the onset of necking is found to be identical to that given by (3), with the stress-strain curve obtained in simple compression. For extremely brittle materials, the fracture mode seems to remain brittle even under a fluid pressure acting on the lateral surface.§

At room temperature, the stress-strain curve of metals is practically independent of the rate of straining attainable in ordinary testing machines. High-speed tensile tests have shown that the yield stress increases with the strain rate, and this effect is more pronounced at elevated temperatures. The true strain rate in simple compression is defined as $\dot{\varepsilon} = -\dot{h}/h$, where h is the current specimen height and $\dot{h}$ its rate of change. To obtain a constant strain rate during a test, it is therefore

---

† P. W. Bridgman, *J. Appl. Phys.*, **18**: 246 (1947). The effect of hydrolastic pressure on the shear properties of metals has been investigated by B. Crossland, *Proc. Inst. Mech. Eng.*, **169**: 935 (1954); B. Crossland and W. H. Dearden, ibid., **172**, 805 (1958). See also M. C. Shaw, *Int. J. Mech. Sci.*, **22**: 673 (1980).

‡ J. Chakrabarty. *Proc. 13th Int. M.T.D.R. Conf.*, p. 565, Pergamon Press, Oxford (1972).

§ P. W. Bridgman, *Phil. Mag.*, July, 63 (1912).

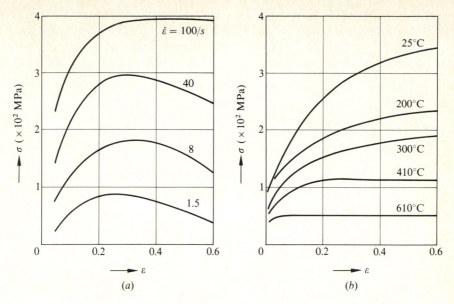

Figure 1.7 Effects of strain rate and temperature on the stress-strain curve of metals. (*a*) EN25 steel at 1000°C (*after Cook*); (*b*) annealed copper at a strain rate of $10^{-3}/s$ (*after Mahtab et al.*).

necessary to increase the platen speed in proportion to the specimen height. This is achieved by using a cam plastometer in which one of the compression platens is actuated by a cam of logarithmic profile.† Maintaining a constant temperature during a test is more difficult, since the heat generated during the test raises the temperature of the specimen adiabatically. Figure 1.7 shows typical stress-strain curves of metals in compression, obtained under constant temperatures and strain rates.‡

For a given value of the strain, the combined effect of strain rate and temperature on the yield stress may be expressed by the functional relationship§

$$\sigma = f\left\{\dot{\varepsilon}\exp\left(\frac{Q}{RT}\right)\right\} \tag{12}$$

† The cam plastometer has been devised by E. Orowan, Brit. Iron and Steel Res. Assoc. Rep., MW/F/22 (1950).

‡ For experimental methods and results on the high-speed compression at elevated temperatures, see P. M. Cook, *Proc. Conf. Properties of Materials at High Rates of Strain, Inst. Mech. Eng.*, **86** (1957); F. U. Mahtab, W. Johnson, and R. A. C. Slater, *Proc. Inst. Mech. Eng.*, **180**: 285 (1965); S. K. Samanta, *Int. J. Mech. Sci.*, **10**: 613 (1968), *J. Mech. Phys. Solids*, **19**: 117 (1971); T. A. Dean and C. E. N. Sturgess, *Proc. Inst. Mech. Eng.*, **187**: 523 (1973). See also R. A. C. Slater, *Engineering Plasticity*, ch. 6, Wiley and Sons, London (1977).

§ C. Zener and J. H. Hollomon, *J. Appl. Phys.*, **15**: 22 (1944); W. F. Hosford and R. M. Caddell, *Metal Forming Mechanics and Metallurgy*, ch. 5, Prentice-Hall, Englewood Cliffs, N.J. (1983). The expression in the curly bracket of (12) is often called the Zener-Hollomon parameter, which is also useful in the theory of high-temperature creep.

where Q is an activation energy for plastic flow, T the absolute testing temperature, and R the universal gas constant equal to 8.314 J/g mol °K. The above relationship has been experimentally confirmed for several metals over wide ranges of strain rate and temperature. When the temperature is held constant, the test results can be fitted by the power law[†]

$$\sigma = C\varepsilon^m \dot{\varepsilon}^n$$

where C, m and n depend on the operating temperature. The exponent n is known as the *strain-rate sensitivity*, which generally increases with both strain and temperature.

The dependence of the flow stress on strain rate and temperature for a given strain is sometimes expressed in the alternative form[‡]

$$\sigma = f\left\{ T\left(1 - m \ln \frac{\dot{\varepsilon}}{\dot{\varepsilon}_0}\right)\right\} \tag{13}$$

where m and $\dot{\varepsilon}_0$ are constants, the quantity in the curly bracket being known as the velocity modified temperature. It is consistent with the fact that an increase in strain rate is in effect equivalent to a decrease in temperature. Equation (14) agrees with test data for a relatively small range of values of $\dot{\varepsilon}$ but over a large temperature range.

Above the recrystallization temperature, the yield stress attains a saturation value after a small amount of strain, as a result of the work-hardening rate being balanced by the rate of thermal softening. The dependence of the saturation stress on strain rate and temperature can be expressed with reasonable accuracy by the empirical equation[§]

$$\sigma = C \sinh^{-1}\left(m\dot{\varepsilon}^n \exp \frac{b}{T}\right) \tag{14}$$

where b, C, m, and n are material constants. The activation energy Q is then independent of the temperature, and is approximately equal to Rb/n. A distinction between cold- and hot-working of metals is usually made on the basis of the recrystallization temperature, whose absolute value is roughly one-half of the absolute melting temperature. Equation (14) has the form of (12), and reduces to a power law when the expression in the parenthesis is sufficiently small.[¶]

[†] K. Inouye, *Tetsu to Hagane* (in Japanese), **41**: 593 (1955).

[‡] C. W. MacGregor and J. C. Fisher, *J. Appl. Mech.*, **13**: 11 (1946).

[§] C. M. Sellars and W. J. McG. Tegart, *Mem. Sci. Rev. Met.*, **63**: 731 (1966); S. K. Samanta, *Proc. 11th Int. M.T.D.R. Conf.*, Pergamon Press, Oxford (1970).

[¶] Large neck-free extensions are possible in certain highly rate-sensitive alloys, called *superplastic alloys*. See W. A. Backofen, I. Turner and H. Avery, *Trans. Q. ASM*, **57**: 981 (1966); J. W. Edington, K. N. Melton, and C. P. Cutler, *Prog. Mater. Sci.*, **21**: 63 (1976); K. A. Padmanabhan and G. J. Davies, *Superplasticity*, Springer-Verlag, Berlin (1980).

1.3 Analysis of Stress

(i) *Stress tensor* When a body is subjected to a set of external forces, internal forces are produced in different parts of the body so that each element of the body is in a state of statical equilibrium. Through any point O within the body, consider a small surface element δS whose orientation is specified by the unit vector $\mathbf{l}$ along the normal drawn on one side of the element (Fig. 1.8a). The material on this side of δS may be regarded as exerting a force $\delta \mathbf{P}$ across the surface element upon the material on the other side. The limit of the ratio $\delta \mathbf{P}/\delta S$ as δS tends to zero is the stress vector $\mathbf{T}$ at O associated with the direction $\mathbf{I}$. For given external loading, the stress acting across any plane passing through a given point O depends on the orientation of the plane. The resolved component of the stress vector along the unit normal $\mathbf{l}$ is called the direct or normal stress denoted by σ, while the component tangential to the plane is known as the shear stress denoted by τ.

Consider now a set of rectangular axes Ox, Oy, and Oz emanating from a typical point O, and imagines a small rectangular parallelepiped at O having its edges parallel to the axes of reference (Fig. 1.8b). The normal stresses across the faces of the block are denoted by σ_x, σ_y, and σ_z, where the subscripts denote the directions of the normal to the faces. The shear stress acting on the faces normal to the x axis is resolved into the components τ_{xy} and τ_{xz} parallel to the y and z axes respectively. The first suffix denotes the direction of the normal to the face and the second suffix the direction of the component. In a similar way, the shear stresses on the faces normal to the y axis are denoted by τ_{yx} and τ_{yz}, and those on the faces normal to the z axis by τ_{zx} and τ_{zy}. The stresses are taken as positive if they are directed as shown in the figure, when the outward normals to the faces are in the positive directions of the coordinate axes. The positive directions are all reversed on the remaining faces of the block where the outward normals are in the negative directions of the axes of reference. The nine components of the stress at any point form a second-order tensor σ_{ij}, known as the stress tensor, where i and j take integral values 1, 2, and 3. The stress components may be displayed as elements of

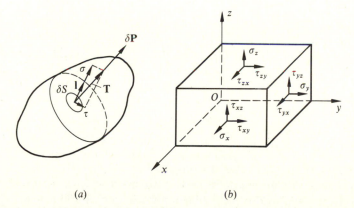

(a) (b)

Figure 1.8 Definition of stress. (a) Normal and shear stresses; (b) components of stress tensor.

the square matrix.

$$
\begin{bmatrix}
\sigma_x & \tau_{xy} & \tau_{xz} \\
\tau_{yx} & \sigma_y & \tau_{yz} \\
\tau_{zx} & \tau_{zy} & \sigma_z
\end{bmatrix}
=
\begin{bmatrix}
\sigma_{11} & \sigma_{12} & \sigma_{13} \\
\sigma_{21} & \sigma_{22} & \sigma_{23} \\
\sigma_{31} & \sigma_{32} & \sigma_{33}
\end{bmatrix}
$$

The forces acting on the faces of the parallelepiped are clearly in equilibrium. To examine the couple equilibrium, let δx, δy, δz denote the lengths of these faces along the respective coordinate axes. Then the resultant couple about the z axis is $(\tau_{xy} - \tau_{yx})\,\delta x\,\delta y\,\delta z$, which must vanish for equilibrium. This gives $\tau_{xy} = \tau_{yx}$. Similarly, the conditions for couple equilibrium about the other two axes give $\tau_{yz} = \tau_{zy}$ and $\tau_{zx} = \tau_{xz}$. These identities may be expressed as $\sigma_{ij} = \sigma_{ji}$, implying that the stress tensor is symmetric with respect to its subscripts. Thus there are six independent stress components, three normal components σ_x, σ_y, σ_z, and three shear components τ_{xy}, τ_{yz}, τ_{zx}, which completely specify the state of stress at each point of the body. The matrix representing the stress tensor is evidently symmetrical.

The mean of the three normal stresses, equal to $(\sigma_x + \sigma_y + \sigma_z)/3$, is known as the *hydrostatic stress* denoted by σ_0. A *deviatoric* or *reduced stress* tensor s_{ij} is defined as that which is obtained from σ_{ij} by reducing the normal stress components by σ_0. This gives the deviatoric normal stresses

$$
s_x = \sigma_x - \sigma_0 = \frac{2\sigma_x - \sigma_y - \sigma_z}{3}
$$

$$
s_y = \sigma_y - \sigma_0 = \frac{2\sigma_y - \sigma_z - \sigma_x}{3}
$$

$$
s_z = \sigma_z - \sigma_0 = \frac{2\sigma_z - \sigma_x - \sigma_y}{3}
$$

The deviatoric shear stresses are the same as the actual shear stresses. Since $s_x + s_y + s_z = 0$, the deviatoric normal stresses cannot all have the same sign. The difference between any two normal components of the deviatoric stress is the same as that between the corresponding components of the actual stress. Expressed in suffix notation, the relationship between s_{ij} and σ_{ij} is

$$
s_{ij} = \sigma_{ij} - \sigma_0\delta_{ij} = \sigma_{ij} - \tfrac{1}{3}\sigma_{kk}\delta_{ij} \tag{15}
$$

where δ_{ij} is the *Kronecker delta* whose value is unity when $i = j$ and zero when $i \neq j$. Evidently, $\delta_{ij} = \delta_{ji}$. Any repeated or dummy suffix indicates a summation of all terms obtainable by assigning the values 1, 2, and 3 to this suffix in succession. Thus $\sigma_{kk} = \sigma_x + \sigma_y + \sigma_z$. It follows from the definition of the delta symbol that $\sigma_{ij}\delta_{jk} = \sigma_{ik}$, where j is a dummy suffix and i, k are free suffixes. Each term of a tensor equation must have the same free suffixes, but a dummy suffix can be replaced by any other letter different from the free suffixes.

(ii) Stresses on an oblique plane Consider the equilibrium of a small tetrahedron $OABC$ of which the edges OA, OB, and OC are along the coordinate axes (Fig. 1.9). Let (l, m, n) be the directions cosines of a straight line drawn along the exterior normal to the oblique plane ABC. These are the components of the unit normal $\mathbf{l}$ with respect to Ox, Oy, and Oz. If the area of the face ABC is denoted by δS, the faces OAB, OBC, and OCA have areas $n\delta S$, $l\delta S$, and $m\delta S$ respectively. The stress vector $\mathbf{T}$ acting across the face ABC has components T_x, T_y, and T_z along the axes of reference. Resolving the forces in the directions Ox, Oy, and Oz, we get

$$T_x = l\sigma_x + m\tau_{xy} + n\tau_{zx}$$
$$T_y = l\tau_{xy} + m\sigma_y + n\tau_{yz} \tag{16}$$
$$T_z = l\tau_{zx} + m\tau_{yz} + n\sigma_z$$

on cancelling out δS from each equation of force equilibrium. When δS tends to zero, these equations give the components of the stress vector at O, associated with the direction (l, m, n), in terms of the components of the stress tensor. Using the suffix notation and the summation convention, (16) can be expressed as

$$T_j = l_i\sigma_{ij}$$

where $l_1 = l$, $l_2 = m$, $l_3 = n$. The above equation is equivalent to three equations corresponding to the three possible values of the free suffix j. A single free suffix therefore characterizes a vector. The normal stress across the plane specified by its normal (l, m, n) is

$$\sigma = lT_x + mT_y + nT_z = l_jT_j = l_il_j\sigma_{ij}$$
$$= l^2\sigma_x + m^2\sigma_y + n^2\sigma_z + 2lm\tau_{xy} + 2mn\tau_{yz} + 2nl\tau_{zx} \tag{17}$$

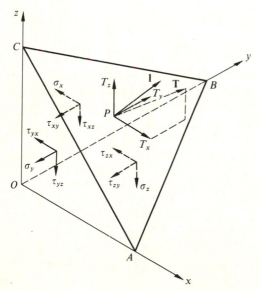

Figure 1.9 Stresses across an oblique plane in a three-dimensional state of stress.

The shear stress across the plane can be resolved into two components in a pair of mutually perpendicular directions in the plane. Denoting one of these directions by (l', m', n'), the corresponding shear component is obtained as

$$\tau' = l'T_x + m'T_y + n'T_z = l'_j T_j = l_i l'_j \sigma_{ij}$$

$$= ll'\sigma_x + mm'\sigma_y + nn'\sigma_z + (lm' + ml')\tau_{xy} + (mn' + nm')\tau_{yz} + (nl' + ln')\tau_{zx} \qquad (18)$$

This evidently is the resolved component of the resultant stress in the direction (l', m', n'). The direction cosines satisfy the well-known geometrical relations

$$l^2 + m^2 + n^2 = 1 \qquad l'^2 + m'^2 + n'^2 = 1$$

$$ll' + mm' + nn' = 0 \qquad (19)$$

The first two equations express the fact (l, m, n) and (l', m', n') represent unit vectors, while the last relation expresses the orthogonality of these vectors. The shear stress is most conveniently found from the fact that its magnitude is $\sqrt{T^2 - \sigma^2}$, and its direction cosines are proportional to its rectangular components

$$T_x - l\sigma \qquad T_y - m\sigma \qquad T_z - n\sigma$$

Let x_i and x'_i represent two sets of rectangular axes through a common origin O, and a_{ij} denote the direction cosine of the x'_i axis with respect to the x_j axis. The direction cosine of the x_i axis with respect to the x'_j axis is then equal to a_{ji}. It follows from geometry that the coordinates of any point in space referred to the two sets of axes are related by the equations

$$x'_i = a_{ij}x_j \qquad x_j = a_{ij}x'_i \qquad (20)$$

The components of any vector transform† according to the same law as (20). Let σ'_{ij} denote the components of the stress tensor when referred to the set of axes x'_i. A defining property of tensors is the transformation law

$$\sigma'_{ij} = a_{ik}a_{jl}\sigma_{kl} \qquad (21)$$

Let us suppose that $a_{11} = l, a_{12} = m, a_{13} = n$, and $a_{21} = l', a_{22} = m', a_{23} = n'$. The normal stress across the plane (l, m, n) is then equal to σ'_{11}, and the corresponding expression (17) can be readily verified from (21). Similarly, the component of the shear stress across the plane resolved in the direction (l', m', n') is equal to σ'_{12} which can be shown to be that given by (18).

(iii) *Principal stresses* The normal stress σ has maximum and minimum values for varying orientations of the oblique plane. Regarding l and m as the independent direction cosines, the conditions for stationary σ may be written as $\partial\sigma/\partial l = 0, \partial\sigma/\partial m = 0$. Differentiating the first equation of (19) partially with

---

† It follows from (20) that $x'_i = a_{ik}x_k = a_{ik}a_{jk}x'_j$, indicating that $a_{ik}a_{jk} = \delta_{ij}$, which furnishes six independent relations of types (19).

respect to l and m, we get $\partial n/\partial l = -l/n$ and $\partial n/\partial m = -m/n$. Inserting these results into the partial derivatives of (17), and using (16), the stationary condition can be expressed as

$$\frac{T_x}{l} = \frac{T_y}{m} = \frac{T_z}{n}$$

This shows that the resultant stress across the plane acts in the direction of the normal when the normal stress has a stationary value. Each of the above ratios is therefore equal to the normal stress σ. The substitution into (16) gives

$$l(\sigma_x - \sigma) + m\tau_{xy} + n\tau_{zx} = 0$$

$$l\tau_{xy} + m(\sigma_y - \sigma) + n\tau_{yz} = 0 \tag{22}$$

$$l\tau_{zx} + m\tau_{yz} + n(\sigma_z - \sigma) = 0$$

In suffix notation, these relations are equivalent to $l_i(\sigma_{ij} - \sigma\delta_{ij}) = 0$, which follows directly from the fact that $T_j = \sigma l_j$ across a principal plane. The set of linear homogeneous equations (22) would have a nonzero solution for l, m, n if the determinant of their coefficients vanishes. Thus

$$\begin{vmatrix} \sigma_x - \sigma & \tau_{xy} & \tau_{zx} \\ \tau_{xy} & \sigma_y - \sigma & \tau_{yz} \\ \tau_{zx} & \tau_{yz} & \sigma_z - \sigma \end{vmatrix} = 0$$

Expanding this determinant, we obtain a cubic equation in σ having three real roots $\sigma_1, \sigma_2, \sigma_3$, which are known as the *principal stresses*. These stresses act across planes on which the shear stresses are zero. The cubic may be expressed in the form

$$\sigma^3 - I_1\sigma^2 - I_2\sigma - I_3 = 0 \tag{23}$$

where

$$I_1 = \sigma_x + \sigma_y + \sigma_z = \sigma_1 + \sigma_2 + \sigma_3 = \sigma_{ii} \tag{24}$$

$$I_2 = -(\sigma_x\sigma_y + \sigma_y\sigma_z + \sigma_z\sigma_x) + \tau_{xy}^2 + \tau_{yz}^2 + \tau_{zx}^2$$

$$= -(\sigma_1\sigma_2 + \sigma_2\sigma_3 + \sigma_3\sigma_1) = \tfrac{1}{2}(\sigma_{ij}\sigma_{ij} - \sigma_{ii}\sigma_{jj}) \tag{25}$$

$$I_3 = \sigma_x\sigma_y\sigma_z + 2\tau_{xy}\tau_{yz}\tau_{zx} - \sigma_x\tau_{yz}^2 - \sigma_y\tau_{zx}^2 - \sigma_z\tau_{xy}^2$$

$$= \begin{vmatrix} \sigma_x & \tau_{xy} & \tau_{zx} \\ \tau_{xy} & \sigma_y & \tau_{yz} \\ \tau_{zx} & \tau_{yz} & \sigma_z \end{vmatrix} = \sigma_1\sigma_2\sigma_3 \tag{26}$$

The expressions for I_1, I_2, I_3 in terms of the principal stresses follow from the fact that (23) is equivalent to the equation $(\sigma - \sigma_1)(\sigma - \sigma_2)(\sigma - \sigma_3) = 0$. Since the stationary values of the normal stress do not depend on the orientation of the coordinate axes, the coefficients of (23) must also be independent of the choice of the axes of references. The quantities I_1, I_2, I_3 are therefore known as the *invariants* of the stress tensor.†

The direction cosines corresponding to each principal stress can be found from the first equation of (19) and any two equations of (22) with the appropriate value of σ. Let (l_1, m_1, n_1) and (l_2, m_2, n_2) represent the directions of σ_1 and σ_2 respectively. If we express (22) in terms of l_1, m_1, n_1, and σ_1, multiply these equations by l_2, m_2, n_2 in order and add them together, and then subtract the resulting equation from that obtained by interchanging the subscripts, we arrive at the result

$$(\sigma_1 - \sigma_2)(l_1 l_2 + m_1 m_2 + n_1 n_2) = 0$$

If $\sigma_1 \neq \sigma_2$, the above equation indicates that the directions (l_1, m_1, n_1) and (l_2, m_2, n_2) are perpendicular to one another. It follows, therefore, that the principal directions corresponding to distinct values of the principal stresses are mutually orthogonal. These directions are known as the *principal axes* of the stress. When two of the principal stresses are equal to one another, the direction of the third principal stress is uniquely determined, but all directions perpendicular to this principal axis are principal directions. When $\sigma_1 = \sigma_2 = \sigma_3$, representing a hydrostatic state of stress, any direction in space is a principal direction.

The invariants of the deviatoric stress tensor are obtained by replacing the actual stress components in (24) to (26) by the corresponding deviatoric components. The first deviatoric stress invariant is

$$J_1 = s_x + s_y + s_z = s_1 + s_2 + s_3 = s_{ii} = 0$$

where s_1, s_2, s_3 are the principal deviatoric stresses. These principal values are the roots of the cubic equation

$$s^3 - J_2 s - J_3 = 0 \tag{27}$$

where

$$J_2 = -(s_x s_y + s_y s_z + s_z s_x) + \tau_{xy}^2 + \tau_{yz}^2 + \tau_{zx}^2$$
$$= \tfrac{1}{2}(s_x^2 + s_y^2 + s_z^2) + \tau_{xy}^2 + \tau_{yz}^2 + \tau_{zx}^2$$
$$= \tfrac{1}{6}[(\sigma_x - \sigma_y)^2 + (\sigma_y - \sigma_z)^2 + (\sigma_z - \sigma_x)^2] + \tau_{xy}^2 + \tau_{yz}^2 + \tau_{zx}^2 \tag{28}$$
$$J_3 = s_x s_y s_z + 2\tau_{xy}\tau_{yz}\tau_{zx} - s_x \tau_{yz}^2 - s_y \tau_{zx}^2 - s_z \tau_{xy}^2$$

$$= \begin{vmatrix} s_x & \tau_{xy} & \tau_{zx} \\ \tau_{xy} & s_y & \tau_{yz} \\ \tau_{zx} & \tau_{yz} & s_z \end{vmatrix} = s_1 s_2 s_3 = \tfrac{1}{3}(s_1^3 + s_2^3 + s_3^3) \tag{29}$$

The last two expressions for J_2 are obtained from the first expression by adding the identically zero terms $\tfrac{1}{2}(s_x + s_y + s_z)^2$ and $\tfrac{1}{3}(s_x + s_y + s_z)^2$ respectively, and noting

---

† Any symmetric tensor of second order has three real principal values, the basic invariants of the tensor being identical in form to those for the stress.

the fact that $s_x - s_y = \sigma_x - \sigma_y$ etc. Similarly, the last expression for J_3 follows from the preceding one on adding the term $\frac{1}{3}(s_1 + s_2 + s_3)^3$. In suffix notation, these invariants can be written as

$$J_2 = \tfrac{1}{2}s_{ij}s_{ij} \qquad J_3 = \tfrac{1}{3}s_{ij}s_{jk}s_{ki} \qquad (30)$$

The repetition of all suffixes is a characteristic of invariants, which are scalars. Substituting $\sigma = s + I_1/3$ in (23) and comparing the coefficients of the resulting equation with those of (27), we obtain

$$J_2 = I_2 + \tfrac{1}{3}I_1^2 \qquad J_3 = I_3 + \tfrac{1}{3}I_1 I_2 + \tfrac{2}{27}I_1^3$$

When J_2 and J_3 have been found, equation (27) may be solved by means of the substitution $s = 2\sqrt{J_2/3} \cos \phi$, which reduces the cubic to

$$\cos 3\phi = \frac{J_3}{2}\left(\frac{3}{J_2}\right)^{3/2} \qquad (31)$$

Since $4J_2^3 \geqslant 27J_3^2$, the right-hand side† of (31) lies between -1 and 1, and one value of ϕ lies between 0 and $\pi/3$. The principal deviatoric stresses may therefore be written as

$$s_1 = 2\sqrt{\frac{J_2}{3}} \cos \phi \qquad s_2, s_3 = -2\sqrt{\frac{J_2}{3}} \cos\left(\frac{\pi}{3} \pm \phi\right) \qquad (32)$$

where $0 \leqslant \phi \leqslant \pi/3$. Any function of these principal components is also a function of the invariants, which play an important part in the mathematical development of the theory of plasticity.

(iv) *Principal shear stresses* When the principal stresses and their directions are known, it is convenient to take the principal axes as the axes of reference. If Ox, Oy, Oz denote the coordinate axes associated with the principal stresses $\sigma_1, \sigma_2, \sigma_3$ respectively, the components of the stress vector across a plane whose normal is in the direction (l, m, n) are

$$T_x = l\sigma_1 \qquad T_y = m\sigma_2 \qquad T_z = n\sigma_3$$

The normal stress across the oblique plane therefore becomes

$$\sigma = l^2\sigma_1 + m^2\sigma_2 + n^2\sigma_3 \qquad (33)$$

If the magnitude of the shear stress across the plane is denoted by τ, then

$$\tau^2 = T^2 - \sigma^2 = (l^2\sigma_1^2 + m^2\sigma_2^2 + n^2\sigma_3^2) - (l^2\sigma_1 + m^2\sigma_2 + n^2\sigma_3)^2$$
$$= (\sigma_1 - \sigma_2)^2 l^2 m^2 + (\sigma_2 - \sigma_3)^2 m^2 n^2 + (\sigma_3 - \sigma_1)^2 n^2 l^2 \qquad (34)$$

in view of the relation $l^2 + m^2 + n^2 = 1$. Since the components of the normal stress along the coordinate axes are $(l\sigma, m\sigma, n\sigma)$, the components of the shear stress are

---

† Using (32) and (31), it can be shown that $4J_2^3 - 27J_3^2 = (\sigma_1 - \sigma_2)^2(\sigma_2 - \sigma_3)^2(\sigma_3 - \sigma_1)^2$, which is a positive quantity for distinct values of the principal stresses.

$l(\sigma_1 - \sigma)$, $m(\sigma_2 - \sigma)$, $n(\sigma_3 - \sigma)$. Hence the direction cosines of the shear stress are

$$l_s = l\left(\frac{\sigma_1 - \sigma}{\tau}\right) \qquad m_s = m\left(\frac{\sigma_2 - \sigma}{\tau}\right) \qquad n_s = n\left(\frac{\sigma_3 - \sigma}{\tau}\right) \tag{35}$$

A plane which is equally inclined to the three principal axes is known as the *octahedral plane*, the direction cosines of its normal being given by $l^2 = m^2 = n^2 = 1/3$. These relations are satisfied by four pairs of parallel planes forming a regular octahedron having its vertices on the principal axes. By (33) and (34), the octahedral normal stress is equal to the hydrostatic stress σ_0, and the octahedral shear stress is of the magnitude

$$\tau_0 = \tfrac{1}{3}\sqrt{(\sigma_1 - \sigma_2)^2 + (\sigma_2 - \sigma_3)^2 + (\sigma_3 - \sigma_1)^2} = \sqrt{\tfrac{2}{3}J_2}$$

The components of the octahedral shear stress along the principal axes are numerically equal to $1/\sqrt{3}$ times the deviatoric principal stresses.

We now proceed to determine the stationary values of the shear stress for varying orientations of the oblique plane. To this end, we put $n^2 = 1 - l^2 - m^2$ in (34), and express it in the form

$$\tau^2 = l^2(\sigma_1^2 - \sigma_3^2) + m^2(\sigma_2^2 - \sigma_3^2) + \sigma_3^2 - \{l^2(\sigma_1 - \sigma_3) + m^2(\sigma_2 - \sigma_3) + \sigma_3\}^2$$

where l and m are treated as the independent variables. We shall follow the convention $\sigma_1 > \sigma_2 > \sigma_3$. Equating to zero the derivatives of τ^2 with respect to l and m, we obtain

$$l(\sigma_1 - \sigma_3)[(1 - 2l^2)(\sigma_1 - \sigma_3) - 2m^2(\sigma_2 - \sigma_3)] = 0$$
$$m(\sigma_2 - \sigma_3)[(1 - 2m^2)(\sigma_2 - \sigma_3) - 2l^2(\sigma_1 - \sigma_3)] = 0 \tag{36}$$

These equations are obviously satisfied for $l = m = 0$, and hence $n = 1$, which corresponds to a principal stress direction for which the shear stress has a minimum value of zero. To obtain a maximum value of the shear stress, we set $l = 0$ satisfying the first equation of (36), and use this value in the second equation to get $1 - 2m^2 = 0$. This gives $l = 0$, $m^2 = n^2 = 1/2$ corresponding to a maximum shear stress equal to $\tfrac{1}{2}(\sigma_2 - \sigma_3)$ according to (34). Similarly, the direction represented by $m = 0$, $n^2 = l^2 = 1/2$ satisfies (36), and furnishes a maximum value of $\tfrac{1}{2}(\sigma_1 - \sigma_3)$ for the shear stress. Finally, setting $n = 0$ and hence $l^2 + m^2 = 1$, we find that τ is a maximum for $l^2 = m^2 = 1/2$, giving a stationary value equal to $\tfrac{1}{2}(\sigma_1 - \sigma_2)$. The three stationary shear stresses, known as the *principal shear stresses*, may therefore be written as

$$\tau_1 = \tfrac{1}{2}(\sigma_2 - \sigma_3) \qquad \tau_2 = \tfrac{1}{2}(\sigma_1 - \sigma_3) \qquad \tau_3 = \tfrac{1}{2}(\sigma_1 - \sigma_2) \tag{37}$$

These stresses act in directions which bisect the angles between the principal axes. By (33), the normal stresses acting on the planes of τ_1, τ_2, τ_3 are immediately found to be, respectively,

$$\tfrac{1}{2}(\sigma_2 + \sigma_3) \qquad \tfrac{1}{2}(\sigma_1 + \sigma_3) \qquad \tfrac{1}{2}(\sigma_1 + \sigma_2)$$

In view of the assumption $\sigma_1 > \sigma_2 > \sigma_3$, the greatest shear stress is of magnitude

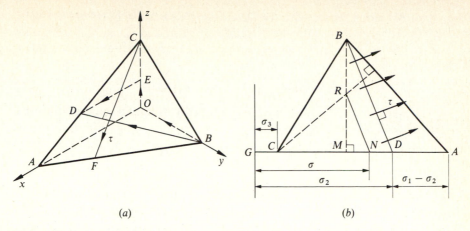

Figure 1.10 Construction for the normal stress and the direction of the shear stress.

$\frac{1}{2}(\sigma_1 - \sigma_3)$, and it acts across a plane whose normal bisects the angle between the directions of σ_1 and σ_3. It follows from (32) that the greatest shear stress is equal to $\sqrt{J_2} \cos(\pi/6 - \phi)$, where ϕ lies between zero and $\pi/3$ satisfying (31).

(v) *Shear stress and the oblique triangle* Consider now the direction of the shear stress on an inclined plane in relation to the true shape of the oblique triangle. It is assumed for simplicity that the direction cosines (l, m, n) are all positive.† Let δh denote the perpendicular distance from the origin O to the oblique plane ABC (Fig. 1.10a). Then the distances of the vertices A, B, C from O are $\delta h/l$, $\delta h/m$, $\delta h/n$ respectively, their ratios being

$$OA:OB:OC = mn:nl:lm \tag{38}$$

The sides of the triangle are readily found from the right-angled triangles AOB, BOC, and COA. The true shape of the oblique triangle ABC is therefore defined by the ratios

$$AB:BC:CA = n\sqrt{1 - n^2}:l\sqrt{1 - l^2}:m\sqrt{1 - m^2} \tag{39}$$

The vertical angles of the triangle follow from (39) and the well-known cosine law. The results can be conveniently put in the form

$$\tan A = \frac{l}{mn} \qquad \tan B = \frac{m}{nl} \qquad \tan C = \frac{n}{lm} \tag{40}$$

The coordinate axes in Fig. 1.10a are in the directions of the principal stresses. A line BD is drawn from the apex B to meet the opposite side of AC at D, such that BD is perpendicular to the direction of the shear stress across the plane. The components

---

† No generality is lost in this assumption, since the expressions for σ and τ involve only the squares of the direction cosines.

of the vector **BD** along the axes Ox, Oy, Oz are equal to $ED, -OB, OE$ respectively. Since BD is orthogonal to both the directions (l, m, n) and (l_s, m_s, n_s), the scalar products of **BD** with the unit vectors representing these directions must vanish. Using (35) and (33), it is easily shown that

$$ED:OB:OE = mn(\sigma_2 - \sigma_3):nl(\sigma_1 - \sigma_3):lm(\sigma_1 - \sigma_2) \tag{41}$$

If $\sigma_1 > \sigma_2 > \sigma_3$, the line BD must meet AC internally as shown. Indeed, from the similar triangles CDE and CAO, we have

$$\frac{CD}{CA} = \frac{ED}{OA} = \frac{ED}{OB}\frac{OB}{OA} = \frac{\sigma_2 - \sigma_3}{\sigma_1 - \sigma_3} \tag{42}$$

in view of (38) and (41). If points A, D, C, and G are located along a straight line, such that $GA = \sigma_1$, $GD = \sigma_2$, and $GC = \sigma_3$, and the true shape triangle ABC is constructed on CA as base (Fig. 1.10b), then in view of (42), the shear stress is directed at right angles to the line joining B and D. Since $n_s < 0$ by (35), the direction of the shear stress vector is obtained by a 90° counterclockwise rotation from the direction BD. If R is the orthocentre of the triangle ABC, and BM is drawn perpendicular to CA, then by Eqs. (40),

$$\frac{CM}{AM} = \frac{\cot C}{\cot A} = \frac{l^2}{n^2} \qquad \frac{MR}{MB} = \frac{\cot A}{\tan C} = m^2 \tag{43}$$

since angle MRC is equal to the vertical angle A. If RN is drawn parallel to BD, meeting CA at N, then $MN/MD = MR/MB = m^2$, which gives

$$GN = GM + MN = (l^2 + m^2 + n^2)GM + m^2MD$$

$$= l^2(GA - MA) + m^2GD + n^2(GC + CM) = l^2GA + m^2GD + n^2GC$$

The expression on the right-hand side is equal to σ in view of (33). Hence GN represents the magnitude of the normal stress transmitted across the plane.[†] It follows from (34) and (41) that if OB represents the quantity $nl(\sigma_1 - \sigma_3)$ to a certain scale, then BD will represent the shear stress τ to the same scale. Hence

$$\frac{OB}{BD} = nl\left(\frac{\sigma_1 - \sigma_3}{\tau}\right) = \frac{nl}{\tau}CA$$

with reference to Fig. 1.10. Since $RN/BD = MR/MB = m^2$ by (43), and $CA = \sqrt{OC^2 + OA^2}$, we have

$$RN = m^2 \cdot BD = \frac{m^2\tau}{nl}\frac{OB}{CA} = \frac{m\tau}{\sqrt{1 - m^2}}$$

in view of (38). It follows that the magnitude of the shear stress on the plane is $\tau = RN \tan \beta$, where β is the angle made by the normal to the plane with the direction of the intermediate principal stress σ_2.

[†] The constructions for the normal stress and the direction of the shear stress are due to H. W. Swift, *Engineering*, **162**: 381 (1946).

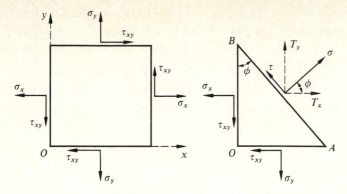

Figure 1.11 An element in a state of plane stress.

(vi) *Plane stress* A state of plane stress is defined by $\sigma_z = \tau_{yz} = \tau_{zx} = 0$. The z axis then coincides with a principal axis, and the corresponding principal stress vanishes.† The orientation of Ox and Oy with respect to the other two principal axes is, however, arbitrary. Consider a plane AB perpendicular to the xy plane, and let ϕ be the counterclockwise angle made by the normal to the plane with the x axis (Fig. 1.11). The shear stress τ will be reckoned positive when it is directed to the left of the exterior normal. Setting $l = \cos\phi$, $m = \sin\phi$, and $n = 0$ in (16), the components of the stress vector across AB are obtained as

$$T_x = \sigma_x \cos\phi + \tau_{xy} \sin\phi \qquad T_y = \tau_{xy} \cos\phi + \sigma_y \sin\phi \qquad (44)$$

The resolved components of the resultant stress along the normal and the tangent to the plane are

$$\sigma = T_x \cos\phi + T_y \sin\phi \qquad \tau = -T_x \sin\phi + T_y \cos\phi$$

Substituting for T_x and T_y in the above equations, the normal and shear stresses across the plane are obtained as

$$\sigma = \sigma_x \cos^2\phi + \sigma_y \sin^2\phi + 2\tau_{xy} \sin\phi \cos\phi$$
$$= \tfrac{1}{2}(\sigma_x + \sigma_y) + \tfrac{1}{2}(\sigma_x - \sigma_y)\cos 2\phi + \tau_{xy}\sin 2\phi \qquad (45)$$

$$\tau = -(\sigma_x - \sigma_y)\sin\phi\cos\phi + \tau_{xy}(\cos^2\phi - \sin^2\phi)$$
$$= -\tfrac{1}{2}(\sigma_x - \sigma_y)\sin 2\phi + \tau_{xy}\cos 2\phi \qquad (46)$$

These results may be directly obtained from (16) and (17) by setting $l = m' = \cos\phi$, $m = -l' = \sin\phi$ and $n = n' = 0$. Since $d\sigma/d\phi = 2\tau$, which is readily verified from above, the shear stress vanishes on the plane for which the normal stress has a stationary value. This corresponds to $\phi = \alpha$, where

$$\tan 2\alpha = \frac{2\tau_{xy}}{\sigma_x - \sigma_y} \qquad (47)$$

---

† The results for plane stress are directly applicable to the more general situation where the z axis coincides with the direction of any nonzero principal stress.

which defines two directions at right angles to one another, giving the principal axes in the plane of Ox and Oy. The principal stresses σ_1, σ_2 are the roots of the equation

$$(\sigma - \sigma_x)(\sigma - \sigma_y) = \tau_{xy}^2$$

which is obtained by writing $T_x = \sigma \cos \phi$ and $T_y = \sigma \sin \phi$ in (44), and then eliminating ϕ between the two equations. The solution is

$$\sigma_1, \sigma_2 = \tfrac{1}{2}(\sigma_x + \sigma_y) \pm \tfrac{1}{2}\sqrt{(\sigma_x - \sigma_y)^2 + 4\tau_{xy}^2} \tag{48}$$

The acute angle made by the direction of the algebraically greater principal stress σ_1 with the x axis is measured in the counterclockwise sense when τ_{xy} is positive, and in the clockwise sense when τ_{xy} is negative. It follows from (48) that

$$\sigma_x + \sigma_y = \sigma_1 + \sigma_2 \qquad \sigma_x\sigma_y - \tau_{xy}^2 = \sigma_1\sigma_2 \tag{49}$$

These are the basic invariants of the stress tensor in a state of plane stress. Evidently, any function of these invariants is also an invariant.

Let $O\xi$, and $O\eta$ represent a new pair of rectangular axes in the (x, y) plane, and let ϕ be the angle of inclination of the ξ axis to the x axis measured in the counterclockwise sense. Then the stress components σ_ξ and $\tau_{\xi\eta}$, referred to the new axes, are directly given by the right-hand sides of (45) and (46) respectively. The remaining stress component σ_η is obtained by writing $\pi/2 + \phi$ for ϕ in (45), resulting in

$$\sigma_\eta = \sigma_x \sin^2 \phi + \sigma_y \cos^2 \phi - 2\tau_{xy} \sin \phi \cos \phi$$

$$= \tfrac{1}{2}(\sigma_x + \sigma_y) - \tfrac{1}{2}(\sigma_x - \sigma_y) \cos 2\phi - \tau_{xy} \sin 2\phi \tag{50}$$

It immediately follows that $\sigma_\xi + \sigma_\eta = \sigma_x + \sigma_y$, which shows the invariance of the first expression of (49). The invariance of the second expression may be similarly verified.

Considering the principal axes as the axes of reference, the shear stress across an inclined plane can be written as $\tau = -\tfrac{1}{2}(\sigma_1 - \sigma_2) \sin 2\phi$, which indicates that the shear stress is directed to the right of the outward normal to the plane when $\sigma_1 > \sigma_2$ and $0 < \phi < \pi/2$. The shear stress has its greatest magnitude when $\phi = \pm\pi/4$, the maximum value of the shear stress being

$$\tau_{\max} = \tfrac{1}{2}|\sigma_1 - \sigma_2| = \tfrac{1}{2}\sqrt{(\sigma_x - \sigma_y)^2 + 4\tau_{xy}^2} \tag{51}$$

There are two other principal shear stresses, having magnitudes $\tfrac{1}{2}|\sigma_1|$ and $\tfrac{1}{2}|\sigma_2|$, and bisecting the angles between the z axis and the directions of σ_1 and σ_2 respectively. A little examination of the three principal values reveals that the numerically greatest shear stress occurs in the plane of the applied stresses when σ_1 and σ_2 have opposite signs, and out of the plane of the applied stresses when they are of the same sign. In view of (49), the former corresponds to $\sigma_x\sigma_y < \tau_{xy}^2$ and the latter to $\sigma_x\sigma_y > \tau_{xy}^2$. A state of pure shear is given by $\sigma_1 = -\sigma_2$, since the normal stress then vanishes on the planes of maximum shear.

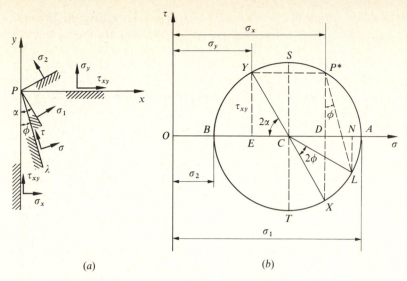

(a)   (b)

Figure 1.12 Mohr's construction for a two-dimensional state of stress. (a) Physical plane; (b) stress plane.

1.4 Mohr's Representation of Stress

(i) *Two-dimensional stress state* A useful graphical method of analyzing the state of stress has been developed by Mohr.† In this method, the normal and shear stresses across any plane are represented by a point on a plane diagram in which σ and τ are taken as rectangular coordinates. For the present purpose, it is necessary to regard the shear stress as positive when it has a clockwise moment about a point within the element. In Fig. 1.12, the stresses acting on planes perpendicular to the x and y axes are represented by the points X and Y on the (σ, τ) plane. The circle drawn on XY as diameter, and having its center C on the σ axis, is called the Mohr circle for the considered state of stress. The points A and B, where the circle is intersected by the σ axis, define the principal stresses, since $OA = \sigma_1$ and $OB = \sigma_2$ in view of (48) and the geometry of Mohr's diagram. By (47), the angle made by CA with CX is twice the angle α which the direction of σ_1 makes with the x axis in the physical plane. The normal and shear stresses transmitted across a plane, whose normal is inclined at a counterclockwise angle ϕ to the x axis, correspond to the point L on the Mohr circle, where CL is inclined to CX at an angle 2ϕ measured in the same sense. The proof of the construction follows from the fact that $CD = CL \cos 2\alpha$ and $XD = CL \sin 2\alpha$, where XD is perpendicular to OA. Then from the geometry of the figure,

$$ON = OC + CL \cos 2(\alpha - \phi) = OC + CD \cos 2\phi + XD \sin 2\phi$$

$$LN = CL \sin 2(\alpha - \phi) = -CD \sin 2\phi + XD \sin 2\phi$$

†O. Mohr, *Zivilingenieur*, **28**: 112 (1882). See also his book, *Technische Mechanik*, 2d ed., p. 192 (1914).

These expressions are equivalent to (45) and (46) in view of the present sign convention. If LC is produced to meet the circle again at M, then the coordinates of M give the stresses across a plane perpendicular to that corresponding to L. The maximum shear stress is evidently equal to the radius of the Mohr circle, and acts on planes that correspond to the extremities of the vertical diameter. The normal stress across these planes is equal to the distance of the center of the circle from the origin of the stress plane.

It is instructive to consider the following alternative construction, also due to Mohr. Let a generic point P, the state of stress at which is being discussed, be taken as the origin of coordinates in the physical plane (Fig. 1.12a). All planes passing through P and containing the z axis are denoted by their traces in the xy plane. The normal and shear stresses corresponding to the points X and Y on the Mohr circle are transmitted across the planes Py and Px respectively. The lines through X and Y drawn parallel to these planes intersect the circle at a common point P^*, which is called the pole of the Mohr circle. When the stress circle and the pole are given, the stresses acting across any plane $P\lambda$ through P are found by locating the point L on the circle such that P^*L is parallel to $P\lambda$, the angle XCL at the center being twice the peripheral angle XP^*L over the arc XL. The planes corresponding to the principal stresses are parallel to P^*A and P^*B, and those corresponding to the maximum shear stress are parallel to P^*S and P^*T. It may be noted that the magnitude of the resultant stress across any plane is equal to the distance of the corresponding stress point on the Mohr circle from the origin of the stress plane.

(ii) Three-dimensional stress state Suppose that the principal stresses $\sigma_1, \sigma_2, \sigma_3$ are known in magnitude and direction for a three-dimensional state of stress. These principal values are assumed as distinct, and so labeled that $\sigma_1 > \sigma_2 > \sigma_3$. A graphical method developed by Mohr can be used to find the variation of normal and shear stresses with the direction (l, m, n). We begin with the relations

$$l^2\sigma_1 + m^2\sigma_2 + n^2\sigma_3 = \sigma$$
$$l^2\sigma_1^2 + m^2\sigma_2^2 + n^2\sigma_3^2 = \sigma^2 + \tau^2 \tag{52}$$
$$l^2 + m^2 + n^2 = 1$$

This is a set of three linear equations for the squares of the direction cosines. The solution is most conveniently obtained by eliminating n^2 from the first two equations by means of the third, resulting in

$$l^2 = \frac{(\sigma - \sigma_2)(\sigma - \sigma_3) + \tau^2}{(\sigma_1 - \sigma_2)(\sigma_1 - \sigma_3)} \tag{53}$$

$$m^2 = \frac{(\sigma - \sigma_3)(\sigma - \sigma_1) + \tau^2}{(\sigma_2 - \sigma_3)(\sigma_2 - \sigma_1)} \tag{54}$$

$$n^2 = \frac{(\sigma - \sigma_1)(\sigma - \sigma_2) + \tau^2}{(\sigma_3 - \sigma_1)(\sigma_3 - \sigma_2)} \tag{55}$$

Let one of the direction cosines, say n, be held constant while the other two are varied. By (55), the normal and shear streses then vary according to the equation

$$\tau^2 + \{\sigma - \tfrac{1}{2}(\sigma_1 + \sigma_2)\}^2 = \tfrac{1}{4}(\sigma_1 - \sigma_2)^2 + n^2(\sigma_1 - \sigma_3)(\sigma_2 - \sigma_3) \qquad (56)$$

In the stress plane, σ and τ therefore lie on a circle whose center is on the σ axis at a distance $\tfrac{1}{2}(\sigma_1 + \sigma_2)$ from the origin. The square of the radius of the circle is given by the right-hand side of (56). The radius varies from $\tfrac{1}{2}(\sigma_1 - \sigma_2)$ for $n = 0$ to $\tfrac{1}{2}(\sigma_1 + \sigma_2) - \sigma_3$ for $n = 1$.

In Fig. 1.13, the points A, B, C with coordinates $(\sigma_1, 0)$, $(\sigma_2, 0)$, $(\sigma_3, 0)$ are the principal points of the Mohr diagram. The centers of the segments AB, BC, and CA are denoted by the points P, Q, and R. The upper semicircle drawn on the diameter AB corresponds to $n = 0$. As n increases from 0 to 1, the radius of the semicircle varies from PB to PC. Similarly, the upper semicircles with BC and CA as diameters correspond to $l = 0$ and $m = 0$ respectively. For constant values of l, (53) defines a family of circles having the equation

$$\tau^2 + \{\sigma - \tfrac{1}{2}(\sigma_2 + \sigma_3)\}^2 = \tfrac{1}{4}(\sigma_2 - \sigma_3)^2 + l^2(\sigma_1 - \sigma_2)(\sigma_1 - \sigma_3) \qquad (57)$$

The center of these circles is at Q, while the radius varies from QB for $l = 0$ to QA for $l = 1$. Finally, considering constant values of m, we have the family of circles

$$\tau^2 + \{\sigma - \tfrac{1}{2}(\sigma_1 + \sigma_3)\}^2 = \tfrac{1}{4}(\sigma_1 - \sigma_3)^2 + m^2(\sigma_1 - \sigma_2)(\sigma_3 - \sigma_2) \qquad (58)$$

with the center at R, and the radius decreasing from RC for $m = 0$ to RB for $m = 1$. For arbitrary values of (l, m, n), the state of stress will correspond to a point in the space between the three semicircles drawn on the diameters AB, BC, and CA.

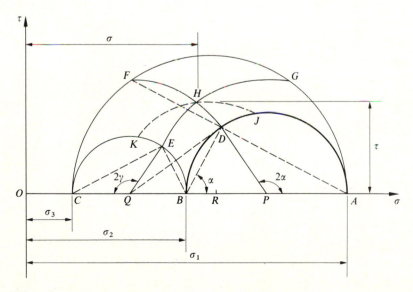

Figure 1.13 Mohr's representation of stress in three dimensions.

To find the values of σ and τ across any given plane, let $\alpha = \cos^{-1} l$ and $\gamma = \cos^{-1} n$ be the angles made by the normal to the plane with the directions of σ_1 and σ_3 respectively. Set off angles APD and CQE equal to 2α and 2γ respectively, by drawing the radii PD and QE to the appropriate semicircles. The circular arcs DHF and EHG, drawn with centers Q and P respectively, intersect one another at H giving the required stress point.[†] If the lines AD and CE are produced, they will meet the outermost semicircle at F and G respectively. Since the angle ABD is equal to α, and $BD = (\sigma_1 - \sigma_2) \cos \alpha$, the triangle BDQ furnishes

$$QD^2 = QB^2 + BD^2 + 2QB \cdot BD \cos \alpha$$

$$= \tfrac{1}{4}(\sigma_2 - \sigma_3)^2 + (\sigma_1 - \sigma_2)(\sigma_1 - \sigma_3) \cos^2 \alpha$$

Hence QD is identical to the radius of the circle (57) corresponding to the given value of l. Similarly, the radius PE is equal to that of the circle (56) corresponding to the given value of n. This completes the proof of the construction for the stress point H. It can be shown that the circular arc drawn through H with center at R cuts the semicircles on AB and BC at J and K respectively, where BJ and BK are each inclined at an angle $\beta = \cos^{-1} m$ to the vertical through B.

The semicircles with centers P, Q, R are in fact one-half of the two-dimensional Mohr circles for the planes perpendicular to the directions of $\sigma_3, \sigma_1, \sigma_2$ respectively. Considering the first semicircle, the coordinates of any point such as D are easily shown to be those given by (45) and (46) with the principal axes taken as the axes of reference. For three-dimensional stress states, there is no graphical construction for finding the principal stresses and their directions from given components of the stress. When one of the axes of reference coincides with a principal axis, the problem of finding the remaining principal stresses and their directions is essentially two-dimensional in character.

1.5 Analysis of Strain Rate

(i) Rates of deformation and rotation A body is said to be deformed or strained when changes occur in the relative positions of the particles forming the body. The instantaneous rate of straining at any point of the body is specified by the velocity field in the neighborhood of this point. Let v_i denote the components (u, v, w) of the velocity of a typical particle P whose instantaneous coordinates are denoted by x_i (Fig. 1.14). Consider a neighboring particle Q situated at an infinitesimal distance from P, the coordinates of Q being $x_i + \delta x_i$. Then the relative velocity of Q with respect to P is given by

$$\delta v_i = \frac{\partial v_i}{\partial x_j} \delta x_j$$

which is equivalent to three equations corresponding to the three components

---

† Numerical examples have been given by J. M. Alexander, *Strength of Materials*, chap. 4, Ellis Horwood Limited, Chichester, U.K. (1981).

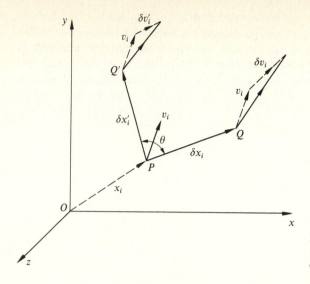

Figure 1.14 Instantaneous velocities of three neighboring particles in a deforming region.

$(\delta u, \delta v, \delta w)$ of the relative velocity. The velocity gradient tensor $\partial v_i/\partial x_j$ may be regarded as the sum of its symmetric part $\dot{\varepsilon}_{ij}$ and antisymmetric part ω_{ij}, where

$$\dot{\varepsilon}_{ij} = \frac{1}{2}\left(\frac{\partial v_i}{\partial x_j} + \frac{\partial v_j}{\partial x_i}\right) \qquad \omega_{ij} = \frac{1}{2}\left(\frac{\partial v_i}{\partial x_j} - \frac{\partial v_j}{\partial x_i}\right) \tag{59}$$

Evidently, $\dot{\varepsilon}_{ij} = \dot{\varepsilon}_{ji}$ and $\omega_{ij} = -\omega_{ji}$, indicating the properties of symmetry and antisymmetry of the respective tensors. The expression for the relative velocity therefore becomes

$$\delta v_i = \dot{\varepsilon}_{ij}\,\delta x_j + \omega_{ij}\,\delta x_j \tag{60}$$

To obtain the physical significance of the decomposed parts of the relative velocity, let $\omega_{21} = -\omega_{12} = \omega_z$, $\omega_{32} = -\omega_{23} = \omega_x$, and $\omega_{13} = -\omega_{31} = \omega_y$, the remaining components of ω_{ij} being identically zero. The second equation of (59) then gives

$$\omega_x = \frac{1}{2}\left(\frac{\partial w}{\partial y} - \frac{\partial v}{\partial z}\right) \qquad \omega_y = \frac{1}{2}\left(\frac{\partial u}{\partial z} - \frac{\partial w}{\partial x}\right) \qquad \omega_z = \frac{1}{2}\left(\frac{\partial v}{\partial x} - \frac{\partial u}{\partial y}\right) \tag{61}$$

It follows from above that the quantities ω_x, ω_y, ω_z form the components of the vector†

$$\boldsymbol{\omega} = \tfrac{1}{2}\,\mathrm{curl}\,\mathbf{v}$$

where $\mathbf{v}$ is the velocity vector of the particle P. The components of the relative velocity given by the second term on the right-hand side of (60) are $\omega_y\,\delta z - \omega_z\,\delta y$, $\omega_z\,\delta x - \omega_x\,\delta z$, $\omega_x\,\delta y - \omega_y\,\delta x$. They form the components of the vector product

---

† The direction of $\boldsymbol{\omega}$ is parallel to the direction of advancement of a right-handed screw turning in the same sense as that of the rigid body rotation.

$\boldsymbol{\omega} \times \delta s$, where δs denotes the infinitesimal vector PQ. The second part of the relative velocity therefore corresponds to an instantaneous rigid body rotation of the neighborhood of P with an angular velocity $\boldsymbol{\omega}$. The antisymmetric tensor ω_{ij} is known as the *spin tensor*. The relationship between the tensor ω_{ij} and the associated *spin vector* ω_k may be written as

$$\omega_{ij} = -e_{ijk}\omega_k \qquad 2\omega_k = -e_{kij}\omega_{ij} = e_{kij}\frac{\partial v_j}{\partial x_i} \tag{62}$$

where e_{ijk} is the *permutation symbol* whose value is $+1$ or -1 according to whether i, j, k form an even or odd permutation† of 1, 2, 3. When two of the suffixes i, j, k are equal, e_{ijk} is identically zero. It follows from the definition that

$$e_{ijk} = e_{jki} = e_{kij} = -e_{ikj} = -e_{kji} = -e_{jik}$$

If the neighborhood of P undergoes an instantaneous deformation, the first term on the right-hand side of (60) must be nonzero. The symmetric tensor $\dot{\varepsilon}_{ij}$ is therefore called the *rate of deformation* or the true strain rate at P at the instant under consideration. The rectangular components of the strain rate are

$$\dot{\varepsilon}_x = \frac{\partial u}{\partial x} \qquad\qquad \dot{\varepsilon}_y = \frac{\partial v}{\partial y} \qquad\qquad \dot{\varepsilon}_z = \frac{\partial w}{\partial z}$$

$$\dot{\gamma}_{xy} = \frac{1}{2}\left(\frac{\partial u}{\partial y} + \frac{\partial v}{\partial x}\right) \qquad \dot{\gamma}_{yz} = \frac{1}{2}\left(\frac{\partial w}{\partial y} + \frac{\partial v}{\partial z}\right) \qquad \dot{\gamma}_{zx} = \frac{1}{2}\left(\frac{\partial w}{\partial x} + \frac{\partial u}{\partial z}\right) \tag{63}$$

The first three are the normal components and the last three are the shear components of the strain rate. When the total deformation is small, the expressions on the right-hand sides of (63) give the components of the strain itself with u, v, w regarded as the components of the displacement of the particle.‡

For the mechanical interpretation of the components of the tensor $\dot{\varepsilon}_{ij}$, consider first the rate of change of the instantaneous length δs of the material line element PQ. The square of this line element is $\delta s^2 = \delta x_i\,\delta x_i$, which gives

$$\delta s(\delta s)^{\cdot} = \delta v_i\,\delta x_i = \frac{\partial v_i}{\partial x_j}\delta x_i\,\delta x_j = \frac{\partial v_j}{\partial x_i}\delta x_i\,\delta x_j$$

where the dot denotes the material derivative, specifying the rate of change following the motion of the particles. Using the expression for $\dot{\varepsilon}_{ij}$ given by (59), the above relation may be written as

$$\delta s(\delta s)^{\cdot} = \frac{1}{2}\left(\frac{\partial v_i}{\partial x_i} + \frac{\partial v_j}{\partial x_i}\right)\delta x_i\,\delta x_j = \dot{\varepsilon}_{ij}\,\delta x_i\,\delta x_j$$

---

† This means that $e_{123} = e_{231} = e_{312} = 1$ and $e_{213} = e_{132} = e_{321} = -1$, while all other components are zero. The permutation tensor has the important property $e_{mij}e_{mkl} = \delta_{ik}\delta_{jl} - \delta_{il}\delta_{jk}$, which may be obtained by eliminating ω_k between the two relations (62), and comparing the result for ω_{ij} with that given by (59).

‡ The expressions for the strain rates and the components of spin in cylindrical and spherical coordinates are given in App. B.

If the unit vector in the direction PQ is denoted by l_i, then $\delta x_i = l_i\, \delta s$. The ratio $(\delta s)^{\cdot}/\delta s$, called the *rate of extension* $\dot{\varepsilon}$ in the direction PQ, is then obtained as

$$\dot{\varepsilon} = l_i l_j \dot{\varepsilon}_{ij}$$

$$= l^2 \dot{\varepsilon}_x + m^2 \dot{\varepsilon}_y + n^2 \dot{\varepsilon}_z + 2lm\dot{\gamma}_{xy} + 2mn\dot{\gamma}_{yz} + 2nl\dot{\gamma}_{zx} \tag{64}$$

where (l, m, n) are the direction cosines of PQ. It follows that the components $\dot{\varepsilon}_x, \dot{\varepsilon}_y, \dot{\varepsilon}_z$ are the rates of extension at the particle P in the coordinate directions.

Consider, now, a second material line element PQ' emanating from P, the instantaneous coordinates of Q' being $x_i + \delta x_i'$. Then the velocity of Q' relative to that of P is

$$\delta v_i' = \frac{\partial v_i}{\partial x_j}\, \delta x_j'$$

The scalar product of the infinitesimal vectors PQ and PQ' is $\delta x_i\, \delta x_i'$, and its material rate of change is

$$(\delta x_i\, \delta x_i')^{\cdot} = \delta x_i\, \delta v_i' + \delta x_i'\, \delta v_i = \left(\frac{\partial v_i}{\partial x_j} + \frac{\partial v_j}{\partial x_i}\right) \delta x_i\, \delta x_j' = 2\dot{\varepsilon}_{ij}\, \delta x_i\, \delta x_j'$$

Let the instantaneous length of the line element PQ' be denoted by $\delta s'$, and the rate of extension in this direction by $\dot{\varepsilon}'$. If the included angle between PQ and PQ' is denoted by θ, the scalar product $\delta x_i\, \delta x_i'$ is equal to $\delta s\, \delta s'\cos\theta$. The above equation therefore becomes

$$[(\dot{\varepsilon} + \dot{\varepsilon}')\cos\theta - \dot{\theta}\sin\theta]\, \delta s\, \delta s' = 2\dot{\varepsilon}_{ij}\, \delta x_i\, \delta x_j' \tag{65}$$

If the neighborhood of P undergoes an instantaneous rigid body motion, the material triangle PQQ' retains its shape following the motion, giving $\dot{\varepsilon} = \dot{\varepsilon}' = \dot{\theta} = 0$. It follows from (65) that $\dot{\varepsilon}_{ij}$ then vanishes identically as expected. The rate at which an instantaneous right angle between a pair of material line elements decreases is twice the rate of shear, denoted by $\dot{\gamma}$. Setting $\delta x_i = l_i\, \delta s$, $\delta x_j' = l_j'\, \delta s'$, and $\theta = \pi/2$ in (65), the *rate of shear* associated with the directions l_i and l_i' is obtained as

$$\dot{\gamma} = l_i l_j' \dot{\varepsilon}_{ij} = ll' \dot{\varepsilon}_x + mm' \dot{\varepsilon}_y + nn' \dot{\varepsilon}_z + (lm' + ml')\dot{\gamma}_{xy}$$

$$+ (mn' + m'n)\dot{\gamma}_{yz} + (nl' + ln')\dot{\gamma}_{zx} \tag{66}$$

where (l', m', n') are the direction cosines of PQ'. It follows from (66) that $\dot{\gamma}_{xy}, \dot{\gamma}_{yz}$, and $\dot{\gamma}_{zx}$ are the rates of shear associated with the appropriate coordinate directions. In the engineering literature, the shear rate is taken as equal to the rate of decrease of the angle formed by an instantaneous pair of orthogonal material line elements. The engineering components of the rate of shear are therefore twice the corresponding tensor components. During a finite deformation, the engineering shear strain associated with a pair of orthogonal line elements in the unstrained state is the tangent of the angle by which the right angle decreases.

(ii) Principal strain rates The relative velocity of Q with respect to P, corresponding to pure deformation in the neighborhood of P, may be resolved into a component along PQ and a component perpendicular to PQ. These resolved components are equal to $\dot{\varepsilon}\,\delta s$ and $\dot{\gamma}\,\delta s$ respectively, as may be seen from (60), (64), and (66), the unit vector l'_i being considered in the appropriate perpendicular direction. The direction PQ represents a principal direction of the rate of deformation, if the relative velocity of pure deformation is directed along PQ. In this case $\dot{\gamma} = 0$, and $\dot{\varepsilon}_{ij}\,\delta x_j$ is equal to $\dot{\varepsilon}\,\delta x_i$, where $\delta x_i = l_i\,\delta s$. Hence

$$\dot{\varepsilon}_{ij}l_j = \dot{\varepsilon}l_i \qquad \text{or} \qquad (\dot{\varepsilon}_{ij} - \dot{\varepsilon}\,\delta_{ij})l_j = 0 \tag{67}$$

This consists of three scalar equations, analogous to (22), for the components of the unit vector l_j. Equating to zero the determinant of the coefficients formed by the expression in the parenthesis of (67), we obtain the cubic equation

$$\dot{\varepsilon}^3 - N_1\dot{\varepsilon}^2 - N_2\dot{\varepsilon} - N_3 = 0$$

whose roots are the *principal strain rates* $\dot{\varepsilon}_1, \dot{\varepsilon}_2, \dot{\varepsilon}_3$. The coefficients N_1, N_2, N_3 are the basic invariants of the strain rate tensor, their expressions in terms of the components of $\dot{\varepsilon}_{ij}$ being

$$N_1 = \dot{\varepsilon}_{ii} \qquad N_2 = \tfrac{1}{2}(\dot{\varepsilon}_{ij}\dot{\varepsilon}_{ij} - \dot{\varepsilon}_{ii}\dot{\varepsilon}_{jj}) \qquad N_3 = |\dot{\varepsilon}_{ij}| \tag{68}$$

where the last expression denotes the determinant of the matrix of the tensor $\dot{\varepsilon}_{ij}$. When the principal values of the strain rate are distinct, each principal strain rate is associated with a unique principal direction. The three principal directions are mutually orthogonal and are known as the *principal axes* of the strain rate. In analogy with (15), a deviatoric strain rate $\dot{e}_{ij}$ is defined as

$$\dot{e}_{ij} = \dot{\varepsilon}_{ij} - \dot{\varepsilon}_0\,\delta_{ij} = \dot{\varepsilon}_{ij} - \tfrac{1}{3}\dot{\varepsilon}_{kk}\,\delta_{ij} \tag{69}$$

The principal axes of $\dot{e}_{ij}$ are the same as those of $\dot{\varepsilon}_{ij}$. The principal components of the deviatoric strain rate are obtained by subtracting the mean extension rate $\dot{\varepsilon}_0$ from the corresponding principal strain rates. The principal shear rates have the values

$$\tfrac{1}{2}|\dot{\varepsilon}_1 - \dot{\varepsilon}_2| \qquad \tfrac{1}{2}|\dot{\varepsilon}_2 - \dot{\varepsilon}_3| \qquad \tfrac{1}{2}|\dot{\varepsilon}_3 - \dot{\varepsilon}_1|$$

These are the maximum values of the magnitude of $\dot{\gamma}$ at the considered particle. Each principal shear rate is associated with directions which bisect the angles between the corresponding pair of principal axes of the rate of deformation.

The first invariant N_1 is equal to the rate of change of volume per unit volume in the neighborhood of a typical particle P. This may be shown by considering a small rectangular parallelepiped at P with its edges parallel to the principal axes. If the instantaneous lengths of the edges are denoted by $\delta a, \delta b, \delta c$, the rates at which these lengths change following the motion are $\dot{\varepsilon}_1\,\delta a, \dot{\varepsilon}_2\,\delta b, \dot{\varepsilon}_3\,\delta c$ respectively. The instantaneous volume $\delta a\,\delta b\,\delta c$ of the parallelepiped therefore changes at the rate $(\dot{\varepsilon}_1 + \dot{\varepsilon}_2 + \dot{\varepsilon}_3)\,\delta a\,\delta b\,\delta c$. If the local density of the material is denoted by ρ, the mass of the parallelepiped is $\rho\,\delta a\,\delta b\,\delta c$, which remains constant following the motion. Setting the rate of change of this mass to zero, we have

$$\dot{\rho}/\rho = -(\dot{\varepsilon}_1 + \dot{\varepsilon}_2 + \dot{\varepsilon}_3) = -(\dot{\varepsilon}_x + \dot{\varepsilon}_y + \dot{\varepsilon}_z) = -\dot{\varepsilon}_{ii}$$

Expressing the rates of extension in terms of the velocity gradients, the above relation can be written as

$$\dot{\rho} + \rho \frac{\partial v_i}{\partial x_i} = \dot{\rho} + \rho \left(\frac{\partial u}{\partial x} + \frac{\partial v}{\partial y} + \frac{\partial w}{\partial z} \right) = 0 \qquad (70)$$

For an incompressible material, the density remains constant, and consequently $\dot{\varepsilon}_{ii}$ must vanish. In this case, the components of the deviatoric strain rate are identical to those of the actual strain rate.

Consider the situation where the principal axes of the strain rate remain fixed with respect to an element as it continues to deform. The axes of reference are assumed to take part in the rotation of the element so that they are parallel to the principal axes at each stage. Let x, y, z denote the coordinates of the center of the element at any instant t, measured in the directions of $\dot{\varepsilon}_1$, $\dot{\varepsilon}_2$, $\dot{\varepsilon}_3$ respectively. If the initial coordinates x_0, y_0, z_0 are taken as independent space variables, which do not change following the motion, the material rate of change of each variable is given by its partial derivative with respect to t. The first principal strain rate may therefore be written as

$$\frac{\partial \varepsilon_1}{\partial t} = \frac{\partial u / \partial x_0}{\partial x / \partial x_0} = \frac{(\partial / \partial t)(\partial x / \partial x_0)}{\partial x / \partial x_0}$$

Similar expressions may be written down for the other two principal strain rates. These equations are immediately integrated to give the total principal strains

$$\varepsilon_1 = \ln \left(\frac{\partial x}{\partial x_0} \right) \qquad \varepsilon_2 = \ln \left(\frac{\partial y}{\partial y_0} \right) \qquad \varepsilon_3 = \ln \left(\frac{\partial z}{\partial z_0} \right)$$

which are the logarithms of the ratios of the final and initial lengths of the material line elements along the principal axes. When the principal axes of the strain rate rotate with respect to the element, the principal components of the successive strain increments cannot be interpreted as increments of principal strains.[†]

(iii) *Instantaneous plane strain* The instantaneous state of strain is called plane if one of the principal strain rates vanishes at each point of the deforming body. If the z axis is taken along this principal axis, we have $\dot{\varepsilon}_z = \dot{\gamma}_{yz} = \dot{\gamma}_{zx} = 0$ for an instantaneous plane strain condition. The velocity field therefore has the form

$$u = u(x, y) \qquad v = v(x, y) \qquad w = 0$$

and the nonzero strain rates $\dot{\varepsilon}_x$, $\dot{\varepsilon}_y$, and $\dot{\gamma}_{xy}$ are all independent of z. A typical material line element PQ in the plane $z = $ const instantaneously extends and rotates in the same plane. A part of the instantaneous rotation corresponds to a local rigid body spin of the material about an axis through P with an angular velocity $\omega_z = \omega$, which is reckoned positive when the rotation is counterclockwise.

[†] For a discussion of finite homogeneous strains, in which all straight lines remain straight during straining, see J. C. Jaeger, *Elasticity, Fracture, and Flow*, chap. 1, Methuen and Company, London (1969).

The condition of compatibility of the components of strain rate is

$$\frac{\partial^2 \dot{\varepsilon}_x}{\partial y^2} + \frac{\partial^2 \dot{\varepsilon}_y}{\partial x^2} = 2 \frac{\partial^2 \dot{\gamma}_{xy}}{\partial x \, \partial y} \tag{71}$$

which is readily verified on direct substitution from (63). It is a consequence of the fact that three strain-rate components are defined by two velocity components.†

Let ϕ denote the counterclockwise orientation of a line element PQ with respect to the x axis. Then the instantaneous coordinate differences between the particles P and Q are $\delta x = \delta s \cos \phi$ and $\delta y = \delta s \sin \phi$, where δs is the length of the element PQ. Let $(\delta u^*, \delta v^*)$ denote the relative velocity of Q with respect to P corresponding to pure deformation. Then

$$\delta u^* = (\dot{\varepsilon}_x \cos \phi + \dot{\gamma}_{xy} \sin \phi) \, \delta s$$

$$\delta v^* = (\dot{\gamma}_{xy} \cos \phi + \dot{\varepsilon}_y \sin \phi) \, \delta s$$

in view of the first term on the right-hand side of (60). The resolved component of this relative velocity in the direction PQ is equal to $\dot{\varepsilon} \, \delta s$, where $\dot{\varepsilon}$ is the rate of extension along PQ. Hence

$$\dot{\varepsilon} = \frac{\delta u^* \cos \phi + \delta v^* \sin \phi}{\delta s}$$

$$= \dot{\varepsilon} \cos^2 \phi + \dot{\varepsilon}_y \sin^2 \phi + 2\dot{\gamma}_{xy} \sin \phi \cos \phi \tag{72}$$

The expression for $\dot{\varepsilon}$ is also obtained from (65) by setting $l = \cos \phi$, $m = \sin \phi$, $n = 0$. The shear strain rate $\dot{\gamma}$ associated with the directions ϕ and $\pi/2 + \phi$ is given by the resolved component of the relative velocity $(\delta u^*, \delta v^*)$ in the direction perpendicular to PQ. Thus

$$\dot{\gamma} = \frac{-\delta u^* \sin \phi + \delta v^* \cos \phi}{\delta s}$$

$$= -(\dot{\varepsilon}_x - \dot{\varepsilon}_y) \sin \phi \cos \phi + \dot{\gamma}_{xy}(\cos^2 \phi - \sin^2 \theta) \tag{73}$$

It follows from the nature of the derivation that $\dot{\gamma}$ is the counterclockwise angular velocity of PQ corresponding to pure deformation of the neighborhood of P. An element PR inclined at $\pi/2 + \phi$ to the x axis has a clockwise angular velocity equal to $\dot{\gamma}$. The right angle between PQ and PR therefore decreases at the rate $2\dot{\gamma}$, which is the engineering shear rate at P associated with these directions. It follows that the total angular velocities of PQ and PR are $\omega + \dot{\gamma}$ and $\omega - \dot{\gamma}$ respectively measured in the counterclockwise sense.

The direction ϕ corresponds to a principal direction of the strain rate if the corresponding shear rate vanishes. Since $d\dot{\varepsilon}/d\phi = 2\dot{\gamma}$ in view of (72) and (73),

---

† For a three-dimensional velocity field, there are six equations of compatibility, three of which are of type (71). They are necessary and sufficient conditions for the existence of single-valued velocities. See, for example, L. E. Malvern, *Introduction to Mechanics of a Continuous Medium*, p. 189, Prentice-Hall, Englewood Cliffs, N.J. (1969).

the longitudinal strain rate has a stationary value in the principal direction. The condition $\dot{\gamma} = 0$ gives

$$\tan 2\phi = \frac{2\dot{\gamma}_{xy}}{\dot{\varepsilon}_x - \dot{\varepsilon}_y} \tag{74}$$

which defines two mutually perpendicular directions representing the principal axes in the xy plane. The principal axes of stress and strain rate coincide if the ratios on the right-hand sides of (47) and (74) are equal to one another. The principal strain rates are expressed as

$$\dot{\varepsilon}_1, \dot{\varepsilon}_2 = \tfrac{1}{2}(\dot{\varepsilon}_x + \dot{\varepsilon}_y) \pm \tfrac{1}{2}\sqrt{(\dot{\varepsilon}_x - \dot{\varepsilon}_y)^2 + 4\dot{\gamma}_{xy}^2} \tag{75}$$

The second term on the right-hand side represents the maximum rate of shear in the plane of the instantaneous motion. Choosing the principal axes in this plane as the new axes of reference, the rate of extension $\dot{\varepsilon}$ and the total angular velocity $\dot{\phi}$ of a material line element PQ may be written from (72) and (73) as

$$
\begin{aligned}
\dot{\varepsilon} &= \tfrac{1}{2}(\dot{\varepsilon}_1 + \dot{\varepsilon}_2) + \tfrac{1}{2}(\dot{\varepsilon}_1 - \dot{\varepsilon}_2)\cos 2\phi \\
\dot{\phi} &= \quad \omega \quad - \tfrac{1}{2}(\dot{\varepsilon}_1 - \dot{\varepsilon}_2)\sin 2\phi
\end{aligned}
\tag{76}
$$

where ω is the component of spin at P. It follows that $\dot{\varepsilon}$ and $\dot{\phi}$ can be represented by a point whose locus is a circle with parametric equations (76). If $\dot{\varepsilon}$ is taken as the ordinate and $\dot{\phi}$ as the abscissa (Fig. 1.15), the coordinates of the center C of the circle are $\omega, \tfrac{1}{2}(\dot{\varepsilon}_1 + \dot{\varepsilon}_2)$, and the radius of the circle is $\tfrac{1}{2}(\dot{\varepsilon}_1 - \dot{\varepsilon}_2)$, where $\dot{\varepsilon}_1 > \dot{\varepsilon}_2$. The highest and lowest points of the circle, denoted by A and B, represent the maximum and minimum rates of extension, together with an angular velocity equal to ω.

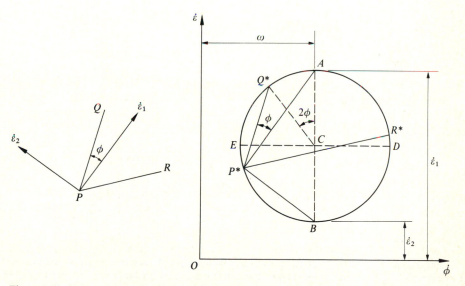

Figure 1.15 Geometrical representation of the rates of extension and rotation in plane strain.

Let a point Q^* on the circle correspond to the direction ϕ with respect to the first principal axis. Then in view of (76), the angle ACQ^* is equal to 2ϕ. The line Q^*P^* drawn parallel to the direction PQ meets the circle again at P^* which may be regarded as the pole of the circle. Since the peripheral angle AP^*Q^* is equal to ϕ, the line P^*A is parallel to the first principal direction at P. To find the point on the circle corresponding to any given direction in the plane of motion, it is only necessary to draw a line in this direction through P^* and locate its second intersection with the circle. Let P^*R^* be drawn parallel to some given direction PR through P. Then the rate of change of the material angle QPR is equal to the difference between the abscissas of the corresponding points R^* and Q^* on the circle. For the considered orientation of the line elements PQ and PR, the angle between them instantaneously decreases. The difference between the abscissas of the points D and E, which correspond to the maximum shear directions at P, is greater than that of any other pair of points on the circle. It follows, therefore, that the right angle formed by the material line elements in the maximum shear directions changes at a rate which is numerically greater than that for any other material angle in the plane of motion.†

(iv) Equilibrium and virtual work Consider a mass of material occupying a finite volume V and bounded by a surface S at a generic instant t. The material is in equilibrium under surface forces distributed over S, and body forces (such as gravitational and centrifugal forces) distributed throughout V. The body force acting on a typical volume element dV is equal to $\rho g_j\, dV$, where g_j denotes the body force per unit mass and ρ the current density. The force T_j acting on a typical surface element dS, specified by its exterior unit normal l_i, is equal to $\sigma_{ij} l_i\, dS$. The condition of force equilibrium requires the resultant of these forces to vanish, leading to

$$\int \sigma_{ij} l_i\, dS + \int \rho g_j\, dV = 0$$

where the surface integral extends over S and the volume integral over V. Using Green's theorem, the surface integral can be transformed into a volume integral, reducing the above expression to

$$\int \left(\frac{\partial \sigma_{ij}}{\partial x_i} + \rho g_j \right) dV = 0$$

The vanishing of the above integral requires that the expression in the parenthesis must vanish identically. The equilibrium condition therefore becomes

$$\frac{\partial \sigma_{ij}}{\partial x_i} + \rho g_j = 0 \tag{77}$$

---

† The above construction is due to W. Prager, *Introduction to Mechanics of Continua*, p. 69, Ginn and Company, Boston (1961). Mohr's construction for the rates of extension and shear associated with any angle ϕ is identical to that for the normal and shear stresses.

which is equivalent to three equations corresponding to the three coordinate directions. In view of the symmetry of the stress tensor, (77) also ensures that the resultant moment of the surface and body forces is identically zero.†

Equation (77) must be satisfied throughout the interior of the body. At the boundary of the body, the force T_j per unit area acting on a typical surface element must be equal to the stress vector across this element. The boundary condition may therefore be written as

$$T_j = l_i \sigma_{ij} \tag{78}$$

where l_i is the unit vector along the exterior normal to the surface at the considered point. In general, (77) and (78) must be supplemented by other equations to determine the stress components uniquely.

Consider, now, a continuous velocity field v_j, which is chosen independently of an equilibrium distribution of stress σ_{ij}. The rate of work done by the distribution of surface traction T_j (in the absence of body forces) is

$$\int T_j v_j \, dS = \int l_i \sigma_{ij} v_j \, dS = \int \frac{\partial}{\partial x_i} (\sigma_{ij} v_j) \, dV = \int \sigma_{ij} \frac{\partial v_j}{\partial x_i} \, dV$$

in view of (77). The transformation of the surface integral into the volume integral follows from Gauss' divergence (or Green's theorem) applied to the vector $\sigma_{ij} v_j$. Since

$$\sigma_{ij} \frac{\partial v_j}{\partial x_i} = \sigma_{ji} \frac{\partial v_j}{\partial x_i} = \sigma_{ij} \frac{\partial v_i}{\partial x_j} = \frac{1}{2} \sigma_{ij} \left(\frac{\partial v_i}{\partial x_j} + \frac{\partial v_j}{\partial x_i} \right) = \sigma_{ij} \dot{\varepsilon}_{ij}$$

by the symmetry of the stress tensor and the interchangeability of dummy suffixes, we obtain the principle of virtual work in the form

$$\int T_j v_j \, dS = \int \sigma_{ij} \dot{\varepsilon}_{ij} \, dV \tag{79}$$

where $\dot{\varepsilon}_{ij}$ is the rate of deformation associated with the velocity v_j. Equation (79) states that the rate of work done by the external forces on any virtual velocity field is equal to the rate of dissipation of internal energy. If the velocity field is discontinuous, the energy dissipated due to shearing across the discontinuities must be included on the right-hand side of (79).

1.6 Concepts of Stress Rate

(i) *Objective stress rates* The rate of deformation of a solid, for a given state of stress, is generally a function of the instantaneous rate of change of the stress. The stress rate tensor used in this relation, known as the constitutive relation, must be defined in such a way that it vanishes in the event of an instantaneous rigid body

---

† See, for example, W. Prager, *Introduction to Mechanics of Continua*, p. 46, Ginn and Company, Boston (1961).

rotation. Such a stress rate is called an *objective stress rate*. From the physical standpoint, it is natural to consider the rate of change of the stress referred to a set of axes that participates in the instantaneous rotation of a typical element. Although the stress components with respect to a fixed coordinate system are changed by the rotation of the element, the components with respect to the rotating system remain unaffected.

Consider two sets of rectangular axes x_i and x_i', which have a common origin O, and which are coincident at an instant t. During a small interval of time dt, the first set of axes is assumed to remain fixed, while the second set of axes takes part in the rigid-body rotation of the given element. An infinitesimal vector PQ drawn in the element from its center P is denoted by δx_j and $\delta x_j'$ with respect to the two coordinate systems at the instant $t + dt$. The difference $\delta x_j - \delta x_j'$ is equal to $\delta v_j \, dt$, where δv_j denotes the relative velocity of rotation of Q with respect to P at the instant t. Recalling that $\delta v_j = \omega_{ji} \, \delta x_i'$, where ω_{ij} denotes the rate of rotation of the neighborhood of P, we have

$$\delta x_j = \delta x_j' + (\omega_{ji} \, dt) \, \delta x_i' = (\delta_{ij} - \omega_{ij} \, dt) \, \delta x_i' \tag{80}$$

The angle which the x_j axis makes with the x_i' axis instantaneously changes at the rate ω_{ij} when $i \neq j$. Let σ_{ij} denote the true (or Cauchy) stress at the particle P when $dt = 0$. The material rates of change of the stress referred to the fixed and the rotating axes are denoted by $\dot\sigma_{ij}$ and $\mathring\sigma_{ij}$ respectively. At the time $t + dt$, the primed and the unprimed stress components become $\sigma_{ij} + \mathring\sigma_{ij} \, dt$ and $\sigma_{ij} + \dot\sigma_{ij} \, dt$ respectively. In view of (20), expressed in the infinitesimal form, a_{ij} is given by the expression in the parenthesis of (80). The transformed stress tensor may therefore be written from (21) as

$$\sigma_{ij} + \mathring\sigma_{ij} \, dt = (\delta_{ik} - \omega_{ik} \, dt)(\delta_{jl} - \omega_{jl} \, dt)(\sigma_{kl} + \dot\sigma_{kl} \, dt)$$

Neglecting the terms containing squares and cubes of dt, and using the symmetry of the stress tensor, the relationship between $\dot\sigma_{ij}$ and $\mathring\sigma_{ij}$, which is due to Jaumann,[†] is obtained as

$$\mathring\sigma_{ij} = \dot\sigma_{ij} - \sigma_{ik}\omega_{jk} - \sigma_{jk}\omega_{ik} \tag{81}$$

The quantity $\mathring\sigma_{ij}$ may be regarded as the rigid body derivative of the true stress at the instant under consideration. It can be significantly different from the material derivative $\dot\sigma_{ij}$ whenever the rate of rotation is important.

Consider now the scalar triple product $p_{ij}\sigma_{jk}\omega_{ik}$ where p_{ij} is an arbitrary tensor having the same principal axes as those of σ_{ij}. Since this expression is an invariant, it is convenient to take the common principal axes as the axes of reference. The tensor $p_{ij}\sigma_{jk}$ then corresponds to a diagonal matrix, while the diagonal components of ω_{ik} are always zero. Consequently, the scalar product of these two tensors is identically

† G. Jaumann, *Sitz. Akad. Wiss. Wien* **120**: 385 (1911). The result has been rederived by H. Fromm, *Ing.-Arch.*, **4**: 452 (1933); S. Zaremba, *Mem. Sci. Math.*, No. 82, Paris (1937); W. Noll, *J. Rat. Mech. Anal.*, **4**: 3 (1955); R. Hill, *J. Mech. Phys. Solids*, 7: 209 (1959). See also W. Prager, *Q. Appl. Math.*, **18**: 403 (1961).

zero. It is similarly shown that the triple product $p_{ij}\sigma_{ik}\omega_{jk}$ also vanishes. It follows, therefore, from (81) that

$$p_{ij}\overset{\circ}{\sigma}_{ij} = p_{ij}\dot{\sigma}_{ij}$$

Thus $\overset{\circ}{\sigma}_{ij}$ and $\dot{\sigma}_{ij}$ have the same scalar product with any second-order tensor whose principal axes coincide with those of σ_{ij}. This property has an important consequence in the theory of plasticity.

Various other definitions of the stress rate, vanishing for an instantaneous rigid-body rotation, have been proposed in the literature. An objective stress rate sometimes used in the literature to replace $\overset{\circ}{\sigma}_{ij}$ is the material rate of change of the modified stress tensor

$$\tau_{ij} = \frac{\partial a_i}{\partial x_k}\frac{\partial a_j}{\partial x_l}\sigma_{kl}$$

where a_i are the initial coordinates of the particle which is currently at x_i. The material derivative of τ_{ij}, when the initial state coincides with that at the generic instant t, is easily shown to be†

$$\dot{\tau}_{ij} = \dot{\sigma}_{ij} - \sigma_{ik}\frac{\partial v_j}{\partial x_k} - \sigma_{jk}\frac{\partial v_i}{\partial x_k} \tag{82}$$

It follows from (81) and (82) that $\dot{\tau}_{ij}$ differs from $\overset{\circ}{\sigma}_{ij}$ by the quantity $\sigma_{ik}\dot{\varepsilon}_{jk} + \sigma_{jk}\dot{\varepsilon}_{ik}$, which is appreciable when the rate of deformation becomes significant. It is important to note that $\dot{\tau}_{ij}$ and $\dot{\sigma}_{ij}$ do not have the same scalar product with any tensor p_{ij} which is coaxial with σ_{ij}.

(ii) Nominal stress rate Through a typical particle P in the deforming material, consider a small surface element represented by the vector δS_i at any instant t. The magnitude of this vector is the current area δS, and the direction of this vector is that of the normal to the surface in the current state. The coordinates of P are denoted by x_i in the instantaneous state, and by a_i in some initial reference state with respect to a fixed set of rectangular axes. The initial area of the surface element is δS_0, the corresponding vector being denoted by δS_i°. Consider now a material line element PQ emanating from the particle P. If the instantaneous components of the vector PQ are denoted by δx_i, the corresponding components in the initial state are given by

$$\delta a_i = \frac{\partial a_i}{\partial x_j}\delta x_j$$

The volume of the material cylinder, specified by the axial vector PQ and having the given surface element as its base, changes from $\delta S_i^\circ\,\delta a_i$ in the initial state to

---

† The derivation of (82) is very similar to that of (87). The tensor $(\rho^\circ/\rho)\tau_{ij}$, where ρ° and ρ are the initial and final densities of the material, is called the Kirchhoff stress. The material rate of change of the Kirchhoff stress at the initial state is $\dot{\tau}_{ij} + \dot{\varepsilon}_{kk}\sigma_{ij}$.

$\delta S_i \, \delta x_i$ in the current state. The conservation of mass requires

$$\rho \, \delta S_j \, \delta x_j = \rho_0 \, \delta S_i^{\circ} \, \delta a_i$$

where ρ_0 and ρ are the initial and current densities of the material at the particle P. Substituting for δa_i, we have

$$\left(\frac{\rho}{\rho_0} \delta S_j - \frac{\partial a_i}{\partial x_j} \delta S_i^{\circ} \right) \delta x_j = 0$$

Since this equation must be satisfied for any arbitrary vector δx_j, the expression in the parenthesis must vanish. Hence

$$\frac{\rho}{\rho_0} \delta S_j = \frac{\partial a_i}{\partial x_j} \delta S_i^{\circ} \tag{83}$$

The infinitesimal force δP_j transmitted in the current state may be referred to the surface element in the initial state through a nominal stress tensor t_{ij}. The true stress tensor σ_{ij}, on the other hand, is associated with the surface element in the current state to give the same infinitesimal force. Expressed mathematically,

$$\delta P_j = t_{ij} \, \delta S_i^{\circ} = \sigma_{kj} \, \delta S_k \tag{84}$$

Thus $t_{ij} \, \delta S_0$ is the jth component of the force currently acting on a surface element which was initially perpendicular to the ith axis. Substitution for δS_k from (83) leads to the relationship between t_{ij} and σ_{ij} as

$$t_{ij} = \frac{\rho_0}{\rho} \frac{\partial a_i}{\partial x_k} \sigma_{kj} \tag{85}$$

This relation shows that the nominal stress tensor t_{ij} is not symmetric. Nevertheless, it is convenient to introduce this tensor for treating the problems of uniqueness and stability. The material derivative of t_{ij} is obtained by applying the operator.

$$\frac{d}{dt} = \frac{\partial}{\partial t} + v_m \frac{\partial}{x_m}$$

where v_m is the instantaneous velocity of the considered particle. The first term on the right-hand side represents the local part and the second term the convective part of the derivative. Since the initial coordinates do not change following the particle, $da_i/dt = 0$, in view of which the material rate of change of the tensor $\partial a_i/\partial x_k$ is obtained as

$$\frac{d}{dt} \left(\frac{\partial a_i}{\partial x_k} \right) = \frac{\partial}{\partial x_k} \left(\frac{\partial a_i}{\partial t} \right) + v_m \frac{\partial}{\partial x_k} \left(\frac{\partial a_i}{\partial x_m} \right) = - \frac{\partial v_m}{\partial x_k} \frac{\partial a_i}{\partial x_m} \tag{86}$$

Considering the material derivative of (85), and using (70) and (86), it is easily shown that

$$\frac{dt_{ij}}{dt} = \frac{\rho_0}{\rho} \left(\frac{d\sigma_{mj}}{dt} + \sigma_{mj} \frac{\partial v_k}{\partial x_k} - \sigma_{kj} \frac{\partial v_m}{\partial x_k} \right) \frac{\partial a_i}{\partial x_m}$$

where a dummy suffix has been replaced by another. If the initial state is now assumed to coincide with the instantaneous state, $\rho_0 = \rho$, $a_i = x_i$, and consequently $\partial a_i/\partial x_m = \delta_{im}$. Denoting the instantaneous rate of change by a dot as usual, we finally obtain†

$$\dot{t}_{ij} = \dot{\sigma}_{ij} + \sigma_{ij}\frac{\partial v_k}{\partial x_k} - \sigma_{jk}\frac{\partial v_i}{\partial x_k} \tag{87}$$

which relates the *nominal stress rate* $\dot{t}_{ij}$ to the true stress rate $\dot{\sigma}_{ij}$ with respect to a fixed set of rectangular axes. It follows from (84) that when $\dot{t}_{ij}$ vanishes, the force transmitted across the surface element instantaneously remains constant, despite the deformation and the rotation of the element.

(iii) *Equilibrium equations and boundary conditions* Let V_0 be the initial volume and S_0 the initial surface of the material which instantaneously fills the volume V with surface S. Denote by l_i° the unit vector along the exterior normal to an initial surface element of area dS_0. The forces currently acting on typical surface and volume elements of the material may be expressed as $t_{ij}l_i^\circ\, dS_0$ and $\rho_0 g_j\, dV_0$ respectively. Equating the resultant of these forces over the entire body to zero, we get

$$\int t_{ij}l_i^\circ\, dS_0 + \int \rho_0 g_j\, dV_0 = 0$$

where the integrands are considered as functions of the initial coordinates a_i and the time t. The transformation by Green's theorem furnishes the result

$$\int \left(\frac{\partial t_{ij}}{\partial a_i} + \rho_0 g_j\right) dV_0 = 0$$

Since this equation holds for any arbitrary region V_0, the integrand must vanish. The equation of equilibrium in terms of the nominal stress therefore becomes

$$\frac{\partial t_{ij}}{\partial a_i} + \rho_0 g_j = 0$$

The material derivative of this equation is simply the partial derivative with respect to t, since the initial coordinates are taken as the space variables. Denoting the material rate of change of the nominal stress by $\dot{t}_{ij}$ when the initial state is assumed as that at the instant t, we obtain

$$\frac{\partial \dot{t}_{ij}}{\partial x_i} + \rho \dot{g}_j = 0 \tag{88}$$

This is the rate equation of equilibrium expressed in its simplest form. Inserting from (87), the equation may be written down in terms of the true stress rate $\dot{\sigma}_{ij}$.

† The expression (87) is due to R. Hill, *J. Mech. Phys. Solids*, **6**: 236 (1958).

When body forces are neglected (as is usual with gravitational forces), the rate equation becomes

$$\frac{\partial \dot{\sigma}_{ij}}{\partial x_i} - \frac{\partial v_i}{\partial x_k}\frac{\partial \sigma_{jk}}{\partial x_i} = 0 \tag{89}$$

This expression is also obtained if we apply the operator d/dt on equation (77). The second term on the left-hand side of (89) represents the effect of the instantaneous motion of the element.

The components of the stress rate must be in equilibrium with the instantaneous rate of change of boundary tractions. Since the future position of a typical surface element is not known in advance, when positional changes are taken into account, it is convenient to express the boundary condition in terms of the traction rate based on the initial configuration. If δP_j denotes the current load vector acting on a surface element of initial area δS_0, then the ratio $\delta P_j / \delta S_0$ as δS_0 tends to zero is the nominal traction F_j. It follows from (84) that

$$F_j = l_i^{\circ} t_{ij}$$

If the material rate of change of the nominal traction is denoted by $\dot{F}_j$ when the initial state is taken as that at the instant considered ($l_i^{\circ} = l_i$), then

$$\dot{F}_j = l_i \dot{t}_{ij} \tag{90}$$

A different situation arises when a part of the boundary surface is subjected to a uniform normal pressure p through an inviscid fluid. In this case, the infinitesimal load vector on the surface element is

$$\delta P_j = -p\,\delta S_j = -p\left(\frac{\rho_0}{\rho}\frac{\partial a_i}{\partial x_j}\right) l_i^{\circ}\,\delta S_0$$

in view of (83). The load per unit initial area of the element therefore becomes

$$F_j = -p\left(\frac{\rho_0}{\rho}\frac{\partial a_i}{\partial x_j}\right) l_i^{\circ} \tag{91}$$

Taking the material derivative of the nominal traction (91), and using the relations (70) and (86), we obtain

$$\frac{dF_j}{dt} = \frac{\rho_0}{\rho}\left\{ -\left(\dot{p} + p\frac{\partial v_k}{\partial x_k}\right)\frac{\partial a_i}{\partial x_j} + p\frac{\partial v_k}{\partial x_j}\frac{\partial a_i}{\partial x_k}\right\} l_i^{\circ}$$

since the unit vector l_i° does not change during the motion of the surface. If the initial state is regarded as identical to the instantaneous state, $\rho_0 = \rho$, $a_i = x_i$, $l_i^{\circ} = l_i$, and the nominal traction rate becomes†

$$\dot{F}_j = -\dot{p}l_j + p\left(l_k\frac{\partial v_k}{\partial x_j} - l_j\frac{\partial v_k}{\partial x_k}\right) \tag{92}$$

† J. Chakrabarty, *Z. angew. Math. Phys.*, **24**: 270 (1973). A more general type of loading has been examined by R. Hill, *J. Mech. Phys. Solids*, **10**: 185 (1962).

It follows that even when the pressure remains constant, the nominal traction changes as a result of the instantaneous distortion of the unconstrained surface. The equilibrium equation and the boundary condition, expressed in the rate form, must be supplemented by the constitutive equation for the particular solid in formulating the boundary value problem of the incremental type.

Problems

1.1 In a certain annealed material, the yield point is taken as that for which the permanent strain is one-quarter of the recoverable elastic strain. The true stress-strain curve for the material in the plastic range may be represented by the empirical equation

$$\sigma = \frac{E}{180} \varepsilon^{0.25}$$

where E is Young's modulus. Determine the stress Y at the yield point as a fraction of E, and compute the true and nominal values of the uniaxial instability stress in terms of Y.

Answer: $Y = E/943$, $\sigma = 3.71Y$, $s = 2.89Y$.

1.2 The true stress/engineering strain curve of a material in simple tension may be represented by the equation $\sigma = Ce^n$, where C and n are empirical constants. Show that the value of e at the onset of necking in uniaxial tension is $n/(1 - n)$. Suppose that a bar of material is axially compressed to a strain of $e < n$, and is subsequently extended to the point of necking. Assuming no buckling, and neglecting Bauschinger effect, show that the ratio of the final and initial lengths of the bar is $(1 - e)^2/(1 - n)$.

1.3 Prove that according to the Voce equation for the stress-strain curve, the true stress and the natural strain at the onset of instability in uniaxial tension are

$$\sigma = \frac{Cn}{1 + n} \qquad \varepsilon = \frac{\ln[m(1 + n)]}{n}$$

What would be the instability strain when $m(1 + n)$ is less than unity? Show that the stress-strain curve can be linearized by introducing a new strain measure ε^* defined as

$$\varepsilon^* = \frac{1}{n}\left\{1 - \left(\frac{l_0}{l}\right)^n\right\}$$

where l_0 and l are the initial and final lengths of a tension bar.

1.4 In the simple compression of a short cylinder, the curve representing the variation of the load with the amount of compression shows a point of inflection. If the true stress-strain curve of the material is expressed by the empirical equation $\sigma = C\varepsilon^n$, show that the natural strain corresponding to the point of inflection is

$$\varepsilon = \tfrac{1}{4}[\sqrt{n(8 + n)} - 3n]$$

For what range of values of n will this strain exceed the instability strain in simple tension?

Answer: $0 < n < 1/6$.

1.5 In the homogeneous compression of a cylindrical specimen, the curve for the nominal stress against the natural strain has a point of inflection. Show that the corresponding point on the true stress-strain curve is given by

$$\left(\frac{d}{d\varepsilon} + 1\right)^2 \sigma = 0$$

Assuming the empirical equation $\sigma = C\varepsilon^n$, show that the true strain is $\sqrt{n} - n$ at the point of inflection. For what values of n will this strain exceed the uniaxial instability strain?

Answer: $0 < n < 0.25$.

1.6 The effect of elastic deformation of the material on the instability strain may be estimated by considering the stress-strain equation in the Ramberg-Osgood form

$$\varepsilon = \frac{\sigma}{E} + \frac{3\sigma_0}{7E}\left(\frac{\sigma}{\sigma_0}\right)^{1/n}$$

where σ_0 is the nominal yield stress and n is the strain-hardening exponent. Show that the true strain at the onset of necking in simple tension becomes

$$\varepsilon \simeq n + \left(\frac{7n}{3}\right)^n \left(\frac{\sigma_0}{E}\right)^{1/n}$$

to a close approximation. Assuming $n = 0.05$ and $\sigma_0/E = 0.002$, compute the percentage error involved in using the simple power law $\sigma = C\varepsilon^n$.

Answer: 4.76%.

1.7 The plane structure shown in Fig. A consists of three bars pin-jointed at their ends. The central bar OB is made of a material whose stress-strain curve is represented by $\sigma = C_1\varepsilon^{n_1}$. The inclined bars OA and OC are made of a different material, having its stress-strain law expressed by $\sigma = C_2\varepsilon^{n_2}$, where $n_2 < n_1$. If the initial angle of inclinations ψ is such that plastic instability occurs simultaneously in the three bars on the application of a vertical load at O, show that

$$\cos\psi = \sqrt{\frac{\exp(2n_2) - 1}{\exp(2n_1) - 1}}$$

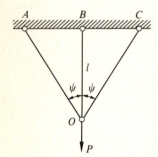

Figure A

1.8 Suppose that the bars of Fig. A have the same cross-sectional area A, and are made of a material that strain-hardens according to the law $\sigma/Y = (E\varepsilon/Y)^n$. Show that the relationship between the applied load P and the deflection δ of point O, for sufficiently small strains in the fully plastic range, is given by

$$\frac{P}{AY} = (1 + 2\cos^{2n+1}\psi)\left(\frac{E\delta}{Yl}\right)^n \qquad \frac{E\delta}{Yl} \geqslant \sec^2\psi$$

How is this equation modified when OB is plastic while OA and OC are still elastic? Obtain a graphical plot of P/AY against $E\delta/Yl$ over the range $0 < E\delta/Yl < 5$, assuming $n = 0.25$ and $\psi = 45°$.

1.9 The stress-strain curve of a rigid/plastic metal can be accurately fitted (except for very small strains) by the Ludwik equation $\sigma = C\varepsilon^n$. It is required to approximate this curve by the straight line $\sigma = Y + H\varepsilon$, giving the same plastic work over a total strain of ε_0 (Fig. B). If the difference between the stresses predicted by the two equations at $\varepsilon = \frac{1}{2}\varepsilon_0$ is exactly one-half of that at $\varepsilon = \varepsilon_0$, show that

$$\frac{Y}{\sigma_0} = \frac{3 - n}{1 + n} - 2^{1-n} \qquad \frac{H\varepsilon_0}{2\sigma_0} = 2^{1-n} - \frac{2 - n}{1 + n}$$

where $\sigma_0 = C\varepsilon_0^n$. Assuming $n = 0.3$, estimate the maximum percentage error in the linear approximation where the straight line falls below the curve.

Answer: 7.8%.

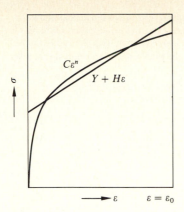

Figure with curves labeled $C\varepsilon^n$, $Y + H\varepsilon$, axes σ and ε, with $\varepsilon = \varepsilon_0$.

Figure B   **Figure C**

1.10 Derive an expression for the hoop stress that exists in a thin circular ring of mean radius r, thickness t, and density ρ, rotating about its own axis with an angular velocity ω. If the deformation is continued in the plastic range, tensile instability would occur when the angular velocity attains a maximum. Representing the true stress-strain curve by the empirical equation $\sigma = C\varepsilon^n$, show that the instability or bursting speed is given by

$$\rho\omega^2 r_0^2 = C\left(\frac{n}{2}\right)^n \exp(-n)$$

where r_0 is the mean radius of the undeformed ring.

1.11 In a thin-walled spherical shell under a uniform internal pressure p, the state of stress is a balanced biaxial tension σ, which is related to the compressive thickness strain ε by the uniaxial stress-strain law for the material. If plastic instability occurs in the shell when the internal pressure attains a maximum, show that $d\sigma/d\varepsilon = \frac{3}{2}\sigma$ at the onset of instability. Assuming the empirical stress-strain equation $\sigma = C\varepsilon^n$, obtain the dimensionless bursting pressure

$$\frac{p}{C} = \frac{2t_0}{r_0}\left(\frac{2}{3}n\right)^n \exp(-n)$$

where t_0 and r_0 are the initial wall thickness and mean radius respectively.

1.12 A compound bar is made up of a solid cylinder which just fits into a hollow one, the two cylinders being firmly bonded at their common interface. The true stress-strain curve is given by $\sigma = C_1\varepsilon^{n_1}$ for the inner cylinder and by $\sigma = C_2\varepsilon^{n_2}$ for the outer cylinder. If the two cylinders carry equal loads at the onset of instability, when the compound bar is subjected to longitudinal tension, show that the ratio of the cross-sectional areas of the outer and inner cylinders is

$$\frac{A_2}{A_1} = \frac{C_1}{C_2}\left(\frac{n_1 + n_2}{2}\right)^{n_1 - n_2}$$

1.13 Fig. C illustrates the perforation of a uniform plate of thickness t_0 by a smooth cylindrical drift of radius a having a conical end. Each element of the raised lip may be assumed to form under a uniaxial tensile hoop stress of varying intensity. Show that the height of the lip is $h = \frac{2}{3}a$, and that its thickness varies as the cube root of the distance from the outer edge. If the material strain-hardens according to the law $\sigma = C\varepsilon^n$, show that the plastic work done during the process is

$$W = \pi t_0 a^2 C \frac{\Gamma(1+n)}{2^{1+n}}$$

where $\Gamma(x)$ is the gamma function of any positive variable x. Find the numerical value of $W/t_0 a^2 C$ when $n = 0.5$.

    Answer: 0.984.

1.14 A plate of uniform thickness t_0 is perforated by a smooth conical drift of semiangle α as shown in Fig. D. The axis of the drift moves perpendicular to the plane of the plate and develops a conical lip of base radius a. Assuming a uniaxial state of stress to exist in each element, show that the radius of the outer cross section of the lip is $b = a(1 - \sin \alpha)^{2/3}$. Show also that the thickness t of an element that was situated at a radius r_0 in the undeformed state is given by

$$\frac{t}{t_0} = \left(\frac{r_0}{b}\right)^{1/2} \left\{1 + \left(\frac{r_0}{b}\right)^{3/2} \sin \alpha\right\}^{-1/3}$$

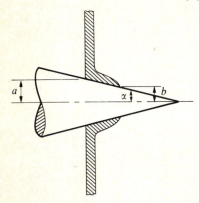

Figure D

1.15 Suppose that the structure shown in Fig. A is loaded in the fully plastic range to produce a small vertical deflection δ of the joint O. The bars are identical in material and cross section, and the strain-hardening of each bar is given by $\sigma/Y = (E\varepsilon/Y)^n$. If the residual stresses left in the vertical and the inclined bars are σ'_1 and σ'_2 respectively on complete unloading from the plastic state, show that

$$\frac{\sigma'_2}{Y} = -\frac{\sigma'_1}{2Y} \sec \psi = \frac{\cos^{2n} \psi - \cos^2 \psi}{1 + 2 \cos^3 \psi} \left(\frac{E\delta}{Yl}\right)^n$$

Assuming δ to be three times that at the initial yielding, calculate σ'_1 and the residual deflection δ' when $n = 0.25$ *and* $\psi = 45°$.

Answer: $\sigma'_1/Y = -0.372$, $E\delta'/Yl = 1.581$.

1.16 Two uniform vertical wires AB and CD, shown in Fig. E, support a load W acting at the free end of an initially horizontal rigid bar hinged at O. The lower ends of the wires are attached to blocks which can slide along a frictionless groove in the rigid bar. The strain-hardening exponents for the wires AB and CD are n and $2n$ respectively. If plastic instability occurs simultaneously in them when the load is increased to a critical value, show that

$$\frac{b}{a} = e^n + \frac{(e^n - 1)(e^{2n} - 1)}{1 - e^n\sqrt{2 - e^{2n}}}$$

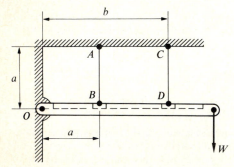

Figure E

1.17 Let $\sigma_1 > \sigma_2 > \sigma_3$ be the principal stresses at any point P in a stressed body, and consider a straight line through P having direction cosines

$$\sqrt{\frac{\sigma_1 - \sigma_2}{\sigma_1 - \sigma_3}}, \quad 0, \quad \sqrt{\frac{\sigma_2 - \sigma_3}{\sigma_1 - \sigma_3}}$$

with respect to the principal axes. Show that the resultant shear stress at P across any plane containing the given straight line is in the direction of this line.

1.18 At a typical point O in a stressed body, the normal stress across a certain plane is equal to the intermediate principal stress, while the shear stress is the geometric mean of the principal shear stresses other than the absolute maximum. Assuming $\sigma_1 > \sigma_2 > \sigma_3$, show that the direction cosines of the normal to the plane with respect to the principal axes are

$$\frac{1}{2}\sqrt{\frac{\sigma_2 - \sigma_3}{\sigma_1 - \sigma_3}}, \quad \frac{\sqrt{3}}{2}, \quad \frac{1}{2}\sqrt{\frac{\sigma_1 - \sigma_2}{\sigma_1 - \sigma_3}}$$

Find the direction of the shear stress across the plane, and show that it it coincides with that of the greatest shear stress at O when $\sigma_1 + \sigma_3 = 2\sigma_2$.

1.19 Referring to the oblique triangle of Fig. 1.10a, where CF is drawn along the shear stress vector to meet the side AB at F, show that

$$\frac{AF}{AB} = \left(\frac{\sigma_2 - \sigma}{\sigma - \sigma_3}\right)\frac{m^2}{n^2}$$

Show also that AB is divided internally or externally at F according as the ratio $(\sigma_2 - \sigma_3)/(\sigma_1 - \sigma_2)$ is greater or less than l^2/n^2.

1.20 A typical point O in a stressed body is taken as the origin of coordinates with rectangular axes in the directions of the principal stresses σ_1, σ_2, and σ_3. If P is any point on the surface of the quadric

$$\sigma_1 x^2 + \sigma_2 y^2 + \sigma_3 z^2 = \pm c^2$$

where c is a constant, show that the normal stress at O acting on the plane perpendicular to OP has the magnitude c^2/r^2, where $OP = r$. Show also that the resultant stress across this plane is directed along the normal to the quadric surface at P, and is of magnitude c^2/hr, where h is the perpendicular distance of O from the tangent plane through P.

1.21 The rectangular components of the stess tensor at a certain point are found to be proportional to the elements of the square matrix

$$\begin{bmatrix} 2 & 3 & 2 \\ 3 & 2 & 1 \\ 2 & 1 & c \end{bmatrix}$$

Find the value of c for which there will be a traction-free plane passing through the given point. Compute the direction cosines of the normal to the traction-free plane.

Answer: $c = 0.4$, $l = 0.154$, $m = -0.617$, $n = 0.772$.

1.22 If OP represents the resultant stress vector across a plane passing through O, show that P will lie on the surface of an ellipsoid, known as stress ellipsoid, whose principal axes coincide with those of the stress at O. Prove that the given plane is parallel to the tangent plane of the stress director surface

$$\frac{x^2}{\sigma_1} + \frac{y^2}{\sigma_2} + \frac{z^2}{\sigma_3} = \text{const}$$

at the point where it is intersected by OP, the coordinate axes being taken through O along the principal stress axes.

1.23 The resultant stress at a given point O across an oblique plane is 135 MPa, acting in the direction $(1/3, 2/3, -1/3)$ with respect to a set of rectangular axes. If the normal to the plane is inclined at $45°$ to the x axis, and makes equal acute angles with the y and z axes, find the normal and shear components of the stress. Assuming the state of stress at O to correspond to $\sigma_x = \sigma_y$, $\tau_{xy} = \tau_{yz}$, and $\tau_{zx} = 0$, determine the nonzero components of the stress tensor.

Answer: In units of MPa, $\sigma = 86.13$, $\tau = 103.95$, $\sigma_x = \sigma_y = 105.44$, $\sigma_z = -120.88$, $\tau_{xy} = 30.88$, $\tau_{yz} = 30.88$.

1.24 The state of stress at a certain point in a material body is defined by the following reactangular components:

$$\sigma_x = 64 \text{ MPa} \qquad \sigma_y = -76 \text{ MPa} \qquad \sigma_z = 48 \text{ MPa}$$

$$\tau_{xy} = 30 \text{ MPa} \qquad \tau_{yz} = -25 \text{ MPa} \qquad \tau_{zx} = 55 \text{ MPa}$$

Determine the normal and shear stresses acting on a plane whose normal in inclined at 40 and 70° to the x and y axes respectively. Find also the direction cosines of the shear stress, assuming an acute angle between the normal and the z axis.

Answer: $\sigma = 95.12$ MPa, $\tau = 52.42$ MPa, $l_s = 0.312$, $m_s = -0.937$, $n_s = 0.152$.

1.25 In a prismatic beam subjected to combined bending and twisting, the components of the stress tensor at a given point are

$$\sigma_x = 72.5 \text{ MPa} \qquad \sigma_y = -12.8 \text{ MPa} \qquad \sigma_z = 0$$

$$\tau_{xy} = 62.3 \text{ MPa} \qquad \tau_{yz} = 0 \qquad \tau_{zx} = -45.4 \text{ MPa}$$

where the x axis is along the centroidal axis of the beam. Find the values of the principal stresses, the greatest shear stress, and the direction cosines of the largest principal stress.

Answer: $\sigma_1 = 119.2$ MPa, $\sigma_2 = -3.4$ MPa, $\sigma_3 = 55.5$ MPa, $l_1 = 0.855$, $m_1 = 0.404$, $n_1 = -0.326$.

1.26 A strain rosette, consisting of three strain gauges OP, OQ, and OR (Fig. F), is constructed to measure simultaneously three extensional small strains ε_p, ε_Q, and ε_R along the surface of a strained body. Using the transformation formula for ε, show that the directions of the principal surface strains make angles α and $\pi/2 + \alpha$ with OQ in the counterclockwise sense, where

$$\tan 2\alpha = \frac{(\varepsilon_p - \varepsilon_R) \tan \psi}{\varepsilon_p + \varepsilon_R - 2\varepsilon_Q}$$

In the special case of an equiangular rosette ($\psi = \pi/3$), show that the principal values of the surface strain are

$$\tfrac{1}{3}(\varepsilon_P + \varepsilon_Q + \varepsilon_R) \pm \tfrac{1}{3}\sqrt{3(\varepsilon_p - \varepsilon_R)^2 + (\varepsilon_p + \varepsilon_R - 2\varepsilon_Q)^2}$$

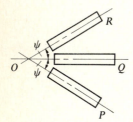

Figure F

1.27 A simple shear is a state of plane strain in which the final coordinates (x, y) of a typical particle are related to the initial coordinates (x_0, y_0) by the transformation

$$x = x_0 + y_0 \tan \phi \qquad y = y_0$$

where $\tan \phi$ is the amount of shear. Show that the straight lines which suffer the maximum extension and contraction are inclined to the x axis at angles $\pm \pi/4 - \alpha/2$ in the strained state, where

$$\alpha = \tan^{-1} (\tfrac{1}{2} \tan \phi)$$

Show also that the logarithms of the length ratios associated with maximum extension and contraction are $\pm \sinh^{-1}(\tan \alpha)$.

1.28 A state of uniform plane strain of arbitrary magnitude is given by the coordinate transformation

$$x = cx_0 \qquad y = dy_0$$

where c and d are positive constants. Assuming $c > 1 > d$, show that the straight lines whose lengths remain unchanged make angles $\pm \beta$ and $\pm \beta_0$ with the x-axis in the final and initial states respectively, where

$$\tan \beta = \frac{d}{c} \sqrt{\frac{c^2 - 1}{1 - d^2}} = \frac{d}{c} \tan \beta_0$$

Prove that the deformation is associated with a change in volume unless $cd = 1$, which corresponds to pure shear. Show also that the maximum engineering shear strain is $(c^2 - d^2)/(2cd)$, associated with lines that are inclined at $\pm \pi/4$ with the x axis in the unstrained state.

1.29 Find the relationship between the constants A, B, and C in the following expressions, which represent a possible deformation rate in a two-dimensional field:

$$\dot{\varepsilon}_x = Ax^2(x^2 + y^2) \qquad \dot{\varepsilon}_y = By^2(x^2 + y^2)$$

$$\dot{\gamma}_{xy} = Cxy(x^2 + y^2)$$

Show that the associated velocity field, to within a rigid-body motion, is given by

$$u = Cx^3(\tfrac{3}{5}x^2 + y^2) + Dy$$

$$v = Cy^3(x^2 + \tfrac{3}{5}y^2) - Dx$$

where D is an arbitrary constant. Obtain an expression for the component of spin in the xy plane.

Answer: $C = 3A = 3B$.

1.30 An element of material deforms in plane strain such that the principal axes of the strain rate remain fixed in the element as it rotates during its motion. The directions of $\dot{\varepsilon}_1$ and $\dot{\varepsilon}_2$ are assumed to be parallel to the x and y axes respectively in the initial state. Show that the principal natural strains produced by an arbitrary small deformation of the element are

$$\varepsilon_1 = \frac{\partial u}{\partial x} + \frac{1}{2}\left\{\left(\frac{\partial u}{\partial y}\right)^2 - \left(\frac{\partial u}{\partial x}\right)^2\right\}$$

$$\varepsilon_2 = \frac{\partial u}{\partial y} + \frac{1}{2}\left\{\left(\frac{\partial v}{\partial x}\right)^2 - \left(\frac{\partial v}{\partial y}\right)^2\right\}$$

to second order, where u and v are the components of the displacement of the center of the element whose initial coordinates are x and y.

1.31 Let a_i and x_i be the initial and final coordinates of a typical particle P with respect to a fixed set of rectangular axes. Show that the ratio of the final and initial squared lengths of the material line elements through P, parallel to the coordinate axes in the initial state, are equal to the diagonal elements of the matrix of the tensor

$$g_{ij} = \frac{\partial x_k}{\partial a_i} \frac{\partial x_k}{\partial a_j}$$

Prove that the ratio of the initial and final densities of the material in the neighborhood of the considered particle is equal to the jacobian $|\partial x_i/\partial a_j|$ of the transformation of coordinates.

1.32 Green's strain tensor γ_{ij} at a typical particle in a finitely deformed body, having initial coordinates a_i, is defined as that whose scalar product with the tensor $2da_i\, da_j$ is equal to the difference between the final and initial squared lengths of a material line element emanating from the particle. Show that

$$\gamma_{ij} = \frac{1}{2}\left(\frac{\partial u_i}{\partial a_j} + \frac{\partial u_i}{\partial a_i} + \frac{\partial u_k}{\partial a_i}\frac{\partial u_k}{\partial a_j}\right)$$

where u_i is the displacement of the principle. Show also that the material rate of change of γ_{ij}, when the initial reference state coincides with the instantaneous state, is identical to the rate of deformation.

FOUNDATIONS OF PLASTICITY

2.1 The Criterion of Yielding

Suppose that an element of material is subjected to a system of stresses of gradually increasing magnitude. The initial deformation of the element is entirely elastic and the original shape of the element is recovered on complete unloading. For certain critical combinations of the applied stresses, plastic deformation first appears in the element. A law defining the limit of elastic behavior under any possible combination of stresses is called *yield criterion*. The law applies not only to loading directly from the annealed state, but also to reloading of an element unloaded from a previous plastic state. In developing a mathematical theory, it is necessary to take into account a number of idealizations at the outset. Firstly, it is assumed that the conditions of loading are such that all strain rate and thermal effects can be neglected. Secondly, the Bauschinger effect and the hysteresis loop, which arise from nonuniformity on the microscope scale, are disregarded. Finally, the material is assumed to be isotropic, so that its properties at each point are the same in all directions. There is a useful and immediate simplification resulting from the experimental fact that yielding is practically unaffected by a uniform hydrostatic tension or compression.† The effects of these restrictions on the nature of the yield criterion will be first examined in geometrical terms.

(i) *A geometrical representation* Consider a system of three mutually perpendicular axes with the principal stresses taken as rectangular coordinates (Fig. 2.1).

---

† P. W. Bridgman, Metals Technology, Tech. Pub. 1782 (1944); B. Crossland, *Proc. Inst. Mech. Eng.*, **168**: 935 (1954).

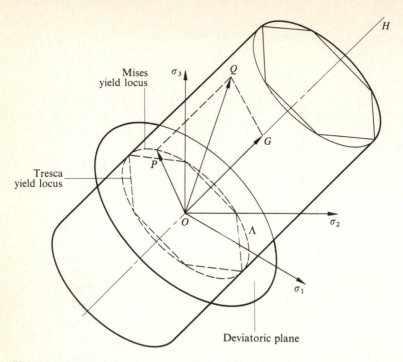

Figure 2.1 Geometrical representation of the yield criterion in the principal stress space.

The state of stress at any point in a body may be represented by a vector† emanating from the origin O. Imagine a line OH equally inclined to the three axes, so that its direction cosines are $(1/\sqrt{3}, 1/\sqrt{3}, 1/\sqrt{3})$. The stress vector **OQ**, whose components are $(\sigma_1, \sigma_2, \sigma_3)$, may be resolved into a vector **OG** along OH and a vector **OP** perpendicular to OH. The vector **OG** is of magnitude $\sqrt{3}\,\sigma_0$ and represents the hydrostatic stress with components $(\sigma_0, \sigma_0, \sigma_0)$. The vector **OP** represents the deviatoric stress with components (s_1, s_2, s_3) and its magnitude is $\sqrt{2J_2}$ by Eq. (28), Chap. 1. For any given state of stress, the deviatoric stress vector will lie in the plane passing through O and perpendicular to OH. This plane is known as the *deviatoric plane* and its equation is $\sigma_1 + \sigma_2 + \sigma_3 = 0$ in the principal stress space. Since a uniform hydrostatic stress has no effect on yielding, it follows that yielding can depend only on the magnitude and direction of the deviatoric stress vector **OP**. The yield surface is therefore a right cylinder whose generators are perpendicular to the deviatoric plane. Any stress state in which the stress point lies on the surface of the cylinder corresponds to a state of yielding. Any point inside the cylinder represents an elastic state of stress. The curve Λ in which the yield surface is intersected by the deviatoric plane is called the yield locus. The

† H. M. Westergaard, J. Franklin Institute, **189**: 627 (1920); W. W. Sokolovsky, *Dok. Akad. Nauk, USSR*, **61**: 223 (1946); R. Hill, *The Mathematical Theory of Plasticity*, p. 17, Clarendon Press, Oxford (1950).

equation to a possible yield locus, which is assumed to be *convex* (i.e., concave to the origin), is a possible yield criterion.

Consider now the yield locus together with the orthogonal projections of the stress axes on the deviatoric plane, which is taken in the plane of the paper (Fig. 2.2). It is evident that the yielding of isotropic materials can depend only on the values of principal stresses and not on their directions. Thus if the stresses (p, q, r) cause yielding, so will the stresses (p, r, q), implying that the yield locus is symmetrical about the projected σ_1 axis. It similarly follows that the yield locus is symmetrical with respect to the projections of the σ_2 and σ_3 axes. This amounts to the fact that the yield criterion is a function of the invariants of the deviatoric stress tensor. A further restriction on the form of the yield locus is imposed by the assumption that the Bauschinger effect is absent. Thus if (p, q, r) is a plastic state, $(-p, -q, -r)$ is also a plastic state. In other words, a radial line drawn from any point on the yield locus must meet the locus again at the same distance from the origin. Hence the yield locus must also be symmetrical about the lines orthogonal to the projected axes. The shape of the yield locus is therefore repeated over the twelve 30° segments formed by the six diameters as shown in the figure. Stated mathematically, the yield criterion is expressible in the form

$$f(J_2, J_3) = \text{const} \tag{1}$$

where f is an even function of J_3, which changes sign with the stresses, The yield criterion is called *regular* if the locus has a continuously turning tangent everywhere, and *singular* when the locus has sharp corners. The yield surface is considered as strictly convex when a straight line joining any two points on the yield locus lies completely inside the locus. A flat on the yield surface can be regarded as a limiting state of convexity.

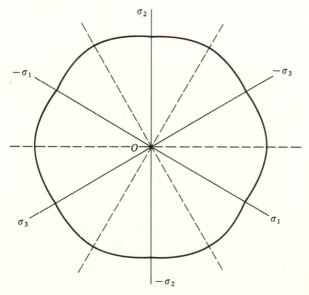

Figure 2.2 General appearance of the deviatoric yield locus having six axes of symmetry.

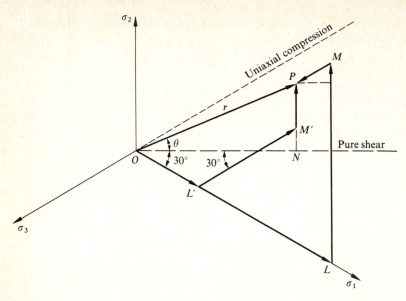

Figure 2.3 Deviatoric stress vector and its components along the projected axes.

(ii) *General considerations* The deviatoric stress vector **OP** may be regarded as the sum of the projections on the deviatoric plane of the component vectors of **OQ** along the axes of reference (Fig. 2.3). Since each stress axis is inclined to the deviatoric plane at an angle $\sin^{-1}(1/\sqrt{3})$ in the original stress space, each projected length along the axes is $\sqrt{\tfrac{2}{3}}$ times the actual length. Hence the lengths of the component vectors in the deviatoric plane are $OL = \sqrt{\tfrac{2}{3}}\,\sigma_1$, $LM = \sqrt{\tfrac{2}{3}}\,\sigma_2$, $MP = \sqrt{\tfrac{2}{3}}\,\sigma_3$. From geometry, the rectangular components of **OP** with respect to the horizontal and vertical through O are

$$ON = \frac{\sigma_1 - \sigma_3}{\sqrt{2}} = r\cos\theta \qquad PN = \frac{2\sigma_2 - \sigma_3 - \sigma_1}{\sqrt{6}} = r\sin\theta \qquad (2)$$

where (r, θ) are the polar coordinates of P. For the experimental determination of the yield criterion, it is convenient to introduce the Lode[†] parameter μ defined as

$$\mu = \frac{2\sigma_2 - \sigma_3 - \sigma_1}{\sigma_3 - \sigma_1} = -\sqrt{3}\tan\theta \qquad \sigma_1 > \sigma_2 > \sigma_3 \qquad (3)$$

To obtain the yield locus, it is only necessary to apply stresses for which θ covers the range from 0 to $\pm\pi/6$, μ varying from 0 to ∓ 1. When $\mu = 0$, $\sigma_2 = \tfrac{1}{2}(\sigma_3 + \sigma_1)$ and we have a state of pure shear denoted by $\tfrac{1}{2}(\sigma_3 - \sigma_1, \sigma_1 - \sigma_3, 0)$ together with a hydrostatic stress $\tfrac{1}{2}(\sigma_3 + \sigma_1)$. When $\mu = -1$, $\sigma_1 = \sigma_2$ and the state of stress is equivalent to a uniaxial compression $(\sigma_3 - \sigma_1, 0, 0)$ and a hydrostatic stress σ_1. Let

† W. Lode, *Z. angew. Math. Mech.*, **5**: 142 (1925), and *Z. Phys.*, **36**: 913 (1926).

the yield stresses in pure shear and uniaxial tension (or compression) be denoted by k and Y respectively. Then it follows from (2) that $r = \sqrt{2}\,k$ at $\theta = 0$ and $r = \sqrt{\frac{2}{3}}\,Y$ at $\theta = \pi/6$. For most metals k lies between $Y/2$ and $Y/\sqrt{3}$. Equations (2) may be written alternatively in the form

$$s_1 - s_3 = \sqrt{2}\,r \cos \theta \qquad s_1 + s_3 = -s_2 = -\sqrt{\frac{2}{3}}\,r \sin \theta$$

from which the deviatoric principal stresses can be expressed as

$$s_1 = \sqrt{\frac{2}{3}}\,r \cos\left(\frac{\pi}{6} + \theta\right) \qquad s_2 = \sqrt{\frac{2}{3}}\,r \sin \theta \qquad s_3 = -\sqrt{\frac{2}{3}}\,r \cos\left(\frac{\pi}{6} - \theta\right) \quad (4)$$

When the yield locus is given, r is a known function of θ, and equations (4) define the yield criterion parametrically through θ.

When the vector **OP** is given, the deviatoric stresses can be obtained graphically by noting the fact that $s_2 = \sqrt{\frac{2}{3}}\,PN$. Hence if a point M' is located on PN such that $PM'/NM' = 2$, then $PM' = \sqrt{\frac{2}{3}}\,s_2$. Let $L'M'$ be drawn parallel to $O\sigma_3$. Since $L'M' = OL' + PM'$ by geometry, it follows that $OL' = \sqrt{\frac{2}{3}}\,s_1$ and $L'M' = -\sqrt{\frac{2}{3}}\,s_3$. The points L' and M' therefore define the deviatoric stresses. The hydrostatic stress is of magnitude $\sqrt{\frac{2}{3}}\,LL'$, but it is not defined by the vector **OP**. The ratios of the deviatoric stresses are

$$s_1 : s_2 : s_3 = (\sqrt{3} - \tan \theta) : 2 \tan \theta : -(\sqrt{3} + \tan \theta) \qquad (5)$$

in view of (4). Multiplying the three equations in (4), and remembering that $s_1 s_2 s_3 = J_3$ and $r = \sqrt{2J_2}$, we obtain

$$J_3^2 = \tfrac{4}{27} J_2^3 \sin^2 3\theta \qquad (6)$$

If the polar equation of the yield locus is given, J_2 is a known function of θ. Eliminating θ between (6) and the given equation for the yield locus, it is possible to obtain the yield criterion in the form (1). Alternatively, if the yield criterion is given in terms of J_2 and J_3, the equation to the yield locus may be derived from it. It may be noted that as θ varies from 0 to $\pi/6$, the value of J_2 varies from k^2 to $Y^2/3$, and that of J_3 varies from 0 to $-\frac{2}{27} Y^3$. Plastic yielding is predominantly influenced by the magnitude of J_2.

(iii) *The Tresca and Mises criteria* Various criteria have been suggested in the past to predict the yielding of metals under complex stresses. Most of them are, however, only of historical interest, because they conflict with the experimental finding that a hydrostatic stress has no effect on yielding. The two entirely satisfactory and widely used criteria are those due to Tresca and von Mises. From a series of experiments on the extrusion of metals, Tresca[†] concluded that yielding occurred

---

[†] H. Tresca, *Comptes Rendus Acad. Sci.*, Paris, **59**: 754 (1864), and *Mem. Sav. Acad. Sci.*, Paris, **18**: 733 (1868). Tresca was probably influenced by a more general criterion for the failure of soils proposed earlier by C. A. Coulomb, *Mem. Math. Phys.*, **7**: 343 (1773).

when the maximum shear stress reached a critical value. If the stress vector lies within the sector $-\pi/6 \leqslant \theta \leqslant \pi/6$ in the plane diagram, the Tresca yield criterion may be written as

$$\sigma_1 - \sigma_3 = 2k \qquad \sigma_1 > \sigma_2 > \sigma_3 \tag{7}$$

The polar equation of the yield criterion in this range, by the first equation of (2), is

$$r \cos \theta = \sqrt{2}k \qquad -\frac{\pi}{6} \leqslant \theta \leqslant \frac{\pi}{6}$$

which means that the yield locus in this sector is a straight line parallel to the σ_2 direction. It is evident that $Y = 2k$ according to Tresca's yield criterion. The consideration of all possible values of the stresses leads to a regular hexagon (Fig. 2.4) as the complete yield locus in the deviatoric plane. The yield criterion is therefore singular, the sharp corners of the locus being at the points representing uniaxial tension or compression. The Tresca criterion may be expressed in terms of the invariants J_2 and J_3 if we observe that the polar equation of the Tresca yield locus is equivalent to $J_2 \cos^2 \theta = k^2$. The elimination of θ between this equation and (6) gives

$$4(J_2 - k^2)(J_2 - 4k^2)^2 = 27J_3^2 \tag{8}$$

Tresca's yield criterion is piecewise linear only when it is expressed in terms of the principal stresses. In a number of important physical problems involving high degrees of symmetry, the directions of the principal stresses are known in advance. The relative values of these principal stresses are also frequently indicated by the nature of the applied loading. In such cases, Tresca's criterion provides considerable simplifications in the theoretical analysis. In general, however, the Tresca criterion would lead to serious mathematical complexities.

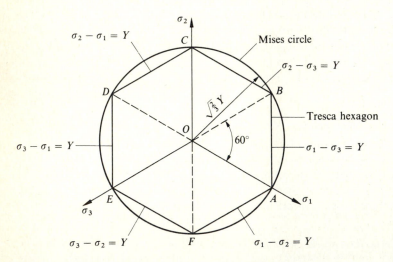

Figure 2.4 Deviatoric yield loci of Tresca and von Mises.

Von Mises[†] suggested, from purely theoretical considerations, that yielding occurred when J_2 attained a critical value. It immediately follows that the Mises yield locus is a circle of radius $\sqrt{2}\,k$ or $\sqrt{\frac{2}{3}}\,Y$, and the yield criterion is $J_2 = k^2$. Evidently, $Y = \sqrt{3}\,k$ according to the von Mises criterion. Using the expressions for J_2 from equations (28) of Chap. 1, the yield criterion proposed by von Mises may be written

$$s_{ij}s_{ij} = s_x^2 + s_y^2 + s_z^2 + 2\tau_{xy}^2 + 2\tau_{yz}^2 + 2\tau_{xz}^2 = 2k^2$$
$$(\sigma_x - \sigma_y)^2 + (\sigma_y - \sigma_z)^2 + (\sigma_z - \sigma_x)^2 + 6\tau_{xy}^2 + 6\tau_{yz}^2 + 6\tau_{xz}^2 = 2Y^2 \tag{9}$$

Isotropy is implied by the symmetry of the yield function with respect to the stress components, while absence of Bauschinger effect is indicated by the fact that only squares of the stresses and stress differences are involved.

Nadai[‡] has pointed out that according to the von Mises criterion, yielding begins when the octahedral shear stress attains a certain value. Indeed, it follows from Sec. 1.3(iv) that the octahedral shear stress has the value $\sqrt{\frac{2}{3}}\,k$ at the yield point. Hencky[§] proposed that the Mises law implies the elastic energy of distortion reaching a certain value at the yield point. Thus a hydrostatic stress, which only produces elastic energy of volume change in isotropic materials, does not cause yielding. Equations (9) may also be regarded as implying that yielding occurs when the root mean square value of either the principal shear stresses or the principal deviatoric stresses becomes critical. In view of (4), the Mises criterion may be written in the parametric form

$$s_1 = \frac{2}{3}\,Y \cos\left(\frac{\pi}{6} + \theta\right) \qquad s_2 = \frac{2}{3}\,Y \sin\theta \qquad s_3 = -\frac{2}{3}\,Y \cos\left(\frac{\pi}{6} - \theta\right) \tag{10}$$

It follows from the first equation of (2) that the maximum shear stress in any plastic state according to the von Mises yield criterion is $\cos\theta$ times that in pure shear. The Tresca criterion, on the other hand, predicts the same maximum shear stress in all plastic states. It is customary to make the two criteria agree with each other in uniaxial tension or compression, so that the Mises circle passes through the corners of the Tresca hexagon (Fig. 2.4). The two yield criteria then have the same value of Y, but the value of k in the Mises criterion is $2/\sqrt{3}$ times that in the Tresca criterion. The two yield loci therefore differ most in a state of pure shear. For most metals, the yield criterion of von Mises defines the yield limit more accurately than does that of Tresca. If the latter is adopted for simplicity, the overall accuracy can be improved by replacing $2k$ in (7) by mY, where m is an empirically assigned number lying between 1.0 and 1.155. Then the error in the calculated stresses can be limited to ± 7.5 per cent.

[†] R. von Mises, *Göttinger Nachrichten Math. Phys. Klasse*, 582 (1913). It was apparently anticipated by M. T. Huber, *Czas. Tech.*, Lemberg, **22**: 81 (1940).

[‡] A. Nadai, *J. Appl. Phys.*, **8**: 205 (1937). See also *The Theory of Flow and Fracture of Solids*, p. 402, McGraw-Hill Book Co., New York (1950).

[§] H. Hencky, *Z. Angew. Math. Mech.*, **4**: 323 (1924).

(iv) *The plane stress yield locus* In a number of important physical problems, one of the principal stresses may be assumed to vanish. The yield criterion may then be represented by a closed curve where the nonzero principal stresses are plotted as rectangular coordinates. According to Tresca's yield criterion, the magnitude of the numerically greater of the two principal stresses is equal to Y when these stresses are of the same sign, while the principal stress difference is of magnitude Y when the stresses have opposite signs. Assuming $\sigma_3 = 0$, the Tresca yield locus in the (σ_1, σ_2) plane is represented by a hexagon defined by the straight lines

$$\sigma_1 = \pm Y \qquad \sigma_2 = \pm Y \qquad \sigma_1 - \sigma_2 = \pm Y \tag{11}$$

When $\sigma_3 = 0$, the von Mises yield criterion (9), expressed in terms of the principal stresses, reduces to

$$\sigma_1^2 - \sigma_1\sigma_2 + \sigma_2^2 = Y^2 \tag{12}$$

which is the equation to an ellipse whose major and minor axes are inclined at an angle of 45° with the σ_1 and σ_2 axes (Fig. 2.5). The Mises ellipse circumscribes the Tresca hexagon for a given uniaxial yield stress Y.

Consider, now, the general plane stress components σ_x, σ_y, and τ_{xy}, referred to a pair of rectangular axes. The numerically greatest shear stress occurs in or out of the plane of the applied stresses according as $\sigma_x\sigma_y$ is less or greater than τ_{xy}^2. In the former situation, the maximum shear stress criterion of Tresca becomes

$$(\sigma_x - \sigma_y)^2 + 4\tau_{xy}^2 = Y^2 \qquad \sigma_x\sigma_y \leqslant \tau_{xy}^2 \tag{13}$$

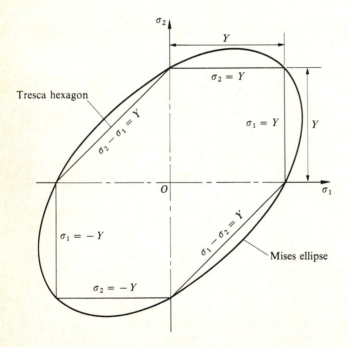

Figure 2.5 Tresca and Mises yield loci on the (σ_1, σ_2) plane when $\sigma_3 = 0$.

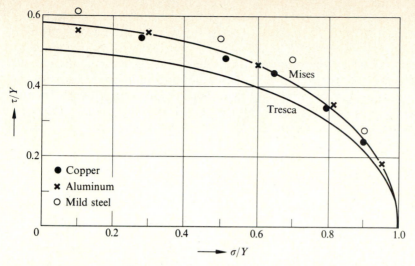

Figure 2.6 Tresca and Mises ellipses on the (σ, τ) plane together with the experimental results of Taylor and Quinney.

When $\sigma_x \sigma_y \geq \tau_{xy}^2$, either of the first two equations of (11) must be satisfied for yielding. Consequently

$$(\sigma_1 \mp Y)(\sigma_2 \mp Y) = 0$$

Expanding this product and expressing it in terms of the general stress components, we obtain the Tresca criterion in the form

$$\tau_{xy}^2 - \sigma_x \sigma_y \pm Y(\sigma_x + \sigma_y) = Y^2 \qquad \sigma_x \sigma_y \geq \tau_{xy}^2$$

where the upper sign holds when σ_x and σ_y are both positive and the lower sign when they are both negative. The Mises criterion, on the other hand, reduces to the unique expression

$$\sigma_x^2 - \sigma_x \sigma_y + \sigma_y^2 + 3\tau_{xy}^2 = Y^2 \tag{14}$$

In the combined tension and torsion of a thin-walled tube, each element of the tube wall is subjected to a longitudinal stress σ and a shear stress τ. In view of (13) and (14), the yield criterion may be expressed as

$$\sigma_2 + \alpha \tau^2 = Y^2$$

where $\alpha = 4$ for the Tresca criterion and $\alpha = 3$ for the von Mises criterion. In the (σ, τ) plane, these criteria are represented by ellipses having the same length of the major axis (Fig. 2.6). Taylor and Quinney† in their classical experiments of this

---

† G. I. Taylor and H. Quinney, *Phil. Trans. Roy. Soc.*, **A230**: 323 (1931). The influence of the intermediate principal stress on the yielding of metals was demonstrated earlier by W. Lode, *Z. angew. Math. Mech.*, **5**: 142 (1925).

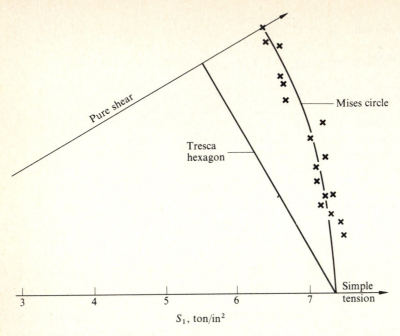

Figure 2.7 Experimental verification of the yield locus considered in the deviatoric plane (*due to Lianis and Ford*).

kind observed that the experimental points fell between the two ellipses, but were more inclined toward the Mises ellipse. For the considered stress state, it is easy to show that

$$\mu = \frac{\sigma}{\sqrt{\sigma^2 + 4\tau^2}}$$

Hence it is only necessary to apply longitudinal stresses in the range $0 \leqslant \sigma \leqslant Y$ in order to cover a 30° segment of the experimental yield locus plotted in the deviatoric plane.

The testing of thin-walled tubes requires complicated apparatus and entails great difficulty in ensuring that the material of the tube is isotropic. Hill,[†] on the other hand, has shown that arbitrary uniform states of combined stresses can be produced by pulling a thin rectangular strip having a pair of asymmetrical notches. The method has the advantage that anisotropy can be either eliminated or effectively controlled. Using such a notched specimen, Lianis and Ford[‡] carried out experiments on the yielding of commercially pure aluminum, specially treated

---

[†] R. Hill, *J. Mech. Phys. Solids*, **1**: 271 (1953).

[‡] G. Lianis and H. Ford, *J. Mech. Phys. Solids*, **5**: 215 (1957). For a similar experimental confirmation, using combined bending and torsion of thin tubes, see M. P. L. Siebel, *J. Mech. Phys. Solids*, **1**: 189 (1953).

to give a sharp yield, to obtain plastic stress states covering the required 30° segment in the deviatoric plane. The results, displayed in Fig. 2.7, provide sufficient evidence for yielding taking place according to the von Mises criterion.

2.2 Strain-Hardening Postulates

(i) *Isotropic hardening* We have seen that an element of material yields when the magnitude of the deviatoric stress vector is increased to a value such that the stress point reaches the yield locus. Unless the locus is a circle (as for the Mises criterion), the magnitude of the stress vector causing yielding depends on its final direction in the deviatoric plane. If the material is nonhardening, the plastic stress state can change in such a way that the stress point always lies on a constant yield locus. For a strain-hardening material, the size and shape of the yield locus depend on the complete history of plastic deformation since the previous annealing. It is assumed that the material is isotropic at the annealed state and that the anisotropy and the Bauschinger effect developed during the cold work may be neglected. The preceding discussion of the yield criterion is then appropriate for any given state of hardening of the material.

A convenient mathematical formulation for strain-hardening is obtained by assuming further that the yield surface uniformly expands without change in shape, as the state of stress changes along a certain path P_0P in the stress space (Fig. 2.8), the amount of hardening being given by the final plastic state. Since the yield locus merely increases in size, any given state of hardening may be defined by the current yield stress in uniaxial tension. It is, therefore, necessary to relate the current yield stress to the amount of plastic deformation following a given initial state of yielding. To this end, we replace Y in the yield criterion by $\bar{\sigma}$, which is known as the *equivalent stress*, *effective stress*, or *generalized stress*. Referring to the von Mises

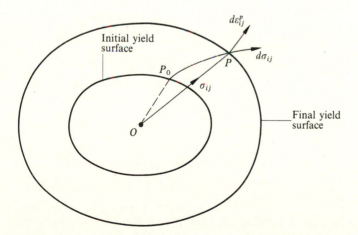

Figure 2.8 Geometrical representation of the isotropic hardening rule.

yield criterion, we write

$$\bar{\sigma} = \sqrt{\tfrac{3}{2}} \, (s_{ij} s_{ij})^{1/2} = \sqrt{\tfrac{3}{2}} \, \{s_x^2 + s_y^2 + s_z^2 + 2\tau_{xy}^2 + 2\tau_{yz}^2 + 2\tau_{zx}^2\}^{1/2}$$

$$= \sqrt{\tfrac{1}{2}} \, \{(\sigma_x - \sigma_y)^2 + (\sigma_y - \sigma_z)^2 + (\sigma_z - \sigma_x)^2 + 6\tau_{xy}^2 + 6\tau_{yz}^2 + 6\tau_{zx}^2\}^{1/2} \qquad (15)$$

Consider first the hypothesis in which the amount of hardening is taken as a function of the total plastic work per unit volume.† This assumption is obviously consistent with the fact that no hardening is produced by purely elastic strains. If the plastic part of the strain increment tensor is denoted by $d\varepsilon_{ij}^p = \dot{\varepsilon}_{ij}^p \, dt$, where $\dot{\varepsilon}_{ij}$ is the rate of deformation and dt the time element, the increment of plastic work per unit volume is

$$dW_p = \sigma_{ij} \, d\varepsilon_{ij}^p = (s_{ij} + \sigma_0 \delta_{ij}) \, d\varepsilon_{ij}^p = s_{ij} \, d\varepsilon_{ij}^p$$

where the last step follows from the condition $d\varepsilon_{ii}^p = 0$, implying that there is no plastic volume change. Plastic incompressibility of metals is in close agreement with experimental observations, and is also consistent with the fact that a uniform hydrostatic stress produces no plastic strain. The work-hardening hypothesis may be stated mathematically as

$$\bar{\sigma} = \Phi\left(\int dW_p \right) = \Phi\left(\int \sigma_{ij} \, d\varepsilon_{ij}^p \right) \qquad (16)$$

where the integral is taken over the actual strain path starting from some initial state. The function Φ can be determined from the true stress-strain curve in uniaxial tension or compression. If the true stress σ is plotted againnst the plastic part of the strain, then W_p is equal to the area under the curve up to the ordinate σ. Since $\bar{\sigma} = \sigma$ in this case, the area directly gives the argument of Φ in (16).

In an alternative hypothesis, more frequently in use, $\bar{\sigma}$ is regarded a function of a certain measure of the total plastic strain. Considering the second invariant of the plastic strain increment tensor, an *equivalent* (or generalized) *plastic strain increment* is defined as

$$\overline{d\varepsilon^p} = \sqrt{\tfrac{2}{3}} \, (d\varepsilon_{ij}^p \, d\varepsilon_{ij}^p)^{1/2}$$

$$= \sqrt{\tfrac{2}{3}} \, \{(d\varepsilon_x^p)^2 + (d\varepsilon_y^p)^2 + (d\varepsilon_z^p)^2 + 2(d\gamma_{xy}^p)^2 + 2(d\gamma_{yz}^p)^2 + 2(d\gamma_{zx}^p)^2\}^{1/2} \qquad (17)$$

where only the positive root is implied. The numerical factor in the above expression is so chosen that in uniaxial tension, $\overline{d\varepsilon}^p$ equals the longitudinal plastic strain increment. This follows from the fact that the magnitude of the lateral compressive plastic strain in the tensile test of an isotropic bar is half the

---

† R. Hill, *The Mathematical Theory of Plasticity*, p. 26, Clarendon Press, Oxford (1950). For rigid/plastic materials, the work-hardening hypothesis was suggested earlier by R. Schmidt, *Ing.-Archiv.*, **3**: 215 (1932).

longitudinal tensile plastic strain. The strain-hardening hypothesis may now be expressed as†

$$\bar{\sigma} = F\left(\int \overline{d\varepsilon}^{\,p}\right) = F\left(\int \sqrt{\tfrac{2}{3} d\varepsilon_{ij}^p \, d\varepsilon_{ij}^p}\right) \qquad (18)$$

where the integral is taken along the strain path as before. The integrated strain, known as the total equivalent plastic strain, provides a suitable measure of the plastic deformation. The function F is given by the relationship between the true stress and the plastic strain in uniaxial tension or compression. Equation (18) implies that the amount of hardening is determined by every infinitesimal plastic distortion leading to the final shape of an element, and not merely by the difference between the initial and final shapes of the element.

Both (16) and (18) imply that the stress-strain curves in tension and compression coincide only when the stresses are plotted against the logarithmic strains, but not when they are plotted against the engineering strain. It is evident that the compressive stress is the same function of the height ratio (h_0/h) as the tensile stress is of the length ratio (l/l_0). The stress-strain curves obtained under different stress systems can be compared on the basis of either (16) or (18). Consider, as an example, a thin-walled circular cylinder under pure torsion. A line on the tube, originally parallel to the axis, becomes a helix making an angle ϕ with the axis when the shear stress is τ. The shear strain in the tube is $\gamma = \tan\phi$ in engineering measure, and the total equivalent plastic strain is $(\tan\phi - \tau/G)/\sqrt{3}$, where G is the shear modulus. Since $\bar{\sigma} = \sqrt{3}\,\tau$, it follows from (18) that $\sqrt{3}\,\tau$ is the same function of $(\tan\phi - \tau/G)/\sqrt{3}$ as the tensile stress σ is of $\ln(l/l_0) - \sigma/E$ in uniaxial tension.‡ In the plastic range, the uniaxial stress-strain curve can be derived from the engineering shear stress-strain curve by plotting $\sigma = \sqrt{3}\,\tau$ against

$$\varepsilon = \frac{1}{\sqrt{3}}\left[\gamma + (1 - 2v)\frac{\tau}{E}\right]$$

where E is Young's modulus and v Poisson's ratio. The correspondence between the uniaxial stress-strain curve and the shear stress-strain curve is indicated in Fig. 2.9. The strain-hardening hypothesis would generally give results different from those obtained from the work-hardening hypothesis. However, the stress-strain relations for metals are usually such that the two hypotheses will lead approximately to the same result.

(ii) *Anistropic hardening* We shall now consider hardening rules that account for

† In the special case where the elastic strains are negligible, this was proposed by F. K. G. Odquist, *Z. angew. Math. Mech.*, **13**: 360 (1933).

‡ This is confirmed by the results of an experimental investigation by W. M. Shepherd, *Proc. Inst. Mech. Eng.*, **159**: 95 (1948). See also C. Zener and J. H. Hollomon, *J. Appl. Phys.*, **17**: 2 (1946).

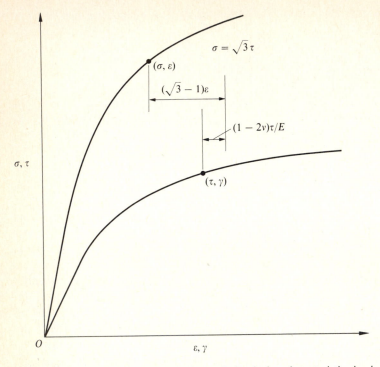

Figure 2.9 Correspondence between the strain-hardening characteristics in simple tension and pure torsion for an isotropic material.

anisotropy and Bauschinger effect exhibited by real materials.† It is assumed that the material is initially isotropic, having identical yield stresses in tension and compression. In the kinematic hardening rule, due to Prager, the yield surface is assumed to undergo translation in a nine-dimensional stress space. The initial yield surface is represented by the equation $f(\sigma_{ij}) = k^2$, where k is a constant. If the resultant displacement of the yield surface at any stage is denoted by a symmetric tensor α_{ij}, the current yield surface is given by

$$f(\sigma_{ij} - \alpha_{ij}) = k^2 \tag{19}$$

Since α_{ij} is not a scalar multiple of the isotropic tensor δ_{ij}, which represents a hydrostatic change in stress, the material becomes anisotropic as a result of the hardening process. It is reasonable to suppose that the incremental translation of the yield surface is in the direction of the plastic strain increment $d\varepsilon_{ij}^p$, considered as

---

† Experimental investigations on subsequent yield surface have been carried out by P. M. Naghdi, F. Essenberg, and W. Koff, *J. Appl. Mech.*, **25**: 201 (1958); H. J. Ivy, *J. Mech. Eng. Sci.*, **3**: 15 (1961); W. M. Mair and H. Ll. D. Pugh, *J. Mech. Eng. Sci.*, **6**: 93 (1964); P. S. Theocaris and C. R. Hazell, *J. Mech. Phys. Solids*, **13**: 281 (1965); J. Rogan and A. Shelton, *J. Strain Anal.*, **4**: 138 (1969).

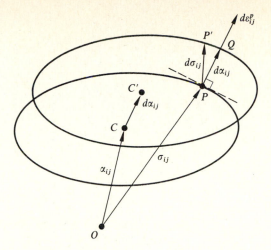

Figure 2.10 Stress-space representation of Prager's hardening rule.

a vector in the 9-space. Then†

$$d\alpha_{ij} = c \, d\varepsilon_{ij}^p \qquad (20)$$

where c is a scalar quantity characterizing the material behavior. The deformation is assumed small, so that the effect of rotation of the element on $d\alpha_{ij}$ may be disregarded. Since $d\alpha_{ii} = 0$, the translation of the yield surface is always parallel to the deviatoric hyperplane. The hardening rule is represented in Fig. 2.10, where O is the origin of the stress space and C the current center of the yield surface. The incremental translation of the yield surface, during a stress increment PP', is represented by CC', which is equal and parallel to PQ.

When c is a constant, (20) immediately integrates to $\alpha_{ij} = c\varepsilon_{ij}^p$, indicating that the total translation of the yield surface is a measure of the total plastic strain. If, in addition, the initial surface is that of von Mises, the yield criterion becomes

$$(s_{ij} - c\varepsilon_{ij}^p)(s_{ij} - c\varepsilon_{ij}^p) = 2k^2 \qquad (21)$$

where k is now the initial yield stress in pure shear. A constant value of c represents linear strain-hardening with a plastic modulus $H = \frac{3}{2}c$, whatever the yield function. If a specimen is loaded in simple tension until the longitudinal plastic strain is ε^p, the current yield stress is $Y + H\varepsilon^p$, but subsequent loading in simple compression will produce yielding when the intensity of the stress becomes $Y - H\varepsilon^p$. The sum of the predicted yield stresses in tension and compression is therefore independent of the plastic strain, and is equal to twice the initial yield stress Y.

According to (20), and the associated flow rule (Sec. 2.3(ii)), the yield surface translates in the direction of the exterior normal at the stress point in a nine-dimensional space. Many practical problems, on the other hand, are conveniently

---

† W. Prager, *Proc. Inst. Mech. Eng.*, **169**: 41 (1955), and *J. Appl. Mech.*, **23**: 93 (1956). See also I. U. Ishlinsky, *Ukr. Mat. Zh.*, **6**: 314 (1954).

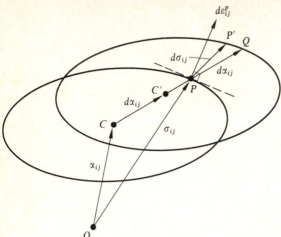

Figure 2.11 Stress-space representation of Ziegler's hardening rule.

treated in appropriate subspaces. In most subspaces the yield surface undergoes translation but in a direction different from the outward normal. If the material obeys Tresca's yield criterion, the yield surface tends to deform in certain subspaces.[†] The difficulty can be avoided if we replace (20) by

$$d\alpha_{ij} = (\sigma_{ij} - \alpha_{ij})\, d\mu \qquad (22)$$

where $d\mu$ is a positive scalar.[‡] This equation states that the yield surface translates in the direction of the line connecting the center of the yield surface to the current stress point P. In Fig. 2.11, the incremental translation of the yield surface is represented by the vector $\mathbf{CC'}$, equal to the vector $\mathbf{PQ}$, where Q lies on CP extended. The multiplying factor $d\mu$ is determined from the condition that the infinitesimal chord QP' is perpendicular to the plastic strain increment vector. Thus

$$(d\sigma_{ij} - d\alpha_{ij})\, d\varepsilon_{ij}^p = 0$$

which expresses the fact that the stress point remains on the yield surface. On substitution from (22), we get

$$d\mu = \frac{d\sigma_{ij}\, d\varepsilon_{ij}^p}{(\sigma_{kl} - \alpha_{kl})\, d\varepsilon_{kl}^p} \qquad (23)$$

A disadvantage of the modified hardening rule is that the path traced by the center of the yield surface does not, in general, represent the strain path. The two hardening rules coincide in uniaxial tension and compression, as well as in simple and pure shears.

† R. T. Shield and H. Zeigler, Z. angew. Math. Phys., **9**: 260 (1958).

‡ The modified hardening rule has been proposed by H. Zeigler, Q. Appl. Math., **17**: 55 (1959). See also C. C. Clavot and H. Zeigler, Ing.-Archiv., **28**: 13 (1959). For a critical review, see P. M. Naghdi, Plasticity, Proc. 2d Symp. Naval Struct. Mech., pp. 121–169, Pergamon Press (1960).

A more realistic hardening process should involve simultaneous translation and expansion of the yield surface.† The simplest yield criterion of this type is (19) with k^2 replaced by $\bar{\sigma}^2/3$, where $\bar{\sigma}$ is a function of the total equivalent plastic strain defined in (18). It should be noted that $\bar{\sigma}$ is only the isotropic part of the current uniaxial yield stress of the material. In the case of a monotonic simple tension σ, we have $d\bar{\sigma} = d\sigma - \frac{3}{2}c\, d\varepsilon^p$ and $\overline{d\varepsilon^p} = d\varepsilon^p$. Denoting $d\bar{\sigma}/d\varepsilon^p$ by $\frac{3}{2}h$, the plastic modulus may be written as

$$H = \tfrac{3}{2}(h + c)$$

When h and c have constant values, implying a linear strain-hardening law, the yield stress in tension is increased by the amount $\frac{3}{2}(h + c)\varepsilon^p$ when the longitudinal strain is ε^p. The yield stress in compression, due to a subsequent reversal of loading, exceeds the initial yield Y by the amount $\frac{3}{2}(h - c)\varepsilon^p$, which can take both positive and negative value. The same conclusion holds when a specimen is loaded in compression with a tensile prestrain in the plastic range. The variation of the parameter c may be allowed for by assuming it to depend only on the total equivalent plastic strain. Suitable empirical relations may be used to express h and c as functions of $\int \overline{d\varepsilon^p}$, involving arbitrary constants that can be determined from experimental measurements.‡

2.3 The Rule of Plastic Flow

When an element of material is unloaded from a certain plastic state, it recovers elasticity and the stress point moves inside the yield locus. If anisotropy is disregarded, the elastic behavior of the material is characterized by two independent elastic constants which retain their initial values. When the element is reloaded along a certain strain-path, yielding will again occur when the stress point reaches the current yield locus. For a work-hardening material, a further plastic flow can be enforced only by increasing the stress to a point outside the yield locus. If the stress increment is such that the stress point remains on the same yield locus, no hardening is produced and the plastic strain increments are zero. Such changes in stress are called neutral since they represent neither loading nor unloading. The elastic part of the strain increment corresponding to any plastic flow is directly related to the stress increment by means of Hooke's law. It is

---

† The combined hardening rule has been discussed by P. G. Hodge, Jr., *J. Appl. Mech.*, **24**: 482 (1957); I. Kadashevich and V. V. Novozhilov, *Prikl. Mat. Mekh.*, **22**: 104 (1959); Z. Mroz, H. P. Shrivastava, and R. N. Dubey, *Acta Mech.*, **25**: 51 (1976). For an extension of the hardening rule to large strains, see E. H. Lee, R. L. Mallett, and T. B. Wertheimer, *J. Appl. Mech.*, **50**: 554 (1983).

‡ More complex hardening rules, predicting rotations of the yield surface in addition to expansion and translation, have been considered by A. Baltov and A. Sawczuk, *Acta Mech.*, **1**: 81 (1965); J. F. Williams and N. L. Svensson, *Meccanica*, **6**: 104 (1971); H. P. Shrivastava, Z. Mroz, and R. N. Dubey, *Z. angew. Math. Mech.* **53**: 625 (1973); M. Tanaka and Y. Miyagawa, *Ing. Archiv.*, **44**: 255 (1975); A. Phillips and G. J. Weng, *J. Appl. Mech., Trans. ASME*, **42**: 315 (1975); D. W. A. Rees, *J. Strain Anal.*, **16**, 85 (1981).

therefore necessary to relate the increment of plastic strain to the stress increment and the current stress.

(i) *The plastic potential* The observed plastic behavior of polycrystalline metals clearly indicates that for isotropic materials, the principal axes of the plastic strain increment coincide with those of the stress.† The plastic strain increment may therefore be regarded as a vector $2G\,(d\varepsilon_1^p, d\varepsilon_2^p, d\varepsilon_3^p)$ in the principal stress space, the factor $2G$ being introduced to obtain the dimension of stress. Since $d\varepsilon_1^p + d\varepsilon_2^p + d\varepsilon_3^p = 0$, the plastic strain-increment vector may be regarded as normal to a right cylinder perpendicular to the deviatoric plane. The curve Γ in which the cylinder is intersected by the deviatoric plane is a level curve of a scalar function of the deviatoric principal stresses (s_1, s_2, s_3). It is reasonable to stipulate that the ratios of the components of the plastic strain increment depend on the current stress and not on the stress increment. The magnitude of the strain increment is, however, determined by the stress increment through the strain-hardening characteristic of the material. The plastic strain-increment vector is therefore parallel to the normal to the curve Γ at the point where it is intersected by the deviatoric stress vector. Since, in an isotropic material, the strain increments are interchanged when the stresses are so, the curve Γ must be symmetrical with respect to the stress axes. Moreover, the reversal of the sign of the applied stresses in our idealized material should merely change the sign of the strain increments. Hence the slope of the curve Γ at the opposite ends of a diameter must be the same. This is possible if the curve is also symmetrical about diameters perpendicular to the three axes. Thus the curve Γ, like the yield locus Λ, is identical in each of the 30° segments marked off by the directions representing the states of uniaxial stress and pure shear. It follows that the equation of the potential surface, whose size is immaterial, may be written in the form

$$g(J_2, J_3) = \text{const} \tag{24}$$

The function g, defining the ratios of the components of the plastic strain increment, is known as the *plastic potential*.‡ In a nine-dimensional space, the plastic strain increment may be expressed by the *flow rule*

$$d\varepsilon_{ij}^p = \frac{\partial g}{\partial \sigma_{ij}}\, d\lambda \tag{25}$$

where $d\lambda$ is a positive scalar that depends on the stress increment, and is generally a function of the space variables as well as the time scale. Since g is independent of the hydrostatic stress, the plastic incompressibility condition $d\varepsilon_{ii}^p = 0$ is identically

---

† This was recognized by B. de Saint-Venant, *Comptes Rendus Acad. Sci., Paris,* **70**: 473 (1870); *J. Math. Pures et Appl.,* **16**: 308 (1871).

‡ The concept of plastic potential is due to R. von Mises, *Z. angew. Math. Mech.,* **8**: 161 (1928). A generalization has been made by W. Prager, *Proc. 8th Int. Congr. Appl. Mech., Istanbul,* **2**: 65 (1952). See also H. Zeigler, *Q. Appl. Math.,* **19**: 39 (1961). The present discussion follows R. Hill, *The Mathematical Theory of Plasticity,* p. 35, Clarendon Press, Oxford (1950).

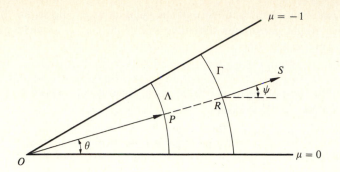

Figure 2.12 Geometrical representation of the plastic flow rule in the deviatoric plane.

satisfied. If g is assumed to be a homogeneous function of degree n, involving the stress components, the increment of plastic work per unit volume may be written as

$$dW_p = \frac{\partial g}{\partial \sigma_{ij}} \sigma_{ij} \, d\lambda = ng \, d\lambda \tag{26}$$

by Euler's theorem. Since plastic deformation is irreversible, $d\lambda$ is necessary positive during plastic flow.

Figure 2.12 shows the yield locus and the plastic potential curve over a typical $30°$ segment bounded by the radial lines $\theta = 0$ and $\theta = \pi/6$. The deviatoric stress vector **OP** meets the curve Γ at R, and the vector **RS** drawn normal to Γ represents the plastic strain increment. If ψ is the angle made by the vector **RS** with the direction $\theta = 0$, the associated Lode parameter† is

$$v = \frac{2d\varepsilon_2^p - d\varepsilon_3^p - d\varepsilon_1^p}{d\varepsilon_3^p - d\varepsilon_1^p} = -\sqrt{3} \tan \psi \tag{27}$$

The new parameter v should not be confused with Poisson's ratio. Now, for an isotropic material in uniaxial tension, $d\varepsilon_2^p = d\varepsilon_3^p = -\frac{1}{2} d\varepsilon_1^p$ from symmetry, giving $v = -1$ when $\mu = -1$. Moreover, a pure shear stress in an ideal element must produce a pure shear strain, for which $d\varepsilon_1^p = -d\varepsilon_2^p$, $d\varepsilon_3^p = 0$; hence $v = 0$ when $\mu = 0$. In other words, the vector **RS** is along the radial line for both uniaxial stress and pure shear. It follows that the curve Γ must intersect the bounding radii of each $30°$ segment orthogonally. The magnitude of the vector **RS** is $2G\sqrt{\frac{3}{2}} \, \overline{d\varepsilon^p}$ in view of (17), while vector **OP** is of magnitude $\sqrt{\frac{2}{3}} \, \bar{\sigma}$ in view of (15). Hence, for the von Mises criterion, the increment of plastic work per unit volume is

$$dW^p = s_{ij} \, d\varepsilon_{ij}^p = \frac{\mathbf{OP} \cdot \mathbf{RS}}{2G} = \bar{\sigma} \, \overline{d\varepsilon}^p \cos (\psi - \theta) \tag{28}$$

For a general yield criterion, $\bar{\sigma}$ in the above expression must be replaced by $\sqrt{3J_2}$.

---

† W. Lode, *Z. Phys.*, **36**: 913 (1936).

Since the material is isotropic, the dependence of the yield locus on the strain history appears through a single parameter that determines its size. The yield criterion may therefore be put in the form

$$f(J_2, J_3) = c$$

where c is assumed to be a function of the total plastic work per unit volume of the considered element. During continued loading,

$$df = dc = h\sigma_{ij}\, d\varepsilon_{ij}^p = ngh\, d\lambda$$

in view of (26), the parameter h being a function of the stress and strain history. The substitution for $d\lambda$ from above into the flow rule (25) furnishes†

$$h\, d\varepsilon_{ij}^p = \frac{1}{ng}\frac{\partial g}{\partial \sigma_{ij}}\, df \qquad (29)$$

This expression is consistent with the fact that no plastic strain can occur during a neutral loading for which $df = 0$.

The plastic stress-strain relation becomes more revealing when expressed in terms of the partial derivatives of g with respect to J_2 and J_3. Using the fact that

$$\frac{\partial s_{kl}}{\partial \sigma_{ij}} = \delta_{ik}\,\delta_{jl} - \frac{1}{3}\delta_{ij}\,\delta_{kl}$$

it is easy to show from Eqs. (30), Chap. 1, that

$$\frac{\partial J_2}{\partial \sigma_{ij}} = s_{kl}\frac{\partial s_{kl}}{\partial \sigma_{ij}} = s_{ij} \qquad \frac{\partial J_3}{\partial \sigma_{ij}} = s_{kl}s_{km}\frac{\partial s_{lm}}{\partial \sigma_{ij}} = p_{ij}$$

where

$$p_{ij} = s_{ik}s_{kj} - \tfrac{2}{3}J_2\,\delta_{ij}$$

Equation (29) may therefore be written as

$$h\, d\varepsilon_{ij}^p = \frac{1}{ng}\left(\frac{\partial g}{\partial J_2}s_{ij} + \frac{\partial g}{\partial J_3}p_{ij}\right) df \qquad (30)$$

which holds for $df \geqslant 0$. Since the principal axes of p_{ij} are identical to those of σ_{ij}, the flow rule implies that the principal axes of stress and plastic strain increment coincide.

(ii) The associated flow rule In view of the similarity of the properties of the yield function and the plastic potential, it may be assumed that they are actually identical. The flow rule obtained on the basis of the identity of g and f is known as the *associated flow rule* for the given yield criterion. The partial derivatives $\partial f/\partial \sigma_{ij}$ correspond to a specified position of the stress point on the yield surface, and are

---

† The flow rule expressed in the form (29) is essentially due to E. Melan, *Ing.-Arch.*, **9**: 116 (1938). See also W. Prager, ibid. **20**: 235 (1949).

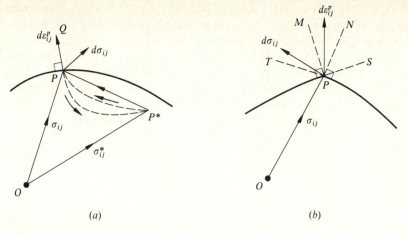

Figure 2.13 Stress increment and strain increment vectors for a given state of stress. (a) Regular yield surface; (b) singular yield surface.

uniquely defined at all points when the yield criterion is regular. The assumption $g = f$ follows from considerations of the plastic deformation of polycrystalline aggregates in which individual crystals deform by slipping over preferred planes.[†] When the curves Λ and Γ are similar, the plastic strain increment vector is directed along the normal to the yield locus at P making an angle ψ with the line $\theta = 0$. The polar equation of the yield locus may therefore be written as

$$dr/d\theta = r \tan(\theta - \psi)$$

or

$$r = r_0 \exp \int_0^\theta \tan(\theta - \psi)\, d\theta \tag{31}$$

where r_0 is the length of the radius at $\theta = 0$. When the (μ, ν) relation is experimentally determined, ψ is a known function of θ, and the yield locus can be derived from (31) using numerical integration.[‡]

Suppose that the plastic strain increment $d\varepsilon_{ij}^p$ is given. The corresponding stress σ_{ij}, determined from the normality rule and the yield criterion, is represented by a point P in the stress space (Fig. 2.13a). If σ_{ij}^* is an arbitrary state of stress represented by a point P^* on or inside the yield surface, the difference between the incremental plastic works done by the two stress states is

$$\delta W_p = (\sigma_{ij} - \sigma_{ij}^*)\, d\varepsilon_{ij}^p$$

which represents the scalar product of the vectors P^*P and PQ. If the yield surface is strictly convex, the angle between these vectors is acute and the scalar product is positive. Hence

$$(\sigma_{ij} - \sigma_{ij}^*)\, d\varepsilon_{ij}^p \geqslant 0 \tag{32}$$

[†] J. F. W. Bishop and R. Hill, *Phil. Mag.*, **42**: Ser. 7: 414 (1951).
[‡] G. I. Taylor, *Proc. Roy. Soc.*, **A191**: 441 (1947).

where the equality holds when P^* coincides with P. This is the maximum work theorem, due to von Mises,† which states that the actual work done in a given plastic strain increment is greater than the fictitious work done by an arbitrary state of stress not exceeding the yield limit. If the yield surface contains a flat, the equality in (32) holds for all stress points P and P^* lying on this flat.

Consider an element of material that has been deformed by taking it along a certain strain path, the current stress in the element being σ_{ij}. For a work-hardening material, a further increment of plastic strain $d\varepsilon_{ij}^p$ requires the stress to be increased in such a way that the infinitesimal vector $d\sigma_{ij}$ lies outside the current yield surface. For a nonhardening material, the stress increment is always tangential to the yield surface. In view of (25) with $g = f$, we have

$$d\sigma_{ij}\, d\varepsilon_{ij}^p = \frac{\partial f}{\partial \sigma_{ij}}\, d\sigma_{ij}\, d\lambda = df\, d\lambda \geqslant 0 \tag{33}$$

during the plastic deformation. The equality holds only for a perfectly plastic material. Inequality (33) states that the plastic strain increment vector makes an acute angle with the stress increment vector during the loading of an element of work-hardening material.

The above inequalities may be obtained from a different standpoint, which also furnishes the associated flow rule.‡ Consider an external agency that applies additional stresses to an element having an initial stress σ_{ij}^*. The element is first brought to the yield point by increasing the stress to σ_{ij}, and this is followed by a plastic strain increment $d\varepsilon_{ij}^p$ produced by a further increment of stress $d\sigma_{ij}$. The agency subsequently releases $d\sigma_{ij}$, and returns the state of stress to σ_{ij}^* along an elastic path (Fig. 2.13a). For a work-hardening material, it is reasonable to assert that the work done by the external agency over the complete cycle is positive. Since all the elastic energy is recovered on closing the cycle, the net work done by the external agency may be written as

$$(\sigma_{ij} - \sigma_{ij}^*)\, d\varepsilon_{ij}^p + \tfrac{1}{2} d\sigma_{ij}\, d\varepsilon_{ij}^p > 0$$

to second order. Inequality (33) follows on setting $\sigma_{ij}^* = \sigma_{ij}$. This brings us to Drucker's first postulate, which states that the plastic work done by an external agency during the application of additional stresses is positive for a work-hardening material and zero for a nonhardening material.

If σ_{ij}^* is distinct from σ_{ij}, the magnitude of the difference $\sigma_{ij} - \sigma_{ij}^*$ may be made as large as we please compared to $d\sigma_{ij}$. The above inequality then reduces to (32). Drucker's second postulate therefore states that the net work done by an external agency during a cycle of addition and removal of stresses is nonnegative. An immediate consequence of this postulate is that the angle between the vectors representing $d\varepsilon_{ij}^p$ and $\sigma_{ij} - \sigma_{ij}^*$ cannot be obtuse, so that all points σ_{ij}^* lie on one side of the plane perpendicular to $d\varepsilon_{ij}^p$ and passing through the point σ_{ij}. The yield

---

† R. von Mises, *Z. angew. Math. Mech.*, **8**: 161 (1928).
‡ D. C. Drucker, *Q. Appl. Math.*, **7**: 411 (1950); *Proc. 1st U.S. Nat. Congr. Appl. Mech.*, 487 (1951). See also D. R. Bland, *J. Mech. Phys. Solids*, **6**: 71 (1957).

surface must therefore be convex. At any point σ_{ij} on the yield surface, with a uniquely defined normal, the only direction of $d\varepsilon_{ij}^p$ that satisfies inequality (32) for all possible points σ_{ij}^* is that of the exterior normal.

When the yield criterion is singular, the normal is not uniquely defined at an edge of the yield surface, but inequalities (32) and (33) continue to hold if Drucker's postulates are accepted. The first inequality indicates that the plastic strain increment vector lies between the exterior normals to the regular faces meeting at the edge (Fig. 2.13b). For a work-hardening material, the second inequality allows the plastic strain increment vector to lie anywhere in the angle NPM formed by the normals at P, so long as the stress increment vector lies in the angle SPT formed by the tangents at P. If, on the other hand, $d\sigma_{ij}$ falls outside the angle SPT, the direction of $d\varepsilon_{ij}^p$ is further restricted by the requirement that the angle between the two incremental vectors must be acute. When the material is nonhardening, $d\varepsilon_{ij}^p$ can lie anywhere between the normals PN and PM so long as the stress point remains on the considered edge of the yield surface.

(iii) *Constitutive equations* For physical reasons, the stress increment that enters into the stress-strain relation must be based on an objective rate of change of the true stress (see Sec. 1.6(i)). In the theory of plasticity, the objective stress rate should be such that the yield function has a stationary value whenever the stress point remains on the same yield surface. It turns out that only the Jaumann stress rate $\overset{\circ}{\sigma}_{ij}$ is capable of satisfying this condition. Indeed, the material rate of change of the yield function is

$$\dot{f} = \frac{\partial f}{\partial \sigma_{ij}} \dot{\sigma}_{ij} = \frac{\partial f}{\partial \sigma_{ij}} \overset{\circ}{\sigma}_{ij}$$

in view of the fact that $\partial f / \partial \sigma_{ij}$ is coaxial with σ_{ij} for an isotropic material. This shows that $\dot{f} = 0$ when $\overset{\circ}{\sigma}_{ij}$ is tangential to the yield surface.[†] In the following discussion, therefore, the stress increment $d\sigma_{ij}$ will denote $\overset{\circ}{\sigma}_{ij} dt$, where dt is the increment of time scale.

When the material work-hardens, the magnitude of the plastic strain increment depends only on the component of the stress increment along the normal to the yield surface. Let n_{ij} denote the outward drawn unit normal to the yield surface at the current stress point σ_{ij} in a nine-dimensional space. Since the vector representing the plastic strain increment is in the direction of n_{ij}, the plastic stress-strain relation may be expressed as[‡]

$$h \, d\varepsilon_{ij}^p = n_{ij} n_{kl} \, d\sigma_{kl} \qquad n_{kl} \, d\sigma_{kl} \geqslant 0 \qquad (34)$$

where h is a positive scalar representing the rate of hardening. For any given yield function $f(\sigma_{ij})$, the unit normal n_{ij} can be found from the relation $n_{ij} n_{ij} = 1$, and the fact that the components of n_{ij} are proportional to those of $\partial f / \partial \sigma_{ij}$. The scalar

---

[†] The yield function has a stationary value when $\overset{\circ}{\sigma}_{ij}$ itself vanishes. No other objective stress rate satisfies this requirement.

[‡] R. Hill, *J. Mech. Phys. Solids*, **4**: 247 (1956).

products of (34) with $d\sigma_{ij}$ and $d\varepsilon_{ij}^p$ lead to the relationship

$$d\sigma_{ij}\, d\varepsilon_{ij}^p = h^{-1}(n_{ij}\, d\sigma_{ij})^2 = h\, d\varepsilon_{ij}^p\, d\varepsilon_{ij}^p \tag{35}$$

during plastic deformation. In the case of a uniaxial tension represented by the principal stresses $(\sigma, 0, 0)$, the unit normal in the principal stress space is $\sqrt{\frac{2}{3}}(1, -\frac{1}{2}, -\frac{1}{2})$, and (34) gives $h = \frac{2}{3}H$, where H is the plastic modulus at the current state of hardening.

For a nonhardening material ($h = 0$), the scalar product $n_{ij}\, d\sigma_{ij}$ vanishes during plastic flow, and the magnitude of the plastic strain increment is indeterminate by (34). The stress-strain relation in this case is more conveniently written as

$$d\varepsilon_{ij}^p = n_{ij}\, d\lambda \qquad n_{kl}\, d\sigma_{kl} = 0 \qquad d\lambda > 0$$

If the stress increment is such that the scalar produce $n_{kl}\, d\sigma_{kl}$ is negative, the element unloads from the plastic state, and $d\varepsilon_{ij}^p$ is identically zero whether the material work-hardens or not.

The stress-strain relation (34) can be modified for a singular yield criterion when the plastic state of stress is represented by a point on an edge common to two or more regular faces. In a nine-dimensional space, the exterior normal to each face at the considered point is conveniently represented by a unit vector $\mathbf{n}_\alpha$ ($\alpha = 1, 2, \ldots$). If the stress increment is denoted by the vector $d\boldsymbol{\sigma}$, the associated plastic strain increment for a work-hardening material is[†]

$$d\boldsymbol{\varepsilon}^p = \sum h_\alpha^{-1}(\mathbf{n}_\alpha \cdot d\boldsymbol{\sigma})\mathbf{n}_\alpha \qquad \mathbf{n}_\alpha \cdot d\boldsymbol{\sigma} \geqslant 0 \tag{36}$$

where h_α is a positive scalar function of the stress and strain history, and the summation includes the contributions from all the operative yield mechanisms. When $\mathbf{n}_\alpha \cdot d\boldsymbol{\sigma} \leqslant 0$ for all α, the plastic strain increment vanishes. For a non-hardening material, $d\boldsymbol{\varepsilon}^p = \sum \mathbf{n}_\alpha\, d\lambda_\alpha$, where the summation is taken over the normal to those surfaces on which $d\boldsymbol{\sigma}$ lies.

The complete stress-strain relation for an elastic/plastic material is obtained by adding the plastic strain increment $d\varepsilon_{ij}^p$ to the elastic strain increment $d\varepsilon_{ij}^e$ corresponding to a stress increment $d\sigma_{ij}$. For an isotropic material, there are two independent elastic constants, the shear modulus G being related to Young's modulus E and Poisson's ratio v by $E = 2(1 + v)G$. The elastic part of the strain increment may be written as

$$d\varepsilon_{ij}^e = \frac{ds_{ij}}{2G} + \frac{1 - 2v}{3E} \delta_{ij}\, d\sigma_{kk} \tag{37}$$

The first term on the right-hand represents the deviatoric part and the second term the hydrostatic part. In terms of the actual stress increment,

$$d\varepsilon_{ij}^e = \frac{1}{2G}\left(d\sigma_{ij} - \frac{v}{1 + v} \delta_{ij}\, d\sigma_{kk}\right) \tag{37a}$$

[†] W. T. Koiter, *Q. Appl. Math.*, **11**: 350 (1953); W. E. Boyce and W. Prager, *J. Mech. Phys. Solids*, **6**: 9 (1957).

The right-hand sides of above equations give the total strain increment when the element unloads from a current plastic state. The incremental change in volume, which is entirely elastic, is expressed by the equation

$$d\varepsilon_{ii} = \frac{1 - 2v}{E} \, d\sigma_{ii}$$

When the flow rule is associated with the yield criterion, which is assumed to be regular, Eqs. (34) and (37a) furnish the complete stress-strain relation

$$d\varepsilon_{ij} = \frac{1}{2G}\left(d\sigma_{ij} - \frac{v}{1 + v}\,\delta_{ij}\,d\sigma_{kk} \right) + \frac{3}{2H}\,n_{ij}n_{kl}\,d\sigma_{kl} \tag{38}$$

for a work-hardening material, whenever $n_{kl}\,d\sigma_{kl} \geqslant 0$. Taking the scalar product of (38) with n_{ij}, and remembering that $n_{ij}n_{ij} = 1$ and $n_{ii} = 0$, we obtain

$$n_{ij}\,d\varepsilon_{ij} = \frac{3G + H}{2GH}\,n_{ij}\,d\sigma_{ij}$$

which indicates that $n_{ij}\,d\varepsilon_{ij} \gtrless 0$ for $n_{ij}\,d\sigma_{ij} \gtrless 0$ when $H > 0$. In view of this, the stress-strain relation may be written in the inverted form

$$d\sigma_{ij} = 2G\left[d\varepsilon_{ij} + \frac{v}{1 - 2v}\,\delta_{ij}\,d\varepsilon_{kk} - \frac{3G}{3G + H}\,n_{ij}n_{kl}\,d\varepsilon_{kl} \right] \tag{39}$$

whenever $n_{kl}\,d\varepsilon_{kl} \geqslant 0$. If there is unloading represented by $n_{kl}\,d\varepsilon_{kl} < 0$, the last term of (39) must be omitted. Equation (39) applies equally well to nonhardening materials ($H = 0$) and strain-softening materials ($H < 0$) during plastic deformation. Since $n_{ij}\,d\sigma_{ij} < 0$ for both loading and unloading when $H < 0$, the restriction $n_{ij}\,d\varepsilon_{ij} \geqslant 0$ must be applied when using (38) for a strain-softening material.†
When $H = 0$, (38) must be modified to

$$d\varepsilon_{ij} = \frac{1}{2G}\left(d\sigma_{ij} - \frac{v}{1 + v}\,\delta_{ij}\,d\sigma_{kk} \right) + \lambda n_{ij} \tag{40}$$

where $\lambda > 0$ for $n_{ij}\,d\sigma_{ij} = 0$ and $\lambda = 0$ for $n_{ij}\,d\sigma_{ij} < 0$. When a strain increment is prescribed such that $n_{ij}\,d\varepsilon_{ij} > 0$, the corresponding stress increment $d\sigma_{ij}$ can be uniquely determined from (39), whether the material work-hardens or not. If, on the other hand, a stress increment satisfying $n_{ij}\,d\sigma_{ij} = 0$ is prescribed, and the material is nonhardening, the plastic part of the strain increment cannot be determined without considering the applied constraints.‡

---

† Evidently, the restriction $n_{ij}\,d\varepsilon_{ij} < 0$ must be used for unloading. With this modification of loading and unloading criteria, the constitutive equations become structurally equivalent to those obtained from a strain space formulation by P. M. Naghdi and J. A. Trapp, *Int. J. Eng. Sci.*, **13**: 785 (1975); J. Casey and P. M. Naghdi, *J. Appl. Mech.*, **48**: 285 (1981), and *Q. J. Mech. Appl. Math.*, **37**: 231 (1984).

‡ A theory of plasticity in which both elastic and plastic strains are of finite magnitude has been developed by E. H. Lee, *J. Appl. Mech.*, **36**: 1 (1969). See also A. E. Green and P. M. Naghdi, *Arch. Ration. Mech. Anal.*, **18**: 251 (1965); P. M. Naghdi and J. A. Trapp, *Q. J. Mech. Appl. Math.*, **28**: 25 (1975).

2.4 Particular Stress-Strain Relations

(i) *Lévy-Mises and Prandtl-Reuss equations* The plastic flow rule corresponding to any particular choice of the plastic potential may be readily obtained. The simplest form of the potential curve, having all the properties described in the preceding section, is obviously a circle. In this case $\theta = \psi$, or equivalently $\mu = v$, whatever the form of the yield locus. Taking $g = J_2 = \frac{1}{2}s_{ij}s_{ij}$, and employing (25), we obtain the corresponding flow rule

$$d\varepsilon_{ij}^p = s_{ij}\, d\lambda$$

or

$$\frac{d\varepsilon_x^p}{s_x} = \frac{d\varepsilon_y^p}{s_y} = \frac{d\varepsilon_z^p}{s_z} = \frac{d\gamma_{xy}^p}{\tau_{xy}} = \frac{d\gamma_{yz}^p}{\tau_{yz}} = \frac{d\gamma_{xz}^p}{\tau_{zx}} = d\lambda \tag{41}$$

The stress-strain relation in this form was suggested independently by Lévy and von Mises, who used the total strain increments instead of the plastic strain increments.[†] The modified equations (41), which allow for the elastic strain increments, were proposed by Prandtl for plane strain and by Reuss for an arbitrary state of strain.[‡] Since the plastic shear strain increments, according to (41), vanish with the corresponding shear stresses, the principal axes of the stress and the plastic strain increment coincide. Equations (41) also indicate that the Mohr circle for the stress can be used for the plastic strain increment, provided the origin is moved in the appropriate direction of the σ axis by an amount equal to the hydrostatic stress.

If, in addition, the yield function is taken to be that of von Mises, then the above flow rule is associated with the yield criterion, which is given by $f = J_2$. Since the increment of plastic work per unit volume is $\bar{\sigma}\, \overline{d\varepsilon^p}$ in view of equation (28), it immediately follows from (16) that $\bar{\sigma}$ is the same function of $\int \overline{d\varepsilon^p}$ as the stress is of the plastic strain in uniaxial tension. Thus the work-hardening hypothesis is in this case equivalent to the strain-hardening hypothesis. Since g is a homogeneous function of degree two having the magnitude $\bar{\sigma}^2/3$, it follows from equations (26) and (28) that

$$dW^p = \tfrac{2}{3}\bar{\sigma}^2\, d\lambda = \bar{\sigma}\, \overline{d\varepsilon^p}$$

which gives

$$d\lambda = \frac{3\,\overline{d\varepsilon^p}}{2\bar{\sigma}} = \frac{3\,d\bar{\sigma}}{2H\bar{\sigma}} \tag{42}$$

The Prandtl-Reuss flow rule is completely defined by (41) and (42). The principal plastic strain increments may be expressed by the right-hand sides of (10) with $2Y/3$

† M. Lévy, *J. Math. Pures et Appl.*, **16**: 369 (1871); R. von Mises, *Göttinger Nachrichten Math. Phys. Klasse*, 582 (1913).

‡ L. Prandtl, *Proc. First Int. Congr. Appl. Mech., Delft*, 43 (1924); A. Reuss, *Z. angew. Math. Mech.*, **10**: 266 (1930).

replaced by $\overline{d\varepsilon}{}^p$ or $d\bar{\sigma}/H$. During continued loading of a plastic element,

$$\bar{\sigma}\,d\bar{\sigma} = \tfrac{3}{2}s_{ij}\,ds_{ij} = \tfrac{3}{2}s_{ij}\,d\sigma_{ij} \geqslant 0$$

where the equality holds for a nonhardening material. Since $d\sigma_{ij}\,d\varepsilon_{ij}^p = d\bar{\sigma}\,\overline{d\varepsilon}{}^p$ in view of (41), (42), and the preceding expression, we have

$$\frac{d\bar{\sigma}}{\bar{\sigma}} = \frac{d\sigma_{ij}\,d\varepsilon_{ij}^p}{\sigma_{kl}\,d\varepsilon_{kl}^p}$$

The flow rule may be derived from (34) by noting the fact that the exterior unit normal to the yield surface is in the direction of the deviatoric stress vector, and the stress increment normal to the yield surface is equal to the increase in radius of the Mises cylinder. Thus

$$n_{ij} = \sqrt{\tfrac{3}{2}}\,(s_{ij}/\bar{\sigma}) \qquad n_{kl}\,d\sigma_{kl} = \sqrt{\tfrac{2}{3}}\,d\bar{\sigma}$$

With these substitutions, the expression on the right-hand side of (34) reduces to $(d\bar{\sigma}/\bar{\sigma})s_{ij}$, which completes the proof.

From (37a) and (41), the complete Prandtl-Reuss equation for an elastic/plastic material may be expressed as

$$d\varepsilon_{ij} = \frac{1}{E}\left[(1+v)\,d\sigma_{ij} - v\,\delta_{ij}\,d\sigma_{kk}\right] + (\sigma_{ij} - \tfrac{1}{3}\sigma_{kk}\,\delta_{ij})\,d\lambda \tag{43}$$

where $d\lambda$ is given by (42). For a nonhardening material, $d\lambda$ may be treated as a basic unknown of the problem. Equation (43) consists of three equations of each of the two types

$$d\varepsilon_x = \frac{1}{E}\left[d\sigma_x - v(d\sigma_y + d\sigma_z)\right] + \tfrac{2}{3}d\lambda[\sigma_x - \tfrac{1}{2}(\sigma_y + \sigma_z)]$$

$$d\gamma_{xy} = \frac{d\tau_{xy}}{2G} + \tau_{xy}\,d\lambda$$

In a number of practical problems, the loading paths are such that the elastic strains are small in comparison with the plastic strains. The Prandtl-Reuss equations may then be replaced by the more tractable Lévy-Mises equations, which correspond to (41) with the superscripts omitted. This is equivalent to assuming the material to be rigid/plastic.[†]

The early experimental investigation by Lode on the combined tension and internal pressure of thin tubes, and by Taylor and Quinney on the combined tension and torsion of thin tubes, verified the relation $\mu = v$ only to a first approximation.[‡] However, since the strain ratios are far more sensitive to

† The rigid/plastic theory has been treated by H. Lippman and O. Mahrenholt, *Plastomechanik der Unformung Metallischer Werkstoffe*, Springer Verlag, Berlin (1967).

‡ W. Lode, *Z. Phys.*, **36**: 913 (1926); G. I. Taylor and H. Quinney, *Phil. Trans. R. Soc.*, **A230**: 323 (1931).

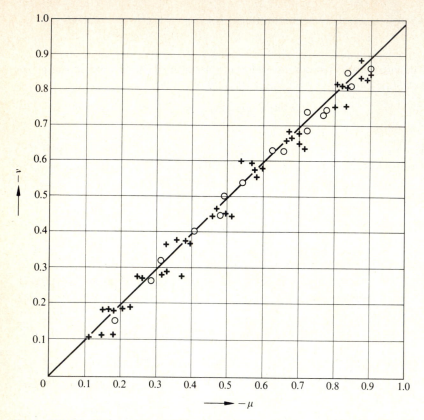

Figure 2.14 Experimental verification of the (μ, v) relationship by Hundy and Green $(+)$, and Lianis and Ford $(\circ)$.

anisotropy than the stress ratios are, the estimation of v is somewhat unreliable in experiments with the thin-walled tubes where some degree of anisotropy is likely to remain undetected. Following a method suggested by Hill, involving the tension of a thin rectangular strip with asymmetrical notches, the identity of μ and v has been experimentally verified† to a close approximation (Fig. 2.14). There is at present sufficient evidence to conclude that the von Mises yield criterion and the Prandtl-Reuss flow rule constitute the most realistic description of the plastic behavior of metals, when anisotropy and Bauschinger effect are of secondary importance.‡

---

† B. B. Hundy and A. P. Green, *J. Mech. Phys. Solids*, **3**: 16 (1954), and G. Lianis and H. Ford, ibid., **5**: 215 (1957), provided the experimental confirmation. Similar agreement has also been found by M. P. L. Siebel, *J. Mech. Phys. Solids*, **1**: 189 (1953), using the combined bending and twisting of thin-walled tubes.

‡ For an account of the various experimental investigations on the verification of the laws of plasticity, see E. Mroz and A. H. Olszak, *Recent Advancements in the Mathematical Theory of Plasticity*, Pergamon Press (1963).

(ii) *A matrix formulation* For the numerical solution of elastic/plastic problems, using the Prandtl-Reuss theory, it is useful to consider the incremental stress-strain relation in the matrix form. To this end, we write $\sqrt{\frac{3}{2}}\, s_{ij}/\bar{\sigma}$ for n_{ij} in (39), and obtain the Prandtl-Reuss stress-strain relation for a work-hardening material in the form

$$d\sigma_{ij} = c_{ijkl}\, d\varepsilon_{kl}$$

where $d\sigma_{ij}$ is considered in the Jaumann sense,

$$c_{ijkl} = 2G\left[\delta_{ik}\,\delta_{jl} + \frac{v}{1-2v}\,\delta_{ij}\,\delta_{kl} - \frac{1}{\alpha\bar{\sigma}^2}\,s_{ij}\,s_{kl}\right] \tag{44}$$

in the loading part of the plastic region ($d\bar{\sigma} > 0$), and

$$\alpha = \frac{2}{3}\left(1 + \frac{H}{3G}\right)$$

Let $\{d\sigma\}$ and $\{d\varepsilon\}$ denote column vectors whose elements are given by the components of $d\sigma_{ij}$ and $d\varepsilon_{ij}$ respectively. The stress-strain equation, which gives a linear relationship between the increments of stress and strain, can be expressed in the matrix notation

$$\{d\sigma\} = [C]\{d\varepsilon\}$$

where $[C]$ is a symmetric square matrix, known as the *constitutive matrix*. In the case of plane strain ($d\varepsilon_z = 0$), the matrix equation becomes

$$\begin{Bmatrix} d\sigma_x \\[2mm] d\sigma_y \\[2mm] d\tau_{xy} \end{Bmatrix} = 2G \begin{bmatrix} \dfrac{1-v}{1-2v} - \dfrac{s_x^2}{\alpha\bar{\sigma}^2} & \dfrac{v}{1-2v} - \dfrac{s_x s_y}{\alpha\bar{\sigma}^2} & -\dfrac{s_x \tau_{xy}}{\alpha\bar{\sigma}^2} \\[3mm] \dfrac{v}{1-2v} - \dfrac{s_x s_y}{\alpha\bar{\sigma}^2} & \dfrac{1-v}{1-2v} - \dfrac{s_y^2}{\alpha\bar{\sigma}^2} & -\dfrac{s_y \tau_{xy}}{\alpha\bar{\sigma}^2} \\[3mm] -\dfrac{s_x \tau_{xy}}{\alpha\bar{\sigma}^2} & -\dfrac{s_y \tau_{xy}}{\alpha\bar{\sigma}^2} & \dfrac{1}{2} - \dfrac{\tau_{xy}^2}{\alpha\bar{\sigma}^2} \end{bmatrix} \begin{Bmatrix} d\varepsilon_x \\[2mm] d\varepsilon_y \\[2mm] d\gamma_{xy} \end{Bmatrix} \tag{45}$$

where $d\gamma_{xy}$ is the engineering component, not the tensor component. The remaining stress-strain equation (for $d\sigma_z$), not included in (45), can be separately written down. In general, when the nonzero components of the strain increment are capable of being independently prescribed, the constitutive matrix is given by (45), expanded as necessary. In the case of plane stress ($\sigma_z = 0$), the strain increments cannot be independently chosen, as they are required to satisfy the equation

$$\left(\frac{v}{1-2v} - \frac{s_x s_z}{\alpha\bar{\sigma}^2}\right) d\varepsilon_x + \left(\frac{v}{1-2v} - \frac{s_y s_z}{\alpha\bar{\sigma}^2}\right) d\varepsilon_y$$

$$+ \left(\frac{1-v}{1-2v} - \frac{s_z^2}{\alpha\bar{\sigma}^2}\right) d\varepsilon_z - \frac{s_z \tau_{xy}}{\alpha\bar{\sigma}^2}\, d\gamma_{xy} = 0$$

which ensures $d\sigma_z = 0$. Using the above relation, $d\varepsilon_z$ can be eliminated from the remaining stress-strain equations. After some algebraic manipulation, using the relation $s_x + s_y + s_z = 0$ and the yield criterion

$$3(s_x^2 + s_x s_y + s_y^2 + \tau_{xy}^2) = \bar{\sigma}^2$$

the matrix equation can be put in the form of (45) with the constitutive matrix modified to[†]

$$[C] = \frac{2G}{N}
\begin{bmatrix}
(1+v)\dfrac{s_y^2}{\bar{\sigma}^2} + 2M & -(1+v)\dfrac{s_x s_y}{\bar{\sigma}^2} + 2vM & -\dfrac{(s_x + vs_y)\tau_{xy}}{\bar{\sigma}^2} \\[4mm]
-(1+v)\dfrac{s_x s_y}{\bar{\sigma}^2} + 2vM & (1+v)\dfrac{s_x^2}{\bar{\sigma}^2} + 2M & -\dfrac{(s_y + vs_x)\tau_{xy}}{\bar{\sigma}^2} \\[4mm]
-\dfrac{(s_x + vs_y)\tau_{xy}}{\bar{\sigma}^2} & -\dfrac{(s_y + vs_x)\tau_{xy}}{\bar{\sigma}^2} & \dfrac{N}{2} - (1-v)\dfrac{\tau_{xy}^2}{\bar{\sigma}^2}
\end{bmatrix} \tag{46}$$

where

$$M = \frac{H}{9G} + \frac{\tau_{xy}^2}{\bar{\sigma}^2} \qquad N = \tfrac{2}{3}(1-v)\left(1 + \frac{H}{3G}\right) - (1-2v)\frac{s_z^2}{\bar{\sigma}^2} \tag{47}$$

and $d\gamma_{xy}$ is again the engineering shear component. If the element remains elastic ($\bar{\sigma} < Y$), or unloads from a plastic state ($d\bar{\sigma} < 0$), the constitutive matrix $[C]$ should correspond to Hooke's law. It may be noted that an elastic material is indistinguishable from an elastic/plastic material for which H is infinitely large.

A plasticity deforming body generally contains one or more boundaries separating elastic and plastic regions. These boundaries are themselves unknown and are usually so complicated in shape that a complete solution of the elastic/plastic problem would involve tedious computations. The determination of the elastic/plastic boundary can be avoided by using the Ramberg-Osgood equation (Sec. 1.2(iii)) for the uniaxial stress-strain curve, the entire body being regarded as plastic during continued loading due to the nature of the empirical equation, which involves a zero initial yield stress.

(iii) *Geometrical representation of stress and strain* We have seen that the principal axes of the plastic strain increment coincide with those of the current stress, while the principal axes of the elastic strain increment coincide with those of the stress increment. If the principal axes of the stress do not rotate with respect to the element while it deforms, the principal components of the stress increment are the same as the increments of the principal stress components. In this special case, the principal axes of the total strain-increment coincide with the principal axes of the stress. The stresses and strains at any stage may therefore be represented by vectors in the principal stress space, a factor of $2G$ being used for the strains to have the dimension of stress. Since the hydrostatic parts of the stress and the strain are related to one another by the equation $\varepsilon_0 = (1 - 2v)\sigma_0/E$, it is only necessary to

---

[†] Y. Yamada, N. Yoshimura, and T. Sakurai, *Int. J. Mech. Sci.*, **10**: 343 (1968). For a somewhat different formulation, see P. V. Marcal and I. P. King, *Int. J. Mech. Sci.* **9**: 143 (1967).

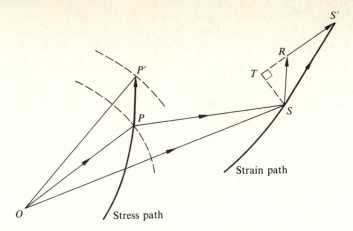

Figure 2.15 Geometrical representation of the Prandtl-Reuss relation for a work-hardening material.

consider the deviatoric stresses and strains on a plane diagram. The Prandtl-Reuss stress-strain relation will be assumed in what follows.†

Let **OP** and **OS** represent the current stress and strain vectors in the deviatoric plane (Fig. 2.15), the coordinates of S in the stress space being $2G$ times the components of the deviatoric strain. In view of (37), the vector **OP** also represents the deviatoric elastic strain. The vector **PS** therefore represents the total plastic strain. Let **PP'** represent a change in the deviatoric stress such that P' is outside the current yield locus, the new yield locus being represented by the dotted curve through P'. The corresponding strain-increment vector **SS'** is the sum of the vectors **SR** and **RS'** representing the elastic and plastic strain increments respectively. The elastic strain-increment vector **SR** is equal to the stress-increment vector **PP'** and the plastic strain-increment vector **RS'** is parallel to the current stress vector **OP**. If ST is drawn perpendicular to $S'R$ produced, the angle SRT is the same as the acute angle between OP and PP'. Since $SR = PP'$, TR is equal to the increment in the radius of the von Mises circle and is therefore of length $\sqrt{\frac{2}{3}}\, d\bar{\sigma}$. Now the length of the plastic strain-increment vector is

$$RS' = 2G\sqrt{\frac{3}{2}}\,\overline{d\varepsilon^{p}} = \sqrt{\frac{3}{2}}\frac{2G}{H}\,d\bar{\sigma}$$

and it is defined on the plane diagram by the ratio

$$\frac{RS'}{RT} = \frac{3G}{H} = \frac{3(E/T - 1)}{2(1 + v)} \tag{48}$$

where T is the tangent modulus to the uniaxial stress-strain curve at $\sigma = \bar{\sigma}$. During the continued plastic flow, the points P and S describe curves representing the

† R. Hill, *The Mathematical Theory of Plasticity*, p. 41, Clarendon Press, Oxford (1950). In general, a five-dimensional space will be necessary for the complete geometrical representation. See A. Ilyushin, *Prikl. Mat. Mekh.*, **18**: 641 (1954).

stress and strain paths respectively, starting from a common initial point P_0. These curves are the projections on the deviatoric plane of the paths traced by the stress and strain points in the principal stress space. The sum of the lengths of the vector **RS'** along the strain path is $\sqrt{6}\,G$ times the total equivalent plastic strain denoted by $\int \overline{d\varepsilon}{}^p$.

When the stress-increment **PP'** representing loading is given, **SR** is drawn equal and parallel to PP', and RS' is drawn parallel to the stress vector **OP**. The strain increment **SS'** is then obtained by making $RS' = (3G/H)RT$, where T is the foot of the perpendicular form S on RS'. The principal components of the deviatoric strain may be determined in the same manner as that described for the deviatoric stress. When the strain increment **SS'** is given, $S'T$ is drawn parallel to OP, and ST is drawn perpendicular to it. By locating R on $S'T$ such that $TS' = (1 + H/3G)RS'$, the stress increment **PP'**, equal to the elastic strain increment **SR**, is obtained. If the strain increment is such that S' lies inside the circle through S, the element unloads and PP' is then equal and parallel to SS'. In the case of no work-hardening, the stress point can only move along the same yield circle and SR is always perpendicular to RS'. Since the scalar product of **OP** and **SR** then vanishes, the increment of the elastic energy of distortion becomes zero.

In the initial stages of plastic deformation of an annealed material, the elastic and plastic strain increments are comparable, since H is then of the order G. Even when H is small compared to G, the elastic and plastic strain increments may be comparable if SS' makes a large angle with OP. In the limiting case of neutral loading the plastic strain increment is zero and SS' is perpendicular to OP, the points S' and R then coinciding with T. In the other extreme case when the stress point moves outward along a radial line, the strain point moves parallel to it and the elastic strain increment is a minimum. When the material is slightly prestrained and the stress path does not depart greatly from the radial direction, the elastic component of the strain is negligible. For a nonhardening material, when the elastic strain is neglected, the work expended per unit volume during the deformation is $k/\sqrt{2}\,G$ times the length of the strain path. Thus for a given final strain, the work is a minimum when the strain path is a straight line obtained by maintaining constant ratios of the principal strain components.

(iv) *Tresca's associated flow rule* If the material obeys Tresca's yield criterion, the plastic potential may be represented by a regular hexagon similar to the yield hexagon. When the stress point lies on one of the sides of the hexagon, the plastic strain-increment vector is directed along the normal to the side. The plastic deformation is therefore a pure shear in the direction of the maximum shear stress. When the stress corresponds to a corner, it is customary to assume that the stress point remains at the corner during a finite strain. The plastic strain-increment vector then lies within the 60° angle formed by the normals to the sides meeting at the corner. The strain increment may, however, be uniquely determined from the conditions of constraint of the problem. It follows that $v = 0$ while μ varies from 0 to -1, and v varies from 0 to -1 when $\mu = -1$. Considering the side AB of the

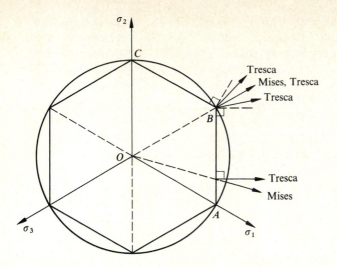

Figure 2.16 Plastic strain increment vectors associated with the Tresca and Mises criteria.

deviatoric hexagon (Fig. 2.16), Tresca's yield criterion and the associated flow rule may be written as

$$\sigma_1 - \sigma_3 = \bar{\sigma}, \quad \sigma_1 > \sigma_2 > \sigma_3 \qquad d\varepsilon_1^p = -d\varepsilon_3^p > 0, \quad d\varepsilon_2^p = 0$$

where $\bar{\sigma}$ is the current uniaxial yield stress. At the corner B, we have $\sigma_1 = \sigma_2$, and the ratios of the plastic strain increments can lie between those corresponding to the sides AB and BC. The yield criterion and the flow rule corresponding to the corner B may therefore be written as†

$$\sigma_1 - \sigma_3 = \sigma_2 - \sigma_3 = \bar{\sigma} \qquad d\varepsilon_1^p > 0, d\varepsilon_2^p > 0, d\varepsilon_3^p < 0$$

which must be supplemented by the incompressibility equation $d\varepsilon_{ii}^p = 0$. Similar results hold for the other sides and corners of the hexagon. An assumed plastic regime (side or corner) will be acceptable if all the inequalities are found to be satisfied. During the continued plastic flow, the stress point may, of course, move from one regime to another. The flow rule can be written in the integrated form when the stress point moves in regular progression. This includes plastic states in which the stress point remains on a side, remains at a corner, or moves from a side to a corner, while the principal axes of the stress remain fixed in the element. If the stress point moves from one side to another or from a corner to a side, only the incompressibility condition may be written in the integrated form.

It may be noted that a strain-increment vector normal to a side of the hexagon does not uniquely define the stress point. The increment of plastic work is, however, uniquely determined, since the projection of the stress vector on the normal to the side is constant along the side. This work is greater than that done by the stress

† W. T. Koiter, Biezeno Anniversary Volume, Stam, Haarlem, 232 (1953).

state represented by any other side of the hexagon. Using the yield criterion and the flow rule, it is easy to show that the increment of plastic work per unit volume corresponding to any plastic regime is $\bar{\sigma}|d\varepsilon^p|$, where $d\varepsilon^p$ is the numerically largest principal plastic strain increment. This suggests that the work-hardening hypothesis is a convenient basis for relating $\bar{\sigma}$ to the degree of cold work. Then $\bar{\sigma}$ is the same function of $\int |d\varepsilon^p|$ as the stress is of the plastic strain in uniaxial tension or compression. Stated mathematically, the work-hardening hypothesis according to the Tresca theory is reduced to

$$\bar{\sigma} = F\left\{ \int |d\varepsilon^p| \right\}$$

where the integral is taken along the strain path. It follows that the magnitude of the plastic strain-increment vector is equal to $d\bar{\sigma}/H$. If the stress path is such that the stress point moves between any two consecutive sides of the hexagon, $d\varepsilon^p$ represents the same principal plastic strain increment throughout the deformation. In this case, the above integral represents the magnitude of the numerically largest principal plastic strain, provided the principal axes do not rotate with respect to the element. For other stress paths, the value of the integral will depend on the complete stress and strain history, not merely on the final state of strain.†

(v) *Anisotropic flow rule* Consider a material that is initially isotropic and obeys the von Mises criterion for yielding. The material is assumed to harden anisotropically by a combined expansion and translation of the yield surface in stress space. After a certain amount of plastic deformation, the yield criterion assumes the form

$$(s_{ij} - \alpha_{ij})(s_{ij} - \alpha_{ij}) = \tfrac{2}{3}\bar{\sigma}^2 \tag{49}$$

where α_{ij} are the components of the total translation of the yield surface. The right-hand side of (49) is the square of the radius of the current deviatoric yield locus.

Adopting the associated flow rule of plasticity, and denoting the left-hand side of (49) by $2f$, the plastic strain increment may be written as

$$d\varepsilon_{ij}^p = \frac{\partial f}{\partial \sigma_{ij}} d\lambda = (s_{ij} - \alpha_{ij})\, d\lambda \tag{50}$$

where $d\lambda$ is a positive factor of proportionality. Let us assume that the physical coordinate axes initially coincide with the principal stress axes, so that $\sigma_{ij} = 0$ ($i \neq j$) when $\alpha_{ij} = 0$. It follows from (50) and (20) that $d\varepsilon_{ij}^p = 0$ ($i \neq j$) and $d\alpha_{ij} = 0$ ($i \neq j$) in the initial state, implying that $d\varepsilon_{ij}^p$ is coaxial with σ_{ij}. The result does not continue to hold unless the principal axes of stress remain fixed in the element. Thus, due to the anistropy caused by strain-hardening, the principal axes of stress and plastic strain increment do not generally coincide.

The quantity $\bar{\sigma}$ represents the isotropic part of the yield stress, and is assumed to depend on the total equivalent strain $\int d\bar{\varepsilon}^p$, where $d\bar{\varepsilon}^p$ is given by (17). The

---

† A theory of piecewise linear isotropic plasticity has also been developed by P. G. Hodge, *J. Franklin Inst.*, **263**: 13 (1957).

relationship between $\overline{d\varepsilon}^p$ and $d\lambda$ can be established by substituting for $d\varepsilon^p_{ij}$ into (17). Since

$$\sqrt{\tfrac{3}{2}}\,\overline{d\varepsilon}^p = d\lambda\sqrt{(s_{ij} - \alpha_{ij})(s_{ij} - \alpha_{ij})} = \sqrt{\tfrac{2}{3}}\,\bar{\sigma}\,d\lambda$$

in view of (50) and (49), we have

$$d\lambda = \frac{3\,\overline{d\varepsilon}^p}{2\bar{\sigma}} = \frac{d\bar{\sigma}}{h\bar{\sigma}}$$

where h is a measure of the isotropic part of the rate of strain-hardening. The anisotropic part of the hardening rate is given by the parameter c in Eq. (20). In view of the above relation, the plastic strain increment becomes

$$d\varepsilon^p_{ij} = (s_{ij} - \alpha_{ij})\frac{d\bar{\sigma}}{h\bar{\sigma}} \tag{51}$$

For the combined hardening process, $d\bar{\sigma}$ must be positive during continued plastic flow. Since $ds_{ii} = d\alpha_{ii} = 0$, Eq. (49) gives

$$\bar{\sigma}\,d\bar{\sigma} = \tfrac{3}{2}(s_{ij} - \alpha_{ij})(d\sigma_{ij} - d\alpha_{ij})$$

Substituting from (20) and (51), and using (49), the above relation is reduced to

$$\left(1 + \frac{c}{h}\right)d\bar{\sigma} = \frac{3}{2\bar{\sigma}}(s_{kl} - \alpha_{kl})\,d\sigma_{kl} > 0 \tag{52}$$

The stress-strain relation defined by (51) and (52) may be obtained directly from (34), whose form is unchanged by the anisotropy of hardening, with $h + c$ written for h. For purely isotropic hardening ($c = 0$), the stress-strain relation reduces to that given by (41) and (42).

When the hardening is purely kinematic ($h = 0$), $\bar{\sigma}$ has a constant value, equal to $\sqrt{3}\,k$, where k is the initial yield stress in pure shear. The loading condition $d\bar{\sigma} = 0$ is equivalent to

$$(s_{ij} - \alpha_{ij})(d\sigma_{ij} - d\alpha_{ij}) = 0$$

where $d\alpha_{ij}$ is given by (20). Substituting from (50) and using (49), whose right-hand side is now equal to $2k^2$, we obtain

$$d\lambda = \frac{1}{ck^2}(s_{kl} - \alpha_{kl})\,d\sigma_{kl} > 0 \tag{53}$$

The stress-strain relation for kinetic hardening is completely defined by (50) and (53). When c is a constant, α_{ij} can be replaced by $c\varepsilon^p_{ij}$ in the above equations.†

We have seen that the principal axes of stress and plastic strain coincide when

---

† Plastic stress-strain relations for cyclic loading have been discussed by D. C. Drucker and L. Palgen, *J. Appl. Mech.*, **48**: 479 (1981), and by J. Casey and P. M. Naghdi, ibid., 285 (1981). A simplified method of analysis for cyclic loading has been proposed by J. Tribout, G. Inglebert, and J. Casier, *J. Pressure Vessel Technol., Trans. ASME*, **105**: 222 (1983).

the principal stress axes remain fixed in the element from the start. In this special case, it is convenient to express the stress-strain relation in terms of the principal components. The translation of the yield surface in the principal stress space is along its exterior normal, and the hardening rule applies without change in form. For a piecewise linear yield condition such as that of Tresca, the analysis is greatly simplified by the fact that there is a limited path-independence of the final plastic strain, similar to that noted for isotropic hardening.†

2.5 The Total Strain Theory

(i) *Hencky's stress-strain relation* Although the Prandtl-Reuss stress-strain relation provides the most satisfactory basis for treating plasticity problems, the theory is incremental and generally leads to mathematical complexities. Considerable simplifications are often achieved by using a system of equations due to Hencky,‡ who postulated a one-to-one correspondence between the stress and the strain. Thus, the components of the total plastic strain are taken to be proportional to the corresponding deviatoric stresses. Assuming small strains, the plastic stress-strain relation proposed by Hencky may be written as

$$\varepsilon_{ij}^p = \lambda s_{ij} \tag{54}$$

where λ is positive during loading and zero during unloading. For a nonhardening material, λ may be treated as an unspecified factor of proportionality. When the material work-hardens, λ depends on the equivalent stress $\bar{\sigma}$, which may be regarded as a function of an equivalent total plastic strain $\bar{\varepsilon}^p$ defined as

$$\bar{\varepsilon}^p = \sqrt{\tfrac{2}{3}\varepsilon_{ij}^p \varepsilon_{ij}^p} \tag{55}$$

the relationship between $\bar{\sigma}$ and $\bar{\varepsilon}^p$ being given by the uniaxial stress–plastic-strain curve. Substituting from (54), and using (15) for $\bar{\sigma}$, which implies the von Mises yield criterion, we have

$$\sqrt{\tfrac{3}{2}}\,\bar{\varepsilon}^p = \lambda\sqrt{s_{ij}s_{ij}} = \lambda\sqrt{\tfrac{2}{3}}\,\bar{\sigma}$$

giving $\lambda = 3\bar{\varepsilon}^p/2\bar{\sigma}$. The stress-strain relation (54) may therefore be expressed in the form

$$\varepsilon_{ij}^p = \frac{3\bar{\varepsilon}^p}{2\bar{\sigma}}\,s_{ij} = \frac{3}{2}\left(\frac{1}{S} - \frac{1}{E}\right)s_{ij} \tag{56}$$

---

† If the stress increment is such that the stress point remains at a corner of the yield locus, which is a regular hexagon in the deviatoric plane, each component of the principal plastic strain increment for purely kinematic hardening is $1/c$ times the corresponding deviatoric stress increment. A general theory of piecewise linear plasticity, based on maximum shear, has been developed by P. G. Hodge, Jr., *J. Mech. Phys. Solids*, **5**: 242 (1957). A more general theory based on an arbitrary coordinate system has been discussed by P. G. Hodge, Jr., *Int. J. Mech. Sci.*, **22**: 21 (1980).

‡ H. Hencky, *Z. angew. Math. Mech.* **4**: 323 (1924). The Hencky stress-strain relation has been extensively used by A. Nadai, *Theory of Flow and Fracture of Solids*, McGraw-Hill Book Co. (1950), and W. W. Sokolovsky, *Theory of Plasticity*, Moscow (1969).

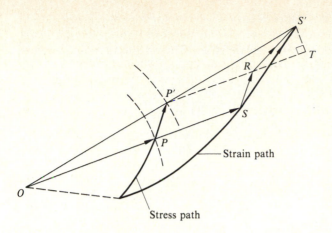

Strain path

Stress path

Figure 2.17 Geometrical representation of the Hencky stress-strain relation in the deviatoric plane.

where S is the secant modulus of the uniaxial stress-strain curve at $\sigma = \bar{\sigma}$. For an incompressible material, $\varepsilon_{ij}^e = 3s_{ij}/2E$ by Hooke's law, and (56) then furnishes $\varepsilon_{ij} = 3s_{ij}/2S$. The incremental form of (56) is

$$d\varepsilon_{ij}^p = \frac{3}{2\bar{\sigma}} \left\{ \left(d\bar{\varepsilon}^p - \frac{\bar{\varepsilon}^p \, d\bar{\sigma}}{\bar{\sigma}} \right) s_{ij} + \bar{\varepsilon}^p \, ds_{ij} \right\} \tag{57}$$

The Hencky equation is equivalent to the Prandtl-Reuss equation when the ratios of the deviatoric stress components are held constant. This may be shown by writing $s_{ij} = (\bar{\sigma}/Y)s_{ij}^0$, where s_{ij}^0 is the deviatoric stress at the initial yielding. Since $ds_{ij} = (d\bar{\sigma}/\bar{\sigma})s_{ij}$, the above equation reduces to that given by (41) and (42), the quantity $\overline{d\varepsilon^p}$ in this case being identical to $d\bar{\varepsilon}^p$. When the stress–plastic-strain relationship in uniaxial tension is represented by a power law, the Hencky equation is initially equivalent to the Prandtl-Reuss equation.† Indeed, if $\bar{\sigma}$ varies as the nth power of $\bar{\varepsilon}^p$, it is easily shown that $d\bar{\sigma}/\bar{\sigma} = n(d\bar{\varepsilon}^p/\bar{\varepsilon}^p)$, while $\bar{\varepsilon}^p/\bar{\sigma} \to 0$ as $\bar{\varepsilon}^p$ tends to zero. The complete equivalence of the two equations is established by setting $\overline{d\varepsilon^p} = (1 - n)\, d\bar{\varepsilon}^p$ at the initial stage.

The Hencky theory can be extended to large strains by using a suitable definition of the strain tensor ε_{ij}, the most natural definition in this context being $\varepsilon_{ij} = \int d\varepsilon_{ij}$, where the integral is taken along the path of the particle. When the principal axes of strain increment remain fixed in the element, it leads to the logarithmic strains in the principal directions.

If the principal axes of the stress do not rotate with respect to the element, a geometrical representation of stresses and strains is again possible. The vectors **OP** and **PS**, representing the deviatoric stress and the plastic strain respectively (Fig. 2.17), are directed along the same radial line whatever the stress path. According to

---

† A. A. Ilyushin, *Prikl. Mat. Mekh.*, **10**: 347 (1946); ibid., **11**: 293 (1947); F. Edelman, *Proc. 1st U.S. Nat. Congr. Appl. Mech.*, 493 (1951); L. M. Kachanov, *Foundations of the Theory of Plasticity*, p. 68, North Holland Pub. Co. (1971).

the Prandtl-Reuss theory, this happens only when the stress path is a straight line through the origin. The vector **OP** also represents the deviatoric elastic strain, and the vector **OS** represents the total deviatoric strain. The magnitudes of the vectors **OP** and **PS** are $\sqrt{\frac{2}{3}}\,\bar{\sigma}$ and $2G\sqrt{\frac{3}{2}}\,\bar{\varepsilon}^p$ respectively. Hence

$$\frac{PS}{OP} = 3G\left(\frac{1}{S} - \frac{1}{E}\right) = \frac{3(E/S - 1)}{2(1 + v)}$$

When the point S describes a certain curved strain path, the stress vector rotates so that it is always along OS. The stress increment **PP′** corresponding to a strain increment **SS′** is obtained by locating P' on OS' by the above relation. The vector **SR**, equal to **PP′**, represents the elastic strain increment, and the vector **RS′** represents the plastic strain increment. This is evidently incompatible with the Prandtl-Reuss flow rule which requires RS' to be parallel to OP. The stress path corresponding to the Hencky equation therefore differs from that given by the Prandtl-Reuss equation. The projection of RS' on a line through R parallel to OS is $RT = 2G\sqrt{\frac{3}{2}}\,d\bar{\varepsilon}^p$, where $d\bar{\varepsilon}^p$ is the increment of the equivalent total plastic strain. The scalar product of the vectors **OP** and **RS′** gives the increment of plastic work $dW_p = \bar{\sigma}\,d\bar{\varepsilon}^p$, which agrees with that obtained from the scalar product of (57) with the deviatoric stress s_{ij}.

The Hencky theory is unsuitable for describing the complete plastic behavior of metals, as can be seen by considering the unloading of the element after a certain amount of plastic deformation. If the element is reloaded to a different stress state on the current yield locus, the ratios of the plastic strain components will be entirely different. This is absurd, since only elastic changes in strain can occur while the stress point is inside the yield locus.[†] Nevertheless, when the loading is continuous and the deviatoric stress ratios do not vary significantly during the loading, the Hencky equation should lead to approximately correct results. Unloading can be permitted, provided the element is not reloaded to a plastic state different from that at the time of unloading.

(ii) *A range of validity* The total strain theory of Hencky is, in general, theoretically unacceptable, since it violates the fundamental rule of plastic flow of metals. Except for proportional loading, the Hencky stress-strain relation cannot be associated with a plastic potential with continously turning normal. Consider now the situation in which a corner is formed in the yield locus at the current stress point during the ensuing plastic deformation.[‡] The Hencky stress-strain relation is then compatible with the associated flow rule at the corner for a certain range of nonproportional loadings. It is assumed that the yield locus changes in such a way that the stress point continues to be at the corner during the deformation. It is

---

† The physical shortcomings of the Hencky theory have been demonstrated by C. H. Handelman, C. C. Lin, and W. Prager, *Q. Appl. Math.*, **4**: 397 (1947).

‡ The analysis presented here is much simpler than that originally given by B. Budiansky, *J. Appl. Mech.*, **26**: 259 (1959). For another approach to the problem, see V. D. Kliushnikov, *Prikl. Mat. Mekh.*, **23**: 405 (1959).

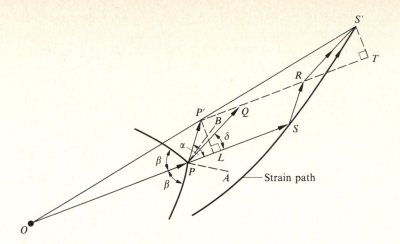

Figure 2.18 Development of a corner on the yield locus and its influence on the Hencky theory.

also supposed that the shape of the yield locus in the neighborhood of the corner at each stage is symmetrical about the current deviatoric stress vector.†

Let β denote the acute angle made by the deviatoric stress vector **OP** with either of the limiting tangents to the yield locus at a given stage (Fig. 2.18). According to the associated flow rule, the plastic strain-increment vector **PQ** = **RS'** can lie anywhere between the exterior normals PA and PB so long as the stress-increment vector **PP'** falls between the tangents produced. If α and δ are the angles made by the stress and plastic strain increments respectively with the current stress vector, the plastic stress-strain relation at the corner requires

$$\alpha < \beta \qquad \delta < \frac{\pi}{2} - \beta \tag{58}$$

Let L denote the foot of the perpendicular drawn from P' on OS. Then from geometry, $PL = \sqrt{\frac{2}{3}}\, d\bar{\sigma}$, and $RT = 2G\sqrt{\frac{3}{2}}\, d\bar{\varepsilon}^p$, giving

$$\frac{RT}{PL} = \frac{3G}{H} = 3G\left(\frac{1}{T} - \frac{1}{E}\right)$$

Also, from the similar triangles $P'OL$ and $S'P'T$, we have

$$\frac{S'T}{P'L} = \frac{P'S'}{OP'} = 3G\left(\frac{1}{S} - \frac{1}{E}\right)$$

† This is a feature of a slip theory of plasticity developed by S. B. Batdorf and B. Budiansky, *NACA Tech. Note*, 1971 (1949), and *J. Appl. Mech.*, **15**: 323 (1954). A similar theory, based on linear loading surfaces, has been put forward by J. L. Sanders, *Proc. 2d U.S. Nat. Cong. Appl. Mech.*, p. 445 (1954). For experimental evidences, see A. Phillips and G. A. Gray, *J. Basic Eng.*, **83**: 275 (1961). The slip theory has been further developed by T. H. Lin and M. Ito, *J. Mech. Phys. Solids*, **13**: 103 (1965); *Int. J. Engng. Sci.*, **1**: 543 (1966). See also T. H. Lin, *Theory of Inelastic Structures*, Chapter 4, John Wiley and Sons, New York (1968).

Since $S'T = RT \tan \delta$ and $P'L = PL \tan \alpha$, the above equations give the relationship between the angles α and δ as

$$\tan \alpha \cot \delta = \frac{E/T - 1}{E/S - 1}$$

Denoting the right-hand side of this equation by N, which is obtained from the given stress-strain curve, inequalities (58) may be expressed as

$$\tan \alpha < \tan \beta < \frac{N}{\tan \alpha}$$

The angle α represents the deviation from proportional loading. It will be assumed that the value of β at each stage is such that it allows the maximum variation of α. For a given stage, as α is increased, $\tan \alpha$ increases and $N/\tan \alpha$ decreases. So long as $\tan \alpha$ is less than $\sqrt{N}$, the above inequalities will be satisfied if $\tan \beta$ has an optimum value equal to $\sqrt{N}$. All other values of β will further restrict the variation of α. Hence the condition for validity of the Hencky theory becomes

$$\alpha < \tan^{-1} \sqrt{N} = \tan^{-1} \sqrt{\frac{E/T - 1}{E/S - 1}} \qquad (59)$$

For a work-hardening material with a definite yield point, β decreases from the initial value of $\pi/2$ as the loading proceeds. When the stresses at each stage of the loading are known, $\bar{\sigma}$ may be plotted as a function of the polar angle θ given by Eq. (3). The acute angle made by the tangent at a generic point of the curve with the radius vector is α, and $\tan \alpha$ is equal to the magnitude of $\bar{\sigma}(d\theta/d\bar{\sigma})$. The loading curve may also be plotted in rectangular coordinates using (2), where a factor of $\sqrt{\frac{3}{2}}$ may be introduced to make the length of the radius vector equal to $\bar{\sigma}$. For practical applications, it is sometimes convenient to express (59) in an alternative form, using the result

$$PP' = PL \sec \alpha < \sqrt{\tfrac{2}{3}} \, d\bar{\sigma} \sqrt{1 + N}$$

But PP' is equal to $\sqrt{ds_{ij} \, ds_{ij}}$, which is the magnitude of the deviatoric stress-increment vector. In terms of the principal components of the actual stress increment, the above inequality may be written as

$$(d\sigma_1 - d\sigma_2)^2 + (d\sigma_2 - d\sigma_3)^2 + (d\sigma_3 - d\sigma_1)^2 < 2(1 + N) \, d\bar{\sigma}^2 \qquad (60)$$

When the material work-hardens, the inequality (60) will be satisfied in a large number of practical problems where the stress ratios vary during the deformation. The parameter N is in fact the ratio of the slopes of the radius vector and the tangent at any point of the uniaxial stress–plastic-strain curve. If the stress-strain curve is represented by the Ramberg-Osgood equation, in which the plastic strain varies as the mth power of the true stress, N has a constant value equal to m. For a nonhardening material, the yield locus does not change during the deformation, and the Hencky theory is strictly valid when the stress point remains fixed during the deformation.

2.6 Theorems of Limit Analysis

(i) *Significance of the yield point* The yield point of a rigid/plastic body is defined as the moment when the deformation first becomes possible as the load is increased beyond the stage of initial yielding. During the loading interval marked by the beginnings of plastic yielding and plastic deformation, the body remains entirely rigid even though partly plastic. The rate at which the load must be increased to produce further distortion following the yield point depends on the rate of strain-hardening and the change in geometry. When the rate of hardening is of the order of the yield stress, and geometry changes are disregarded, only a slight increase in load is sufficient to produce strains of appreciable magnitude. In a slowly hardening elastic/plastic body, such strains generally occur under loads differing marginally from the rigid/plastic *yield point load*. More precisely, in a nonhardening deformable body, the load rapidly approaches an asymptotic value which is identical to the rigid/plastic yield point load.

Consider the state of stress at the yield point of a rigid/plastic body under given tractions T_j over a part S_F of the surface, and given velocities v_j over the remainder S_v. Let (σ_{ij}, v_j) and (σ_{ij}^*, v_j^*) be any two consistent solutions for the stress and associated velocity distributions corresponding to the same boundary conditions. Across certain internal surfaces, the velocity may be tangentially discontinuous. Such a surface may be considered as the limit of a narrow layer through which the velocity changes rapidly but continuously. Since the material is isotropic, the state of stress at the discontinuity is a pure shear k tangential to the surface, together with a hydrostatic stress. Let τ be the shear component of σ_{ij} along a discontinuity of v_j^*, and τ^* the shear component of σ_{ij}^* along a discontinuity of v_j. Then, by the principle of virtual work, expressed by (79), Chap. 1, with the field quantities replaced by the corresponding differences, we have

$$\int (T_j - T_j^*)(v_j - v_j^*) \, dS = \int (\sigma_{ij} - \sigma_{ij}^*)(\dot{\varepsilon}_{ij} - \dot{\varepsilon}_{ij}^*) \, dV$$

$$+ \int (k - \tau^*)[v] \, dS_D + \int (k - \tau)[v^*] \, dS_D^* \quad (61)$$

where the square brackets denote the magnitudes of the velocity discontinuities across the respective internal surfaces S_D and S_D^*. Since no shear-stress component can be numerically greater than k, the surface integrals on the right-hand side are nonnegative. By the maximum work inequality, and the fact that the strain rate vanishes in a rigid region,

$$(\sigma_{ij} - \sigma_{ij}^*)(\dot{\varepsilon}_{ij} - \dot{\varepsilon}_{ij}^*) = (\sigma_{ij} - \sigma_{ij}^*)\dot{\varepsilon}_{ij} + (\sigma_{ij}^* - \sigma_{ij})\dot{\varepsilon}_{ij}^* \geqslant 0$$

which indicates that the volume integral in (61) is also nonnegative. Since $T_j = T_j^*$ on S_F and $v_j = v_j^*$ on S_v, the integrand on the left-hand side of (61) vanishes at each point of the boundary. Hence the integrals on the right-hand side must separately vanish. It follows that for a strictly convex yield surface, wherever $\dot{\varepsilon}_{ij}$ or $\dot{\varepsilon}_{ij}^*$ is nonzero, σ_{ij} and σ_{ij}^* can only differ from one another by a hydrostatic stress which must be uniform for equilibrium. In other words, the state of stress at the yield point is uniquely defined in a region where the material can deform under given boundary conditions.† The present analysis says nothing, however, about the possible modes of deformation, nor the extent of the deforming zone.

In a nonhardening rigid/plastic body, the deformation following the yield point occurs under constant loads so long as the changes in geometry are negligible. This may be shown by considering the equilibrium equations $\partial \dot{\sigma}_{ij}/\partial x_i = 0$ and the boundary conditions $\dot{T}_j = l_i \dot{\sigma}_{ij}$ satisfied by the stress rate $\dot{\sigma}_{ij}$. The transformation by Green's theorem furnishes

$$\int \dot{T}_j v_j \, dS = \int \frac{\partial}{\partial x_i} (\dot{\sigma}_{ij} v_j) \, dV = \int \dot{\sigma}_{ij} \frac{\partial v_j}{\partial x_i} \, dV = \int \dot{\sigma}_{ij} \dot{\varepsilon}_{ij} \, dV$$

in view of the symmetry of the tensor $\dot{\sigma}_{ij}$. Since $\dot{\sigma}_{ij}$ and $\dot{\varepsilon}_{ij}$ are mutually orthogonal in the vector representation wherever $\dot{\varepsilon}_{ij} \neq 0$, the scalar product $\dot{\sigma}_{ij} \dot{\varepsilon}_{ij}$ is zero at each point. It follows that each component of the traction rate $\dot{T}_j$ must also vanish unless the corresponding component of the surface velocity is zero. In an elastic/plastic solid, on the other hand, the loads do not generally attain constant values for any finite strain, but the rate of change of the loads becomes insignificant after a permanent strain which is still of an elastic order of magnitude. As the asymptotic values of the loads are approached, the deformation increases so rapidly that in an actual structure the result is tantamount to plastic collapse. The elastic part of the strain rate then becomes negligible compared to the plastic part.

(ii) Lower and upper bound theorems We now proceed to develop the extremum principles associated with the uniqueness theorem. An assumed stress field will be called *statically admissible* if it satisfies the equilibrium equations and the stress boundary conditions without violating the yield criterion. Let $(\sigma_{ij}, \dot{\varepsilon}_{ij})$ now refer to the actual stress and the associated strain rate corresponding to any yield point state. If σ_{ij}^* is any other statically admissible state of stress,‡ then by the principle of virtual work,

$$\int (T_j - T_j^*) v_j \, dS = \int (\sigma_{ij} - \sigma_{ij}^*) \dot{\varepsilon}_{ij} \, dV + \int (k - \tau^*)[v] \, dS_D \tag{62}$$

The second integral on the right-hand side, representing the contribution from any velocity discontinuity in the actual state, is definitely nonnegative since $\tau^* \leqslant k$. The first integral on the right-hand side is also nonnegative by the maximum work inequality and the fact that $\dot{\varepsilon}_{ij}$ vanishes in any rigid region. If some or all of the traction components are prescribed on S_F, while the complementary velocity

---

† The uniqueness theorem is due to R. Hill, *Phil. Mag.*, **42**: 868 (1951). It is a generalization of an earlier result by R. Hill, *Q. J. Mech. Appl. Math.*, **1**: 18 (1948). See also E. H. Lee, *Phil. Mag.*, Ser. 7, **43**: 549 (1952).

‡ The assumed stress field may include surfaces of stress discontinuity provided the tractions across them are continuous.

components are zero, the left-hand side of (62) vanishes over S_F. Then

$$\int T_j v_j \, dS_v \geqslant \int T_j^* v_j \, dS_v = \int l_i \sigma_{ij}^* v_j \, dS_v \qquad (63)$$

The result provides the *lower bound theorem*, which states that the rate of work done by the actual surface tractions on S_v is greater than or equal to that done by the surface tractions in any statically admissible stress field.† If the prescribed velocity on S_v is uniform, a knowledge of the rate of work is equivalent to a knowledge of the overall load acting on S_v. The theorem then gives a lower bound on the load itself at the yield point. Similarly, when the velocity on S_v corresponds to a rigid-body rotation, a lower bound on the yield couple can be estimated.

For the complementary minimum principle, let v_j^* denote any piecewise continuous velocity field satisfying the incompressibility condition. Let σ_{ij}^* be the stress that would be compatible with the strain rate $\dot\varepsilon_{ij}^*$ according to the normality rule. If the actual stress at the yield point is σ_{ij} as before, then by the virtual work principle,

$$\int T_j v_j^* \, dS = \int \sigma_{ij} \dot\varepsilon_{ij}^* \, dV + \int \tau[v^*] \, dS_D^* \qquad (64)$$

which includes the contribution to the rate of energy dissipation from any discontinuity S_D^* in the virtual velocity field. The distribution σ_{ij}^* is not necessarily in equilibrium, and is undefined where $\dot\varepsilon_{ij}^* = 0$. Since $\tau \leqslant k$, and $\sigma_{ij} \dot\varepsilon_{ij}^* \leqslant \sigma_{ij}^* \dot\varepsilon_{ij}^*$ by the maximum work inequality, we have

$$\int T_j v_j^* \, dS_v \leqslant \int \sigma_{ij}^* \dot\varepsilon_{ij}^* \, dV + \int k[v^*] \, dS_D^* - \int T_j v_j^* \, dS_F \qquad (65)$$

The virtual velocity field is regarded as *kinematically admissible* if it also satisfies the velocity boundary conditions on S_v. For a kinematically admissible field, therefore, we may write $v_j^* = v_j$ on the left-hand side of (65). The result is generally known as the *upper bound theorem* which states that the rate of work done by the unknown surface tractions on S_v is less than or equal to the rate of internal energy dissipated in any kinematically admissible velocity field.‡ The theorem gives an upper bound on the yield point load itself in the special case when the velocity on S_v is constant in magnitude and direction.¶

---

† R. Hill, op. cit. A heuristic principle of maximum plastic resistance was proposed earlier by M. A. Sadowsky, *J. Appl. Mech.*, **10**: A-65 (1943), on the basis of its success in a few special cases. The theorem also holds for an incompressible material when the Hencky equations are used, and the body is assumed entirely plastic. See A. H. Philippidis, *J. Appl. Mech.*, **15**: 241 (1948), and H. J. Greenberg, ibid., **16**: 103 (1949).

‡ R. Hill, *Phil. Mag.*, **42**: 868 (1951). For a Mises material, the theorem was proved earlier by A. A. Markov, *Prikl. Mat. Mekh.*, **11**: 339 (1947), and R. Hill. *J. Appl. Mech.*, **17**: 64 (1950), under the restriction that the entire body is deforming plastically.

¶ Assuming that the load remains constant at the moment of collapse in a non-hardening elastic/plastic material, while geometry changes are still negligible, the theorems of limit analysis have been proved by D. C. Drucker, W. Prager, and H. J. Greenberg, *Q. Appl. Math.*, **9**: 381 (1952).

A slightly different situation arises when several loads are applied to a structure, their ratios at the yield point being given. These ratios may be regarded as maintained constant as the loads are gradually increased to their collapse values (proportional loading). The absolute magnitude of the loads can be specified by a single parameter T at the yield point. Each load corresponds to a surface traction $T_j = Tr_j$, where r_j is a given vector specifying the load. Let S_R denote the part of the surface where the load ratios are given, the remaining surface being assumed partly traction-free and partly attached with rigid constraints ($v_j = 0$). If an assumed stress field is such that $T_j^* = T^* r_j$ on S_R, then (62) immediately shows that $T \geqslant T^*$, providing a lower bound on the loading parameter. Equation (64), together with the maximum work inequality, furnishes

$$T \int r_j v_j^* \, dS_R \leqslant \int \sigma_{ij}^* \dot{\varepsilon}_{ij}^* \, dV + \int k[v^*] \, dS_{\overline{D}} \tag{66}$$

from which an upper bound on T can be estimated by using a trial velocity field v_j^* vanishing at the specified constraints. With a judicious choice, these bounds can be brought close enough to provide a useful approximation to the collapse load.

The magnitude of the yield point load evidently depends on the particular choice of the yield function and the plastic potential. Suppose that the von Mises yield and the associated Lévy-Mises flow rule constitute a standard for comparison. When Tresca's yield function and potential are used for an approximation, the corresponding hexagonal yield prism may be regarded as either inscribed or circumscribed with respect to the Mises cylinder. Considering a mean of the two corresponding load values, which are in the ratio $2/\sqrt{3}$, the error in the approximation can be limited to ± 7.5 percent.† The Tresca theory has been extensively used in the literature for the estimation of limit loads of structures.

(iii) *Influence of Coulomb friction* Suppose, now, that the surface S_v represents the interface between a workpiece and a rigid die. The boundary conditions on S_v require the normal component of velocity to be compatible with the rigid motion of the die, and the tangential traction to satisfy a prescribed frictional law. We begin with the situation where the tangential traction has a constant value along the interface. If (v_n^*, v_t^*) are the normal and tangential components of the trial velocity at the interface, and (T_n, T_t) the corresponding components of the actual traction, then

$$\int T_j v_j^* \, dS_v = \int T_n v_n^* \, dS_v + \int T_t v_t^* \, dS_v$$

where $T_t = 0$ for a perfectly smooth die and $T_t = -k$ for a perfectly rough die, both v_n^* and v_t^* being considered positive. If the interface is planar, a velocity field for which v_n^* is equal to the prescribed normal velocity v_n provides an upper bound on the total normal load. The trial velocity field is then kinematically admissible.

---

† R. Hill, *Phil. Mag.*, Ser. 7, **43**: 353 (1952).

In the case of Coulomb friction existing at the interface, the resultant traction on S_v, having a magnitude T_r, acts in a direction making an angle $\lambda = \tan^{-1} \mu$ with the inward normal to the surface, where μ is the coefficient of friction along the interface. If v_r^* denotes the component of the trial velocity vector in the direction of the resultant traction, then $T_j v_j^* = T_r v_r^*$. When S_v is planar, and the trial velocity field is such that v_r^* is constant on S_v, (65) furnishes an upper bound on the resultant force transmitted across the interface.[†] The assumed velocity field is not kinematically admissible, unless the corresponding resultant velocity on S_v is constant in magnitude and direction.

Suppose, now, that the interface S_v is nonplanar, and we are required to find an upper bound on the component of the resultant force in a fixed direction $\mathbf{v}$. Then the trial velocity field should be such that its component in the direction of the resultant traction is $v_r^* = V_0 \cos \theta$, where V_0 is a constant, and θ is the acute angle between $\mathbf{v}$ and the direction of the resultant traction. Since

$$\int T_j v_j^* \, dS_v = \int T_r v_r^* \, dS_v = V_0 \int T_r \cos \theta \, dS_v$$

the integral on the left-hand side of (65) is equal to V_0 times the required load at the yield point. The trial velocity field is not kinematically admissible, but it does provide an upper bound.

In actual practice, Coulomb's law may be operative over a part of the interface between the material and the die. The remaining part of the interface S_v generally consists of two sections, with sticking friction existing on one and zero relative motion on the other. The magnitude of the tangential traction is less than μT_n on both these sections. If S_v is a single surface, an upper bound on the total normal load can be obtained by setting $v_r^* = V_0 \cos \lambda$ over the Coulomb part, and $v_n^* = V_0$ over the remainder of the interface.

The limit theorems are equally relevant to steady state problems, involving nonhardening materials, where the external configuration of the body is specified. Since the deformed state of the body may be regarded as the sum total of incremental distortions due to a succession of yield points, lower and upper bounds on the load required to maintain the steady state can be found from (63) and (65). When the problem is nonsteady, the continually changing external configuration is not known in advance, and the present approximation methods are not directly applicable for a finitely deformed body.

(iv) *A shakedown theorem* An elastic/plastic structure subjected to a cyclic loading program may fail due to excessive deformation, even though the extreme values of the load may be lower than that corresponding to plastic collapse. The plastic strains produced in a critical load cycle progressively increase as the cycle is repeated, and the structure eventually becomes unserviceable. In some cases, cycles of plastic flow may occur in alternating directions causing an early fatigue failure.

[†] I. F. Collins, *J. Mech. Phys. Solids*, **17**: 323 (1969).

The structure will be safe, however, if the load varies in such a way that the stress distribution remains within the yield limit after an initial period of plastic flow. The structure is then said to *shake down* to a state of residual stress, the change in strain being entirely elastic as the load is subsequently varied between the prescribed limits.

Consider a nonhardening elastic/plastic body in which the deformation is small enough for changes in geometry to be negligible. Let $(\sigma_{ij}, \varepsilon_{ij})$ denote the actual stress and strain distributions at any instant, and $(\sigma'_{ij}, \varepsilon'_{ij})$ those which would exist under the same boundary conditions if the body behaved elastically. It is convenient to write

$$\sigma_{ij} = \sigma'_{ij} + \rho_{ij} \qquad \dot{\varepsilon}^e_{ij} = \dot{\varepsilon}'_{ij} + \dot{\gamma}_{ij}$$

where ρ_{ij} denotes a self-equilibrilating distribution of residual stress, and γ_{ij} the corresponding elastic strain. The residual stress may be regarded as that remaining in the body on complete removal of the external load by an elastic process. The plastic strain ε^p_{ij} produced by ρ_{ij} is identical to that by σ_{ij}. It can be shown that not more than one residual stress distribution can exist for a given plastic strain distribution with zero displacement at the constraints.†

Suppose it is possible to find a time-independent residual stress ρ^*_{ij} such that the superposition of this stress on σ'_{ij} gives a safe state of stress σ^*_{ij} that is always below the yield limit. The elastic strain energy of the body associated with the stress difference $\rho_{ij} - \rho^*_{ij}$ may be written as

$$W = \frac{1}{2} \int (\rho_{ij} - \rho^*_{ij})(\gamma_{ij} - \gamma^*_{ij}) dV = \frac{1}{2} \int c_{ijkl}(\rho_{ij} - \rho^*_{ij})(\rho_{kl} - \rho^*_{kl}) \, dV$$

where c_{ijkl} is the tensor of elastic constants having the symmetry properties $c_{ijkl} = c_{jikl} = c_{ijlk} = c_{klij}$. Since ρ^*_{ij} does not vary with time, the rate of change of W is

$$\dot{W} = \int c_{ijkl}(\rho_{ij} - \rho^*_{ij})\dot{\rho}_{kl} \, dV = \int (\rho_{ij} - \rho^*_{ij})\dot{\gamma}_{ij} \, dV \qquad (67)$$

Let T_j and T^*_j denote the surface tractions associated with the stresses σ_{ij} and σ^*_{ij} respectively. If v_j and v'_j are continuous velocity distributions associated with the strain rates $\dot{\varepsilon}_{ij}$ and $\dot{\varepsilon}'_{ij}$, it follows from the virtual work principle that

$$\int (\sigma_{ij} - \sigma^*_{ij})(\dot{\varepsilon}_{ij} - \dot{\varepsilon}'_{ij}) \, dV = \int (T_j - T^*_j)(v_j - v'_j) \, dV = 0$$

in view of the fact that $T_j = T^*_j$ on S_F and $v_j = v'_j$ on S_v. Since $\dot{\varepsilon}_{ij} - \dot{\varepsilon}'_{ij} = \dot{\gamma}_{ij} + \dot{\varepsilon}^p_{ij}$, the above equation becomes

$$\int (\sigma_{ij} - \sigma^*_{ij})\dot{\gamma}_{ij} \, dV = -\int (\sigma_{ij} - \sigma^*_{ij})\dot{\varepsilon}^p_{ij} \, dV$$

---

† The proof is based on the differences $\Delta\rho_{ij}$ and $\Delta\gamma_{ij}$ between the residual stresses and strains in two possible solutions. The integral of the scalar product $\Delta\rho_{ij} \, \Delta\gamma$ taken over the entire volume of the body vanishes by the principle of virtual work. Since $\Delta\gamma_{ij} = \Delta\gamma^e_{ij}$, the integrand is positive unless $\Delta\rho_{ij} = 0$, which completes the proof.

The integral on the right-hand side is positive by the maximum work inequality, whenever plastic flow occurs in the actual loading program. Hence (67) gives

$$\dot{W} = \int (\rho_{ij} - \rho_{ij}^*)\dot{\gamma}_{ij}\,dV = \int (\sigma_{ij} - \sigma_{ij}^*)\dot{\gamma}_{ij}\,dV \leqslant 0 \tag{68}$$

with equality only when there is no plastic flow. Since W can never become negative, (68) indicates that plastic flow cannot continue indefinitely, and the body will ultimately shake down to a steady distribution of residual stress. This completes the proof of the *shakedown theorem* which states that an elastic/perfectly plastic structure will shake down for given extreme values of the load if there exists a time-independent distribution of residual stress that nowhere leads to stresses beyond the yield limit when superimposed on the stress distribution corresponding to the elastic response of the structure.† The theorem says nothing about the amount of plastic deformation that may occur before the shakedown state is attained.

2.7 Uniqueness Theorems

(i) *Small elastic/plastic deformation* Consider an elastic/plastic body undergoing deformation under prescribed surface tractions and surface displacements. In general, a part of the body will be plastic and a part will be elastic. The current distribution of stresses and strains in the entire body is assumed to be known. Consider now a further infinitesimal distortion produced by prescribed traction rates on a part of the boundary, and prescribed velocities on the remainder. We propose to examine the conditions under which the stress and strain rates throughout the body are uniquely determined from the given boundary conditions. When positional changes and rotations of all material elements are disregarded, and the material is work-hardening, the bilinear constitutive equation may be written as

$$\dot{\varepsilon}_{ij} = \frac{1}{2G}\left(\dot{\sigma}_{ij} - \frac{v}{1+v}\dot{\sigma}_{kk}\,\delta_{ij}\right) + \frac{3}{2H}\dot{\sigma}_{kl}n_{kl}n_{ij} \qquad \dot{\sigma}_{kl}n_{kl} \geqslant 0$$

$$\dot{\varepsilon}_{ij} = \frac{1}{2G}\left(\dot{\sigma}_{ij} - \frac{v}{1+v}\dot{\sigma}_{kk}\,\delta_{ij}\right) \qquad\qquad \dot{\sigma}_{kl}n_{kl} \leqslant 0 \tag{69}$$

whenever an element is currently plastic. The unit normal n_{ij} to the yield surface, which is assumed regular, corresponds to the current state of stress σ_{ij}. The conditions $\dot{\sigma}_{ij}n_{ij} > 0$, $\dot{\sigma}_{ij}n_{ij} = 0$, and $\dot{\sigma}_{ij}n_{ij} < 0$ represent loading, neutral loading, and unloading respectively. The second relation of (69) also applies to an elastic element which may or may not have been previously plastic.

---

† The static shakedown theorem is due to E. Melan, *Ing.-Arch.*, **9**: 116 (1938). See also P. S. Symonds, *J. Appl. Mech.*, **18**: 85 (1951). For a kinematic counterpart of Melan's theorem, see W. T. Koiter, *Proc. Kon. Ned. Ak. Wet.*, **B59**: 24 (1956), *Progress in Solid Mechanics* (Eds. I. N. Sneddon and R. Hill), **1**: 167, North-Holland Publishing Co. (1960). For a finite element formulation of the problem, see G. Maier, *Meccanica*, **6**: 250 (1969), and **7**: 51 (1970).

Let $(\dot{\sigma}_{ij}, \dot{\varepsilon}_{ij})$ and $(\dot{\sigma}_{ij}^*, \dot{\varepsilon}_{ij}^*)$ represent two distinct distributions of stress and strain rates under prescribed traction rates and surface velocities. The strain rates $\dot{\varepsilon}_{ij}$ and $\dot{\varepsilon}_{ij}^*$ are derivable from continuous velocity distributions v_j and v_j^* respectively. The stress rates $\dot{\sigma}_{ij}$ and $\dot{\sigma}_{ij}^*$ separately satisfy the equation of equilibrium, and give rise to traction rates $\dot{T}_j$ and $\dot{T}_j^*$ respectively. The field equations satisfied by the differences $\Delta\dot{\sigma}_{ij} = \dot{\sigma}_{ij} - \dot{\sigma}_{ij}^*$ and $\Delta\dot{\varepsilon}_{ij} = \dot{\varepsilon}_{ij} - \dot{\varepsilon}_{ij}^*$ are

$$\frac{\partial}{\partial x_i}(\Delta\dot{\sigma}_{ij}) = 0 \qquad \Delta\dot{T}_j = l_i\,\Delta\dot{\sigma}_{ij}$$

$$2\Delta\dot{\varepsilon}_{ij} = \frac{\partial}{\partial x_j}(\Delta v_i) + \frac{\partial}{\partial x_i}(\Delta v_j)$$

where $\Delta\dot{T}_j$ and Δv_j denote the differences $\dot{T}_j - \dot{T}_j^*$ and $v_j - v_j^*$ respectively, while l_i is the exterior unit normal to a typical surface element. Using Green's theorem for the transformation of a surface integral into a volume integral, we get

$$\int \Delta\dot{T}_j\,\Delta v_j\,dS = \int \frac{\partial}{\partial x_i}(\Delta\dot{\sigma}_{ij}\,\Delta v_j)\,dV = \int \Delta\dot{\sigma}_{ij}\frac{\partial}{\partial x_i}(\Delta v_j)\,dV$$

Since $\Delta\dot{T}_j = 0$ on the part of the boundary where the traction rate is prescribed, and $\Delta v_j = 0$ on the part of the boundary where the velocity is prescribed, the surface integral is identically zero. In view of the symmetry of the stress-rate tensor, the integrand on the right-hand side is equal to $\Delta\dot{\sigma}_{ij}\,\Delta\dot{\varepsilon}_{ij}$. Hence

$$\int \Delta\dot{\sigma}_{ij}\,\Delta\dot{\varepsilon}_{ij}\,dV = \int (\dot{\sigma}_{ij} - \dot{\sigma}_{ij}^*)(\dot{\varepsilon}_{ij} - \dot{\varepsilon}_{ij}^*)\,dV = 0 \qquad (70)$$

The relationship between $\Delta\dot{\varepsilon}_{ij}$ and $\Delta\dot{\sigma}_{ij}$ is identical in form to the first equation of (69) when $\dot{\sigma}_{ij}n_{ij} \geqslant 0$ and $\dot{\sigma}_{ij}^*n_{ij} \geqslant 0$, and to the second equation of (69) when $\dot{\sigma}_{ij}n_{ij} \leqslant 0$ and $\dot{\sigma}_{ij}^*n_{ij} \leqslant 0$. Under these conditions, it is immediate that

$$\Delta\dot{\sigma}_{ij}\,\Delta\dot{\varepsilon}_{ij} \geqslant \frac{1}{2G}\left(\Delta\dot{\sigma}_{ij}\,\Delta\dot{\sigma}_{ij} - \frac{v}{1+v}\Delta\dot{\sigma}_{ii}\,\Delta\dot{\sigma}_{jj}\right) \qquad (71)$$

since $(n_{ij}\,\Delta\dot{\sigma}_{ij})^2 \geqslant 0$. If, on the other hand, the stress changes in such a way that either $\dot{\sigma}_{ij}n_{ij} \geqslant 0$ and $\dot{\sigma}_{ij}^*n_{ij} \leqslant 0$, or $\dot{\sigma}_{ij}n_{ij} \leqslant 0$ and $\dot{\sigma}_{ij}^*n_{ij} \geqslant 0$, the scalar products of (69) with $\dot{\sigma}_{ij}$ and $\dot{\sigma}_{ij}^*$ indicate that

$$\dot{\sigma}_{ij}\dot{\varepsilon}_{ij} \geqslant \frac{1}{2G}\left(\dot{\sigma}_{ij}\dot{\sigma}_{ij} - \frac{v}{1+v}\dot{\sigma}_{ii}\dot{\sigma}_{jj}\right)$$

$$\dot{\sigma}_{ij}^*\dot{\varepsilon}_{ij} \leqslant \frac{1}{2G}\left(\dot{\sigma}_{ij}^*\dot{\sigma}_{ij} - \frac{v}{1+v}\dot{\sigma}_{ii}^*\dot{\sigma}_{jj}\right) \qquad (72)$$

The first inequality, in fact, applies in all cases. The inequality satisfied by $\dot{\sigma}_{ij}\dot{\varepsilon}_{ij}^*$ is identical to the second of (72), while that satisfied by $\dot{\sigma}_{ij}^*\dot{\varepsilon}_{ij}$ is similar to the first of

(72). Thus

$$(\dot{\sigma}_{ij} - \dot{\sigma}_{ij}^*)(\dot{\varepsilon}_{ij} - \dot{\varepsilon}_{ij}^*) = (\dot{\sigma}_{ij}\dot{\varepsilon}_{ij} + \dot{\sigma}_{ij}^*\dot{\varepsilon}_{ij}^*) - (\dot{\sigma}_{ij}^*\dot{\varepsilon}_{ij} + \dot{\sigma}_{ij}\dot{\varepsilon}_{ij}^*)$$

$$\geqslant \frac{1}{2G}\left((\dot{\sigma}_{ij} - \dot{\sigma}_{ij}^*)(\dot{\sigma}_{ij} - \dot{\sigma}_{ij}^*) - \frac{v}{1+v}(\dot{\sigma}_{ii} - \dot{\sigma}_{ii}^*)^2\right)$$

Since the equality always holds in an element that is currently elastic, (71) is satisfied at each point of the body. The right-hand side of (71) is definitely positive unless $\dot{\sigma}_{ij}^* = \dot{\sigma}_{ij}$, as is evident from the fact that the expression in the parenthesis is equal to

$$\Delta\dot{s}_{ij}\,\Delta\dot{s}_{ij} + \frac{1}{3}\left(\frac{1-2v}{1+v}\right)\Delta\dot{\sigma}_{ii}\,\Delta\dot{\sigma}_{jj}$$

where $\Delta\dot{s}_{ij}$ is the deviatoric part of $\Delta\dot{\sigma}_{ij}$. It follows that the integrand of (70) is positive except when $\dot{\sigma}_{ij}^* = \dot{\sigma}_{ij}$, indicating that the distribution of stress rate is unique when the strains are small.† The uniqueness theorem holds equally good for a nonhardening material, since (70) and (71) are independent of the rate of hardening. Whereas in a work-hardening material uniqueness of the stress rate also ensures uniqueness of the strain rate, the plastic part of the strain rate in a nonhardening material is indeterminate from the constitutive equation alone. The distribution of velocities in a nonhardening material is, therefore, not necessarily unique.

(ii) *Large elastic/plastic deformation* When large plastic strains are possible under given boundary conditions, the preceding analysis must be modified by including changes in geometry of the deforming element. As explained earlier, the stress rate appearing in the constitutive equation must be considered in Jaumann's sense, the instantaneous rotation of the axes of reference being associated with the local spin tensor $\omega_{ij} = -e_{ijk}\omega_k$, where ω_k are the components of the vector $\frac{1}{2}$ curl $\mathbf{v}$. The relationship between the stress rate and the strain rate is

$$\overset{\circ}{\sigma}_{ij} = 2G\left(\dot{\varepsilon}_{ij} + \frac{v}{1-2v}\dot{\varepsilon}_{kk}\delta_{ij} - \frac{3G}{3G+H}\dot{\varepsilon}_{kl}n_{kl}n_{ij}\right) \qquad \dot{\varepsilon}_{kl}n_{kl} \geqslant 0$$

$$\qquad\qquad\qquad\qquad\qquad\qquad\qquad\qquad\qquad\qquad\qquad\qquad\qquad (73)$$

$$\overset{\circ}{\sigma}_{ij} = 2G\left(\dot{\varepsilon}_{ij} + \frac{v}{1-2v}\dot{\varepsilon}_{kk}\delta_{ij}\right) \qquad\qquad\qquad \dot{\varepsilon}_{kl}n_{kl} \leqslant 0$$

whenever an element is currently plastic. The second expression also holds in an element that is stressed below the yield limit. The Jaumann stress rate $\overset{\circ}{\sigma}_{ij}$ is related to the material derivative $\dot{\sigma}_{ij}$ by Eq. (81), Chap. 1. Thus

$$\dot{\sigma}_{ij} = \overset{\circ}{\sigma}_{ij} + \sigma_{ik}\omega_{jk} + \sigma_{jk}\omega_{ik} \qquad\qquad\qquad\qquad\qquad\qquad (74)$$

---

† This uniqueness theorem is due to E. Melan, *Ing.-Arch.*, **9**: 116 (1938). For a nonhardening Prandtl-Reuss material, the uniqueness of the stress-rate distribution has been proved by H. J. Greenberg, *Q. Appl. Math.*, **7**: 85 (1949). See also W. T. Koiter, *Q. Appl. Math.*, **11**: 350 (1953).

When geometry changes are taken into account, it is convenient to formulate the boundary-value problem in terms of the rate of change of the nominal traction, based on the current configuration of the body. Denoting the nominal stress rate by $\dot{t}_{ij}$, and the nominal traction rate by $\dot{F}_j$, the equilibrium equations and the boundary conditions may be written as

$$\frac{\partial \dot{t}_{ij}}{\partial x_i} + \dot{g}_j = 0 \qquad \dot{F}_j = l_i \dot{t}_{ij} \tag{75}$$

where $\dot{g}_j$ is the body force rate per unit current volume, x_i the current position vector, and l_i the unit outward normal to the surface at the current configuration. If the instantaneous velocity is denoted by v_j, the relationship between $\dot{t}_{ij}$ and $\dot{\sigma}_{ij}$ is

$$\dot{t}_{ij} = \dot{\sigma}_{ij} + \sigma_{ij}\frac{\partial v_k}{\partial x_k} - \sigma_{jk}\frac{\partial v_i}{\partial x_k} \tag{76}$$

Suppose, now, that there could be two distinct solutions to the boundary-value problem for given nominal traction rates $\dot{F}_j$ on a part S_F of the boundary, given velocities v_j on the remainder S_v, and given body force rates $\dot{g}_j$ throughout the volume V. If the difference between the corresponding quantities in the two possible solutions is denoted by a prefix Δ, Eqs. (75) and (76) provide

$$\frac{\partial}{\partial x_i}(\Delta \dot{t}_{ij}) = 0 \qquad \Delta \dot{F}_j = l_i \, \Delta \dot{t}_{ij}$$

$$\Delta \dot{t}_{ij} = \Delta \dot{\sigma}_{ij} + \sigma_{ij}\frac{\partial}{\partial x_k}(\Delta v_k) - \sigma_{jk}\frac{\partial}{\partial x_k}(\Delta v_i) \tag{77}$$

In view of the given boundary conditions, $\Delta \dot{F}_j = 0$ on S_F and $\Delta v_j = 0$ on S_v. The application of Green's theorem to integrals involving the surface S and the volume V in the current state gives

$$\int \Delta \dot{F}_j \, \Delta v_j \, dS = \int \frac{\partial}{\partial x_i}(\Delta \dot{t}_{ij} \, \Delta v_j) \, dV = \int \Delta \dot{t}_{ij} \frac{\partial}{\partial x_i}(\Delta v_j) \, dV$$

by (77). Since the surface integral vanishes throughout the boundary, the condition for having two possible solutions, constituting the phenomenon of bifurcation, becomes[†]

$$\int \Delta \dot{t}_{ij} \frac{\partial}{\partial x_i}(\Delta v_j) \, dV = 0$$

Substituting for $\Delta \dot{t}_{ij}$, and using the symmetry of the stress tensor, the above condition can be expressed as

$$\int \left\{ \Delta \dot{\sigma}_{ij} \, \Delta \dot{\varepsilon}_{ij} + \sigma_{ij}\frac{\partial}{\partial x_k}(\Delta v_k)\frac{\partial}{\partial x_i}(\Delta v_j) - \sigma_{ij}\frac{\partial}{\partial x_j}(\Delta v_k) \right\} dV = 0 \tag{78}$$

[†] The principles of uniqueness and stability in elastic/plastic solids, using a generalized constitutive law, have been discussed by R. Hill, *J. Mech. Phys. Solids*, **6**: 637 (1958). More general boundary conditions have been examined by R. Hill, *J. Mech. Phys. Solids*, **10**: 185 (1962).

after suitable interchanging of dummy suffixes. It follows from (74) that

$$\Delta \dot{\sigma}_{ij} = \Delta \overset{\circ}{\sigma}_{ij} + \sigma_{ik} \Delta \omega_{jk} + \sigma_{jk} \Delta \omega_{ik}$$

From the symmetry of σ_{ij} and $\dot{\varepsilon}_{ij}$, and the antisymmetry of ω_{ij}, it is easy to show that

$$\Delta \dot{\sigma}_{ij} \Delta \dot{\varepsilon}_{ij} = \Delta \overset{\circ}{\sigma}_{ij} \Delta \dot{\varepsilon}_{ij} - 2\sigma_{ij} \Delta \dot{\varepsilon}_{jk} \Delta \omega_{ik}$$

which gives the leading term of the integrand in terms of the stress rates that appear in the constitutive equations. A sufficient condition for uniqueness of the boundary-value problem may therefore be written as

$$\int \left\{ \Delta \overset{\circ}{\sigma}_{ij} \Delta \dot{\varepsilon}_{ij} + \sigma_{ij} \left[\Delta \dot{\varepsilon}_{kk} \Delta \dot{\varepsilon}_{ij} - 2\Delta \dot{\varepsilon}_{jk} \Delta \omega_{ik} - \frac{\partial}{\partial x_k} (\Delta v_i) \frac{\partial}{\partial x_j} (\Delta v_k) \right] \right\} dV > 0$$

for every distinct pair of continuous differentiable velocity fields taking prescribed values on S_v.

Consider, now, a hypothetical solid whose constitutive law is given by the first equation of (73) regardless of the sign of $\dot{\varepsilon}_{ij} n_{ij}$, whenever the element is currently plastic. Since the stress rate is then a unique linear function of the strain rate with identical loading and unloading responses, the new solid may be regarded as a linearized elastic/plastic solid.[†] If the true stress rate for the linearized solid is denoted by $\hat{\tau}_{ij}$, corresponding to a given strain rate $\dot{\varepsilon}_{ij}$, it follows from (73) that

$$\overset{\circ}{\sigma}_{ij} \dot{\varepsilon}_{ij} \geqslant \hat{\tau}_{ij} \dot{\varepsilon}_{ij} = 2G \left\{ \dot{\varepsilon}_{ij} \dot{\varepsilon}_{ij} + \frac{v}{1 - 2v} (\dot{\varepsilon}_{kk})^2 - \frac{3G}{3G + H} (\dot{\varepsilon}_{ij} n_{ij})^2 \right\} \tag{79}$$

where the equality holds only in the loading part of the plastic region. In the elastic region, the linearized solid behaves identically to the actual elastic/plastic solid, giving

$$\overset{\circ}{\sigma}_{ij} \dot{\varepsilon}_{ij} = \hat{\tau}_{ij} \dot{\varepsilon}_{ij} = 2G \left\{ \dot{\varepsilon}_{ij} \dot{\varepsilon}_{ij} + \frac{v}{1 - 2v} (\dot{\varepsilon}_{kk})^2 \right\} \tag{79a}$$

Let $(\dot{\varepsilon}_{ij}, \dot{\varepsilon}_{ij}^*)$ denote two distinct strain rates and $(\overset{\circ}{\sigma}_{ij}, \overset{\circ}{\sigma}_{ij}^*)$ the corresponding stress rates in the actual elastic/plastic solid. If the element is currently plastic, each of these strain rates will produce either unloading or further loading. The stress rates in the linearized solid corresponding to $\dot{\varepsilon}_{ij}$ and $\dot{\varepsilon}_{ij}^*$ may be denoted by $\hat{\tau}_{ij}$ and $\hat{\tau}_{ij}^*$ respectively. Considering the various combinations of loading and unloading, it is easily shown from the scalar product of (73) with $\dot{\varepsilon}_{ij}^*$ that

$$\overset{\circ}{\sigma}_{ij} \dot{\varepsilon}_{ij}^* \leqslant \hat{\tau}_{ij} \dot{\varepsilon}_{ij}^* = 2G \left\{ \dot{\varepsilon}_{ij} \dot{\varepsilon}_{ij}^* + \frac{v}{1 - 2v} \dot{\varepsilon}_{ii} \dot{\varepsilon}_{kk}^* - \frac{3G}{3G + H} \dot{\varepsilon}_{ij} \dot{\varepsilon}_{kl}^* n_{ij} n_{kl} \right\} \tag{80}$$

except when both $\dot{\varepsilon}_{ij}$ and $\dot{\varepsilon}_{ij}^*$ correspond to instantaneous unloading. The equality in (80) holds when $\dot{\varepsilon}_{ij}$ calls for additional loading, whatever the nature of $\dot{\varepsilon}_{ij}^*$. It

[†] The concept of linearization of a general nonlinear solid has been discussed by R. Hill, *J. Mech. Phys. Solids*, 7: 209 (1959).

follows from (79) and (80), and similar inequalities satisfied by $\mathring{\sigma}_{ij}^{*}\dot{\varepsilon}_{ij}^{*}$ and $\mathring{\sigma}_{ij}^{*}\dot{\varepsilon}_{ij}$, that

$$(\mathring{\sigma}_{ij} - \mathring{\sigma}_{ij}^{*})(\dot{\varepsilon}_{ij} - \dot{\varepsilon}_{ij}^{*}) \geqslant (\mathring{\tau}_{ij} - \mathring{\tau}_{ij}^{*})(\dot{\varepsilon}_{ij} - \dot{\varepsilon}_{ij}^{*})$$

or

$$\Delta\mathring{\sigma}_{ij}\,\Delta\dot{\varepsilon}_{ij} \geqslant \Delta\mathring{\tau}_{ij}\,\Delta\dot{\varepsilon}_{ij} = 2G\left\{\Delta\dot{\varepsilon}_{ij}\,\Delta\dot{\varepsilon}_{ij} + \frac{v}{1-2v}(\Delta\dot{\varepsilon}_{kk})^{2} - \frac{3G}{3G+H}(n_{ij}\,\Delta\dot{\varepsilon}_{ij})^{2}\right\} \quad (81)$$

for a plastic element, with equality for loading caused by both $\dot{\varepsilon}_{ij}$ and $\dot{\varepsilon}_{ij}^{*}$. When both the states call for unloading, $\Delta\mathring{\tau}_{ij}$ and $\Delta\mathring{\sigma}_{ij}$ are given by the first and second equations respectively of (73) with $\dot{\varepsilon}_{ij}$ replaced by $\Delta\dot{\varepsilon}_{ij}$, leading to the inequality $\Delta\mathring{\sigma}_{ij}\,\Delta\dot{\varepsilon}_{ij} > \Delta\mathring{\tau}_{ij}\,\Delta\dot{\varepsilon}_{ij}$. For an elastic element, we have the immediate identity $\Delta\mathring{\sigma}_{ij}\,\Delta\dot{\varepsilon}_{ij} = \Delta\mathring{\tau}_{ij}\,\Delta\dot{\varepsilon}_{ij}$, the expression for which is obtained from (81) by omitting the last term. Since $\Delta\mathring{\sigma}_{ij}\,\Delta\dot{\varepsilon}_{ij} \geqslant \Delta\mathring{\tau}_{ij}\,\Delta\dot{\varepsilon}_{ij}$ throughout the body, uniqueness is ensured by the slightly oversufficient criterion

$$\int\left\{\Delta\mathring{\tau}_{ij}\,\Delta\dot{\varepsilon}_{ij} + \sigma_{ij}\left[\Delta\dot{\varepsilon}_{kk}\,\Delta\dot{\varepsilon}_{ij} - 2\Delta\dot{\varepsilon}_{jk}\,\Delta\omega_{ik} - \frac{\partial}{\partial x_{k}}(\Delta v_{i})\frac{\partial}{\partial x_{j}}(\Delta v_{k})\right]\right\}dV > 0 \quad (82)$$

which is more useful for practical applications. It follows that uniqueness of the linearized solid also ensures uniqueness for the nonlinear elastic/plastic solid. If the constraints are rigid, so that $v_{j} = 0$ on S_{v}, then every difference field Δv_{j} is a member of the admissible field v_{j} for the linearized solid. The uniqueness criterion then becomes

$$\int\left\{\mathring{\tau}_{ij}\dot{\varepsilon}_{ij} + \sigma_{ij}\left(\dot{\varepsilon}_{kk}\dot{\varepsilon}_{ij} - 2\dot{\varepsilon}_{jk}\omega_{ik} - \frac{\partial v_{i}}{\partial x_{k}}\frac{\partial v_{k}}{\partial x_{j}}\right)\right\}dV > 0$$

for all continuous differentiable fields vanishing on S_{v}. If the constraints are not rigid, the inequality still holds, but v_{j} is no longer an admissible field for the actual boundary-value problem. Splitting the tensor $\partial v_{i}/\partial x_{k}$ into its symmetric and antisymmetric parts, it can be shown that

$$\sigma_{ij}\frac{\partial v_{i}}{\partial x_{k}}\frac{\partial v_{k}}{\partial x_{j}} = \sigma_{ij}(\dot{\varepsilon}_{jk}\dot{\varepsilon}_{ik} + \omega_{jk}\omega_{ik})$$

The remaining triple products cancel one another on account of the symmetry and antisymmetry properties of their factors. Discarding the term in $\dot{\varepsilon}_{kk}$, which always makes an insignificant contribution, we have

$$\int\left[\mathring{\tau}_{ij}\dot{\varepsilon}_{ij} - \sigma_{ij}(2\dot{\varepsilon}_{jk}\omega_{ik} + \dot{\varepsilon}_{jk}\dot{\varepsilon}_{ik} + \omega_{jk}\omega_{ik})\right]dV > 0 \quad (83)$$

The leading term in the square bracket is given by (79) over the plastic part and (79a) over the elastic part. When curvilinear coordinates are employed, it is only necessary to interpret $\dot{\varepsilon}_{ij}$, etc., in (83) as the curvilinear components. If the integral in (83) vanishes for some nonzero field v_{j}, bifurcation in the linearized solid can occur for any value of the traction rate on S_{F}. In the actual elastic/plastic solid, however, bifurcation will occur under those traction rates for which there is no unloading of the current plastic region.

In a class of problems, such as those involving elastic/plastic buckling, the elastic and plastic parts of the strain rate are generally of comparable magnitudes. A useful simplification then results from the fact that the quantity $\sigma_{ij}\dot{\varepsilon}_{ik}\dot{\varepsilon}_{jk}$ is negligible compared to similar terms in the expression for $\mathring{\tau}_{ij}\dot{\varepsilon}_{ij}$. The necking type problems, on the other hand, generally involve strain rates whose elastic parts are negligible. The strain rate vector in such cases is nearly normal to the yield surface, giving $\mathring{\tau}_{ij}\dot{\varepsilon}_{ij} \simeq \frac{2}{3}H\dot{\varepsilon}_{ij}\dot{\varepsilon}_{ij}$ to a close approximation.

As an elementary application of the uniqueness criterion, consider the longitudinal tension of a cylindrical specimen of uniform cross section. The existing state of stress is a uniaxial tension defined by the principal components $\sigma_1 = \sigma$ and $\sigma_2 = \sigma_3 = 0$. If the elastic part of the strain rate is disregarded, then for an isotropic material the principal strain rates are $\dot{\varepsilon}_1 = \dot{\varepsilon}$ and $\dot{\varepsilon}_2 = \dot{\varepsilon}_3 = -\frac{1}{2}\dot{\varepsilon}$. Choosing coordinate axes along the common principal axes of stress and strain rate, the integrand of (83) is found to be $(H - \sigma)\dot{\varepsilon}^2 + \sigma(\omega_2^2 + \omega_3^2)$, where ω_2 and ω_3 are two of the components of the spin vector. The uniqueness functional is certainly positive when $H > \sigma$, irrespective of the possible spin. At the critical value $H = \sigma$, the load attains its maximum, and either a further uniform extension or local necking can occur in principle.[†]

(iii) *Rigid/plastic materials* A material is considered as rigid/plastic when the modulus of elasticity is assigned an indefinitely large value. The elastic strains are zero in the limit, and consequently the change in volume also disappears. If the plastic zone is fully constrained by the nonplastic material, the entire body is rigid in the limit. The distribution of stress rate is then uniquely determined for a prescribed distribution of traction rates, provided G is assumed to be large but finite. If, on the other hand, a distribution of surface velocities is prescribed, the behavior of the rigid/plastic material must be distinguished from that of the elastic/plastic material. When the existing state of stress in the rigid/plastic body is regarded as given, the prescribed velocity distribution will be consistent if the corresponding strain rate vector is normal to the yield surface at the given stress point. In general, the existing stress would change discontinuously so as to become consistent with the strain rate at all points.

The material is assumed to be isotropic, so that the unit normal n_{ij} to the yield surface (which is regular) has principal axes coinciding with those of the current stress σ_{ij}. Since $\mathring{\sigma}_{ij}n_{ij} = \dot{\sigma}_{ij}n_{ij}$, the constitutive equations for a work-hardening rigid/plastic solid become

$$\dot{\varepsilon}_{ij} = (3/2H)\dot{\sigma}_{kl}n_{kl}n_{ij} \qquad \dot{\sigma}_{kl}n_{kl} \geqslant 0$$
$$\dot{\varepsilon}_{ij} = 0, \qquad\qquad \dot{\sigma}_{kl}n_{kl} \leqslant 0 \tag{84}$$

[†] Detailed investigations of the bifurcation in uniaxial tension have been carried out by S. Y. Cheng, S. T. Ariaratnam, and R. N. Dubey, *Q. Appl. Math.*, **29**: 41 (1971); A. Needleman, *J. Mech. Phys. Solids*, **20**: 111 (1972); J. W. Hutchinson and J. P. Miles, ibid., **22**: 61 (1974); L. G. Chen, *Int. J. Mech. Sci.*, **25**: 47 (1983). The bifurcation problem in plane strain tension has been treated by S. T. Ariaratnam and R. N. Dubey, *Q. Appl. Math.*, **27**: 349 (1969); R. Hill and J. W. Hutchinson, *J. Mech. Phys. Solids*, **23**: 239 (1975).

whenever an element is stressed to the yield point. Taking scalar products of the first relation of (84) with $\dot{\sigma}_{ij}$ and $\dot{\varepsilon}_{ij}$ in turn, and remembering that $n_{ij}n_{ij} = 1$, it is easily shown that

$$\dot{\sigma}_{ij}\dot{\varepsilon}_{ij} = \frac{3}{2H}(\dot{\sigma}_{ij}n_{ij})^2 = \frac{2}{3}H\dot{\varepsilon}_{ij}\dot{\varepsilon}_{ij} \tag{85}$$

The last expression also holds in a plastic element that instantaneously unloads, as well as in an element that is currently nonplastic, the strain rate being identically zero in both cases.

Suppose there could be two physically possible modes $\dot{\varepsilon}_{ij}$ and $\dot{\varepsilon}_{ij}^*$ associated with equilibrium distribution of stress rates $\dot{\sigma}_{ij}$ and $\dot{\sigma}_{ij}^*$, under prescribed nominal traction rates on S_F and velocities on S_v. When $\dot{\varepsilon}_{ij}$ is nonzero, the scalar product of (84) with $\dot{\sigma}_{ij}^*$ gives

$$\dot{\sigma}_{ij}^*\dot{\varepsilon}_{ij} = \frac{3}{2H}(\dot{\sigma}_{ij}^*n_{ij})(\dot{\sigma}_{kl}n_{kl}) \leqslant \tfrac{2}{3}H\dot{\varepsilon}_{ij}\dot{\varepsilon}_{ij}^* \tag{86}$$

in view of the constitutive relations for $\dot{\varepsilon}_{ij}^*$, with equality if and only if $\dot{\varepsilon}_{ij}^*$ is also nonzero. The result follows from the fact that $\dot{\sigma}_{ij}^*\dot{\varepsilon}_{ij} \leqslant 0$ when $\dot{\varepsilon}_{ij} \neq 0$ and $\dot{\varepsilon}_{ij}^* = 0$, implied by $\dot{\sigma}_{ij}n_{ij} > 0$ and $\dot{\sigma}_{ij}^*n_{ij} \leqslant 0$. Starting with the equation for $\dot{\varepsilon}_{ij}^* \neq 0$, it is easily shown that $\dot{\sigma}_{ij}\dot{\varepsilon}_{ij}^*$ satisfies the same inequality as (86), while $\dot{\sigma}_{ij}^*\dot{\varepsilon}_{ij}^*$ is given by an expression similar to (85). Hence

$$\Delta\dot{\sigma}_{ij}\,\Delta\dot{\varepsilon}_{ij} = (\dot{\sigma}_{ij} - \dot{\sigma}_{ij}^*)(\dot{\varepsilon}_{ij} - \dot{\varepsilon}_{ij}^*) \geqslant \tfrac{2}{3}H\,\Delta\dot{\varepsilon}_{ij}\,\Delta\dot{\varepsilon}_{ij} \tag{87}$$

Since the equality is identically satisfied in those regions where $\dot{\varepsilon}_{ij}$ and $\dot{\varepsilon}_{ij}^*$ are both zero, (87) holds throughout the body. The existing stress is known in principle wherever $\dot{\varepsilon}_{ij}$ or $\dot{\varepsilon}_{ij}^*$ is nonzero.

The analysis for the occurrence of bifurcation for the rigid/plastic material follows that for the elastic/plastic material. In view of the incompressibility condition $\partial v_k/\partial x_k = 0$, the condition for having two possible solutions can be written down directly from (78) as

$$\int \left\{ \Delta\dot{\sigma}_{ij}\,\Delta\dot{\varepsilon}_{ij} - \sigma_{ij}\frac{\partial}{\partial x_k}(\Delta v_i)\frac{\partial}{\partial x_j}(\Delta v_k) \right\} dV = 0 \tag{88}$$

The integral really extends over the region where the existing stress is uniquely known, but may be formally extended over the entire body on the understanding that the motion of the rigid zones is known in advance ($\Delta v_j = 0$). In view of (87), a sufficient condition for uniqueness of the deformation mode is†

$$\int \left\{ \tfrac{2}{3}H\,\Delta\varepsilon_{ij}\,\Delta\dot{\varepsilon}_{ij} - \sigma_{ij}\frac{\partial}{\partial x_k}(\Delta v_i)\frac{\partial}{\partial x_j}(\Delta v_k) \right\} dV > 0 \tag{89}$$

for all incompressible difference fields Δv_j vanishing on S_v. Setting $\Delta v_j = w_j$ and

† R. Hill, *J. Mech. Phys. Solids*, **5**: 153 and 302 (1957).

$\Delta\dot{\varepsilon}_{ij} = \dot{\eta}_{ij}$, the uniqueness criterion may be written as

$$\int \left\{ \frac{2}{3} H\dot{\eta}_{ij}\dot{\eta}_{ij} - \sigma_{ij}\frac{\partial w_i}{\partial x_k}\frac{\partial w_k}{\partial x_j} \right\} dV > 0$$

for all incompressible velocity fields w_j vanishing on the constraints.† The corresponding strain rate η_{ij} is nonzero in the uniquely stressed zone where it is parallel to n_{ij}, though not necessarily in the same sense. When the constraints are rigid, w_j may be regarded as an actual field for a linearized rigid/plastic solid in which the constitutive law is given by the first equation of (84) whenever the existing stress is uniquely defined. If the second term of the above integral is negative for all nonzero fields w_j, the inequality is certainly satisfied for all $H \geqslant 0$, indicating that there is not more than one physically possible mode whatever the rate of hardening. When all positional changes and rotations of material elements are disregarded (the second term discarded), there is a unique deformation mode for a work-hardening material, although the corresponding stress rate is not necessarily unique.‡

(iv) *Pressure type loading* The preceding discussion of uniqueness is based on the assumption that the change in the load vector on an infinitesimal surface element is assigned, irrespective of changes in its area and orientation. An important special case is that of dead loading where the load remains constant during an infinitesimal distortion of the deforming body. In a variety of practical situations, the actual traction is given to be a uniform fluid pressure acting over a part of the boundary, where the nominal traction rate is given by the equation

$$\dot{F}_j = \dot{p}l_j + p\left(l_k\frac{\partial v_k}{\partial x_j} - l_j\frac{\partial v_k}{\partial x_k} \right) \tag{90}$$

where l_j is the unit vector along the outward normal to the surface at the instant when the applied fluid pressure is p. It may be noted that the part of the nominal traction rate given by only the first term can be prescribed. Consider now the possibility of two distinct solutions of the boundary value problem under given nominal traction rate $\dot{F}_j$ on S_F, velocity v_j on S_v, and pressure rate $\dot{p}$ on the remainder S_f. Then

$$\Delta\dot{F}_j = p\left[l_k\frac{\partial}{\partial x_j}(\Delta v_k) - l_j\frac{\partial}{\partial x_k}(\Delta v_k) \right] \qquad \text{on } S_f \tag{91}$$

since $\Delta\dot{p} = 0$ in view of the boundary condition. The natural starting point is the

---

† The most general velocity field for a uniform plastic state has been given by W. Prager, *Rev. Fac. Sci., Univ. Istanbul*, **19**: 23 (1954).

‡ A detailed analysis for the bifurcation in a rigid/plastic bar under plane strain tension has been presented by G. R. Cowper and E. T. Onat, *Proc. 4th U.S. Nat. Congr. Appl. Mech.*, 1023 (1962). An approximate solution for the plane stress buckling has been given by J. Chakrabarty, *Int. J. Mech. Sci.*, **11**: 659 (1969).

transformation

$$\int \Delta \dot{F}_j \, \Delta v_j \, dS = \int l_i \, \Delta \dot{t}_{ij} \, \Delta v_j \, dS = \int \Delta \dot{t}_{ij} \frac{\partial}{\partial x_i} (\Delta v_j) \, dV$$

where $\Delta \dot{F}_j$ vanishes on S_F and Δv_j vanishes on S_v. In view of (91), the above equation becomes

$$\int \Delta \dot{t}_{ij} \frac{\partial}{\partial x_i} (\Delta v_j) \, dV - p \int \Delta v_j \left[l_k \frac{\partial}{\partial x_j} (\Delta v_k) - l_j \frac{\partial}{\partial x_k} (\Delta v_k) \right] dS_f = 0$$

Inserting the expression for $\Delta \dot{t}_{ij}$ from (77) gives the condition for bifurcation in terms of $\Delta \dot{\sigma}_{ij}$. Proceeding as before, a sufficient condition for uniqueness in the elastic/plastic material, when a part of the boundary is subjected to uniform normal pressure p, may be expressed as[†]

$$\int \left\{ \overset{\circ}{\tau}_{ij} \dot{\varepsilon}_{ij} + \sigma_{ij} \left(\dot{\varepsilon}_{kk} \dot{\varepsilon}_{ij} + 2 \dot{\varepsilon}_{jk} \omega_{ki} - \frac{\partial v_i}{\partial x_k} \frac{\partial v_k}{\partial x_j} \right) \right\} dV$$

$$- p \int \left(l_k \frac{\partial v_k}{\partial x_j} - l_j \frac{\partial v_k}{\partial x_k} \right) v_j \, dS_f > 0 \qquad (92)$$

where v_j is any continuous differentiable velocity field vanishing on S_v. The leading term in the volume integral is given by (79) in the plastic region and (79a) in the elastic region. When $S_F = 0$, the surface integral in (92) may be formally extended over the entire surface of the body, since $v_j = 0$ on the remainder S_v. Transformation of the surface into volume integral by Green's theorem then furnishes

$$\int \left\{ \overset{\circ}{\tau}_{ij} \dot{\varepsilon}_{ij} + 2 \sigma_{ij} \dot{\varepsilon}_{jk} \omega_{ki} + (\sigma_{ij} + p \delta_{ij}) \left(\dot{\varepsilon}_{kk} \dot{\varepsilon}_{ij} - \frac{\partial v_i}{\partial x_k} \frac{\partial v_k}{\partial x_j} \right) \right\} dV > 0 \qquad (93)$$

The second term of the integrand is unaffected by the addition of a quantity $p \delta_{ij}$ to the current stress tensor σ_{ij}. It follows from (93) that in a body partly constrained and having a uniform fluid pressure p acting on the remaining surface, the condition for the occurrence of bifurcation is the same as that without the pressure provided the actual normal stresses are augmented by the amount p. Thus, in the simple tensile test, if the specimen begins to neck when the applied tensile stress is σ, the application of a uniform fluid pressure p on the lateral surface would reduce the tensile stress to $\sigma - p$ at the onset of necking, but the critical rate of hardening and hence the amount of uniform strain is unaffected by the pressure.

In the case of rigid/plastics solids, $\dot{\varepsilon}_{kk}$ is identically zero, and the scalar product $\sigma_{ij} \dot{\varepsilon}_{jk} \omega_{ki}$ also vanishes on account of the coincidence of the principal axes of stress and strain rate. Since $\overset{\circ}{\tau}_{ij} \dot{\varepsilon}_{ij}$ then reduces to $\frac{2}{3} H \dot{\varepsilon}_{ij} \dot{\varepsilon}_{ij}$, the deformation mode will be unique if[‡]

$$\int \left(\frac{2}{3} H \dot{\varepsilon}_{ij} \dot{\varepsilon}_{ij} - \sigma_{ij} \frac{\partial v_i}{\partial x_k} \frac{\partial v_k}{\partial x_j} \right) dV - p \int \left(l_k \frac{\partial v_k}{\partial x_j} v_j \right) dS_f > 0 \qquad (94)$$

[†] J. Chakrabarty, Z. angew. Math. Phys., **24**: 270 (1973).
[‡] J. Chakrabarty, Int. J. Mech. Sci., **11**: 723 (1969); M. Miles, J. Mech. Phys. Solids, **17**: 303 (1969). Inequality (93) is an extension of one given earlier by R. Hill, J. Mech. Phys. Solids, **5**: 153 (1957).

where v_j is any incompressible velocity field vanishing on S_v, and producing a strain rate $\dot{\varepsilon}_{ij}$ that is parallel to n_{ij} in the plastically deforming region. Although the rigid/plastic assumption is a useful approximation in problems involving large plastic strains, the limitation imposed by the normality rule on the choice of admissible velocity fields is sometimes too severe to permit an effective analysis. A more realistic approximation would be accomplished, when the entire body is plastic, by considering (92) for an incompressible velocity field v_j such that the associated strain rate has a small elastic part. Then (92) approximately reduces to (94) with the addition of the term $\sigma_{ij}\dot{\varepsilon}_{jk}\omega_{ki}$, which is not necessarily negligible in the volume integral.

The problem of uniqueness is closely related to that of stability of a solid under given boundary conditions. In the dynamic sense, a sufficient condition of stability is that the internal energy dissipated in any geometrically possible small displacement from the position of equilibrium must exceed the work done by the external forces. Calculating the internal energy and the external work to the second order for an additional infinitesimal displacement, it can be shown that the criterion for stability for an elastic/plastic solid under pressure-type loading is (92) with $\dot{\tau}_{ij}$ replaced by $\mathring{\sigma}_{ij}$. Since $\mathring{\sigma}_{ij}\dot{\varepsilon}_{ij} \geqslant \dot{\tau}_{ij}\dot{\varepsilon}_{ij}$, a boundary-value problem that has a unique solution is certainly stable.† For a rigid/plastic solid, the stability functional is identical to (94), but the velocity field v_j is subject to the further restriction that the associated strain rate has the same sense as that of n_{ij}. Since the class of fields for uniqueness then contains that for stability, a point of bifurcation is again possible before an actual loss of stability. Moreover, at such a stable bifurcation, the load must increase with continuing deformation.‡

2.8 Extremum Principles

(i) *Elastic/plastic solids for small strains* It is assumed at the outset that the existing state of stress is such that the increment of the plastic strain is constrained to be of the elastic order of magnitude. Let $(\dot{\sigma}_{ij}, \dot{\varepsilon}_{ij})$ be the actual stress and strain rates in an elastic/plastic body which is subjected to prescribed traction rates $\dot{T}_j$ on a part S_F of the surface, and prescribed velocities v_j on the remainder S_v. When geometry changes are disregarded, the equilibrium equations (in the absence of body forces) and the boundary conditions in terms of $\dot{\sigma}_{ij}$ are

$$\frac{\partial \dot{\sigma}_{ij}}{\partial x_i} = 0 \qquad \dot{T}_j = l_i \dot{\sigma}_{ij}$$

The strain rate $\dot{\varepsilon}_{ij}$ must be derivable from a continuous velocity field v_j satisfying the boundary conditions.

Let $\dot{\sigma}_{ij}^*$ denote any statically admissible state of stress rate that satisfies the equilibrium equations and the stresses boundary conditions, the corresponding

---

† The stability of rigid/plastic solids has been discussed by R. Hill, *J. Mech. Phys. Solids*, **6**: 1 (1957), and also by J. Chakrabarty, *Int. J. Mech. Sci.* **11**: 723 (1969).

‡ For a different approach to uniqueness, see D. C. Drucker, *Q. Appl. Math.*, **14**: 35 (1956). The material stability has been examined by D. C. Drucker, *J. Appl. Mech.*, **26**: 101 (1959).

strain rate $\dot{\varepsilon}_{ij}^*$ being given by the stress-strain relations. The application of Green's theorem furnishes

$$\int (\dot{T}_j^* - \dot{T}_j)v_j \, dS = \int l_i(\dot{\sigma}_{ij}^* - \dot{\sigma}_{ij})v_j \, dS = \int (\dot{\sigma}_{ij}^* - \dot{\sigma}_{ij})\frac{\partial v_j}{\partial x_i} \, dV$$

The surface integral vanishes on S_F where $\dot{T}_j^* = \dot{T}_j$. In view of the symmetry of $\dot{\sigma}_{ij}$ and $\dot{\sigma}_{ij}^*$, we get

$$\int (\dot{\sigma}_{ij}^* - \dot{\sigma}_{ij})\dot{\varepsilon}_{ij} \, dV = \int (\dot{T}_j^* - \dot{T}_j)v_j \, dS \tag{95}$$

When the material is nonhardening, the constitutive equation for an element stressed to the yield point is

$$\dot{\varepsilon}_{ij} = \frac{1}{2G}\left(\dot{\sigma}_{ij} - \frac{v}{1+v}\dot{\sigma}_{kk}\delta_{ij}\right) + \dot{\lambda}n_{ij} \tag{96}$$

where $\dot{\lambda} > 0$ for loading ($\dot{\sigma}_{ij}n_{ij} = 0$) and $\dot{\lambda} = 0$ for unloading ($\dot{\sigma}_{ij}n_{ij} < 0$). The scalar product of (96) with $\dot{\sigma}_{ij}$ and $\dot{\sigma}_{ij}^*$ indicates that (72) is satisfied under all conditions of loading and unloading. It follows from (72), and a similar inequality for $\dot{\sigma}_{ij}^*\dot{\varepsilon}_{ij}^*$, that

$$\dot{\sigma}_{ij}^*\dot{\varepsilon}_{ij}^* - 2\dot{\sigma}_{ij}^*\dot{\varepsilon}_{ij} + \dot{\sigma}_{ij}\dot{\varepsilon}_{ij} \geqslant \frac{1}{2G}\left\{(\dot{\sigma}_{ij}^* - \dot{\sigma}_{ij})(\dot{\sigma}_{ij}^* - \dot{\sigma}_{ij}) - \frac{v}{1+v}(\dot{\sigma}_{kk}^* - \dot{\sigma}_{kk})^2\right\} \tag{97}$$

For a work-hardening material, the second inequality of (72) does not hold when both $\dot{\sigma}_{ij}$ and $\dot{\sigma}_{ij}^*$ correspond to further loading, but an independent calculation shows that (97) still holds with a strict inequality. Since the equality always holds in any nonplastic element, (97) is satisfied throughout the elastic/plastic body. The expression on the right-hand side is definitely positive unless $\sigma_{ij}^* = \sigma_{ij}$. Hence

$$\int (\dot{\sigma}_{ij}^*\dot{\varepsilon}_{ij}^* - \dot{\sigma}_{ij}\dot{\varepsilon}_{ij}) \, dV > 2\int (\dot{\sigma}_{ij}^* - \dot{\sigma}_{ij})\dot{\varepsilon}_{ij} \, dV$$

or

$$\int \dot{\sigma}_{ij}^*\dot{\varepsilon}_{ij}^* \, dV - 2\int \dot{T}_j^*v_j \, dS_v > \int \dot{\sigma}_{ij}\dot{\varepsilon}_{ij} \, dV - 2\int \dot{T}_j v_j \, dS_v$$

in view of (95). Transformaton of the volume integral on the right-hand side by Green's theorem then furnishes

$$\int \dot{\sigma}_{ij}^*\dot{\varepsilon}_{ij}^* \, dV - 2\int \dot{T}_j^*v_j \, dS_v > \int \dot{T}_j v_j \, dS_F - \int \dot{T}_j v_j \, dS_v \tag{98}$$

unless $\dot{\sigma}_{ij}^* = \dot{\sigma}_{ij}$. Thus, among all statically admissible distributions of stress rate, the actual one minimizes† the left-hand side of (98). The left-hand side has in fact a

---

† The minimum principle for the stress rate is substantially due to P. G. Hodge, Jr., and W. Prager, *J. Math. Phys.*, **27**: 1 (1948). See also H. J. Greenberg, *Q. Appl. Math.*, **7**: 85 (1949).

stationary value in the actual state,† as may be shown by writing this expression in the form

$$\int (\dot\sigma_{ij}^*\dot\varepsilon_{ij}^* - 2\dot\sigma_{ij}^*\dot\varepsilon_{ij})\,dV + 2\int \dot{T}_j^* v_j\,dS_F$$

Since $\dot{T}_j^* = \dot{T}_j$ on S_F, the last integral is independent of $\dot\sigma_{ij}^*$. The expression will therefore have a stationary value if a variation of the first integrand vanishes at every point of the body. Considering equations (69) and (96), it can be shown in a straightforward manner that

$$\delta(\dot\sigma_{ij}^*\dot\varepsilon_{ij}^*) = 2\dot\varepsilon_{ij}^*\delta\dot\sigma_{ij}^* \qquad \delta(\dot\sigma_{ij}^*\dot\varepsilon_{ij}) = \dot\varepsilon_{ij}\delta\dot\sigma_{ij}^*$$

in both elastic and plastic regions, under all conditions of loading and unloading, and for all $H \geqslant 0$. Hence

$$\delta(\dot\sigma_{ij}^*\dot\varepsilon_{ij}^* - 2\dot\sigma_{ij}^*\dot\varepsilon_{ij}) = 2(\dot\varepsilon_{ij}^* - \dot\varepsilon_{ij})\delta\dot\sigma_{ij}^*$$

which evidently vanishes when $\dot\varepsilon_{ij}^* = \dot\varepsilon_{ij}$, or $\dot\sigma_{ij}^* = \dot\sigma_{ij}$, thus establishing the condition of a stationary value in the actual state.‡

A second extremum principle is associated with any distribution of kinematically admissible strain rate $\dot\varepsilon_{ij}^*$, obtained from a velocity field v_j^* satisfying the boundary conditions on S_v. The corresponding stress rate $\dot\sigma_{ij}^*$ satisfies the stress-strain relations, but it is not necessarily in equilibrium. The usual transformation of integrals furnishes

$$\int (v_j^* - v_j)\dot{T}_j\,dS = \int l_i(v_j^* - v_j)\dot\sigma_{ij}\,dS = \int \dot\sigma_{ij}\frac{\partial}{\partial x_i}(v_j^* - v_j)\,dV$$

Since $v_j^* = v_j$ on S_v, the above result may be written as

$$\int (\dot\varepsilon_{ij}^* - \dot\varepsilon_{ij})\dot\sigma_{ij}\,dV = \int (v_j^* - v_j)\dot{T}_j\,dS_F \tag{99}$$

on account of the symmetry of $\dot\sigma_{ij}$. Using inequalities similar to (72), and proceeding as before, it can be shown that

$$\dot\sigma_{ij}^*\dot\varepsilon_{ij}^* - 2\dot\sigma_{ij}\dot\varepsilon_{ij}^* + \dot\sigma_{ij}\dot\varepsilon_{ij} > 0$$

for both nonhardening and work-hardening materials, except when $\dot\sigma_{ij}^* = \dot\sigma_{ij}$. This

---

† For a Prandtl-Reuss material, this was proved by W. Prager, *Proc. 6th Int. Congr. Appl. Mech.*, Paris (1946).

‡ The Hencky stress-strain relations for a nonhardening material are associated with a variational principle suggested by A. Haar and Th. von Karman, *Göttinger Nachrichten, Math. Phys. Klasse*, 204 (1909). The Haar-Karman principle involves the minimization of the left-hand side of (98) with $\dot\sigma_{ij}^*$, $\dot\varepsilon_{ij}^*$, and $\dot{T}_j^*$ replaced by σ_{ij}^*, ε_{ij}^*, and T_j^* respectively. The theorem was proved by H. J. Greenberg, Report A11-S4, *Grad. Div. Appl. Math.*, Brown University (1949), under the assumption that the state of stress remains constant at any point where the yield limit has once been reached.

leads to the inequality

$$2 \int (\dot{\varepsilon}_{ij}^* - \dot{\varepsilon}_{ij})\dot{\sigma}_{ij} \, dV < \int (\dot{\sigma}_{ij}^* \dot{\varepsilon}_{ij}^* - \dot{\sigma}_{ij}\dot{\varepsilon}_{ij}) \, dV$$

or

$$2 \int \dot{T}_j v_j^* \, dS_F - \int \dot{\sigma}_{ij}^* \dot{\varepsilon}_{ij}^* \, dV < \int \dot{T}_j v_j \, dS_F - \int \dot{\sigma}_{ij}\dot{\varepsilon}_{ij} \, dV$$

by (99). Transforming the volume integral on the right-hand side into surface integral, the result may be expressed as

$$2 \int \dot{T}_j v_j^* \, dS_F - \int \dot{\sigma}_{ij}^* \dot{\varepsilon}_{ij}^* \, dV < \int \dot{T}_j v_j \, dS_F - \int \dot{T}_j v_j \, dS_v \qquad (100)$$

unless $\dot{\sigma}_{ij}^* = \sigma_{ij}$. Thus, among all kinematically admissible distributions of strain rate, the actual one maximizes† the left-hand side of (100). The right-hand side is the same in both (98) and (100), which provide the means of obtaining upper and lower bounds in the approximate solution of elastic/plastic problems.‡

(ii) *Rigid/plastic solids for small strains* The following discussion is restricted to the situation where changes in geometry are again negligible. The material is assumed to be work-hardening and rigid/plastic with identical yield function and plastic potential. For given surface tractions T_j over a part S_F and velocities v_j over the remainder S_v, the stress is uniquely defined in the deforming part of the plastic zone. The mode of deformation, on the other hand, may not be uniquely determined by these boundary conditions alone. In order to define the deformation mode uniquely, it is also necessary to specify the traction rate on the boundary S_F. The unique mode is then singled out by the existence of an equilibrium distribution of stress rate compatible with the rate of hardening. All other modes are only kinematically possible virtual modes of deformation.§

Let $\dot{\varepsilon}_{ij}$ denote the actual mode derivable from a continuous velocity v_j, and $\dot{\varepsilon}_{ij}^*$ any virtual mode derivable from a velocity v_j^*, which is equal to v_j on S_v. If $\dot{\sigma}_{ij}$ is the actual stress rate satisfying the equilibrium equations and the boundary conditions, then

$$\int \dot{T}_j v_j \, dS = \int \dot{\sigma}_{ij}\dot{\varepsilon}_{ij} \, dV = \int h\dot{\varepsilon}_{ij}\dot{\varepsilon}_{ij} \, dV = \int h^{-1}(\dot{\sigma}_{ij}n_{ij})^2 \, dV \qquad (101)$$

† This theorem has been proved by H. J. Greenberg, *Q. Appl. Math.*, **7**: 85 (1949). A weaker variational principle, asserting the existence of a stationary value of the left-hand side of (100), was given earlier by W. Prager, op. cit. The generalization of both (98) and (100) to materials with a singular yield surface is due to W. T. Koiter, *Q. Appl. Math.*, **11**: 350 (1953).

‡ For other bounding theorems related to both incremental and total strain theories of plasticity, see P. G. Hodge, Jr., *Engineering Plasticity* (Eds J. Heyman and F. A. Leckie), 237, Cambridge University Press (1968); G. Maier, *J. Mech.*, **8**: 5 (1969); J. F. Soechting and R. H. Lance, *J. Appl. Mech.*, **36**: 228 (1969); A. R. Ponter and J. B. Martin, *J. Mech. Phys. Solids*, **20**: 281 (1972).

§ The extremum principles of this subsection, corresponding to a regular yield surface, have been obtained by R. Hill, *J. Mech. Phys. Solids*, **4**: 247 (1956). An extension of these results to a singular yield surface has been made by W. E. Boyce and W. Prager, ibid., **6**: 9 (1957).

in view of (85) with $h = \frac{2}{3}H$. The last integral should extend only over the region where $\dot{\varepsilon}_{ij}$ is nonzero. From (85) and (86), it follows that

$$(\dot{\varepsilon}_{ij}^* - \dot{\varepsilon}_{ij})\dot{\sigma}_{ij} \leqslant h(\dot{\varepsilon}_{ij}^* - \dot{\varepsilon}_{ij})\dot{\varepsilon}_{ij} \leqslant \tfrac{1}{2}h(\dot{\varepsilon}_{ij}^*\dot{\varepsilon}_{ij}^* - \dot{\varepsilon}_{ij}\dot{\varepsilon}_{ij})$$

with equality if and only if $\dot{\varepsilon}_{ij}^* = \dot{\varepsilon}_{ij}$. The last inequality follows from the fact that the expression on the right-hand side exceeds the preceding one by the amount $\tfrac{1}{2}h(\dot{\varepsilon}_{ij}^* - \dot{\varepsilon}_{ij})(\dot{\varepsilon}_{ij}^* - \dot{\varepsilon}_{ij})$, which is always positive unless $\dot{\varepsilon}_{ij}^* = \dot{\varepsilon}_{ij}$. Equation (99) therefore gives

$$2\int (v_j^* - v_j)\dot{T}_j \, dS_F \leqslant \int h\dot{\varepsilon}_{ij}^*\dot{\varepsilon}_{ij}^* \, dV - \int h\dot{\varepsilon}_{ij}\dot{\varepsilon}_{ij} \, dV$$

Expressing the second volume integral as surface integral using (101), we finally obtain

$$\int h\dot{\varepsilon}_{ij}^*\dot{\varepsilon}_{ij}^* \, dV - 2\int \dot{T}_j v_j^* \, dS_F \geqslant \int \dot{T}_j v_j \, dS_v - \int T_j v_j \, dS_F \tag{102}$$

with equality if and only if $\dot{\varepsilon}_{ij}^*$ coincides with the actual mode $\dot{\varepsilon}_{ij}$. The inequality (102) gives a minimum characterization of the unique strain rate and velocity fields of the solution.

Consider now an equilibrium distribution of stress rate $\dot{\sigma}_{ij}^*$ such that $\dot{T}_j^* = T_j$ on S_F. In the region where $\dot{\varepsilon}_{ij}$ is nonzero, the constitutive equation (84), and the fact that $[(\sigma_{ij}^* - \sigma_{ij}) \cdot n_{ij}]^2 \geqslant 0$, furnish

$$(\dot{\sigma}_{ij}^* - \dot{\sigma}_{ij})\dot{\varepsilon}_{ij} = h^{-1}\dot{\sigma}_{kl}n_{kl}(\dot{\sigma}_{ij}^* - \dot{\sigma}_{ij})n_{ij} \leqslant \tfrac{1}{2}h^{-1}[(\dot{\sigma}_{ij}^*n_{ij})^2 - (\dot{\sigma}_{ij}n_{ij})^2]$$

The equality holds when $(\dot{\sigma}_{ij}^* - \dot{\sigma}_{ij})n_{ij} = 0$, and this could be satisfied with $\dot{\sigma}_{ij}^* \neq \dot{\sigma}_{ij}$. In view of the above inequality, (95) may be written as

$$2\int \dot{T}_j^* v_j \, dS_v - \int h^{-1}(\dot{\sigma}_{ij}^*n_{ij})^2 \, dV \leqslant 2\int \dot{T}_j v_j \, dS_v - \int h^{-1}(\dot{\sigma}_{ij}n_{ij})^2 \, dV$$

where the volume integrals are taken over the zone where $\dot{\varepsilon}_{ij} \neq 0$. Using (101), the inequality can be expressed as

$$2\int \dot{T}_j^* v_j \, dS_v - \int h^{-1}(\dot{\sigma}_{ij}^*n_{ij})^2 \, dV \leqslant \int \dot{T}_j v_j \, dS_v - \int \dot{T}_j v_j \, dS_F \tag{103}$$

Since the actual deforming zone is usually not known in advance the volume integral would have to be taken over the deformable zone defined by T_j on S_F and v_j on S_v, thereby weakening the inequality to some extent. The inequality (103) gives a maximum characterization of the stress rate field of the solution. Combining (102) and (103), we have a means of obtaining upper and lower bound approximations for the expression on the right-hand side of each of these inequalities.

If the plastic state of stress is represented by a point on an edge of a singular yield surface, it is only necessary to replace the expressions in the volume integrals of (102) and (103) by $\sum h_\alpha \varepsilon_\alpha^2$ and $\sum h_\alpha^{-1}(\dot{\sigma} \cdot \mathbf{n}_\alpha)^2$ respectively, in view of (36), where the summation includes the contributions of the yield mechanisms associated with all the faces meeting in the considered edge of the yield surface.

(iii) Elastic/plastic solids with large strains At any stage of the deformation of an elastic/plastic body, the internal distribution of stress and the current external configuration are supposed to be known. A sufficient condition for uniqueness of the solution, when changes in geometry are duly allowed for, has already been established. The extremum principle that characterizes the unique solution is now considered for the typical boundary-value problem involving a regular yield surface. Let v_j denote the actual velocity field in the elastic/plastic body, and v_j^* any other distinct field satisfying the velocity boundary conditions. If the differences between starred and unstarred quantities are denoted by Δ, so that $\Delta v_j = v_j^* - v_j$, etc., the usual transformation of integrals for the surface S and volume V in the current state furnishes

$$\int \dot{F}_j \, \Delta v_j \, dS = \int \frac{\partial}{\partial x_i} (\dot{t}_{ij} \, \Delta v_j) \, dV = \int \dot{t}_{ij} \, \Delta\left(\frac{\partial v_j}{\partial x_i}\right) dV \tag{104}$$

in view of (75), body forces being disregarded. The nominal traction rate $\dot{F}_j$ is given on a part S_F of the boundary, and the velocity v_j is given on the remainder S_v. By (74) and (76), the relationship between the nominal stress rate $\dot{t}_{ij}$ and the Jaumann stress rate $\mathring{\sigma}_{ij}$ is

$$\dot{t}_{ij} = \mathring{\sigma}_{ij} + \sigma_{ij} \frac{\partial v_k}{\partial x_k} - (\sigma_{ij} \dot{\varepsilon}_{jk} + \sigma_{jk} \dot{\varepsilon}_{ik}) + \sigma_{ik} \frac{\partial v_j}{\partial x_k} \tag{105}$$

in view of the identity $\omega_{ij} = \partial v_i / \partial x_j - \dot{\varepsilon}_{ij}$. Consider now the scalar product of (105) with $\Delta(\partial v_j / \partial x_i)$. Using the symmetry of the stress tensor, and the standard method of interchanging dummy suffixes, it can be shown that†

$$\dot{t}_{ij} \, \Delta\left(\frac{\partial v_j}{\partial x_i}\right) = \mathring{\sigma}_{ij} \, \Delta\dot{\varepsilon}_{ij} + \sigma_{ij} \left\{ \dot{\varepsilon}_{kk} \, \Delta\dot{\varepsilon}_{ij} - 2\dot{\varepsilon}_{ik} \, \Delta\dot{\varepsilon}_{jk} + \frac{\partial v_k}{\partial x_i} \, \Delta\left(\frac{\partial v_k}{\partial x_j}\right) \right\}$$

Since the elastic moduli are always large compared to the stress supportable by the material, there is hardly any point in retaining the first term in the curly bracket. Inserting into (104), and noting that $\Delta v_j = 0$ on S_v, while $\dot{F}_j$ is prescribed on S_F, we get

$$\Delta \int \dot{F}_j v_j \, dS_F = \int \left\{ \mathring{\sigma}_{ij} \, \Delta\dot{\varepsilon}_{ij} - \sigma_{ij} \left[2\dot{\varepsilon}_{ik} \, \Delta\dot{\varepsilon}_{jk} - \frac{\partial v_k}{\partial x_i} \, \Delta\left(\frac{\partial v_k}{\partial x_j}\right) \right] \right\} dV \tag{106}$$

From inequalities (79) and (80), and the fact that $\dot{t}_{ij}\dot{\varepsilon}_{ij}^* = \dot{t}_{ij}^*\dot{\varepsilon}_{ij}$, we have the immediate result

$$\mathring{\sigma}_{ij}^*\dot{\varepsilon}_{ij}^* - 2\mathring{\sigma}_{ij}\dot{\varepsilon}_{ij}^* + \mathring{\sigma}_{ij}\dot{\varepsilon}_{ij} \geqslant (\dot{t}_{ij}^* - \dot{t}_{ij})(\dot{\varepsilon}_{ij}^* - \dot{\varepsilon}_{ij})$$

The equality arises in the elastic region and the loading part of the plastic region

---

† The analysis given here is essentially due to R. Hill, *J. Mech. Phys. Solids*, **6**: 236 (1958).

common to both the actual and varied fields. Then

$$\Delta \int \mathring{\sigma}_{ij} \mathring{\varepsilon}_{ij} \, dV \geqslant 2 \int \mathring{\sigma}_{ij} \, \Delta \mathring{\varepsilon}_{ij} \, dV + \int \Delta \mathring{\tau}_{ij} \, \Delta \mathring{\varepsilon}_{ij} \, dV \tag{107}$$

Writing down the starred and unstarred terms explicitly, it can be shown in a straightforward manner that

$$\sigma_{ij} \Delta \left(2\mathring{\varepsilon}_{ik}\varepsilon_{jk} - \frac{\partial v_k}{\partial x_i} \frac{\partial v_k}{\partial x_j} \right)$$

$$= 2\sigma_{ij} \left[2\mathring{\varepsilon}_{ik} \, \Delta \mathring{\varepsilon}_{jk} - \frac{\partial v_k}{\partial x_i} \Delta \left(\frac{\partial v_k}{\partial x_j} \right) \right] + \sigma_{ij} \left[2\Delta \mathring{\varepsilon}_{ik} \, \Delta \mathring{\varepsilon}_{jk} - \Delta \left(\frac{\partial v_k}{\partial x_i} \right) \Delta \left(\frac{\partial v_k}{\partial x_j} \right) \right]$$

Since the existing stress is given, the prefix Δ on the left-hand side can be equally applied to the product of σ_{ij} and the expression in the curved bracket. Combining the last result with (106) and (107) gives

$$\Delta \left\{ \int \left[\mathring{\sigma}_{ij}\mathring{\varepsilon}_{ij} - \sigma_{ij} \left(2\mathring{\varepsilon}_{ik}\mathring{\varepsilon}_{jk} - \frac{\partial v_k}{\partial x_i} \frac{\partial v_k}{\partial x_j} \right) \right] dV - 2 \int \mathring{F}_j v_j \, dS_F \right\} > 0 \tag{108}$$

in view of the uniqueness criterion (82), which is expressible in the alternative form

$$\int \left\{ \Delta \mathring{\tau}_{ij} \, \Delta \mathring{\varepsilon}_{ij} - \sigma_{ij} \left[2\Delta \mathring{\varepsilon}_{ik} \, \Delta \mathring{\varepsilon}_{jk} - \Delta \left(\frac{\partial v_k}{\partial x_i} \right) \Delta \left(\frac{\partial v_k}{\partial x_j} \right) \right] \right\} dV > 0$$

when the insignificant term $\sigma_{ij} \, \Delta \mathring{\varepsilon}_{ij} \, \Delta \mathring{\varepsilon}_{kk}$ is neglected. The functional in the curly bracket of (108) evidently has a minimum value in the actual state. The minimum is also analytic and absolute in the sense that the first variation of the functional vanishes when the velocity field is actual. Thus

$$\delta \left\{ \int \left[\mathring{\sigma}_{ij}\mathring{\varepsilon}_{ij} - \sigma_{ij} \left(2\mathring{\varepsilon}_{ik}\mathring{\varepsilon}_{jk} - \frac{\partial v_k}{\partial x_i} \frac{\partial v_k}{\partial x_j} \right) \right] dV - 2 \int \mathring{F}_j v_j \, dS_F \right\} = 0 \tag{109}$$

Even when the solution is not unique, the variational principle holds for each solution, though the extremum principle may not exist. It is important to remember that $\mathring{\sigma}_{ij}\mathring{\varepsilon}_{ij}$ differs from $\mathring{\tau}_{ij}\mathring{\varepsilon}_{ij}$ in that part of the plastic zone which instantaneously unloads from the plastic state. When geometry changes are disregarded, so that the terms in σ_{ij} are considered negligible, the extremum principle reduces to that implied by (100).

(iv) *Rigid/plastic solids with large strains* It is instructive to treat the rigid/plastic solid purely on its own merits rather than as a limiting case of the elastic/plastic solid. Since the constitutive law (84) cannot be written in the inverted form, the boundary-value problem does not, in general, define the stress rate uniquely. Consider the typical problem in which the nominal traction rate $\mathring{F}_j$ is given on a part S_F of the external surface, and the velocity v_j on the remainder S_v. Broadly speaking, a unique solution exists for the deformation mode only when the rate of hardening exceeds a certain critical value. To obtain the associated extremum principle, the natural starting point is (104), where

$$\dot{t}_{ij} = \dot{\sigma}_{ij} - \sigma_{jk}\frac{\partial v_i}{\partial x_k}$$

and Δv_j is the difference $v_j^* - v_j$ between the virtual and actual velocities of a typical particle. The substitution for $\dot{t}_{ij}$ into (104) gives

$$\int \dot{F}_j(v_j^* - v_j)\,dV = \int \left\{ \dot{\sigma}_{ij}(\dot{\varepsilon}_{ij}^* - \dot{\varepsilon}_{ij}) - \sigma_{ij}\frac{\partial v_k}{\partial x_j}\frac{\partial}{\partial x_k}(v_i^* - v_i) \right\}dV \qquad (110)$$

in view of the symmetry of the stress and stress-rate tensors, and the interchangeability of dummy suffixes. From (85) and (86), we get

$$2\dot{\sigma}_{ij}(\dot{\varepsilon}_{ij}^* - \dot{\varepsilon}_{ij}) + \tfrac{2}{3}H(\dot{\varepsilon}_{ij}^* - \dot{\varepsilon}_{ij})(\dot{\varepsilon}_{ij}^* - \dot{\varepsilon}_{ij}) \leqslant \tfrac{2}{3}H(\dot{\varepsilon}_{ij}^*\dot{\varepsilon}_{ij}^* - \dot{\varepsilon}_{ij}\dot{\varepsilon}_{ij}) \qquad (111)$$

for $H \geqslant 0$. The inequality arises in those elements which are deforming in the virtual mode but unloading in the actual mode. To the right-hand side of equation (110), multiplied by 2, we add the positive uniqueness functional (89), and use (111), the result being

$$2\int \dot{F}_j(v_j^* - v_j)\,dS < \int \left\{ \frac{2}{3}H(\dot{\varepsilon}_{ij}^*\dot{\varepsilon}_{ij}^* - \dot{\varepsilon}_{ij}\dot{\varepsilon}_{ij}) - \sigma_{ij}\frac{\partial}{\partial x_j}(v_k^* + v_k)\frac{\partial}{\partial x_k}(v_i^* - v_i) \right\}dV \qquad (112)$$

The second integrand on the right-hand side can be simplified by splitting the tensors $\partial v_i/\partial x_j$ and $\partial v_i^*/\partial x_j$ into their symmetric and antisymmetric parts. Then

$$\sigma_{ij}\left(\frac{\partial v_i^*}{\partial x_k}\frac{\partial v_k}{\partial x_j} - \frac{\partial v_i}{\partial x_k}\frac{\partial v_k^*}{\partial x_j} \right) = 2\sigma_{ij}(\dot{\varepsilon}_{ij}^*\omega_{jk} - \dot{\varepsilon}_{ik}\omega_{jk}^*)$$

The other triple products vanish in pairs due to symmetry and antisymmetry properties of their factors. Since the tensors σ_{ij} and $\dot{\varepsilon}_{ij}$ are coaxial (the material being isotropic), the triple products $\sigma_{ij}\dot{\varepsilon}_{ik}\omega_{jk}^*$ and $\sigma_{ij}\dot{\varepsilon}_{ik}^*\omega_{jk}$ individually vanish (see Sec. 1.6(i)). Inequality (112) therefore reduces to

$$\int \left\{ \frac{2}{3}H(\dot{\varepsilon}_{ij}^*\dot{\varepsilon}_{ij}^* - \dot{\varepsilon}_{ij}\dot{\varepsilon}_{ij}) - \sigma_{ij}\left(\frac{\partial v_i^*}{\partial x_k}\frac{\partial v_k^*}{\partial x_j} - \frac{\partial v_i}{\partial x_k}\frac{\partial v_k}{\partial x_j} \right) \right\}dV > 2\int \dot{F}_j(v_j^* - v_j)\,dS_F$$

since $v_j^* = v_j$ on S_v. Now, for the actual field v_j, the transformation of integrals by Green's theorem gives

$$\int \dot{F}_j v_j\,dS = \int \dot{t}_{ij}\frac{\partial v_j}{\partial x_i}\,dV = \int \left(\frac{2}{3}H\dot{\varepsilon}_{ij}\dot{\varepsilon}_{ij} - \sigma_{ij}\frac{\partial v_i}{\partial x_k}\frac{\partial v_k}{\partial x_j} \right)dV$$

in view of (85). Using this result to eliminate the unstarred part of the volume integral in the preceding inequality, we finally obtain

$$\int \left(\frac{2}{3}H\dot{\varepsilon}_{ij}^*\dot{\varepsilon}_{ij}^* - \sigma_{ij}\frac{\partial v_i^*}{\partial x_k}\frac{\partial v_k^*}{\partial x_j} \right)dV - 2\int \dot{F}_j v_j^*\,dS_F > \int \dot{F}_j v_j\,dS_v - \int \dot{F}_j v_j\,dS_F \qquad (113)$$

The expression on the left-hand side of (113) therefore has a minimum value in the actual state.† It can be shown that the minimum is analytic only when no unloading occurs in the actual state. The term in σ_{ij} gives a finite contribution at a velocity discontinuity, which can be allowed only when $H = 0$. The minimum principle is directly obtainable from (108) by letting the shear modulus tend to infinity, and using the coaxiality of the stress and strain rate tensors.

Problems

2.1 An isotropic material exhibiting no Bauschinger effect is found to yield under biaxial stresses of p and q. Show that the plane stress yield locus must pass through the stress points (p, q), $(-p, -q)$, $(p - q, p)$, $(p - q, -q)$, $(q - p, -p)$, $(q - p, q)$, as well as those obtained by interchanging each pair of coordinates, whatever the form of the yield criterion. Assuming $p > 0 > q$, find the ratio of the uniaxial yield stress predicted by the Tresca criterion to that by the Mises criterion.

    Answer: $(p - q)/\sqrt{p^2 - pq + q^2}$.

2.2 Show that the strain energy per unit volume for an isotropic elastic solid consists of a volumetric resilience that depends on the bulk modulus K, and a shear resilience that depends on the shear modulus G. Establish the fact that yielding occurs according to the von Mises criterion when the shear resilience attains the value $Y^2/6G$. If the Mises criterion is approximated by the Tresca criterion with Y replaced by mY, what value of m will make the percentage error in uniaxial tension identical to that in pure shear?

    Answer: $m = 1.071$.

2.3 A closed-ended thin-walled tube of thickness t and mean radius r is subjected to an axial tensile force P, which is less than the value P_0 necessary to cause yielding. If a gradually increasing internal pressure p is now applied, show that the tube will yield according to the Tresca criterion when

$$\frac{pr}{Yt} = \begin{cases} 1 & \frac{P}{P_0} \leqslant \frac{1}{2} \\[2mm] 2\left(1 - \frac{P}{P_0}\right) & \frac{P}{P_0} \geqslant \frac{1}{2} \end{cases}$$

and according to the von Mises criterion when

$$\frac{pr}{Yt} = \frac{2}{\sqrt{3}}\left\{1 - \left(\frac{P}{P_0}\right)^2\right\}^{1/2}$$

2.4 A closed-ended thin-walled tube of initial mean radius r_0 is subjected to an internal pressure p, and an external pressure αp on the cylindrical surface. The loading is continued into the plastic range by maintaining a constant value of $\alpha > 0$. Assuming the deformation to be uniform, and using the Lévy-Mises flow rate, show that the total equivalent strain at any stage is

$$\bar{\varepsilon} = \frac{2\sqrt{3 - 6\alpha + 4\alpha^2}}{3 - 4\alpha} \ln\left(\frac{r}{r_0}\right)$$

where r is the current mean radius. For what range of values of α is this expression valid?

    Answer: $0 \leqslant \alpha < \frac{3}{4}$.

2.5 A uniform cylindrical bar is rendered plastic by the application of a longitudinal tensile force. The material is isotropic rigid/plastic, obeying an arbitrary regular yield criterion. Assuming the

---

† R. Hill, *J. Mech. Phys. Solids,* **5**: 229 (1957). This paper includes an extremum principle for the modified boundary-value problem in which S_F is submitted to a uniform fluid pressure.

deformation mode to be axially symmetrical, show that the distribution of radial and axial velocities has the general form

$$v_r = -r(Az + B) \qquad v_z = A(\tfrac{1}{2}r^2 + z^2) + 2Bz$$

where A and B are arbitrary constants, the z axis being aken along the longitudinal axis of the bar.

2.6 A thin-walled tube, subjected to combined tension and torsion in the plastic range, hardens kinematically according to Prager's hardening rule. The tube is initially isotropic and yields according to the von Mises criterion. Assuming a linear strain-hardening law with a plastic modulus H, show that the yield criterion at any stage of the loading is

$$(\sigma - H\varepsilon^p)^2 + 3(\tau - \tfrac{2}{3}H\gamma^p)^2 = Y^2$$

where (σ, τ) are the applied tensile and shear stresses and $(\varepsilon^p, \gamma^p)$ the corresponding plastic strain components at a generic stage.

2.7 A solid circular cylinder of radius a and height h is axially compressed by a pair of partially rough platens. The stresses acting on an element, bounded by radial planes and concentric cylindrical surfaces, are shown in Fig. A. Assuming $\sigma_\theta = \sigma_r$, obtain the radial equilibrium equation

$$\frac{d\sigma_r}{dr} = -\frac{2\mu p}{h}$$

where μ is the coefficient of friction between the cylinder and the platens. Neglecting the effect of friction on the yield criterion, show that the load P required to bring the cylinder to the yield point is given by

$$\frac{P}{\pi a^2} \simeq Y\left(1 + \frac{2\mu a}{3h}\right)$$

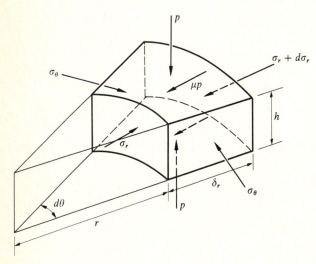

Figure A

2.8 Figure B shows the stresses acting on an element in the meridian plane of a necked cylindrical bar subjected to an axial force P. The element is bounded by principal stress trajectories in the neighborhood of a typical point in the minimum section. Show that the condition for radial equilibrium is

$$\frac{d\sigma_r}{dr} + \frac{\sigma_z - \sigma_r}{\rho} = 0$$

where ρ is the local radius of curvature of the longitudinal trajectory. Assuming $\rho = R(a/r)$, where R is the radius of curvature of the neck, and setting $\sigma_\theta = \sigma_r$ in the yield criterion, show that

$$\frac{P}{\pi a^2} = \sigma\left(1 + \frac{a}{4R}\right)$$

where σ is the current uniaxial yield stress of the material in the minimum section.

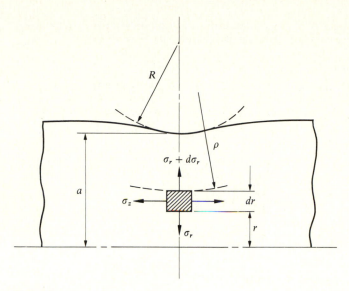

Figure B

2.9 As a possible yield criterion for isotropic metals, it is postulated that the numerically largest deviatoric principal stress attains a critical value at yielding. Show that the yield locus is a regular hexagon whose sides are inclined at 30° to those of the Tresca hexagon. If the new hexagon is made to circumscribe the Mises circle, prove that the new yield criterion is expressible in the form

$$k^2(J_2 - k^2)^2 = J_3^2$$

where k is the yield stress in pure shear. What is the corresponding relationship between k and Y?
Answer: $Y = 1.5k$.

2.10 From the observed experimental results of Taylor and Quinney, Prager suggested the plastic potential

$$g = J_2\left(1 - \frac{3J_3^2}{4J_2^3}\right)$$

Show that the polar equation of its level curve in the deviatoric plane is

$$r^2(9 - \sin^2 3\theta) = \text{const}$$

Assuming the yield locus to be a geometrically similar curve, find the ratio of the yield stress in pure shear to that in uniaxial tension or compression.
Answer: 0.544.

2.11 The experimentally determined (μ, ν) relationship for a certain metal may be reasonably expressed by the equation

$$\nu = \frac{3\mu(2 + \mu^2)}{9 + \mu^2(1 - \mu^2)}$$

Assuming the yield function and the plastic potential to be identical, show that the polar equation of the deviatoric yield locus is

$$r = r_0 \sec^{4/3} \theta \exp\left(-\tfrac{1}{2} \tan^2 \theta\right)$$

where r_0 is a constant. What value of k/Y is predicted by this equation?

Answer: 0.563.

2.12 A material yields according to the von Mises yield criterion and hardens isotropically. If the plastic potential is the same as that of Prob. 2.10, show that the flow rule may be written as

$$d\varepsilon_{ij}^p = \tfrac{9}{8}\{(1 + 2\alpha^2)\sqrt{\tfrac{3}{2}}\, n_{ij} + \alpha(\delta_{ij} - 3n_{ik}n_{kj})\} \frac{d\bar{\sigma}}{H}$$

where $\alpha = 9J_3/2\bar{\sigma}^3$, and n_{ij} is the unit vector in the direction of s_{ij}. Find the predicted slope of the engineering shear stress-strain curve, excluding the elastic strain, in terms of the current plastic modulus H.

Answer: $8H/27$.

2.13 In the combined bending and twisting of a thin-walled tube, which involves longitudinal and shearing stresses σ and τ respectively, show that the scalar parameter α of the preceding problem is

$$\alpha = \frac{1 + 9\tau^2/2\sigma^2}{3(1 + 3\tau^2/\sigma^2)^{3/2}}$$

Using the above flow rule, compute the plastic strain increment ratio and the associated Lode parameter v when the stress ratio τ/σ is 0.5, 1.0, and 2.0. Obtain a graphical plot for the relationship between the Lode parameters μ and v.

2.14 A circular cylindrical bar is subjected to a uniform fluid pressure p on the lateral surface. The ends of the bar are supported in such a way that they are free to move axially during the deformation. To determine the condition for necking, it is convenient to use the incompressible trial velocity field

$$v_r = -\tfrac{1}{2}rf'(z) \qquad v_\theta = 0 \qquad v_z = f(z)$$

in cylindrical coordinates (r, θ, z). Assuming the uniqueness functional to have a stationary value at the bifurcation, obtain the condition $H > p$ for uniqueness, whatever the form of f.

2.15 A rigid/plastic bar of arbitrary cross section is subjected to a uniaxial plastic state of stress in the z direction which coincides with the longitudinal axis. If the stress is uniform throughout the bar, show that the rectangular components of the associated velocity have the general expressions

$$u = -\tfrac{1}{4}A(x^2 - y^2 + 2z^2) - \tfrac{1}{2}x(By + Cz + D)$$

$$v = -\tfrac{1}{4}B(y^2 - x^2 + 2z^2) - \tfrac{1}{2}y(Ax + Cz + D)$$

$$w = \tfrac{1}{4}C(x^2 + y^2 + 2z^2) + z(Ax + By + D)$$

for any regular yield surface, where A, B, C, and D are arbitrary constants.

2.16 A long slender column of uniform cross section, made of a rigid work-hardening material, is built-in at one end and loaded axially at the other by an increasing compressive load P. Using the velocity field of the preceding problem, show that the condition for uniqueness of the deformation mode is $H > Pl^2/3I$ to a close approximation, where H is the current plastic modulus, l the length of the column, and I the least moment of inertia of the cross section. Verify that the velocity field corresponding to the critical rate of hardening reduces to that for pure bending.

2.17 A uniform rigid/plastic bar of rectangular cross-section is made of a von Mises material that hardens isotropically. The ends of the bar, defined at any instant by $x = \pm l$, are moved apart with equal and opposite velocities of magnitude U, while zero tangential traction is maintained on these faces. The initiation of necking of the bar, whose instantaneous thickness is $2h$, may be represented by the velocity field

$$u = U\left(\frac{x}{l} + \frac{c}{\pi} \sin\frac{\pi x}{l} \cos\frac{\pi y}{l}\right)$$

$$v = -U\left(\frac{y}{l} + \frac{c}{\pi} \cos\frac{\pi x}{l} \sin\frac{\pi y}{l}\right)$$

where c is a constant. Considering the field equations for the rate problem, and assuming the value of c that corresponds to neutral loading at $x = \pm l$, $y = 0$, show that the condition for necking is

$$\sqrt{3}\,\frac{\bar{\sigma}}{H} = 1 + \frac{2\pi h}{l} \operatorname{cosec} \frac{2\pi h}{l} \qquad \frac{h}{l} < \frac{1}{2}$$

2.18 Suppose that the bar of the preceding problem is axially compressed by means of a pair of frictionless rigid platens in contact with the end faces. The velocity field at the incipient buckling, when the bar is of length $2l$ and thickness $2h$, may be taken as

$$u = U\left(-\frac{x}{l} + \frac{c}{\pi} \sin\frac{\pi x}{l} \sin\frac{\pi y}{l}\right)$$

$$v = U\left(\frac{y}{l} + \frac{c}{\pi} \cos\frac{\pi x}{l} \cos\frac{\pi y}{l}\right)$$

where c is a constant. Choosing the value of c for which there is neutral loading at $x = 0$, $y = h$, show that the critical stress is given by

$$\sqrt{3}\,\frac{\bar{\sigma}}{H} = \frac{2\pi h}{l} \operatorname{cosec} \frac{2\pi h}{l} - 1 \qquad \frac{h}{l} < \frac{1}{2}$$

Verify that for a sufficiently small h/l, the critical compressive stress reduces to that given by the tangent modulus theory (Section 4.7).

THREE

ELASTOPLASTIC BENDING AND TORSION

In a deformable body subjected to external loads of gradually increasing magnitude, plastic flow begins at a stage when the yield criterion is first satisfied in the most critically stressed element. Further increase in loads causes spreading of the plastic zone which is separated from the elastic material by an elastic/plastic boundary. The position of this boundary is an unknown of the problem, and is generally so complicated in shape that the solution of the boundary-value problem often involves numerical methods. The solution must be carried out in a succession of small increments of strain even when the deformation is restricted to an elastic order of magnitude. It is necessary to ensure at each stage that the calculated stresses and displacements in the elastic and plastic regions satisfy the conditions of continuity across the elastic/plastic boundary. In this chapter, we shall be concerned mainly with the problems of bending and torsion in the elastic/plastic range, assuming the deformation to be sufficiently small. The related problems of limit analysis will be discussed in the next chapter.

3.1 Plane Strain Compression and Bending

(i) *Plane strain compression of a block* As a simple application of the Prandtl-Reuss theory, consider the frictionless compression of a rectangular block of metal between a pair of rigid overlapping platens (Fig. 3.1). The edges of the block are parallel to the rectangular axes, with the x axis taken in the direction of compression. A condition of plane strain is achieved by suppressing lateral

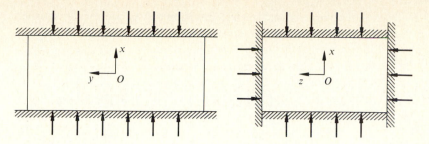

Figure 3.1 Plastic compression of a block between smooth rigid platens under conditions of plane strain.

expansion in the z direction with the help of rigid dies.† It is therefore a case of homogeneous compression in which $\sigma_y = 0$ throughout the deformation, and $\sigma_z = v\sigma_x$ while the block is still elastic. If Tresca's yield criterion is adopted, yielding begins when $\sigma_x = -Y$ in each element of the block. The relevant stress-strain equations in the plastic range are

$$d\varepsilon_x = \frac{1}{E}(d\sigma_x - v\, d\sigma_z) + \frac{1}{3}(2\sigma_x - \sigma_z)\, d\lambda$$

$$d\varepsilon_z = \frac{1}{E}(d\sigma_z - v\, d\sigma_x) + \frac{1}{3}(2\sigma_z - \sigma_x)\, d\lambda = 0$$

$$(1)$$

If the material is nonhardening, $\sigma_x = -Y$ throughout the plastic compression. The elimination of $d\lambda$ from (1) then gives

$$E\, d\varepsilon_x = \left(\frac{1}{2} - v\right) d\sigma_z + \frac{3\, d\sigma_z}{2(2\sigma_z + Y)}$$

At the initial yielding, $\sigma_z = -vY$ and $\varepsilon_x = -(1 - v^2)Y/E$. Under these initial conditions, the above equation integrates to

$$\frac{E}{Y}\varepsilon_x = \left(\frac{1}{2} - v\right)\left(\frac{\sigma_z}{Y} + v\right) - \frac{3}{4}\ln\left(\frac{1 - 2v}{1 + 2\sigma_z/Y}\right) - (1 - v^2) \qquad (2)$$

giving the variation of σ_z with the amount of compression. As the deformation proceeds, the first term becomes increasingly unimportant, while σ_z rapidly approaches the limiting value $-\frac{1}{2}Y$. Taking $v = 0.3$, for instance, σ_z is found to have the value $-0.49Y$ when ε_x is only 3.5 times that at the initial yielding.

When the material yields according to the von Mises yield criterion $\sigma_x^2 - \sigma_x\sigma_z + \sigma_z^2 = Y^2$, the initial yielding of the block corresponds to $\sigma_x = \sigma_x^0$ and $\sigma_z = v\sigma_x^0$, where

$$\sigma_x^0 = -\frac{Y}{\sqrt{1 - v + v^2}}$$

† An experimental set up of this kind has been employed by P. W. Bridgman, *J. Appl. Phys.*, **17**: 225 (1946). Since no dies are absolutely rigid, a direct experimental verification of the theory is difficult.

During the subsequent compression, the yield criterion can be identically satisfied by writing the stresses in terms of a parameter θ as

$$\sigma_x = -\frac{2Y}{\sqrt{3}} \cos \theta \qquad \sigma_z = -\frac{2Y}{\sqrt{3}} \sin\left(\frac{\pi}{6} - \theta\right) \qquad (3)$$

The condition $\sigma_z = v\sigma_x$ at the initial yielding furnishes the initial value of θ as

$$\theta_0 = \tan^{-1} \frac{1 - 2v}{\sqrt{3}} \qquad (4)$$

When $v = 0.3$, we get $\sigma_x^0 \simeq -1.127Y$ and $\theta_0 \simeq 13°$. Substitution from (3) into the stress-strain relations (1) gives

$$d\varepsilon_x = \frac{2Y}{\sqrt{3}\,E} \left\{ \sin\theta - v\cos\left(\frac{\pi}{6} - \theta\right) \right\} d\theta - \frac{2Y}{\sqrt{3}} \cos\left(\frac{\pi}{6} - \theta\right) d\lambda$$

$$0 = \frac{2Y}{\sqrt{3}\,E} \left\{ \cos\left(\frac{\pi}{6} - \theta\right) - v\sin\theta \right\} d\theta + \frac{2Y}{\sqrt{3}} \sin\theta\, d\lambda$$

Since $d\lambda$ must be positive, the second equation indicates that θ decreases as the compression proceeds. Eliminating $d\lambda$ from the above equations, we get

$$E\,d\varepsilon_x = \frac{2Y}{\sqrt{3}} \left\{ (1 - 2v)\cos\left(\frac{\pi}{6} - \theta\right) + \frac{3}{4}\operatorname{cosec}\theta \right\} d\theta$$

Using the initial condition $\varepsilon_x = -(1 - v^2)\sigma_x^0/E$ when $\theta = \theta_0$, the above equation is readily integrated to obtain†

$$-\frac{E}{Y}\varepsilon_x = \frac{2}{\sqrt{3}}(1 - 2v)\sin\left(\frac{\pi}{6} - \theta\right) + \frac{\sqrt{3}}{2}\ln\left(\cot\frac{\theta}{2}\tan\frac{\theta_0}{2}\right) + \sqrt{1 - v + v^2} \qquad (5)$$

As the deformation continues, the first term on the right-hand side soon becomes negligible. The angle θ rapidly approaches the limiting value zero, the corresponding values of σ_x and σ_z being $-2Y/\sqrt{3}$ and $-Y/\sqrt{3}$ respectively. It is found that σ_z is within 1 percent of its limiting value when ε_x is only four times that at the initial yielding, for $v = 0.3$. Owing to the rapid initial change in stress, the elastic and plastic strain increments are comparable up to a total strain which is three to four times that at the elastic limit. A graphical comparison of the solutions based on the Tresca and Mises criteria is made in Fig. 3.2.

(ii) *Plane strain bending of a beam* A related problem is the bending of a uniform rectangular beam by terminal couples under conditions of plane strain (Fig. 3.3). The radius of curvature of the bent beam is assumed large compared to its depth $2h$, so that transverse stresses may be neglected. The neutral fibre coincides with

---

† The solution given here is essentially due to R. Hill, *J. Appl. Mech.*, **16**: 295 (1949). A graphical solution of this problem has been discussed by J. M. Alexander, *Proc. Inst. Mech. Eng.*, **173**: 73 (1959).

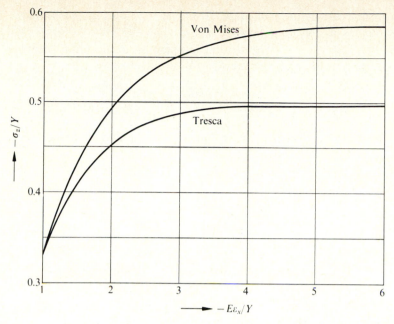

Figure 3.2 Effect of elastic compressibility on the plane strain compression of a block in the plastic range.

Ox, and is bent into a circular arc of radius R. All fibres above this line are extended and those below this line are compressed during the bending. So long as the beam remains elastic, the longitudinal stress σ_x is distributed linearly across the depth of the beam according to the relationship

$$\sigma_x = \frac{Ey}{(1 - v^2)R} = \frac{My}{I_z}$$

where M is the bending couple per unit width of the beam, and $I_z = \frac{2}{3}h^3$ the moment of inertia of the cross section per unit width about the z axis. The factor

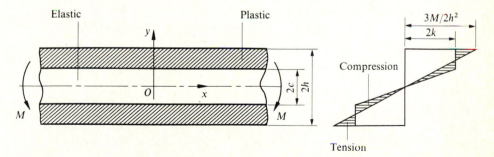

Figure 3.3 Geometry and stress distribution (approximate) in the plane strain bending of a beam. The residual stress is given by the shaded triangles.

$(1 - v^2)$ arises from the condition of plane strain ($\varepsilon_z = 0$) during the bending. Plastic yielding begins at the boundaries $y = \pm h$ when the longitudinal stress attains the value $\pm Y/\sqrt{1 - v + v^2}$. The bending moment M_e at the initial yielding, and the corresponding radius of curvature R_e of the neutral surface, are

$$M_e = \frac{2h^2 Y}{3\sqrt{1 - v + v^2}} \qquad R_e = \frac{Eh\sqrt{1 - v + v^2}}{Y(1 - v^2)} \tag{6}$$

where subscript e represents the elastic limit. For $v = 0.3$, the numerical values of M_e and R_e are $0.751\,Yh^2$ and $0.934Eh/Y$ respectively.

If the bending moment is increased further, plastic zones spread inward from the outer surfaces, the depth of the elastic part at any stage being denoted by $2c$. The stresses in the elastic region are

$$\sigma_x = \frac{Yy}{c\sqrt{1 - v + v^2}} \qquad \sigma_z = v\sigma_x \qquad -c \leqslant y \leqslant c \tag{7}$$

The longitudinal strain at a generic point of the cross section is y/R throughout the bending. The application of Hooke's law to the elastic part of the beam gives

$$R = \frac{Ec\sqrt{1 - v + v^2}}{Y(1 - v^2)} = \left(\frac{c}{h}\right)R_e$$

during the elastic/plastic bending. In the lower plastic region, the stresses are given by (3), where θ depends on y according to (5) with $\varepsilon_x = y/R$. The stresses and strains in the upper plastic region are identical in magnitude but opposite in sign. The applied couple per unit width is

$$M = 2\int_0^h \sigma_x y\, dy = \frac{2Yc^2}{3\sqrt{1 - v + v^2}} + \frac{2Y}{\sqrt{3}}\int_c^h y\cos\theta\, dy \tag{8}$$

Using equation (5), with $-\varepsilon_x$ replaced by y/R, the above integral can be evaluated numerically to obtain $M/h^2 Y$ for any assumed value of c/h.

For practical purposes, it is sufficiently accurate to replace the von Mises criterion by the modified Tresca criterion $\sigma_x = \pm 2Y/\sqrt{3}$. Then the magnitude of the longitudinal stress increases from zero at the neutral surface to $2Y/\sqrt{3}$ at the elastic/plastic boundary. The integration in this case is straightforward, and the result is

$$M \simeq \frac{2Y}{\sqrt{3}}\left(h^2 - \frac{1}{3}c^2\right) = \frac{1}{2}M_e\left\{3 - \left(\frac{R}{R_e}\right)^2\right\} \tag{9}$$

where

$$M_e \simeq \frac{4Yh^2}{3\sqrt{3}} \qquad R_e \simeq \frac{\sqrt{3}\,Eh}{2Y(1 - v^2)}$$

The maximum error in this approximation is about 2 percent, occurring at the initial yielding. The bending moment M rapidly approaches the asymptotic value

$(2/\sqrt{3})h^2 Y$ or $\frac{3}{2}M_e$, which is the fully plastic or collapse moment per unit width of the beam. The limiting plastic state involves a stress discontinuity of amount $4Y/\sqrt{3}$ across the neutral surface.

If the beam is unloaded from the partly plastic state, there is a certain distribution of residual stress left in the beam. The residual stress can be calculated on the assumption that the change in stress during the unloading is purely elastic. It is therefore necessary to superpose an elastic stress distribution due to an opposite moment equal in magnitude to that which is released. Subtracting My/I_z from the existing stress in the elastic/plastic beam, where M is given by (9), we obtain the residual stress on complete unloading as

$$\frac{\sigma_x}{Y} = \frac{2}{\sqrt{3}} \left\{ \frac{y}{c} - \frac{y}{2h}\left(3 - \frac{c^2}{h^2}\right)\right\} \qquad |y| \leqslant c$$

$$\frac{\sigma_x}{Y} = \frac{2}{\sqrt{3}} \left\{ 1 - \frac{y}{2h}\left(3 - \frac{c^2}{h^2}\right)\right\} \qquad |y| \geqslant c$$

(10)

The distribution is shown diagrammatically by the shaded triangles in Fig. 3.3. The residual stress changes sign in the region $c < |y| < h$, vanishing at a distance $2h/(3 - c^2/h^2)$ from the neutral surface. The stress attains its greatest magnitude at the outer surface for $c/h \geqslant \sqrt{2} - 1$, and at the plastic boundary for $c/h \leqslant \sqrt{2} - 1$. As the beam is rendered increasingly plastic, the residual stress at $y = \pm h$ approaches the limiting value $\mp Y/\sqrt{3}$.

The curvature of the unloaded beam is obtained by subtracting from the elastic/plastic curvature h/cR_e the amount of elastic spring-back equal to $(1 - v^2)M/EI_z$. Substituting for R_e, M, and I_z, the residual curvature may be expressed as

$$\frac{1}{R} = \frac{2}{\sqrt{3}}(1 - v^2)\frac{Y}{Ec}\left(1 - \frac{3c}{2h} + \frac{c^3}{2h^3}\right)$$

(11)

The factor outside the bracket is the curvature of the beam at the moment of unloading. The expression within the bracket is the ratio of the residual stress at $y = c$ to the plane strain yield stress $2Y/\sqrt{3}$. For small elastic/plastic bending, the residual curvature is comparable to the amount of elastic spring-back.

3.2 Cylindrical Bars Under Torsion and Tension

(i) *Pure torsion of a bar* We begin with a solid cylindrical bar of radius a, subjected to a twisting moment T. So long as the bar is elastic, the shear stress acting over any cross section is proportional to the radial distance r from the central axis. The applied torque T is the resultant moment of the stress distribution about this axis. If the angle of twist per unit length of the bar is denoted by θ, the elastic shear stress may be written as

$$\tau = Gr\theta = \frac{2Tr}{\pi a^4}$$

Since the shear stress has its greatest value at $r = a$, the bar begins to yield at this radius when the torque is increased to T_e, the corresponding twist being θ_e. Setting $\tau = k$ at $r = a$, we get

$$T_e = \frac{1}{2} \pi k a^3 \qquad \theta_e = \frac{k}{Ga}$$

If the torque is increased further, a plastic annulus forms near the boundary, leaving a central zone of elastic material within a radius c (Fig. 3.4a). The stress distribution in the elastic region is linear, with the shear stress reaching the value k at $r = c$. For a non-hardening material, the shear stress has the constant value k throughout the plastic region, and the stress distribution becomes

$$\tau = k \frac{r}{c} \qquad 0 \leqslant r \leqslant c$$

$$\tau = k \qquad c \leqslant r \leqslant a$$

Since the shear stress within the elastic zone is also equal to $Gr\theta$, we have $\theta = k/Gc$. The twisting moment is

$$T = 2\pi \int_0^a \tau r^2 \, dr = \frac{2}{3} \pi k \left(a^3 - \frac{1}{4} c^3 \right) = \frac{1}{3} T_e \left\{ 4 - \left(\frac{\theta_e}{\theta} \right)^3 \right\} \tag{12}$$

As the elastic/plastic torsion continues, the torque rapidly approaches the fully plastic value $\frac{2}{3} \pi k a^3$. Since θ tends to infinity as c tends to zero, an elastic core of material must exist for all finite values of the angle of twist.

In the case of an annealed material, there is no well-defined yield point, and the elastic/plastic boundary is therefore absent. Since the engineering shear strain at

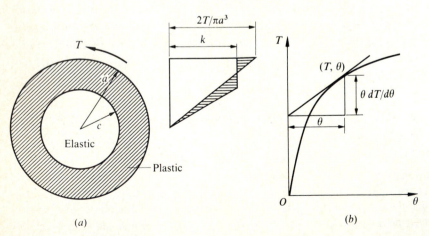

Figure 3.4 Torsion of a solid cylindrical bar. (a) Plastic annulus and stress distribution for $H = 0$; (b) Nadai's construction for an annealed bar.

any radius r is $\gamma = r\theta$, the torque may be expressed as

$$T = 2\pi \int_0^a \tau r^2 \, dr = \frac{2\pi}{\theta^3} \int_0^{a\theta} \tau\gamma^2 \, d\gamma$$

When the shear stress-strain curve of the material is given, the torque can be calculated from above, using the known (τ, γ) relationship. Conversely, if the torque-twist relationship for a solid bar has been experimentally determined, the shear stress-strain curve can be easily derived from it. The differentiation of the above equation with respect to θ gives

$$\frac{d}{d\theta}(T\theta^3) = 2\pi a^3 \theta^2 \tau_0$$

where τ_0 is the value of τ at $r = a$ where the shear strain is $\gamma_0 = a\theta$. The relationship between τ_0 and γ_0 is therefore given by†

$$\tau_0 = \frac{1}{2\pi a^3} \left(\theta \frac{dT}{d\theta} + 3T \right) \qquad \gamma_0 = a\theta \qquad (13)$$

The geometrical significance of the first term in the bracket is indicated in Fig. 3.4b. Since $dT/d\theta$ must be obtained numerically or graphically from the measured (T, θ) curve, the computation based on (13) is not very accurate for the initial part of the curve. The accuracy may, however, be improved by rewriting the shear stress as‡

$$\tau_0 = \frac{1}{2\pi a^3} \left\{ \theta^2 \frac{d}{d\theta} \left(\frac{T}{\theta} \right) + 4T \right\}$$

The ratio T/θ is constant in the elastic range, and decreases slowly over the initial part of the plastic range. The contribution of the first term in the bracket is therefore small over this part. For the latter part of the curve, where the strain-hardening is small, the formula (13) should give more satisfactory results.

Suppose, now, that a bar of external radius a has a concentric circular hole of radius b. The material is assumed to work-harden, the uniaxial stress–plastic strain curve being represented by a straight line of slope H over the relevant range. If the bar is subjected to pure torsion, yielding begins at $r = a$ when the torque and the specific angle of twist become

$$T_e = \frac{1}{2} \pi k a^3 \left(1 - \frac{b^4}{a^4} \right) \qquad \theta_e = \frac{k}{Ga}$$

During the elastic/plastic torsion, the shear stress increases with the radius in both elastic and plastic regions. Since the total equivalent strain in any plastic element is equal to $(1/\sqrt{3})(r\theta - \tau/G)$, the assumed strain-hardening law gives

$$\tau = k + \frac{H}{3} \left(r\theta - \frac{\tau}{G} \right) \qquad \theta = \frac{k}{Gc}$$

† A. Nadai, *Theory of Flow and Fracture of Solids*, vol. I, p. 349, McGraw-Hill Book Company, New York (1950).

‡ R. Hill, *The Mathematical Theory of Plasticity*, p. 94, Clarendon Press, Oxford (1950).

where c is the radius to the elastic/plastic boundary. Substituting for θ, the shear stress in the plastic region is obtained as

$$\tau = k\left(\frac{1 + Hr/3Gc}{1 + H/3G}\right) \qquad c \leqslant r \leqslant a$$

The stress in the elastic region $(b \leqslant r \leqslant c)$ is $\tau = kr/c$ as before. The stress distribution over the entire cross section furnishes the applied torque T. A straightforward integration results in

$$\left(1 + \frac{H}{3G}\right)T = \frac{2}{3}\pi ka^3\left\{1 + \frac{1}{4}\left[-\frac{c^3}{a^3}\left(1 + \frac{3b^4}{c^4}\right) + \frac{Ha}{Gc}\left(1 - \frac{b^4}{a^4}\right)\right]\right\} \qquad (14)$$

which reduces to (12) when $H = 0$ and $b = 0$. The fully plastic torque T_0 for a work-hardening hollow bar is given by

$$\left(1 + \frac{H}{3G}\right)T_0 = \frac{2}{3}\pi ka^3\left\{1 - \frac{b^3}{a^3} + \frac{H}{4G}\left(\frac{a}{b} - \frac{b^3}{a^3}\right)\right\}$$

The variation of T/T_e with θ/θ_e for $H = 0$ and $H = 0.3G$ is shown in Fig. 3.5. The

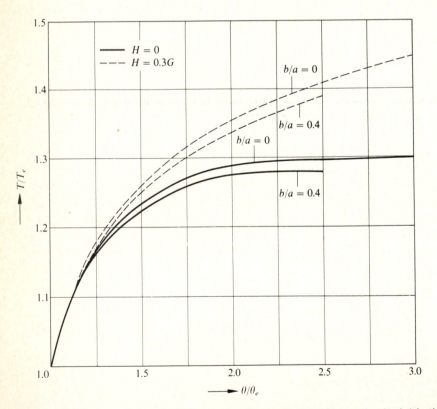

Figure 3.5 Torque-twist relationship in pure torsion of cylindrical bars in the elastic/plastic range.

fully plastic angle of twist per unit length has the finite value $\theta_0 = k/Gb$, which is independent of the rate of hardening of the material.

The residual stress left in the bar on unloading from an elastic/plastic state can be determined as in the case of bending. It is only necessary to superpose an elastic distribution of stress produced by an opposite torque equal in magnitude to that which is released. For a completely unloaded bar, we therefore have to subtract the quantity $2Tr/\pi(a^4 - b^4)$ from the stress existing at the moment of unloading. Using (14) for T, the residual stress distribution for $H = 0$ may be written as

$$\frac{\tau}{k} = \frac{r}{c} - \frac{r}{3a}\frac{4 - (c^3/a^3)(1 + 3b^4/c^4)}{1 - b^4/a^4} \qquad b \leqslant r \leqslant c$$

$$\frac{\tau}{k} = 1 - \frac{r}{3a}\frac{4 - (c^3/a^3)(1 + 3b^4/c^4)}{1 - b^4/a^4} \qquad c \leqslant r \leqslant a \tag{15}$$

The residual stress is negative in an outer part of the plastic annulus and positive over the remainder of the cross section. When $b = 0$, the numerically greatest residual stress occurs at $r = c$ for $c/a \leqslant 0.576$ and at $r = a$ for $c/a \geqslant 0.576$.

The angle of elastic untwist per unit length on complete unloading is equal to $2T/\pi G(a^4 - b^4)$, where T is given by (14). Assuming $H = 0$, the residual angle of twist per unit length may be expressed as

$$\theta = \frac{k}{Gc}\left\{1 - \frac{c}{3a}\frac{4 - (c^3/a^3)(1 + 3b^4/c^4)}{1 - b^4/a^4}\right\} \tag{16}$$

The factor outside the curly bracket is the value of θ at the moment of unloading, while the expression inside the bracket is the residual value of τ/k at the elastic/plastic boundary. For a given elastic/plastic twist, the residual angle of twist decreases as the rate of hardening increases.

(ii) Combined torsion and tension—I A solid cylindrical bar of radius a and length l is subjected to any combination of twist and axial extension. While the deformation is elastic, the longitudinal stress σ is constant over the cross section, and the shear stress τ is directly proportional to the radial distance r from the axis. It follows that yielding first occurs at $r = a$ when the stresses satisfy the von Mises yield criterion

$$\sigma^2 + 3\tau^2 = Y^2 \tag{17}$$

When the loading is continued into the plastic range, so that the radius to the elastic/plastic boundary is c, the stresses in the elastic region for an incompressible material are

$$\sigma = 3G\varepsilon \qquad \tau = \frac{Gr\phi}{l} \qquad 0 \leqslant r \leqslant c$$

where ε is the total longitudinal strain and ϕ the total angle of twist. In the plastic region $(c \leqslant r \leqslant a)$, the Prandtl-Reuss stress-strain equations give

$$
d\varepsilon = \frac{dl}{l} = \frac{d\sigma}{3G} + \frac{2}{3}\sigma \, d\lambda
$$

$$
d\gamma = \frac{r \, d\phi}{2l} = \frac{d\tau}{2G} + \tau \, d\lambda
$$

(18)

The stresses in the plastic region are also required to satisfy the yield criterion (17), work-hardening being neglected. Eliminating $d\lambda$ from (18), and using (17), it is easily shown that

$$
\frac{3}{r}\frac{dl}{d\phi} = \frac{\sigma}{\tau} + \frac{l}{Gr}\left(\frac{d\sigma}{d\phi} - \frac{\sigma}{\tau}\frac{d\tau}{d\phi}\right) = \frac{\sigma}{\tau} - \frac{Y^2 l}{Gor}\left(\frac{1}{\tau}\frac{d\tau}{d\phi}\right)
$$

Suppose that the ratio of the rate of extension to the rate of twist is held constant during the elastic/plastic loading.† The constant value of $dl/d\phi$, denoted by $a\alpha/3$, would be compatible with the above equation and the yield criterion, if σ and τ are both constant in any given plastic element. Then

$$
\frac{\sigma}{\tau} = \frac{3}{r}\frac{dl}{d\phi} = \frac{a\alpha}{r} \qquad c \leqslant r \leqslant a
$$

(19)

Since $\varepsilon/\phi = a\alpha/3l$ according to this strain path, σ/τ is continuous across the elastic/plastic boundary. Combining (17) and (19), we have

$$
\sigma = \frac{\alpha Y}{\sqrt{\alpha^2 + 3r^2/a^2}} \qquad \tau = \frac{(r/a)Y}{\sqrt{\alpha^2 + 3r^2/a^2}} \qquad c \leqslant r \leqslant a
$$

(20)

Thus, the axial stress decreases and the shear stress increases as we move outward from $r = c$. The continuity of the axial stress across the plastic boundary $r = c$ requires

$$
\varepsilon \sqrt{\alpha^2 + \frac{3c^2}{a^2}} = \frac{\alpha Y}{3G}
$$

giving the relationship between c/a and ε during the loading. Using the expressions for σ and τ in the elastic and plastic regions, the axial load N and the torque T are found by integration as

$$
\frac{N}{\pi a^2 Y} = \frac{\alpha c^2}{a^2}\left(\alpha^2 + \frac{3c^2}{a^2}\right)^{-1/2} + \frac{2}{3}\alpha\left\{\sqrt{\alpha^2 + 3} - \sqrt{\alpha^2 + \frac{3c^2}{a^2}}\right\}
$$

$$
\frac{T}{\pi a^3 Y} = \frac{c^4}{2a^4}\left(\alpha^2 + \frac{3c^2}{a^2}\right)^{-1/2} - \frac{2}{9}\alpha^2\left\{\sqrt{\alpha^2 + 3} - \sqrt{\alpha^2 + \frac{3c^2}{a^2}}\right\}
$$

$$
+ \frac{2}{27}\left\{(\alpha^2 + 3)^{3/2} - \left(\alpha^2 + \frac{3c^2}{a^2}\right)^{3/2}\right\}
$$

(21)

---

† This solution, due to R. Hill, has been presented by F. A. Gaydon, *Q. J. Mech. Appl. Math.*, **5**: 29 (1952). The fully plastic stress distribution was given earlier by A. Nadai, *Trans. A.S.M.E.*, **52**: 93 (1930).

For any given value of α, (21) defines the relationship between N and T parametrically through c/a. When $\alpha = 0$, the axial force vanishes, and the torque reduces to that given by (12). The fully plastic stress distribution over the cross section for an arbitrary strain path under combined loading is given by (20) in terms of the final value of α.

(iii) *Combined torsion and tension–II* Suppose that a cylindrical bar of radius a is first twisted elastically and then extended into the elastic/plastic range by an increasing axial load.† The angle of twist of the bar is maintained at a constant value θ_0 per unit length during the extension. Yielding begins at the outer radius when the longitudinal strain is ε_0, the corresponding axial stress being $3G\varepsilon_0$ for an incompressible material. Since the shear stress is $Ga\theta_0$ at $r = a$, the relationship between θ_0 and ε_0 is

$$a^2\theta_0^2 + 3\varepsilon_0^2 = \frac{Y^2}{3G^2} \tag{22}$$

in view of the yield criterion (17). Subsequently, when the bar is plastic to a radius c, the stresses in the elastic zone corresponding to an axial strain ε are

$$\sigma = 3G\varepsilon \qquad \tau = Gr\theta_0 \qquad 0 \leqslant r \leqslant c$$

Since the element at $r = c$ must be at the point of yielding, the radius to the elastic/plastic boundary is given by

$$c^2\theta_0^2 + 3\varepsilon^2 = \frac{Y^2}{3G^2}$$

In the plastic region, the stresses must satisfy the yield criterion (17) and the stress-strain equations (18), where $d\gamma = 0$. Eliminating $d\lambda$, and substituting for $d\tau/\tau$ using (17), we obtain

$$3G\,d\varepsilon = \frac{Y^2\,d\sigma}{Y^2 - \sigma^2}$$

which is readily integrated to

$$\frac{3G}{Y}\varepsilon = \tanh^{-1}\left(\frac{\sigma}{Y}\right) + \text{const} \qquad c \leqslant r \leqslant a$$

The constant of integration must be determined from the condition that

$$\sigma = 3G\varepsilon = \sqrt{Y^2 - 3G^2r^2\theta_0^2}$$

when an element at radius r first becomes plastic. Hence the tensile stress in the

---

† The solutions discussed here are due to F. A. Gaydon, *Q. J. Mech. Appl. Math.*, **5**: 29 (1952). See also W. Prager and P. G. Hodge, Jr., *Theory of Perfectly Plastic Solids*, chap. 3, Wiley and Sons (1951). Numerical solutions for strain-hardening materials with arbitrary values of v have been discussed by D. S. Brooks, *Int. J. Mech. Sci.*, **11**: 75 (1969).

plastic region ($c \leqslant r \leqslant a$) is given by

$$\frac{\sigma}{Y} = \tanh\left(\frac{3G}{Y}\varepsilon - \sqrt{1 - \frac{3G^2}{Y^2}r^2\theta_0^2} + \tanh^{-1}\sqrt{1 - \frac{3G^2}{Y^2}r^2\theta_0^2}\right) \quad (23)$$

The shear stress in the plastic region follows from (23) and the yield criterion (17). The variations of load and torque with extension can be calculated numerically if required.

If the bar is initially twisted to an extent that makes it just plastic at $r = a$, then $Ga\theta_0 = Y/\sqrt{3}$ and $\varepsilon_0 = 0$. Substituting in (23), the stress distribution in the plastic region is obtained as

$$\frac{\sigma}{Y} = \tanh\left(\frac{3G}{Y}\varepsilon - \sqrt{1 - \frac{r^2}{a^2}} + \tanh^{-1}\sqrt{1 - \frac{r^2}{a^2}}\right) \quad (24)$$

The bar becomes completely plastic ($c = 0$) when $\varepsilon = Y/3G$, giving $\sigma/Y = \tanh 1 \simeq 0.762$ at $r = a$. If the extension is continued in the fully plastic range, (24) holds over the entire cross section of the bar. The stresses σ and τ at the boundary $r = a$ approach their asymptotic values Y and zero respectively as the strain is increased. The approach is so rapid that σ is within 0.5 percent of Y when ε is only equal to Y/G.

Consider now the situation where the bar is first extended to produce an axial strain ε_0 elastically, and then twisted by a gradually increasing torque while the extension is held constant. The bar begins to yield at the outer radius again when the angle of twist per unit length is θ_0, given by (22). When the specific angle of twist θ is large enough to render the bar plastic to a radius c, the stresses in the elastic region are

$$\sigma = 3G\varepsilon_0 \qquad \tau = Gr\theta \qquad 0 \leqslant r \leqslant c$$

Since the material at $r = c$ is at the point yielding,

$$c^2\theta^2 + 3\varepsilon_0^2 = \frac{Y^2}{3G^2}$$

Setting $d\varepsilon = 0$ and $d\phi = l\,d\theta$ in the Prandtl-Reuss equations (18), and eliminating $d\lambda$, we obtain the differential equation

$$Gr\,d\theta = \frac{Y^2\,d\tau}{Y^2 - 3\tau^2}$$

in view of (17). The integration of the above equation gives

$$\frac{\sqrt{3}\,G}{Y}r\theta = \tanh^{-1}\left(\frac{\sqrt{3}\,\tau}{Y}\right) + \text{const} \qquad c \leqslant r \leqslant a$$

When an element first becomes plastic, its tensile stress is $\sigma_0 = G\varepsilon_0$, the corresponding shear stress being given by

$$\sqrt{3}\,\tau = \sqrt{Y^2 - \sigma_0^2} = \sqrt{3}\,Gr\theta$$

The constant of integration follows from this initial condition, and the shear stress in the plastic region ($c \leqslant r \leqslant a$) finally becomes

$$\frac{\sqrt{3}\,\tau}{Y} = \tanh\left(\frac{\sqrt{3}\,G}{Y}r\theta - \sqrt{1 - \frac{\sigma_0^2}{Y^2}} + \tanh^{-1}\sqrt{1 - \frac{\sigma_0^2}{Y^2}}\right) \tag{25}$$

The tensile stress in the plastic region then follows from the yield criterion. If the bar is initially extended just to the yield point before the torque is applied, $\sigma_0 = Y$ and $\theta_0 = 0$, giving

$$\frac{\sqrt{3}\,\tau}{Y} = \tanh\left(\frac{\sqrt{3}\,G}{Y}r\theta\right) \qquad \frac{\sigma}{Y} = \operatorname{sech}\left(\frac{\sqrt{3}\,G}{Y}r\theta\right) \tag{26}$$

These expressions hold throughout the cross section of the bar, which is now completely plastic. When $a\theta$ is equal to $\sqrt{3}\,Y/G$, the value of $\sqrt{3}\,\tau$ at $r = a$ is already within 0.5 percent of Y. The torque T and the axial load N are given by

$$\frac{\sqrt{3}\,T}{\pi a^3 Y} = 2\int_0^1 \xi^2 \tanh\left(\frac{\sqrt{3}\,G}{Y}\xi a\theta\right) d\xi$$

$$\frac{N}{\pi a^2 Y} = 2\int_0^1 \xi \operatorname{sech}\left(\frac{\sqrt{3}\,G}{Y}\xi a\theta\right) d\xi \tag{27}$$

where $\xi = r/a$. The variations of the dimensionless load and torque for the initially plastic bar are shown in Fig. 3.6, which indicates that N and T rapidly approach their asymptotic values of zero and $2\pi Y a^3/3\sqrt{3}$ respectively.

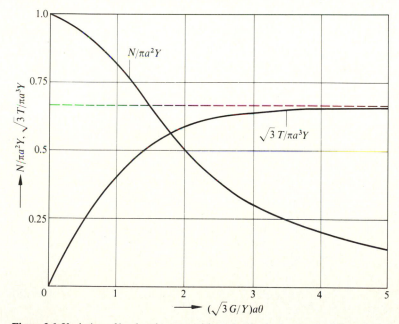

Figure 3.6 Variation of load and torque with angle of twist in a bar initially extended to the yield point.

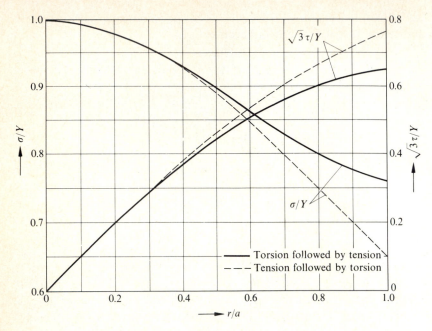

Figure 3.7 Stress distribution in a circular cylindrical bar subjected to combined tension and torsion ($\varepsilon = Y/3G$ and $a\theta = Y/\sqrt{3}\,G$).

The distributions of σ/Y and $\sqrt{3}\,\tau/Y$ for both the strain paths, expressed by equations (24) and (26), are plotted in Fig. 3.7, assuming $\varepsilon = Y/3G$ and $a\theta = Y/\sqrt{3}\,G$ respectively. The first strain path is represented by solid curves and the second strain path is represented by broken curves. Although the final states of deformation in the two cases are the same, the stress distribution differs as a consequence of its path dependence.

3.3 Thin-walled Tubes Under Combined Loading

(i) Combined torsion and tension A thin-walled tube is first twisted to the point of yielding and then extended longitudinally while holding the angle of twist constant. The material is prestrained to obtain a sharp yield point, and the subsequent rate of hardening is assumed negligible compared to the elastic modulus. The loading path is such that the elastic and plastic strain increments are initially of comparable magnitudes, even though the tube is fully plastic. The longitudinal stress σ and the shear stress τ at any stage are approximately uniform through the thickness of the tube, the relevant stress-strain equations being

$$d\varepsilon = \frac{d\sigma}{E} + \tfrac{2}{3}\sigma\,d\lambda \qquad d\gamma = \frac{d\tau}{2G} + \tau\,d\lambda \tag{28}$$

where $d\gamma = 0$ during the extension. The elimination of $d\lambda$ between the two equations gives

$$d\varepsilon = \frac{d\sigma}{E} - \frac{\sigma}{3G}\frac{d\tau}{\tau} = \frac{d\sigma}{E} + \frac{\sigma^2}{Y^2 - \sigma^2}\frac{d\sigma}{3G}$$

where the last step follows from the yield criterion (17), the material being assumed nonhardening. Since the shear stress decreases from its initial value $Y/\sqrt{3}$ as the tensile stress increases from zero, the integration of the above equation results in

$$\frac{6G}{Y}\varepsilon = \left(\frac{1 - 2v}{1 + v}\right)\frac{\sigma}{Y} + \ln\left(\frac{Y + \sigma}{Y - \sigma}\right) \tag{29}$$

in view of the relation $E = 2(1 + v)G$. As the longitudinal strain is increased from zero, σ rapidly increases to approach the value Y asymptotically. The first term on the right-hand side of (29) soon becomes negligible. For $v = 0.3$, the longitudinal stress is within one percent of Y when ε is only equal to $3Y/E$. The above theory forms the basis of an experimental verification of the Prandtl-Reuss theory. The experiment has been carried out by Hohenemser using prestrained tubes of mild steel,[†] and reasonable agreement has been found with the theoretical prediction.

Consider the related problem in which a thin-walled tube is brought to the yield point in simple tension, and subsequently twisted in the plastic range holding the tensile stress constant at the value Y. The loading can be continued only if the material work-hardens, having an effective stress $\bar{\sigma}$ at any stage. Then

$$\bar{\sigma}^2 = Y^2 + 3\tau^2 \qquad \bar{\sigma}\,d\bar{\sigma} = 3\tau\,d\tau$$

The stress-strain relations (28) still hold with $d\sigma = 0$, while $d\lambda$ is given by

$$\frac{2}{3}d\lambda = \frac{d\bar{\sigma}}{H\bar{\sigma}} = \frac{3\tau\,d\tau}{H(Y^2 + 3\tau^2)}$$

where H is the current slope of the stress-plastic strain curve. Equations (28) therefore become

$$d\varepsilon = \frac{3Y\tau\,d\tau}{H(Y^2 + 3\tau^2)} \qquad d\gamma = \frac{d\tau}{2G} + \frac{9\tau^2\,d\tau}{2H(Y^2 + 3\tau^2)}$$

It follows from the second of these relations that the torsional rigidity is equal to G when $\tau = 0$, although the tube is plastic.[‡] Assuming H to be a constant, the above equations can be integrated to

$$\varepsilon = \frac{Y}{E} + \frac{Y}{2H}\ln\left(1 + \frac{3\tau^2}{Y^2}\right)$$

$$\gamma = \frac{\tau}{2G} + \frac{3}{2H}\left[\tau - \frac{Y}{\sqrt{3}}\tan^{-1}\left(\frac{\sqrt{3}\,\tau}{Y}\right)\right] \tag{30}$$

[†] K. Hohenemser, *Z. angew. Math. Mech.*, **11**: 15 (1931). See also K. Hohenemser and W. Prager, ibid., **12**: 1 (1932).

[‡] This has been experimentally confirmed by J. L. M. Morrison and W. M. Shepherd, *Proc. Inst. Mech. Eng.*, **163**: 1 (1950), who subjected thin-walled tubes to various combinations of tension and torsion. The results were in substantial agreement with the Prandtl-Reuss theory.

The incremental shear strain $d\gamma$ at any stage is equal to $r\,d\phi/2l$, where ϕ is the total angle of twist, l the current length of the tube, and r the current mean radius. Since r and l are varying from stage to stage, 2γ cannot be interpreted as being the tangent of the angle of the helix formed by an original generator of the tube.

Suppose, now, that a thin-walled tube is subjected to simultaneous torsion and tension following a strain path which is such that the elastic part of the strain is negligible. The Lévy-Mises flow rule furnishes the relations

$$d\varepsilon = \frac{dl}{l} = \frac{\sigma\,d\bar{\sigma}}{H\bar{\sigma}} \qquad d\gamma = \frac{r\,d\phi}{2l} = \frac{3\tau\,d\bar{\sigma}}{2H\bar{\sigma}} \qquad \frac{dr}{r} = \frac{dt}{t} = -\frac{dl}{2l} \tag{31}$$

where l, r, and t denote the current length, mean radius, and thickness of the tube, their initial values being l_0, r_0, and t_0 respectively. The last set of equations immediately furnish

$$\frac{r}{r_0} = \frac{t}{t_0} = \sqrt{\frac{l_0}{l}}$$

Since $r^2 l = r_0^2 l_0 = \text{const}$, it follows that the internal volume of the tube remains unchanged during the deformation. This is really a consequence of the isotropy of the material and the equality of the Lode variables μ and ν. From (31), we now obtain

$$\frac{3\tau}{\sigma} = r\frac{d\phi}{dl} = r_0\sqrt{\frac{l_0}{l}}\frac{d\phi}{dl}$$

$$\bar{\sigma} = \sqrt{\sigma^2 + 3\tau^2} = \sigma\left\{1 + \frac{r_0^2 l_0}{3l}\left(\frac{d\phi}{dl}\right)^2\right\}^{1/2} \tag{32}$$

The elimination of σ between the first equation of (31) and the second equation of (32), and the integration of the resulting differential equation, result in

$$\int_Y^{\bar{\sigma}}\frac{d\bar{\sigma}}{H} = \int_{l_0}^{l}\left\{1 + \frac{r_0^2 l_0}{3l}\left(\frac{d\phi}{dl}\right)^2\right\}^{1/2}\frac{dl}{l} \tag{33}$$

If $dl/d\phi$ is considered as a given function of the length of the tube, the integral on the right-hand side can be evaluated numerically or otherwise to obtain the relationship between $\bar{\sigma}$ and l. Since σ and τ then follow from (32), the axial load $2\pi r t\sigma$ and the torque $2\pi r^2 t\tau$ can be calculated at any stage of the deformation.†

(ii) **Combined bending and twisting** A thin-walled tube of thickness t and mean radius a is rendered partly plastic by the simultaneous application of bending and twisting couples (Fig. 3.8). Lateral forces being absent, the stresses and strains do

---

† The combined torsion and tension of thin-walled cylinders formed the basis of an experimental verification of the laws of plasticity by G. I. Taylor and H. Quinney, *Philos. Trans. R. Soc. London*, **A230**: 323 (1931).

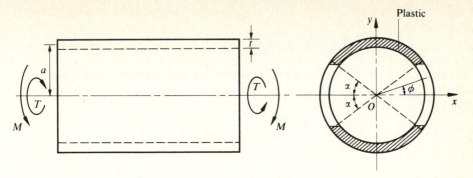

Figure 3.8 Combined bending and twisting of a thin-walled tube in the plastic range.

not vary along the length of the tube.† The shear stress τ is uniformly distributed over the section, while the longitudinal stress σ varies along the circumference at each stage. Let ψ denote the angle of bend and θ the angle of twist, per unit length of the tube. While the tube is entirely elastic, the stresses are

$$\sigma = Ey\psi \qquad \tau = Ga\theta$$

where y is reckoned positive in the direction of convexity of the bent tube. There is no warping of the cross section in the elastic range, the bending couple M and the twisting couple T being

$$M = \pi t a^3 E\psi \qquad T = 2\pi t a^3 G\theta$$

It is evident that plastic yielding first occurs at the extremities of the vertical diameter when the yield criterion (17) is satisfied. Assuming $\psi = \psi_e$ and $\theta = \theta_e$ at the initial yielding, we get

$$\frac{1}{3}\left(\frac{Ea\psi_e}{k}\right)^2 + \left(\frac{Ga\theta_e}{k}\right)^2 = 1$$

Since $E\psi/G\theta$ is equal to $2M/T$ throughout the elastic deformation, the values of ψ and θ at the initial yielding may be expressed in terms of the couple ratio as

$$\frac{Ea\psi_e}{2k} = \left(\frac{4}{3} + \frac{T^2}{M^2}\right)^{-1/2} \qquad \frac{Ga\theta_e}{k} = \left(1 + \frac{4M^2}{3T^2}\right)^{-1/2} \tag{34}$$

If the deformation is continued, two identical plastic zones spread toward the neutral axis. Let the position of the elastic/plastic boundary at any stage be specified by its angular distance α from the neutral axis Ox. Let ϕ be the counterclockwise angle defining the position of a generic point on the circumference with

---

† Thin-walled tubes of more general shape of the cross section have been considered by R. Hill and M. P. L. Siebel, *Phil. Mag.*, **42**: 722 (1951). For experimental confirmations, see M. P. L. Siebel, *J. Mech. Phys. Solids*, **1**: 149 (1953). The effects of work-hardening have been examined by R. T. Shield and E. T. Onat, *J. Appl. Mech.*, **20**: 345 (1953).

respect to Ox. Neglecting work-hardening, the distribution of the axial stress may be written as

$$\sigma = Ea\psi \sin \phi(|\phi| \leqslant \alpha) \qquad \sigma = \pm Ea\psi \sin \alpha(|\phi| \geqslant \alpha) \tag{35a}$$

where the upper sign holds for positive values of ϕ and the lower sign for negative values of ϕ. In view of (17), the shear stress at any stage is†

$$\tau = k\sqrt{1 - \frac{1}{3}\left(\frac{Ea\psi}{k}\right)^2 \sin^2 \alpha} \qquad -\frac{\pi}{2} \leqslant \phi \leqslant \frac{\pi}{2} \tag{35b}$$

Using the normal stress distribution, the bending couple is obtained in the nondimensional form

$$\frac{M}{T_0} = \frac{2}{\pi k} \int_0^{\pi/2} \sigma \sin \phi \, d\phi = \frac{Ea\psi}{\pi k}(\alpha + \sin \alpha \cos \alpha) \tag{36}$$

where $T_0 = 2\pi ka^2 t$ is the yield point torque in pure torsion. Using the value of $Ea\psi/k$ given by (36), the stresses in the plastic region can be expressed in terms of the applied couples as

$$\frac{\sigma}{k} = \pm \frac{\pi \sin \alpha}{\alpha + \sin \alpha \cos \alpha} \frac{M}{T_0} \qquad \frac{\tau}{k} = \frac{T}{T_0}$$

As the cross section is rendered increasingly plastic, the magnitude of σ/k rapidly approaches the limiting value $\pi M/2T_0$. Inserting from above into (17) gives

$$\frac{\pi^2}{3}\left(\frac{\sin \alpha}{\alpha + \sin \alpha \cos \alpha}\right)^2 \left(\frac{M}{T_0}\right)^2 + \left(\frac{T}{T_0}\right)^2 = 1 \tag{37}$$

This is the relationship between M and T for any given position of the elastic/plastic boundary. As α decreases from $\pi/2$, the coefficient of $(M/T_0)^2$ decreases from its initial value of $4/3$, and approaches the limiting value $\pi^2/12$ corresponding to the fully plastic state. From (36) and (37), the angle of bend at any stage may be expressed as

$$\frac{Ea\psi}{k} = \left\{\frac{1}{3}\sin^2 \alpha + \frac{T^2}{\pi^2 M^2}(\alpha + \sin \alpha \cos \alpha)^2\right\}^{-1/2} \tag{38}$$

which reduces to that given by (34) when $\alpha = \pi/2$. If the ratio of the applied couples is given at each stage, (38) defines the relationship between ψ and α. Equations (35) and (36) then uniquely determine σ, τ, and M. The variation of the bending moment with the angle of bend is shown graphically in Fig. 3.9 for constant values of M/T. The limiting value of M/T_0, which is equal to $(\pi^2/12 + T^2/M^2)^{-1/2}$ in view

---

† The use of Hencky equations and the assumption of elastic incompressibility lead to the expression $\tau = \pi Ga\theta/(2\alpha + 2\cot \alpha)$. The corresponding values of M and T, when θ/ψ is held constant, are found to be in close agreement with those given by the Prandtl-Reuss theory.

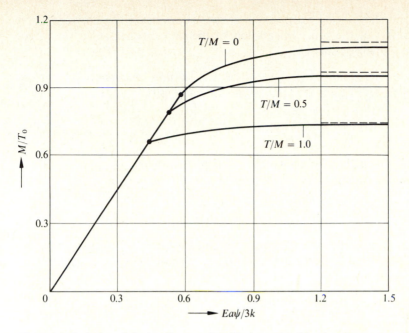

Figure 3.9 Variation of the bending couple with the angle of bend under constant T/M ratios. The asymptotes are shown by broken lines.

of (37), is indicated in each case by a broken line. The approach to the limit is seen to be extremely rapid.†

Plane transverse sections become warped when the tube becomes partly plastic, the warping displacement at any point on the circumference being denoted by w. The longitudinal and shear strain increments, denoted $d\varepsilon$ and $d\gamma$, are related to $d\psi$ and $d\theta$ by the equations

$$d\varepsilon = a \sin \phi \, d\psi \qquad 2d\gamma = a \, d\theta + \frac{1}{a} \frac{\partial}{\partial \phi} (dw)$$

If geometry changes are disregarded, the incremental warp dw is equal to $(\partial w/\partial \psi) \, d\psi$. In the plastic regions, the Prandtl-Reuss stress-strain relations give

$$\frac{d\gamma - d\tau/2G}{d\varepsilon - d\sigma/E} = \frac{3\tau}{2\sigma} \qquad (|\phi| \geqslant \alpha)$$

Substituting for the strain increments, and using (17), we obtain

$$\frac{1}{a} \frac{\partial}{\partial \phi} \left(\frac{\partial w}{\partial \psi} \right) + a \frac{d\theta}{d\psi} = \frac{1}{E} \left\{ 2(1 + v) + \frac{3\tau^2}{k^2 - \tau^2} \right\} \frac{d\tau}{d\psi} \pm \frac{\sqrt{3} \, \tau a \sin \phi}{\sqrt{k^2 - \tau^2}}$$

† When the tube is deformed under a constant value of θ/ψ, the applied couples approach their limiting values in an oscillatory manner. See R. Hill and M. P. L. Seibel, *Phil. Mag.*, **42**: 722 (1951).

Because of symmetry, $\partial w/\partial \psi$ must vanish at the extremities of the horizontal and vertical diameters of the tube. Integration of the above equation in the upper plastic region therefore gives

$$\frac{\partial}{\partial \psi}\left(\frac{w}{a}\right) = \left(\frac{\pi}{2} - \phi\right)\left\{a\frac{d\theta}{d\psi} - \left[2(1 + v) + \frac{3\tau^2}{k^2 - \tau^2}\right]\frac{1}{E}\frac{d\tau}{d\psi}\right\} - \frac{\sqrt{3}\,\tau a \cos \phi}{\sqrt{k^2 - \tau^2}}$$

$$\alpha \leqslant \phi \leqslant \frac{\pi}{2} \quad (39a)$$

In the lower plastic region, $\partial w/\partial \psi$ has the same magnitude but an opposite sign. In the elastic region, the stress-strain relation $d\gamma = d\tau/2G$ leads to

$$\frac{\partial}{\partial \psi}\left(\frac{w}{a}\right) = \left\{\frac{2}{E}(1 + v)\frac{d\tau}{d\psi} - a\frac{d\theta}{d\psi}\right\}\phi \qquad -\alpha \leqslant \phi \leqslant \alpha \qquad (39b)$$

Since w should be a continuous function of position round the periphery, $\partial w/\partial \psi$ is necessarily continuous at each point. In view of equations (39), the continuity condition at $\phi = \alpha$ furnishes

$$\left\{2(1 + v) - \left[\frac{6\alpha}{\pi} - (1 - 2v)\right]\frac{\tau^2}{k^2}\right\}\frac{1}{Ea}\frac{d\tau}{d\psi} = \left(1 - \frac{\tau^2}{k^2}\right)\frac{d\theta}{d\psi} - \frac{2\sqrt{3}\,\tau}{\pi k}\sqrt{1 - \frac{\tau^2}{k^2}}\cos \alpha$$

$$(40)$$

If the loading path is prescribed, (40) gives $d\theta/d\psi$ as a function of ψ or α, and θ can be obtained by integration using the initial yield values (34). If, on the other hand, the strain path is given, so that θ is a prescribed function of ψ, the differential equation (40) can be solved numerically with the help of (35b) to determine τ and α. The corresponding bending couple then follows from (36).

When θ is a known function of ψ, the warping displacement w can be determined by the integration of (39). In the elastic region ($|\phi| \leqslant \alpha$), the warping function immediately follows from (39b) and the fact that $w = 0$ for $\psi = 0$. Substituting for $d\theta/d\psi$ into equation (39a), the differential equation for the warping in the plastic region is reduced to

$$\frac{\partial}{\partial \psi}\left(\frac{w}{a}\right) = -\frac{\alpha}{E}\left(1 - \frac{2\phi}{\pi}\right)\frac{3\tau^2}{k^2 - \tau^2}\frac{d\tau}{d\psi} + \left[\left(1 - \frac{2\phi}{\pi}\right)\cos \alpha - \cos \phi\right]\frac{\sqrt{3}\,a\tau}{\sqrt{k^2 - \tau^2}}$$

Let ψ_0 and τ_0 denote the values of ψ and τ when a given element, specified by the angle ϕ, first becomes plastic. Then

$$\Delta\left(\frac{w}{a}\right) = -\frac{3}{E}\left(1 - \frac{2\phi}{\pi}\right)\int_{\tau_0}^{\tau}\frac{\alpha\tau^2 d\tau}{k^2 - \tau^2}$$

$$+ \sqrt{3}\,a\int_{\psi_0}^{\psi}\left[\left(1 - \frac{2\phi}{\pi}\right)\cos \alpha - \cos \phi\right]\frac{\tau\,d\psi}{\sqrt{k^2 - \tau^2}} \qquad \alpha \leqslant \phi \leqslant \frac{\pi}{2} \quad (41)$$

For a given value of M/T, the ratio $Ea\psi_0/k$ is given by the right-hand side of (38) with α replaced by ϕ. The value of τ_0/k is obtained from (35b), where ψ_0 and ϕ must be written for ψ and α respectively. Figure 3.10 shows the variation of w/a with ϕ for $\nu = 0.3$, when the ratio M/T has a constant value of unity, and the plastic boundary corresponds to $\alpha = 25°$.

For the preceding analysis to be valid, it is essential that no plastic element unloads during the deformation. This condition is satisfied if the rate of plastic work is nowhere negative. In view of the flow rule, the validity of the solution requires

$$\sigma(d\varepsilon - d\sigma/E) \geqslant 0 \qquad |\phi| \geqslant \alpha$$

Since $\varepsilon = a\psi \sin \phi$ and $\sigma = \pm E a \psi \sin \alpha$, the inequality becomes

$$(|\sin \phi| - \sin \alpha) \, d\psi - \psi \cos \alpha \, d\alpha \geqslant 0 \qquad |\phi| \geqslant \alpha$$

The above restriction is certainly satisfied when the bending is monotonic ($d\psi > 0$) and the plastic zone steadily increases in size ($d\alpha < 0$). These conditions are fulfilled in the numerical solutions presented in Figs. 3.9 and 3.10.

(iii) *Combined tension and internal pressure* A thin-walled tube with closed ends is loaded in the plastic range by the combined action of an internal pressure p and an independent axial tensile force N. The elastic strains are negligible compared to the plastic strains, so that a rigid/plastic model would be justified for the analysis. So long as the rate of hardening exceeds a certain critical value, the deformation of the tube is uniquely determined by the applied loads which are varied in a prescribed manner. If r is the current mean radius of the tube and t the current wall

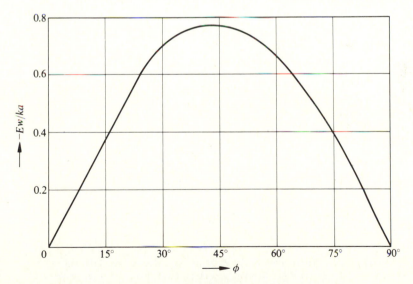

Figure 3.10 Distribution of axial displacement, representing warping, when $\alpha = 25°$ and $M/T = 1$.

thickness, the circumferential stress σ_θ and the axial stress σ_z at any stage are given by

$$\sigma_\theta = \frac{pr}{t} \qquad \sigma_z = (1 + \alpha)\frac{pr}{2t} \qquad (42)$$

where α denotes the ratio $N/\pi r^2 p$ between the independent axial force and the end load due to the internal pressure. According to the Lévy-Mises flow rule, the components of the strain increment are given by

$$\frac{d\varepsilon_\theta}{2\sigma_\theta - \sigma_z} = \frac{d\varepsilon_z}{2\sigma_z - \sigma_\theta} = -\frac{d\varepsilon_r}{\sigma_\theta + \sigma_z} = \frac{\overline{d\varepsilon}}{2\bar{\sigma}}$$

where $d\varepsilon_r$ is the thickness strain increment dt/t, and $\overline{d\varepsilon}$ the equivalent strain increment. Since $2\sigma_z/\sigma_\theta = 1 + \alpha$, the equivalent stress is

$$\bar{\sigma} = \sqrt{\sigma_\theta^2 - \sigma_\theta\sigma_z + \sigma_z^2} = \tfrac{1}{2}\sigma_\theta\sqrt{3 + \alpha^2}$$

in view of which, the principal strain increments become

$$d\varepsilon_\theta = \frac{(3 - \alpha)\,\overline{d\varepsilon}}{2\sqrt{3 + \alpha^2}} \qquad d\varepsilon_r = -\frac{(3 + \alpha)\,\overline{d\varepsilon}}{2\sqrt{3 + \alpha^2}} \qquad d\varepsilon_z = \frac{\alpha\,\overline{d\varepsilon}}{\sqrt{3 + \alpha^2}} \qquad (43)$$

Since the equivalent strain must be positive, it follows from (43) that $d\varepsilon_\theta \gtrless 0$ for $\alpha \lessgtr 3$, $d\varepsilon_z \gtrless 0$ for $\alpha \gtrless 0$, and $d\varepsilon_r < 0$ for all α.

The deformation of the tube ceases to be uniform when the rate of hardening decreases to a critical value. At this stage, the internal energy dissipated during a further infinitesimal deformation, computed to the second order of smallness, equals the work done by the external forces. It is reasonable to suppose that both N and p attain stationary values at the onset of instability.† The corresponding change in α is given by

$$\frac{d\alpha}{\alpha} = d\left(\ln\frac{N}{r^2 p}\right) = -2\frac{dr}{r} = -2d\varepsilon_\theta$$

since $dN = dp = 0$ when instability sets in. The logarithmic differentiation of (42) then gives

$$\frac{d\sigma_\theta}{\sigma_\theta} = d\varepsilon_\theta - d\varepsilon_r = \frac{3\,\overline{d\varepsilon}}{\sqrt{3 + \alpha^2}}$$

$$\frac{d\sigma_z}{\sigma_z} = \left(\frac{1 - \alpha}{1 + \alpha}\right)d\varepsilon_\theta - d\varepsilon_r = \frac{\sqrt{3 + \alpha^2}}{1 + \alpha}\,\overline{d\varepsilon}$$

Differentiating the von Mises yield criterion, and eliminating $2\sigma_\theta - \sigma_r$ and $2\sigma_r - \sigma_\theta$

† H. W. Swift, *J. Mech. Phys. Solids*, **1**: 1 (1952); Z. Marciniak, *Rozpr. Inz.*, **110**: 529 (1958); B. Storakers, *Int. J. Mech. Sci.*, **10**: 519 (1968). Useful experimental results have been reported by B. H. Jones and P. B. Mellor, *J. Strain Anal.*, **2**: 62 (1967); E. A. Davis, *J. Appl. Mech.*, **12**: 13 (1945).

by means of the flow rule, it is easy to show that

$$d\bar{\sigma}\,\overline{d\varepsilon} = d\sigma_\theta\,d\varepsilon_\theta + d\sigma_z\,d\varepsilon_z$$

Substituting for $d\sigma_\theta$ and $d\sigma_z$ from the preceding relations, and using the values of $d\varepsilon_\theta$ and $d\varepsilon_z$ given by (43), we get

$$\frac{1}{\bar{\sigma}}\frac{d\bar{\sigma}}{d\varepsilon} = \left(\frac{9+\alpha^3}{3+\alpha^2}\right)\frac{\sigma_\theta}{2\bar{\sigma}} = \frac{9+\alpha^3}{(3+\alpha^2)^{3/2}} \tag{44}$$

as the condition for plasticity instability when $\alpha \geqslant 0$. The quantity on the left-hand side of (44) is the reciprocal of the critical subtangent to the generalized stress-strain curve. For a given material, the instability strain therefore depends on the final stress ratio, which must be determined from the prescribed loading path. Since the stress ratio remains constant for $\alpha = 0$ (internal pressure alone), $\alpha = 3$, and $\alpha = \infty$ (axial force alone), equation (44) provides the complete solution in these cases.

To obtain the solution for a variable stress ratio, consider the particular loading path in which the ratio N/p is held constant. Then $d\alpha = -2\alpha\,d\varepsilon_\theta$ throughout the loading, and (43) gives

$$\overline{d\varepsilon} = -\frac{\sqrt{3+\alpha^2}}{\alpha(3-\alpha)}\,d\alpha$$

It may be noted that $d\alpha \gtrless 0$ for $\alpha \gtrless 3$. If α_0 denotes the initial value of α, integration of the above equation results in[†]

$$\bar{\varepsilon} = -\int_{\alpha_0}^{\alpha}\frac{\sqrt{3+\alpha^2}}{\alpha(3-\alpha)}\,d\alpha = f(\alpha)-f(\alpha_0)$$

where

$$f(\alpha) = \ln\left(\alpha + \sqrt{3+\alpha^2}\right) + \frac{1}{\sqrt{3}}\ln\left(\frac{\sqrt{3}+\sqrt{3+\alpha^2}}{\alpha}\right)$$

$$-\frac{2}{\sqrt{3}}\ln\left(\pm\frac{\sqrt{3}}{2}\mp\frac{2\sqrt{3}+\sqrt{3+\alpha^2}}{3-\alpha}\right) \tag{45}$$

The upper sign holds when $\alpha > 3$ and the lower sign when $\alpha < 3$. Since $\bar{\varepsilon}$ is positive, $f(\alpha)$ increases with the deformation of all variable stress ratios.[‡]

The solution is most conveniently obtained by representing the strain-hardening characteristic by the power law $\bar{\sigma} = C\bar{\varepsilon}^n$, which makes the left-hand side of (44) equal to $n/\bar{\varepsilon}$. Then $\bar{\varepsilon}$ and $f(\alpha)-f(\alpha_0)$ can be plotted against α for various values of n and α_0 respectively, as shown in Fig. 3.11. The points of intersection

† J. Chakrabarty, *METU J. Pure Appl. Sci.*, **3**: 29 (1970).

‡ For an analysis of the instability problem based on the assumption of constant stress ratio associated with either the internal pressure or the axial force attaining a maximum, see W. T. Lankford and E. Saibel, *Metals Tehnology*, T. P. 2238 (1947), and P. B. Mellor, *J. Mech. Eng. Sci.*, **1**: 251 (1962).

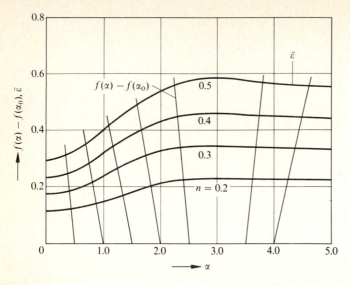

Figure 3.11 Plastic instability of closed-ended tubes under combined internal pressure and axial tension.

define the required values of $\bar{\varepsilon}$ and α at the onset of instability. The circumferential strain at instability is then obtained from the fact that the ratio of the final to the initial radius of the tube is equal to $\sqrt{\alpha_0/\alpha}$ for nonzero values of N and p. The ratio of the final and initial thicknesses is obtained by the integration of the equation

$$\frac{dt}{t} = -\left(\frac{3+\alpha}{3-\alpha}\right) d\varepsilon_\theta = \left(\frac{3+\alpha}{3-\alpha}\right)\frac{d\alpha}{2\alpha}$$

which follows from the first two relations of (43). Thus†

$$\frac{t}{t_0} = \left(\frac{3-\alpha_0}{3-\alpha}\right)\sqrt{\frac{\alpha}{\alpha_0}} \qquad \frac{r}{r_0} = \sqrt{\frac{\alpha_0}{\alpha}}$$

When $\alpha = \alpha_0 = 0$, an independent analysis gives $r/r_0 = t_0/t = \exp(n/2)$. Similarly, when $\alpha = \alpha_0 = 3$, it is easily shown that $r/r_0 = 1$ and $t/t_0 = \exp(-n)$. The instability or bursting pressure is finally obtained from the first relation of (42), using the fact that σ_θ is equal to $2C\bar{\varepsilon}^n/\sqrt{3+\alpha^2}$, the values of $\bar{\varepsilon}$ and α at instability being found from Fig. 3.11 for any given α_0.

3.4 Pure Bending of Prismatic Beams

(i) *Symmetrical bending* Consider a uniform prismatic bar bent by two equal and opposite couples M applied at its ends (Fig. 3.12). The cross section of the beam has

---

† Plastic instability in tubes of finite length has been examined by W. A. Weil, *Int. J. Mech. Sci.*, **5**: 487 (1963), and by J. K. Banerjee, ibid., **17**: 659 (1975).

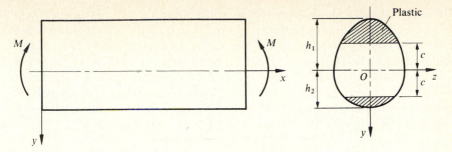

Figure 3.12 Bending of a prismatic beam under terminal couples.

an axis of symmetry Oy, and the axis of the bending couple is parallel to Oz, where O is taken on the neutral plane. The plane of bending then coincides with the xy plane, the neutral fiber Ox being bent to a circular arc of radius R. During the elastic bending, O is situated at the centroid of the cross section, and the only nonzero stress $\sigma_x = \sigma$ is given by

$$\sigma = \frac{Ey}{R} = \frac{My}{I_z}$$

where E is Young's modulus for the material, and I_z the moment of inertia of the cross section about the neutral axis Oz.

The longitudinal strain in the elastic beam is $\varepsilon_x = y/R$, and this is accompanied by the transverse strains $\varepsilon_y = \varepsilon_z = -vy/R$, where v is Poisson's ratio. If the components of the displacement are denoted by u, v, and w with respect to the coordinate axes, then

$$\frac{\partial u}{\partial x} = \frac{y}{R} \qquad \frac{\partial v}{\partial y} = \frac{\partial w}{\partial z} = -\frac{vy}{R}$$

$$\frac{\partial u}{\partial y} + \frac{\partial v}{\partial x} = \frac{\partial u}{\partial z} + \frac{\partial w}{\partial x} = \frac{\partial v}{\partial z} + \frac{\partial w}{\partial y} = 0$$

Assuming an element of the x axis and an element of the yz plane to be fixed in space at $x = y = z = 0$, the solution is obtained from the conditions $u = v = w = 0$ and $\partial v/\partial x = \partial w/\partial x = \partial w/\partial y = 0$ at the origin of coordinates. The result is†

$$u = \frac{xy}{R} \qquad v = -\frac{x^2 + v(y^2 - z^2)}{2R} \qquad w = -\frac{vyz}{R} \tag{46}$$

The deformation is such that transverse planes remain plane during the bending. The neutral plane xz, and every parallel plane, is deformed into an anticlastic surface having a transverse curvature v/R with an upward convexity.

---

† See, for example, S. Timoshenko and J. N. Goodier, *Theory of Elasticity*, 3rd ed., chap. 10, McGraw-Hill Book Co., New York (1970).

Yielding first occurs in the fiber that is farthest from the neutral surface, when the longitudinal stress becomes numerically equal to Y. If the cross section is not symmetrical about the neutral axis Oz, the plastic zone spreads inward from this side before the other side begins to yield. The subsequent bending of the beam involves two separate plastic zones, with the elastic/plastic boundaries situated at equal distances $c = (Y/E)/R$ from the neutral surface. The position of the neutral surface varies with the amount of bending, and is determined from the condition of zero resultant longitudinal force across any transverse section, namely

$$\int \sigma b(y)\, dy = 0$$

where b is the width of the cross section at any distance y from Oz. If Oz is an axis of symmetry of the cross section, the neutral axis coincides with the centroidal axis in both elastic and plastic ranges of the bending.

It is customary to assume that the state of stress is uniaxial even when the beam is partly plastic. This, however, is not strictly correct, as may be seen by considering the deformation of the beam. During a small incremental distortion, the anticlastic curvature changes by the amount $v\, d(1/R)$ in the elastic region, and by the amount $\eta\, d(1/R)$ in the plastic region, where η denotes the contraction ratio for the material beyond the yield point. The elements would not therefore fit together at the elastic/plastic interface except in the special case $\eta = v = \frac{1}{2}$. It follows that the preceding theory of elastic/plastic bending would be strictly valid only if the material is incompressible. For $v < \frac{1}{2}$, the necessary continuity restriction cannot be maintained without introducing transverse stresses, which affect the shape of the plastic boundaries.† The problem then becomes extremely complicated. According to the simplified treatment,‡ the displacement in both elastic and plastic regions is given by (46) with the appropriate value $v = \frac{1}{2}$, so long as the deformation is sufficiently small.

The stress σ in the elastic region varies linearly from zero on the neutral axis to a magnitude Y on the elastic/plastic boundary. In a plastic fibre, the stress has the local yield value in tension or compression, and is a given function of the strain $|y/R|$. The bending moment at any stage can be calculated from the expression

$$M = \int \sigma y b(y)\, dy$$

For an annealed material, the elastic/plastic interface disappears, but the integral can still be evaluated from a given stress-strain law holding over the entire cross section. For a nonhardening material, the ratio of the fully plastic moment to the initial yield moment of a given cross section is called the *shape factor*.

---

† R. Hill, *The Mathematical Theory of Plasticity*, p. 82, Clarendon Press, Oxford (1950).

‡ J. W. Roderick, *Phil. Mag.*, **39**: 529 (1948); J. W. Roderick and J. Heyman, *Proc. Inst. Mech. Eng.*, **165**: 189 (1951). For experimental support, see J. W. Roderick and I. H. Phillipps, *Research (Eng. Struct. Suppl)*, *Colston Papers*, **2**: 9 (1949). The effect of upper yield point on the bending of beams has been discussed by C. Leblois and C. Massonet, *Int. J. Mech. Sci.*, **14**: 95 (1972).

(ii) *Rectangular and circular cross section* As a first example, consider the bending of a beam whose cross section is a rectangle of depth $2h$ and width b, the bending couple being applied in the vertical plane (Fig. 3.13a). In view of the symmetry of the cross section, the neutral axis always passes through its centroid, the moment of inertia about this axis being $I_z = \frac{2}{3}bh^3$. Plastic yielding begins at $y = \pm h$ when the bending moment and the radius of curvature become

$$M_e = \frac{2}{3}bh^2 Y \qquad R_e = \frac{Eh}{Y}$$

The radius of curvature at any stage during the elastic/plastic bending is $R = Ec/Y$, where c is the semidepth of the elastic core. It is supposed that the material strain-hardens according to the law

$$\frac{\sigma}{Y} = \left(\frac{E\varepsilon}{Y}\right)^n \qquad \varepsilon \geqslant \frac{Y}{E}$$

where $0 \leqslant n < 1$. Evidently, σ and ε are equal to the magnitudes of the longitudinal stress and strain in the plastic regions. Since $\varepsilon = |y/R|$, the stress distribution on the tension side of the cross section may be written as

$$
\begin{aligned}
\sigma &= Y\left(\frac{y}{c}\right) & 0 \leqslant y \leqslant c \\[2mm]
\sigma &= Y\left(\frac{y}{c}\right)^n & c \leqslant y \leqslant h
\end{aligned}
\tag{47}
$$

In view of the symmetry of the cross section, the bending moment at any stage is given by

$$M = 2b \int_0^h \sigma y \, dy$$

Substituting from (47) and integrating, the relationship between the bending

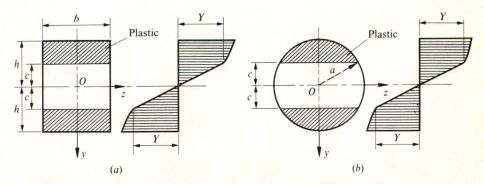

Figure 3.13 Geometry and stress distribution for the elastic/plastic bending of beams of rectangular and circular cross sections for work-hardening materials.

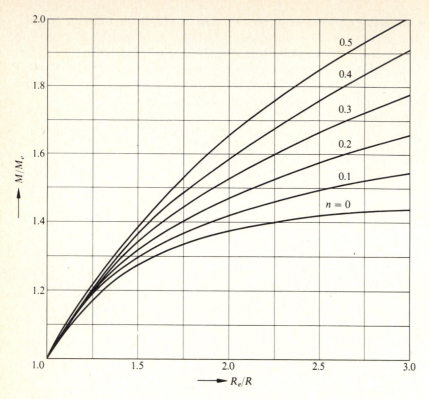

Figure 3.14 Moment-curvature relationship for work-hardening beams of rectangular cross section in pure bending.

moment and the curvature may be expressed as†

$$\frac{M}{M_e} = \frac{1}{2+n}\left\{3\left(\frac{R_e}{R}\right)^n - (1-n)\left(\frac{R}{R_e}\right)^2\right\} \tag{48}$$

For a nonhardening material ($n = 0$), the moment-curvature relationship reduces to that given by (9). The variation of M/M_e with R_e/R is shown graphically in Fig. 3.14 for several values of n. The bending moment increases steadily with the curvature, except when $n = 0$, for which there is a limiting moment equal to 1.5 M_e. The shape factor for a beam of rectangular cross section is therefore 1.5.

If the bending moment of an elastic-plastic beam is released by an amount M', a purely elastic stress equal to $-M'y/I_z$ is superposed on (47). The residual curvature of the beam is obtained by subtracting the spring-back curvature M'/EI_z from the elastic/plastic curvature $1/R$. If M' is continued to increase in the opposite

---

† Numerical solutions for work-hardening beams of rectangular, circular and hexagonal cross sections, based on the Ramberg-Osgood equation for the stress-strain curve, have been obtained by J. Betten, *Ing.-Archiv.*, **44**: 199 (1975).

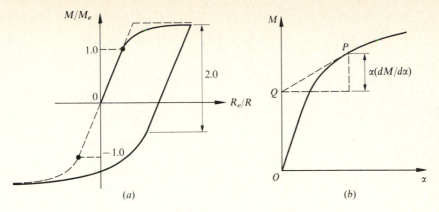

Figure 3.15 Moment-curvature diagrams for rectangular sections. (*a*) Effects of unloading and reversed loading without hardening; the broken lines at the top represent an idealized behavior; (*b*) Nadai's construction for an annealed beam.

sense, yielding will occur in compression at $y = h$ and tension at $y = -h$ when

$$\frac{M'}{M_e} = 1 + \left(\frac{h}{c}\right)^n = 1 + \left(\frac{R_e}{R}\right)^n \tag{49}$$

provided there is no Bauschinger effect. Thus, for a nonhardening material ($n = 0$), the elastic range of bending moment† is always equal to $2M_e$, as shown in Fig. 3.15*a*. As the bending is continued in the negative sense, the resultant bending moment of the nonhardening beam approaches the value $-1.5\,M_e$ in an asymptotic manner. When the magnitude of the negative curvature becomes equal to or greater than that at the instant of unloading, the bending moment is identical to that required to bend the beam monotonically to this curvature from the unstrained state.

The stress-strain curve of a material in uniaxial tension or compression may be derived from an experimentally determined relationship between the bending moment M and the angle of bend α measured over a length l. We begin by writing the moment in the form

$$M = 2bR^2 \int_0^{\varepsilon_0} \sigma\varepsilon\, d\varepsilon \qquad \varepsilon_0 = \frac{h}{R}$$

Multiplying the expression for M by α^2, and using the fact that $R^2\alpha^2 = l^2$, we obtain the derivative

$$\frac{d}{d\alpha}(M\alpha^2) = 2bh^2\alpha\sigma_0$$

where σ_0 is the tensile stress corresponding to the strain ε_0 occurring at the

---

† This result is independent of the shape of the cross section, provided the material is ideally plastic.

boundary $y = h$. The relationship between σ_0 and ε_0 may therefore be written as

$$\sigma_0 = \frac{1}{2bh^2}\left(\alpha\,\frac{dM}{d\alpha} + 2M\right) \qquad \varepsilon_0 = \frac{h\alpha}{l} \tag{50}$$

If a tangent is drawn at any point P on the (M, α) curve to meet the M axis at Q, then the projection of PQ parallel to the M axis represents the first term in the parenthesis of (50), as indicated in Fig. 3.15b. A graphical construction of a $(\sigma_0, \varepsilon_0)$ curve on the basis of a given (M, α) curve is therefore possible.

Consider now a beam of circular cross section subjected to pure bending in the vertical diametral plane. The moment of inertia of the cross section about the neutral axis, which coincides with the horizontal diameter, is equal to $\pi a^4/4$, where a is the radius of the circular boundary. Yielding first occurs at the extremities of the vertical diameter when M and R attain the values

$$M_e = \frac{\pi}{4}a^3 Y \qquad R_e = \frac{Ea}{Y}$$

A typical elastic/plastic stage is specified by the distance c of the plastic boundary from the neutral axis (Fig. 3.13b). The elastic/plastic bending moment is given by

$$M = 2\int_0^a \sigma yb(y)\,dy = 4\int_0^a \sigma y\sqrt{a^2 - y^2}\,dy$$

where $\sigma = (y/c)Y$ for $y \leqslant c$, and $\sigma_x = Y$ for $y \geqslant c$, work-hardening being neglected. Carrying out the integration, the result may be expressed as

$$\frac{M}{M_e} = \frac{2}{\pi}\left\{\frac{1}{3}\left(5 - \frac{2c^2}{a^2}\right)\sqrt{1 - \frac{c^2}{a^2}} + \frac{a}{c}\sin^{-1}\frac{c}{a}\right\} \tag{51}$$

The radius of curvature of the elastic/plastic beam is c/a times that at the initial yielding. As the cross section is rendered increasingly plastic, the ratio M/M_e rapidly approaches the asymptotic value $16/3\pi \simeq 1.698$, which is the shape factor for a circular cross section.

(iii) *Solution for a triangular cross section* Consider, now, the pure bending of a beam whose cross section is an isosceles triangle of height h and base width b (Fig. 3.16). While the beam is purely elastic, the stress varies linearly across the depth of the cross section, with the neutral axis situated at a distance $\frac{2}{3}h$ below the apex A of the triangle. The moment of inertia of the cross section about the neutral axis is then equal to $bh^3/36$. The yield point is first reached in compression at the outermost point A, the values of M and R at the initial yielding being

$$M_e = \frac{1}{24}bh^2 Y \qquad R_e = \frac{2Eh}{3Y}$$

For $M > M_e$, there is at first a single plastic zone whose inner boundary is at some distance c from the instantaneous neutral axis Oz. If the material is nonhardening, the stress in the plastic triangle of height d is a uniform compression of amount Y.

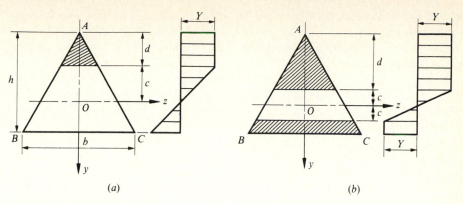

Figure 3.16 Elastoplastic bending of a beam of triangular cross section. (a) $1 \leqslant M/M_e \leqslant 1.804$; (b) $1.804 \leqslant M/M_e \leqslant 2.344$.

The stress in a typical elastic element is $(y/c)Y$, and the local width of the cross section is $b(d + c + y)/h$. The resultant force across the section would vanish if

$$\int_{-c}^{h-c-d} (c + d + y)y \, dy = \tfrac{1}{2}cd^2$$

which leads to the relationship

$$\frac{c}{h} = \frac{1}{3}\left(2 + \frac{d^3}{h^3}\right) - \frac{d}{h} \tag{52}$$

The neutral axis progressively moves downward till the beam begins to yield also on its tension side. The first elastic/plastic phase therefore continues so long as $2c \geqslant h - d$, which condition is equivalent to

$$1 - \frac{3d}{h} + \frac{2d^3}{h^3} \geqslant 0 \qquad \text{or} \qquad \frac{d}{h} \leqslant \frac{\sqrt{3}-1}{2} \simeq 0.366$$

The bending moment at any stage of the first elastic/plastic phase is[†]

$$M = \frac{bd^2}{2h}\left(\frac{d}{3} + c\right)Y + \frac{bY}{ch}\int_{-c}^{h-c-d} (c + d + y)y^2 \, dy$$

obtained by taking moment about the instantaneous neutral axis Oz. Evaluating the integral, and using (52), we obtain

$$\frac{M}{M_e} = 2\left(\frac{1 - 4d^3/h^3 + 3d^4/h^4}{2 - 3d/h + d^3/h^3}\right) \qquad \frac{d}{h} \leqslant 0.366 \tag{53}$$

At the end of the first elastic/plastic phase, $M/M_e \simeq 1.804$ and $c/h \simeq 0.317$. If the

[†] A. Nadai, *Theory of Flow and Fracture of Solids*, vol. I, p. 358, McGraw-Hill Book Company, New York (1950).

bending moment is increased further, a second plastic zone is formed near the base of the triangle. The neutral axis at each stage is equidistant from the two elastic/plastic boundaries. The condition of zero resultant force across the section then becomes

$$\int_{-c}^{c} (c + d + y)y \, dy = \tfrac{1}{2}c[d^2 - h^2 + (d + 2c)^2]$$

or

$$\frac{d}{h} = \sqrt{\frac{1}{2} - \frac{c^2}{h^2}} - \frac{c}{h} \tag{54}$$

The neutral axis slowly moves upward as the two plastic boundaries approach one another. The bending moment during the second elastic/plastic phase is given by

$$M = \frac{bd^2}{2h}\left(\frac{d}{3} + c\right)Y + \frac{bY}{ch}\int_{-c}^{c}(c + d + y)y^2 \, dy + \frac{bY}{h}\int_{c}^{h-d-2c}(c + d + y)y \, dy$$

in view of the stress distributions in the elastic and plastic regions. Using (54), the moment can be expressed in the form

$$\frac{M}{M_e} = 8\left\{1 - \left(1 + \frac{4c^2}{3h^2}\right)\sqrt{\frac{1}{2} - \frac{c^2}{3h^2}}\right\} \qquad \frac{c}{h} \leqslant 0.317 \tag{55}$$

The radius of curvature of the neutral surface is Ec/Y throughout the elastic/plastic bending. As the cross section becomes fully plastic, the two plastic boundaries coincide with the neutral axis, giving $d/h = 0.707$ and $M/M_e \simeq 2.344$ in the limiting state. The limiting value of the moment is very closely attained while the curvature of the beam is still of the elastic order of magnitude.[†]

(iv) *Unsymmetrical bending* Consider a prismatic beam of arbitrary cross section whose principal axes of inertia are Oy and Oz passing through the centroid O. Suppose that the bending couple M is applied in an axial plane passing through the line mm, inclined at a counterclockwise angle α to the y axis (Fig. 3.17a). The vector representing the moment M can be resolved into two components of magnitudes $M \cos \alpha$ and $M \sin \alpha$ along the principal axes. So long as the beam is elastic, the resultant longitudinal stress at any point P of the cross section may be written as

$$\sigma = \frac{My}{I_z}\cos \alpha + \frac{Mz}{I_y}\sin \alpha \tag{56a}$$

where I_y and I_z are the principal moments of inertia of the cross section. Since the stress vanishes along the neutral axis, the equation of this axis is

$$\frac{y}{I_z}\cos \alpha + \frac{z}{I_y}\sin \alpha = 0$$

[†] The elastic/plastic bending of a circular plate under an all-round couple has been treated by F. A. Gaydon and H. Nuttall, *J. Mech. Phys. Solids*, **5**: 62 (1956).

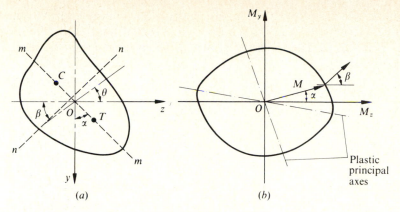

Figure 3.17 Unsymmetrical bending of a beam. (*a*) An arbitrary cross section; (*b*) a typical yield locus.

If θ denotes the angle which the elastic neutral axis makes with the z axis in the counterclockwise sense, then

$$\tan \theta = -\frac{y}{z} = \frac{I_z}{I_y} \tan \alpha \qquad (56b)$$

When the principal axes are distinct, θ is different from α, except when α is either zero or $\pi/2$. It follows that the plane of bending does not coincide with the plane of the applied couple unless the latter coincides with one of the principal planes of the beam. If the cross section has an axis of symmetry, which coincides with one of the principal axes of intertia, ($56b$) immediately furnishes the orientation of the neutral axis corresponding to any given axis of the applied moment.

As the beam is rendered increasingly plastic by the bending couple applied in a given plane, which is not a plane of symmetry of the beam, the neutral axis generally rotates during the bending. The complete elastic/plastic analysis for unsymmetrical bending leads to complicated expressions even for relatively simple cross sections. The determination of the fully plastic moment is, however, fairly straightforward for any cross section when the direction of the neutral axis is given. It will be assumed, for simplicity, that the material of the beam is ideally plastic. Referring to Fig. 3.17*a*, suppose the cross section is brought to the fully plastic state by a bending couple such that the neutral axis makes an angle β with Oz. Since both sides of the neutral axis are uniformly stressed to the yield point, the line *nn* must divide the cross section into equal areas so as to give zero resultant force across the section. The centroid O of the entire cross section therefore bisects the line TC joining the centroids of the tension and compression sides. The magnitude of the fully plastic moment is $M_0 = YAr$, where A is the total area of the cross section and $2r$ the length of the line TC, the direction of which defines the plane of the applied moment.† The components of the bending moment and the direction

---

† E. H. Brown, *Int. J. Mech. Sci.*, **9**: 77 (1967); J. Heyman, *Proc. Inst. Civ. Eng.*, **41**: 751 (1968).

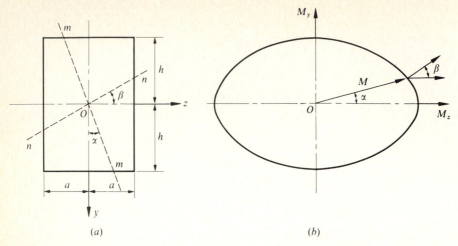

(a) (b)

Figure 3.18 Unsymmetrical bending of a beam of rectangular cross section. (a) Geometry; (b) yield locus.

of their resultant are given by[†]

$$M_z = YAy_T \qquad M_y = YAz_T \qquad \tan \alpha = \frac{z_T}{y_T}$$

where (y_T, z_T) are the rectangular coordinates of point T. A graphical plot of M_y against M_z defines a yield locus of the type shown in Fig. 3.17(b). The radius vector to a generic point P on the yield locus defines the plastic moment in direction and magnitude. The normal to the yield locus at P is in the direction of the corresponding neutral axis. Since M_y and M_z are merely reversed in sign when α is increased by π, the yield locus is generally skew-symmetric around the origin of the moment plane. The directions of the maximum and minimum diameters of the yield locus correspond to $\alpha = \beta$, and represent the plastic principal axes which are not generally orthogonal.

We shall now derive an explicit solution for the unsymmetrical bending of a beam of rectangular cross section[‡] of width $2a$ and height $2h$. In view of the symmetry of the cross section, the neutral axis passes through the centroid in both the elastic and plastic range of bending (Fig. 3.18a). The initial yielding of the cross sections occurs at the opposite corners farthest from the neutral axis when the longitudinal stress is of magnitude Y. Equations (56) therefore furnish

$$M_e = \frac{\frac{4}{3}ah^2 Y}{\cos \alpha + (h/a)\sin \alpha} \qquad \tan \theta = \frac{h^2}{a^2}\tan \alpha \qquad (57)$$

[†] The components M_y and M_z are taken as positive when they produce clockwise and counterclockwise moments respectively about the positive y and z axes.

[‡] This problem has been discussed by A. J. Barrett, *J. R. Aeronaut. Soc.*, **57**: 503 (1953), and H. B. Harrison, *Struct. Eng.*, **41**: 231 (1963). For an elastic/plastic analysis of the bending problem, see M. S. Aghbabian and E. P. Popov, *Proc. First U.S. Nat. Congr. Appl. Mech.* **1**: 579 (1951).

The solution for the fully plastic state is most conveniently obtained by finding the position of the centroid of the area on either side of the neutral axis, which is inclined at an angle β to the horizontal axis of symmetry. When the neutral axis meets the vertical sides of the cross section as shown, the coordinates of the centroid T of the tension side are found to be

$$y_T = \frac{h}{2}\left(1 - \frac{a^2}{3h^2}\tan^2\beta\right) \qquad z_T = \frac{a^2}{3h}\tan\beta \qquad \tan\beta \leqslant \frac{h}{a}$$

For higher values of β, the neutral axis meets the rectangle on its horizontal sides, and the rectangular coordinates of T become

$$y_T = \frac{h^2}{3a}\cot\beta \qquad z_T = \frac{a}{2}\left(1 - \frac{h^2}{3a^2}\cot^2\beta\right) \qquad \tan\beta \geqslant \frac{h}{a}$$

The two sets of equations for y_T and z_T completely define a centroidal locus which is symmetrical about the coordinate axes. The direction of the applied moment is given by

$$\tan\alpha = \frac{2\tan\beta}{3h^2/a^2 - \tan^2\beta} \qquad \tan\beta \leqslant \frac{h}{a}$$

$$\tan\alpha = \frac{\frac{1}{2}\tan\beta}{3a^2/h^2 - \cot^2\beta} \qquad \tan\beta \geqslant \frac{h}{a} \tag{58}$$

It may be noted that $\tan\alpha \gtrless a/h$ for $\tan\beta \gtrless h/a$. The neutral axis is therefore along a diagonal of the cross section when the plane of the applied moment passes through the opposite diagonal. Since $r = \sqrt{y_T^2 + z_T^2}$, the value of the fully plastic moment is given by

$$M_0 = \frac{2}{3}a^3 Y\left\{\left(\frac{3h^2}{a^2} - \tan^2\beta\right)^2 + 4\tan^2\beta\right\}^{1/2} \qquad \tan\beta \leqslant \frac{h}{a}$$

$$M_0 = \frac{2}{3}h^3 Y\left\{\left(\frac{3a^2}{h^2} - \cot^2\beta\right)^2 + 4\cot^2\beta\right\}^{1/2} \qquad \tan\beta \geqslant \frac{h}{a} \tag{59}$$

Equations (58) and (59) give the relationship between M_0 and α parametrically through the angle β. From (57), (58), and (59), the shape factor M_0/M_e can be expressed as

$$\frac{M_0}{M_e} = \frac{1}{2}\left(1 + \frac{a}{h}\tan\beta\right)\left(3 - \frac{a}{h}\tan\beta\right) \qquad \tan\beta \leqslant \frac{h}{a}$$

$$\frac{M_0}{M_e} = \frac{1}{2}\left(1 + \frac{h}{a}\cot\beta\right)\left(3 - \frac{h}{a}\cot\beta\right) \qquad \tan\beta \geqslant \frac{h}{a} \tag{60}$$

The shape factor has a maximum value equal to 2.0 corresponding to $\tan\beta = h/a$. Setting $\beta = 0$ or $\pi/2$ in (60) furnishes a shape factor of 1.5, in agreement with the result for symmetrical bending. From (59), it is possible to express β in terms of α as

$$\tan \beta = -\cot \alpha + \sqrt{\frac{3h^2}{a^2} + \cot^2 \alpha} \qquad \tan \alpha \leqslant \frac{a}{h}$$

$$\tan \beta = -\tan \alpha + \sqrt{\frac{3a^2}{h^2} + \tan^2 \alpha} \qquad \tan \alpha \geqslant \frac{a}{h}$$

The total rotation of the neutral axis during bending from the elastic to the fully plastic state is given by the difference between the angles θ and β for any given α. The rotation is counterclockwise for $0 < \alpha < \tan^{-1}(a/h)$, and clockwise for $\tan^{-1}(a/h) < \alpha < \pi/2$. The rotation vanishes when the axis of bending coincides with one of the axes of symmetry or through one of the diagonals of the rectangle.

The components M_z and M_y of the fully plastic moment are $4ahY$ times y_T and z_T respectively. Eliminating β, and setting $m_z = M_z/2ah^2Y$ and $m_y = M_y/2ha^2Y$, we obtain the dimensionless *interaction relations*

$$m_z + \tfrac{3}{4}m_y^2 = 1(m_y \leqslant \tfrac{2}{3}) \qquad m_y + \tfrac{3}{4}m_z^2 = 1(m_y \geqslant \tfrac{2}{3}) \tag{61}$$

which define the first quadrant of the yield locus shown in Fig. 3.18b. As expected, the yield locus is symmetrical about its axes of reference. The ratio of the partial derivatives of the yield function with respect to M_y and M_x is equal to $\tan \beta$, which is in agreement with the normality rule for the direction of the neutral axis. The principal axes of bending coincide with the axes of symmetry of the cross section in both the elastic and plastic ranges.†

3.5 Bending of Beams Under Transverse Loads

(i) *Basic principles* When a beam is bent by transverse loads, which are assumed here to act in a plane of symmetry, the bending moment varies along the length of the beam. The corresponding variation of curvature along the beam usually produces a complicated shape of the bent axis, which is known as the deflection curve. It is assumed, as usual, that deflections due to shearing and axial forces are negligible compared to those due to bending. The relationship between the bending moment and the curvature is therefore identical to that for pure bending in both the elastic and plastic ranges. If the material is nonhardening, and the cross section rectangular, the moment-curvature relationship for continued loading may be written as

$$\frac{R_e}{R} = \frac{M}{M_e} \qquad |M| \leqslant M_e$$

$$\frac{R_e}{R} = \pm \left(3 - 2 \left| \frac{M}{M_e} \right| \right)^{-1/2} \qquad |M| \geqslant M_e \tag{62}$$

---

† Equations (61) approximately hold, with appropriate normalizing moments, for most other doubly symmetric solid and closed hollow sections.

where M_e and R_e correspond to the elastic limit. The sign of the curvature $1/R$ is identical to that of the bending moment M, which is considered positive when it has a sagging effect on the beam. Since R/R_e is numerically equal to c/h for an elastic/plastic section, where $2h$ is the depth of the beam and $2c$ that of the elastic core, we have

$$\left|\frac{M}{M_e}\right| = \frac{1}{2}\left(3 - \frac{c^2}{h^2}\right) \qquad 0 \leqslant c \leqslant h \qquad\qquad (62a)$$

The slope of the *deflection curve* changes discontinuously at a fully plastic cross section where the curvature becomes infinitely large. In reality, the discontinuity is the limit of a narrow region through which the slope changes rapidly in a continuous manner.

The shearing force at any elastic/plastic cross section is carried entirely by the elastic core, as may be seen by considering the equation of longitudinal equilibrium and the yield criterion. If the longitudinal and shear stresses in a typical plastic element are denoted by σ and τ respectively, then

$$\frac{\partial\sigma}{\partial x} + \frac{\partial\tau}{\partial y} = 0 \qquad \sigma^2 + 3\tau^2 = Y^2 \qquad\qquad (63)$$

where the x axis is taken along the central axis of the beam, and the y and z axes parallel to the vertical and horizontal sides of the cross section. For a beam loaded by normal forces, $\tau = \partial\tau/\partial x = 0$ along $y = \pm h$, and it follows from (63) that

$$\frac{\partial\sigma}{\partial x} = \frac{\partial\tau}{\partial y} = \frac{\partial\sigma}{\partial y} = 0 \qquad \text{at} \qquad y = \pm h$$

By successive differentiation of (63), it is easily shown that all the higher-order derivatives of σ and τ also vanish at $y = \pm h$. The shear stress therefore vanishes everywhere in the plastic regions, where the longitudinal stress has the constant magnitude Y. Since the depth of the elastic core decreases as the bending proceeds, the maximum shear stress on a given elastic/plastic cross section must increase with progressive bending.

The downward displacement v of any particle on the longitudinal axis of the beam is assumed to be small compared with the dimensions of its cross section. Then the local curvature of the bent axis is numerically equal to $\partial^2 v/x^2$ to a close approximation. If ψ denotes the counterclockwise angle which the tangent to the deflection curve makes with the x axis, then

$$\psi \simeq -\frac{\partial v}{\partial x} \qquad \frac{1}{R} \simeq \frac{\partial\psi}{\partial x} = -\frac{\partial^2 v}{\partial x^2}$$

The second expression is consistent with the fact the curvature is positive when the bent beam is concave upward. If the problem is statically determined, M is a known function of x, and the shape of the deflection curve can be determined, in view of (62), by the direct integration of the differential equation. For a statically indeterminate problem, the bending moment distribution cannot be determined without recourse to the displacement.

Consider a pair of neighboring points on the deflection curve situated at horizontal distances x and $x + dx$ from a given point A as shown in Fig. 3.19a. The counterclockwise angle turned through by this infinitesimal segment is $d\psi \simeq dx/R$. The distance between the two points where the tangents at the ends of the segment meet the vertical drawn through a selected point P is $d\delta = (a - x) \, d\psi$. The total angle $\Delta\psi$ turned through by AP in the counterclockwise sense, and the vertically upward distance $\Delta\delta$ of P away from the tangent at A, are†

$$\Delta\psi = \int_0^a \frac{dx}{R} \qquad \Delta\delta = \int_0^a (a - x) \frac{dx}{R} \tag{64}$$

The deflection of P away from the tangent at some other point B, at a distance l from A, may be obtained from above by an appropriate change of the limits of integration. Assuming A and B to be situated at the same horizontal level, the deflection of P below AB may be written as

$$\delta = -a\psi_A - \int_0^a (a - x) \frac{dx}{R} = (l - a)\psi_B - \int_a^l (x - a) \frac{dx}{R} \tag{65}$$

where ψ_A and ψ_B are the slopes of the deflection curve at A and B respectively. In an elastic beam, the slope is everywhere continuous, but local discontinuities may exist in a beam that is elastic/plastic.

The deflection analysis for an elastic/plastic beam is greatly simplified if the spread of the plastic zone is disregarded. This approximation is equivalent to the assumption that the beam is made of an idealized I section of negligible web thickness and having the same fully plastic moment M_0 as the actual beam. The idealized beam obviously behaves elastically except at those sections where the bending moment attains the fully plastic value. The curvature can grow indefinitely at these sections, permitting the formation of plastic hinges, while the bending moment remains constant at the magnitude M_0. The moment-curvature relationship for the idealized beam is represented by the dashed lines of Fig. 3.15a.

Figure 3.19b shows a uniformly loaded member of length $2l$, carrying a total load W. It is supposed that the bending moments at the ends A and B, as well as at the center C, are known. Let the maximum bending moment $\bar{M}$ occur at a section D, whose distance from C is denoted by d. Since the shearing force must vanish at D, the conditions of overall moment equilibrium of the segments of AD and DB furnish

$$\bar{M} = M_A + \frac{Wl}{4} \left(1 - \frac{d}{l}\right)^2 = M_B + \frac{Wl}{4} \left(1 + \frac{d}{l}\right)^2$$

The elimination of $\bar{M}$ between these two relations immediately gives d. The

---

† This is the basis of the well-known moment-area method used in structural mechanics for the analysis of elastic beams.

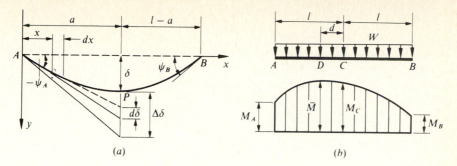

Figure 3.19 Bending of a transversely loaded beam: (a) Geometry of deflected axis; (b) a uniformly loaded segment.

maximum moment $\bar{M}$ is most conveniently obtained from the condition of overall vertical equilibrium of DC. The results are

$$d = \frac{M_A - M_B}{W} \qquad \bar{M} = M_C + \frac{Wd^2}{4l} \tag{66}$$

The maximum bending moment occurs to the left of the central section when $M_A > M_B$, and to the right of the central section when $M_A < M_B$.

Consider now the deflection of the uniformly loaded member at the point of maximum bending moment. It is easy to show that the bending moment M at any distance x from A, in terms of $a = l - d$, is given by

$$M - M_A = (\bar{M} - M_A)\frac{x}{a}\left(2 - \frac{x}{a}\right)$$

For an ideal beam section with a unit shape factor, the curvature is equal to M/EI throughout the bending, where EI is the flexural rigidity. Then the height of D above the tangent at A is

$$\Delta\delta = \frac{I}{EI}\int_0^a (a - x)M \, dx = (\bar{M} + M_A)\frac{a^2}{4EI} \tag{66a}$$

It follows that $\Delta\delta$ vanishes when $M_A = -\bar{M}$, implying that the tangent to the deflection curve at A in this case passes through D. This conclusion remains valid if the loading is uniform only over a part of the member, such as AC, provided the maximum bending moment occurs in the same part.

(ii) Cantilever with a terminal load A uniform cantilever of length l, having a rectangular cross section of width b and height $2h$, is loaded by a concentrated load W at the free end (Fig. 3.20a). Yielding first occurs at the top and bottom corners of the built-in end, where the bending moment has its greatest value, the corresponding load W_e being such that $M_e = W_e l$. When the load W exceeds W_e, the plastic zones spread symmetrically inward from the corners, and the yield moment M_e is attained at some distance a from the free end. If M is the bending moment at any

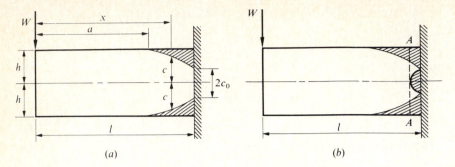

Figure 3.20 Cantilever loaded by a concentrated end load. (*a*) Any elastic/plastic state; (*b*) collapse state.

section, then

$$\frac{W_e}{W} = \frac{a}{l} \qquad \frac{M}{M_e} = -\frac{x}{a}$$

where x is measured from the free end of the cantilever. If c denotes the semiheight of the elastic core at any partially plastic section, then $c/h = R/R_e$. The bending moment relationship (62a) for the elastic/plastic region therefore furnishes

$$\left(\frac{c}{h}\right)^2 = 3 - \frac{2x}{a} = 3 - \frac{2x}{l}\left(\frac{W}{W_e}\right) \tag{67}$$

This shows that the plastic boundary is part of a parabola having its vertex on the central axis at a distance $\frac{3}{2}a$ from the free end. The beam eventually collapses when the bending moment at the built-in cross section attains the fully plastic value $\frac{3}{2}M_e$. The collapse load is therefore equal to $\frac{3}{2}W_e$, and the corresponding value of a is equal to $\frac{2}{3}l$.

Since the vertical deflection v decreases as x increases, $\partial v/\partial x$ is negative, and its value algebraically increases with x. Consequently, $\partial^2 v/\partial x^2$ is everywhere positive, and the differential equations for the deflection curve become

$$R_e \frac{\partial^2 v}{\partial x^2} = \frac{x}{a} \qquad (x \leqslant a)$$

$$R_e \frac{\partial^2 v}{\partial x^2} = \left(3 - \frac{2x}{a}\right)^{-1/2} \qquad (x \geqslant a)$$

in view of (62). Using the facts that $\partial v/\partial x$ vanishes at $x = l$, and is continuous across $x = a$, the above equations can be integrated to obtain the slope

$$\frac{\partial v}{\partial x} = -\frac{a}{R_e}\left[\sqrt{3 - \frac{2x}{a}} - \sqrt{3 - \frac{2l}{a}}\right] \qquad a \leqslant x \leqslant l$$

$$\frac{\partial v}{\partial x} = -\frac{a}{R_e}\left[\frac{1}{2}\left(3 - \frac{x^2}{a^2}\right) - \sqrt{3 - \frac{2l}{a}}\right] \qquad 0 \leqslant x \leqslant a$$

Since the displacement v vanishes at the built-in end $x = l$, the integration of the above equations results in

$$v = \frac{a^2}{3R_e}\left[\left(3 - \frac{2x}{a}\right)^{3/2} - \left(3 + \frac{l-3x}{a}\right)\sqrt{3 - \frac{2l}{a}}\right] \qquad a \leqslant x \leqslant l$$

$$v = \frac{a^2}{3R_e}\left[5 - \frac{x}{2a}\left(9 - \frac{x^2}{a^2}\right) - \left(3 + \frac{l-3x}{a}\right)\sqrt{3 - \frac{2l}{a}}\right] \qquad 0 \leqslant x \leqslant a$$

(68)

Setting $x = 0$ in the second relations for $\partial v/\partial x$ and v, the angle of rotation ψ and the deflection δ of the free end of the cantilever may be written as†

$$\frac{\psi}{\psi_e} = \frac{W_e}{W}\left[3 - 2\sqrt{3 - \frac{2W}{W_e}}\right]$$

$$\frac{\delta}{\delta_e} = \left(\frac{W_e}{W}\right)^2\left[5 - \left(3 + \frac{W}{W_e}\right)\sqrt{3 - \frac{2W}{W_e}}\right]$$

(69)

where ψ_e and δ_e are the slope and the deflection at the initial yielding, their values being

$$\psi_e = \frac{l}{2R_e} = \frac{Yl}{2Eh} \qquad \delta_e = \frac{l^2}{3R_e} = \frac{Yl^2}{3Eh}$$

The results (69) may be directly obtained by using (64), with the integrals taken over the entire length of the beam. When the load W just reaches the collapse value $1.5W_e$, the quantities ψ and δ are seen to have the values $2\psi_e$ and $2.22\delta_e$ respectively.‡ The distortion of the beam at the moment of collapse is therefore only of an elastic order of magnitude. Equations (69) also hold for the terminal slope and the central deflection of a simply supported beam of length $2l$, carrying a concentrated load $2W$ at the midspan.

The longitudinal stress is distributed linearly in the elastic region of the cantilever. This is accompanied by a parabolic distribution of shear stress satisfying the equation of longitudinal equilibrium. Since the shear stress vanishes in the plastic region, the longitudinal and shear stress distributions in the elastic core of a partially plastic cross section may be written as

$$\sigma = -\left(\frac{y}{c}\right)Y \qquad \tau = -\frac{3W}{4bc}\left(1 - \frac{y^2}{c^2}\right) \qquad -c \leqslant y \leqslant c$$

(70)

where y is the vertical distance below the central axis. It may be noted that

$$\frac{dc}{dx} = -\frac{h^2}{ac} = -\frac{h^2 W}{cM_e} = -\frac{3W}{2bcY}$$

† The load-deflection relationship expressed by the second equation of (69) is due to J. Fritsche, *Bauingenieur*, **11**: 851 (1930).

‡ For a statically determinate beam, the ratio of the collapse load to the elastic limit load is always equal to the shape factor.

in view of (67). The maximum shear stress at any cross section occurs on $y = 0$, and its magnitude is such that the resultant shearing force over the section is equal to W.

The transverse normal stress σ_y vanishes in the plastic region where $\tau = 0$. Equilibrium requires, however, that σ_y must be nonzero in the elastic region. Indeed, it follows from the condition of vertical equilibrium that

$$\frac{\partial \sigma_y}{\partial y} = -\frac{\partial \tau}{\partial x} = -\frac{3W}{4bc^2}\left(1 - \frac{3y^2}{c^2}\right)\frac{dc}{dx}$$

Substituting for dc/dx and integrating, we get

$$Y\sigma_y = \left(\frac{3W}{2bc}\right)^2 \frac{y}{2c}\left(1 - \frac{y^2}{c^2}\right) \qquad -c \leqslant y \leqslant c$$

An inner plastic zone begins to form at the center of the built-in cross section when the shear stress at this point attains the value $-Y/\sqrt{3}$ according to the von Mises yield criterion. If c_0 is the value of c at $x = l$, it follows from (70) and (67) that yielding occurs at $x = l$ and $y = 0$ when

$$\frac{2}{\sqrt{3}} = \frac{3W}{2bc_0 Y} = \frac{hW}{lW_e}\left(3 - \frac{2W}{W_e}\right)^{-1/2} \tag{71}$$

The stress distribution over the elastic part of the built-in cross section at this stage is given by

$$\sigma = -\frac{Yy}{c_0} \qquad \tau = -\frac{Y}{\sqrt{3}}\left(1 - \frac{y^2}{c_0^2}\right) \qquad \sigma_y = \frac{2Yy}{3c_0}\left(1 - \frac{y^2}{c_0^2}\right)$$

Substitution into the von Mises yield function $\sigma^2 - \sigma\sigma_y + \sigma_y^2 + 3\tau^2$ shows that this has a maximum value of $1.006 Y^2$ for $y = \pm 0.68 c_0$. Since the extent of violation of the yield criterion is only marginal, the stress distribution may be regarded as statistically admissible.

The shear force that corresponds to $\tau = -Y/\sqrt{3}$ attained at each point of the cross section is $F_0 = 2bhY/\sqrt{3}$. Since the actual shear force F is equal to the load W, we have $F/F_0 = 2c_0/3h$ in view of (71). The magnitude of the bending moment at the built-in end, when its center is at the point of yielding, is very closely equal to the fully plastic moment M_0' in the presence of shear. It follows from (67) that[†]

$$\frac{M_0'}{M_0} = \frac{2W}{3W_e} = 1 - \frac{c_0^2}{3h^2} = 1 - \frac{3}{4}\left(\frac{F}{F_0}\right)^2 \qquad \frac{F}{F_0} \leqslant \frac{2}{3} \tag{72}$$

The restriction $F/F_0 \leqslant 2/3$ ensures that the shear stress in the region $0 \leqslant x \leqslant a$ nowhere exceeds the yield stress in shear. If the growth of the inner plastic zone

---

† B. G. Neal, *The Plastic Methods of Structural Analysis*, 3d ed., p. 138, Chapman and Hall, London (1977).

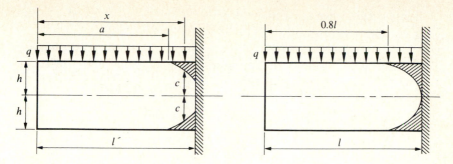

Figure 3.21 Cantilever having a uniformly distributed load. (a) $q_e < q < 1.5q_e$; (b) $q = 1.5 \, q_e$ (shear neglected).

(Fig. 3.20b) is taken into account,† the coefficient of $(F/F_0)^2$ in (72) is found to be modified to 0.44 approximately, the restriction on the equation being modified to $F/F_0 \leqslant 0.79$.

(iii) *Cantilever with a distributed load* Suppose that the load carried by the cantilever is uniformly distributed over its length, the intensity of the normal pressure at any stage being denoted by q (Fig. 3.21). From statical considerations, the bending moment at any distance x from the free end is $M = -\frac{1}{2}qbx^2$, where b is the width of the cross section. Yielding begins at the built-in end when the pressure becomes q_e, such that

$$q_e = \frac{2M_e}{bl^2} = \frac{4}{3} Y \left(\frac{h}{l}\right)^2$$

The effect of normal pressure on the nature of the stress distribution across the beam is neglected. For $q > q_e$, the plastic zones would spread symmetrically from top and bottom to reach some distance a from the free end of the cantilever. Since the bending moment at $x = a$ must be of magnitude M_e, we have

$$\frac{M}{M_e} = -\frac{x^2}{a^2} \qquad \frac{q_e}{q} = \frac{a^2}{l^2}$$

In view of the assumed moment-curvature relationship, the semi-depth c of the elastic core of any partially plastic cross section varies according to the equation

$$\left(\frac{c}{h}\right)^2 + 2\left(\frac{x}{a}\right)^2 = 3 \tag{73}$$

The elastic/plastic boundary is therefore part of an ellipse with centre at O, and having its semiminor and semimajor axes equal to $\sqrt{3}\,h$ and $\sqrt{\frac{3}{2}}\,a$ respectively.

† See M. R. Horne, *Proc. R. Soc.*, **A207**: 216 (1951). For a rigorous rigid/plastic solution, taking account of the conditions at the support, see A. P. Green, *J. Mech. Phys. Solids*, **3**: 143 (1954). Lower and upper bound solutions have been given by D. C. Drucker, *J. Appl. Mech.*, **23**: 509 (1956).

Plastic collapse occurs when the semimajor axis is l, the values of q and a at this stage being $\frac{3}{2}q_e$ and $\sqrt{\frac{2}{3}}\,l$ respectively.

The slope and the deflection of the cantilever during the elastic/plastic loading can be determined as before. Since the curvature $1/R$ at any point of the deflection curve is equal to $-\partial^2 v/\partial x^2$, Eqs. (62) give

$$R_e \frac{\partial^2 v}{\partial x^2} = \frac{x^2}{a^2} \qquad (x \leqslant a)$$

$$R_e \frac{\partial^2 v}{\partial x^2} = \left(3 - \frac{2x^2}{a^2}\right)^{-1/2} \qquad (x \geqslant a)$$

These equations can be integrated with the boundary conditions $v = \partial v/\partial x = 0$ at $x = l$. Using the conditions of continuity of v and $\partial v/\partial x$ across $x = a$, it is easy to show that

$$v = \frac{a^2}{2R_e} \left\{ \sqrt{3 - \frac{2x^2}{a^2}} - \sqrt{3 - \frac{2l^2}{a^2}} - \sqrt{2}\frac{x}{a}\left(\sin^{-1}\frac{l}{a}\sqrt{\frac{2}{3}} - \sin^{-1}\frac{x}{a}\sqrt{\frac{2}{3}}\right) \right\}$$

$$a \leqslant x \leqslant l$$

$$v = \frac{a^2}{2R_e} \left\{ \frac{3}{2} - \frac{2x}{3a}\left(1 - \frac{x^3}{4a^3}\right) - \sqrt{3 - \frac{2l^2}{a^2}} - \sqrt{2}\frac{x}{a}\left(\sin^{-1}\frac{l}{a}\sqrt{\frac{2}{3}} - \sin^{-1}\sqrt{\frac{2}{3}}\right) \right\}$$

$$0 \leqslant x \leqslant a$$

The slope of the deflection curve is given by the derivative of v with respect to x. At the free end $x = 0$, the amount of slope ψ and the deflection δ may be expressed nondimensionally as

$$\frac{\psi}{\psi_e} = \sqrt{\frac{q_e}{q}}\left[1 + \frac{3}{\sqrt{2}}\left(\sin^{-1}\sqrt{\frac{2q}{3q_e}} - \sin^{-1}\sqrt{\frac{2}{3}}\right)\right]$$

$$\frac{\delta}{\delta_e} = \frac{q_e}{q}\left[3 - 2\sqrt{3 - \frac{2q}{q_e}}\right]$$

(74)

where $\psi_e = l/3R_e$ and $\delta_e = l^2/4R_e$ are the values ψ and δ at the initial yielding. At the moment of incipient collapse, $\psi \simeq 1.88\psi_e$ and $\delta = 2\delta_e$, indicating that the deformation of the beam is still small.

The influence of shear is slightly more significant for the distributed load than for the concentrated end load. Since the total shearing force across any section is equal to qbx, the shear stress distribution on an elastic/plastic section is given by

$$\tau = -\frac{3qx}{4c}\left(1 - \frac{y^2}{c^2}\right) \qquad a \leqslant x \leqslant l \qquad -c \leqslant y \leqslant c$$

The greatest value of the shear stress occurs at the center of the built-in section

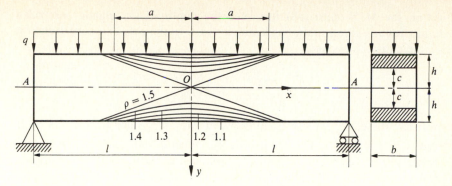

Figure 3.22 Simply supported beam carrying a uniformly distributed load q per unit area. The growth of the plastic zone is shown by its boundaries for constant values of $\rho = q/q_e$.

$x = l$. Using (73), the greatest shear stress of magnitude τ_0 may be written as

$$\frac{\tau_0}{Y} = \frac{3ql}{4hY}\left(3 - \frac{2l^2}{a^2}\right)^{-1/2} = \frac{hq}{lq_e}\left(3 - \frac{2q}{q_e}\right)^{-1/2}$$

A central plastic zone will begin at $x = l$ when τ_0 attains the value $Y/\sqrt{3}$, the corresponding load ratio being readily found from the formula

$$\frac{q_e}{q} = \frac{1}{3}\left\{1 + \sqrt{1 + \frac{9h^2}{l^2}}\right\} \tag{75}$$

If the length of the beam is greater than about three times its depth, the value of q/q_e given by (75) differs very little from that at the instant of collapse. For shorter beams, the additional deflection due to shear, which also affects the magnitude of the collapse pressure, may be appreciable.†

(iv) *Simply supported beam under uniform loading* A simply supported beam of rectangular cross section carries a uniformly distributed load of intensity q per unit area on the top surface (Fig. 3.22). The bending moment is a maximum at the central cross section $x = 0$, and the magnitude of this moment is $\frac{1}{2}qbl^2$, where $2l$ denotes the total length of the beam. Yielding first occurs at $y = \pm h$ on the central section when the maximum bending moment reaches the value M_e, the corresponding pressure being $q_e = 2M_e/bl^2$. For some value of q greater than q_e, there will be two symmetrical plastic zones covering a length $2a$ along $y = \pm h$ of the beam. Since the statical moment at any section is $M = \frac{1}{2}qb(l^2 - x^2)$, and the sections $x = \pm a$ are just at the point of yielding, we have

$$\frac{M}{M_e} = \frac{q}{q_e}\left(1 - \frac{x^2}{l^2}\right) \qquad \frac{a}{l} = \sqrt{1 - \frac{q_e}{q}} \tag{76}$$

† The effect of shear on the fully plastic moment of an I section beam has been discussed by A. P. Green, *J. Mech. Phys. Solids*, **3**: 143 (1954). See also J. Heyman and V. L. Dutton, *Weld. Metal Fabr.*, **22**: 265 (1954). A lower-bound solution has been given B. G. Neal, *J. Mech. Eng. Sci.*, **3**: 258 (1961).

Substitution for M/M_e into (48) furnishes the depth of the elastic core as a function of x for any given strain-hardening index n. In the case of a nonhardening material, the equation becomes

$$\left(\frac{c}{h}\right)^2 - \frac{2q}{q_e}\left(\frac{x}{l}\right)^2 = 3 - \frac{2q}{q_e} \tag{77}$$

It follows that the elastic/plastic boundaries at each stage are hyperbolas with asymptotes $y = \pm\sqrt{3}(h/l)x$. The beam collapses when $q = \frac{3}{2}q_e$, for which the two plastic zones join at the center O. The positions of the elastic/plastic boundaries for several values of q/q_e are shown in Fig. 3.22. The plastic boundaries at the incipient collapse coincide with the asymptotes.

Since the deflection v is an even function of x, it is necessary to consider only one-half of the beam. Setting $q/q_e = \rho$, the differential equation for the elastic/plastic portion of the beam may be written as

$$\frac{\partial^2 v}{\partial x^2} = -\frac{1}{R_e}\left\{3 - 2\rho\left(1 - \frac{x^2}{l^2}\right)\right\}^{-1/2} \qquad 0 \leqslant x \leqslant a$$

in view of (62) and (76). Integrating, and using the fact that $\partial v/\partial x$ vanishing at $x = 0$, we find

$$\frac{\partial v}{\partial x} = -\frac{l}{R_e\sqrt{2\rho}} \sinh^{-1}\frac{x}{l}\sqrt{\frac{2\rho}{3 - 2\rho}} \qquad 0 \leqslant x \leqslant a$$

and

$$v = \delta - \frac{l^2}{R_e\sqrt{2\rho}}\left\{\frac{x}{l}\sinh^{-1}\frac{x}{l}\sqrt{\frac{2\rho}{3 - 2\rho}} - \sqrt{\frac{3 - 2\rho}{2\rho} + \frac{x^2}{l^2}} + \sqrt{\frac{3 - 2\rho}{2\rho}}\right\}$$

$$0 \leqslant x \leqslant a \tag{78}$$

where δ denotes the deflection at $x = 0$. The slope and the deflection for the elastic portion of the beam are governed by the differential equation

$$\frac{\partial^2 v}{\partial x^2} = -\frac{\rho}{R_e}\left(1 - \frac{x^2}{l^2}\right) \qquad a \leqslant x \leqslant l$$

Taking account of the continuity of $\partial v/\partial x$ at $x = a$, the integral of this equation may be written as

$$\frac{\partial v}{\partial x} = -\frac{l}{R_e}\left\{\frac{\rho x}{l}\left(1 - \frac{x^2}{3l^2}\right) - \frac{2\rho + 1}{3}\sqrt{1 - \frac{1}{\rho}} + \frac{1}{\sqrt{2\rho}}\sinh^{-1}\sqrt{\frac{2\rho - 2}{3 - 2\rho}}\right\}$$

$$a \leqslant x \leqslant l$$

Integrating again, and using the boundary condition $v = 0$ at $x = l$, we obtain

$$v = \frac{l^2}{R_e}\left\{\left(1 - \frac{x}{l}\right)\left[\frac{1}{\sqrt{2\rho}}\sinh^{-1}\sqrt{\frac{2\rho - 2}{3 - 2\rho}} - \frac{2\rho + 1}{3}\sqrt{1 - \frac{1}{\rho}}\right]\right.$$

$$\left. + \frac{\rho}{2}\left[\frac{5}{6} - \frac{x^2}{l^2}\left(1 - \frac{x^2}{6l^2}\right)\right]\right\} \qquad a \leqslant x \leqslant l \quad (79)$$

The central deflection δ is now obtained from (78) and (79), using the fact that v is continuous at $x = a$. Thus†

$$\delta = \frac{l^2}{R_e}\left\{\frac{1}{\sqrt{2\rho}}\sinh^{-1}\sqrt{\frac{2\rho - 2}{3 - 2\rho}} + \frac{\sqrt{3 - 2\rho}}{2\rho} - \frac{2\rho + 1}{3}\sqrt{1 - \frac{1}{\rho}} + \frac{2\rho}{3} - \frac{3}{4\rho}\right\} \quad (80)$$

This result is also obtained by direct integration, using (64) with a replace by l. Setting $\rho = 1$, the terminal slope and the central deflection at the initial yielding are found as

$$\psi_e = \frac{2l}{3R_e} = \frac{2Yl}{3Eh} \qquad \delta_e = \frac{5l^2}{12R_e} = \frac{5Yl^2}{12Eh}$$

As ρ approaches the collapse value 1.5, the deflection according to (80) tends to infinity.‡ However, δ is found to have an elastic order of magnitude so long as ρ is only 3 to 4 percent smaller than the collapse value. The load-deflection curves for the three beams analyzed so far are plotted nondimensionally in Fig. 3.23. It may be noted that whereas the cantilever collapses for a finite value of the deflection, the collapse load for the simply supported beam is approached in an asymptotic manner. For all practical purposes, the beam may be regarded as having reached the collapse state when the deflection is only a few times that at the initial yielding.

The shear stress is distributed parabolically over the elastic part of each cross section of the beam, the resultant shearing force to the right of any section being of amount qbx in the upward sense. Hence

$$\tau = -\frac{3qx}{4c}\left(1 - \frac{y^2}{c^2}\right) \qquad 0 \leqslant x \leqslant a$$

$$\tau = -\frac{3qx}{4h}\left(1 - \frac{y^2}{h^2}\right) \qquad a \leqslant x \leqslant l$$

The numerically greatest shear stress occurs at $x = l$, but its value for usual l/h ratios is not sufficient to cause yielding. In the limiting state of $q = 1.5q_e$, the maximum shear stress becomes identical for all elastic/plastic cross sections.

It is not difficult to establish the load-deflection relationship when the material

---

† This solution has been given by W. Prager and P. G. Hodge, Jr., *Theory of Perfectly Plastic Solids*, p. 49, Wiley and Sons, New York (1951).

‡ The deflection tends to infinity as the load approaches the collapse value when the plastic moment is attained at a local maximum or over a finite length of the beam.

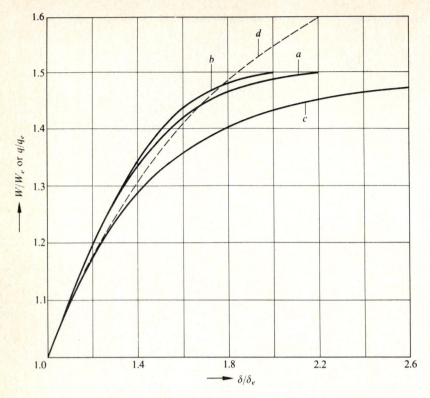

Figure 3.23 Load-deflection curves for statically determinate beams. (*a*) Cantilever with a terminal load; (*b*) uniformly loaded cantilever; (*c*) simply supported and uniformly loaded beam; (*d*) effect of strain-hardening on case (*c*) when $n = 0.2$.

strain-hardens. For a given intensity of loading, the depth of penetration of the plastic zone is decreased by strain-hardening, but the length of the plastic zone is unaffected. In view of (76), the moment-curvature relationship (48) for the elastic/plastic portion of the beam becomes

$$3\left(\frac{R_e}{R}\right)^n - (1 - n)\left(\frac{R}{R_e}\right)^2 = (2 + n)\rho\left(1 - \frac{x^2}{l^2}\right) \qquad 0 \leqslant x \leqslant a \qquad (81)$$

Thus, R_e/R can be computed for any given x/l and ρ. Since the slope of the deflection curve vanishes at $x = 0$, the central deflection of the beam is given by

$$\delta = \frac{l^2}{R_e}\int_0^l \frac{R_e}{R}\left(1 - \frac{x}{l}\right)\frac{dx}{l}$$

irrespective of the material property. The integral is readily evaluated over the elastic part $a \leqslant x \leqslant l$, for which R_e/R is equal to $\rho(1 - x^2/l^2)$. Inserting the value of

a/l from (76), we obtain

$$\delta = \frac{l^2}{R_e} \left\{ \frac{2\rho}{3} - \frac{1}{4\rho} - \frac{2\rho+1}{3} \sqrt{1 - \frac{1}{\rho}} + \int_0^{\sqrt{(\rho-1)/\rho}} \frac{R_e}{R}\left(1 - \frac{x}{l}\right) d\left(\frac{x}{l}\right) \right\} \quad (82)$$

In view of (81), the integral in (82) can be obtained numerically for any assumed ρ, which is not limited to the ideally plastic collapse value. The variation of δ/δ_e with q/q_e for a strain-hardening index $n = 0.2$ is shown by the broken curve in Fig. 3.23. Over the considered range, the load intensity for the work-hardening beam steadily increases with the amount of deflection.

(v) Beams fixed at both ends We begin with the simple example where a concentrated load W is applied at the midspan of a beam which is built-in at both ends. The terminal sections of the beam are prevented from rotation by the existence of reactant moments M_r due to the fixed-end condition. By symmetry, each half of the beam is under equal and opposite forces $W/2$ and identical couples M_r acting at its ends. The condition of statical equilibrium for each half requires $M_r = Wl/4$ throughout the bending, where l is the semilength of the beam. The bending moment varies linearly along the beam, vanishing at the midpoint of each half. The shape of the deflected axis of the beam and the conditions of symmetry then indicate that the central deflection of the beam is twice the deflection at the free end of a cantilever of length $l/2$ and carrying an end load $W/2$. The variation of the terminal slope ψ and the central deflection δ for the fixed-ended beam is therefore given by (69), where

$$W_e = \frac{4M_e}{l} \qquad \psi_e = \frac{l}{4R_e} \qquad \delta_e = \frac{l^2}{6R_e}$$

These are the values of W, ψ, and δ at the initial yielding of the built-in beam. The collapse load is $\frac{3}{2}W_e$, and the corresponding deflection is $\frac{20}{9}\delta_e$. The effect of fixing the ends of the beam is to double the load-carrying capacity and halve the limiting deflection.†

Suppose, now, that a uniformly distributed load of intensity w per unit length is applied to a fixed-ended beam of length $2l$ as shown in Fig. 3.24. The bending moment diagram and the shape of the deflection curve are symmetrical as before about the central section. If the magnitude of the end moments is denoted by M_r, the bending moment at any cross section may be written as

$$M = \tfrac{1}{2}w(l^2 - x^2) - M_r \quad (83)$$

where x is measured from the central cross section of the beam. The unknown moment M_r can be determined from the condition that the tangent to the deflection curve at the middle is parallel to that at either end. While the beam is entirely

† The bending of fixed-ended beams under symmetrical three-point loading has been investigated by W. Prager, *Bauingenieur*, **14**: 65 (1933).

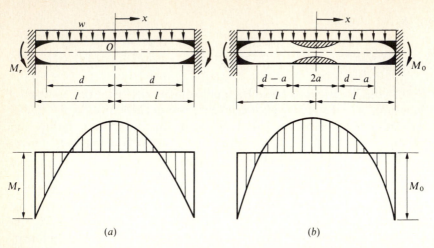

Figure 3.24 Geometry and bending moment distribution for a uniformly loaded elastic/plastic beam with fixed ends. (a) $w \leqslant 1.60w_e$; (b) $w \geqslant 1.67w_e$.

elastic, its curvature is proportional to the bending moment, and (64) and (83) give

$$\int_0^l M \, dx = 0 \quad \text{or} \quad M_r = \tfrac{1}{3}wl^2$$

The bending moment has a maximum value of $\tfrac{1}{2}M_r$ occurring at $x = 0$. The deflection curve has inflections at $x = \pm l/\sqrt{3}$, where the bending moment vanishes. The central deflection of the elastically loaded beam is obtained from (64) and (83) as

$$\delta = -\frac{l}{EI} \int_0^l Mx \, dx = -\frac{wl^2}{6EI} \int_0^l \left(1 - \frac{3x^2}{l^2}\right) x \, dx = \frac{wl^4}{24EI}$$

The negative sign arises before the integral because δ is measured at $x = 0$. Since the bending moment has its greatest magnitude at the built-in ends, yielding begins at these sections when the load intensity and the corresponding deflection become

$$w_e = \frac{3M_e}{l^2} \qquad \delta_e = \frac{l^2}{8R_e}$$

For $w > w_e$, plastic zones spread out from the ends of the beam to reach the sections $x = \pm d$ where $M = -M_e$. The bending moment M and the distance d may be expressed in the nondimensional forms

$$\frac{M}{M_e} = \frac{3w}{2w_e}\left(1 - \frac{x^2}{l^2}\right) - \frac{M_r}{M_e} \qquad \frac{d}{l} = \sqrt{1 - \frac{2w_e}{3w}\left(\frac{M_r}{M_e} - 1\right)} \tag{84}$$

The relationship between M_r/M_e and w/w_e during the elastic/plastic loading is given by

$$\int_0^d \left(\frac{M}{M_e}\right) dx - \int_d^l \left(3 - \frac{2M}{M_e}\right)^{-1/2} dx = 0$$

in view of the conditions of zero slope at $x = 0$ and $x = l$. Denoting M_r/M_e by m, the complicated relationship resulting from the integration can be written approximately in the more convenient form†

$$\frac{w}{w_e} \simeq \frac{2m+1}{3} + \frac{1}{\sqrt{2+m}} \left(\sin^{-1}\sqrt{\frac{3m}{3+m}} - \sin^{-1}\sqrt{\frac{2+m}{3+m}}\right) \qquad m \leqslant \frac{3}{2} \quad (85)$$

which is sufficiently accurate for practical purposes. The above expression is based on the first approximation $w/w_e \simeq m$, which is reasonable for $m \leqslant 1.5$. The built-in sections become fully plastic when $m = 1.5$, and this corresponds to $w/w_e \simeq 1.60$. In the absence of work-hardening, m has the constant value 1.5 for all higher values of w/w_e. The problem then becomes statically determined, giving

$$\frac{M}{M_e} = \frac{3}{2}\left\{\frac{w}{w_e}\left(1 - \frac{x^2}{l^2}\right) - 1\right\} \qquad \frac{d}{l} = \sqrt{1 - \frac{w_e}{3w}} \qquad \frac{w}{w_e} \geqslant 1.60 \quad (86)$$

The ends of the beam progressively rotate about the plastic hinges as the loading is continued in this range. A new plastic zone begins to form at the central cross section when the bending moment at $x = 0$ becomes equal to M_e. By (86), the value of w/w_e at this stage is $\frac{5}{3}$, and the corresponding value of d/l is $2/\sqrt{5}$. A further increase in load causes the central plastic zone to extend over a length $2a$, where

$$\frac{a}{l} = \sqrt{1 - \frac{5w_e}{3w}} \qquad \frac{w}{w_e} \geqslant \frac{5}{3}$$

The central cross section becomes fully plastic when $w/w_e = 2$, representing a state of plastic collapse caused by simultaneous hinge actions at $x = 0$ and $x = \pm l$. The ratios a/l and d/l attain their limiting values $\frac{1}{6}$ and $\frac{5}{6}$ respectively at the incipient collapse.

Consider now the deflection of the beam as w is increased from w_e. For $w \leqslant 1.60w_e$, the slope of the deflection curve vanishes at $x = \pm l$, and the central deflection is given by

$$\delta = -\frac{1}{R_e}\left\{\int_0^d \left(\frac{M}{M_e}\right) x\, dx - \int_d^l \left(3 + \frac{2M}{M_e}\right)^{-1/2} x\, dx\right\}$$

in view of the moment-curvature relations (62). Inserting from (84) and integrating, we obtain

$$\delta = \frac{l^2}{R_e}\left\{\frac{1}{2}\left(m - \frac{3}{4}\rho\right) + \frac{1}{3\rho}\left[\frac{1}{2}(3 - m^2) - \sqrt{3 - 2m}\right]\right\} \qquad m \leqslant \frac{3}{2} \quad (87)$$

† J. Chakrabarty, unpublished work (1980).

where ρ denotes the ratio w/w_e. When ρ is increased to 1.60, the deflection becomes $\delta \simeq 1.89\delta_e$. Since the slope at the ends of the beam is nonzero when $\rho > 1.60$, the central deflection must be found from the modified expression

$$\delta = \frac{l^2}{R_e} \left\{ \int_0^d \frac{M}{M_e} \left(1 - \frac{x}{l}\right) \frac{dx}{l} - \int_d^l \left(3 + \frac{2M}{M_e}\right)^{-1/2} \left(1 - \frac{x}{l}\right) \frac{dx}{l} \right\}$$

where M/M_e is given by (86). The above expression follows from (62) and (64). Performing the integration, and substituting for d/l, we have

$$\delta = \frac{l^2}{R_e} \left\{ \left(\rho - \frac{4}{3}\right) \sqrt{1 - \frac{1}{3\rho}} + \frac{3}{8} \left(\frac{5}{3} - \rho\right) \left(1 - \frac{1}{3\rho}\right) + \frac{1}{3\rho} \right.$$

$$\left. - \frac{1}{\sqrt{3\rho}} \left[\frac{\pi}{2} - \sin^{-1} \sqrt{1 - \frac{1}{3\rho}}\right] \right\} \qquad 1.6 \leqslant \rho \leqslant \frac{5}{3} \quad (88)$$

The beginning of yielding at $x = 0$ corresponds to $\delta \simeq 2.33\delta_e$, obtained by setting $\rho = \frac{5}{3}$ in (88). For higher values of ρ, the central deflecting is given by

$$\delta = \frac{l^2}{R_e} \left\{ \int_0^a \frac{R_e}{R} \left(1 - \frac{x}{l}\right) \frac{dx}{l} + \int_a^d \frac{M}{M_e} \left(1 - \frac{x}{l}\right) \frac{dx}{l} + \int_d^l \frac{R_e}{R} \left(1 - \frac{x}{l}\right) \frac{dx}{l} \right\}$$

where R_e/R is equal to $(3 - 2M/M_e)^{-1/2}$ for the first integral and $-(3 + 2M/M_e)^{-1/2}$ for the last integral, M/M_e being given by (86). Using the values of a/l and d/l on integration, we get

$$\delta = \frac{l^2}{R_e} \left\{ \left(\rho - \frac{4}{3}\right) \sqrt{1 - \frac{1}{3\rho}} - \left(\rho - \frac{2}{3}\right) \sqrt{1 - \frac{5}{3\rho}} + \frac{1}{\rho} \sqrt{\frac{2 - \rho}{3}} \right.$$

$$\left. + \frac{1}{\sqrt{3\rho}} \left[\sinh^{-1} \sqrt{\frac{3\rho - 5}{6 - 3\rho}} + \sin^{-1} \sqrt{1 - \frac{1}{3\rho}} - \frac{\pi}{2}\right] \right\} \qquad \rho \geqslant \frac{5}{3} \quad (89)$$

The deflection tends to infinity as ρ tends to the collapse value 2, but δ is found to be only four times that at the initial yielding when the load intensity is within 5 percent of the limiting value. The variation of δ/δ_e with w/w_e is represented by the lower solid curve in Fig. 3.25. If the spread of the plastic zones is disregarded, the load-deflection curve reduces to that shown by the broken lines, predicting a finite deflection at the instant of collapse.†

(vi) *Analysis of a propped cantilever* A cantilever of length l is simply supported at its free end and loaded by a uniformly distributed load w per unit length, as shown in Fig. 3.26. The vertical reaction at the prop and the reactant moment at the built-

---

† The effect of strain-hardening on the deflection of fixed-ended beams has been considered by M. R. Horne, *Weld. Res.*, **5**: 147 (1951). The increase in carrying capacity due to the prevention of axial movement of the ends of the beam has been examined by R. M. Haythornwaite, *Engineering*, **183**: 110 (1957).

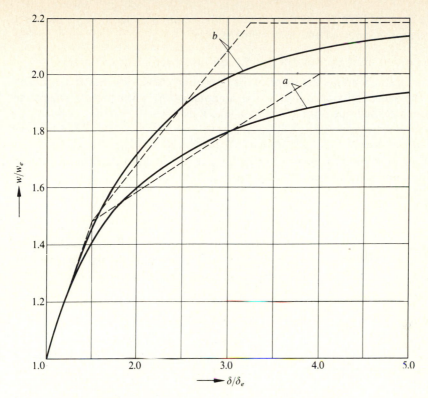

Figure 3.25 Load-deflection relationship for statically indeterminate beams under uniform loading. (*a*) Fixed-ended beam; (*b*) propped cantilever. The broken lines correspond to the ideal beam section.

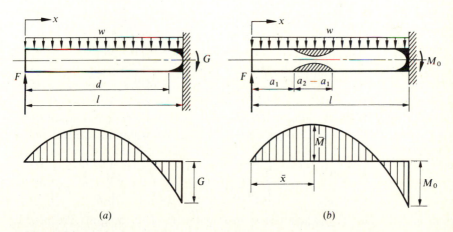

Figure 3.26 Elastic/plastic bending of a uniformly loaded propped cantilever. (*a*) $w \leqslant 1.64 w_e$; (*b*) $w \geqslant 1.67 w_e$.

in end are denoted by F and G respectively. The bending moment at a distance x from the simply supported end may be written as

$$M = Fx - \tfrac{1}{2}wx^2$$

During the elastic bending, the slope of the deflection curve vanishes at $x = l$. The fact that the deflection vanishes at $x = 0$ is expressed by $\int Mx\,dx = 0$, where the integration is carried out from $x = 0$ to $x = l$. This gives

$$F = \tfrac{3}{8}wl \qquad G = \tfrac{1}{8}wl^2$$

The bending moment in the beam vanishes at $x = \tfrac{3}{4}l$, and has a maximum value at $x = \tfrac{3}{8}l$. The central deflection of the elastic beam is

$$\delta = -\frac{1}{EI}\int_{l/2}^{l}\left(x - \frac{l}{2}\right)M\,dx = \frac{wl^4}{192EI}$$

This is marginally lower than the maximum deflection which occurs at $x \simeq 0.58l$. Yielding first occurs at the built-in end when G becomes equal to M_e, the corresponding values of w and δ being

$$w_e = \frac{8M_e}{l} \qquad \delta_e = \frac{l^2}{24R_e}$$

As the load intensity is steadily increased from w_e, the bending moment at the built-in end also increases, until G attains the fully plastic value $\tfrac{3}{2}M_e$. The bending moment distribution and the spread of the plastic zone at this stage are given by

$$\frac{M}{M_e} = \frac{x}{l}\left[4\rho\left(1 - \frac{x}{l}\right) - \frac{3}{2}\right]$$

$$\frac{2d}{l} = \left(1 - \frac{3}{8\rho}\right) + \left[\left(1 - \frac{3}{8\rho}\right)^2 + \frac{1}{\rho}\right]^{1/2} \qquad \rho \geqslant 1.64 \qquad (90)$$

where ρ denotes the ratio w/w_e. The last expression follows from the fact that $M = -M_e$ at $x = d$. The value of ρ at this stage is found from the conditions of zero slope and deflection at $x = l$ and $x = 0$ respectively. By (62) and (65), we have

$$\int_{0}^{d}\left(\frac{M}{M_e}\right)x\,dx - \int_{d}^{l}\left(3 + \frac{2M}{M_e}\right)^{-1/2}x\,dx = 0 \qquad (91)$$

Substituting from (90) and integrating, we obtain a transcendental equation of which the solution is $\rho \simeq 1.64$, giving $d/l \simeq 0.93$. The corresponding central deflection is given by

$$\delta = \frac{1}{R_e}\left\{\int_{0}^{l/2}\left(\frac{M}{M_e}\right)x\,dx + \frac{l}{2}\int_{l/2}^{d}\left(\frac{M}{M_e}\right)dx - \frac{l}{2}\int_{d}^{l}\left(3 + \frac{2M}{M_e}\right)^{-1/2}dx\right\}$$

in view of (65) and (62), when use is made of (91). The integrals are readily evaluated on inserting the expression for M/M_e, the result being $\delta \simeq 1.82\delta_e$ corresponding to $\rho = 1.64$.

For $\rho < 1.64$, the problem is not statically determined, but the preceding expression still holds for the estimation of the central deflection. The bending moment distribution and the extent of the plastic zone may be written as

$$\frac{M}{M_e} = \frac{x}{l}\left(f - \frac{4\rho x}{l}\right) \qquad \frac{d}{l} = \frac{f + \sqrt{f^2 + 16\rho}}{8\rho} \qquad 1 \leqslant \rho \leqslant 1.64 \qquad (92)$$

where $f = FL/M_e$. Using these relations in the relevant integrals, the deflection is obtained in the dimensionless form

$$\frac{\delta}{\delta_e} = 2[2f - \sqrt{f^2 + 16\rho}]\frac{d^2}{l^2} - \frac{1}{2}(f - \rho)$$

$$-3\sqrt{\frac{2}{\rho}}\left\{\sin^{-1}\frac{8\rho - f}{\sqrt{f^2 + 24\rho}} - \sin^{-1}\sqrt{\frac{f^2 + 16\rho}{f^2 + 24\rho}}\right\} \qquad 1 \leqslant \rho \leqslant 1.64 \quad (93)$$

The relationship between f and ρ can be found from (91) and (92), but the result is rather complicated. However, for $1 \leqslant \rho \leqslant 1.5$, a close approximation is achieved by assuming $f \simeq 3\rho$ in (93) for the computation of the deflection.

When $\rho > 1.64$, the built-in end undergoes rotation under a constant bending moment of magnitude $1.5M_e$. Equations (90) remain valid for this range of loading, the reaction at $x = 0$ being given by $f = 4\rho - 1.5$. The maximum bending moment $\bar{M}$ at any stage occurs at a distance $\bar{x}$ from the simply supported end, where

$$\frac{\bar{x}}{l} = \frac{1}{2}\left(1 - \frac{3}{8\rho}\right) \qquad \frac{\bar{M}}{M_e} = \rho\left(1 - \frac{3}{8\rho}\right)^2 \qquad \rho \geqslant 1.64 \qquad (94)$$

Both $\bar{x}$ and $\bar{M}$ steadily increase as ρ increases, and a second plastic zone begins to form at $x = \bar{x}$ when $\bar{M} = M_e$. The load intensity ratio at this stage is $\rho \simeq 1.67$, which corresponds to $\bar{x} \simeq 0.39l$. For a somewhat higher value of ρ, the inner plastic zone would extend from $x = a_1$ to $x = a_2$, the bending moments at these two sections being equal to M_e. It follows from (90) that

$$a_1, a_2 = \frac{l}{2}\left\{\left(1 - \frac{3}{8\rho}\right) \mp \left[\left(1 - \frac{3}{8\rho}\right)^2 - \frac{1}{\rho}\right]^{1/2}\right\} \qquad \rho \geqslant 1.67 \qquad (95)$$

The central section of the beam begins to yield when $a_2 = l/2$, giving $\rho = 1.75$. The section $x = l/2$ is therefore purely elastic for $\rho \leqslant 1.75$ and elastic/plastic for $\rho \geqslant 1.75$. When $\bar{M}$ attains the value $1.5M_e$, the beam collapses due to the formation of plastic hinges at $x = l$ and $x = \bar{x}$. The load intensity and the position of the sagging hinge at the incipient collapse are given by

$$\rho = \frac{3}{8}(3 + 2\sqrt{2}) \simeq 2.185 \qquad \frac{\bar{x}}{l} = \sqrt{2} - 1 \simeq 0.414$$

in view of (94). The ratio Fl/M_e rises to the value $3(\sqrt{2} + 1)$ at collapse. The lengths of the plastic zones near $x = \bar{x}$ and $x = l$ have the limiting values $0.478l$ and $0.051l$ respectively.

The deflection analysis for $\rho > 1.64$ is complicated by the fact that the plastic zone near the fixed end of the beam slowly unloads as ρ increases. However, for $\rho \geqslant 1.75$, it is a good approximation to neglect the small longitudinal spread of the plastic zone near $x = l$. Using (65), and the condition of zero displacement at $x = 0$, the central deflection of the beam for $\rho \geqslant 1.75$ may then be written as

$$\delta = \frac{l^2}{2R_e} \left\{ \int_0^{a_1} \frac{M}{M_e}\left(\frac{x}{l}\right)\frac{dx}{l} + \int_{a_1}^{l/2} \left(3 - \frac{2M}{M_e}\right)^{-1/2}\left(\frac{x}{l}\right)\frac{dx}{l} \right.$$
$$\left. + \int_{l/2}^{a_2} \left(3 - \frac{2M}{M_e}\right)^{-1/2}\left(1 - \frac{x}{l}\right)\frac{dx}{l} + \int_{a_2}^{l} \frac{M}{M_e}\left(1 - \frac{x}{l}\right)\frac{dx}{l} \right\} \quad (96)$$

The integrals are easily evaluated for any given value of ρ using (90) and (95). It is found that $\delta \simeq 2.13\delta_e$ when $\rho = 1.75$. The complete load-deflection relationship for the propped cantilever is represented by the upper solid curve in Fig. 3.25. The associated broken lines are based on the neglect of the spread of the plastic zones.

3.6 Torsion of Prismatic Bars

(i) *Elastic torsion* A prismatic bar of arbitrary cross section is twisted by equal and opposite terminal couples T about an axis parallel to the generators. These are deformed into helical curves as the cross sections rotate during the torsion. Except for a circular cylindrical bar, the cross sections become warped, and the amount of warping depends on the angle of twist θ per unit length of the bar. We choose rectangular coordinates (x, y, z) whose origin is at an end section of the bar, the z axis being taken parallel to the generators. The rectangular components of the displacement, representing the rotation and the warping of the cross sections, are given by

$$u = -\theta yz \qquad v = \theta xz \qquad w = w(x, y, \theta) \qquad (97)$$

The first two expressions are obtained from the coordinate transformation $x = r\cos\phi$ and $y = r\sin\phi$ for a typical particle (Fig. 3.27a), and the fact that $dx = u$, $dy = v$ and $dr = 0$ during an infinitesimal rotation $d\phi = \theta z$. In the elastic range of torsion,† the axial displacement w is proportional to θ, but the proportionality ceases to hold in the elastic/plastic range. The nonzero components of the strain are γ_{xz} and γ_{yz} given by

$$2\gamma_{xz} = \frac{\partial w}{\partial x} + \frac{\partial u}{\partial z} = \frac{\partial w}{\partial x} - \theta y$$

$$2\gamma_{yz} = \frac{\partial w}{\partial y} + \frac{\partial v}{\partial z} = \frac{\partial w}{\partial y} + \theta x$$

It follows from Hooke's law that the only nonzero stress components are

---

† The theory of elastic torsion of prismatic bars is due to B. Saint-Venant, *Mem. savants etrangers, Sci. Math. Phys.*, **14**: 233 (1855).

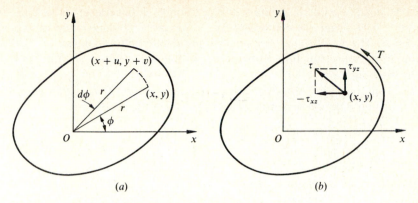

Figure 3.27 Stresses and displacements in the torsion of a prismatic bar of arbitrary cross section.

$\tau_{xz} = 2G\gamma_{xz}$ and $\tau_{yz} = 2G\gamma_{yz}$. The elimination of w from the above equations and the use of stress-strain relations furnish

$$\frac{\partial \tau_{yz}}{\partial x} - \frac{\partial \tau_{xz}}{\partial y} = 2G\theta \tag{98}$$

which is the equation of strain-compatibility expressed in terms of the stresses. The equation of equilibrium for the shear stresses is

$$\frac{\partial \tau_{xz}}{\partial x} + \frac{\partial \tau_{yz}}{\partial y} = 0$$

This equation is identically satisfied if we introduce a stress function $\phi(x, y)$ such that the stress components are

$$\tau_{xz} = \frac{\partial \phi}{\partial y} \qquad \tau_{yz} = -\frac{\partial \phi}{\partial x} \tag{99}$$

Substituting these derivatives into (98), we obtain the governing differential equation

$$\frac{\partial^2 \phi}{\partial x^2} + \frac{\partial^2 \phi}{\partial y^2} = -2G\theta \tag{100}$$

It follows from (99) that the resultant shear stress at any point on a transverse section is tangential to the curve of constant ϕ considered through the point.[†] The lines of shearing stress therefore coincide with the contour lines of ϕ. Since the shear stress at each point of the boundary of the section must be directed along the tangent to the boundary, the lateral surface of the bar being stress free, the

---

† The components of the unit exterior normal to the curve $\phi(x, y) = $ const with respect to the x and y axes are proportional to $\partial\phi/\partial x$ and $\partial\phi/\partial y$ respectively. The scalar product of the normal vector with the shear-stress vector therefore vanishes.

boundary curve must be a line of constant ϕ. For a simply connected cross section, we may take $\phi = 0$ along the boundary, since we are interested only in the derivatives of ϕ. The solution of the torsion problem is thus reduced to the determination of the stress function ϕ satisfying the differential equation (100) and vanishing on the boundary.

When the components of the shear stress have been found for a given angle of twist, the warping function w can be determined from the relations

$$\frac{\partial w}{\partial x} = \frac{\tau_{xz}}{G} + \theta y \qquad \frac{\partial w}{\partial y} = \frac{\tau_{yz}}{G} - \theta x \tag{101}$$

The stresses transmitted across any transverse section is statically equivalent to a twisting moment T about the z axis, the resultant shearing force being easily shown to vanish. Referring to Fig. 3.27b, we have

$$T = \iint (x\tau_{yz} - y\tau_{xz})\, dx\, dy = -\iint \left(x\frac{\partial \phi}{\partial x} + y\frac{\partial \phi}{\partial y} \right) dx\, dy$$

where the integration is taken over the entire cross section of the bar. Integrating by parts, we get

$$T = -\int \phi(x\, dy - y\, dx) + 2\iint \phi\, dx\, dy$$

the first integral being taken round the boundary. In view of the boundary condition $\phi = 0$, the expression for the torque reduces to

$$T = 2\iint \phi\, dx\, dy \tag{102}$$

When ϕ is known for a given cross section, the integral can be evaluated in a straightforward manner. It is apparent from the above derivation that one half of the torque is due to the stress component τ_{xz} and the other half to the stress component τ_{yz}.

Consider, as an example, the elastic torsion of a bar whose cross section is an ellipse with semiaxes a and b. If the origin of coordinates is taken at the center O of the ellipse (Fig. 3.28a), the stress function may be written as

$$\phi = C\left(1 - \frac{x^2}{a^2} - \frac{y^2}{b^2} \right)$$

where C is a constant. The expression in the bracket vanishes along the boundary of the ellipse, while (100) is satisfied if

$$C = \frac{G\theta a^2 b^2}{a^2 + b^2}$$

The lines of shearing stress are therefore concentric ellipses with center at O.

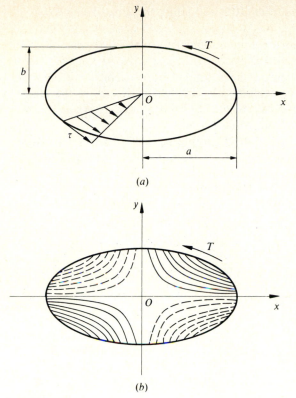

(a)

(b)

Figure 3.28 Elastic torsion of a bar of elliptical cross section. (a) Geometry of the section and distribution of resultant shear stress, (b) lines of constant warping. The alternate concave and convex areas are shown by full and broken lines respectively.

Substituting for ϕ into (99), the shear stresses are obtained as

$$\tau_{xz} = -\frac{2G\theta a^2 y}{a^2 + b^2} \qquad \tau_{yz} = \frac{2G\theta b^2 x}{a^2 + b^2} \qquad (103)$$

These expressions indicate that the resultant shear stress has a constant direction for all points on a radial line through O. By (102), the applied torque is

$$T = 2C \iint \left(1 - \frac{x^2}{a^2} - \frac{y^2}{b^2}\right) dx\, dy = \frac{\pi G\theta a^3 b^3}{a^2 + b^2} \qquad (104)$$

It follows from (103) that the resultant shear stress has its greatest magnitude τ_0 at the extremities of the minor axis of the ellipse, where

$$\tau_0 = \frac{2G\theta a^2 b}{a^2 + b^2} = \frac{2T}{\pi a b^2}$$

Plastic yielding would therefore occur at these points of the cross section when $\tau_0 = k$, the corresponding torque being $\frac{1}{2}\pi k a b^2$. The warping of the cross section during the elastic torsion follows from (101) and (103), the result being

$$w = -\left(\frac{a^2 - b^2}{a^2 + b^2}\right)\theta xy$$

since $w = 0$ at $x = y = 0$. The contour lines defined by $w = \text{const}$ are therefore rectangular hyperbolas whose asymptotes are the principal axes of the ellipse. These curves are shown in Fig. 3.28b, where solid lines indicate concavity and broken lines indicate convexity of the cross-section.†

Suppose, now, that the bar has a central elliptic hole of semiaxes ρa and ρb, where $\rho < 1$. The boundary of the hole then coincides with a shearing line of the solid bar. The same stress function therefore holds for the hollow bar, giving the stress distribution (103) in terms of θ. The torque required by the hollow bar is smaller than that for the solid bar by an amount which is carried by the portion replaced by the hole. The torque-twist relationship for the hollow bar is therefore obtained by multiplying the right-hand side of (104) by the factor $(1 - \rho^4)$.

(ii) *Elastic/plastic torsion* When the torque is increased to a critical value, the resultant shear stress attains the yield value k at one or more points of the cross section. In view of (99), the shear stress has the magnitude $|\text{grad }\phi|$, which attains its greatest value at the boundary as a consequence of the differential equation (100). Plastic yielding therefore begins somewhere on the boundary of the cross section, where the yield criterion

$$\tau_{xz}^2 + \tau_{yz}^2 = \left(\frac{\partial \phi}{\partial x}\right)^2 + \left(\frac{\partial \phi}{\partial y}\right)^2 = k^2 \tag{105}$$

is first satisfied. As the torque is increased further, plastic zones spread inward from such points on the boundary. If the material is nonhardening, the shear stress vector has the constant magnitude k throughout the plastic region. The stress function ϕ in the plastic region satisfies the differential equation (105), while in the elastic region ϕ continues to be given by (100). The elastic/plastic boundary at each stage is determined from the condition that the shear stress is continuous across the interface. It is evident that the stress distribution in an elastic/plastic bar can be found without recourse to the deformation.

Figure 3.29a shows a typical partially plastic cross section in which the elastic/plastic interface is represented by Γ. On the plastic side of this boundary, the stress is uniquely determined by the shape of the external boundary. In view of the yield condition

$$|\text{grad }\phi| = k$$

the derivative of ϕ along the inward normal to the contour curve is equal to k. Since ϕ vanishes along the external contour, the lines of shearing stress in the plastic

---

† For a detailed discussion of the elastic torsion of bars, reference may be made to S. Timoshenko and J. N. Goodier, *Theory of Elasticity*, 3rd ed., McGraw-Hill Book Co., New York (1970), and I. S. Sokolnikoff, *Mathematical Theory of Elasticity*, 2nd ed., McGraw-Hill Book Co., New York (1956).

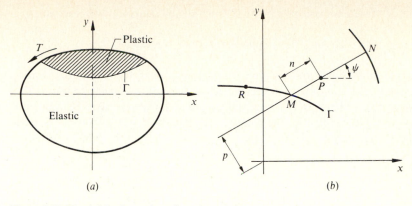

Figure 3.29 Elastic/plastic torsion of a prismatic bar.

region are parallel curves spaced at constant distances from the boundary. The stress function at any point in the plastic region is therefore k times the distance from the boundary along the normal through the point. If the boundary has a reentrant corner, the vertex is a singularity of stress, and the contour lines in the neighborhood of the corner are circular arcs within an angular span defined by the terminal normals at the vertex.

Consider a typical point P in the cross section of a bar which is subjected to a gradually increasing torque. So long as the neighborhood of P remains elastic, the strain ratio γ_{xz}/γ_{yz} is equal to the stress ratio τ_{xz}/τ_{yz} by Hooke's law. When the advancing plastic boundary reaches P, the resultant shear stress has the magnitude k and is tangential to the curve through P drawn parallel to the external boundary of the plastic region. If geometry changes are disregarded, the shear-stress vector remains unchanged in magnitude and direction at all subsequent stages. The elastic strain increments therefore disappear, and the Prandtl-Reuss relation reduces to $d\gamma_{xz}/\tau_{xz} = d\gamma_{yz}/\tau_{yz}$. Since the stress ratio remains constant, we get

$$\frac{\gamma_{xz}}{\gamma_{yz}} = \frac{\partial w/\partial x - \theta y}{\partial w/\partial y + \theta x} = \frac{\tau_{xz}}{\tau_{vz}} \tag{106}$$

This equation holds in both elastic and plastic regions, and is identical to the corresponding relation given by the Hencky theory. The displacement system (97) is evidently compatible with the plastic state of stress with τ_{xz} and τ_{yz} as the only nonzero stress components.

Let ψ denote the counterclockwise angle made with the x axis by a typical normal MN to the external boundary at N, intersecting the elastic/plastic boundary Γ at M (Fig. 3.29b). At each point of MN, the resultant shear stress acts in the perpendicular direction with components

$$\tau_{xz} = -k \sin \psi \qquad \tau_{yz} = k \cos \psi \tag{107}$$

Substitution of these expressions into the stress-strain relation (106) furnishes

$$\frac{\partial w}{\partial x} \cos \psi + \frac{\partial w}{\partial y} \sin \psi = \theta(y \cos \psi - x \sin \psi)$$

This linear differential equation is hyperbolic, and its characteristics† are given by $dy/dx = \tan \psi$. The characteristics therefore coincide with the family of normals to the external contour. The expression on the left-hand side of the above equation is equal to the partial derivative $\partial w/\partial n$, where n denotes the distance MP of a generic point P on the considered normal. Hence

$$\frac{\partial w}{\partial n} = \theta(y \cos \psi - x \sin \psi) = \theta p$$

where p denotes the perpendicular distance of the characteristic MN from the origin O. The sign of p will be taken as positive when the vector represented by MN has a clockwise moment about O. Since $n \cos \psi = x - \xi$ and $n \sin \psi = y - \eta$, where (ξ, η) are the coordinates of M, the above equation furnishes‡

$$w(x, y) - w(\xi, \eta) = \theta(x\eta - y\xi) = \theta np \tag{108}$$

In view of the continuity of w across Γ, the value of $w(\xi, \eta)$ is known from the integration of (101) in the elastic region. Hence the warping displacement at any point in the plastic region can be readily calculated. If the cross section has an axis of symmetry, w vanishes along this line in both the elastic and plastic regions. In this case, $w(\xi, \eta)$ can be found by direct integration along the plastic boundary Γ. The variation of w along this boundary is

$$dw = \frac{\partial w}{\partial x} dx + \frac{\partial w}{\partial y} dy = \theta(y \, dx - x \, dy) + \frac{1}{G}(\tau_{xz} \, dx + \tau_{yz} \, dy)$$

in view of (101). Let R denote the point where Γ is intersected by the axis of symmetry. Since $w = 0$ at R, the integration of the above equation along Γ results in

$$w(\xi, \eta) = \theta \int_R^M (y \, dx - x \, dy) + \frac{1}{G} \int_R^M (\tau_{xz} \, dx + \tau_{yz} \, dy) \tag{109}$$

The expression in the second integral is equal to $\tau_s \, ds$, where τ_s is the resolved component of the shear-stress vector along the tangent to Γ, and ds the corresponding arc element. When the angle of twist is large, the contribution from the second integral in (109) is negligible.§

---

† For a general first-order differential equation $P(\partial z/\partial x) + Q(\partial z/\partial y) = R$, where z is the dependent variable and P, Q, R are continuous functions of (x, y, z), the characteristics are the family of curves $dx/P = dy/Q$. The substitution of (107) into the equilibrium equation shows that the stress equation is also hyperbolic having the same characteristics.

‡ See, for example, P. G. Hodge, Jr., *J. Appl. Mech.*, **16**: 399 (1949), who applied the equation to the calculation of the warping of an I beam with fillets.

§ Computer methods for elastic/plastic torsion based on finite difference and finite element models have been developed by P. G. Hodge, Jr., C. T. Herakovich, and R. B. Stout, *J. Appl. Mech.*, **35**: 454 (1968); C. T. Herakovich and P. G. Hodge, Jr., *Int. J. Mech. Sci.*, **11**: 53 (1969).

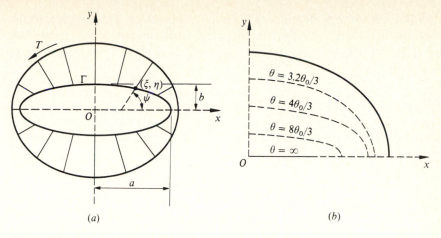

Figure 3.30 Characteristics and plastic boundaries for Sokolovsky's oval when $A = 7B$.

Consider a point Q on the external boundary which is the first to become plastic during the loading. Since such a point must be on the boundary, the direction of the shear stress remains constant for all angles of twist. The magnitude of this shear stress has a constant value k when the torque T exceeds the initial yield value T_e. If the bar is completely unloaded from the partly plastic state, a negative shear stress of magnitude greater than k is superimposed at Q, leaving a residual shear stress acting in the opposite sense. For $T = 2T_e$ at the instant of unloading, the magnitude of the residual stress attains the value $-k$, causing yielding to restart at Q. Evidently, a secondary yielding of this type cannot occur in a bar for which the fully plastic torque is less than twice the torque at the initial yielding.

(iii) Solution for an oval cross section An inverse method of solution, in which the external contour is determined from an assumed stress distribution in the elastic region, has been applied by Sokolovsky† to the elastic/plastic torsion of a bar having an oval cross section (Fig. 3.30a). Assuming that the stress components τ_{zx} and τ_{yz} in the elastic region are proportional to $-y/b$ and x/a respectively, where a and b are constants, we write

$$\tau_{xz} = \frac{\partial \phi}{\partial y} = -\frac{ky}{b} \qquad \tau_{yz} = -\frac{\partial \phi}{\partial x} = \frac{kx}{a} \tag{110}$$

so that the elastic/plastic boundary, where $\tau_{xz}^2 + \tau_{yz}^2 = k^2$, is an ellipse with semiaxes a and b. The stress function in the elastic region is

$$\phi = C - \frac{k}{2}\left(\frac{x^2}{a} + \frac{y^2}{b}\right)$$

† W. W. Sokolovsky, *Prikl. Mat. Mekh.*, **6**: 241 (1942). See also R. von Mises, *Reissner Anniversary Volume*, p. 241, Ann Arbor, Mich. (1949). An alternative inverse method of solution has been described by L. A. Galin, *Prikl. Mat. Mekh.*, **13**: 285 (1949).

where C is a constant to be determined from the condition of continuity across the elastic/plastic boundary. The lines of shearing stress in the elastic region are concentric ellipses with semiaxes proportional to $\sqrt{a}$ and $\sqrt{b}$. The substitution of (110) into (98) yields

$$\theta = \frac{k(a + b)}{2Gab}$$

Using Eqs. (101), the warping displacement in the elastic region is easily found to be

$$w = -(a - b)\frac{kxy}{2Gab} = -\left(\frac{a - b}{a + b}\right)\theta xy \tag{111}$$

The stresses in the plastic region are given by (107). Since the stresses must be continuous across the elastic/plastic boundary, the characteristic angle ψ at any point (ξ, η) of the ellipse is such that

$$\xi = a \cos \psi \qquad \eta = b \sin \psi$$

Hence ψ represents the eccentric angle of the ellipse. The equation of the characteristic through (ξ, η) is

$$y = x \tan \psi - (a - b) \sin \psi \tag{112}$$

which defines the angle ψ corresponding to any given point (x, y). The shearing lines in the plastic region are the orthogonal trajectories to the family of straight lines. Since $dy/dx = -\cot \psi$ along any such trajectory, the differentiation of (112) with respect to ψ gives

$$\frac{dx}{d\psi} + x \tan \psi = (a - b) \sin \psi \cos^2 \psi$$

on a trajectory. The integration of this equation leads to the equation of the trajectory in the parametric form

$$x = \cos \psi [A + B(2 - \cos 2\psi)]$$
$$y = \sin \psi [A - B(2 + \cos 2\psi)] \tag{113}$$

where A is an arbitrary constant and $4B = a - b$. For $A > 3B$, the curve is a closed oval with semiaxes $A + B$ and $A - B$. The oval is very nearly an ellipse. When a and b are given, (113) defines the external contour with any suitable value of A. If, on the other hand, the external contour is specified, the position of the elastic/plastic boundary for a given angle of twist can be determined from the relations

$$\frac{a + b}{ab} = \frac{2G\theta}{k} \qquad a - b = 4B$$

Solving for a and b in terms of the parameters B and $k/G\theta$, we get

$$a = \frac{k}{2G\theta} + 2B + \sqrt{4B^2 + \frac{k^2}{4G^2\theta^2}}$$

$$b = \frac{k}{2G\theta} - 2B + \sqrt{4B^2 + \frac{k^2}{4G^2\theta^2}}$$

$$(114)$$

For each increment of the angle of twist, a and b decrease by equal amounts. The smallest angle of twist for which the elliptic interface lies within the external contour corresponds to $a = A + B$. The solution is therefore valid only in the range

$$\theta \geqslant \theta_0 = \frac{(A - B)k}{(A + B)(A - 3B)G}$$

The shape of the elastic/plastic boundary is not known for smaller angles of twist. Figure 3.30b indicates the spread of the plastic zone with increasing twist when $A/B = 7$ for the external boundary.

To obtain the functions ϕ and w in the plastic region, consider a generic point P at a distance n from (ξ, η) measured along the characteristic of inclination ψ. The coordinates of P are

$$x = (a + n) \cos\psi \qquad y = (b + n) \sin\psi \qquad (115)$$

It follows from (113) and (115) that the length of the characteristic between the elliptic interface and the external boundary is

$$n_0 = A - \tfrac{1}{2}(a + b) - B \cos 2\psi$$

Hence the stress function in the plastic region is

$$\phi = k(n_0 - n) = k\left(A - B \cos 2\psi - \frac{k}{2G\theta} - \sqrt{4B^2 + \frac{k^2}{4G^2\theta^2}} - n \right) \qquad (116)$$

where ψ and n are given by (115) as functions of (x, y). Setting $x = a \cos\psi$ and $y = b \sin\psi$ in the elastic stress function, and using (116), it is found that ϕ would be continuous across the plastic boundary $n = 0$ if

$$C = A - \tfrac{1}{4}(a + b) = A - \frac{k}{4G\theta} - \sqrt{B^2 + \frac{k^2}{16G^2\theta^2}}$$

To obtain the warping in the plastic region, it is only necessary to insert the expression for $w(\xi, \eta)$ into (108). Since $p = -(a - b) \sin\psi \cos\psi$ by (112), and

$$w(\xi, \eta) = -(a - b) \frac{k}{2G} \sin\psi \cos\psi$$

in view of (111), the warping displacement at any point in the plastic region may be written as

$$w = -2B\left(\theta n + \frac{k}{2G} \right) \sin 2\psi \qquad 0 \leqslant n \leqslant n_0 \qquad (117)$$

The contours of constant w in the plastic region can be found by calculating a series of values of ψ and n for selected values w, the coordinates of the corresponding points on the curves being then obtained from (115). These curves can be continued into the elastic region employing (111). Figure 3.31 shows the contour lines in one quadrant of the cross section when $A = 7B$ and $\theta = \frac{4}{3}\theta_0$. Using (114), the plastic warping function (117) may be expressed alternatively as

$$w = \theta\left(-2B + \sqrt{4B^2 + \frac{k^2}{4G^2\theta^2}}\right)x \sin\psi - \theta\left(2B + \sqrt{2B^2 + \frac{k^2}{2G^2\theta^2}}\right)y \cos\psi$$

where ψ is given by Eq. (112). For sufficiently large values of θ, the warping displacement is approximately equal to $-4B\theta y \cos\psi$ in most of the plastic region.

The applied torque can be obtained directly from the condition of statical equilibrium, using the stress distributions in the elastic and plastic regions. The torque carried by the elastic core is

$$T_1 = \iint (x\tau_{yz} - y\tau_{xz})\, dx\, dy = k \iint \left(\frac{x^2}{a} + \frac{y^2}{b}\right) dx\, dy$$

in view of (110), the integration being taken over the central ellipse of semiaxes a and b. It is easily shown that

$$T_1 = \tfrac{1}{4}\pi kab(a + b)$$

Using (107), the contribution to torque from the plastic part of the cross section may be written as

$$T_2 = k \iint (x \cos\psi + y \sin\psi)\, dx\, dy$$

The integral extends over the region between the elliptic interface and the external

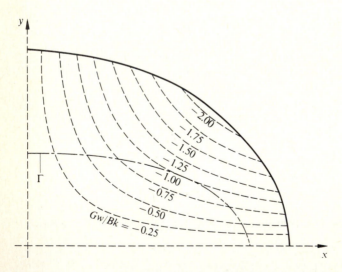

Figure 3.31 Contours of constant warping for Sokolovsky's oval when $A = 7B$, $\theta = 4\theta_0/3$.

boundary. It is convenient to change the variables to n and ψ using the transformation (115), the result being

$$T_2 = 4k \int_0^{\pi/2} \int_0^{n_0} (a \cos^2 \psi + b \sin^2 \psi + n)(a \sin^2 \psi + b \cos^2 \psi + n) \, dn \, d\psi$$

The expression in the second parenthesis is the jacobian of the transformation. The integration is now carried out in a straightforward manner, and the resultant torque T is found as

$$T = T_1 + T_2 = \pi k \{\tfrac{2}{3}[A^3 - \tfrac{1}{8}(a^3 + b^3)] - B^2[3A - 2(a + b)]\} \qquad (118)$$

Equations (114) and (118) give the torque-twist relationship for $\theta \geqslant \theta_0$. The magnitude of the torque corresponding to $\theta = \theta_0$ is obtained by setting $a = A + B$ and $b = A - 3B$ in (118). As the cross section approaches the fully plastic state, the elastic region tends to be a line of stress discontinuity of length $8B$. The fully plastic torque, which follows from (118) with $a = 4B$ and $b = 0$, is rapidly approached in an asymptotic manner as θ is increased.

In principle, we could always start with a suitable elastic distribution of stress, and determine the elastic/plastic boundary as the curve along which the resultant shear stress is equal to k. The continuity of the stress components across this boundary gives the inclination of the straight characteristics, of which any orthogonal trajectory represents a suitable external boundary of the cross section. In general, the solution would correspond to a particular contour for a given angle of twist, while we have no information regarding the stress distribution for the same contour with other angles of twist. The success of Sokolovsky's solution lies in the fact that the elastic/plastic boundary and the associated stress distribution are similar for a range of values of the angle of twist.†

(iv) *The membrane analogy* An analogy between the elastic torsion of a bar and the small deflection of a laterally loaded membrane has been pointed out by Prandtl.‡ The membrane is stretched by a uniform tension F per unit length of its boundary, and is attached to a die whose edge plane is of the same shape as the cross section of the twisted bar. A uniform lateral pressure q per unit area is then applied to the membrane to produce a deflection ω at a generic point (Fig. 3.32a). Since the curvatures of the deflected membrane in the xz and yz planes are $-\partial^2\omega/\partial x^2$ and $-\partial^2\omega/\partial y^2$ respectively for small deflections, the equation of lateral equilibrium is

$$\frac{\partial^2 \omega}{\partial x^2} + \frac{\partial^2 \omega}{\partial y^2} = -\frac{q}{F}$$

† A conformal transformation method has been applied by E. Trefftz, *Z. angew. Math. Mech.*, **5**: 64 (1925), to determine the plastic region at the reentrant corner of an L beam. The shape of the plastic region at the tip of a sharp notch has been investigated by J. A. H. Hult, *J. Mech. Phys. Solids*, **6**: 79 (1957).

‡ L. Prandtl, *Physik. Z.* **4**: 758 (1903). For experimental methods based on this analogy, see A. A. Griffith and G. I. Taylor, *Proc. Inst. Mech. Eng.*, Oct.–Dec. (*1917*). See also T. J. Higgins, *Proc. Soc. Exp. Stress Anal.*, **2**: 17 (1945).

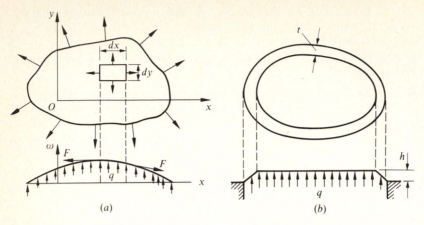

Figure 3.32 The deflection of stretched membranes under uniform lateral pressure.

The boundary condition is $\omega = 0$ along the edge of the die. A comparison with (100) indicates that ω satisfies the same differential equation and boundary condition as the ratio ϕ/G, provided we choose $q = 2\theta F$. The contours of constant deflection therefore correspond with the lines of shearing of the twisted bar, while the resultant shear stress at any point is G times the greatest slope at the corresponding point of the deflected membrane. It follows from (102) that the applied torque is equal to $2G$ times the volume bounded by the deflected membrane and the xy plane. Since the membrane is everywhere concave to the applied pressure, the greatest value of the shear stress must occur somewhere on the boundary.

An extension of the membrane analogy to elastic/plastic torsion has also been suggested by Prandtl. It is necessary to erect a roof of constant slope equal to k/G and having its base identical to the boundary of the cross section. The base is filled with a stretched membrane which is loaded by a uniform surface pressure as before. When the intensity of the pressure becomes sufficiently high, the membrane begins to touch the roof, and this corresponds to a state of initial yielding of the twisted bar. As the pressure is further increased, certain parts of the membrane come in contact with the roof, while the remainder is still free to deflect. The supported and unsupported parts of the membrane correspond to the plastic and elastic zones respectively of the twisted bar.† The twisting moment is again equal to the volume under the entire membrane multiplied by $2G$. The membrane analogy provides a useful experimental means of finding the stress distribution in elastic and partly plastic bars.

It is possible to obtain simple approximate solutions to certain problems using the membrane analogy. Consider, for example, the torsion of a bar whose cross

---

† See A. Nadai, *Theory of Flow and Fracture of Solids*, vol. I, chaps. 23 and 25, McGraw-Hill Book Co., New York (1950).

section is a narrow rectangle of width a and length b. It is intuitively obvious that the surface of the deflected membrane would be cylindrical except near the shorter sides of the rectangle (Fig. 3.33). Neglecting the end effect, the deflection at any point may be written as

$$\omega \simeq \delta\left(1 - \frac{4x^2}{a^2}\right)$$

while the bar is still elastic. The central deflection δ is obtained from the condition of equilibrium of the membrane under a normal pressure q, which is 2θ times the surface tension F. Since the slope of the membrane with respect to the x axis is of amount $4\delta/a$ along $x = \pm a$, we find $\delta = \frac{1}{4}a^2\theta$. The elastic stress function therefore becomes

$$\phi = G\omega = G\theta\left(\frac{a^2}{4} - x^2\right)$$

and the elastic stress distribution is given by

$$\tau_{xz} = \frac{\partial\phi}{\partial y} = 0 \qquad \tau_{yz} = -\frac{\partial\phi}{\partial x} = 2G\theta x$$

except near the ends $y = \pm b/2$ where τ_{xz} would be appreciable. The volume of the region between the deflected membrane and the xy plane is equal to $\frac{2}{3}ab\delta$, which gives

$$T = \tfrac{4}{3}Gab\delta = \tfrac{1}{3}G\theta a^3 b \tag{119}$$

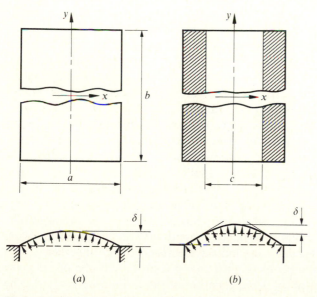

(a) (b)

Figure 3.33 Application of membrane analogy to the torsion of a bar of narrow rectangular cross section. (a) Elastic torsion, (b) elastic/plastic torsion.

The greatest shear stress occurs along the longer sides of the rectangle, and has a magnitude equal to $G\theta a$. The initial yielding of the bar therefore corresponds to

$$\theta_e = \frac{k}{Ga} \qquad T_e = \tfrac{1}{3}ka^2b$$

For a higher value of the torque, the plastic zone spreads inward from either side, leaving an elastic core of width c. The membrane is in contact with the roof over the portions representing the plastic zones. The stress component τ_{xz} continues to vanish, while

$$\tau_{yz} = \frac{2kx}{c} \qquad 0 \leqslant x \leqslant \frac{c}{2}$$

$$\tau_{yz} = k \qquad \frac{c}{2} \leqslant x \leqslant \frac{a}{2}$$

The equilibrium of the free portion of the membrane requires $\theta = k/Gc$, while the parabolic shape of the deflection gives $\delta = kc/4G$. The total torque is

$$T = \frac{4}{3} Gbc\delta + \frac{1}{2} kb(a^2 - c^2)$$

substituting for δ, this is reduced to

$$T = \frac{1}{2} kb\left(a^2 - \frac{c^2}{3}\right) = \frac{1}{2} kb\left(a^2 - \frac{k^2}{3G^2\theta^2}\right) \tag{120}$$

The solution obtained for the narrow rectangle may be employed to derive approximate torque-twist relations for rolled sections (such as angles, channels, and I sections) by dividing the cross section into a number of rectangles. For a given angle of twist, the torque shared by each rectangular part is obtained from either (119) or (120), depending on whether this part is elastic or partially plastic. There is, however, a considerable stress concentration at the reentrant corners, and the intensity of the stress depends on the radius of the fillet.

The membrane analogy is also applicable with some modifications to the torsion of hollow bars. A rigid flat plate having the same shape as the hole is attached to the inner boundary of the stretched membrane. The plate is constrained to move vertically during the application of the pressure, using a counterweight to balance the weight of the plate. The height attained by the horizontal plate, giving a constant stress function at the boundary of the hole, is governed by the condition of equilibrium of the plate under the action of the uniform membrane forces F and the applied pressure $q = 2\theta F$. The torque is $2G$ times the total volume under the plate and the membrane taken together.

A solution to the torsion problem for a closed thin-walled tube of arbitrary cross section may be obtained by the application of the membrane analogy. The resultant shear stress τ in this case may be assumed constant across the thickness, since the variation of the slope of the membrane across the wall is negligible (Fig.

3.32*b*). If *h* denotes the difference in level of the two boundaries of the membrane, its slope at any point is h/t, where *t* is the local thickness of the tube. Since the slope is equal to τ/G by the membrane analogy, the local shear stress is $\tau = Gh/t$. If the mean area enclosed by the outer and inner boundaries of the cross section is denoted by *A*, the torque is $T = 2GAh$, giving

$$\tau = \frac{Gh}{t} = \frac{T}{2At}$$

The shear stress in the elastic tube is therefore inversely proportional to the thickness. From the condition of vertical equilibrium of the plate,

$$\int \tau \, ds = Gh \int \frac{ds}{t} = 2G\theta A$$

where *ds* is an arc element of the center line of the cross section, and the integral extends over the whole perimeter. Hence, for an elastic tube,[†]

$$\theta = \frac{h}{2A} \int \frac{ds}{t} = \frac{T}{4GA^2} \int \frac{ds}{t} \qquad (121)$$

As the torque is increased, the shear yield stress *k* is first attained in the element whose thickness is the least. Suppose that the thickness of the tube varies between a maximum and a minimum round the periphery. The height of the rigid plate corresponding to an elastic/plastic torsion would be $h = kt^*/G$, where t^* is the wall thickness at the extremities of the plastic arc length s^* of the mean periphery. Then the applied torque is $T = 2kAt^*$, and the angle of twist per unit length is

$$\theta = \frac{k}{2GA} \left(s^* + t^* \int \frac{ds}{t} \right) \qquad (122)$$

where the integral is taken over the elastic part of the cross section. The analysis can be easily extended to the tension of tubular members with multicellular cross sections.

(v) *Fully plastic torsion* In a solid bar made of nonhardening material, the fully plastic stress distribution represents a limiting state which is approached in an asymptotic manner as the angle of twist increases. The fully plastic value of the torque has a physical significance, since it is very closely attained while the deformation is still of the elastic order of magnitude. The stress surface for a fully plastic cross section can be obtained experimentally by piling dry sand on a horizontal base whose shape is identical to that of the cross section of the bar.[‡] The *sandhill* so formed represents a roof of constant slope determined by the internal friction of the sand. The analogy is completed by introducing a factor of

[†] This formula is due to R. Bredt, *Ver. Dtsch. Ing.*, **40**: 815 (1896).
[‡] A. Nadai, *Z. angew. Math. Mech.*, **3**: 442 (1923). The influence of work-hardening on the fully plastic stress distribution has been discussed by R. Hill, *J. Mech. Phys. Solids*, **5**: 1 (1956).

proportionality that makes the slope of the sandhill equal to k/G. The applied torque is $2G$ times the volume of the sand forming the hill. It is important to note that the shear stress in a fully plastic bar is discontinuous across lines which are projections of ridges on the sandhill surface. The component of the shear stress normal to such a line must of course be continuous for equilibrium. A line of discontinuity therefore bisects the angle formed by the intersecting shearing lines. The shear strain rate vanishes on a line of stress discontinuity.

The warping of the cross section of a fully plastic bar may be obtained to a close approximation by assuming the material to be rigid/plastic. For such a material, no twist is at all possible before the torque attains the fully plastic value. Once the limiting torque has been reached, the bar is free to twist in an unrestricted manner. The warping displacement in a rigid/plastic bar is therefore obtained from (108) and (109) by letting G tend to infinity, bearing in mind that Γ has now become a line of stress discontinuity. Thus

$$w(x, y) = \theta np + \theta \int_R^M (y \, dx - x \, dy) \tag{123}$$

where the integral is taken along the discontinuity, and w is assumed to vanish at some point R on the discontinuity.[†] The expression in the parenthesis is equal to the moment of a line element of Γ about the origin. When there are characteristics that do not intersect a stress discontinuity, the warping function cannot be determined uniquely without consideration of the work-hardening.[‡]

Consider the torsion of a bar whose cross section is a rectangle of width $2a$ and length $2b$. During the elastic torsion, the maximum shear stress occurs at the midpoints of the longer sides.[§] Yielding begins at these points when the applied torque becomes

$$T_e = 8ka^2b \left\{ \frac{\frac{1}{3} - (64a/\pi^5 b) \tanh (\pi b/2a)}{1 - (8/\pi^2) \operatorname{sech} (\pi b/2a)} \right\} \tag{124}$$

to a close approximation. The expression in the denominator is approximately equal to $k/2G\theta_e a$. With further increase in torque, plastic zones spread inward and along the longer sides until the shorter sides begin to yield at their centers. The cross section becomes fully plastic when the elastic zones shrink into lines of discontinuity as shown in Fig. 3.34a. There are four discontinuities bisecting the angles at the corners, and another one joining their meeting points C. In each of the four regions separated by the discontinuities, the lines of shearing stress are parallel to the corresponding side of the rectangle.

The discontinuity CC corresponds to the central ridge of the sandhill whose height is $h = ka/G$ to within a scale factor. The total volume of the heaped sand is the sum of the volume of a square pyramid of height h and base area $4a^2$, and the

---

† This formula is due to J. Mandel, *C. R. Acad. Sci., Paris*, **222**: 1205 (1946).

‡ This question has been discussed by R. Hill, op. cit., in relation to a hollow bar.

§ See, for example, S. Timoshenko and J. N. Goodier, *Theory of Elasticity*, chap. 11, 3d ed., McGraw-Hill Book Co., New York (1970).

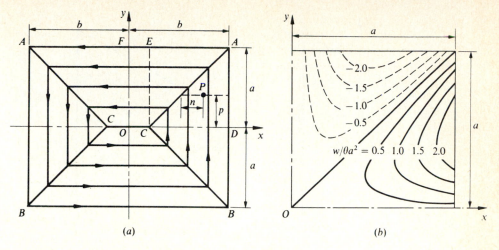

Figure 3.34 Contour lines of ϕ and w for fully plastic torsion of bars of rectangular and square cross sections.

volume of a triangular prism of length $2(b - a)$ and cross-sectional area ah. Hence, the fully plastic torque for the rectangular cross section is

$$T_0 = 2G[\tfrac{4}{3}a^2h + 2(b - a)ah] = 4ka^2\left(b - \frac{a}{3}\right) \tag{125}$$

In the case of a square cross section ($b = a$), the fully plastic torque is $\tfrac{8}{3}ka^3$, which is about 1.60 times that at the initial yielding.

The warping of the cross section during the fully plastic torsion can be easily calculated from (123) using rectangular axes as shown. Due to the symmetry of the cross section, it is only necessary to consider one quadrant of the rectangle. In the region ACD, the straight characteristic drawn through a generic point P gives $p = y$ and $n = x - (b - a) - y$. Since $x - y = b - a$ and $dx = dy$ along the discontinuity CA, it follows from (123) that

$$w = \theta y[x - y - 2(b - a)] \qquad x - y \geqslant b - a \tag{126}$$

The value of w in this region is entirely positive when $b = a$, and entirely negative when $b \geqslant 2a$. Since $p = -x$ in the remainder of the quadrant, while $n = y - x + (b - a)$ in ACE and $n = y$ in ECOF, we have

$$w = -\theta xy \qquad\qquad 0 \leqslant x \leqslant b - a$$
$$w = -\theta[xy - (x - b + a)^2] \qquad b - a \leqslant x \leqslant b - a + y \tag{127}$$

The contour lines of w in the first quadrant for a square cross section are shown in Fig. 3.34b, the elevated and depressed areas of the cross section being indicated by the solid and broken lines respectively.

As a second example, consider a prismatic bar whose cross section is an equilateral triangle, Fig. 3.35a, the length of each side of the triangle being denoted

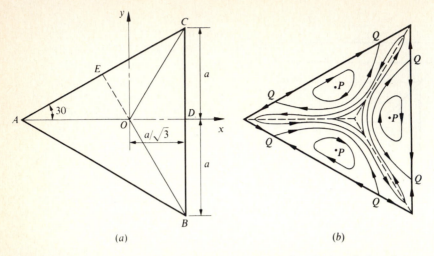

Figure 3.35 Geometry and residual shearing lines in the torsion of a bar of triangular cross section.

by $2a$. An elastic analysis of this problem reveals that the resultant shear stress has the maximum intensity $(\sqrt{3}/2)G\theta a$, occurring at the midpoint of each side. Plastic yielding therefore begins at these points when the specific angle of twist is $\theta_e = 2k/\sqrt{3}\,Ga$. The corresponding torque is found to be[†]

$$T_e = \frac{\sqrt{3}}{5}\,G\theta_e a^4 = \frac{2}{5}\,ka^3 \tag{128}$$

The fully plastic stress distribution involves stress discontinuities along the internal bisectors of the angles of the triangle, the lines of shearing stress being parallel to the sides of the triangle. The height of the sandhill that corresponds with the plastic state of stress is $h = ka/\sqrt{3}\,G$, and the fully plastic torque is

$$T_0 = 2G(\tfrac{1}{3}h)(\sqrt{3}\,a^2) = \tfrac{2}{3}ka^3$$

By Eq. (123), the warping displacement w corresponding to a specific angle of twist θ in the fully plastic range is easily shown to be

$$w = \theta y\left(x - \frac{y}{\sqrt{3}}\right) \qquad \text{in} \qquad COD$$

$$w = \frac{\sqrt{3}}{2}\,\theta\left(x^2 - \frac{y^2}{3}\right) \qquad \text{in} \qquad COE$$

(129)

[†] Relaxation methods have been used for the elastic/plastic torsion of bars of triangular cross section by D. G. Christopherson and R. V. Southwell, *Proc. R. Soc.*, A, **168**: 317 (1938), and I sections by D. G. Christopherson, *J. Appl. Mech.*, **7**: 1 (1940). The torsion of hollow bars has been investigated by F. S. Shaw, Australian Council for Aeronautics, Rep. A.C.A. 11 (1944). See also R. V. Southwell, *Q. J. Mech. Appl. Math.*, **2**: 385 (1949).

when the axes of reference are as shown. The expression for w in region AOE is obtained by changing the sign of y in the first equation of (129). The warping in the remaining half of the triangle follows from the condition of antisymmetry of w with respect to the x axis.

The distribution of residual stresses on complete unloading approach a limiting state as the cross section tends to become fully plastic during the loading. The residual stresses can be calculated by using the expressions for the elastic stresses in the bar corresponding to a specific angle of twist θ. For the triangular cross section shown in Fig. 3.35a, the elastic stresses are found to be

$$\tau_{xz} = -G\theta y\left(1 - \frac{\sqrt{3}x}{a}\right) \qquad \tau_{yz} = G\theta\left[x + \frac{\sqrt{3}}{2a}(x^2 - y^2)\right] \qquad (130)$$

By (128), the angle of elastic untwist per unit length due to unloading from the fully plastic state is

$$\theta = \frac{5T_0}{\sqrt{3}\,Ga^4} = \frac{10k}{3\sqrt{3}\,Ga}$$

Inserting the value of θ into the elastic stresses (130), and subtracting them from the fully plastic stresses, we obtain the residual stress distribution. Considering the region BOC, where $\tau_{xz} = 0$ and $\tau_{yz} = k$ in the fully plastic state, the residual stress function is easily shown to be

$$\phi = k\left(\frac{a}{\sqrt{3}} - x\right)\left\{1 - \frac{5}{9a^2}\left[\left(x + \frac{2a}{\sqrt{3}}\right)^2 - 3y^2\right]\right\} \qquad (131)$$

The residual stress vanishes in this region at $x = 0.389a$, $y = 0$ and at $x = 0.577a$, $y = \pm0.632a$. The stress distribution is, of course, symmetrical about the lines of discontinuity. The lines of shearing stress in the unloaded bar,[†] which are the level curves of ϕ, are plotted in Fig. 3.35b, the points of zero stress being denoted by P and Q.

In an experimental determination of the fully plastic torque, the problem of measuring the volume of the sand heap may be avoided by measuring its weight, and using the fact that the ratio of the weight of the sand heap for any given cross section to that for a circular cross section is equal to the ratio of the corresponding fully plastic torques. When the prismatic bar has a symmetrical longitudinal hole, the sandhill analogy may be extended by heaping the sand around a fixed cylinder whose cross section has the shape of the hole, and which stands perpendicular to the base plate from an identical hole in it.The sandhill analogy is useful for visualizing the nature of the fully plastic stress distribution in a twisted bar, the difficulties arising from sharp corners being overcome by considering the contour as a limiting case of one with rounded corners. This is illustrated in Fig. 3.36 for an L-shaped contour in which the curved discontinuity is found to remain curved

† P. G. Hodge, Jr., *J. Appl. Mech.*, **16**: 399 (1949).

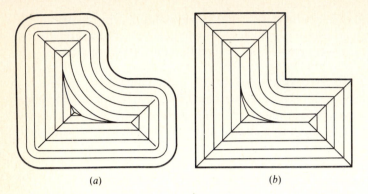

(a) $\qquad\qquad$ (b)

Figure 3.36 Stress discontinuities and shearing lines for fully plastic L sections.

when the limiting process is completed.† The rounding of the shearing lines at a reentrant corner is due to the fact that the corner is the origin of a fan of characteristics whose orthogonal trajectories are circular arcs.

(vi) *Inclusion of work-hardening* To analyze the problem of elastic/plastic torsion of a bar made of work-hardening material, it is convenient to employ a stress-strain equation that corresponds to no well-defined yield point. The problem is then simplified by the absence of an elastic/plastic boundary, which permits the same equations to apply throughout the cross section, although the governing differential equation becomes much more complicated. We shall use the Ramberg-Osgood equation

$$\varepsilon = \frac{\sigma}{E}\left\{1 + m\left(\frac{\sigma}{\sqrt{3}\,k}\right)^{2n}\right\} \tag{132}$$

for the uniaxial stress-strain curve, where m and n are dimensionless constants, and k a nominal yield stress in simple or pure shear, the slope of the stress-strain curve being equal to E when $\sigma = 0$. As the value of n increases, the stress-strain curve approaches that for a nonhardening material having a uniaxial yield stress $\sqrt{3}\,k$.

Consider, first, the formulation of the torsion problem according to the Prandtl-Reuss theory.‡ Since the velocity field in the twisted bar is given by the partial derivative of (97) with respect to θ, which is taken as the time scale, the components of the shear-strain increment are expressed as

$$2d\gamma_{xz} = \left(\frac{\partial^2 w}{\partial x\,\partial\theta} - y\right)d\theta \qquad 2d\gamma_{yz} = \left(\frac{\partial^2 w}{\partial y\,\partial\theta} + x\right)d\theta$$

---

† W. Prager and P. G. Hodge, Jr., *Theory of Perfectly Plastic Solids*, p. 66, John Wiley and Sons, New York (1951).

‡ W. Prager, *J. Appl. Phys.*, **18**: 375 (1947). For a work-hardening bar, the Prandtl-Reuss and Hencky theories lead to identical results only when the cross section is circular.

The elimination of w between these two relations leads to the strain compatibility equation

$$\frac{\partial}{\partial x}(d\gamma_{yz}) - \frac{\partial}{\partial y}(d\gamma_{xz}) = d\theta \tag{133}$$

The strain increment depends on the stress increment and the current stress according to the relations

$$2G\,d\gamma_{xz} = d\tau_{xz} + \tau_{xz}\,d\lambda \qquad 2G\,d\gamma_{yz} = d\tau_{yz} + \tau_{yz}\,d\lambda$$

where $d\lambda$ is a positive scalar depending on the plastic modulus H. Thus

$$d\lambda = \frac{3G\,d\bar{\sigma}}{H\bar{\sigma}} = (1 + 2n)\frac{3Gm}{E}\left(\frac{\tau}{k}\right)^{2n}\frac{d\tau}{\tau}$$

in view of (132), the resultant shear stress being denoted by τ. The above relation is readily integrated to

$$\lambda = \int_0^{\bar{\sigma}} \frac{3G\,d\bar{\sigma}}{H\bar{\sigma}} = \frac{3m}{4n}\left(\frac{1 + 2n}{1 + v}\right)\left\{\left(\frac{\partial\phi}{\partial x}\right)^2 + \left(\frac{\partial\phi}{\partial y}\right)^2\right\}^n \tag{134}$$

on using (99), modified by including a factor k on the right-hand side for convenience. Substituting into (133) from the stress-strain relations, and introducing the modified stress function, we finally obtain

$$\frac{\partial}{\partial x}\left(\frac{\partial\dot{\phi}}{\partial x} + \dot{\lambda}\frac{\partial\phi}{\partial x}\right) + \frac{\partial}{\partial y}\left(\frac{\partial\dot{\phi}}{\partial y} + \dot{\lambda}\frac{\partial\phi}{\partial y}\right) = -\frac{2G}{k} \tag{135}$$

where the dot denotes differentiation with respect to θ. This is a quasi-linear partial differential equation in three independent variables (x, y, θ), the quantity $\dot{\lambda}$ being obtained from (134). The solution of (135) must be carried out numerically under the initial condition $\phi = 0$ and the boundary condition ϕ = constant. For a small initial angle of twist per unit length, we may take the elastic solution for ϕ, and use the corresponding values of $\dot{\phi}$ and $\dot{\lambda}$ to continue the solution.

When the loading is monotonic, a good approximation to the Prandtl-Reuss solution for the torsion problem would be obtained by using the Hencky stress-strain relations, which may be written as

$$2G\gamma_{xz} = k(1 + \lambda)\frac{\partial\phi}{\partial y} \qquad 2G\gamma_{yz} = -k(1 + \lambda)\frac{\partial\phi}{\partial x}$$

where ϕ is the stress function, and λ is a positive quantity given by

$$\lambda = \frac{3Gm}{E}\left(\frac{\tau}{k}\right)^{2n} = \frac{3m}{2(1 + v)}\left\{\left(\frac{\partial\phi}{\partial x}\right)^2 + \left(\frac{\partial\phi}{\partial y}\right)^2\right\}^n \tag{136}$$

The compatibility equation (98) holds in the plastic range with τ_{xz} and τ_{yz} replaced by $2G\gamma_{xz}$ and $2G\gamma_{yz}$ respectively. On substitution from the Hencky equations, the compatibility equation becomes

$$\frac{\partial}{\partial x}\left[(1 + \lambda)\frac{\partial \phi}{\partial x}\right] + \frac{\partial}{\partial y}\left[(1 + \lambda)\frac{\partial \phi}{\partial y}\right] = -\frac{2G\theta}{k}$$

Inserting the expression for λ and rearranging, we obtain the governing differential equation

$$(1 + cF^n)(\phi_{xx} + \phi_{yy}) + 2ncF^{n-1}(\phi_x^2\phi_{xx} + 2\phi_x\phi_y\phi_{xy} + \phi_y^2\phi_{yy}) = -\frac{2G\theta}{k} \quad (137)$$

where $c = 3m/2(1 + v)$, and F denotes the quantity in the curly bracket of (136). The subscripts refer to partial differentiation with respect to the independent variables. Equation (137) can be solved numerically starting with the elastic solution corresponding to a small initial angle of twist.

The numerical solution is carried out by superimposing a finite square mesh upon the cross section of the bar and replacing the partial derivatives by the corresponding central differences.† Denoting by $\phi(i, j)$ the value of the stress function at $x = ih$ and $y = jh$, where h is the distance between two successive mesh lines, the finite difference form of the relevant partial derivatives may be written as

$2h\phi_x(i, j) = \phi(i + 1, j) - \phi(i - 1, j)$

$2h\phi_y(i, j) = \phi(i, j + 1) - \phi(i, j - 1)$

$h^2\phi_{xx}(i, j) = \phi(i + 1, j) - 2\phi(i, j) + \phi(i - 1, j)$

$h^2\phi_{yy}(i, j) = \phi(i, j + 1) - 2\phi(i, j) + \phi(i, j + 1)$

$4h^2\phi_{xy}(i, j) = \phi(i + 1, j + 1) - \phi(i - 1, j + 1) - \phi(i + 1, j - 1) + \phi(i - 1, j - 1)$

The finite difference equation obtained by replacing the derivatives in (137) by the above expressions can be solved explicitly for $\phi(i, j)$, using the most recently computed values of ϕ at the eight neighboring mesh points. After a sufficient number of iterations from a suitable trial solution for any given θ, the values of $\phi(i, j)$ should converge to the solution of the difference equation. The associated torque may then be calculated from (102) by a repeated use of the two-dimensional form of Simpson's rule.

Some results of the computation for a square cross section of side $2a$, based on $h = a/24$ and $c = 1$ are shown in Fig. 3.37. Choosing $n = 9$, which very nearly represents a perfectly plastic material, the positions of the elastic/plastic boundary for increasing angles of twist are obtained as contours of $\tau = k$, and plotted in (a). The influence of work-hardening on the torque-twist relationship is indicated in (b), where T/ka^3 is plotted against $G\theta a/k$ for $n = 1$ and $n = 9$. The validity of the

† H. J. Greenberg, W. S. Dorn, and E. H. Wetherell, *Plasticity, Proc. 2nd Symp. Naval Struct. Mech.*, E. H. Lee and P. S. Symonds (eds.), pp. 279–297 (1960). For another method of numerical solution, see A. Mendelson, *Plasticity: Theory and Applications*, chap. 11, Macmillan Pub. Co., New York (1968).

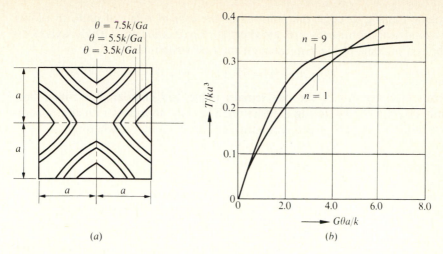

$$\theta = 7.5k/Ga$$
$$\theta = 5.5k/Ga$$
$$\theta = 3.5k/Ga$$

(a)

(b)

Figure 3.37 Torsion of a work-hardening bar of square cross section. (a) Contours of $\tau = k$ when $n = 9$; (b) torque-twist relationship.

Hencky theory requires (Sec. 2.5) that the counterclockwise angle α which the radius vector to a generic point on the loading curve, obtained by plotting τ_{yz} against τ_{xz}, makes with the local tangent to the curve must be less than $\tan^{-1}\sqrt{1 + 2n}$. This condition is indeed found to be satisfied by the solution for representative values of n.

3.7 Torsion of Bars of Variable Diameter

(i) *Elastic torsion* Consider the pure torsion of a nonuniform bar of circular cross section having its radius varying along the longitudinal axis (Fig. 3.38). We begin with the assumption that the displacement at any point is perpendicular to the

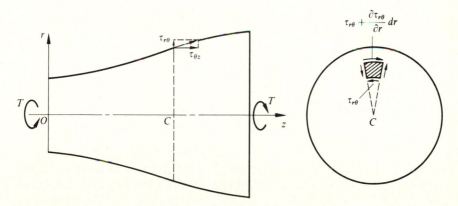

Figure 3.38 Torsion of a shaft of variable diameter.

axial plane passing through the point as in the case of a bar of uniform radius. Using cylindrical coordinates (r, θ, z), where the z axis coincides with the axis of the bar, the components of the displacement may be written as

$$u = 0 \qquad v = v(r, z) \qquad w = 0$$

The transverse sections of the bar remain plane and circular, but radial lines generally become curved during the torsion. The only nonzero strain components corresponding to the above displacement field are $\gamma_{r\theta}$ and $\gamma_{\theta z}$, where

$$2\gamma_{r\theta} = \frac{\partial v}{\partial r} - \frac{v}{r} \qquad 2\gamma_{\theta z} = \frac{\partial v}{\partial z}$$

By Hooke's law, all stress components are identically zero except $\tau_{r\theta}$ and $\tau_{\theta z}$, which are equal to $2G\gamma_{r\theta}$ and $2G\gamma_{\theta z}$ respectively. It follows from the above relations that

$$\frac{\partial}{\partial r}\left(\frac{v}{r}\right) = \frac{\tau_{r\theta}}{Gr} \qquad \frac{\partial}{\partial z}\left(\frac{v}{r}\right) = \frac{\tau_{\theta z}}{Gr} \tag{138}$$

These equations must be supplemented by the condition of statical equilibrium which requires

$$\frac{\partial \tau_{r\theta}}{\partial r} + \frac{\partial \tau_{\theta z}}{\partial z} + \frac{2\tau_{r\theta}}{r} = 0 \tag{139}$$

This is identically satisfied by introducing a stress function $\phi(r, z)$ such that

$$r^2 \tau_{r\theta} = -\frac{\partial \phi}{\partial z} \qquad r^2 \tau_{\theta z} = \frac{\partial \phi}{\partial r} \tag{140}$$

Substituting into (138) and eliminating v/r, we obtain the governing differential equation†

$$\frac{\partial^2 \phi}{\partial r^2} - \frac{3}{r}\frac{\partial \phi}{\partial r} + \frac{\partial^2 \phi}{\partial z^2} = 0 \tag{141}$$

Since the lateral surface of the bar is free from external forces, the resultant shear stress across an axial section at the boundary must be directed along the tangent to the boundary. Hence

$$\tau_{r\theta}\, dz - \tau_{\theta z}\, dr = 0$$

along the boundary. Substituting from (140), we obtain the boundary condition

$$\frac{\partial \phi}{\partial z}\, dz + \frac{\partial \phi}{\partial r}\, dr = d\phi = 0$$

which shows that ϕ has a constant value ϕ_0 along the boundary of the twisted bar. Since $\tau_{r\theta}$ must vanish along the axis of symmetry, ϕ is also constant along $r = 0$. We may assume $\phi = 0$ along $r = 0$ without loss of generality. The torque is then found

† This theory is due to J. H. Michell, *Proc. London Math. Soc.*, **31**: 141 (1899).

from the formula

$$T = 2\pi \int_0^a r^2 \tau_{\theta z} \, dr = 2\pi \int_0^a \frac{\partial \phi}{\partial r} \, dr = 2\pi \phi_0 \tag{142}$$

where a denotes the external radius of any cross section. When ϕ has been determined from the differential equation (141) and the boundary condition $\phi = \phi_0$, the stresses follow from (140) and the displacement from (138). The quantity v/r represents the angle of rotation of an elemental ring of radius r in a transverse section. Since v is not proportional to r, the angle of twist varies over any given cross section of the bar. The proportionality holds only in the special case of a circular cross section, which corresponds to $\phi = Tr^4/2\pi a^4$.

Consider, as an example, the torsion of a bar in the form of a truncated cone whose virtual apex is taken as the origin of coordinates. By geometry, the quantity $z/\sqrt{r^2 + z^2}$ has the constant value $\cos \alpha$ at the boundary, where α is the semiangle of the cone. Any function of this parameter would therefore satisfy the boundary condition. It is easily verified that the stress function

$$\phi = c \left\{ 2 - \frac{3z}{\sqrt{r^2 + z^2}} + \frac{z^3}{(r^2 + z^2)^{3/2}} \right\}$$

where c is a constant, satisfies the differential equation (141) and vanishes on $r = 0$. Substituting into (142) the boundary value of ϕ, we get

$$c = \frac{T}{2\pi(2 - 3\cos \alpha + \cos^3 \alpha)}$$

The lines of shearing in the elastic bar are radial lines drawn through the apex. By (140), the components of the shear stress are

$$\tau_{r\theta} = \frac{3cr^2}{(r^2 + z^2)^{5/2}} \qquad \tau_{\theta z} = \frac{3crz}{(r^2 + z^2)^{5/2}} \tag{143}$$

The displacement is readily obtained by the integration of (138). Assuming that an element at the centre of the smaller end $z = h$ is prevented from rotation, we find

$$v = \frac{cr}{Gh^3} \left\{ 1 - \frac{h^3}{(r^2 + z^2)^{3/2}} \right\}$$

The angle of twist v/r is constant over any spherical surface having its centre at the apex of the cone. The greatest resultant shear stress occurs at the pheriphery of the smaller end of the bar, the magnitude of this stress being $(3c/h^3) \sin \alpha \cos^3 \alpha$ in view of (143). Hence, plastic yielding begins at $r = h \tan \alpha$ and $z = h$ when the torque becomes

$$T_e = 2\pi k h^3 \left[\frac{2 - 3\cos \alpha + \cos^3 \alpha}{3 \sin \alpha \cos^3 \alpha} \right] \tag{144}$$

which is obtained by setting $c = kh^3/(3 \sin \alpha \cos^3 \alpha)$. Taking for instance $\alpha = 30°$, the value of T_e is found to be $0.322kh^3$.

(ii) *Elastic/plastic torsion* When the bar is twisted beyond the elastic limit, a plastic zone spreads inward from somewhere on the boundary. The stresses in the plastic region must satisfy the yield criterion $\tau_{r\theta}^2 + \tau_{\theta z}^2 = k^2$, which gives

$$|\text{grad } \phi| = \sqrt{\left(\frac{\partial \phi}{\partial r}\right)^2 + \left(\frac{\partial \phi}{\partial z}\right)^2} = kr^2$$

in view of (140). Let ψ denote the counterclockwise angle which the resultant shear stress vector over an axial plane makes with the z axis. Then the stress components in the plastic region may be written as

$$\tau_{r\theta} = k \sin \psi \qquad \tau_{\theta z} = k \cos \psi \tag{145}$$

Inserting these expressions into the equilibrium equation (139), we get

$$\frac{\partial \psi}{\partial r} - \tan \psi \frac{\partial \psi}{\partial z} + \frac{2}{r} \tan \psi = 0$$

This is a hyperbolic equation having characteristics in the directions $dz/dr = -\tan \psi$, which are orthogonal to the lines of shearing stress in the axial plane. The variation of ψ along a characteristic is

$$d\psi = \frac{\partial \psi}{\partial r} dr + \frac{\partial \psi}{\partial z} dz = \left(\frac{\partial \psi}{\partial r} - \tan \psi \frac{\partial \psi}{\partial z}\right) dr = -2 \tan \psi \frac{dr}{r}$$

Let (r_0, z_0) be the coordinates of the point where a given characteristic meets the external boundary, and α the angle of inclination of the local tangent to the boundary with respect to the z axis. The integration of the above equation then gives[†]

$$r^2 \sin \psi = r_0^2 \sin \alpha = \rho \quad \text{(say)} \tag{146}$$

along a characteristic, which is generally curved. In view of (145) and (146), the stress component $\tau_{r\theta}$ varies inversely as r^2 along a characteristic. The shape of the characteristic through (r_0, z_0) is given by

$$z - z_0 = \int_r^{r_0} \tan \psi \, dr = \rho \int_r^r (r^4 - \rho^2)^{-1/2} \, dr \tag{147}$$

To obtain the stress function in the plastic region, we combine (140) and (145) to write the derivatives of ϕ as

$$\frac{\partial \phi}{\partial z} = -k\rho \qquad \frac{\partial \phi}{\partial r} = k\sqrt{r^4 - \rho^2}$$

on eliminating ψ by means of (146). The variation of ϕ along a characteristic therefore becomes

$$d\phi = k(-\rho \, dz + \sqrt{r^4 - \rho^2} \, dr)$$

[†] W. W. Sokolovsky, *Prikl. Mat. Mekh.*, **9**: 343 (1945).

Since the boundary value of ϕ is $\phi_0 = T/2\pi$ in view of (142), the integration of the above equation results in

$$\phi = \frac{T}{2\pi} - k\left\{\rho(z - z_0) + \int_r^{r_0} \sqrt{r^4 - \rho^2}\, dr\right\} \tag{148}$$

In the elastic region, the stress function is obtained from the solution of (141), using the boundary condition $\phi = \phi_0$ along the meridian, and the condition of continuity of ϕ across the elastic/plastic boundary. The position of this unknown boundary is determined from the additional requirement that the first derivatives of ϕ must also be continuous. If the plastic zone is assumed to spread far enough, the characteristics emanating from the boundary will generally form an envelope, beyond which they cannot be continued. Consequently, the bar will not become fully plastic, no matter how large the angle of twist. For instance, in a conical bar of $15°$ semiangle, a region of $7°$ around the axis will never become plastic.†

The shear stress in any plastic element remains constant in magnitude and direction so long as changes in geometry are negligible. The only non-zero strain increments are $d\gamma_{r\theta}$ and $d\gamma_{\theta z}$ whose elastic parts are identically zero. The Prandtl-Reuss flow rule is therefore expressible in the integrated form

$$\frac{\gamma_{r\theta}}{\gamma_{\theta z}} = \frac{\tau_{r\theta}}{\tau_{\theta z}} = \tan \psi$$

Substituting for the stress and strain components, the above relationship may be expressed in the form

$$\frac{\partial v}{\partial r} - \tan \psi \frac{\partial v}{\partial z} - \frac{v}{r} = 0$$

which is a hyperbolic equation having the same characteristics as those for the stress. The variation of v along a characteristic is

$$dv = \frac{\partial v}{\partial r}\, dr + \frac{\partial v}{\partial z}\, dz = \left(\frac{\partial v}{\partial r} - \tan \psi \frac{\partial v}{\partial z}\right) dr = v \frac{dr}{r} \tag{149}$$

Hence $v/r = \text{const}$ along a characteristic. It follows that the surfaces of revolution formed by the characteristics in the plastic region undergo rotation without distortion during the twisting. The surfaces of constant angle of twist v/r are therefore orthogonal to the lines of shearing stress. This is equally true in the elastic region as may be seen by setting the variation of v/r to zero, using (138) for the partial derivatives.

For any given shape of the external boundary, the purely elastic distribution of stress may be determined numerically using the finite difference form of equation (141). This gives the magnitude and position of the maximum resultant shear stress

---

† The plastic stress distribution in a conical bar and a stepped bar has been discussed by W. W. Sokolovsky, op. cit. The elastic/plastic torsion of cylindrical bars with a circumferential notch has been investigated by J. B. Walsh and A. C. Mackenzie, *J. Mech. Phys. Solids*, **7**: 247 (1959).

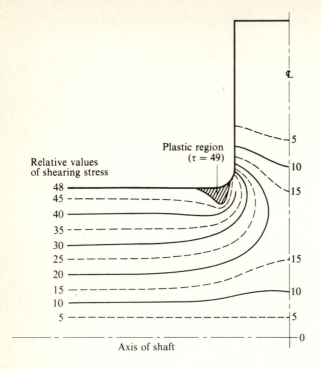

Relative values
of shearing stress

Plastic region
($\tau = 49$)

48	
45	
40	
35	
30	
25	
20	
15	
10	
5	

Axis of shaft

Figure 3.39 Yielding near a fillet in a cylindrical shaft provided with a collar (*after Eddy and Shaw*).

which must be equal to k at the commencement of plastic yielding. Choosing a suitable value of T exceeding the torque T_e at the elastic limit, the stress function in an assumed plastic region is evaluated numerically from (147) and (148). Using a trial-and-error procedure, the elastic/plastic boundary is determined from the conditions of continuity of ϕ and its first derivatives. Since the solution in the elastic region depends on the shape of the elastic/plastic boundary, an adjustment of this boundary must be accompanied by a recalculation of the elastic stress function.

Figure 3.39 shows contours of the resultant shear stress τ around the plastic zone when a cylindrical shaft with a collar† is subjected to a torque $T = 1.5T_e$. The width of the collar and the radius of the fillet are taken as $\frac{3}{4}a$ and $\frac{1}{8}a$ respectively, where a is the radius of the parallel portion of the shaft. Due to an elastic stress concentration factor of 1.53 existing at the fillet, yielding beings at a torque $T_e \simeq 0.327\pi ka^3$, which is 0.654 times the torque required for a uniform shaft of radius a. The plastic zone is found to grow rapidly when $T > 1.53T_e$, the state of full plasticity being attained when the torque becomes $\frac{2}{3}\pi ka^3$, which is equal to $2.04T_e$. Due to the large stress gradient in the neighborhood of the stress concentration, the extent of the plastic zone is fairly small even for a torque that is 1.5 times that at the initial yielding.

† R. P. Eddy and F. S. Shaw, *J. Appl. Mech.*, **16**: 139 (1949).

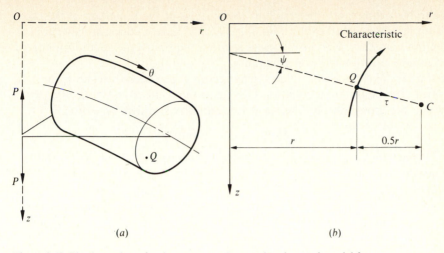

(a)   *(b)*

Figure 3.40 Plastic torsion of a ring sector under equal and opposite axial forces.

(iii) *Torsion of a ring sector* A somewhat similar problem arises when a circular ring sector of uniform cross section is twisted by two equal and opposite forces through the center of curvature and along the perpendicular to the plane of the ring. The problem is of some importance in the design of close-coiled helical springs.† Using cylindrical coordinates (r, θ, z) with the z axis taken in the direction of the applied force as indicated in Fig. 3.40*a*, the components of the velocity may be written as

$$u = 0 \qquad v = v(r, z) \qquad w = c\theta$$

where c is a constant. The circumferential velocity v represents the warping of the cross section. The nonzero components of strain rate associated with the velocity field are given by

$$2\dot{\gamma}_{r\theta} = \frac{\partial v}{\partial r} - \frac{v}{r} \qquad 2\dot{\gamma}_{\theta z} = \frac{\partial v}{\partial z} + \frac{c}{r} \tag{150}$$

The torsion problem for the elastic ring sector can be solved by expressing the shear stresses $\tau_{r\theta}$ and $\tau_{\theta z}$ by (140), so that the equilibrium equation (139) is identically satisfied. It is easy to show that the stress function ϕ satisfies the differential equation

$$\frac{\partial^2 \phi}{\partial r^2} - \frac{3}{r}\frac{\partial \phi}{\partial r} + \frac{\partial^2 \phi}{\partial z^2} = -2G\delta$$

where $\dot{\delta}$ is equal to c, which specifies the axial velocity of one end of the ring sector

† The elastic torsion of a ring sector of circular cross section has been investigated by W. Freiberger, *Aust. J. Sci. Res.*, A, **2**: 351 (1949). The elastic/plastic torsion has been discussed by W. Freiberger, *Q. Appl. Math.*, **14**: 259 (1956).

relative to the other. Since the resultant shear stress at the boundary of the cross section must be tangential to the boundary, ϕ must have a constant value at each point of the boundary. For a solid cross section, the boundary condition may be written as $\phi = 0$ without loss of generality.

In the case of a completely plastic ring sector, the nonzero stresses $\tau_{r\theta}$ and $\tau_{\theta z}$ must satisfy the equilibrium equation (139) and the yield criterion $\tau_{r\theta}^2 + \tau_{\theta z}^2 = k^2$ throughout the cross section. The latter is identically satisfied by writing

$$\tau_{r\theta} = k \cos \psi \qquad \tau_{\theta z} = k \sin \psi$$

where ψ is the clockwise angle made by the shear stress vector with the positive r axis. The substitution into (139) shows that the stress equation is hyperbolic with the characteristics $dz/dr = -\cot \psi$. The characteristic direction at a generic point of the cross section is therefore normal to the local stress vector. In view of the equilibrium equation expressed in terms of ψ, the variation of ψ along a characteristic is given by

$$\frac{d\psi}{ds} = \left(\frac{\partial \psi}{\partial r} - \cot \psi \frac{\partial \psi}{\partial z} \right) \frac{dr}{ds} = \frac{2}{r} \cos \psi \tag{151}$$

where ds is the line element of the characteristic, whose positive direction is obtained by rotating the stress vector through $90°$ in the counterclockwise sense. It follows from (151) that the radius of curvature of the characteristic is $(r/2) \sec \psi$. This means that the center of curvature C of the characteristic at any point Q is located on the line of action of the shear stress vector at Q, the radial distance of C from the z axis being 1.5 times that of Q from the z axis (Fig. 3.40b). A graphical construction of the characteristics and shearing lines is therefore possible by using a small arc process starting from the boundary of the cross section.

Using (140) for the stress components, and the boundary condition for ϕ, it is easily shown that the resultant radial shearing force over the cross section and the resultant moment of the shearing stresses about the origin O are identically zero. The stress distribution in the ring is therefore statically equivalent to an axial force

$$P = \iint \tau_{\theta z} \, dr \, dz = k \iint \sin \psi \, dr \, dz \tag{152}$$

The double integral can be evaluated numerically by summing the contribution from each curvilinear element of the orthogonal network formed by the characteristics and the shearing lines.

Since the direction of the shear stress vector does not change once an element becomes plastic, the ratio $\dot{\gamma}_{r\theta}/\dot{\gamma}_{\theta z}$ is equal to $\cot \psi$ by the stress-strain relation. Equations (150) therefore give

$$\frac{\partial v}{\partial r} - \cot \psi \left(\frac{\partial v}{\partial z} + \frac{c}{r} \right) - \frac{v}{r} = 0$$

This is a hyperbolic equation having the same characteristics as those of the stress

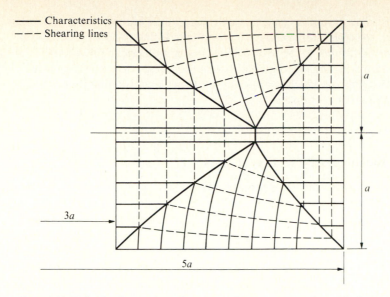

Figure 3.41 Fully plastic state for a ring sector of square cross section with characteristics and shearing lines (*after Wang and Prager*).

equation. The variation of v along a typical characteristic is given by

$$\frac{dv}{ds} = \left(\frac{\partial v}{\partial r} - \cot \psi \frac{\partial v}{\partial z}\right)\frac{dr}{ds} = \frac{1}{r}(v \sin \psi + c \cos \psi) \qquad (153)$$

which can be integrated numerically using the condition that v is continuous across a line of stress discontinuity. If the material is assumed rigid/plastic, the shear strain rate vanishes along a stress discontinuity, resulting in

$$dv = \frac{\partial v}{\partial r} dr + \frac{\partial v}{\partial z} dz = \frac{1}{r}(v \, dr - c \, dz) \qquad (154)$$

along this line. Since the warping displacement is determined to within an arbitrary constant, v may be assumed to vanish on an arbitrarily chosen point on the line of discontinuity, which bisects the angle between the intersecting shear lines.

As an example, we consider the fully plastic torsion of a ring sector whose cross section is a square of side $2a$, the radius of the center line being taken as $4a$. Figure 3.41 shows the characteristics and the shearing lines over the cross section.[†] Starting from a number of equidistant points on the sides of the square, the characteristics are gradually extended into the interior of the cross section by a succession of small circular arcs of appropriate radii. The characteristics emanating from the sides intersect in points which define the lines of discontinuity of the stress

† This solution is due to A. J. Wang and W. Prager, *J. Mech. Phys. Solids*, **8**: 169 (1955). The solution for a hollow cross section has been given by W. Freiberger and W. Prager, *J. Appl. Mech.* **23**, 461 (1956).

field. Two of these discontinuities pass through the corners of the square, bisecting the angle between the characteristics meeting on them. Using the distribution of ψ, the external force is found to be $P = 0.66ka^2$ by a numerical procedure. The contours of constant warping may be obtained by the integration of (153), starting from the discontinuities on which v/c is given by (154), and assuming v to vanish at the center of the cross section.

3.8 Combined Bending and Twisting of Bars

(i) *The exact yield point state* A prismatic bar whose cross section has two orthogonal axes of symmetry is rendered fully plastic by the combined action of bending couples M and twisting couples T applied at the ends. We choose a set of rectangular axes in which the x and y axes are along the axes of symmetry and z axis along the centroidal axis of the bar, the axis of the bending couple being taken to coincide with the x axis. The material is assumed to be rigid/plastic obeying the von Mises yield criterion and the Lévy-Mises flow rule. If the rates of bending and twisting per unit length are denoted by $\dot\psi$ and $\dot\theta$ respectively at the incipient plastic flow, the incompressible velocity field may be expressed as

$$u = -\tfrac{1}{2}xy\dot\psi - yz\dot\theta$$
$$v = \tfrac{1}{4}(x^2 - y^2 - 2z^2)\dot\psi + xz\dot\theta \tag{155}$$
$$w = yz\dot\psi + f(x, y)$$

so that the strain rates are independent of z. The unknown function f specifies the rate of warping of the cross section. The velocity field (155) is the effect of superposition of those corresponding to pure bending and pure torsion, with the neutral plane coinciding with the xz plane

The only nonzero stress components are σ_z, τ_{xz}, and τ_{yz}, which are required to satisfy the equilibrium equation $\partial\tau_{xz}/\partial x + \partial\tau_{yz}/\partial y = 0$, the yield criterion

$$\sigma_z^2 + 3(\tau_{xz}^2 + \tau_{yz}^2) = 3k^2$$

and the flow rule. The first two equations are identically satisfied by the existence of a stress function $\phi(x, y)$ such that

$$\tau_{xz} = k\phi_y \qquad \tau_{yz} = -k\phi_x \qquad \sigma_z = \pm\sqrt{3}\,k(1 - \phi_x^2 - \phi_y^2)^{1/2} \tag{156}$$

where ϕ_x and ϕ_y denote the partial derivatives $\partial\phi/\partial x$ and $\partial\phi/\partial y$ respectively. The nonzero components of the strain rate corresponding to (155) are

$$\dot\varepsilon_x = \dot\varepsilon_y = -\tfrac{1}{2}y\dot\psi \qquad \dot\varepsilon_z = y\dot\psi$$
$$\dot\gamma_{xz} = \frac{1}{2}\left(\frac{\partial f}{\partial x} - y\dot\theta\right) \qquad \dot\gamma_{yz} = \frac{1}{2}\left(\frac{\partial f}{\partial y} + x\dot\theta\right)$$

The substitution for the stresses and the strain rates into the Lévy-Mises flow rule furnishes the relationship between the kinematical and statical variables as

$$y\dot{\psi} = \pm\frac{2}{\sqrt{3}}k\dot{\lambda}(1 - \phi_x^2 - \phi_y^2)^{1/2}$$

$$\frac{\partial f}{\partial x} = 2k\dot{\lambda}\phi_y + y\dot{\theta} \qquad \frac{\partial f}{\partial y} = -2k\dot{\lambda}\phi_x - x\dot{\theta} \tag{157}$$

where $\dot{\lambda}$ is a nonnegative factor of proportionality. The first of these relations indicates that the upper sign corresponds to $y > 0$ and the lower sign to $y < 0$. The last two relations of (157) will be compatible if

$$k\frac{\partial}{\partial x}(\dot{\lambda}\phi_x) + k\frac{\partial}{\partial y}(\dot{\lambda}\phi_y) + \dot{\theta} = 0$$

Inserting the expression for $\dot{\lambda}$ given by the first equation of (157), and evaluating the partial derivatives, we obtain the nonlinear partial differential equation†

$$y\dot{\psi}[(1 - \phi_y^2)\phi_{xx} + 2\phi_x\phi_y\phi_{xy} + (1 - \phi_x^2)\phi_{yy}]$$

$$+ \dot{\psi}\phi_y(1 - \phi_x^2 - \phi_y^2) \pm \frac{2}{\sqrt{3}}\dot{\theta}(1 - \phi_x^2 - \phi_y^2)^{3/2} = 0 \tag{158}$$

For a given value of $\dot{\psi}/\dot{\theta}$, this elliptic equation can be solved numerically under the boundary condition $\phi = 0$ to obtain the stress function ϕ, which specifies the stress distribution. When $\dot{\psi} = 0$, the equation reduces to $\phi_x^2 + \phi_y^2 = 1$ obtained earlier for pure torsion.

The stress distribution in the bar is statically equivalent to a twisting couple T and a bending couple M. Using (156), the external couples are easily shown to be

$$T = \iint (x\tau_{yz} - y\tau_{xz})\,dx\,dy = 2k\iint \phi\,dx\,dx$$

$$M = \iint y\sigma_z\,dx\,dy = \sqrt{3}\,k\iint \pm y\sqrt{1 - \phi_x^2 - \phi_y^2}\,dx\,dy \tag{159}$$

These expressions define the relationship between T and M at the yield point parametrically through the ratio $\dot{\psi}/\dot{\theta}$.

Equation (158) may also be derived by a simple application of the principle of maximum plastic work. We begin with the fact that the rate of work done by the applied couples is

$$\dot{W} = l(M\dot{\psi} + T\dot{\theta})$$

where l is the total length of the bar. Substituting from (159), we get

$$\dot{W} = kl\iint [\pm\sqrt{3}\,y\dot{\psi}\sqrt{1 - \phi_x^2 - \phi_y^2} + 2\dot{\theta}\phi]\,dx\,dy$$

† G. H. Handelman, *Q. Appl. Mech.*, **1**: 351 (1944), and R. Hill, *Q. J. Mech. Appl. Math.*, **1**: 18 (1948). A numerical solution of the equation for a bar of square section has been presented by M. C. Steele, *J. Mech. Phys. Solids*, **3**: 156 (1955).

The function ϕ must be such that the integral becomes a maximum. By the Euler-Lagrange equation of the calculus of variations, the stationary value of the integral corresponds to

$$\frac{\partial F}{\partial \phi} - \frac{\partial}{\partial x}\left(\frac{\partial F}{\partial \phi_x}\right) - \frac{\partial}{\partial y}\left(\frac{\partial F}{\partial \phi_y}\right) = 0$$

where F represents the integrand, and the result then reduces to (158). If the variational method is adopted, it is necessary to verify that the velocity field is associated with the stress field and that the rate of plastic work is nowhere negative.

The analysis is readily extended to include an axial force N applied to the bar. The velocity field then contains additional terms $-\frac{1}{2}x\dot{\varepsilon}$, $-\frac{1}{2}y\dot{\varepsilon}$ and $z\dot{\varepsilon}$ for u, v, and w respectively, where $\dot{\varepsilon}$ denotes a uniform rate of extension. The neutral plane is given by $y\dot{\psi} + \dot{\varepsilon} = 0$, and Eq. (158) is modified by writing $y\dot{\psi} + \dot{\varepsilon}$ in place of $y\dot{\psi}$. The applied force N is equal to $\iint \sigma_z \, dx \, dy$, where σ_z is given by (156) with the upper sign holding for $y\dot{\psi} + \dot{\varepsilon} > 0$ and the lower sign for $y\dot{\psi} + \dot{\varepsilon} < 0$. The relationship between N, T, and M at the yield point depends on the ratios $\dot{\psi}/\dot{\theta}$ and $\dot{\varepsilon}/\dot{\theta}$.

(ii) *Lower bound approximations* Let T_0 denote the fully plastic torque under pure torsion, and M_0 the fully plastic moment under pure bending about the considered axes of symmetry. The former is associated with a shear stress of magnitude k throughout the cross section, while the latter involves normal stresses $\pm\sqrt{3}\,k$ on opposite sides of the neutral plane. To obtain a lower bound on the yield point couples under combined loading, we assume a distribution of constant shear stress $\tau < k$ similar to that in pure torsion, and combine this with a distribution of normal stress of constant magnitude $\sigma < \sqrt{3}\,k$ similar to that in pure bending.[†] Then

$$\frac{T}{T_0} = \frac{\tau}{k} \qquad \frac{M}{M_0} = \frac{\sigma}{\sqrt{3}\,k}$$

Since the fictitious stress state must not violate the yield criterion, the best approximation corresponds to $\sigma^2 + 3\tau^2 = 3k^2$, which gives[‡]

$$\left(\frac{T}{T_0}\right)^2 + \left(\frac{M}{M_0}\right)^2 = 1 \tag{160}$$

Consider now the effect of an axial force N, whose yield point value is N_0 when the force is acting alone. In the absence of torsion, the stress state at the yield point consists of longitudinal stresses $\pm\sqrt{3}\,k$ on opposite sides of the neutral plane,

---

[†] The lower and upper bound approximations discussed here are due to R. Hill and M. P. L. Siebel, *J. Mech. Phys. Solids*, **1**: 207 (1953).

[‡] For applications of this equation to the plastic analysis of grillages, see J. Heyman, *J. Appl. Mech.*, **18**: 157 (1951), **19**: 153 (1952). See also R. Sankaranarayanan and P. G. Hodge, Jr., *J. Mech. Phys. Solids*, **7**: 22 (1959). The plastic torsion of I beams with warping restraint has been investigated by N. S. Boulton, *Int. J. Mech. Sci.*, **4**: 491 (1962), K. S. Dinno and S. S. Gill, ibid., **6**: 27 (1964), and G. Augusti, ibid., **8**: 541 (1966).

which is no longer a plane of symmetry of the cross section. The resultant force and moment acting over the section can be calculated for any assumed position of the neutral axis. The relationship between M and N at the yield point may be put in the form

$$f\left(\frac{M}{M_0}, \frac{N}{N_0}\right) = 0$$

where f is a dimensionless function depending on the shape of the cross section. When a torque T is present, the magnitude of the longitudinal stress is reduced to σ without changing the neutral axis. The relationship between the corresponding bending moment and axial force becomes

$$f\left(\frac{\sqrt{3}\,kM}{\sigma M_0}, \frac{\sqrt{3}\,kN}{\sigma N_0}\right) = 0 \tag{161}$$

which must be combined with the result $T/T_0 = \tau/k$. The elimination of σ and τ by means of the yield criterion then leads to a lower bound relationship between M/M_0, T/T_0, and N/N_0 at the yield point. It represents a surface lying inside that corresponding to the actual yield point.

Consider the yield point state of a bar, whose cross section is a circle of radius a, under the action of a bending moment M and an axial tension N. If $a \sin \beta$ denotes the distance of the neutral plane and $a \sin \phi$ that of any parallel plane from the longitudinal axis, it is easily shown that

$$M = 4\sqrt{3}\,ka^2 \int_0^\beta \cos^2 \phi \, d\phi = \frac{N_0}{\pi}(2\beta + \sin 2\beta)$$

$$M = 4\sqrt{3}\,ka^3 \int_\beta^{\pi/2} \cos^2 \phi \sin \phi \, d\phi = M_0 \cos^3 \beta$$

where $N_0 = \sqrt{3}\,\pi ka^2$ and $M_0 = 4ka^3/\sqrt{3}$. The relationship between N/N_0 and M/M_0 defined by the above expressions may be written approximately as

$$\frac{M}{M_0} + \left(\frac{N}{N_0}\right)^2 = 1 \tag{162}$$

which underestimates the magnitudes of M and N, for a given value of their ratio, by less than 1.5 percent. The above formula is in fact exact for a rectangular cross section with the appropriate values of M_0 and N_0 (see Sec. 4.7(i)). When a bending moment M, a twisting moment T, and an axial force N act simultaneously, it follows from (161) and (162) that

$$\left(\frac{N}{N_0}\right)^2 + \frac{\sigma}{\sqrt{3}\,k}\left(\frac{M}{M_0}\right) = \frac{\sigma^2}{3k^2} \qquad \left(\frac{T}{T_0}\right)^2 = \frac{\tau^2}{k^2} = 1 - \frac{\sigma^2}{3k^2}$$

where $T_0 = 2\pi ka^3/3$ in the case of a circular cross section. Eliminating $\sigma/\sqrt{3}\,k$

between these two equations, we obtain the lower bound approximation

$$\left(\frac{N}{N_0}\right)^2 + \left(\frac{T}{T_0}\right)^2 + \frac{M}{M_0}\sqrt{1 - \frac{T^2}{T_0^2}} = 1 \tag{163}$$

This equation reduces to (160) when $N = 0$. In the case of combined tension and torsion ($M = 0$) of a circular cylindrical bar, the exact yield point solution is obtained from (21) by setting $c = 0$. The lower bound solution given by (163) with $M = 0$ differs from the exact solution by less than 1.5 percent. For an arbitrary shape of the cross section, (163) may be regarded as a reasonable approximation (not necessarily a lower bound) to the actual yield point state under combined loading.

(iii) *Estimation of upper bounds* To obtain an upper bound solution for the yield point state, we assume a fictitious deformation consisting of a bending rate $\dot{\psi}$ and a twisting rate $\dot{\theta}$ per unit length of the bar without any warping of the cross section.[†] The velocity distribution is then given by (155) with $f = 0$, and the corresponding components of the strain rate at any point are

$$\dot{\varepsilon}_x = \dot{\varepsilon}_y = -\tfrac{1}{2}y\dot{\psi} \qquad \dot{\varepsilon}_z = y\dot{\psi} \qquad \dot{\gamma}_{xz} = -\tfrac{1}{2}y\dot{\theta} \qquad \dot{\gamma}_{yz} = \tfrac{1}{2}x\dot{\theta}$$

The rate of dissipation of internal energy per unit volume is $\sqrt{3}\,k\dot{\bar{\varepsilon}}$, where $\dot{\bar{\varepsilon}}$ is the equivalent strain rate expressed as

$$\dot{\bar{\varepsilon}}^2 = \tfrac{2}{3}(\dot{\varepsilon}_x^2 + \dot{\varepsilon}_y^2 + \dot{\varepsilon}_z^2 + 2\dot{\gamma}_{xz}^2 + 2\dot{\gamma}_{yz}^2) = y^2\dot{\psi}^2 + \tfrac{1}{3}(x^2 + y^2)\dot{\theta}^2 \tag{164}$$

Since the rate of external work done per unit length of the bar is equal to $M\dot{\psi} + T\dot{\theta}$, upper bounds on the applied couples are given by

$$M\dot{\psi} + T\dot{\theta} = \sqrt{3}\,k\int \dot{\bar{\varepsilon}}\,dA = \sqrt{3}\,k\int \sqrt{y^2\dot{\psi}^2 + \tfrac{1}{3}(x^2 + y^2)\dot{\theta}^2}\;dA$$

where the integral is taken over the whole cross section of the bar. The partial differentiation of this equation with respect to $\dot{\psi}$ and $\dot{\theta}$, corresponding to a given M/T ratio, furnishes the best upper bound solution

$$M = 3k\iint \frac{\alpha y^2\,dx\,dy}{\sqrt{(1 + 3\alpha^2)y^2 + x^2}} \qquad T = k\iint \frac{(x^2 + y^2)\,dx\,dy}{\sqrt{(1 + 3\alpha^2)y^2 + x^2}} \tag{165}$$

where $\alpha = \dot{\psi}/\dot{\theta}$. The integrals can be evaluated for any given cross section to obtain M and T in terms of the parameter α. The above expressions are precisely those obtained as the moment resultants of the stresses associated with the fictitious strain rates. According to the Lévy-Mises flow rule and the von Mises yield criterion, the nonzero stresses are

$$\sigma_z = \mu y\dot{\psi} \qquad \tau_{xz} = -\tfrac{1}{3}\mu y\dot{\theta} \qquad \tau_{yz} = \tfrac{1}{3}\mu x\dot{\theta}$$

---

† Improved upper bounds can be obtained by including suitable warping functions. See F. A. Gaydon and H. Nuttall, *J. Mech. Phys. Solids*, **6**: 17 (1957).

where μ is a positive multiplier given by

$$\mu^2[y^2\dot{\psi}^2 + \tfrac{1}{3}(x^2 + y^2)\dot{\theta}^2] = 1$$

These stresses do not satisfy the conditions of equilibrium, and give rise to transverse shearing forces and tangential surface tractions which violate the boundary conditions. Nevertheless, the resultant moments produced by these stresses coincide with (165).

When the cross section of the bar is a circle of radius a, it is convenient to use polar coordinates (r, ϕ) with the transformation $x = r \cos \phi$, $y = r \sin \phi$. Equations (165) then reduce to

$$\frac{M}{M_0} = \int_0^{\pi/2} \frac{\sqrt{3}\,\alpha \sin^2 \phi \, d\phi}{\sqrt{1 + 3\alpha^2 \sin^2 \phi}} \qquad \frac{T}{T_0} = \frac{2}{\pi} \int_0^{\pi/2} \frac{d\phi}{\sqrt{1 + 3\alpha^2 \sin^2 \phi}} \qquad (166)$$

These are incomplete elliptic integrals, which are more convenient to evaluate numerically than from available tables. The results given in the following table indicate that the sum of the squares of M/M_0 and T/T_0 exceeds the lower bound value of unity by less than 4 percent.

$\dot{\psi}/\dot{\theta}$	0.000	0.408	0.577	1.000	1.826	5.774	∞
M/M_0	0.000	0.476	0.602	0.776	0.900	0.982	1.000
T/T_0	1.000	0.904	0.835	0.688	0.502	0.235	0.000

For a square cross section with sides $x = \pm a$ and $y = \pm a$, the values of M_0 and T_0 are $2\sqrt{3}\,ka^3$ and $8ka^3/3$ respectively. The integrals in (165) are exactly evaluated, and the relationship between M and T is obtained in the parameteric form

$$\frac{M}{M_0} = \frac{\alpha}{\sqrt{3}} \left\{ \frac{\sqrt{1+\omega}}{\omega} + 2 \ln \left(\frac{1 + \sqrt{1+\omega}}{\sqrt{\omega}} \right) - \frac{\sinh^{-1} \sqrt{\omega}}{\omega\sqrt{\omega}} \right\}$$

$$\frac{T}{T_0} = \frac{1}{4} \left\{ \frac{(1+\omega)^{3/2}}{\omega} + (2-\omega) \ln \left(\frac{1 + \sqrt{1+\omega}}{\sqrt{\omega}} \right) + (2\omega - 1) \frac{\sinh^{-1} \sqrt{\omega}}{\omega\sqrt{\omega}} \right\}$$

$$(167)$$

where $\omega = 1 + 3\alpha^2$. The upper bound approximations for the circular and square cross sections are compared with the lower bound approximation in Fig. 3.42. The solid circles are based on the numerical solution of Eq. (158) for the square cross section. The maximum divergence between the upper and lower bounds occurs in pure torsion for a bar of square cross section, the corresponding error in the upper bound being about 14 percent. In the case of a rectangular cross section with

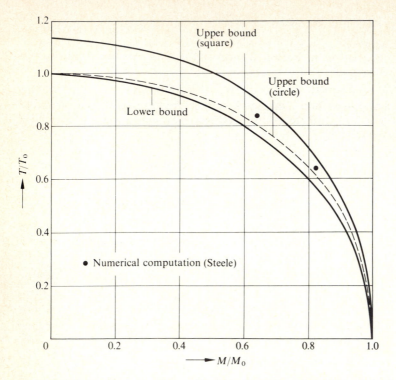

Figure 3.42 Interaction curves for combined bending and torsion of bars of circular and square cross sections.

arbitrary ratio of its sides, the maximum error in the upper bound based on (165) would be less than 14 percent.

According to the maximum work principle, $\dot{\psi}/\dot{\theta}$ is equal to the ratio of $\partial g/\partial M$ to $\partial g/\partial T$, where $g(M, T)$ defines the relationship between M and T at the yield point. In geometrical terms, the vector $(\dot{\psi}, \dot{\theta})$ is directed along the exterior normal to the interaction curve obtained by plotting T against M corresponding to the yield point state. If (160) is taken as an approximation to $g(M, T)$, we immediately get $\dot{\psi}/\dot{\theta} = \pi^2 M/12T$ in the case of a circular cross section. This incidentally is the exact value of $\dot{\psi}/\dot{\theta}$ for a thin-walled cylindrical tube stressed to the yield point.

Problems

3.1 A solid cylindrical bar made of a work-hardening material with an initial shear yield stress k is subjected to pure torsion until the torque has the fully plastic value corresponding to no work-hardening. Assuming a constant plastic modulus $H = G/3$, find the associated twist ratio θ/θ_e. If the bar is fully unloaded from the partly plastic state, compute the residual shear stress at the external boundary in terms of the yield stress k.

Answer: $\theta/\theta_e = 1.825$, $\tau/k = -0.251$.

3.2 A hollow cylindrical bar of external radius a and internal radius $0.5a$ is twisted to full plasticity and the applied torque is subsequently released. The material is nonhardening and obeys the von Mises yield criterion. If the unloaded bar is subjected to a sufficiently large axial tension, show that yielding will restart at the inner radius. Compute the tensile stress σ at the yield point, and the residual twist θ per unit length after the unloading.

Answer: $\sigma = 0.926\,Y$, $\theta = 0.756k/Ga$.

3.3 A solid circular cylinder of radius a is rendered partially plastic to a radius b in pure torsion. An increasing axial tension is then applied to the bar while the angle of twist is held constant. Assuming an incompressible and nonhardening Prandtl-Reuss material, show that the stress distribution in the region $b \leqslant r \leqslant a$ is given by

$$\frac{\sigma}{Y} = \tanh\left(\frac{3G}{Y}\varepsilon\right) \qquad \frac{\sqrt{3}\,\tau}{Y} = \text{sech}\left(\frac{3G}{Y}\varepsilon\right)$$

where ε is the longitudinal strain. Show that the axial stress in the additional plastic region is given by Eq. (24) with a replaced by b.

3.4 A thin-walled tube made of a linearly work-hardening Prandtl-Reuss material is initially twisted to the yield point, and subsequently extended with the angle of twist held constant. Show that the relationship between the longitudinal strain ε and the axial stress σ is given by

$$\frac{6G}{Y}\varepsilon = \frac{3\sigma}{(1+v)Y} + n\int_Y^{\bar{\sigma}}\frac{\sigma\,d\bar{\sigma}}{Y\bar{\sigma}} \qquad \frac{\sigma}{Y} = \left\{\left(\frac{\bar{\sigma}}{Y}\right)^2 - \left(\frac{Y}{\bar{\sigma}}\right)^n\right\}^{1/2}$$

where $\bar{\sigma}$ is the equivalent stress and $n = 6G/H$. Obtain a graphical plot of σ/Y against $6G\varepsilon/Y$ over the range $0 \leqslant \sigma/Y \leqslant 0.99$, and compare it with that for a nonhardening material of yield stress Y.

3.5 A nonhardening thin-walled tube of mean radius a, made of a Prandtl-Reuss material, is subjected to pure bending until the extreme fibers are just stressed to the yield point. The tube is then twisted with a gradually increasing torque T holding the angle of bend constant. Show that the bending couple M decreases according to the relation

$$\frac{M}{T_0} = \frac{\sqrt{3}}{\pi}\left(\cos^{-1}\frac{T}{T_0} + \frac{T}{T_0}\sqrt{1 - \frac{T^2}{T_0^2}}\right)$$

where T_0 denotes the yield torque. Prove that the warping w in the plastic region, whose angular span is specified by the core angle 2α, is given by

$$\Delta\left(\frac{Ew}{ka}\right) = 3\left(1 - \frac{2\phi}{\pi}\right)\left\{\left[\sin\alpha - \alpha\cos\alpha\right]_\alpha^\phi - \int_\alpha^\phi \alpha\,\text{cosec}\,\alpha\,d\alpha\right\} \qquad \alpha \leqslant \phi \leqslant \frac{\pi}{2}$$

3.6 A thin-walled tube, whose cross section is a square of side $2a$, is bent to an angle $\psi = 3k/Ea$ per unit length about an axis passing through the midpoints of two opposite sides. The tube is then twisted in the plastic range while the angle of bend is kept constant. If the material is ideally plastic, show that

$$\frac{M}{T_0} = \frac{1}{6\sqrt{3}}\left(13 + \frac{T^2}{2T_0^2}\right)\sqrt{1 - \frac{T^2}{T_0^2}}$$

Assuming the material to be incompressible, and using the Prandtl-Reuss flow rule, show that the torque-twist relationship is

$$\frac{Ga\theta}{k} = \tanh^{-1}\frac{T}{T_0} + \frac{1}{2\sqrt{3}}\left(\frac{T}{T_0}\sqrt{1 - \frac{T^2}{T_0^2}} - \sin^{-1}\frac{T}{T_0}\right)$$

3.7 Consider the plastic instability of a thin-walled circular cylinder under combined internal pressure and axial load. It may be assumed that the instability is caused by either the internal pressure or the resultant axial force attaining a maximum. Assuming further that the stress ratio remains constant at the onset of instability, show that the critical subtangent to the generalized stress-strain curve for a Mises material is

$$s = 2\left\{1 - \frac{\sigma_\theta}{\sigma_z} + \left(\frac{\sigma_\theta}{\sigma_z}\right)^2\right\}^{1/2} \Big/ \left(2 - \frac{\sigma_\theta}{\sigma_z}\right) \qquad \frac{\sigma_\theta}{\sigma_z} \leqslant 0.5$$

$$s = \frac{2}{3}\left\{1 - \frac{\sigma_z}{\sigma_\theta} + \left(\frac{\sigma_z}{\sigma_\theta}\right)^2\right\}^{1/2} \qquad \frac{\sigma_z}{\sigma_\theta} \leqslant 2.0$$

where σ_z and σ_θ are the axial and circumferential stresses respectively at the point of instability. Show a graphical comparison of this solution with that given by Eq. (44).

3.8 An open-ended thin-walled tube is subjected to an internal pressure p and an axial tensile load N in the plastic range. Plastic instability is assumed to occur when p and N simultaneously attain their maximum values. Using the empirical equation $\bar{\sigma} = C\bar{\varepsilon}^n$, show that the total equivalent strain at the onset of instability is

$$\bar{\varepsilon} = \frac{2n(1 - \beta + \beta^2)^{3/2}}{(1 + \beta)(3 - 4\beta + 2\beta^2)}$$

where β is the final ratio of the axial stress to the hoop stress. Use the von Mises yield criterion and the Lévy-Mises flow rule to derive the result.

3.9 In the pure bending of an elastic/plastic beam, the moment-curvature relationship approaches a limit, which is obtained by assuming the plastic zone to spread over the whole cross section. For a work-hardening material, it is convenient to use the empirical equation $\sigma = C\varepsilon^n$ for the derivation of the limiting moment. Show that in the case of a circular cross section of radius a, the limiting moment is

$$M^* = \frac{2Ca^3\sqrt{\pi}}{3 + n}\left(\frac{a}{R}\right)^n \frac{\Gamma[(2 + n)/2]}{\Gamma[(3 + n)/2]}$$

where $\Gamma(x)$ is the gamma function of a positive quantity x. Note that for $n = 0$, the above expression reduces to the fully plastic moment of a nonhardening beam.

3.10 A nonhardening beam of rectangular cross section is bent about an axis of symmetry to an elastic/plastic curvature equal to κ_0. The beam is then unloaded and reloaded in the opposite sense until plastic deformation again occurs. Show that the new elastic/plastic phase involves a curvature $\kappa \leqslant \kappa_0 - 2\kappa_e$, where κ_e corresponds to the initial yielding, and that the moment-curvature relationship becomes

$$\frac{M}{M_e} = -\frac{1}{2}\left[3 + \left(\frac{\kappa_e}{\kappa_0}\right)^2\right] + \left(\frac{2\kappa_e}{\kappa_0 - \kappa}\right)^2$$

Verify that for $\kappa \leqslant -\kappa_0$, the beam behaves as though the initial positive loading had never taken place.

3.11 A beam of rectangular cross section having a width b and depth $2h$ is bent about an axis parallel to the width. The material is linearly work-hardening with an initial yield stress Y and a tangent modulus T. Show that the moment-curvature relationship during an elastic/plastic bending may be written as

$$\frac{M}{M_e} = \frac{1}{2}\left(1 - \frac{T}{E}\right)\left[3 - \left(\frac{R}{R_e}\right)^2\right] + \frac{TR_e}{ER}$$

If the beam is completely unloaded from a state in which half the cross section is rendered plastic, show that the residual curvature is $5(E - T)/8E$ times the curvature of the beam at the initial yielding.

3.12 A prismatic beam of square cross section, made of an ideally plastic material, is bent about a diagonal in the elastic/plastic range. Show that the initial yield moment is $M_e = 2\sqrt{2}\,a^3 Y/3$, where $2a$ is the length of each side of the square, and that the moment-curvature relationship for the partially plastic beam is

$$\frac{M}{M_e} = 2 - 2\left(\frac{R}{R_e}\right)^2 + \left(\frac{R}{R_e}\right)^3$$

If the applied moment is released after half the area of the cross section has become plastic, find the limits between which a reloading moment M can vary without causing further plastic flow.
Answer: $-0.147M_e \leqslant M \leqslant 1.853M_e$.

3.13 An I section may be regarded as a combination of three rectangles with an overall depth $2h$, flange width b, flange thickness t_f and web thickness t_w. Show that the fully plastic moment for bending of the I section beam about its strong axis is

$$M_0 = bt_f(2h - t_f)Y + t_w(h - t_f)^2 Y$$

The influence of a transverse shear force F on the plastic moment may be approximately estimated by assuming the shear stress to be zero in the flanges and uniformly distributed in the web. Show that the reduced plastic moment is

$$M'_0 = M_f + \left\{M_w\left(M_w - \frac{F^2}{Yt_w}\right)\right\}^{1/2} \qquad F \leqslant Yt_w(h - t_f)$$

where M_f and M_w are the contributions of the flanges and the web to the plastic moment without shear.

3.14 A T section consists of two equal rectangles of length h and thickness t. The beam is made of a material whose yield stresses in tension and compression are Y and $1.5Y$ respectively. If the beam is bent about an axis perpendicular to the web, show that the fully plastic moment is $Yht(0.5h + 0.7t)$ when the tip of the web is in tension and to $Yht(0.7h + 0.5t)$ when it is in compression. Assuming $h/t = 9$, determine the shape factors in the two cases.
Answer: 1.88, 1.64.

3.15 A beam of Z section shown in Fig. A is folded from a thin uniform sheet where t/a is small compared to unity. Determine the angles α_1 and α_2 which define the directions of the principal axes of plastic bending. Find also the angle α which the strong principal axis of elastic bending makes with the z axis. Compute the angle of inclination β of the neutral axis to the y axis when the fully plastic moment is about the z axis. Draw the yield locus and show the directions of the plastic principal axes.
Answer: $\alpha_1 = 18.4°$, $\alpha_2 = 17.4°$, $\alpha = 22.5°$, $\beta = 35.3°$.

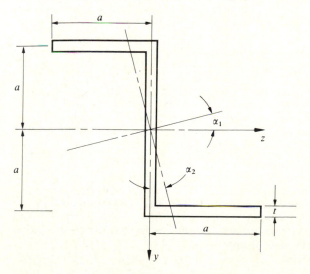

Figure A

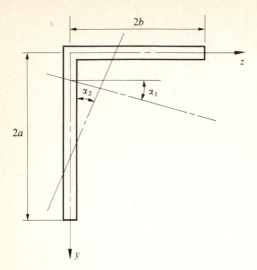

Figure B

3.16 Figure B shows an idealized angle section composed of two rectangles of uniform wall thickness t, which is small compared to the leg lengths $2a$ and $2b$. Show that the direction of the strong principal axis of plastic bending is given by $\tan \alpha_1 = 2b^2/(a^2 + 2ab - b^2)$, while that of the weak principal axis is given by

$$\tan \alpha_2 = \frac{2a^2(1 + \tan \alpha_2)^2 - (a + b)^2}{2b^2(1 + \tan \alpha_2)^2 - (a + b)^2 \tan^2 \alpha_2}$$

Assuming $b/a = 0.75$, compute the values of α_1 and α_2. Sketch the yield locus, indicating the directions of the plastic principal axes.

Answer: $\alpha_1 = 30.1°$, $\alpha_2 = 19.4°$.

3.17 A uniform beam of length $2l$ is simply supported at both ends and symmetrically loaded by two equal loads W at distances $l/2$ from the center. The material of the beam is nonhardening and its cross section is rectangular. Show that the central deflection of the beam at the commencement of plastic yielding is $\delta_e = 11l^2/24R_e$. Sketch the shape of the elastic/plastic boundaries when W exceeds the value W_e at the initial yielding, and obtain the load-deflection relationship

$$\frac{\delta}{\delta_e} = \frac{2}{11}\left\{5 - \left(3 + \frac{W}{W_e}\right)\left(3 - \frac{2W}{W_e}\right)^{1/2}\right\}\left(\frac{W_e}{W}\right)^2 + \frac{9}{11}\left(3 - \frac{2W}{W_e}\right)^{-1/2}$$

Note that δ tends to infinity as W approaches the limiting value $1.5W_e$ corresponding to plastic collapse.

3.18 A propped cantilever of length l, having a rectangular cross section, carries a concentrated load W at the midspan. Show that yielding begins at the built-in end when the load and the central deflection become

$$W_e = \frac{16M_e}{3l} \qquad \delta_e = \frac{7l^2}{144R_e}$$

Assuming that the reactant moment at the built-in end varies directly as the load in the initial stages of elastic/plastic bending, find the value of W when the central cross section of the beam begins to yield. Neglecting work-hardening, calculate the load for which the built-in section just becomes fully plastic, as well as the load that corresponds to plastic collapse.

Answer: $W/W_e = 1.20, 1.62, 1.69$.

3.19 A uniform beam, built-in at both ends, carries a load W uniformly distributed overs its length $2l$. The beam is assumed to have an idealized section whose shape factor is unity and plastic moment is M_0. Show that the central deflection of the beam is equal to $M_0 l^2/8EI$ when plastic hinges form at the ends of the beam, and to $M_0 l^2/3EI$ when it is at the point of collapse. If the beam is completely unloaded from the state of incipient collapse, show that there will be a residual deflection of amount $M_0 l^2/6EI$ at the center, and a uniform residual moment $M_0/3$ throughout the beam.

3.20 A nonhardening beam of length $3l$ has built-in ends and carries a concentrated load W at a distance l from each end. Assuming a shape factor of unity, show that yield hinges begin to form at the ends of the beam when the load is $W = 3M_0/2l$ and the central deflection is $\delta = 5M_0 l^2/16EI$, where M_0 is the plastic moment. Show also that the carrying capacity of the beam is exhausted when the load and the central deflection become $W = 2M_0/l$ and $\delta = 19M_0 l^2/24EI$ respectively. Verify that the residual deflection is $\delta = 9M_0 l^2/24EI$ when the beam is fully unloaded from the point of collapse.

3.21 A circular plate of radius a and thickness $2h$ is bent by an allround couple M per unit circumference. Show that the middle plane of the plate is bent into a spherical surface with a radius of curvature R, which is equal to $2Eh^3/3(1 - v)M$ in the elastic range. If the depth of the elastic core is denoted by $2c$ when the plate is partially plastic, show that

$$R = \frac{Ec}{(1 - v)Y} \qquad M = Y\left(h^2 - \frac{c^2}{3}\right)$$

Using cylindrical coordinates (r, z), where $z = 0$ represents the middle surface, show that the axial displacement w in the plastic region is given by

$$\frac{Ew}{Y} = (1 - v)\left(\frac{a^2 - r^2}{2c} - \frac{z^2}{c}\right) + (1 - 2v)(2|z| - c)$$

if the boundary condition is assumed as $w = 0$ at $z = 0$ and $r = a$.

3.22 Consider the elastic torsion of a circular cylindrical bar of radius a, containing a semicircular longitudinal groove of radius b as shown in Fig. C. Show that the torsion problem is solved by the stress function

$$\phi = -\frac{1}{2}G\theta(r^2 - b^2)\left(1 - \frac{2a}{r}\cos\psi\right)$$

expressed in plane polar coordinates (r, ψ). Find the distribution of shear stress in the bar, and show that the greatest resultant shear stress is of magnitude $G\theta(2a - b)$ occurring at the bottom of the groove.

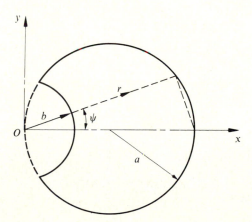

Figure C

3.23 The cross section of a prismatic bar is an equilateral triangle, the length of each side being $2a$. Using the coordinate system of Fig. 3.35, show that the stress function

$$\phi = \frac{1}{2} G\theta \left\{ \frac{4}{9} a^2 - (x^2 + y^2) - \frac{x}{\sqrt{3}\,a} (x^2 - 3y^2) \right\}$$

vanishes along the boundary of the cross section, and provides a valid solution for elastic torsion. Find the stress distribution in the bar and hence derive the warping function

$$w = \frac{\sqrt{3}\,\theta}{2a} \left(x^2 - \frac{y^2}{3} \right) y$$

Show that the applied torque T is $3G\theta/5$ times the polar moment of inertia of the cross section.

3.24 The stresses in the plastic region of a twisted bar may be written as $\tau_{xz} = k \cos \alpha$ and $\tau_{yz} = k \sin \alpha$, where α is the angle made by the shear stress vector with the x axis in the counterclockwise sense. Show that the fully plastic stress distribution in a bar of elliptic cross section of semiaxes a and b is given implicitly by

$$x \cos \alpha + y \sin \alpha = \frac{(a^2 - b^2) \sin \alpha \cos \alpha}{\sqrt{a^2 \sin^2 \alpha + b^2 \cos^2 \alpha}}$$

Verify that the fully plastic stress state involves a straight discontinuity extending between the centers of curvature for the ends of the major axis. Obtain an expression for the warping of the cross section.

3.25 The problem of elastic/plastic torsion of a bar, whose cross section is approximately an equilateral triangle, can be solved by taking the stress function in the elastic region to be identical to that in Prob. 3.23, with a replaced by an arbitrary constant $\sqrt{3}\,c$. Using polar coordinates (r, α), show that the elastic/plastic boundary Γ is given by the equation

$$r^2 \left(1 + \frac{r}{c} \cos 3\alpha + \frac{r^2}{4c^2} \right) = \frac{k^2}{G^2 \theta^2}$$

Assuming $\theta = k/Gc$, draw the characteristics through suitable points on Γ to obtain the shape of the external contour shown in Fig. D. What is the physical significance of c?

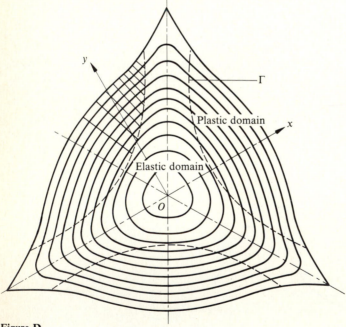

Figure D

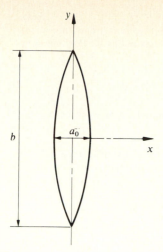

Figure E

3.26 A prismatic bar has a thin symmetrical cross section bounded by a pair of parabolas as shown in Fig. E, the width of the section at a generic height y being

$$a = a_0\left(1 - \frac{4y^2}{b^2}\right)$$

The stress function for purely elastic torsion of the bar may be written as $\phi = G\theta(a^2/4 - x^2)$ to a close approximation. Show that the applied torque is $T = 0.152G\theta a_0^3 b$. Find the ratio of the torque necessary to initiate plastic yielding in the bar to that required for a rectangular section of length b and width a_0.

3.27 The cross section of a thin-walled tube of uniform wall thickness t is shown in Fig. F. Using the membrane analogy for torsion, find the distribution of shear stress when the tube is purely elastic. Hence show that the twisting moment and the specific angle of twist at the initial yielding are given by

$$T_e = \tfrac{16}{3}ka^2t \qquad Ga\theta_e = \tfrac{7}{6}k$$

Show also that the shear stress vanishes in the internal webs at the incipient collapse, and determine the corresponding values of the torque and the twist.

   Answer: $T_0 = 6ka^2t$, $\theta_0 = 1.5k/Ga$.

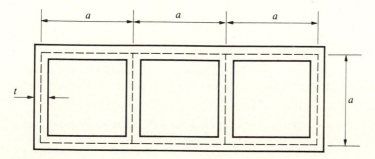

Figure F

3.28 A beam of rectangular cross section, whose sides are defined by $x = \pm b$ and $y = \pm a$, is subjected to a bending moment M about the x axis and a twisting moment T about the z axis. For b/a greater than about 3, a good upper bound may be obtained by using the velocity field of Eqs. (155) with $w = xy\theta$. Show that the interaction relationship is

$$\left(\frac{M}{M_0}\right)^2 + \left(1 - \frac{a}{3b}\right)^2\left(\frac{T}{T_0}\right)^2 = 1$$

Note that this solution approaches the lower bound approximation as a/b tends to be negligible compared to unity.

3.29 A cantilever I beam is built-in at the end $z = 0$, where it is prevented from warping, and is subjected to a twisting couple T at the free end $z = l$ (Fig. G). Each flange is then under a twisting moment T_f, a bending moment M_f, and a shearing force Q. Neglecting the effect of shear, Eq. (160) may be assumed for the yield condition of each flange with fully plastic moments T_0 and M_0. Denoting the horizontal displacement of the upper flange by $u(z)$, and using the upper-bound theorem, show that

$$(T - T_w)u(l) = \int_0^l \sqrt{4(T_0u')^2 + (M_0hu'')^2}\, dz + M_0hu'(0)$$

where T_w is the torque shared by the web, and the primes indicate differentiation with respect to z. Verify that for a linear displacement function, T exceeds the sandhill torque by M_0h/l.

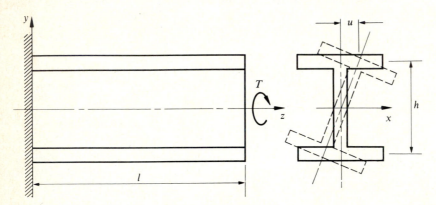

Figure G

3.30 In the plastic torsion of a uniform rigid/plastic bar, made of a work-hardening material, let ds and dc denote the arc elements of the shearing line and the characteristic respectively through a generic point of the cross section. If ds is taken along the shear stress vector, and the direction of dc is obtained from that of ds by a $90°$ counterclockwise rotation, show that the shear stress k and the engineering shear strain rate λ must satisfy the equations

$$\frac{\partial k}{\partial s} + k\frac{\partial \psi}{\partial c} = 0 \qquad \lambda\frac{\partial \psi}{\partial s} - \frac{\partial \lambda}{\partial c} = 2$$

where ψ is the angle of inclination of the tangent to the characteristic. Denoting the local radius of curvature of the shearing line by ρ, show that $\rho(\rho - \lambda)$ is constant along a characteristic when the bar is just fully plastic.

3.31 A rigid/plastic prismatic bar of arbitrary cross section is subjected to combined torsion and axial tension until it is fully plastic. Choosing a set of rectangular coordinates, where the z axis is parallel to the generators, write down the velocity field in terms of the rate of extension $\dot{\varepsilon}$, and the rate of twist $\dot{\theta}$ per unit length. Adopting the Mises theory, show that the stress function ϕ must satisfy the differential equation

$$\frac{\partial}{\partial x}\left(F\frac{\partial\phi}{\partial x}\right) + \frac{\partial}{\partial y}\left(F\frac{\partial\phi}{\partial y}\right) = -\frac{2\theta}{\sqrt{3}\,\dot{\varepsilon}}$$

where

$$F = \left\{1 - \left(\frac{\partial\phi}{\partial x}\right)^2 - \left(\frac{\partial\phi}{\partial y}\right)^2\right\}^{-1/2}$$

Assuming the cross section to be a circle of radius a, show that the differential equation is identically satisfied by the stress function associated with equation (20) for the resultant shear stress.

FOUR

PLASTIC ANALYSIS OF BEAMS AND FRAMES

4.1 Introduction

The conventional methods of analysis and design of engineering structures are based on a permissible working stress whose value is well within the elastic limit. The concentrations of stress that occur at rivet holes and sudden changes in cross section are usually disregarded in the elastic analysis. Since the results of the elastic analysis cease to hold when the yield limit is exceeded at the most critical cross section, the elastic design of a structure requires a margin of safety that ensures a fully elastic response. A limitation of structural designs based on the elastic analysis is evident from the fact that minor structural imperfections, which are without effect on the overall strength of the structure, have a marked influence on the elastic behavior.

The load-carrying capacity of a structure made of a ductile material is rarely exhausted at the onset of plastic yielding, since excessive deflections do not occur before the load is appreciably higher than that at the elastic limit. This effect is more pronounced in statically indeterminate structures, where there is a redistribution of stress beyond the elastic limit, resulting in a marked increase in the carrying capacity. It follows that an economical design of a structure can be based on a suitable safety factor applied to the load for which the overall deflection begins to increase in a more or less unrestricted manner.† Such a load is called the collapse load, which can be determined by the methods of plastic analysis without having to

---

† See, for example, J. F. Baker, *J. Inst. Civ. Eng.*, **31**: 188 (1949).

consider the intervening elastic/plastic range of deformation. The calculations involved in the plastic analysis are much simpler than those required in the corresponding elastic analysis. The influence of work-hardening is usually neglected in the plastic analysis so that the estimated carrying capacity is always conservative.

The strength of a structure is characterized by its collapse load which is obtained on the basis of certain idealizations. Considering a nonhardening elastic/plastic structure, a state of plastic collapse is defined as one for which the ✓ deflections, regarded as small, continue to increase under constant external loads. Since the bending moment distribution remains unchanged during the collapse, the change in curvature vanishes everywhere except at certain critical cross sections where the bending moment attains the fully plastic value. Infinitely large curvatures can theoretically occur at these cross sections, permitting hinge rotations which give rise to a link-type mechanism for plastic collapse.† If the material is assumed as rigid/plastic, the collapse load is precisely that for which the deformation of the structure first becomes possible as the intensity of the load is progressively increased from zero. The ratio of the collapse load to the working load, known as the *load factor*, represents the margin of safety under service conditions.

According to the *lower bound theorem* of limit analysis,‡ an external load in equilibrium with a distribution of bending moment which nowhere exceeds the fully plastic value is less than or equal to the collapse load. Such a distribution of bending moment is referred to as statically admissible. The *upper bound theorem*, on the other hand, states that the load obtained by equating the external work done by it to the internal work absorbed at the plastic hinges in any assumed collapse mechanism is greater than or equal to the collapse load. The deformation mode represented by a collapse mechanism is said to be kinematically admissible. The two limit theorems can be combined to form a *uniqueness theorem* which states that if any statically admissible distribution of bending moment can be found in a structure that has sufficient number of yield hinges to produce a mechanism, the corresponding load is equal to the collapse load. When a structure is subjected to a number of loads which may or may not increase in strict proportion to one another, plastic collapse will occur at the first combination of loads for which a statically admissible bending moment distribution that satisfies the mechanism condition can be found. The load-carrying capacity of the structure can therefore be determined for any given ratios of the applied loads in the state of collapse, without any reference to the loading history. It follows that the collapse load is unaffected by initial internal stresses,§ as well as by any flexibility of support and imperfect fit of members. If the problem is not statically determined at collapse,

---

† The concept of plastic hinges has been introduced by G. Kazinczy, *Betonszemle*, **2**: 6 (1914).

‡ Formal proofs of the limit theorems are given in Sec. 2.6 from the point of view of continuum mechanics. In the context of beams and frames, the theorems have been proved by A. A. Gvozdev, *Akad. Nauk U.S.S.R.*, Moscow-Leningrad, 19 (1938); M. R. Horne, *J. Inst. Civ. Eng.* **34**: 174 (1950); H. J. Greenberg and W. Prager, *Proc. ASCE*, **77**: 59 (1951).

§ This has been pointed out by G. V. Kazinczy, *Bauingenieur*, **19**: 236 (1938). Direct experimental confirmation was provided earlier by H. Maier-Leibnitz, *Bautechnik*, **6**: 11 (1928).

the distribution of bending moment will depend, however, on such factors as the history of loading, initial stresses, and settlement of supports.

An important corollary of the lower bound theorem is that the collapse load cannot be decreased by increasing the strength of any part of the structure.[†] Indeed, the bending moment distribution corresponding to the state of collapse will remain statically admissible for the modified structure in which the fully plastic moment is increased at one or more cross sections. Hence, the load-carrying capacity of the modified structure is at least as high as that of the unmodified structure. The upper-bound theorem provides the corollary that the collapse load cannot be increased by decreasing the strength of any part of the structure. This conclusion follows from the fact that the mechanism corresponding to the state of collapse will produce in the modified structure an internal work that is less than or equal to that in the unmodified structure. The resulting upper bound obtained for the weakened structure, therefore, cannot exceed the collapse load for the original structure.

4.2 Limit Analysis of Beams

(i) *Plastic collapse of simple beams* A statical method of evaluating the collapse load for single-span beams is suggested by the fact that the problem is statically determinate at collapse. The load is associated with an equilibrium distribution of bending moment which attains the fully plastic value at a sufficient number of cross sections. It is usually more convenient, however, to use a kinematical method based on the principle of virtual work. To illustrate the procedure, we consider a uniform fixed-ended beam of length $2l$, Fig. 4.1a, carrying a load W uniformly distributed over its left-hand half. The only possible collapse mechanism, based on the rigid/plastic model, is shown in Fig. 4.1b, the plastic hinge in the beam being assumed at a distance ξl from the left-hand end. The hinge rotations θ and α at the ends of the beam are related to one another by

$$\xi l\theta = (2 - \xi)l\alpha \qquad \text{or} \qquad \alpha = \frac{\theta\xi}{2 - \xi}$$

The work done by the load during the small mechanism motion is W/l times the area of the polygon $ADCE$. The work absorbed at each plastic hinge is necessarily positive, and is equal to the plastic moment M_0 times the amount of hinge rotation. Equating the work done to the work absorbed, we have

$$Wl\left(\xi\theta - \frac{\alpha}{2}\right) = M_0\theta + M_0(\theta + \alpha) + M_0\alpha$$

and the substitution for α gives

$$Wl = \frac{8M_0}{\xi(3 - 2\xi)} \tag{1}$$

---

† This was first stated as an axiom without a formal proof by S. M. Feinberg, *Prikl. Mat. Mekh.*, **17**: 63 (1948).

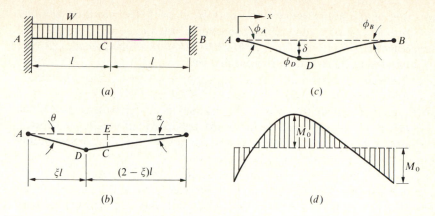

Figure 4.1 Plastic collapse of a fixed-ended beam. (*a*) Loaded beam; (*b*) collapse mechanism; (*c*) deflection curve; (*d*) bending moment diagram.

Since the deformation mode is kinematically admissible, (1) provides an upper bound on the collapse load for any assumed value of ξ. The right-hand side of (1) has a minimum value when $\xi = \frac{3}{4}$, giving the collapse load parameter $Wl = \frac{64}{9}M_0$. The overall equilibrium of the beam requires the vertical reactions at A and B to be $\frac{3}{4}W$ and $\frac{1}{4}W$ respectively at the incipient collapse.

The bending moment distribution shown in Fig. 4.1*d* is parabolic over the left-hand half and linear over the right-hand half of the beam. The plastic hinge conditions $M = -M_0$ at $x = 0$, $M = M_0$ at $x = \frac{3}{4}l$, and $M = -M_0$ at $x = 2l$ furnish

$$M = M_0\left[-1 + \frac{16x}{3l}\left(1 - \frac{2x}{3l}\right)\right] \qquad 0 \leqslant x \leqslant l$$

$$M = M_0\left(\frac{23}{9} - \frac{16x}{9l}\right) \qquad l \leqslant x \leqslant 2l \tag{2}$$

in view of the continuity of M at $x = l$. Since the maximum bending moment in the beam is M_0, occurring at $x = \frac{3}{4}l$, the yield limit is nowhere exceeded. The condition for moment equilibrium of the segment AD provides a check on the collapse load $W = 64M_0/9l$, which is the actual collapse load since it is both an upper and a lower bound.

Suppose that the plastic hinge in the beam is assumed to occur at the midpoint of the loaded segment. The corresponding upper bound is $W = 8M_0/l$, obtained by setting $\xi = 0.5$ in (1). The associated bending moment distribution over the loaded part of the beam is easily shown to be

$$M = M_0\left[-1 + \frac{2x}{l}\left(3 - \frac{2x}{l}\right)\right] \qquad 0 \leqslant x \leqslant l$$

The maximum bending moment again occurs at $x = \frac{3}{4}l$, but its magnitude is $\frac{5}{4}M_0$, which renders the moment distribution statically inadmissible. However, if the right-hand side of the above equation is multiplied by $\frac{4}{5}$, the resulting bending moment would be statically admissible. The external load in equilibrium with the modified moment distribution is $W = \frac{4}{5}(8M_0/l) = 6.4M_0/l$, which provides a lower bound. The collapse load is therefore bounded by

$$6.4M_0 < Wl < 8.0M_0$$

If the position of the yield hinge is now assumed at $\xi = \frac{3}{4}$, suggested by the preceding approximation, the exact value of the collapse load results. In general, the second approximation would lead to an upper bound, but the corresponding lower bound would be sufficiently close.

Under certain simplifying assumptions, the deflection of the beam at collapse can be estimated without having to follow the development of the plastic hinges. Firstly, the beam is assumed to be perfectly elastic everywhere except at the hinges, where the magnitude of the bending moment is M_0. Secondly, a plastic hinge that develops at some stage is assumed to be operative at all subsequent stages, so that the deflection curve has a continuous slope except at the hinges. The deflections and hinge rotations at the instant of collapse may be computed from (64) and (65), Chap. 3. Let ϕ_A, ϕ_B, and ϕ_D denote the magnitudes of the hinge rotations at A, B, and D respectively during the plastic collapse, Fig. 4.1c. Since the sum of the angles turned through by the segments AD and DB in the counterclockwise sense is equal to the sum of the hinge rotations at A and B less the hinge rotation at D, we have

$$\phi_A - \phi_D + \phi_B = \int_0^{2l} \frac{M}{EI}\, dx = \frac{10}{27} \frac{M_0 l}{EI} \tag{3}$$

in view of Eqs. (2). The tangent to the bent axis at A passes through the hinge at D, the sum of the bending moments at these two sections being zero (see Sec. 3.5(i)). Since the terminal slopes of the beam are $\psi_A = -\phi_A$ and $\psi_B = \phi_B$, Eqs. (65), Chap. 3 furnish

$$\delta = \frac{3}{4} l\phi_A = \frac{5}{4} l\phi_B + \frac{1}{EI} \int_0^{2l} M\left(x - \frac{3}{4}l\right) dx$$

where δ is the deflection of the beam at D. The lower limit of integration has been extended to $x = 0$ for convenience without affecting the result. Substituting from (2), and integrating, we have

$$\frac{\delta}{l} = \frac{3}{4}\phi_A = \frac{5}{4}\phi_B + \frac{11}{54}\frac{M_0 l}{EI}$$

These relations may be combined with (3) to express ϕ_A, ϕ_B and δ in the form

$$\phi_A = \frac{M_0 l}{3EI} + \frac{5}{8}\phi_D \qquad \phi_B = \frac{M_0 l}{27EI} + \frac{3}{8}\phi_D$$

$$\frac{\delta}{l} = \frac{M_0 l}{4EI} + \frac{15}{32}\phi_D \tag{4}$$

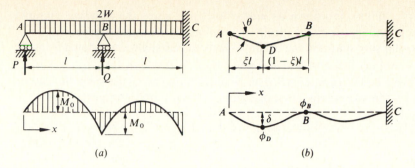

Figure 4.2 Plastic collapse of a continuous beam. (*a*) Loaded beam and bending moment diagram; (*b*) collapse mechanism and deflection curve.

Equations (4) define the motion of the collapse mechanism with superimposed effects of elastic deformation. At the instant of collapse, the rotation must vanish at the hinge that forms last. This hinge is readily identified as that at D, since ϕ_A and ϕ_B are seen to be positive, as required, when ϕ_D is set equal to zero. That the hinge at D is indeed the last to form is further supported by the fact that ϕ_D would be negative if ϕ_A or ϕ_B is assumed to vanish at the point of collapse.

(ii) *Plastic collapse of continuous beams* The kinematical method of analysis is readily applicable to multispan beams which generally remain statically indeterminate at collapse. Possible upper bounds are obtained from the fact that any one of the loaded spans may collapse as a beam. The correct mechanism is that which corresponds to the lowest value of the upper bound. Consider, for example, a two-span beam of uniform plastic moment M_0, loaded and supported as shown in Fig. 4.2*a*. The left-hand span requires two yield hinges and the right-hand span three yield hinges for the plastic collapse.† Since the two spans are of equal lengths, a smaller load is associated with the mechanism of Fig. 4.2*b*, which represents the correct mode of collapse. If θ denotes the angle of rotation of the beam at the left-hand support, the hinge discontinuity at the central support must be $\theta\xi/(1 - \xi)$, where ξl is the distance of the sagging hinge from A. The load W on this span moves through an average distance $\frac{1}{2}\xi l\theta$, leading to the work equation

$$\frac{1}{2} W l \xi \theta = M_0\left(\frac{\xi\theta}{1 - \xi} + \frac{\theta}{1 - \xi}\right) \quad \text{or} \quad W l = \frac{2M_0}{\xi}\left(\frac{1 + \xi}{1 - \xi}\right)$$

The minimum value of W corresponds to $\xi = \sqrt{2} - 1$, and the collapse load is therefore given by

$$W l = 2(3 + 2\sqrt{2})M_0 \simeq 11.65M_0$$

---

† For a complete elastic/plastic analysis of this problem, based on a unit shape factor, see B. Venkatraman and S. Patel, *Structural Mechanics with Introduction to Elasticity and Plasticity*, pp. 589–594, McGraw-Hill Book Co., New York (1970).

This kind of collapse is known as *partial collapse*, since the bending moment distribution at collapse is not statically determined throughout the beam.

In the collapsing span AB, the bending moment is easily written down from the fact that the support reaction P at A is given by $Pl = 2(\sqrt{2} + 1)M_0$ in view of the overall moment equilibrium of AB. The bending moment in the span BC can be expressed in terms of the unknown reaction Q at the central support. Thus

$$\frac{M}{M_0} = (\sqrt{2} + 1)\frac{x}{l}\left[2 - (\sqrt{2} + 1)\frac{x}{l}\right] \qquad 0 \leqslant x \leqslant l$$

$$\frac{M}{M_0} = (\sqrt{2} + 1)\frac{x}{l}\left[2 - (\sqrt{2} + 1)\frac{x}{l}\right] + \frac{Q}{M_0}(x - l) \qquad l \leqslant x \leqslant 2l$$

(5)

A statically admissible bending moment distribution in BC may be constructed by assuming $Ql = (7 + 4\sqrt{2})M_0$, so that $M = -M_0$ at $x = 2l$. The maximum bending moment then occurs at the middle of this span, and has the value $(2\sqrt{2} - 1)M_0/4 \simeq 0.457M_0$. Since the yield moment is nowhere exceeded, the collapse load found above is confirmed as actual.

The reaction Q can be determined exactly by using the kinematical conditions that the deflection vanishes at B and C, while the slope vanishes at C. These conditions are expressed by

$$\int_l^{2l} M(x - l)\, dx = 0$$

where M is given by the second equation of (5). A straightforward integration furnishes

$$Ql = \tfrac{1}{4}(31 + 14\sqrt{2})M_0 \simeq 12.70M_0$$

The bending moment distribution at collapse is now completely determined, and is shown in Fig. 4.2a. At the built-in end $x = 2l$, the bending moment attains the value $-(2\sqrt{2} + 1)M_0/4 \simeq -0.957M_0$, ensuring the static admissibility.

The slope of the beam is discontinuous at B and D during the collapse. Let ψ_B denote the slope of the right-hand span of the beam at the central support. If the spread of the plastic zone is disregarded, it follows from (64), Chap. 3 that

$$\psi_B = -\int_l^{2l} \frac{M}{EI}\, dx = (12 + 5\sqrt{2})\frac{M_0 l}{3EI} - \frac{Ql^2}{2EI} = (3 - 2\sqrt{2})\frac{M_0 l}{24EI} \qquad (6)$$

in view of (5) and the expression for Q. The slope of the left-hand span of the beam at B is $\psi_B + \phi_B$, where ϕ_B is the magnitude of the hinge rotation at this section. The tangent to the corresponding centerline at B makes a counterclockwise angle $\psi_B + \phi_B - \phi_D - \psi_A$ with the tangent at A, where ψ_A is the slope at A, and ϕ_D the hinge rotation at D. By equations (5), we have

$$\psi_B + \phi_B - \phi_D - \psi_A = \int_0^l \frac{M}{EI}\, dx = \frac{\sqrt{2}\, M_0 l}{3EI} \qquad (7)$$

Since the bending moments at B and D are equal and opposite, the deflection δ of the beam at D must be $(2 - \sqrt{2})(\psi_B + \phi_B)l$. By (65), Chap. 3, the deflection may be written alternatively as

$$\delta = -(\sqrt{2} - 1)l\psi_A - \frac{1}{EI}\int_0^{(\sqrt{2}-1)l} M[(\sqrt{2} - 1)l - x]\,dx$$

where M is given by the first equation of (5). Carrying out the integration,

$$\frac{\delta}{l} = -(\sqrt{2} - 1)\psi_A - (3 - 2\sqrt{2})\frac{M_0 l}{4EI} = (2 - \sqrt{2})(\psi_B + \phi_B)$$

Eliminating ψ_A by means of (7), the quantities $\psi_B + \phi_B$ and δ/l can be expressed in terms of $M_0 l/EI$ and ϕ_D. Since ψ_B is given by (6), the results finally become

$$\phi_B = (6\sqrt{2} - 5)\frac{M_0 l}{24EI} + (\sqrt{2} - 1)\phi_D$$

$$\frac{\delta}{l} = (5\sqrt{2} - 6)\frac{M_0 l}{12EI} + (3\sqrt{2} - 4)\phi_D$$

(8)

It follows from (8) that ϕ_D vanishes at the instant of collapse, which means that the first hinge forms at B and the second hinge forms at D, the deflection of the sagging hinge at the incipient collapse being $0.089 M_0 l^2/EI$. This deflection is identical to that in a uniformly loaded cantilever of length l having a vertical support at the free end. The hinge rotation in the latter beam at the point of collapse is slightly higher than that given by (8) due to the complete fixity of the section where the rotation occurs.

(iii) Influence of deflection on collapse In the case of partial collapse, as we have seen, the collapse load can be determined without reference to the noncollapsing part of the beam. This may lead to paradoxical results in certain cases, unless a displacement analysis is carried out to examine the behavior of the complete beam. Consider a uniform beam of plastic moment M_0, resting on four simple supports, and carrying a point load W as shown in Fig. 4.3a. The central span is of length $2l$, and the two outer spans are each of length kl, where k may have any value between zero and infinity. The kinematical method applied to the only possible collapse mechanism, shown in Fig. 4.3b, immediately gives the collapse load

$$W = \frac{4M_0\theta}{l\theta} = \frac{4M_0}{l}$$

The distribution of bending moment at collapse is shown by shaded areas in Fig. 4.3c. The static equilibrium of each half of the central span requires $Wl/2 = 2M_0$, giving $W = 4M_0/l$, which agrees with the kinematical result. The independence of the collapse load of the value of k seems surprising. Indeed, when k is very large, the outer supports have negligible effect, and the beam effectively becomes simply supported, the corresponding collapse load being $2M_0/l$. On the other hand, when

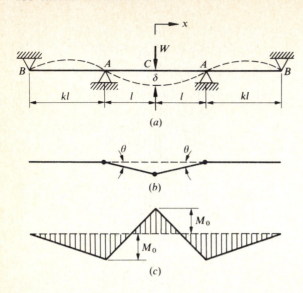

(a)

(b)

(c)

Figure 4.3 Plastic collapse of a two-span beam. (a) Loading and deflection; (b) collapse mechanism; (c) bending moment distribution.

k is very small, the beam is effectively clamped between the two supports, and the collapse load $4M_0/l$ should apply.†

The paradox can be explained by the consideration of displacements. To simplify the analysis, we assume an ideal beam section for which plastic deformation is confined at the yield hinges. Denoting the reaction at the inner supports by P, the distribution of bending moment at any stage may be written as

$$M = -Pkl + \tfrac{1}{2}W[(1 + k)l - x] \qquad 0 \leqslant x \leqslant l$$

$$M = \left(\frac{W}{2} - P\right)[(1 + k)l - x] \qquad l \leqslant x \leqslant (1 + k)l$$

$$(9)$$

where x is measured from the loading point C. In view of the symmetry of the problem, only the right-hand half of the beam will be considered.

When the applied load is sufficiently small, the beam is entirely elastic, and the slope of the beam vanishes at $x = 0$. Since the deflection of the beam vanishes at $x = l$ and $x = (1 + k)l$, the central deflection δ is given by (64), Chap. 3. Thus

$$EI\delta = \int_0^l M(l - x)\,dx = \int_0^{(1 + k)l} M[(1 + k)l - x]\,dx$$

where EI is the flexural rigidity of the beam. The integration is easily carried out on inserting the appropriate expression for M, and the result is

$$\frac{EI\delta}{l^3} = \frac{W}{6}\left(1 + \frac{3}{2}k\right) - \frac{kP}{2} = \frac{W}{6}(1 + k)^3 - \frac{kP}{6}(3 + 6k + 2k^2)$$

† The paradox was pointed out by F. Stüssi and C. F. Kollbrunner, *Bautechnik*, **13**: 264 (1935), and was resolved with the consideration of displacements by P. S. Symonds and B. G. Neal, *J. Aeronaut. Sci.*, **19**: 15 (1952).

These relations furnish the reaction P and the deflection δ as

$$P = \frac{W}{2}\left[1 + \frac{3}{2k(3 + k)}\right] \qquad \delta = \left(\frac{3 + 4k}{3 + k}\right)\frac{Wl^3}{24EI} \tag{10}$$

The above solution will hold so long as the bending moments at the critical sections A and C are less than M_0 in magnitude. It follows from (9) and (10) that the yield moment is first attained at $x = 0$ for $W = W_e$ and $\delta = \delta_e$, where

$$W_e = \frac{4M_0}{l}\left(\frac{3 + k}{3 + 2k}\right) \qquad \delta_e = \frac{M_0 l^2}{6EI}\left(\frac{3 + 4k}{3 + 2k}\right)$$

When the load exceeds W_e, a plastic hinge forms at the central section C, and the slope of the beam becomes discontinuous at $x = 0$. The yield condition $M = M_0$ at $x = 0$ gives the modified moment distribution

$$M = M_0 - \tfrac{1}{2}Wx \qquad\qquad 0 \leqslant x \leqslant l$$

$$\tag{11}$$

$$M = \left(\frac{W}{2k} - \frac{M_0}{kl}\right)[(1 + k)l - x] \qquad l \leqslant x \leqslant (1 + k)l$$

In view of (65), Chap. 3, the condition of zero deflections at A and B can be expressed as

$$kl\psi_A - \frac{1}{EI}\int_l^{(1+k)l} M[(1 + k)l - x]\,dx = 0$$

where ψ_A denotes the continuous slope at A. Using the second equation of (11), the above integral is readily evaluated, the result being

$$\psi_A = \left(\frac{Wl}{2} - M_0\right)\frac{kl}{3EI}$$

The slope is therefore proportional to k. The deflection at the center of the beam is

$$\delta = l\psi_A - \int_0^l \frac{M}{EI}x\,dx = \frac{M_0 l^2}{6EI}\left[(1 + k)\frac{Wl}{M_0} - (3 + 2k)\right] \tag{12}$$

in view of the first equation of (11). When the load reaches the collapse value $4M_0/l$, the bending moment at $x = l$ attains the value $-M_0$, forming a plastic hinge at section A. The corresponding value of δ is

$$\delta_0 = (1 + 2k)\frac{M_0 l^2}{6EI}$$

The load-deflection relationship given by (10) and (12) is shown graphically in Fig. 4.4 for selected values of k. The dashed curve is based on experimental results†

---

† The experimental results are due to F. Stüssi and C. F. Kollbrunner, op. cit. The effect of partial end fixity has also been investigated by G. V. Kazinczy, *Proc. Int. Assoc. Bridge Struct. Eng.*, **2**: 249 (1934).

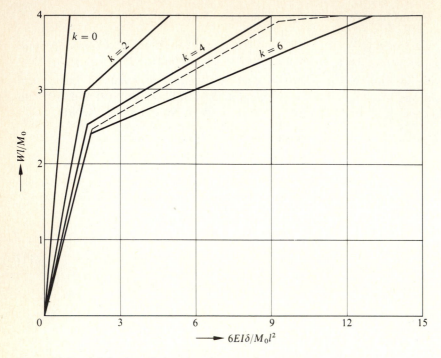

Figure 4.4 Load-deflection relations for a beam on four supports.

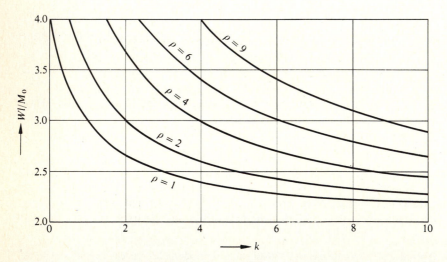

Figure 4.5 Loading-carrying capacity of a two-span beam based on the criterion of permissible deflection.

corresponding to $k = 4$. When $k = 0$ (fixed-end condition), the elastic limit load coincides with the collapse load.

For sufficiently small values of k, the deflection at the point of collapse is of the same order as that at the elastic limit. In this case, the collapse load $4M_0/l$ provides a realistic measure of the carrying capacity of the beam. When k exceeds about 6, unacceptably large deflections occur before the theoretical collapse load is reached, the load $4M_0/l$ is therefore of little practical significance for large values of k. A quantitative measure of the load-carrying capacity may be obtained on the basis of a permissible value of δ. To be specific, we take the greatest permissible deflection as ρ times the value of δ_0 corresponding to $k = 0$. By Eq. (12), the highest allowable load is then given by

$$\frac{Wl}{M_0} = \frac{3 + 2k + \rho}{1 + k} \leqslant 4$$

The variation of the limiting load with k is shown in Fig. 4.5 for various values of ρ. When k is sufficiently small, the limiting load is identical to the collapse load. When k is sufficiently large, the limiting load decreases as k increases, approaching asymptotically the collapse load for a simply supported beam.

4.3 Limit Analysis of Plane Frames

(i) *Interaction diagrams* We begin with the principle of virtual work which states that if a body is in statical equilibrium, the work done by the external loads on an arbitrary set of small external displacements is equal to the work done by the internal forces on the corresponding internal displacements. It is sufficient for the present purpose to consider mechanism type of deformation modes involving rigid-body rotation of members about a set of appropriate hinges. It is assumed at the outset that the external loads are supported by the bending resistance of the members, and that axial and shearing forces produce negligible secondary effects.

Consider a fixed-base rectangular portal frame of uniform cross section, loaded as shown in Fig. 4.6a, the plastic moment for each member of the frame being denoted by M_0. The frame has three degrees of redundancy, because if a cut is made at any section, and the bending moment, shearing force, and axial force are specified at this section, the problem becomes statically determinate. Since the bending moment varies linearly along each segment of the frame, the shearing force being constant, the distribution of bending moment is specified by its values at the five numbered sections. Sagging bending moments will be always taken as positive when viewed from the interior of the frame. Thus, positive values of moments, curvature, and hinge rotation will correspond to tensile stresses and strains in the fibers adjacent to the dashed lines.†

---

† A comprehensive account of the plastic methods for portal frames has been presented by J. Heyman, *Plastic Design of Portal Frames*, Cambridge University Press (1957). Reinforced concrete frames have been treated by A. L. L. Baker, *Limit State Design of Reinforced Concrete*, 2d ed., Cement and Concrete Association, London (1970).

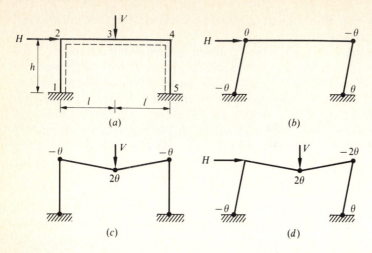

Figure 4.6 Plastic collapse of a portal frame. (a) Loaded frame; (b) panel mechanism; (c) beam mechanism; (d) combined mechanism.

There are three possible mechanisms, shown in Fig. 4.6b to d, each representing a virtual displacement pattern defined by a hinge rotation θ. In the panel mechanism (b), the horizontal load H moves through a distance $h\theta$ producing a virtual work of amount $Hh\theta$. Equating this to the virtual work absorbed at the hinges, we have

$$M_1(-\theta) + M_2(\theta) + M_4(-\theta) + M_5(\theta) = hH\theta$$

or

$$-M_1 + M_2 - M_4 + M_5 = Hh \tag{13}$$

In the beam mechanism (c), the vertical load V does work of amount $Vl\theta$, while no work is done by the horizontal load H. The principle of virtual work gives

$$M_2(-\theta) + M_3(2\theta) + M_4(-\theta) = Vl\theta$$

or

$$-M_2 + 2M_3 - M_4 = Vl \tag{14}$$

The combined panel and beam mechanism (d) involves virtual works $Hh\theta$ and $Vl\theta$ done by the horizontal and vertical loads respectively, and the work equation becomes

$$M_1(-\theta) + M_3(2\theta) + M_4(-2\theta) + M_5(\theta) = Hh\theta + Vl\theta$$

or

$$-M_1 + 2M_3 - 2M_4 + M_5 = Hh + Vl \tag{15}$$

Since (15) may be obtained by adding together (13) and (14), only two of these equations are independent. This is a consequence of the fact that the frame has three redundancies and five unknown critical moments.

We are concerned here with only positive values of H and V. If the frame actually collapses in the mode of Fig. 4.6b, the magnitude of the bending moment at each of the four plastic hinges must be equal to M_0. Since the sign of the bending moment must be the same as that of the corresponding hinge rotation, $M_1 = -M_0$, $M_2 = M_0$, $M_4 = -M_0$, and $M_5 = M_0$ for this mode of collapse. The bending moment distribution, which is linear in each segment of the frame (the shearing force being constant), will be statically admissible if $-M_0 \leqslant M_3 \leqslant M_0$. Equations (13) and (14) therefore furnish

$$Hh = 4M_0 \qquad 0 \leqslant Vl \leqslant 2M_0 \tag{16a}$$

When the mechanism of Fig. 4.6c represents the actual mode of collapse, it is necessary to set $M_1 = -M_0$, $M_3 = M_0$, and $M_4 = -M_0$ in equations (13) and (14). Using the restrictions $-M_0 \leqslant M_1 \leqslant M_0$ and $-M_0 \leqslant M_5 \leqslant M_0$ required by the condition of static admissibility, we have

$$Vl = 4M_0 \qquad 0 \leqslant Hh \leqslant 2M_0 \tag{16b}$$

Finally, regarding the mechanism of Fig. 4.6d as actual for the state of collapse, and setting $M_1 = -M_0$, $M_3 = M_0$, $M_4 = -M_0$ and $M_5 = M_0$ in Eqs. (14) and (15), the relationship between H and V under the restriction $-M_0 \leqslant M_2 \leqslant M_0$ is obtained as

$$Hh + Vl = 6M_0 \qquad 2M_0 \leqslant Hh \leqslant 4M_0 \tag{16c}$$

The relationship between H and V producing plastic collapse is shown graphically in Fig. 4.7, the mode of collapse associated with each linear segment of the diagram being as indicated. Such a diagram, known as an *interaction diagram*, is always a convex locus enclosing the origin. Any combination of H and V represented by a point inside the diagram constitutes a safe state of external loading.

The collapse equation corresponding to any assumed mechanism of collapse may be directly obtained by a kinematical analysis in which the work done by the external loads is equated to the work absorbed at the plastic hinges. Since plastic work is always positive, the sign convention may be dispensed with in writing down the work equation, from which an upper bound can be derived. The kinematical method is equally suitable when one of the loads is uniformly distributed as shown in Fig. 4.8a, the total vertical load in this case being denoted by $2V$ for convenience. When the frame collapses by the combined mechanism shown in Fig. 4.8b, the position of the plastic hinge within the length of the beam may be specified by a distance ξl from the left-hand corner. The motion of the mechanism is defined by a clockwise rotation θ of each column. Since the vertical deflection of the plastic hinge in the beam is equal to $\xi l\theta$, the right-hand portion of the beam of length $(2 - \xi)l$ rotates counterclockwise through an angle $\phi = \xi\theta/(2 - \xi)$. The work done by the vertical and horizontal loads are $Vl\xi\theta$ and $Hh\theta$ respectively, so that the work equation for this mode of collapse is

$$Hh\theta + Vl\xi\theta = 2M_0\theta + 2M_0(\theta + \phi)$$

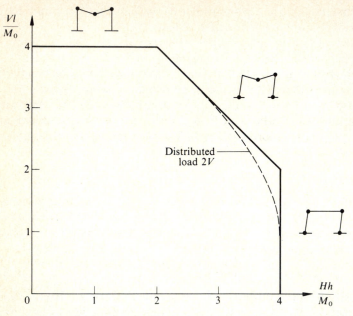

Figure 4.7 Interaction diagram for plastic collapse of a fixed-base rectangular portal frame.

Substituting for ϕ, and simplifying, we get

$$Hh + Vl\xi = 2M_0\left(\frac{4 - \xi}{2 - \xi}\right)$$

For a given ratio H/V, the value of ξ must be that which minimizes V. It is easily shown that

$$\xi = 4 - \sqrt{8 + \frac{2Hh}{Vl}}$$

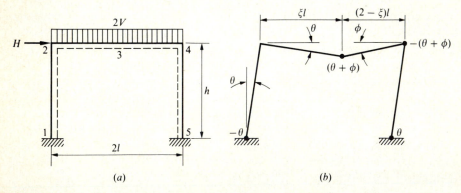

Figure 4.8 Influence of distributed loads. (*a*) Dimensions and loading; (*b*) collapse mechanism.

and the relationship between H and V is expressed parametrically as

$$\frac{Vl}{2M_0} = \frac{2}{(2-\xi)^2} \qquad \frac{Hh}{2M_0} = 1 + \frac{4(1-\xi)}{(2-\xi)^2} \qquad (17)$$

The interaction curve defined by (17) is shown by the dashed line in Fig. 4.7. As ξ increases from zero, Vl/M_0 increases from unity, reaching the value 4 when $\xi = 1$, the ratio Hh/M_0 at this point being equal to 2. For lower values of the horizontal load, Vl/M_0 given by (17) exceeds the value 4 obtained for the beam mechanism, in which case a plastic hinge always forms at $\xi = 1$. The panel mechanism applies for $Vl/M_0 \leqslant 1$, giving $Hh/M_0 = 4$ at the instant of collapse.

(ii) *Combination of mechanisms* When several loads are applied to a structure, and their ratios at the incipient collapse are given, the limiting values of these loads can be specified by a single parameter which represents the collapse load. The actual value of the collapse load is the lowest of all the upper bounds associated with the possible modes of collapse. For complex structures involving a large number of possible collapse modes, it is convenient to identify a specified number of elementary mechanisms and consider all others as suitable combinations of the elementary ones.† Suppose there are n critical sections at which plastic hinges may form under a given loading system, and let r denote the degree of redundancy of the structure. Since the bending moments at the critical sections would be completely determined if the values of these redundancies were known, there must be $n - r$ independent relations connecting the n critical moments. Each of these relations is an equation of statical equilibrium that can be associated with an elementary mechanism through the virtual work principle. It follows, therefore, that the number of elementary mechanisms, from which all other mechanisms of collapse can be deduced, is equal to $n - r$.

As a relatively simple example illustrating the method of combining mechanisms, consider a two-bay rectangular portal frame whose dimensions and loading are as shown in Fig. 4.9a. The fully plastic moments for the columns and the beams are denoted by M_0 and $2M_0$ respectively. It is convenient to begin with the assumption that the plastic hinge in the uniformly loaded beam occurs at the midpoint of this member. The frame has six redundancies, and there are ten possible plastic hinge positions numbered in the figure, so that the number of independent mechanisms is $10 - 6 = 4$. Three of these basic mechanisms may be identified as the panel mechanism of Fig. 4.9b and the beam mechanisms of Fig. 4.9c and d, while the fourth is taken as a simple rotation of the central joint. The last mechanism does not represent an actual mode of collapse, but is clearly independent of the other three, and can be combined with them to form possible collapse mechanisms.

---

† The method of combining mechanisms is due to B. G. Neal and P. S. Symonds, *Proc. Inst. Civ. Eng.*, **1**: 58 (1952). A trial-and-error method was proposed earlier by J. F. Baker, *Struct. Eng.*, **27**: 397 (1949). A moment distribution method has been proposed by M. R. Horne, ibid, **3**: 51 (1954), and J. M. English, *Trans. ASCE*, **119**: 1143 (1954).

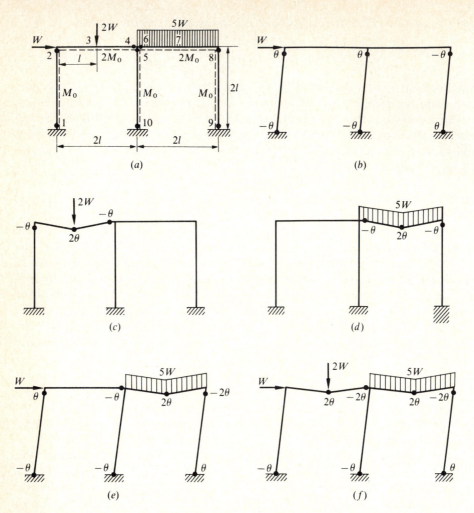

Figure 4.9 Plastic collapse of a two-bay frame. (a) Loaded frame; (b) panel mechanism; (c, d) beam mechanisms; (e, f) combinations of mechanisms.

The upper bounds associated with the elementary panel and beam mechanisms are easily computed from the corresponding work equations, remembering the fact that the work absorbed at the plastic hinges is always positive. Thus

$$2Wl\theta = 2M_0\theta + 2M_0\theta + 2M_0\theta \qquad \text{or} \qquad Wl = 3.0M_0 \qquad (18b)$$

$$2Wl\theta = M_0\theta + 2M_0(2\theta) + 2M_0\theta \qquad \text{or} \qquad Wl = 3.5M_0 \qquad (18c)$$

$$2.5Wl\theta = 2M_0\theta + 2M_0(2\theta) + M_0\theta \qquad \text{or} \qquad Wl = 2.8M_0 \qquad (18d)$$

The letters in the parentheses correspond with those used for the mechanisms in Fig. 4.9. The obvious next step is to combine the two mechanisms which predict the lowest upper bounds, these being the right-hand beam and panel mechanisms. Since the addition of displacements and hinge rotations of the two mechanisms does not lead to any elimination of hinges, a clockwise rotation θ of the central joint is added. The hinge rotations at sections 5 and 6 are then canceled, while a negative hinge rotation of amount θ is introduced at section 4, the resulting mechanism being that shown in Fig. 4.9e. The work absorbed at the central joint is thereby reduced by the amount $M_0\theta$, and the work equation for this combined mechanism becomes

$$4.5Wl\theta = 6M_0\theta + 7M_0\theta - M_0\theta$$

or

$$Wl = \tfrac{24}{9}M_0 \simeq 2.667M_0 \tag{18e}$$

To explore the possibility of obtaining a still lower value of the load, the left-hand beam mechanism is now combined with the mechanism just derived to produce the mechanism shown in Fig. 4.9f. Since the plastic hinge at section 2 is cancelled by this combination, resulting in a saving of internal work by an amount $2M_0\theta$, the work equation for the new mechanism is

$$6.5Wl\theta = 12M_0\theta + 7M_0\theta - 2M_0\theta$$

or

$$Wl = \tfrac{34}{13}M_0 \simeq 2.615M_0 \tag{18f}$$

This gives the lowest value of the upper bound load, and Fig. 4.9f presumably represents the actual collapse mechanism, subject to any adjustment of the position of the plastic hinge under the distributed load.

The conclusion of the above kinematical analysis will now be checked by a statical analysis, for which the equations of equilibrium can be written down from the virtual work principle applied to the four elementary mechanisms. These equations are easily shown to be

$$-M_4 - M_5 + M_6 = 0 \tag{19a}$$

$$-M_1 + M_2 + M_5 - M_8 + M_9 - M_{10} = 2W \tag{19b}$$

$$-M_2 + 2M_3 - M_4 = 2Wl \tag{19c}$$

$$-M_6 + 2M_7 - M_8 = 2.5Wl \tag{19d}$$

For the collapse mechanism of Fig. 4.9f, the bending moments at the plastic hinges are $M_1 = -M_0$, $M_3 = 2M_0$, $M_4 = -2M_0$, $M_7 = 2M_0$, $M_8 = -M_0$, $M_9 = M_0$, and $M_{10} = -M_0$. Substitution into the equilibrium equations (19) then furnishes

$$M_2 = \tfrac{10}{13}M_0 \qquad M_5 = \tfrac{6}{13}M_0 \qquad M_6 = -\tfrac{20}{13}M_0 \qquad Wl = \tfrac{34}{13}M_0$$

This confirms the value of Wl, and indicates that the magnitudes of the three remaining bending moments are less than the corresponding fully plastic moments. By (66), Chap. 3, the maximum bending moment in the right-hand beam occurs at a distance $0.041l$ to the right of the center of this member, the value of the maximum moment being $2.004M_0$. Hence, if the load corresponding to the mechanism of Fig. 4.9f is divided by the factor 1.002, a statically admissible distribution of bending moment would result. The collapse load is therefore bounded by

$$2.610M_0 < Wl < 2.615M_0$$

(iii) *Analysis of a two-story frame* In the preceding example, the statical analysis was straightforward, since the collapse mechanism was complete, the number of plastic hinges being one more than the number of redundancies. When the mode of collapse involves fewer than $(r + 1)$ plastic hinges, where r is the degree of redundancy of the frame, we have a state of *partial collapse* for which the statical check is intrinsically more difficult. Consider, as an example, a two-story rectangular portal frame loaded as shown in Fig. 4.10a, the dimensions and relative plastic moments of the various members being as indicated. The degree of redundancy of the frame is six, and the number of possible plastic hinge positions is twelve, the number of independent mechanisms being $12 - 6 = 6$. Four of these elementary mechanisms are chosen to be the panel and beam mechanisms shown in Fig. 4.10c to f, while the remaining two are taken as the joint mechanisms shown together in Fig. 4.10b. There are a large number of possible mechanisms, and the method of combining mechanisms is useful for identifying the actual mode of collapse.

The collapse equations for the elementary panel and beam mechanisms will be derived here from the corresponding equations of statical equilibrium obtained on the basis of virtual displacements. The virtual work equations are

$$2Wl\theta = -M_4\theta + M_5\theta - M_7\theta + M_8\theta \qquad (20c)$$

$$6Wl\theta = -M_1\theta + M_2\theta - M_{10}\theta + M_{11}\theta \qquad (20d)$$

$$2Wl\theta = -M_5\theta + 2M_6\theta - M_7\theta \qquad (20e)$$

$$6Wl\theta = -M_3\theta + 2M_{12}\theta - M_9\theta \qquad (20f)$$

The upper bounds associated with these mechanisms are readily obtained from the facts that each term of the work equations is positive and the bending moments at the hinges have the appropriate fully plastic values. The results are

$$2Wl\theta = M_0\theta + M_0\theta + M_0\theta + M_0\theta \qquad \text{or} \qquad Wl = 2M_0 \qquad (21c)$$

$$6Wl\theta = 2M_0\theta + 2M_0\theta + 2M_0\theta + 2M_0\theta \qquad \text{or} \qquad Wl = \tfrac{4}{3}M_0 \qquad (21d)$$

$$2Wl\theta = M_0\theta + 3M_0\theta + M_0\theta \qquad \text{or} \qquad Wl = \tfrac{5}{2}M_0 \qquad (21e)$$

$$6Wl\theta = 2M_0\theta + 4M_0\theta + 2M_0\theta \qquad \text{or} \qquad Wl = \tfrac{4}{3}M_0 \qquad (21f)$$

It is natural to start with the combination of the lower beam and panel mechanisms, since these two elementary mechanisms give the lowest value of Wl. By rotating the joint A clockwise through an angle θ, the plastic hinges at sections 2

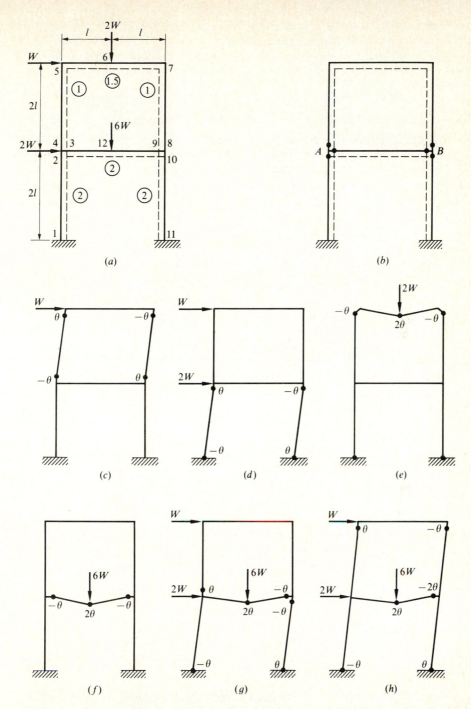

Figure 4.10 Plastic collapse of a two-story frame. (*a*) Dimensions and loading; (*b*) joint mechanisms; (*c*, *d*) elementary panel mechanisms; (*e*, *f*) elementary beam mechanisms; (*g*, *h*) combined mechanisms.

and 3 are eliminated, and replaced by a positive hinge rotation of amount θ at section 4. This produces the mechanism of Fig. 4.10g, and involves a reduction of work absorbed at joint A from $4M_0\theta$ to $M_0\theta$, the work equation for this combination being

$$12Wl\theta = 8M_0\theta + 8M_0\theta - 3M_0\theta \qquad \text{or} \qquad Wl = \tfrac{13}{12}M_0 \qquad (21g)$$

The upper panel mechanism is now combined with the mechanism just obtained, resulting in the cancellation of the hinge at section 4, and a saving of internal work by an amount $2M_0\theta$. A clockwise rotation of the joint B through an angle θ further reduces the internal work by an amount $M_0\theta$, since the hinges at sections 8 and 10 are eliminated while the hinge rotation at section 9 is increased in the negative sense. The resulting mechanism is that shown in Fig. 4.10h, and the associated work equation is

$$14Wl\theta = 13M_0\theta + 4M_0\theta - 3M_0\theta \qquad \text{or} \qquad Wl = M_0 \qquad (21h)$$

The combined panel mechanism (not shown) furnishes the upper bound $Wl = 1.25M_0$, and its combination with the lower beam mechanism again gives $Wl = M_0$. Combinations of mechanisms involving the upper beam mechanism will not be examined, because they are expected to give values of Wl greater than M_0. It is reasonable to conclude that the mechanism of Fig. 4.10h represents the actual mode of collapse, but a statical check is necessary to confirm this.

Since the number of plastic hinges in the collapse mechanism is the same as the degree of redundancy, the frame is statically indeterminate at collapse. However, it is still possible to find a set of statically admissible bending moments that are compatible with the collapse mechanism. Substituting the values $M_1 = -2M_0$, $M_5 = M_0$, $M_7 = -M_0$, $M_9 = -2M_0$, $M_{11} = 2M_0$, $M_{12} = 2M_0$, and $Wl = M_0$ into equations (20c) to (20f), and considering the condition of equilibrium of the joint A, we obtain a set of five independent equations for the six unknown moments. These equations are

$$-M_4 + M_8 = 0 \qquad M_2 - M_{10} = 3M_0 \qquad M_6 = M_0 \qquad M_3 = 0$$
$$-M_2 + M_3 + M_4 = 0$$

Taking guidance from the mechanism of Fig. 4.10g, which gives the next lowest value of the load, the bending moment at section 4 may be arbitrarily set equal to M_0. Then the solution becomes

$$M_2 = M_0 \qquad M_3 = 0 \qquad M_4 = M_0 \qquad M_6 = M_0 \qquad M_8 = M_0$$
$$M_{10} = -2M_0$$

Since none of these bending moments exceeds the corresponding fully plastic moment in magnitude, the distribution of bending moment in the frame is statically admissible. The value $W = M_0/l$ therefore represents the actual collapse load of the considered frame.

(iv) Frames with inclined members As an example for the analysis of nonrectangular frames, consider the symmetrical gable frame shown in Fig. 4.11a. For simplicity, the same plastic moment M_0 is assumed for each member of the frame.

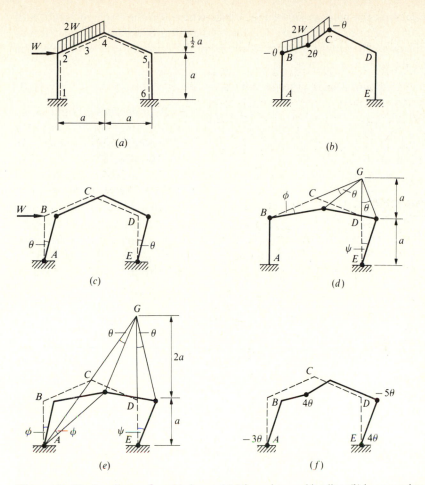

Figure 4.11 Plastic collapse of a gable frame. (*a*) Dimensions and loading; (*b*) beam mechanism; (*c, d, e*) panel mechanisms; (*f*) combined mechanism.

There are six critical bending moments and three degrees of redundancy, so that the number of independent mechanisms must be three. One of these elementary mechanisms is taken as the beam mechanism shown in Fig. 4.11*b*, the plastic hinge being initially assumed at the mid-point of the loaded rafter. The other two basic mechanisms are chosen to be the panel mechanisms of Fig. 4.11*c* and *d*, of which the first one is similar to that previously encountered. All the other possible mechanisms can be obtained by appropriate combinations of these three basic mechanisms.†

† The analysis of gable frames has been discussed by B. G. Neal and P. S. Symonds, *Engineer*, **194**: 315, 363 (1952). Vierendeel girders have been treated by P. S. Symonds and B. G. Neal, *J. Franklin Inst.*, **252**: 469 (1951). See also A. W. Hendry, *Struct. Eng.*, **33**: 213 (1955).

Consider first the mechanism of Fig. 4.11b, where θ is the angle of rotation of each half of the rafter BC. Since the midpoint of the rafter is at a horizontal distance $a/2$ from the ends, the downward displacement of the central hinge is $\frac{1}{2}a\theta$, and the corresponding work equation is

$$W(\tfrac{1}{2}a\theta) = M_0\theta + 2M_0\theta + M_0\theta \qquad \text{or} \qquad Wa = 8M_0 \qquad (22b)$$

In the panel mechanism of Fig. 4.11c, the roof BCD undergoes a rigid body translation by a distance $a\theta$, permitted by equal amounts of hinge rotation at joints A, B, D, and E. The panel mechanism of Fig. 4.11d, on the other hand, involves rigid body rotation of the rafters BC and CD about B and G respectively. Since the velocities of C and D are perpendicular to the members BC and ED respectively, the instantaneous center G of CD is the meeting point of BC and ED, the height of G above D being twice the height of C above D. From geometry, the vertical displacement of the apex C is $a\phi = a\theta$, and the horizontal displacement of the joint D is $a\psi = a\theta$, indicating that $\phi = \theta = \psi$ during collapse. It follows that the hinge rotations at B, C, D, and E in this mechanism are $-\theta$, 2θ, -2θ, and θ respectively. The work equations for the panel mechanisms (c) and (d) are easily obtained as

$$Wa\theta = M_0\theta + M_0\theta + M_0\theta + M_0\theta \qquad \text{or} \qquad Wa = 4M_0 \qquad (22c)$$

$$Wa\theta = M_0\theta + 2M_0\theta + 2M_0\theta + M_0\theta \qquad \text{or} \qquad Wa = 6M_0 \qquad (22d)$$

A useful combination of the elementary panel mechanisms is shown in Fig. 4.11e. It is instructive to deduce the work equation directly from the kinematics of the mechanism, in which joint C moves in a direction perpendicular to AC. The instantaneous center of rotation of the right-hand rafter is therefore at G, whose vertical height above the base of the frame is twice that of the apex C. The vertical motion of C gives $a\phi = a\theta$, and the horizontal motion of D gives $a\psi = 2a\theta$, where θ is the counterclockwise rotation of the rafter CD. The hinge rotations at A, C, D, and E are therefore equal to $-\theta$, 2θ, -3θ, and 2θ respectively, giving

$$2Wa\theta = M_0\theta + 2M_0\theta + 3M_0\theta + 2M_0\theta \qquad \text{or} \qquad Wa = 4M_0 \qquad (22e)$$

We may now combine the mechanisms of Fig. 4.11b, c, and e in such a way that the plastic hinges at B and C are both eliminated. First of all, the motion of (e) is added to twice the motion of (c) to obtain hinge rotations equal to 2θ at both B and C. Twice the rigid body motion of (b) is next added to the resulting motion, which involves five plastic hinges, to arrive at the mechanism shown in Fig. 4.11f. Since the net effect of cancellation of the hinges at B and C is a reduction of internal work by an amount $8M_0\theta$, the work equation becomes

$$2Wa\theta + 2Wa\theta + Wa\theta = 8M_0\theta + 8M_0\theta + 8M_0\theta - 8M_0\theta$$

which gives the least upper bound

$$Wa = \tfrac{16}{5}M_0 = 3.2M_0 \qquad (22f)$$

It is apparent that Fig. 4.11f represents the actual collapse mechanism, subject to a possible adjustment of the location of the plastic hinge under the distributed load.

To perform the necessary statical check, we write down the three independent equations of equilibrium, derived from the virtual work principle applied to the mechanisms of Fig. 4.11b, c, and d. Thus

$$-M_2 + 2M_3 - M_4 = 0.5\,Wa$$

$$-M_1 + M_2 - M_5 + M_6 = Wa \qquad (23)$$

$$-M_2 + 2M_4 - 2M_5 + M_6 = Wa$$

The collapse mechanism of Fig. 4.11f requires $M_1 = -M_0$, $M_3 = M_0$, $M_5 = -M_0$, and $M_6 = M_0$. The above equations therefore furnish

$$M_2 = M_4 = 0.2M_0 \qquad Wa = 3.2M_0$$

It follows therefore that the plastic hinge in the left-hand rafter does in fact form at the central section where the bending moment attains a maximum. Since the distribution of bending moment in the frame is statically admissible, the load $W = 3.2M_0/a$ represents the actual collapse load for the given frame.[†]

(v) *Method of inequalities* The kinematical method of analysis, used in the preceding solutions, involves the estimation of the lowest of all possible upper bounds as the true collapse load. An alternative method, in which the collapse load is obtained as the highest possible lower bound, may also be employed. The method is based on the conditions of statical equilibrium, which must be satisfied under the restriction that none of the critical bending moments can exceed the local plastic moment in magnitude. For the gable frame considered above, the problem is to determine the highest value of Wa under the three constraints (23) imposed by statical equilibrium, and the twelve constraints

$$-M_0 \leqslant M_i \leqslant M_0 \qquad (i = 1, 2, \ldots, 6)$$

imposed by the yield condition. Mathematically, it represents a standard problem of linear programming, and efficient computer techniques are available for its solution. We shall describe here a simple method that brings out the basic features of a system of linear inequalities.[‡]

Consider, as an example, a rectangular portal frame whose dimensions and loading are shown in Fig. 4.12a. For simplicity, the right-hand column foot is assumed to be hinged, so that the degree of redundancy is reduced to two. All members of the frame are assumed to have the same fully plastic moment M_0. The

[†] Various experimental results on the plastic collapse of structures have been reported by J. F. Baker and J. Heyman, *Struct. Eng.*, **28**: 139 (1950); J. F. Baker and J. W. Roderick, *Proc. Inst. Civ. Eng.*, **1**: 71 (1952); C. G. Schilling, F. W. Schutz, and L. S. Beedle, *Welding J.*, Easton, Pa., **35**: 234-S (1956); G. C. Driscoll and L. S. Beedle, *Welding J.*, Easton, Pa., **36**: 275-S (1957). See also J. F. Baker, M. R. Horne, and J. Heyman, *Steel Skeleton*, vol. 2, Cambridge University Press (1956).

[‡] The method of inequalities was first used by B. G. Neal and P. S. Symonds, *J. Inst. Civ. Eng.*, **35**: 21 (1951). The problem has been identified as one of linear programming by A. Charnes and H. J. Greenberg, *Bull. Am. Math. Soc.*, **57**: 480 (1951), and further discussed by J. Heyman, *Proc. Inst. Civ. Eng.*, **12**: 39 (1959). See also J. Munro, *Civ. Eng. Public Works Rev.*, **60** (1965).

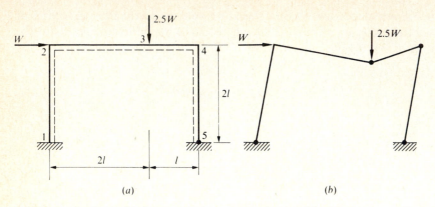

Figure 4.12 Loading and collapse mode for a rectangular portal frame.

two independent equations of equilibrium, obtained from the virtual works associated with the panel and beam mechanisms, are

$$-M_1 + M_2 - M_4 = 2Wl$$
$$-M_2 + 3M_3 - 2M_4 = 5Wl$$

(a)

The critical bending moments must also satisfy the set of linear inequalities $-M_0 \leqslant M_i \leqslant M_0$ (where $i = 1, 2, 3, 4$). Expressing M_1 and M_3 in terms of M_2 and M_4 by means of Eqs. (a), these inequalities may be written as

$$-M_0 \leqslant M_2 - M_4 - 2Wl \leqslant M_0 \qquad -M_0 \leqslant M_2 \leqslant M_0$$
$$-3M_0 \leqslant M_2 + 2M_4 + 5Wl \leqslant 3M_0 \qquad -M_0 \leqslant M_4 \leqslant M_0$$

We consider, next, those inequalities which contain M_2, and rewrite them in the form

$$-M_0 \leqslant M_2 \leqslant M_0$$
$$-M_0 + M_4 + 2Wl \leqslant M_2 \leqslant M_0 + M_4 + 2Wl \qquad (b)$$
$$-3M_0 + 2M_4 - 5Wl \leqslant M_2 \leqslant 3M_0 + 2M_4 - 5Wl$$

The necessary and sufficient condition for these continued inequalities to be simultaneously satisfied is that each left-hand side must be less than or equal to each right-hand side. Three of the resulting inequalities are identically satisfied, while the remaining six are

$$-M_0 \leqslant M_0 + M_4 + 2Wl \qquad\qquad -M_0 \leqslant 3M_0 - 2M_4 - 5Wl$$
$$-M_0 + M_4 + 2Wl \leqslant M_0 \qquad\qquad -M_0 + M_4 + 2Wl \leqslant 3M_0 - 2M_4 - 5Wl$$
$$-3M_0 + 2M_4 - 5Wl \leqslant M_0 \qquad\qquad -3M_0 - 2M_4 - 5Wl \leqslant M_0 + M_4 + 2Wl$$

Solving for M_4, these inequalities may be conveniently expressed in the form

$$-2M_0 - 2Wl \leqslant M_4 \leqslant 2M_0 - 2Wl$$

$$-2M_0 - \tfrac{5}{2}Wl \leqslant M_4 \leqslant 2M_0 - \tfrac{5}{2}Wl$$

$$-\tfrac{4}{3}M_0 - \tfrac{7}{3}Wl \leqslant M_4 \leqslant \tfrac{4}{3}M_0 - \tfrac{7}{3}Wl \tag{c}$$

$$-M_0 \leqslant M_4 \leqslant M_0$$

The last expression has been added to complete the set. We are not concerned here with the actual value of M_4 satisfying (c). Setting each left-hand side less than or equal to each right-hand side, the non-trivial inequities are obtained as

$$-8M_0 \leqslant Wl \leqslant 8M_0 \qquad -10M_0 \leqslant Wl \leqslant 10M_0$$

$$-20M_0 \leqslant Wl \leqslant 20M_0 \qquad -3M_0 \leqslant 2Wl \leqslant 3M_0$$

$$-6M_0 \leqslant 5Wl \leqslant 6M_0 \qquad -M_0 \leqslant Wl \leqslant M_0$$

The largest value of Wl satisfying all these inequalities is evidently equal to M_0. Hence the collapse load is $W = M_0/l$, in view of which inequalities (c) become

$$-4M_0 \leqslant M_4 \leqslant 0 \qquad -\tfrac{9}{2}M_0 \leqslant M_4 \leqslant -\tfrac{1}{2}M_0$$

$$-\tfrac{10}{3}M_0 \leqslant M_4 \leqslant -M_0 \qquad -M_0 \leqslant M_4 \leqslant M_0$$

and it follows that $M_4 = -M_0$. Substituting the values of Wl and M_4 into (b), these inequalities are found to be satisfied if $M_2 = 0$. Equations (a) then give $M_1 = -M_0$ and $M_3 = M_0$. These bending moments correspond to a state of collapse which occurs in the combined panel and beam mechanism, Fig. 4.12b. The work equation for this mode is $Wl = M_0$, in agreement with the statical analysis.

4.4 Displacements In Plane Frames

(i) *Formulation of the problem* The plastic analysis of structures is based on the assumption that changes in geometry prior to collapse are negligible, so that the equilibrium equations are essentially those for the undistorted frame. It is often desirable to estimate the displacements at the point of collapse to ensure that they are not large enough for the theoretical collapse load to be unrealistic. The analysis for the deflection is greatly simplified, without introducing serious errors, if the effects of strain-hardening and the spread of plastic zones are disregarded.[†] Each cross section of the beam therefore remains elastic until the local bending moment attains the fully plastic value. It is further assumed that the loads are increased in strict proportion to their collapse values, so that the rotation at a plastic hinge may never cease once it is formed.[‡] In general, one or more hinges would unload

[†] See P. S. Symonds and B. G. Neal, *J. Franklin Inst.*, **252**: 383 (1951), *J. Aeronaut. Sci.*, **19**: 15 (1952). For experimental confirmation, see P. S. Symonds, *Welding J.*, Easton, Pa., **31**: 33-S (1952).

[‡] This is not necessarily true even for proportional loading, as has been shown by L. Finzi, *9th Int. Congr. Appl. Mech.*, Brussels (1957).

during the loading program, but the deflection based on the above assumption may still be regarded as a reasonable approximation.[†]

The most convenient method of calculating the deflection involves a virtual work approach, in which the actual deformation of the frame is combined with arbitrary states of statical equilibrium.[‡] Since the deformation is elastic/plastic, the effect of hinge discontinuities must be included in the virtual work equation for the distortion of the members. The actual bending moment M at any section produces an elastic curvature $\kappa = M/EI$, so long as the magnitude of M is less than the plastic moment M_0. At a section where $|M| = M_0$, the hinge rotation that actually occurs is ϕ, having the same sign as that of M. In a statically indeterminate frame, there can be a distribution of bending moment m in equilibrium with zero external load. Since the work done by any virtual residual moment m^* on the actual deformation is $m^*\kappa$ per unit length of a typical member, the equation of virtual work may be written as

$$\sum m^*\phi + \int \frac{m^*M}{EI}\, ds = 0 \qquad (24)$$

where the summation extends over all the hinge discontinuities, and the integral extends over the lengths of all the members in the frame. The number of independent distributions of self-stressing moment m^* that can be constructed for a given frame is equal to the degree of redundancy r. The virtual work relation (24) therefore represents r independent equations of compatibility.

When the frame is entirely elastic, the hinge discontinuities are identically zero, and the compatibility equations in conjunction with the equations of equilibrium are sufficient to determine the actual bending moments. In an elastic/plastic frame prior to the instant of collapse, some of the statical unknowns are replaced by kinematical ones representing hinge rotations at the appropriate sections. The total number of unknowns at each stage of the deformation is therefore identical to the number of basic equations of equilibrium and compatibility. For a frame with straight members, the virtual moment m^* varies linearly along the members, and in the case of only concentrated loads acting on the frame, the actual moment M is also linearly distributed. Considering a typical straight segment AB of length l, the distributions of m^* and M may be written as

$$m^* = m_A^* + (m_B^* - m_A^*)\frac{s}{l} \qquad M = M_A + (M_B - M_A)\frac{s}{l}$$

where s is the distance measured from A, and the subscripts refer to the end moments. For a uniform segment AB of flexural rigidity EI, the integral in (24) is

---

[†] The effects of flexibility of joints on the deflection at collapse have been examined by B. G. Neal, *Struct. Eng.*, **38**: 224 (1960). The influence of geometry changes on the post-collapse behavior has been discussed by E. T. Onat, *J. Aeronaut. Sci.*, **22**: 681 (1955).

[‡] The virtual work approach has been proposed by J. Heyman, *Proc. Inst. Civ. Eng.*, **19**: 39 (1961). See also H. Tanaka, *Rep. Inst. Ind. Sci., Univ. Tokyo* (1961).

readily evaluated as

$$\int_A^B \frac{m^*M}{EI} \, ds = \frac{l}{6EI} \left[m_A^*(2M_A + M_B) + m_B^*(2M_B + M_A) \right] \tag{25}$$

Similar expressions can be developed for distributed loads acting along a given segment of the frame, but (25) will be sufficient for our purpose.

When all bending moments and hinge rotations have been found, the deflection at any given point of the frame can be determined from the virtual work principle. The actual deformation of the frame is combined with a distribution of virtual moment M^*, in equilibrium with a unit load applied at the point in the direction of the deflection δ, all other external loads being assumed zero. The virtual work equation then becomes

$$\delta = \sum M^*\phi + \int \frac{M^*M}{EI} \, ds \tag{26}$$

which includes the contributions from all the members and plastic hinges. The value of the integral for a typical uniform straight segment of length l, under a constant shearing force, is given by the right-hand side of (25) with M_A^* and M_B^* replacing m_A^* and m_B^* respectively.

As the load is gradually increased beyond the elastic limit, appropriate plastic hinges form at different stages and continue to undergo rotation. When the collapse load is just attained, the bending moment at the position of the last plastic hinge reaches the fully plastic value. The rotation at the last hinge is zero at the incipient collapse, while those at all other hinges conform in sign with the corresponding bending moments. This condition enables us to estimate the deflection at the point of collapse without having to carry out the complete elastic/plastic analysis. When the frame is collapsing, the distribution of bending moment is uniquely determined, but the hinge rotations are calculated to within one arbitrary quantity.

It is possible to carry out the analysis by assuming at the outset that a particular hinge is the last one to form. If the assumption is incorrect, the sign of at least one of the computed hinge rotations will be opposite to that of the corresponding bending moment. Since any incorrect solution can be derived from the correct one by the superposition of a backward motion of the collapse mechanism, the predicted deflection will be smaller than the actual deflection. The *displacement theorem* therefore states that if the deflection is calculated in turn by assuming each plastic hinge to form last, the largest of the predicted deflections is the actual one, provided no plastic hinge has ever unloaded during the process.†

(ii) *Deflections of simple frames* To illustrate the procedure, we consider the fixed base rectangular portal frame shown in Fig. 4.13a. All the members are of uniform

---

† The displacement theorem is due to P. G. Hodge, Jr., *Plastic Analysis of Structures*, p. 109, McGraw-Hill Book Co. (1959). See also C. E. Massonet and M. A. Save, *Plastic Analysis and Design*, p. 223, Blaisdell Publishing Co., Boston (1965).

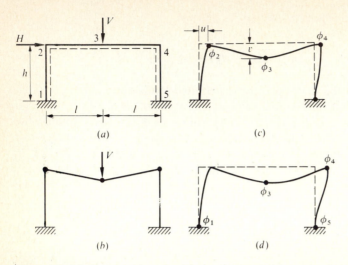

Figure 4.13 Displacements in a portal frame. (*a*) Loaded frame; (*b*) collapse mechanism for $Hh \leqslant 2M_0$; (*c*) deflected shape for $M_0 \leqslant Hh \leqslant 2M_0$; (*d*) deflected shape for $2M_0 \leqslant Hh \leqslant 3.1M_0$.

cross section, having a plastic moment M_0 and flexural rigidity EI. For sufficiently small values of H, the frame collapses in the manner shown in Fig. 4.13*b*, with plastic hinges appearing at sections 2, 3, and 4. It is a case of partial collapse in which one redundancy remains, the vertical load at collapse being given by $Vl = 4M_0$. The virtual residual moments at the critical sections must satisfy two independent equations of equilibrium, which can be immediately written down from (13) and (14). Thus

$$-m_1^* + m_2^* - m_4^* + m_5^* = 0$$

$$-m_2^* + 2m_3^* - m_4^* = 0$$

Since the frame has three redundancies, three independent distributions of m^* are needed, the particular cases shown in Table 4.1 being suitable for the purpose. Using Eq. (25), the integral in (24) corresponding to the distribution (*i*) for m^* is found as

$$\int \frac{m^* M}{EI}\, ds = \frac{h}{6EI}(3M_1 - 3M_0) + \frac{l}{6EI}(-M_0 + M_0) = \frac{h}{2EI}(M_1 - M_0)$$

Similar results are obtained for the other two distributions of m^*. The substitution of the three sets of values into (24) leads to the compatibility relations

$$\phi_2 + 0.5\phi_3 + \frac{h}{2EI}(M_1 - M_0) = 0 \qquad 0.5\phi_3 + \phi_4 + \frac{h}{2EI}(M_5 - M_0) = 0$$

$$\phi_2 + \phi_3 + \phi_4 + \frac{h}{6EI}(M_1 + M_5 - 4M_0) = 0$$

(27)

Table 4.1 Distributions of actual and virtual moments in a rectangular portal frame

Bending moment and hinge rotation	Critical section				
	1	2	3	4	5
M	M_1	$-M_0$	M_0	$-M_0$	M_5
ϕ	0	ϕ_2	ϕ_3	ϕ_4	0
m^* (i)	1	1	0.5	0	0
(ii)	0	0	0.5	1	1
(iii)	0	1	1	1	0
M^* (i)	$-h$	0	0	0	0
(ii)	$-l$	$-l$	0	0	0

These equations must be supplemented by the independent equation of equilibrium which may be taken as (13). To determine the unknown quantities, it is necessary to identify the last hinge where the rotation vanishes at the instant of collapse.

One of the two equations necessary for the unknown moments M_1 and M_5 is obtained by setting the values $M_2 = -M_0$, and $M_4 = -M_0$ in (13). The remaining equation is deduced by eliminating the hinge rotations from (27). The results are

$$M_5 - M_1 = Hh \qquad M_5 + M_1 = M_0$$

yielding

$$M_1 = \tfrac{1}{2}(M_0 - Hh) \qquad M_5 = \tfrac{1}{2}(M_0 + Hh)$$

As H increases from zero, M_1 decreases and M_5 increases, until Hh becomes equal to M_0, causing M_5 to reach the fully plastic value. The present analysis will therefore hold only for $Hh \leqslant M_0$. Inserting the values of M_1 and M_5, Eqs. (27) may be rearranged to give

$$\phi_2 = -\psi \qquad \phi_3 = \frac{M_0 h}{2EI}\left(1 + \frac{Hh}{M_0}\right) + 2\psi \qquad \phi_4 = -\frac{Hh^2}{2EI} - \psi \qquad (28)$$

where ψ is positive during collapse. If we assume $\psi = 0$ at the incipient collapse, ϕ_3 and ϕ_4 are seen to be positive and negative respectively, in accord with the signs of the bending moments at these sections. Since ϕ_2 vanishes at the point of collapse, the plastic hinge at this section is the last one to form.

The virtual unit load moment M^* is most conveniently obtained by introducing a cut at section 2 when $H = 1$ (with $V = 0$), and between sections 3 and 4 when $V = 1$ (with $H = 0$). The values of M^* at the critical sections for the two cases, given in Table 4.1, enable us to calculate the integral in (26) using Eq. (25). If u and v denote the horizontal and vertical displacements of the points of application of the loads H and V respectively, it is easily shown that

$$u = \frac{h^2}{6EI}(M_0 - 2M_1) \qquad v = l\psi + \frac{hl}{2EI}(M_0 - M_1) + \frac{M_0 l^2}{6EI}$$

Substituting for M_1, and remembering that $\psi = 0$ at the instant of collapse, the deflections at collapse are obtained as

$$u = \frac{Hh^3}{6EI} \qquad v = \frac{M_0 l^2}{6EI}\left[1 + \frac{3h}{2l}\left(1 + \frac{Hh}{M_0}\right)\right] \tag{29}$$

For $M_0 \leqslant Hh \leqslant 2M_0$, a plastic hinge with rotation ϕ_5 develops at the right-hand column foot, as shown in Fig. 4.13c, while the collapse load is still given by $Vl = 4M_0$. Since M_5 is now equal to M_0, the condition $M_5 - M_1 = Hh$ required by statical equilibrium gives $M_1 = M_0 - Hh$. Using these values of M_1 and M_5 in Eqs. (27), and introducing the modification due to the rotation at section 5, we obtain the compatibility relations

$$\phi_2 + 0.5\phi_3 - \frac{Hh^2}{2EI} = 0 \qquad 0.5\phi_3 + \phi_4 + \phi_5 = 0$$

$$\phi_2 + \phi_3 + \phi_4 - \frac{h}{6EI}(2M_0 + Hh) = 0$$

In terms of a positive angle ψ, representing the motion of the collapse mechanism, the solution of the above equations may be expressed as

$$\phi_2 = -\psi \qquad \phi_3 = \frac{Hh^2}{EI} + 2\psi \qquad \phi_5 = \frac{M_0 h}{3EI}\left(\frac{Hh}{M_0} - \frac{1}{2}\right)$$

$$\phi_4 = -\frac{M_0 h}{3EI}\left(\frac{5Hh}{2M_0} - 1\right) - \psi \tag{30}$$

Since $Hh \geqslant M_0$, the signs of the hinge rotations conform with those of the corresponding plastic moments if ψ vanishes at the point of collapse. Evidently, the last hinge to form is again at the left-hand end of the beam. The deflections at collapse are readily obtained by setting $\psi = 0$ and $M_1 = M_0 - Hh$ in the previous results. Thus

$$u = \frac{M_0 h^2}{6EI}\left(\frac{2Hh}{M_0} - 1\right) \qquad v = \frac{M_0 l^2}{6EI}\left(1 + \frac{3Hh^2}{M_0 l}\right) \tag{31}$$

It is important to note that the expression for ϕ_5 during collapse does not involve ψ, so that the hinge rotation at the right-hand column foot remains of the elastic order of magnitude.†

When $Hh = 2M_0$, the bending moment at the left-hand column foot reaches the value $-M_0$. For still higher values of Hh, the deformation mode at collapse involves a hinge rotation at this section, as indicated in Fig. 4.13d. The horizontal and vertical loads at collapse depend on one another according to (16c). The frame

---

† For the elastic/plastic analysis of portal frames made of work-hardening materials, see J. W. Roderick, *Struct. Eng.*, **38**: 245 (1960).

is statically determinate at collapse, the bending moments and hinge rotations at the numbered sections being as shown below

Section	1	2	3	4	5
M	$-M_0$	$Hh - 3M_0$	M_0	$-M_0$	M_0
ϕ	ϕ_1	0	ϕ_3	ϕ_4	ϕ_5

The compatibility equations for $Hh \geqslant 2M_0$ may be established as before, using the same distributions of the virtual moment m^* as those given in Table 4.1. Considering the special case $l = h$, the three relevant equations are easily shown to be

$$\phi_1 + 0.5\phi_3 - \frac{h}{6EI}(17M_0 - 5.5\,Hh) = 0$$

$$0.5\phi_3 + \phi_4 + \phi_5 - \frac{h}{6EI}(M_0 - 0.5\,Hh) = 0$$

$$\phi_3 + \phi_4 - \frac{h}{6EI}(14M_0 - 5\,Hh) = 0$$

It is convenient to express the hinge rotations, satisfying the above relations, in terms of an arbitrary angle ψ, the results being

$$\phi_1 = -\psi \qquad\qquad \phi_3 = \frac{M_0 h}{3EI}\left(17 - 5.5\frac{Hh}{M_0}\right) + 2\psi$$

$$\phi_4 = -\frac{M_0 h}{3EI}\left(10 - 3\frac{Hh}{M_0}\right) - 2\psi \qquad \phi_5 = \frac{M_0 h}{6EI}\left(4 - \frac{Hh}{M_0}\right) + \psi \tag{32}$$

For $2 < Hh/M_0 < \frac{34}{11} \simeq 3.09$, the values of ϕ_3 and ϕ_5 will be positive and that of ϕ_4 negative at the instant of collapse, if the last hinge forms at the left-hand column foot requiring $\psi = 0$.

Using Table 4.1 again for the distributions of the unit load moment M^*, in conjunction with Eq. (26), the displacements of the points of application of the loads H and V in their directions are found as

$$u = h\psi + \frac{M_0 h^2}{6EI}\left(5 - \frac{Hh}{M_0}\right) \qquad v = h\psi + \frac{M_0 h^2}{6EI}\left(17 - 5\frac{Hh}{M_0}\right) \tag{33}$$

When $2 \leqslant Hh/M_0 \leqslant 3.09$, the deflections at the incipient collapse are given by (33) with $\psi = 0$. For higher values of $Hh/M_0 \leqslant 4$, the last hinge will form at the center of the beam. The hinge rotations at the instant of collapse are then obtained by setting $\phi_3 = 0$ in (32), and the deflections at this point are obtained by using the corresponding value of ψ in (33). Figure 4.14 indicates how the deflections at collapse vary with the load ratio as H/V increases from zero to unity.

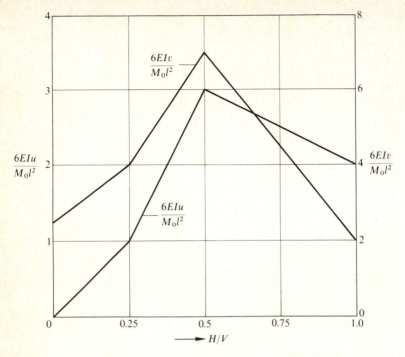

Figure 4.14 Deflections at collapse in a portal frame as functions of the load ratio.

(iii) *Deflections of complex frames* The method of analysis used for simple frames is equally applicable to complex frames. Consider, for example, the two-story rectangular frame of Fig. 4.15a, having a uniform plastic moment M_0 and flexural rigidity EI. The actual mode of collapse, shown in Fig. 4.15b, is most easily found by the method of combining mechanisms, the collapse load W being given by

$$12Wl\theta = 10M_0\theta \quad \text{or} \quad Wl = \tfrac{5}{6}M_0$$

The frame has six degrees of redundancy, and the collapse mechanism involves six plastic hinges, so that one redundancy remains in the state of collapse. The equations of equilibrium, based on the virtual work principle applied to the elementary panel, beam, and joint mechanisms, are easily shown to be

$$-M_4 + M_5 - M_7 + M_8 = Wl$$

$$-M_1 + M_2 - M_{10} + M_{11} = 3Wl$$

$$-M_5 + 2M_6 - M_7 = 4Wl$$

$$-M_3 + 2M_{12} - M_9 = 4Wl \tag{34}$$

$$M_2 - M_3 - M_4 = 0$$

$$M_8 + M_9 - M_{10} = 0$$

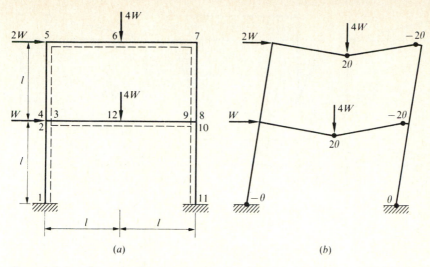

Figure 4.15 Loading and collapse mechanism for a two-story rectangular frame.

Only five of these equations are independent when the value of the collapse load is inserted. Substituting $M_1 = M_7 = M_9 = -M_0$, $M_6 = M_{11} = M_{12} = M_0$ and $Wl = 5M_0/6$ into (34), we get

$$M_3 = M_5 = -\tfrac{1}{3}M_0 \qquad M_4 = M_2 + \tfrac{1}{3}M_0$$
$$M_8 = M_2 + \tfrac{1}{2}M_0 \qquad M_{10} = M_2 - \tfrac{1}{2}M_0$$

These equations are supplemented by the compatibility equation

$$2(M_2 + M_4 + M_8 + M_{10}) + M_5 = M_0$$

which is obtained from (24) by using a distribution of residual moment with the only nonzero values $m_2^* = m_4^* = m_8^* = m_{10}^* = 1$. The assumed residual moments are self-equilibrating since they satisfy (34) with $W = 0$. The actual bending moment distribution, obtained by solving the above equations, is given by

$$M_1 = -M_6 = -M_0 \quad M_2 = \tfrac{1}{12}M_0 \quad M_3 = M_5 = -\tfrac{1}{3}M_0 \quad M_4 = \tfrac{5}{12}M_0$$

$$M_7 = M_9 = -M_0 \qquad M_8 = \tfrac{7}{12}M_0 \quad M_{10} = -\tfrac{5}{12}M_0 \qquad M_{11} = M_{12} = M_0 \tag{35}$$

To obtain the compatibility equations in terms of the hinge rotations, it is convenient to use the set of residual moment distributions† shown in Fig. 4.16. The patterns of residual moment are similar to those used before for the single-story frame. In view of (35), and the assumed distributions for m^*, the calculation

---

† Similar patterns of residual moments may be used for the analysis of multistory frames. See, for example, J. Heyman, *Proc. Inst. Civ. Eng.*, **19**: 39 (1961).

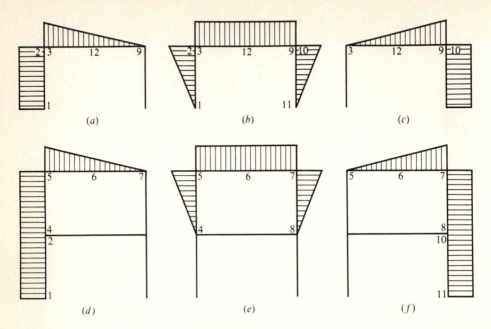

Figure 4.16 Distributions of virtual residual moments for the elastic/plastic analysis of rectangular frames.

based on equations (24) and (25) furnishes

$$\phi_1 + 0.5\phi_{12} = \frac{13}{72}\frac{M_0l}{EI} \qquad \phi_9 + \phi_{11} + 0.5\phi_{12} = -\frac{25}{72}\frac{M_0l}{EI} \qquad (a, b)$$

$$\phi_9 + \phi_{12} = -\frac{2}{9}\frac{M_0l}{EI} \qquad \phi_1 + 0.5\phi_6 = \frac{5}{36}\frac{M_0l}{EI} \qquad (c, d)$$

$$0.5\phi_6 + \phi_7 + \phi_{11} = -\frac{5}{36}\frac{M_0l}{EI} \qquad \phi_6 + \phi_7 = \frac{1}{18}\frac{M_0l}{EI} \qquad (e, f)$$

Since the redundant moment has already been found, only five of these equations are independent. The plastic hinge that forms last is identified by the fact that when the corresponding hinge rotation is set equal to zero, the remaining hinge rotations are found to have the same signs as those of the bending moments. It turns out that the last hinge forms at section 1, and this corresponds to

$$\phi_1 = 0 \qquad \phi_6 = \frac{5}{18}\frac{M_0l}{EI} \qquad \phi_7 = -\frac{1}{3}\frac{M_0l}{EI}$$

$$\phi_9 = -\frac{7}{12}\frac{M_0l}{EI} \qquad \phi_{11} = \frac{1}{18}\frac{M_0l}{EI} \qquad \phi_{12} = \frac{13}{36}\frac{M_0l}{EI}$$

(36)

at the incipient collapse. The rotation has the greatest magnitude at the right-hand hinge of the lower beam.

Consider, now, the horizontal deflection u of the top of the frame at the incipient collapse. The bending moment distribution for a unit horizontal load acting at joint 5 is readily obtained for a modified frame in which the left-hand columns can deform as a single cantilever fixed at the base. Since $\phi_1 = 0$, $M_1^* = -2l$ and $M_2^* = M_4^* = -l$, while all other virtual moments are identically zero, it follows from (26) and (25) that

$$u = \left(\frac{14}{3} - \frac{1}{2}\right)\frac{M_0 l^2}{6EI} = \frac{25}{36}\frac{M_0 l^2}{EI}$$

The vertical deflection v of the midpoint of the upper beam is obtained by considering a unit vertical load acting at this point (all other loads being zero). If we introduce two arbitrary cuts in the frame, one in the right-hand half of the upper beam, and the other anywhere in the lower beam, the nonzero virtual moments become $M_1^* = M_2^* = M_4^* = M_5^* = -l$. The deflection is then computed from (26) and (25) as

$$v = \left(\frac{11}{4} - \frac{7}{12}\right)\frac{M_0 l^2}{6EI} = \frac{13}{36}\frac{M_0 l^2}{EI}$$

If any of the other plastic hinges is assumed to be the last, the value of ϕ_1 would be found as positive at the incipient collapse, and the corresponding deflection would consequently be smaller than that obtained above. This conclusion is in accord with the postulate of the displacement theorem stated in Sec. 4.4(i).

4.5 Variable Repeated Loading

(i) *General considerations* When a structure, made of a nonhardening material, is loaded beyond the elastic limit, the bending moment in a part of the structure lies between the initial yield moment M_e and the fully plastic moment M_0. The change in bending moment that occurs on unloading the structure from the elastic/plastic state will be purely elastic so long as the magnitude of this change nowhere exceeds $2M_e$ (Sec. 3.4(ii)). If a structure is subjected to loads that are alternating in character, yielding may occur alternately in tension and compression in one or more critical sections. This phenomenon, known as *alternating plasticity*, may eventually cause failure by low-cycle fatigue. Another type of failure, due to repeated loading, is caused by certain critical combinations of loads being attained at definite intervals of time. As the number of loading cycles is increased, there is a progressive buildup of plastic deformation, and the structure eventually becomes unserviceable. This phenomenon is known as *incremental collapse* which is marked by a rapid increase in deflections with each loading cycle.

At any stage of the loading program, let M denote the actual bending moment at a given cross section, and $\mathcal{M}$ the bending moment which would exist at this section if the entire structure behaved elastically. If m is the residual moment produced on complete removal of the loads, the unloading process being regarded as wholly elastic, the relationship

$$M = m + \mathcal{M}$$

will hold at each cross section. Any distribution of residual moment must be in statical equilibrium with zero external loading. A structure is said to shake down under a given set of variable loads if a condition is reached at some instant such that all subsequent load applications produce only elastic changes of moments. When such a state is attained, the residual moments remain unchanged throughout the structure.

The algebraically greatest and least elastic moments that can be induced at a given cross section during the loading program will be denoted by $\mathcal{M}^+$ and $\mathcal{M}^-$ respectively. According to the *shakedown theorem* (Sec. 2.6(iv)), a framed structure will shake down under a set of loads varying between prescribed limits if it is possible to postulate a distribution of residual moment m such that the magnitude of the resultant bending moment $m + \mathcal{M}$ nowhere exceeds the local fully plastic moment M_0. The necessary and sufficient conditions for shakedown are expressed by the inequalities†

$$m_j + \mathcal{M}_j^+ \leqslant M_0 \qquad m_j + \mathcal{M}_j^- \geqslant -M_0 \qquad \mathcal{M}_j^+ - \mathcal{M}_j^- \leqslant 2M_e \qquad (37)$$

which must be satisfied at every cross section j. The last inequality is necessary to avoid the danger of alternating plasticity. The first two inequalities control the phenomenon of incremental collapse, and imply the last inequality only when the shape factor is unity. The residual moment in (37) need not be that which would actually exist in the frame after it has shaken down. The shakedown state is generally approached in an asymptotic manner, as the number of loading cycles is increased, but the total plastic strains in the frame are usually restricted to finite values.‡

The statical conditions (37) can be used to obtain a lower bound on the shakedown limit for any assumed distribution of the residual moment. An upper bound on the shakedown limit, from the standpoint of alternating plasticity, can be obtained by assuming a single plastic hinge to form at a section where the range of elastic bending moment $(\mathcal{M}^+ - \mathcal{M}^-)$ has its greatest value $2M_e$. To obtain an upper bound for incremental collapse, we assume a collapse mechanism consisting of sufficient number of yield hinges. Let θ_j denote the hinge rotation at a typical cross section j during a small motion of the collapse mechanism, and let M_{0j} be the plastic moment at this section. If m_j denotes the local residual moment at the

† A general proof of the shakedown theorem in the context of beams and frames has been given by E. Melan, *Prelim. Publ. 2nd Congr. Int. Assoc. Bridge Struct. Eng.*, **43**: Berlin (1936). For a simplified proof, see P. S. Symonds and W. Prager, *J. Appl. Mech.*, **17**: 315 (1950). See also B. G. Neal, *Q. J. Mech. Appl. Math.*, **4**: 78 (1951).

‡ The condition for shakedown of structures in the presence of thermal stresses has been discussed by W. Prager, *Br. Weld. J.*, **3**: 355 (1956).

incremental collapse, then

$$m_j + \mathcal{M}_j^+ = M_{0j} \qquad \text{if} \qquad \theta_j > 0$$
$$m_j + \mathcal{M}_j^- = -M_{0j} \qquad \text{if} \qquad \theta_j < 0$$

(38)

Since the unknown moments m_j are in equilibrium with zero external loads, $\sum m_j \theta_j = 0$ by the principle of virtual work, where the summation includes all the assumed hinges. This relation may be used to eliminate the residual moments from the equation obtained by summing the product of (38) with θ_j over all the plastic hinges of the collapse mechanism. The result may be put in the form

$$\sum \begin{Bmatrix} \mathcal{M}_j^+ \\ \mathcal{M}_j^- \end{Bmatrix} \theta_j = \sum M_{0j} |\theta_j|$$

(39)

where $\mathcal{M}_j^+$ is taken when θ_j is positive, and $\mathcal{M}_j^-$ when θ_j is negative. The actual load defining the shakedown limit is the smallest of all possible upper bound loads based on the considerations of alternating plasticity and incremental collapse.[†]

Since the conditions of the shakedown theorem include those of the lower bound theorem of plastic collapse, the limiting load for shakedown cannot exceed the load corresponding to static collapse. If the limiting value of the load is exceeded, failure would occur due to progressive hinge rotations or alternating plastic flow. The presence of initial stresses has no effect on the conditions for shakedown under a given set of loads. The order in which these loads are applied are also without effect on the occurrence of shakedown. However, the shakedown limit would depend on joint and support stiffnesses, since the elastic bending moment distribution is influenced by these factors.

(ii) *Shakedown in a portal frame* Consider a fixed-base rectangular portal frame of uniform cross section, shown in Fig. 4.17a, having a fully plastic moment equal to M_0. The frame is subjected to horizontal and vertical loads H and V respectively, which vary independently between prescribed limits.[‡] The sign convention for bending moments and hinge rotations will be taken as that used for static loading. The bending moment $\mathcal{M}$ at the five numbered sections, when the response of the frame is purely elastic, can be determined from the equations of equilibrium and compatibility. The two independent equations of equilibrium, associated with the virtual displacement systems of Fig. 4.17b and c, are

$$-\mathcal{M}_1 + \mathcal{M}_2 - \mathcal{M}_4 + \mathcal{M}_5 = 2Hl$$
$$-\mathcal{M}_2 + 2\mathcal{M}_3 - \mathcal{M}_4 = Vl$$

(40)

[†] Methods of estimation of the shakedown limit have been discussed by B. G. Neal and P. S. Symonds, *J. Inst. Civ. Eng.*, **35**: 186 (1950); P. S. Symonds and B. G. Neal, ibid., 41 (1950), *J. Franklin Inst.*, **252**: 283 (1951). See also B. G. Neal and P. S. Symonds, *Proc. Int. Assoc. Bridge Struct. Eng.*, **18**: 171 (1958).

[‡] This solution has been discussed by M. R. Horne, *Plastic Theory of Structures*, 2d ed., pp. 136–140, Pergamon Press, Oxford (1979)

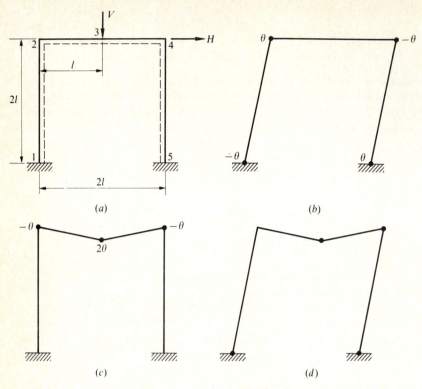

Figure 4.17 Incremental collapse of a portal frame under variable repeated loads $H(0, H_0)$ and $V(0, V_0)$.

The three independent equations of compatibility are easily obtained from (24) and (25), with M replaced by $\mathscr{M}$, and using the virtual residual moments of Table 4.1. Since the hinge discontinuities are zero in an elastic frame, the results are

$$6\mathscr{M}_1 + 8.5\mathscr{M}_2 + 3\mathscr{M}_3 + 0.5\mathscr{M}_4 = 0$$

$$0.5\mathscr{M}_2 + 3\mathscr{M}_3 + 8.5\mathscr{M}_4 + 6\mathscr{M}_5 = 0 \qquad (41)$$

$$2\mathscr{M}_1 + 7\mathscr{M}_2 + 6\mathscr{M}_3 + 7\mathscr{M}_4 + 2\mathscr{M}_5 = 0$$

Adding together the first two equations of (41), and combining the result with the third, we get

$$\mathscr{M}_2 + \mathscr{M}_3 + \mathscr{M}_4 = 0$$

Subtracting the second equation of (41) from the first gives

$$\mathscr{M}_1 - \mathscr{M}_5 + \tfrac{4}{3}(\mathscr{M}_2 - \mathscr{M}_4) = 0$$

The last two relations may be combined with (40) to obtain $\mathscr{M}_2$, $\mathscr{M}_3$, $\mathscr{M}_4$, and $\mathscr{M}_1 - \mathscr{M}_5$ in terms of Hl and Vl. The solution is then completed by substituting into any one of Eqs. (41). Thus

$$\mathcal{M}_1 = -\tfrac{4}{7}Hl + \tfrac{1}{12}Vl \qquad \mathcal{M}_2 = \tfrac{3}{7}Hl - \tfrac{1}{6}Vl \qquad \mathcal{M}_3 = \tfrac{1}{3}Vl$$

$$\mathcal{M}_4 = -\tfrac{3}{7}Hl - \tfrac{1}{6}Vl \qquad \mathcal{M}_5 = \tfrac{4}{7}Hl + \tfrac{1}{12}Vl \tag{42}$$

The algebraically greatest and least values of the elastic moments at the critical sections depend on the limits between which the loads are varied. Assuming, as a first example, $0 \leqslant H \leqslant H_0$ and $0 \leqslant V \leqslant V_0$, the values of $\mathcal{M}^+$ and $\mathcal{M}^-$ are readily found as those shown in the first two rows of Table 4.2.

Table 4.2 Maximum and minimum elastic moments for a portal frame

Cross section		1	2	3	4	5
$0 < H < H_0$	$\mathcal{M}^+$	$\tfrac{1}{12}V_0l$	$\tfrac{3}{7}H_0l$	$\tfrac{1}{3}V_0l$	0	$\tfrac{4}{7}H_0l + \tfrac{1}{12}V_0l$
$0 \leqslant V \leqslant V_0$	$\mathcal{M}^-$	$-\tfrac{4}{7}H_0l$	$-\tfrac{1}{6}V_0l$	0	$-\tfrac{3}{7}H_0l - \tfrac{1}{6}V_0l$	0
$-H_0 < H < H_0$	$\mathcal{M}^+$	$\tfrac{4}{7}H_0l + \tfrac{1}{12}V_0l$	$\tfrac{3}{7}H_0l$	$\tfrac{1}{3}V_0l$	$\tfrac{3}{7}H_0l$	$\tfrac{4}{7}H_0l + \tfrac{1}{12}V_0l$
$0 \leqslant V \leqslant V_0$	$\mathcal{M}^-$	$-\tfrac{4}{7}H_0l$	$-\tfrac{3}{7}H_0l - \tfrac{1}{6}V_0l$	0	$-\tfrac{3}{7}H_0l - \tfrac{1}{6}V_0l$	$-\tfrac{4}{7}H_0l$

Consider first the possibility of incremental collapse represented by the mechanism of Fig. 4.17b. The corresponding work equation, by (39), is found as

$$\tfrac{4}{7}H_0l\theta + \tfrac{4}{7}H_0l\theta + (\tfrac{3}{7}H_0l + \tfrac{1}{6}V_0l)\theta + (\tfrac{4}{7}H_0l + \tfrac{1}{12}V_0l)\theta = 4M_0\theta$$

or

$$H_0l + \tfrac{1}{8}V_0l = 2M_0 \tag{43a}$$

This relationship holds so long as the statical conditions (37) are satisfied at each section. The residual moments associated with this collapse mechanism must satisfy the yield conditions

$$m_1 - \tfrac{4}{7}H_0l = -M_0 \qquad m_2 + \tfrac{3}{7}H_0l = M_0$$

$$m_4 - \tfrac{3}{7}H_0l - \tfrac{1}{6}V_0l = -M_0 \qquad m_5 + \tfrac{4}{7}H_0l + \tfrac{1}{12}V_0l = M_0$$

In view of the interaction relation (43a) and the equilibrium equations

$$-m_1 + m_2 - m_4 + m_5 = 0$$

$$-m_2 + 2m_3 - m_4 = 0 \tag{44}$$

of which the first one is identically satisfied, the residual moments at the cardinal sections may be written as

$$m_1 = -M_0 + \tfrac{4}{7}H_0l \qquad m_2 = M_0 - \tfrac{3}{7}H_0l \qquad m_3 = \tfrac{1}{12}V_0l$$

$$m_4 = \tfrac{5}{3}M_0 - \tfrac{19}{21}H_0l \qquad m_5 = -\tfrac{1}{3}M_0 + \tfrac{2}{21}H_0l \tag{45}$$

The maximum and minimum values of the resultant bending moment at section 3 are

$$m_3 + \mathcal{M}_3^+ = \tfrac{5}{12}V_0l \qquad m_3 + \mathcal{M}_3^- = \tfrac{1}{12}V_0l$$

in view of (45) and Table 4.2. Since the magnitude of the resultant moment cannot exceed M_0, the validity of (43a) requires $0 \leqslant V_0 l \leqslant 2.4 M_0$, which is equivalent to $1.7 M_0 \leqslant H_0 l \leqslant 2.0 M_0$.

Consider now the incremental collapse mechanism of Fig. 4.17c. The work equation for this mechanism is obtained from (39) and the tabulated elastic moments. Thus

$$\tfrac{1}{6} V_0 l\theta + \tfrac{2}{3} V_0 l\theta + (\tfrac{3}{7} H_0 l + \tfrac{1}{6} V_0 l)\theta = 4 M_0\theta$$

or

$$\tfrac{3}{7} H_0 l + V_0 l = 4 M_0 \tag{43b}$$

Since this is a case of partial collapse, the distribution of the residual moment cannot be determined uniquely. The relevant equations are

$$m_2 + \mathcal{M}_2^- = -M_0 \qquad m_3 + \mathcal{M}_3^+ = M_0 \qquad m_4 + \mathcal{M}_4^- = -M_0$$

representing the conditions of yielding, and the first equation of (44) expressing the condition of statical equilibrium. Using Table 4.2, the residual moments may be written as

$$m_2 = -M_0 + \tfrac{1}{6} V_0 l \qquad m_3 = M_0 - \tfrac{1}{3} V_0 l \qquad m_4 = 3 M_0 - \tfrac{5}{6} V_0 l$$
$$m_5 - m_1 = 4 M_0 - V_0 l \tag{46}$$

A limit of applicability of (43b) is imposed by the condition that the resultant bending moments at sections 1 and 5 must lie between $-M_0$ and M_0. Since $\mathcal{M}_5^+ > \mathcal{M}_1^+$, $\mathcal{M}_5^- > \mathcal{M}_1^-$, and $m_5 > m_1$, the relevant inequalities are

$$m_5 + \mathcal{M}_5^+ \leqslant M_0 \qquad m_1 + \mathcal{M}_1^- \geqslant -M_0$$

giving

$$(m_5 - m_1) + (\mathcal{M}_5^+ - \mathcal{M}_1^-) \leqslant 2 M_0$$

Inserting the values $\mathcal{M}_5^+$, $\mathcal{M}_1^-$, and $m_5 - m_1$, and using (43b), we get $V_0 l \geqslant 3.535 M_0$, or $H_0 l \leqslant 1.085 M_0$. The largest permissible value of $V_0 l$ is evidently $4 M_0$, corresponding to $H_0 = 0$.

Assuming the mechanism of Fig. 4.17d for incremental collapse, the work equation is obtained as

$$\tfrac{4}{7} H_0 l\theta + \tfrac{2}{3} V_0 l\theta + 2(\tfrac{3}{7} H_0 l + \tfrac{1}{6} V_0 l)\theta + (\tfrac{4}{7} H_0 l + \tfrac{1}{12} V_0 l)\theta = 6 M_0\theta$$

or

$$H_0 l + \tfrac{13}{24} V_0 l = 3 M_0 \tag{43c}$$

Using the yield conditions and one of the equations of equilibrium (44), the residual moments associated with this collapse mechanism are found to be

$$m_1 = -M_0 + \tfrac{4}{7} H_0 l \qquad m_2 = 3 M_0 - \tfrac{3}{7} H_0 l - \tfrac{5}{6} V_0 l \qquad m_3 = M_0 - \tfrac{1}{3} V_0 l$$
$$m_4 = -M_0 + \tfrac{3}{7} H_0 l + \tfrac{1}{6} V_0 l \qquad m_5 = M_0 - \tfrac{4}{7} H_0 l - \tfrac{1}{12} V_0 l \tag{47}$$

The statical conditions of shakedown require $m_2 + \mathcal{M}_2^- \geqslant -M_0$ and $m_2 + \mathcal{M}_2^+ \leqslant M_0$. Inserting the values of the residual and elastic moments, these inequalities are reduced to $2.4M_0 \leqslant V_0l \leqslant 3.535M_0$, or $1.7M_0 \geqslant H_0l \geqslant 1.085M_0$. The interaction diagram for incremental collapse of the portal frame is compared with that for statical collapse (under the loads $H = H_0$ and $V = V_0$) in Fig. 4.18. It would be noted that for a given ratio H_0/V_0, the mode of collapse under statical loading is not necessarily the same as that under variable loading.

The conditions for alternating plasticity are easily derived by considering the values of $\mathcal{M}^+$ and $\mathcal{M}^-$ at each critical section of the frame. If the shape factor is denoted by α, alternating plasticity can occur simultaneously at sections 1 and 5 when H_0 and V_0 satisfy the relation

$$\alpha(\tfrac{4}{7}H_0l + \tfrac{1}{12}V_0l) = 2M_0$$

Similarly, the conditions for alternating plasticity to occur at sections 2 and 4 is

$$\alpha(\tfrac{3}{7}H_0l + \tfrac{1}{6}V_0l) = 2M_0$$

The occurrence of alternating plasticity at section 3 requires $\alpha V_0l = 6M_0$. The interaction diagram defined by these conditions lies outside that for incremental collapse for usual values of α. The shakedown limit in this case is entirely controlled by incremental collapse.

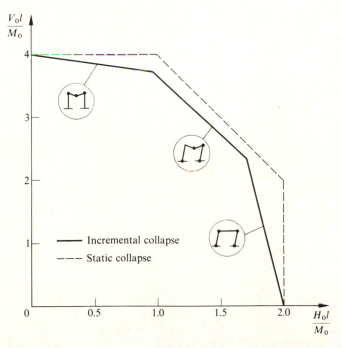

Figure 4.18 Interaction diagrams for a fixed-base portal frame under static and cyclic loading.

(iii) *Cyclic loading in an alternating sense* As a second example, consider the loading of the same frame with $-H_0 \leqslant H \leqslant H_0$ and $0 \leqslant V \leqslant V_0$. The horizontal force therefore acts in either direction in an alternating manner. The modified values of $\mathscr{M}^+$ and $\mathscr{M}^-$ at the critical sections are those given in the last two rows of Table 4.2. In this case, it is natural to expect that the shakedown limit would be controlled by alternating plasticity over a range of values of H_0/V_0. Using Table 4.2, the conditions of alternating plasticity may be written as

$$\alpha(\tfrac{8}{7}H_0 l + \tfrac{1}{12}V_0 l) = 2M_0 \qquad \text{sections 1 and 5} \qquad (48a)$$

$$\alpha(\tfrac{6}{7}H_0 l + \tfrac{1}{6}V_0 l) = 2M_0 \qquad \text{sections 2 and 4} \qquad (48b)$$

$$\tfrac{1}{3}\alpha V_0 l = 2M_0 \qquad \text{section 3} \qquad (48c)$$

where α is the shape factor for each member of the frame. These relations are represented graphically in Fig. 4.19, assuming $\alpha = 1.15$.

Since the horizontal force assumes both positive and negative values, there are two possible modes of incremental collapse associated with the panel mechanism. Either one of these modes may be used to establish the collapse relationship

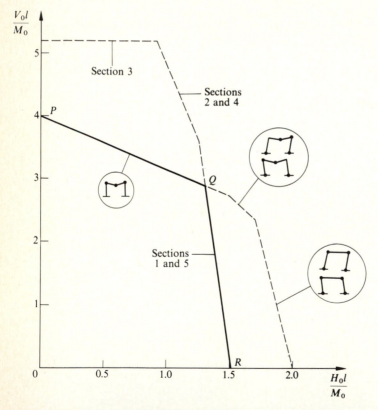

Figure 4.19 Lines of alternating plasticity and incremental collapse for a portal frame with alternating horizontal load and unidirectional vertical load.

between H_0 and V_0, and the result is identical to (43a). Considering the beam mechanism, the incremental collapse equation is obtained as

$$(\tfrac{3}{7}H_0l + \tfrac{1}{6}V_0l)\theta + \tfrac{2}{3}V_0l\theta + (\tfrac{3}{7}H_0l + \tfrac{1}{6}V_0l)\theta = 4M_0\theta$$

or

$$\tfrac{3}{7}H_0l + \tfrac{1}{2}V_0l = 2M_0 \tag{49}$$

The combined panel and beam mechanism gives two possible modes of collapse, and the corresponding equation relating H_0 and V_0 is the same as (43c). The graphical representation of the three collapse equations is shown in Fig. 4.19. The solid lines define the boundary PQR of a permissible region representing possible states of shakedown. The point Q corresponds to $H_0l = 1.312M_0$ and $V_0l = 2.875M_0$, obtained by solving equations (48b) and (49) with $\alpha = 1.15$.

A statical check for the relevant collapse mechanism, which happens to be the beam mechanism, can be carried out as before. It is easily shown that the distribution of residual moment is given by

$$m_2 = m_3 = m_4 = M_0 - \tfrac{1}{3}V_0l \qquad m_1 = m_5$$

Since $\mathscr{M}_5^+ = \mathscr{M}_1^+$ and $\mathscr{M}_5^- = \mathscr{M}_1^-$, all the conditions of the shakedown theorem will be fulfilled if $\mathscr{M}_5^+ - \mathscr{M}_5^- \leqslant 2M_0$. Since the expression on the left-hand side lies between $0.333M_0$ and $1.739M_0$, the inequality is indeed satisfied. Figure 4.19 indicates that the shakedown limit is controlled by alternating plasticity for sufficiently large values of H_0, while incremental collapse becomes the controlling factor when H_0 is sufficiently small.

(iv) *Analysis of a two-story frame* As a final example illustrating the basic principles, we consider the two-story rectangular frame shown in Fig. 4.20a. The frame has uniform cross section throughout with a plastic moment M_0 and flexural rigidity EI. The applied loads H and V are assumed to vary between zero and the maximum values H_0 and V_0 respectively. Since the frame has six redundancies, there must be five equations of equilibrium relating the eleven critical moments. These equations are found in the usual way by considering the virtual modes associated with the sway of lower and upper panels, the lower beam mechanism, and the rotation of joints. Thus, the equilibrium of the elastic moments requires

$$-\mathscr{M}_1 + \mathscr{M}_2 - \mathscr{M}_9 + \mathscr{M}_{10} = 3Hl$$

$$-\mathscr{M}_4 + \mathscr{M}_5 - \mathscr{M}_6 + \mathscr{M}_7 = Hl$$

$$-\mathscr{M}_3 + 2\mathscr{M}_{11} - \mathscr{M}_8 = Vl \tag{50}$$

$$\mathscr{M}_2 - \mathscr{M}_3 - \mathscr{M}_4 = 0$$

$$-\mathscr{M}_7 - \mathscr{M}_8 + \mathscr{M}_9 = 0$$

These equations are supplemented by six equations of compatibility, which are most conveniently obtained from (24) and (25) using the virtual residual moment

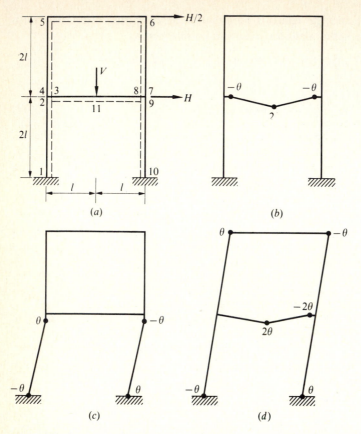

Figure 4.20 A two-story frame with possible collapse mechanisms. The loads H and V vary in the ranges $(0, H_0)$ and $(0, V_0)$ respectively.

patterns of Fig. 4.16. Taking account of the changes in the assigned numbers, the resulting equations are easily shown to be

$$6(\mathcal{M}_1 + \mathcal{M}_2) + 2.5\mathcal{M}_3 + 0.5\mathcal{M}_8 + 3\mathcal{M}_{11} = 0 \qquad (51a)$$

$$2(\mathcal{M}_1 + \mathcal{M}_{10}) + 4(\mathcal{M}_2 + \mathcal{M}_9) + 3(\mathcal{M}_3 + \mathcal{M}_8) + 6\mathcal{M}_{11} = 0 \qquad (51b)$$

$$0.5\mathcal{M}_3 + 2.5\mathcal{M}_8 + 6(\mathcal{M}_9 + \mathcal{M}_{10}) + 3\mathcal{M}_{11} = 0 \qquad (51c)$$

$$3(\mathcal{M}_1 + \mathcal{M}_2 + \mathcal{M}_4) + 5\mathcal{M}_5 + \mathcal{M}_6 = 0 \qquad (51d)$$

$$\mathcal{M}_4 + \mathcal{M}_7 + 5(\mathcal{M}_5 + \mathcal{M}_6) = 0 \qquad (51e)$$

$$\mathcal{M}_5 + 5\mathcal{M}_6 + 3(\mathcal{M}_7 + \mathcal{M}_8 + \mathcal{M}_{10}) = 0 \qquad (51f)$$

This completes the formulation of the problem of finding the distribution of elastic moment in the frame.

It is convenient to work out the solution in two parts. First, we add together in pairs Eqs. (51a) and (51c), Eqs. (51d) and (51f), and the last two equations of (50).

The resulting equations, in conjunction with (51b), (51e), and the third equation of (50), form a set, having the solution

$$\mathcal{M}_1 + \mathcal{M}_{10} = \tfrac{3}{28}Vl \qquad \mathcal{M}_2 + \mathcal{M}_9 = -\tfrac{3}{14}Vl \qquad \mathcal{M}_3 + \mathcal{M}_8 = -\tfrac{11}{28}Vl$$

$$\mathcal{M}_4 + \mathcal{M}_7 = \tfrac{5}{28}Vl \qquad \mathcal{M}_5 + \mathcal{M}_6 = -\tfrac{1}{28}Vl \qquad \mathcal{M}_{11} = \tfrac{17}{56}Vl \tag{52}$$

Next, we consider the differences of the same pairs of equations, and complete the set with the first two equations of (50). The solution for this set of equations is found as

$$\mathcal{M}_1 - \mathcal{M}_{10} = -\tfrac{97}{55}Hl \qquad \mathcal{M}_2 - \mathcal{M}_9 = \tfrac{68}{55}Hl \qquad \mathcal{M}_3 - \mathcal{M}_8 = \tfrac{87}{55}Hl$$

$$\mathcal{M}_4 - \mathcal{M}_7 = -\tfrac{19}{55}Hl \qquad \mathcal{M}_5 - \mathcal{M}_6 = \tfrac{36}{55}Hl \tag{53}$$

Values of the elastic bending moments follow immediately from (52) and (53). The algebraic maximum and minimum elastic moments which occur at the critical sections, as the loads vary between the prescribed limits, are given in the second and third columns of Table 4.3.

Table 4.3 Data for shakedown analysis for the frame of Fig. 4.20

Section	Arbitrary H_0 and V_0			$H_0 l = 1.169 M_0$ $V_0 l = 3.076 M_0$	
	$\mathcal{M}^+$	$\mathcal{M}^-$		$\mathcal{M}^+/M_0$	$\mathcal{M}^-/M_0$
1	$0.0536 V_0 l$	$-0.8818 H_0 l$		0.165	-1.031
2	$0.6182 H_0 l$		$-0.1071 V_0 l$	0.723	-0.329
3	$0.7909 H_0 l$		$-0.1964 V_0 l$	0.924	-0.604
4	$0.0893 V_0 l$	$-0.1727 H_0 l$		0.275	-0.202
5	$0.3273 H_0 l$		$-0.0179 V_0 l$	0.383	-0.055
6	0	$-0.3273 H_0 l$	$-0.0179 V_0 l$	0	-0.438
7	$0.1727 H_0 l + 0.0893 V_0 l$	0		0.476	0
8	0	$-0.7909 H_0 l$	$-0.1964 V_0 l$	0	-1.528
9	0	$-0.6182 H_0 l$	$-0.1071 V_0 l$	0	-1.052
10	$0.8818 H_0 l + 0.0536 V_0 l$	0		1.196	0
11	$0.3036 V_0 l$	0		0.934	0

The static collapse of the frame, produced by the applied loads monotonically increased to their maximum values, is easily investigated by the method of combined mechanisms. Among the large number of possible modes of collapse, those shown in Fig. 4.20b, c, and d are found to be most critical for static collapse, the corresponding collapse equations being

$$V_0 l = 4M_0 \qquad 3H_0 l = 4M_0 \qquad 4H_0 l + V_0 l = 8M_0$$

The interaction diagram defined by these relations is shown by the dashed lines of Fig. 4.21. That no other collapse mode is critical may be confirmed by a statical analysis for each of the load ratios represented by points A and B in the diagram.

It is reasonable to consider the same three mechanisms for the possibility of

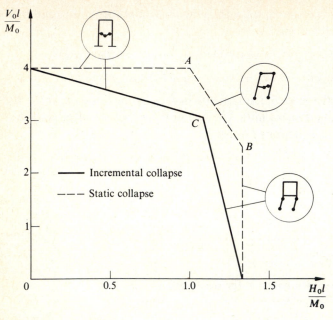

Figure 4.21 Interaction diagrams for a two-story rectangular frame.

incremental collapse. The work equations corresponding to mechanisms (b), (c), and (d) are found as before, using Table 4.3, and the results are†

$$0.7909 H_0 l + V_0 l = 4M_0$$

$$3H_0 l + 0.1607 V_0 l = 4M_0 \qquad (54)$$

$$4H_0 l + 1.0715 V_0 l = 8M_0$$

respectively. The interaction diagram for incremental collapse, shown by solid lines in Fig. 4.21 indicates that the combined panel and beam mechanism does not occur at all for the present range of loading.

To confirm that no other mode is critical, a statical analysis will be carried out for the loads corresponding to the vertex C. The solution of the first two equations of (54) gives $H_0 l = 1.169 M_0$ and $V_0 l = 3.076 M_0$ at C, leading to the maximum and minimum elastic moments shown in the last two columns of Table 4.3. Since seven plastic hinges are simultaneously formed by mechanisms (b) and (c), we have

$$m_1 + \mathscr{M}_1^- = -M_0 \qquad m_2 + \mathscr{M}_2^+ = M_0 \qquad m_3 + \mathscr{M}_3^- = -M_0$$

$$m_8 + \mathscr{M}_8^- = -M_0 \qquad m_9 + \mathscr{M}_9^- = -M_0 \qquad m_{10} + \mathscr{M}_{10}^+ = M_0$$

$$m_{11} + \mathscr{M}_{11}^+ = M_0$$

---

† J. Heyman, *Plastic Design of Frames*, vol. 2, pp. 153–158, Cambridge University Press (1971). The solution is identical to that in which all the loads vary independently between their limiting values.

Substituting the values of the elastic moments from Table 4.3, we obtain the corresponding residual moments

$$m_1 = 0.031M_0 \qquad m_2 = 0.277M_0 \qquad m_3 = -0.396M_0$$

$$m_8 = 0.528M_0 \qquad m_9 = 0.052M_0 \qquad m_{10} = -0.196M_0 \qquad m_{11} = 0.066M_0$$

The equations of equilibrium for the residual moments are given by (50) with $H = V = 0$ and m written for $\mathcal{M}$. The first and third equations are identically satisfied, while the remaining equations furnish

$$m_4 = 0.673M_0 \qquad m_7 = -0.476M_0 \qquad m_5 - m_6 = 1.149M_0$$

The shakedown theorem requires the resultant bending moments at sections 5 and 6 to lie between $-M_0$ and M_0. It follows from Table 4.3 that any value of m_5 in the range $0.587M_0 \leqslant m_5 \leqslant 0.617M_0$ will satisfy this condition. The greatest value of $\mathcal{M}^+ - \mathcal{M}^-$ is $1.528M_0$, occurring at sections 3 and 8 at the ends of the lower beam. Hence, to avoid the danger of alternating plasticity, the shape factor must be less than 1.31 approximately.

4.6 Minimum Weight Design

(i) *Basic concepts* The problem of plastic design is one in which the loads acting on the structure are given, and we are required to find the plastic moments of the various members so that the structure is at the point of collapse when the loads are multiplied by a specified load factor. If the ratios of the plastic moments of the members are chosen at the outset, the methods of plastic analysis are directly applicable for the estimation of the characteristic plastic moment corresponding to the state of collapse. In general, however, the design problem involves the determination of all the plastic moments, whose relative values are not preassigned, so that either the total weight of the structure or its total cost has a minimum value. We shall be concerned here with the plastic design of structures for minimum weight without consideration of the economic and other factors.

Suppose that a frame of given overall dimensions is composed of uniform prismatic members whose plastic moments are taken to be unaffected by shearing and axial forces. When the members have geometrically similar cross sections, and are made of the same material, the weight g per unit length is proportional to h^2, where h is a typical dimension such as the depth of the cross section. Since the fully plastic moment M_0 is proportional to h^3, the relationship $g \propto M_0^{2/3}$ must hold for any given series of similar sections. The empirical relation $g \propto M_0^{0.6}$ is found to agree closely with the tabulated values of g and M_0 for standard I sections. More generally, when geometrical similarities do not hold, the relationship between the specific weight and the plastic moment may be approximated by $g \propto M_0^n$, where n is a physical constant. For the range of sections appropriate to a given problem, it is reasonable to use the linear relationship

$$g = \alpha + \beta M_0$$

where α and β are constants. If l denotes the length of a typical member having a plastic moment M_0, the total structural weight is

$$\alpha \sum l + \beta \sum M_0 l$$

where the summation includes all the members of the frame. The weight of the frame will therefore be a minimum if the parameter

$$G = \sum M_0 l$$

is minimized. The problem of minimum weight design is reduced to that of finding the minimum value of G, which is known as the *weight function*.

It is possible to achieve an absolute minimum weight design if the cross section of each member is allowed to vary continuously in such a way that the fully plastic moment at any section is equal to the magnitude of the local bending moment. The problem is to find an equilibrium distribution of bending moment M that minimizes the linearized weight function

$$G = \int |M| \, ds$$

where the integral extends over all the members. The minimum weight frame then deflects in a form involving no discontinuities in slope, and hence localized hinges do not occur in the collapse mechanism.†

For given external loads acting on a frame, an equilibrium distribution of moment M will correspond to an absolute minimum weight if the deflected shape of the frame at the incipient collapse has a constant absolute curvature, the sign of the curvature being in conformity with that of the bending moment. To prove this theorem, let M^* denote any other distribution of bending moment in equilibrium with the same external loads. By the principle of virtual work, the work done by the external loads on the deflection corresponding to a constant absolute curvature κ_0 is equal to the internal work, which must be the same in both cases. Hence

$$\int |M| \kappa_0 \, ds = \int M^* \kappa \, ds$$

where $|\kappa| = \kappa_0$, but the sign of M^* need not be the same as that as κ. Consequently:

$$\int |M| \, ds = \int M^* \left(\frac{\kappa}{\kappa_0} \right) ds \leqslant \int |M^*| \, ds$$

indicating that the weight of the frame corresponding to the design $|M|$ cannot exceed that corresponding to $|M^*|$. The curvature of the minimum weight frame changes discontinuously from κ_0 to $-\kappa_0$ at a point of inflection where M changes sign from positive to negative.

Once the inflection points have been located from the appropriate kinematical and statical conditions, it is easy to construct the bending moment diagram giving the required variation of the yield moment. In practice, the beam section cannot be reduced to zero at the points of inflection, since some material must be provided to

---

† J. Heyman, *Q. J. Mech. Appl. Math.*, **12**: 314 (1959), *Int. J. Mech. Sci.*, **1**: 121 (1960); M. A. Save and W. Prager, *J. Mech. Phys. Solids*, **11**: 255 (1963).

carry the shearing forces that are neglected in the theory. The theoretical design furnishes valuable information on the maximum saving of material, and provides a basis of comparison for the practical design of frames.

(ii) *Geometrical approach to limit design* Consider the minimum weight design of a fixed-base rectangular portal frame, Fig. 4.22a, made of prismatic members which may be chosen from a wide range of available sections. Both the columns are

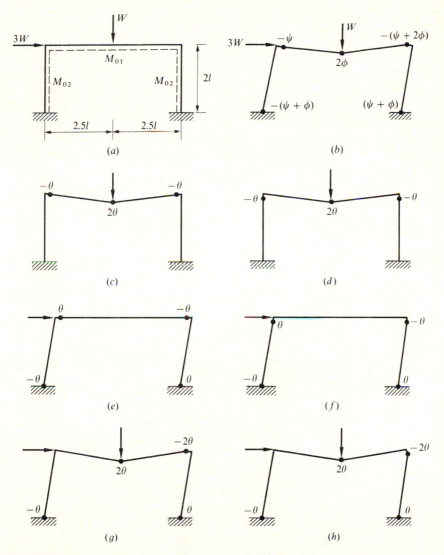

Figure 4.22 Limit design of a rectangular portal frame. (*a*) Loading and dimensions; (*b*) collapse mechanism with two degrees of freedom; (*c–h*) single-degree-of-freedom mechanisms.

assumed to have the same plastic moment M_{02}, which could be different from the plastic moment M_{01} of the beam, so that the weight function becomes

$$G = (5M_{01} + 4M_{02})l \qquad (55)$$

Since it is not known in advance whether M_{01} is greater or less than M_{02}, any plastic hinge occurring at the joints may appear either in the beam or in the column. Each of the three possible modes of collapse is therefore associated with two mechanisms of the same type, resulting in six possible collapse mechanisms as shown in Fig. 4.22c to h. The corresponding work equations are easily found as

$$4M_{01} = 2.5 \ Wl \qquad 2M_{01} + 2M_{02} = 2.5 \ Wl \qquad (c, d)$$

$$2M_{01} + 2M_{02} = 6 \ Wl \qquad 2M_{02} = 3.0 \ Wl \qquad (e, f)$$

$$4M_{01} + 2M_{02} = 8.5 \ Wl \qquad 2M_{01} + 4M_{02} = 8.5 \ Wl \qquad (g, h)$$

The equations on the left-hand side correspond to $M_{01} \leqslant M_{02}$ and those on the right-hand side correspond to $M_{01} \geqslant M_{02}$.

These equations are represented by three pairs of straight lines on a plane diagram, Fig. 4.23, the meeting point of each pair being on the bisector of the coordinate axes. Any given ratio of the plastic moments is represented by a ray, such as ON, interesting some of the mechanism lines. In view of the upper bound theorem of plastic collapse, the values of M_{01} and M_{02} at the incipient collapse are given by the coordinates of the point N, which is the farthest point of intersection relative to the origin. It follows that all possible combinations of M_{01} and M_{02}

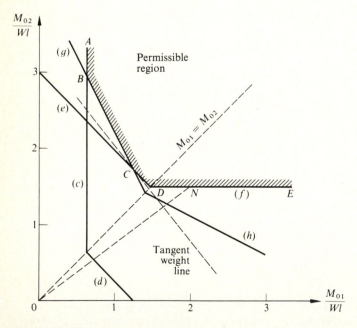

Figure 4.23 Geometrical analog for the plastic design of a rectangular portal frame.

required by the frame under the given loads are represented by the line segments *ABCDE* which is convex toward the origin. The shaded region is called the *permissible region*, since plastic collapse does not occur for any design represented by a point lying in this region. Points on the origin side of the boundary *ABCDE* represent designs which cannot support the given loads.†

The problem of minimum weight design reduces to that of locating a point on the boundary of the permissible region such that the weight function (55) is a minimum. Any point on the straight line

$$5M_{01} + 4M_{02} = \text{const}$$

represents a design of constant weight, the value of G being proportional to the distance of the line from the origin. The minimum weight design is obtained by finding the point where such a straight line is just in contact with the boundary of the permissible region. The appropriate *weight line* is shown broken in Fig. 4.23, the point of contact being the apex C which corresponds to

$$M_{01} = \tfrac{5}{4}Wl \qquad M_{02} = \tfrac{7}{4}Wl$$

giving the minimum weight design for the given frame. By (55), the minimum value of the weight function G is $13.25Wl^2$. For certain dimensions of the frame, the tangent weight line may coincide with one of the inclined faces of the boundary of the permissible domain. In this case, there are infinitely many designs, giving the same minimum weight, two of which are represented by the extremities of the associated line segment.

Since the minimum weight point C is the meeting point of the lines corresponding to mechanisms (*e*) and (*g*), the minimum weight frame may collapse by either of these two modes. A collapse mechanism having two degrees of freedom can actually occur with hinge rotations as shown in Fig. 4.22b. This is formed by a combination of the mechanisms (*e*) and (*g*), after replacing θ in the two cases by ψ and ϕ respectively. Such a mechanism represents an *overcomplete collapse*, since there are more plastic hinges than are necessary for the frame to be statically determinate. The work equation for the new mechanism is easily shown to be

$$2(\psi + 2\phi)M_{01} + 2(\psi + \phi)M_{02} = 6(\psi + \phi)Wl \qquad (56)$$

which is valid for $\psi \geqslant 0$ and $\phi \geqslant 0$. For any given positive values of ψ and ϕ, Eq. (56) represents a straight line on the design plane of Fig. 4.23. The slope of this line lies between the slopes of the mechanism lines (*e*) and (*g*), which correspond to $\phi = 0$ and $\psi = 0$ respectively. Hence, there exists a definite ratio of ψ to ϕ for which the straight line given by (56) has the same slope as the weight line given by (55). Equating the ratios of the coefficients of M_{01} and M_{02}, we have

$$\frac{\psi + 2\phi}{\psi + \phi} = \frac{5}{4} \qquad \text{or} \qquad \psi = 3\phi$$

† The geometrical method has been put forward by J. Foulkes, *Q. Appl. Math.*, **10**: 347 (1953). A trial-and-error method has been discussed by J. Heyman, *Struct. Eng.*, **31**: 125 (1953).

Substituting this value of ψ into (56), the work equation is reduced to

$$5M_{01} + 4M_{02} = 13.25Wl$$

which coincides with the equation for the tangent weight line. The collapse mechanism obtained by setting $\psi = 3\phi$ is regarded as *weight compatible*, since the ratio of the coefficients of the plastic moments in the associated work equation is identical to the ratio of the corresponding coefficients in the weight function. It follows that the sum of the magnitudes of the hinge rotations in an individual member corresponding to a weight compatible mechanism is proportional to the length of this member.

(iii) *Minimum weight theorems* The two-dimensional geometrical analog can be extended to n dimensions for the minimum weight design of a frame having n independent plastic moments. Since each work equation is linear in the plastic moments, it will represent a hyperplane in an n-dimensional design space. The hyperplanes corresponding to all possible mechanisms of collapse will define a permissible region whose boundary will be convex with respect to the origin. The linearized weight function G also represents a hyperplane in n dimensions, and the weight flat will generally intersect the permissible region. As the weight of the frame is reduced, the weight flat moves toward the origin, and the minimum value of G makes the weight flat just touch the boundary of the permissible region. The weight flat then passes through a vertex formed by the intersection of at least n mechanism flats. The collapse mechanism for the minimum weight frame is obtainable by a linear combination of n or more mechanisms, and has at least n degrees of freedom.

Consider a minimum weight frame whose prismatic members are of length l_j and plastic moment M_{0j} corresponding to a set of loads W_i. Suppose it is possible to postulate a weight compatible mechanism of collapse, so that the sum of the magnitudes of the hinge rotations associated with a typical member is $\phi_j = cl_j$, where c is a constant. If the displacements corresponding with the loads W_i are denoted by δ_i, the work equation may be written as

$$\sum_i W_i\delta_i = \sum_j M_{0j}\phi_j = c\sum_j M_{0j}l_j$$

Such a mechanism may be regarded as virtual for any other design of the frame specified by plastic moments M_{0j}^* of its members. Since the bending moment at a hinge occurring in a typical member cannot exceed M_{0j}^* in magnitude, $\sum W_i\delta_i$ must be less than or equal to $c\sum M_{0j}^*l_j$. If the weight functions for the two designs are denoted by G and G^*, then

$$G^* = \sum_j M_{0j}^*l_j \geqslant \sum_j M_{0j}l_j = G$$

This brings us to *Foulkes' theorem* which states that the design of a frame will be of minimum weight for a given set of loads if a weight compatible mechanism can be formulated for plastic collapse and the corresponding distribution of bending

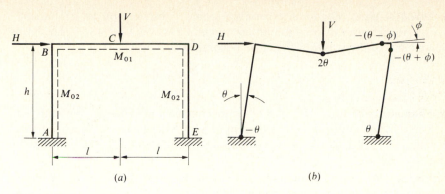

Figure 4.24 Rectangular portal frame with weight compatible collapse mechanism for a range of values of H/V and h/l.

moment is found to be statically admissible.† The theorem includes the possibility of the existence of a range of designs all of which have the same minimum weight.

As an interesting example illustrating the use of Foulkes' theorem, consider the minimum weight design of the fixed-based rectangular portal frame shown in Fig. 4.24a. The required plastic moments M_{01} and M_{02} for the beam and the columns respectively will obviously depend on the ratios h/l and H/V, which specify the geometry and loading of the frame. A design chart can be constructed by assuming a particular collapse mechanism with two degrees of freedom and finding the conditions under which the corresponding design will be valid. Taking, for instance, the mechanism of Fig. 4.24b, the work equation is easily shown to be

$$Vl\theta + Hh\theta = (3\theta - \phi)M_{01} + (3\theta + \phi)M_{02} \tag{57}$$

where θ and ϕ are positive hinge rotations. Equating the coefficients of θ and ϕ gives

$$M_{01} = M_{02} = \tfrac{1}{6}(Vl + Hh)$$

The linearized weight function is $G = 2(M_{01}l + M_{02}h)$, which shows that the mechanism will be weight compatible if

$$\frac{3\theta + \phi}{3\theta - \phi} = \frac{h}{l} \qquad \text{or} \qquad \phi = 3\left(\frac{h - l}{h + l}\right)\theta$$

in view of (57). Since the hinge rotations must accord in sign with the corresponding bending moments, we have the restriction $-\theta < \phi < \theta$, giving $l < 2h < 4l$.

The bending moment at B can be determined from the virtual work equation corresponding to either the panel or the beam mechanism. The latter furnishes the equation

$$-M_B + 2M_C - M_D = Vl \qquad \text{or} \qquad M_B = 3M_{01} - Vl$$

† See J. Foulkes, *Proc. R. Soc.*, A, **233**: 482 (1954).

Since $-M_{01} \leqslant M_B \leqslant M_{01}$ for the bending moment distribution to be statically admissible, the design will be of minimum weight if $2M_{01} \leqslant Vl \leqslant 4M_{01}$. The limits of applicability of the solution therefore become

$$\frac{1}{2} \leqslant \frac{Hh}{Vl} \leqslant 2 \qquad \frac{1}{2} \leqslant \frac{h}{l} \leqslant 2$$

Similar limits can be found for the other mechanisms of collapse, the complete results being shown in Fig. 4.25. The minimum weight design corresponding to any given values of h/l and Hh/Vl is represented by one of seven regions bounded by straight lines parallel to the axes.

Upper and lower bound theorems, analogous to those for plastic analysis, can be established in relation to the design for minimum weight. Any design of a frame which just collapses in a certain mechanism, while admitting a statically admissible distribution of bending moment, provides an upper bound on the minimum weight.

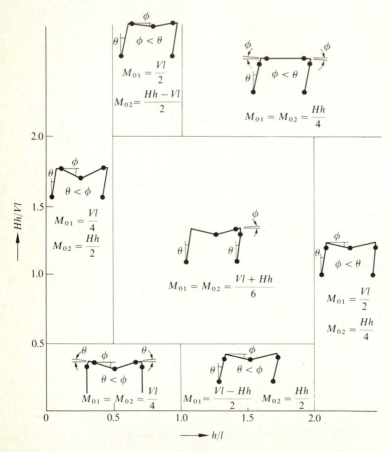

Figure 4.25 Minimum weight solutions for a portal frame loaded as shown in Fig. 4.24a.

In geometrical terms, the point representing the design of the frame will then lie on the boundary of the permissible region. The mechanism condition could be omitted from this theorem without affecting its validity, the design point being then considered within the permissible region. Any design of a frame which is based on a weight-compatible mechanism, without satisfying the condition of static admissibility, provides a lower bound on the minimum weight. In the geometrical representation, the weight hyperplane can always be passed through a vertex that is exterior to the permissible region (see Fig. 4.23). Using suitable positive values of the associated mechanism parameters, the weight flat can be represented by a combined mechanism with appropriate degrees of freedom.

(iv) *Limit design of complex frames* The minimum weight problem can be easily set up for solution by digital computers using the standard techniques of linear programming. To illustrate the basic principles, consider a two-story rectangular frame loaded as shown in Fig. 4.26a. The upper and lower beams have plastic moments M_{01} and M_{03} respectively, the two upper stanchions each have a plastic moment M_{02}, and the two lower stanchions each have a plastic moment M_{04}. These moments must be such that the frame just collapses under the given loads, while the weight function

$$G = 2M_{01}l + 3M_{02}l + 2M_{03}l + 3M_{04}l \tag{58}$$

has its least possible value. Since the frame has six redundancies and twelve potentially critical bending moments, there must be six independent equations of equilibrium. By the principle of virtual work applied to the six elementary

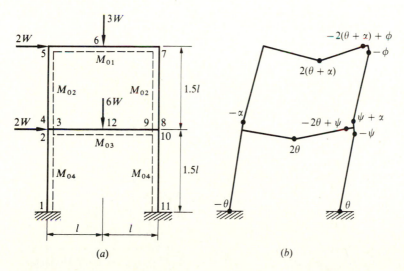

(a)   (b)

Figure 4.26 Minimum weight design of a two-story frame. (a) Loading and dimensions; (b) collapse mechanism with four degrees of freedom.

mechanisms, the equilibrium equations become

$$-M_4 + M_5 - M_7 + M_8 = 3Wl$$
$$-M_1 + M_2 - M_{10} + M_{11} = 6Wl$$
$$-M_5 + 2M_6 - M_7 = 3Wl$$
$$-M_3 + 2M_{12} - M_9 = 6Wl \tag{59}$$
$$M_2 - M_3 - M_4 = 0$$
$$M_8 + M_9 - M_{10} = 0$$

Since the magnitude of the bending moment of any section cannot exceed the local plastic moment, the following inequalities must be satisfied:

$$-M_{01} \leqslant M_i \leqslant M_{01} \qquad (i = 5, 6, 7)$$
$$-M_{02} \leqslant M_j \leqslant M_{02} \qquad (j = 4, 5, 7, 8)$$
$$-M_{03} \leqslant M_k \leqslant M_{03} \qquad (k = 3, 12, 9) \tag{60}$$
$$-M_{04} \leqslant M_r \leqslant M_{04} \qquad (r = 1, 2, 10, 11)$$

The problem is therefore to find the plastic moments M_{01}, M_{02}, M_{03}, and M_{04} that minimize the weight function (58), subject to the six constraints (59) required for equilibrium, and the 28 constraints (60) imposed by the yield condition. The problem formulated in this way represents one of linear programming, which provides a rapid method of locating and testing the vertices of the permissible region.†

For the two-story frame considered here, the required fully plastic moments of its members corresponding to the minimum weight design are obtained as

$$M_{01} = M_{02} = Wl \qquad M_{03} = \tfrac{8}{3}Wl \qquad M_{04} = \tfrac{5}{3}Wl$$

and the linearized weight function then becomes $G = 15.33 Wl^2$. The bending moment distribution in the minimum weight frame at the instant of collapse is given by

$$M_1 = -\tfrac{5}{3}Wl \qquad M_2 = Wl \qquad M_3 = 2Wl \qquad M_4 = -Wl \qquad M_5 = 0$$
$$M_6 = Wl \qquad M_7 = -Wl \qquad M_8 = Wl \qquad M_9 = -\tfrac{8}{3}Wl \qquad M_{10} = -\tfrac{5}{3}Wl$$
$$M_{11} = \tfrac{5}{3}Wl \qquad M_{12} = \tfrac{8}{3}Wl$$

This distribution is statically admissible, and is associated with plastic hinges developed at ten different sections leading to the collapse mechanism of Fig. 4.26b. The four degrees of freedom of the mechanism are specified by the angles θ, α, ϕ,

---

† This formulation was first used by J. Heyman, *Q. Appl. Math.*, **8**: 373 (1951). A computer program has been developed by R. K. Livesley, *Q. J. Mech. Appl. Math.*, **9**: 257 (1956). See also J. Heyman and W. Prager, *J. Franklin Inst.*, **266**: 339 (1958); R. K. Livesley, *Civ. Eng. Public Works. Rev.*, **54**: 737 (1959); H. S. Y. Chan, *J. Appl. Mech.*, **36**: 73 (1969).

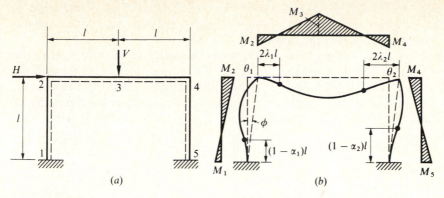

Figure 4.27 Plastic design of a portal frame for absolute minimum weight. (*a*) Dimensions and loading; (*b*) deflected shaped and bending moment diagram.

and ψ, denoting the magnitudes of hinge rotations at the indicated sections. The work equation for this mechanism is

$$(18\theta + 6\alpha)Wl = (4\theta + 4\alpha - \phi)M_{01} + (\phi + \psi + 2\alpha)M_{02}$$

$$+ (4\theta - \psi)M_{03} + (2\theta + \psi)M_{04} \quad (61)$$

The plastic moments are recovered on equating the coefficients of θ, α, ϕ, and ψ. For the mechanism to be weight compatible, the coefficients of the plastic moments in (61) must be proportional to those in (58). Hence

$$\tfrac{1}{2}(4\theta + 4\alpha - \phi) = \tfrac{1}{3}(\theta + \psi + 2\alpha) = \tfrac{1}{2}(4\theta - \psi) = \tfrac{1}{3}(2\theta + \psi)$$

yielding the solution

$$\phi = \tfrac{28}{15}\theta \qquad \alpha = \tfrac{1}{15}\theta \qquad \psi = \tfrac{8}{5}\theta$$

The magnitudes of the hinge rotations at the right-hand ends of the upper and lower beams are $2\theta + 2\alpha - \phi = \tfrac{4}{15}\theta$ and $2\theta - \psi = \tfrac{2}{5}\theta$ respectively. It follows that all the hinge rotations accord in sign with the corresponding bending moments, and the conditions of weight compatibility are therefore satisfied.†

(v) *Absolute minimum weight* The underlying principles for the design of a frame to obtain an absolute minimum weight have been discussed in (i). To illustrate the procedure, consider a fixed-base rectangular portal frame loaded horizontally and vertically as shown in Fig. 4.27*a*. If the load ratio H/V does not exceed a critical value, the deformation mode at collapse involves four inflection points as indicated in Fig. 4.27*b*. Each segment of the deflected frame is a curve of constant absolute

---

† The minimum weight design of frames under shakedown loading has been considered by J. Heyman, *J. Eng. Mech. Div., Proc. ASCE*, **84**: 1790 (1958). The design of minimum weight frames from a finite range of available sections has been discussed by A. R. Toakley, *J. Struct. Div., Proc. ASCE*, **94**: 1219 (1968). The minimum cost design has been considered by P. V. Marcal and W. Prager, *J. de Mec.*, **3**: 509 (1964). See also W. Prager and G. I. N. Rozvani, *Int. J. Mech. Sci.*, **17**: 627 (1975).

curvature, which will be taken as unity for convenience. The deformation is specified by rotations θ_1 and θ_2 of joints 2 and 4 respectively, and an angular displacement ϕ measuring the sidesway.

The slope of the deflected frame vanishes at the column feet 1 and 5, while the angle of the joints 2 and 4 remains unaltered. These conditions, together with the fact that the vertical displacement vanishes at these joints, enable us to express θ_1, θ_2, and ϕ in terms of the parameters α_1, α_2, λ_1, and λ_2, which define the points of inflection. Using (64), Chap. 3, we have

$$\theta_1 = (2\alpha_1 - 1)l \qquad \theta_2 = (2\alpha_2 - 1)l \qquad \phi = (\alpha_1^2 - \tfrac{1}{2})l = (\tfrac{1}{2} - \alpha_2^2)l$$

$$\theta_1 = [2(1 - \lambda_1)^2 - 2\lambda_2^2 - 1]l \qquad \theta_2 = [2(1 - \lambda_2)^2 - 2\lambda_1^2 - 1]l$$

in view of the continuity of slope and deflection at the points of inflection. The first row of results correspond to the columns, while the second row correspond to the beam. The elimination of θ_1, θ_2, and ϕ from the above relations furnishes

$$\alpha_1 = (1 - \lambda_1)^2 - \lambda_2^2 \qquad \alpha_2 = (1 - \lambda_2)^2 - \lambda_1^2$$

$$[(1 - \lambda_1)^2 - \lambda_2^2]^2 + [(1 - \lambda_2)^2 - \lambda_1^2]^2 = 1 \tag{62}$$

The fourth equation necessary for locating the inflections must be found from statical considerations. It follows from the geometry of the bending moment diagram that

$$M_2 = \frac{2\lambda_1}{1 - 2\lambda_1} M_3 \qquad M_4 = \frac{2\lambda_2}{1 - 2\lambda_2} M_3$$

$$M_1 = \frac{1 - \alpha_1}{\alpha_1} M_2 \qquad M_5 = \frac{1 - \alpha_2}{\alpha_2} M_4$$

These are the yield moments at the cardinal sections, and the conditions of equilibrium satisfied by them are obtained from the principle of virtual work as

$$M_2 + 2M_3 + M_4 = Vl \qquad -M_1 - M_2 + M_4 + M_5 = Hl$$

The substitution for M_2 and M_4 into the first equilibrium equation leads to an expression for M_3. The second equilibrium equation may then be used to obtain the relationship between H/V and the inflection parameters. The results may be written as

$$M_3 = \frac{(1 - 2\lambda_1)(1 - 2\lambda_2)}{2(1 - \lambda_1 - \lambda_2)} Vl$$

$$\frac{\lambda_2(1 - 2\lambda_1)}{(1 - \lambda_2)^2 - \lambda_1^2} - \frac{\lambda_1(1 - 2\lambda_2)}{(1 - \lambda_1)^2 - \lambda_2^2} = (1 - \lambda_1 - \lambda_2)\frac{H}{V} \tag{63}$$

Equations (62) and (63) are sufficient to determine the positions of the inflections, and hence to fix the distribution of the yield moment. The validity of the solution requires $\lambda_1 \geqslant 0$, which is equivalent to

$$2(1 + \lambda_2^2)(1 - \lambda_2)^2 \geqslant 1 \qquad \text{or} \qquad \lambda_2 \leqslant 0.328$$

in view of the third equation of (62). It follows from (63) that the corresponding restriction on the load ratio is $H/V \leqslant 1.081$. In the limiting case $\lambda_1 = 0$, the cross section of the left-hand column becomes vanishingly small, while the remaining plastic moments become $M_3 = 0.744Vl$, $M_4 = 1.419Vl$, and $M_5 = 1.723Vl$. In the other extreme case $H = 0$, symmetry requires

$$\alpha_1 = \alpha_2 = 1 - 2\lambda_1 = 1 - 2\lambda_2 = 1/\sqrt{2}$$

Then the plastic moments at sections 2 and 4 are each equal to $\sqrt{2} - 1\, M_3$, and those at sections 1 and 5 are each equal to $(\sqrt{2} - 1)^2 M_3$, where $M_3 = Vl/2\sqrt{2}$.

If the ratio H/V exceeds 1.08, a second inflection appears in the left-hand column,† whose plastic moment continues to vanish. The sagging part of the column near its base decreases as H/V increases, becoming zero when $H/V \geqslant 4.0$. The deformation mode then involves one inflection in each column, at a distance $3l/4$ from the base, and one inflection in the beam occurring at its center. The plastic moments at sections 1 and 5 are therefore three times those at sections 2 and 4 respectively. These moments depend on H and V, and are easily found from the equations of equilibrium.

4.7 Influence of Axial Forces

(i) Combined bending and tension Consider an arbitrary beam section under the combined action of a bending moment M and an axial tensile force N. As the bending moment is gradually increased, with the axial force held constant, a plastic zone forms on one side of the neutral axis, followed by a second plastic zone spreading from the opposite side. During the course of loading, the neutral axis progressively moves inward to assume a final position in the fully plastic state. The values of N and M for full plasticity corresponding to any given position of the neutral axis can be found by integration. Suppose that the cross section has an axis of symmetry Oy which lies in the plane of bending, Fig. 4.28a. Let ξh denote the final distance of the neutral axis from the centroidal axis Oz, where $2h$ is the total depth of the section. Using the stress distribution as shown, the resultant force and moment for a nonhardening beam may be written as

$$N = -Y \int_{-h_1}^{\xi h} b(y)\, dy + Y \int_{\xi h}^{h_2} b(y)\, dy$$

$$M = -Y \int_{-h_1}^{\xi h} b(y) y\, dy + Y \int_{\xi h}^{h_2} b(y) y\, dy$$

where $b(y)$ is the width of the section at any distance y from the centroidal axis. These relations define an interaction curve in the (N, M) plane in terms of the parameter ξ. The differentiation of the above equations with respect to ξ gives

$$\frac{dN}{d\xi} = -2Yhb(\xi h) \qquad \frac{dM}{d\xi} = -2Yh^2 \xi b(\xi h)$$

† See J. Heyman, *Int. J. Mech. Sci.*, **1**: 121 (1960).

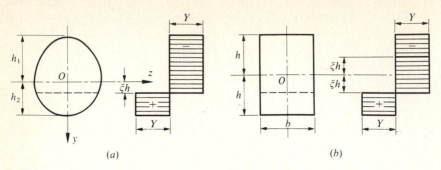

Figure 4.28 Effect of normal force on the fully plastic stress distribution in bars under symmetrical bending.

Since plane sections remain plane, the deformation consists of an arbitrary rotation of the cross section about the neutral axis. The deformation is characterized by a unit longitudinal extension ε and an angle of flexure ψ per unit length. For a rigid/plastic material, we have the immediate relationship

$$\varepsilon = -\xi h \psi$$

The critical values of N and M determine only the ratio of ε and ψ, not their individual values, since the material can flow plastically under constant stress. It follows from the preceding relations that

$$\frac{dM}{dN} = \xi h = -\frac{\varepsilon}{\psi} \tag{64}$$

This result establishes the fact that the vector representing the strain variables (ε, ψ), when superposed on the stress variables in the (N, M) plane, is perpendicular to the local tangent to the interaction curve. The direction of the strain vector is in fact along the exterior normal to the interaction curve at the appropriate stress point.

Consider now the special case of a rectangular cross section† of width b and depth $2h$ as shown in Fig. 4.28b. To express N and M in terms of the parameter ξ, which specifies the position of the neutral axis, it is convenient to decompose the stress distribution into three parts as shown. The central block is responsible for the axial force N, while the outer blocks produce the bending moment M. Consequently,

$$N = -2bhY\xi \qquad M = bh^2Y(1 - \xi^2)$$

The limiting axial tension in the absence of bending is $N_0 = 2bhY$, and the limiting bending moment in the absence of axial force is $M_0 = bh^2Y$. The elimination of ξ

---

† Beams of rectangular cross section subjected to bending moments about both axes of symmetry in the presence of axial forces have been considered by H. Shakir-Khalil and G. S. Tadros, *Struct. Eng.*, **51**: 239 (1973).

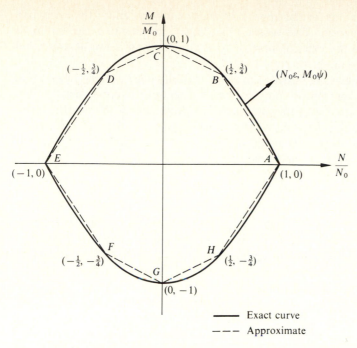

Figure 4.29 Interaction curve for combined bending and axial force in a bar of rectangular cross section.

from the preceding equations therefore gives the interaction relationship†

$$\frac{M}{M_0} + \left(\frac{N}{N_0}\right)^2 = 1 \qquad 0 \leqslant \frac{M}{M_0} \leqslant 1 \qquad (65a)$$

which defines a parabola with vertex at $(0, 1)$ in terms of the dimensionless coordinates N/N_0 and M/M_0. When the fully plastic stress distribution is such that the regions of tension and compression are interchanged, the relationship between N/N and M/M_0 is modified to

$$-\frac{M}{M_0} + \left(\frac{N}{N_0}\right)^2 = 1 \qquad -1 \leqslant \frac{M}{M_0} \leqslant 0 \qquad (65b)$$

Equations (65) define the upper and lower halves of the interaction diagram shown by solid lines in Fig. 4.29. It is symmetrical about the axes of M/M_0 and N/N_0 due to the symmetry of the cross section. It follows from (64) and (65) that

$$\frac{N_0 \varepsilon}{M_0 \psi} = \pm \frac{2N}{N_0}$$

† The interaction formulas for rectangular and I-section beams have been given by K. Girkmann, S. B. Akad. Wiss. Wien (Abt. 2a), **140**: 679 (1931). For experimental confirmations, see L. S. Beedle, J. A. Ready, and B. G. Johnston, Proc. Soc. Exp. Stress Anal., **8**: 109 (1950).

where the upper sign holds for the upper half and the lower sign for the lower half of the interaction diagram. The vector $(N_0\varepsilon, M_0\psi)$, representing the strain, is always directed along the exterior normal to the interaction curve at any point $(N/N_0, M/M_0)$, representing the state of stress.[†]

When the neutral axis lies outside the beam, the fully plastic stress distribution consists of a uniform tension or compression throughout the cross section. This corresponds to point A or E of the interaction diagram, where the normal is not uniquely defined. Indeed, ξ may take on any value between 1 and ∞ or between -1 and $-\infty$ at these two points. The slope $M_0\psi/N_0\varepsilon$ of the strain vector therefore takes on any value between $-\frac{1}{2}$ and $\frac{1}{2}$, which correspond to the limiting normals at the two singular points.

In the solution of special problems, it is often convenient to approximate the actual interaction curve by a polygon as shown by broken lines in Fig. 4.29. The coordinates of the vertices B and D are $(\pm\frac{1}{2}, \frac{3}{4})$, while those of the vertices H and F are $(\pm\frac{1}{2}, -\frac{3}{4})$. The normality rule for the strain vector must be retained to preserve the validity of the principle of maximum plastic work. Thus, the strain vector must have a slope equal to $\frac{2}{3}$ when the stress point is on the side AB, and equal to 2 when the stress point is on the side BC. At the corner B, the slope of the strain vector may lie between $\frac{2}{3}$ and 2, and at the corner A, the slope may have any value less than $\frac{2}{3}$ in magnitude. Similar relations hold for the remaining sides and corners of the yield polygon.

(ii) Limit analysis of arches As in the case of beams, the plastic collapse of arches can be adequately described in terms of rigid segments joined by yield hinges. Due to the presence of axial forces, a yield hinge in an arch must permit localized extension or contraction as well as relative rotation of adjacent cross sections. The mechanical behavior of a yield hinge in an arch can be simplified by using the polygonal approximation to the interaction curve. Since the polygon nowhere extends beyond the actual interaction curve, the collapse load based on this approximation cannot exceed the true collapse load. It is readily shown that an expansion of the polygon of Fig. 4.29 by a factor of $\frac{17}{16}$ yields a similar octagon which is tangential to the parabolic boundaries at the points $(\pm\frac{1}{4}, \pm\frac{15}{16})$ and is elsewhere exterior. It follows that the collapse load obtained from the piecewise linear approximation cannot be lower than the actual one by more than $6\frac{1}{4}$ percent.[‡]

Consider, as an example, a two-hinged circular arch[§] of radius R, subtending an angle $2\phi_0$ at its center of curvature, and carrying a single vertical load $2P$ at the midspan, Fig. 4.30a. Because of symmetry, the vertical component of the reaction at each support is equal to P. The redundant horizontal reaction is equal to

---

[†] Equations (65) approximately hold for many other doubly symmetric cross sections.

[‡] P. G. Hodge, *Plastic Analysis of Structures*, chap. 7, Krieger Publishing Co. (1981).

[§] The limit analysis of arches has been discussed by E. T. Onat and W. Prager, *J. Mech. Phys. Solids*, **1**: 77 (1953). See also A. W. Hendry, *Civ. Eng.*, London, **47**: 38 (1952); L. K. Stevens, *Proc. Inst. Civ. Eng.*, **6**: 493 (1957).

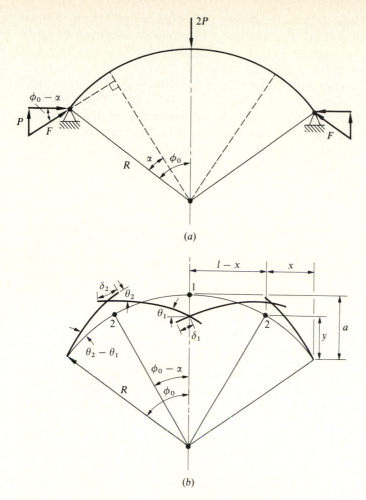

Figure 4.30 Collapse of a circular arch under a concentrated load. (*a*) Dimensions and loading; (*b*) collapse mechanism.

$P \cot(\phi_0 - \alpha)$, where α is the unknown angle between the resultant reaction F and the tangent to the arch at its support. The normal force N and the bending moment M at any section, specified by its angular distance ϕ from the apex, are

$$N = -F \cos(\phi_0 - \alpha - \phi)$$

$$M = FR \left[\cos \alpha - \cos(\phi_0 - \alpha - \phi) \right]$$

(66)

The most critical sections are those corresponding to $\phi = 0$ and $\phi = \phi_0 - \alpha$, where the bending moment has extreme positive and negative values. For sufficiently large values of ϕ_0, the influence of the bending moment would predominate, and the magnitude of the normal force would be less than $\frac{1}{2}N_0$

throughout the arch. Consequently, the stress point would be on CD for $\phi = 0$ and on FG for $\phi = \phi_0 - \alpha$ (Fig. 4.29) at the instant of collapse, satisfying the yield conditions

$$\frac{M}{M_0} - \frac{N}{2N_0} = 1(\phi = 0) \qquad \frac{M}{M_0} + \frac{N}{2N_0} = -1(\phi = \phi_0 - \alpha) \qquad (67)$$

Substituting from (66) into the above relations, we have a pair of equations for α and F. Since $F = P \operatorname{cosec}(\phi_0 - \alpha)$ and $M_0/N_0 = h/2$, the cross section being assumed rectangular, the solution may be written as

$$\cos(\phi_0 - \alpha) = 1 - \frac{2(1 - \cos \alpha)}{1 - h/4R}$$

$$(68)$$

$$\frac{PR}{M_0} = \frac{\sin(\phi_0 - \alpha)}{1 - \cos \alpha + h/4R}$$

giving the relationship between P and ϕ_0 parametrically through α. The normal force has its greatest magnitude equal to F when $\phi = \phi_0 - \alpha$. The validity of the solution therefore requires $F \leqslant N_0/2$, which is equivalent to

$$P \leqslant \frac{N_0}{2} \sin(\phi_0 - \alpha) \qquad \text{or} \qquad \cos \alpha \leqslant 1 - \frac{3h}{4R}$$

in view of (68). For somewhat smaller values of α, the yield condition at $\phi = 0$ remains unchanged, but that at $\phi = \phi_0 - \alpha$ corresponds to the side EF of the yield polygon (Fig. 4.29). When α is sufficiently small, the sides ED and EF of the polygon become appropriate for $\phi = 0$ and $\phi = \phi_0 - \alpha$ respectively. The yield conditions therefore become

$$\frac{2M}{3M_0} - \frac{N}{N_0} = 1(\phi = 0) \qquad \frac{2M}{3M_0} - \frac{N}{N_0} = -1(\phi = \phi_0 - \alpha)$$

Using equations (66), and proceeding as before, the solution for this range is obtained as

$$\cos(\phi_0 - \alpha) = 1 - \frac{2(1 - \cos \alpha)}{1 - 3h/4R}$$

$$(69)$$

$$\frac{PR}{M_0} = \frac{\frac{3}{2}\sin(\phi_0 - \alpha)}{1 - \cos \alpha + 3h/4R}$$

This solution will be valid so long as the magnitude of N at $\phi = 0$ exceeds $N_0/2$. In view of (66) and (69), this restriction is equivalent to

$$P \geqslant \frac{N_0}{2} \tan(\phi_0 - \alpha) \qquad \text{or} \qquad \cos \alpha \geqslant \frac{(1 + 3h/4R)^2}{1 + 9h/4R}$$

The collapse loads given by (68) and (69) are exact for the linearized yield conditions used in this solution. Figure 4.31 shows the variation of the collapse

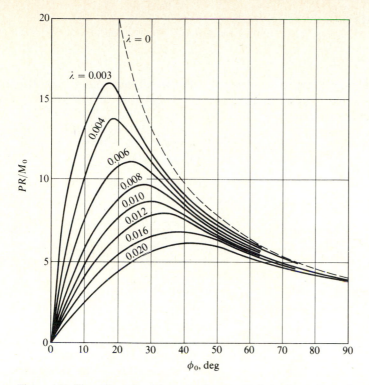

Figure 4.31 The collapse load for a simply supported circular arch as a function of semiangle ϕ_0 and thickness ratio $\lambda = h/4R$.

load with the arch semiangle ϕ_0 for a number of values of $\lambda = h/4R$. When ϕ_0 is sufficiently large, the influence of normal forces becomes negligible, and the collapse load is closely approximated by setting $h/R \simeq 0$. This limiting case is represented by the broken curve.

The problem can also be analyzed by a kinematical approach based on a collapse mechanism. For the pin-supported arch considered here, three yield hinges are necessary to transform the structure into a mechanism. At each of the hinges, there will be a contraction and a rotation as indicated in Fig. 4.30b. The motion of each half of the mechanism is made up of an outward rotation $\theta_2 - \theta_1$ about the end support, an inward rotation θ_2 about the hinge 2, and contractions δ_1 and δ_2 at the hinges 1 and 2 respectively. Considering the effect of each partial motion, the condition of zero horizontal displacement of the center of the arch may be written from simple geometry as

$$a(\theta_2 - \theta_1) - (a - y)\theta_2 + \delta_1 + \delta_2 \cos(\phi_0 - \alpha) = 0$$

We examine the situation where the states of stress in the hinges 1 and 2 correspond to the sides CD and FG of the yield polygon (Fig. 4.29). In view of the normality rule and the adopted sign convention, we have

$$\frac{\delta_1}{\theta_1} = -\frac{\varepsilon_1}{\psi_1} = c \qquad \frac{\delta_2}{\theta_2} = \frac{\varepsilon_2}{\psi_2} = c$$

where $(2\varepsilon_1, 2\psi_1)$ and (ε_2, ψ_2) denote the unit extension and rotation at the respective hinges, and c is a constant equal to $M_0/2N_0 = h/4$. Inserting these values into the preceding equation furnishes

$$\frac{\theta_1}{\theta_2} = \frac{y + c\cos(\phi_0 - \alpha)}{a - c} = \frac{(1 + h/4R)\cos(\phi_0 - \alpha) - \cos\phi_0}{1 - \cos\phi_0 - h/4R} \tag{70}$$

The work done by the applied load is $W = 2PV$, where V is the downward displacement of the center of the arch. Referring to Fig. 4.30b, it is easily shown that

$$V = -l(\theta_2 - \theta_1) + (l - x)\theta_2 + \delta_2\sin(\phi_0 - \alpha)$$

Substituting for the contractions δ_1 and δ_2, and using the geometry of the arch, it is easily shown that

$$V = R\theta_2\left[\left(1 + \frac{h}{4R}\right)\sin(\phi_0 - \alpha) + \left(\frac{\theta_1}{\theta_2} - 1\right)\sin\phi_0\right] \tag{71}$$

Let (N_1, M_1) be the normal force and bending moment at section 1, and (N_2, M_2) those at section 2, satisfying the yield conditions (67). The specific energy dissipated in the yield hinges is

$$D = 2(N_1\varepsilon_1 + M_1\psi_1) + 2(N_2\varepsilon_2 + M_2\psi_2) = 2M_0(\psi_1 - \psi_2)$$

where the last step follows from (67) and the fact that $\varepsilon_1 = -M_0\psi_1/2N_0$ and $\varepsilon_2 = M_0\psi_2/2N_0$. Equating the internal work $2M_0(\theta_1 + \theta_2)$ to the external work $2PV$, and using (71) and (70), we finally obtain

$$\frac{PR}{M_0} = \frac{(1 - \lambda) - 2\cos\phi_0 + (1 + \lambda)\cos(\phi_0 - \alpha)}{(1 + \lambda)\sin\alpha + (1 - \lambda^2)\sin(\phi_0 - \alpha) - (1 - \lambda)\sin\phi_0} \tag{72}$$

where $\lambda = h/4R$. Equation (72) provides an upper bound on the collapse load for any assumed value of α. To obtain the exact solution it is necessary to minimize the right-hand side of (72) with respect to α. The corresponding relationship between ϕ_0 and α is found to be identical to that given by (68).

If the load is uniformly distributed over the entire span, the same collapse mechanism may be used for carrying out the kinematical analysis. Due to the preponderance of axial forces, the sides ED and EF of the yield polygon would be appropriate in this case for a wide range of values of ϕ_0. This means that in equations (70) and (71), the ratio $h/4R$ must be replaced by $3h/4R$. The external work W must be found by integration, using the fact that the contribution from a typical arch element is $pRv\cos\phi\, d\phi$, where p is the intensity of the vertical load and v the downward displacement at an angular distance ϕ from the apex.

(iii) Stability of columns If a straight column is axially compressed by a gradually increasing load, failure may occur by buckling sideways when a critical value of the load is reached. For given material and end conditions, the magnitude of the critical load depends on the *slenderness ratio*, which is the ratio of the length of the column to the least radius of gyration of the cross section. If the slenderness ratio is sufficiently large, buckling will occur in the elastic range, for which an equilibrium configuration infinitesimally near to the straight form is possible under the same axial load. By Euler's well-known formula, the critical stress for elastic buckling is inversely proportional to the square of the slenderness ratio.

The elastic buckling theory does not hold for sufficiently short columns for which the critical stress given by the Euler formula exceeds the stress at the elastic limit. Due to the bilinear character of the incremental material response in the plastic range, a point of bifurcation of the equilibrium path is possible before an actual loss of stability. This means that the column can buckle under an increasing axial load when the rate of work-hardening of the material is decreased to a critical value. The smallest value of the load for which such a bifurcation can occur corresponds to no unloading of the plastically compressed material (Sec. 2.7 (ii)).

The smallest possible buckling load, known as the *tangent modulus* load, is the same as that for a linearized solid, which behaves identically to both loading and unloading with a modulus T equal to the current slope of the actual stress-strain curve. Considering a straight column of length l, which is pin-jointed at its ends as shown in Fig. 4.31a, the differential equation of deflection for a slightly bent form of the column under a load P may be written as

$$\frac{d^2v}{dx^2} + \frac{Pv}{TI} = 0$$

where I is the least moment of inertia of the cross section and v the lateral deflection of any point on the axis. In view of the condition $v = 0$ at $x = 0$, the solution becomes

$$v = v_0 \sin\left(\sqrt{\frac{P}{TI}}\, x\right)$$

where v_0 is the maximum deflection. Since the expression in the parenthesis must be equal to $\pi/2$ when $x = l/2$, the critical load P and the critical compressive stress σ are given by†

$$P = \frac{\pi^2 TI}{l^2} \qquad \sigma = \frac{P}{A} = \pi^2 T\left(\frac{r}{l}\right)^2 \tag{73}$$

since $I = Ar^2$ for a cross section of area A and radius of gyration r. The above relations will hold only when σ exceeds the uniaxial compressive stress Y at the elastic limit. The validity of (73) therefore requires $l/r \leqslant \pi\sqrt{E/Y}$, the tangent modulus being taken as E at the elastic limit.

---

† The tangent modulus formula was first suggested by F. Engesser, *Z. Ver. deut. Ing.*, **42**: 927 (1898). It has been established with physical reasonings by F. R. Shanley, *J. Aero. Sci.*, **14**: 251 (1947). See also R. Hill and M. J. Sewell, *J. Mech. Phys. Solids*, **8**: 105, 112 (1960).

The largest possible buckling load is that which remains constant during the transition from the straight form of equilibrium to a slightly bent form. The induced normal stress due to bending is compressive on the concave side and tensile on the convex side. Since the unloading on the convex side depends on $E > T$, the neutral axis must be situated at some distance to the convex side from the centroid of the cross section. Consider a rectangular cross section of width b and depth $2h$ as shown in Fig. 4.32b, and let h_1 and h_2 denote the distances of the neutral axis from the concave and convex surfaces respectively. The magnitudes of the greatest compressive and tensile stresses induced by bending are σ_1 and σ_2 respectively, where

$$\sigma_1 = \frac{Th_1}{R} \qquad \sigma_2 = \frac{Eh_2}{R}$$

where R is the radius of curvature of the neutral surface. Since the resultant of the bending stress distribution must vanish for equilibrium, $\sigma_1 h_1$ must be equal to $\sigma_2 h_2$. Consequently

$$Th_1^2 = Eh_2^2 \qquad h_1 + h_2 = 2h$$

These equations are easily solved for h_1 and h_2, the result being

$$h_1 = \frac{2h}{1 + \sqrt{T/E}} \qquad h_2 = \frac{2h}{1 + \sqrt{E/T}}$$

The distribution of the additional stresses over the cross section gives rise to the bending moment

$$M = \frac{b}{3}(\sigma_1 h_1^2 + \sigma_2 h_2^2) = \frac{b}{3R}(Th_1^3 + Eh_2^3)$$

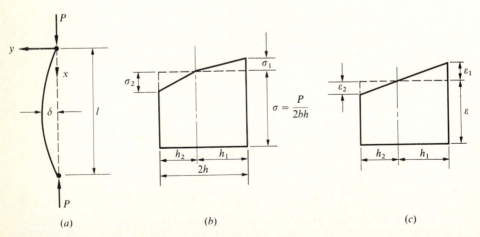

Figure 4.32 Buckling of a straight column. (a) Deflected column; (b, c) stress and strain increments under constant load.

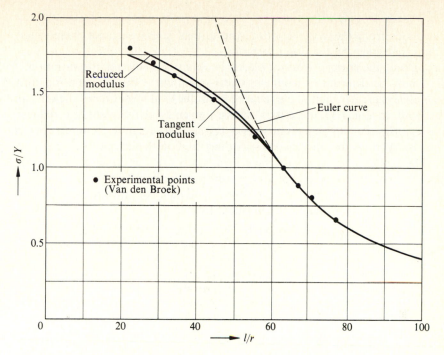

Figure 4.33 Variation of the critical stress with slenderness ratio for an ideal steel column of square cross section.

Substituting for h_1 and h_2, and introducing the moment of inertia $I = \frac{2}{3}bh^3$ about the centroidal axis, we obtain

$$M = \frac{4EI}{R}\left(1 + \sqrt{\frac{E}{T}}\right)^{-2} = \frac{\bar{E}I}{R} \tag{74}$$

where $\bar{E}$ is the *reduced modulus* that depends on E and T as well as on the shape of the cross section. Since $M = Pv$ and $1/R = -d^2v/dx^2$, the differential equation for buckling is

$$\frac{d^2v}{dx^2} + \frac{Pv}{\bar{E}I} = 0$$

which is identical in form to that for elastic buckling. The critical load and the critical stress corresponding to the reduced or double modulus theory therefore become†

$$P = \frac{\pi^2 \bar{E}I}{l^2} \qquad \sigma = \pi^2 \bar{E}\left(\frac{r}{l}\right)^2 \tag{75}$$

† The double modulus formula is due to Th. von Karman, *Z. Ver. deut. Ing.*, **81** (1910).

where $\bar{E}$ is given by (74) when the cross section is rectangular. As the slenderness ratio decreases from the value $\pi\sqrt{E/Y}$, the critical stress becomes increasingly lower than that given by the Euler formula.

To compute the critical stress for plastic buckling of a column, the variation of T with σ must be determined from the stress-strain curve of the given material. Since $\bar{E}$ is always greater than T, the reduced modulus load must exceed the tangent modulus load. In Fig. 4.33, the critical stresses given by the tangent and reduced modulus theories are plotted as functions of the slenderness ratio for a steel bar of square cross section, the curve corresponding to Euler's formula being shown broken. The experimental values of the critical stress are seen to agree better with the tangent modulus theory, except for very short columns.[†] The tangent modulus formula is always used in the design of centrally loaded columns, because it gives the more conservative estimate of the critical load.[‡]

Problems

4.1 A uniform beam of length $2l$ and plastic moment M_0 carries a uniformly distributed load W over the left-hand half, and also a concentrated load W at the midspan. If the beam is built-in at both ends, find by a statical analysis the value of W corresponding to the plastic collapse. Assuming the cross section to be rectangular, find the lengths of the plastic zones in the central part and near the loaded end of the beam at the incipient collapse.

Answer: $8M_0/3l$, $0.477l$, $0.104l$.

4.2 A continuous beam rests on three simple supports at A, B, and C, the spans of AB and BC being l and $1.5l$ respectively. The member AB has a fully plastic moment M_0 and carries a central concentrated load W. The member BC having a plastic moment αM_0 carries a uniformly distributed load W. Using a statical analysis find the value of α for which plastic collapse would occur simultaneously in the two spans. Find the length of the plastic zone near the central hinge at the instant of collapse.

Answer: 0.693, $0.085l$.

4.3 A propped cantilever of length $2l$, plastic moment M_0, and flexural rigidity EI, is subjected to a load W which is uniformly distributed between the center of the beam and the simply supported end. Carry out a kinematical analysis to determine the value of W that will cause plastic collapse. Assuming the moment-curvature relationship corresponding to an idealized beam section, find the deflection δ at the section where the final hinge is formed.

Answer: $W = 4.80M_0/l$, $\delta = 0.286M_0l^2/EI$.

4.4 In a rectangular portal frame $ABCD$ of uniform plastic moment M_0, the columns AB and CD are of heights $2l$ and l respectively. The base A is rigidly fixed and is lower than the pinned base D so that the beam BC, of length $2l$, is horizontal. The frame is subjected to a vertical load $3W$ at the center of the beam and a horizontal load W at C in the direction BC. Determine the upper and lower bounds associated with each collapse mechanism and hence estimate the true collapse load.

Answer: $W = 1.20M_0/l$.

† J. A. Van den Broek, *Eng. J.* (*Canada*), **28**: 772 (1945). See also F. Bleich, *Buckling Strength of Metal Structures*, p. 20, McGraw-Hill Book Co., New York (1952).

‡ The buckling of an eccentric column has been considered by M. R. Horne, *J. Mech. Phys. Solids*, **4**: 104 (1956). The elastic/plastic behavior of beam columns has been discussed by W. F. Chen, *J. Eng. Mech. Div., Proc. ASCE*, **96**: 421 (1970). The stability of frames in the plastic range has been investigated by M. R. Horne, *Proc. R. Soc., London*, A **274**: 343 (1963); K. I. Majid and D. Anderson, *Struct. Eng.*, **46**: 357 (1968); R. H. Wood, *Struct. Eng.* **52**: 235, 295, 341 (1974). See also M. R. Horne and L. J. Morris, *Plastic Design of Low Rise Frames*, chap. 4, Granada Publishing Limited, London (1981).

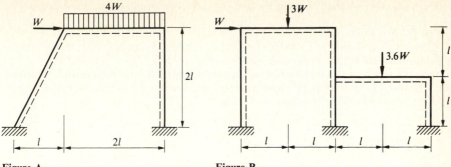

Figure A   **Figure B**

4.5 The rigid frame shown in Fig. A is of uniform cross section, having a plastic moment M_0, and carries a uniformly distributed vertical load $4W$ and a concentrated horizontal load W. Illustrate diagrammatically all the possible modes of collapse, and identify the one that furnishes the best upper bound. Locate the plastic hinge in the beam to find the collapse value of W in terms of M_0 and l.

Answer: $W = 1.08 M_0/l$.

4.6 A two-bay fixed-base rectangular frame $ABCDEF$ is designed to have a uniform section of fully plastic moment M_0. The columns AB, CD, and EF are each of height h, while the beams BC and CE are each of length $2l$. A horizontal load H acts at B in the direction BC, and two equal vertical loads V act at the centers of the beams. Construct an interaction diagram showing the relationship between Hh/M_0 and Vl/M_0 for all positive values of their ratio. Find the required plastic moment M_0 when $h = 2l$ and $V = 2H$.

Answer: $M_0 = 6Hl/11$.

4.7 The rigidly jointed frame shown in Fig. B is composed of prismatic members with plastic moment M_0 for the columns and $1.6M_0$ for the beams. The frame is built-in at the base, and carries concentrated horizontal and vertical loads as indicated. Using the method of combining mechanisms, determine the actual mode of collapse and the associated limit load W. Carry out a statical analysis to confirm the kinematical result.

Answer: $W = 1.44 M_0/l$.

4.8 A two-story rectangular frame, fixed at the base, has uniform beams and columns with a constant plastic moment M_0. The frame is subjected to vertical loads W at the centers of the beams, and horizontal loads $0.9W$ at the ends of the beams as shown in Fig. C. By considering suitable combinations of the

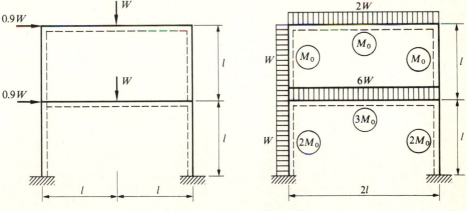

Figure C   **Figure D**

elementary mechanisms, obtain the collapse mechanism, and hence find the collapse load for the given frame.

Answer: $W = 2.13M_0/l$.

4.9 The two-story rectangular frame shown in Fig. D has uniformly distributed vertical loads acting on the beams, and a uniformly distributed horizontal load acting on the left-hand columns. The fully plastic moments of the various members of the frame are as indicated by circles. It may be assumed that plastic hinges can occur only at the ends and midpoints of the loaded members. Use the method of combining mechanisms to estimate the collapse load W in terms of M_0 and l.

Answer: $W = 32M_0/17l$.

4.10 A fixed-base gable frame $ABCDE$ consists of two vertical columns AB and DE, each of length h, and two equally inclined rafters BC and CD, each having a horizontal span l. The roof is subjected to a uniformly distributed vertical load W. If the height of the apex above the base is denoted by $(1 + \lambda)h$, show that the frame would collapse when

$$Wl = 2M_0(1 + \sqrt{1 + 2\lambda})^2$$

where M_0 is the fully plastic moment of each member of the frame.

4.11 A symmetrical gable frame with fixed feet carries a horizontal load W at the top of a column, and a vertical load $3W$ at the midpoint of each rafter. The loading and dimensions of the frame, which is made of prismatic members with a uniform plastic moment M_0, are shown in Fig. E. Neglecting the influence of axial and shearing forces, determine the collapse mechanism and the associated collapse load.

Answer: $W = 5M_0/3l$.

4.12 The dimensions and loading of a fixed-base gable frame are shown in Fig. F. All members of the frame are of uniform cross section having the same plastic moment M_0. Estimate an upper bound on the collapse load W on the assumption that the plastic hinge under the distributed load forms at the midpoint of the rafter. Carry out a statical analysis for the assumed mode of collapse to obtain the corresponding lower bound on the collapse load.

Answer: $2.18M_0 > Wl > 2.12M_0$.

4.13 A fixed-ended beam AB of length $3l$, having a uniform plastic moment M_0 and flexural rigidity EI, is subjected to a concentrated load W at C, where $BC = l$. Assuming an ideal beam section, show that the first plastic hinge forms at B when $W = 9M_0/4l$, the second plastic hinge forms at C when $W = 81M_0/28l$, and the third plastic hinge forms at A when $W = 3M_0/l$. Show also that the deflection δ of the loading point at the three instants are $2M_0l^2/9EI$, $8M_0l^2/21EI$, and $2M_0l^2/3EI$ respectively. Use the virtual work method to obtain the results.

4.14 For the continuous beam shown in Fig. 4.2a, having a unit shape factor, show that the first plastic hinge forms at the central support when $W = 28M_0/3l$, and that the corresponding deflection of the position of the last plastic hinge is $\delta = 0.06M_0l^2/EI$. If the beam is completely unloaded from the state of incipient collapse, show that the residual bending moment at the central support is $0.488M_0$, while the residual deflection of the last plastic hinge is $0.014M_0l^2/EI$.

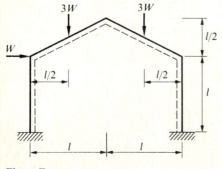

Figure E

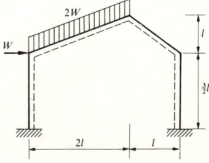

Figure F

4.15 Consider the elastic/plastic behavior of the fixed-base rectangular portal frame shown in Fig. 4.13*a* in the special case when $H = V = W$ and $h = l$. As the load is gradually increased from zero, show that plastic hinges are successively formed at sections 5, 4, 3, and 1 when Wl/M_0 becomes equal to 2.424, 2.567, 2.957, and 3.0 respectively. If δ denotes the horizontal deflection of section 2, show that $3EI\delta/M_0l^2$ attains the values 0.530, 0.637, 0.891, and 1.0 at the instants when these hinges are formed.

4.16 A fixed-base rectangular portal frame, having a height l and span $2l$, carries a horizontal load W at the top of the left-hand column and a vertical load $3W$ at the center of the beam. Each column has a plastic moment M_0 and flexural rigidity EI, the corresponding quantities for the beam being $2M_0$ and $2.5EI$ respectively. Show that the frame collapses in the combined mechanism when $W = 7M_0/4l$. Compute the horizontal and vertical deflections of the center of the beam at the incipient collapse.
    Answer: $u = 0.708M_0l^2/EI$, $v = 1.792M_0l^2/EI$.

4.17 The two-story rectangular frame shown in Fig. C has a uniform plastic moment M_0 and flexural rigidity EI, and remains statically indeterminate at collapse. Using the equations of equilibrium and compatibility, find the bending moment distribution and the hinge rotations at the instant of collapse. Hence obtain the horizontal deflection of the center of the lower beam at the point of plastic collapse.
    Answer: $\delta = 1.362M_0l^2/EI$.

4.18 The horizontal and vertical loads acting on the two-bay rectangular frame of Prob. 4.6 are taken as $H = W$ and $V = 2W$. All the members are of uniform cross section with plastic moment M_0 and flexural rigidity EI. Using the virtual work method, determine the hinge rotations at the instant of collapse when $h = 2l$, and estimate the horizontal and vertical components of the displacement of the center of the left-hand beam at the incipient collapse.
    Answer: $u = 3.67M_0l^2/EI$, $v = 4.06M_0l^2/EI$.

4.19 A uniform beam AB of length l and plastic moment M_0 is built-in at both ends and carries a concentrated load W which rolls back and forth along the beam. Show that the elastic bending moments in the beam when the load acts at a point C, distance λl from A, are

$$\mathcal{M}_A = -Wl\lambda(1 - \lambda)^2 \qquad \mathcal{M}_B = -Wl\lambda^2(1 - \lambda) \qquad \mathcal{M}_C = 2Wl\lambda^2(1 - \lambda)^2$$

Hence show that the beam will collapse incrementally under repeated passage of the rolling load when $Wl = 7.32M_0$.

4.20 A uniform continuous beam ABC, having a length $2l$ and plastic moment M_0, is simply supported at the points A, B, and C, the two spans AB and BC being of equal lengths. Concentrated vertical load P and Q act at the midpoints of AB and BC respectively, and vary independently in the ranges $0 \leqslant P \leqslant W$, and $0 \leqslant Q \leqslant nW$, where n is a positive constant less than unity. Show that the intensity of W required for incremental collapse is $(1 + \frac{3}{16}n)^{-1}$ times that for static collapse under maximum values of the applied loads.

4.21 A pinned-base rectangular portal frame, whose dimensions and loading are shown in Fig. G, has a uniform section with plastic moment M_0. The applied loads H and V can vary independently within the

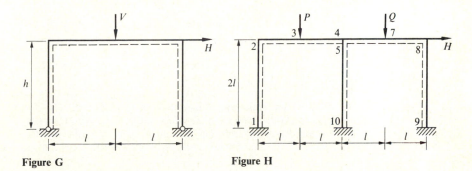

Figure G   **Figure H**

ranges $0 \leqslant H \leqslant H_0$ and $0 \leqslant V \leqslant V_0$. Show that the relationship between H_0 and V_0 for incremental collapse is identical to that for static collapse when

$$\frac{V_0}{H_0} \geqslant \frac{2h}{l} \left(\frac{3l + h}{3l + 2h} \right)$$

Note that the beam mechanism is not critical in this case for incremental collapse over the whole range of values of V_0/H_0.

4.22 In the rectangular portal frame shown in Fig. 4.17a, let the length of the columns be decreased from $2l$ to l, the same plastic moment M_0 being taken for all the members. Show that the elastic bending moments due to the simultaneous action of the horizontal load H and the vertical load V are

$$M_1 = -\tfrac{5}{16}Hl + \tfrac{1}{10}Vl \qquad M_2 = \tfrac{3}{16}Hl - \tfrac{1}{5}Vl \qquad M_3 = \tfrac{3}{10}Vl$$

$$M_4 = -\tfrac{3}{16}Hl - \tfrac{1}{5}Vl \qquad M_5 = \tfrac{5}{16}Hl + \tfrac{1}{10}Vl$$

If the loads vary independently within the ranges $0 \leqslant H \leqslant W$ and $0 \leqslant V \leqslant 2W$, find the value of W that corresponds to the shakedown limit, assuming a unit shape factor.

Answer: $W = 1.83M_0/l$.

4.23 The two-bay rectangular frame shown in Fig. H has a uniform section throughout with a fully plastic moment M_0. The loads P, Q, and H vary independently between the prescribed limits

$$0 \leqslant P \leqslant 2W \qquad 0 \leqslant Q \leqslant 2W \qquad 0 \leqslant H \leqslant W$$

The elastic bending moments at the cardinal sections due to a unit load replacing P, and $Q = H = 0$, are given by

sec.	1	2	3	4	5	6	7	8	9	10
$2M/l$	0.117	-0.281	0.637	-0.445	0.265	-0.180	-0.075	0.031	0.008	-0.156

Using the virtual work method, find the elastic bending moments due to a unit value of H, and compute the greatest value of W that can be permitted without causing incremental collapse.

Answer: $W = 1.61M_0/l$.

4.24 A continuous beam ABC, which is simply supported at A, B, and C, carries concentrated vertical loads $2W$ and W at the centers of the spans AB and BC respectively. The fully plastic moments of the members AB and BC, each having a length l, are denoted by M_1 and M_2 respectively. Using the linearized weight function and the graphical procedure, show that $M_1 = 5Wl/12$ and $M_2 = Wl/6$ for minimum weight consumption of the beam.

4.25 A three-span beam $ABCD$ of length $3l$ rests on four equidistant simple supports at A, B, C, and D. A load $2W$ is uniformly distributed over the central span BC, and a load W is uniformly distributed over each of the two outer spans AB and CD. The beam is designed to have plastic moment M_1 for the outer spans and M_2 for the central span. Assuming the linearized theory, find the values of M_1 and M_2 corresponding to the minimum weight.

Answer: $M_1 = 0.086Wl$, $M_2 = 0.164Wl$.

4.26 In a fixed-base rectangular frame $ABCD$, the columns AB and DC are of equal length $3l$, and the beam BC is of length $4l$. The frame carries a horizontal load W at B in the direction BC, and a vertical load $3W$ at the center of the beam. The columns are required to have a plastic moment M_1 which differs from the plastic moment M_2 of the beam. Using the geometrical method, find the values of M_1 and M_2 in the minimum weight design of the frame.

Answer: $M_1 = M_2 = 7Wl/6$.

4.27 In a pinned-base symmetrical gable frame of horizontal span $2l$, the columns are each of height l, and the apex is $1.5l$ above the base. The plastic moments of the rafters and columns are M_1 and M_2 respectively. The frame carries a horizontal load $2W$ at the top of a column, and a vertical load nW at the apex. Show that the minimum weight design of the frame requires $M_1 = M_2 = Wl$ for $0 \leqslant n \leqslant 2$, and $M_1 = M_2 = (n + 3)Wl/5$ for $n \geqslant 2$.

4.28 The built-in rectangular portal frame shown in Fig. H is to be designed for minimum weight when the applied loads are $P = 2W$, $Q = 3W$, and $H = W$. The columns are required to have a common plastic moment M_1 and the beams to have a common plastic moment M_2. Considering the possible collapse mechanisms with a single degree of freedom, and using the graphical method of solution, determine the values of M_1 and M_2 in terms of W and l.

Answer: $M_1 = 0.40Wl$, $M_2 = 0.87Wl$.

4.29 Suppose that the rectangular frame of the preceding problem is designed to have fully plastic moments M_2 and M_3 for the left-hand and right-hand beams respectively. The columns still have a plastic moment M_1, and the load ratios are the same as before. Starting with a reasonable assumption for the relative values of the plastic moments, and choosing three possible collapse mechanisms at a time to define the vertex of the design space, obtain the correct solution to the minimum weight problem.

4.30 A two-span beam ABC of length $2l$ is simply supported at A, B, and C, where B is the midpoint of AC. The beam is designed for an absolute minimum weight with continuously varying cross section. Show that the points of contraflexure occur at distances $\lambda_1 l$ and $\lambda_2 l$ from A and C respectively, where $\lambda_1^2 + \lambda_2^2 = 1$. If the beam carries concentrated loads W_1 and W_2 at the centers of AB and BC respectively, show that the plastic moment M at the central support is given by

$$\left(1 + \frac{M}{W_1 l}\right)\left(1 + \frac{M}{W_2 l}\right) = \frac{W_1 W_2 l^2}{16 M^2}$$

4.31 In a fixed-base rectangular portal frame $ABCDE$, the columns AB and DE are of height h, and the beam BD is of length $2l$. The frame is to be designed for an absolute minimum weight under a horizontal thrust H at B and a vertical load V at the beam center C. If the deformation mode at collapse involves an inflection in each member with a negative curvature at A, show that the distribution of yield moment is given by $M_C = 0$ and

$$M_B = \frac{H}{4}\left(h - \frac{l}{2}\right) - \frac{Vl}{2} \qquad M_D = \frac{H}{4}\left(h - \frac{l}{2}\right) + \frac{Vl}{2}$$

$$M_A = \left(\frac{2h + l}{2h - l}\right) M_B \qquad M_E = \left(\frac{2h + l}{2h - l}\right) M_D$$

Find the relationship between H/V and h/l required by the validity of the solution.

4.32 An I section consists of two equal rectangular flanges of width b, and a rectangular web of area A_w and thickness t. The total area of the section is A and its overall depth is $2h$. The section is rendered fully plastic by the combined action of a bending moment M about the strong axis, and a normal force N. Show that the interaction curve is given by

$$M = M_0 - Y\left(\frac{A^2}{4t}\right) n^2 \qquad 0 \leqslant n \leqslant \frac{A_w}{A}$$

$$M = Y(1 - n)\left(\frac{4bh}{A} - 1 + n\right) \frac{A^2}{4b} \qquad \frac{A_w}{A} \leqslant n \leqslant 1$$

where M_0 is the plastic moment in the absence of normal force, and n the ratio of the applied force N and its yield value N_0.

4.33 A T section consists of a flange of width b and thickness $b/8$, and a web of depth b and thickness $b/8$. A bending moment M is applied about an axis parallel to the flange in the presence of a normal force. Referring the moment to the equal area axis, and assuming the web tip to be in tension, show that the interaction relationship is

$$M = M_0(1 - \tfrac{2}{9}n^2) \qquad 0 \leqslant n \leqslant 1$$

$$M = M_0(1 - \tfrac{16}{9}n^2) \qquad -1 \leqslant n \leqslant 0$$

where nY is the mean normal stress, and M_0 the plastic moment when $n = 0$. Draw the complete interaction diagram for M/M_0 against n.

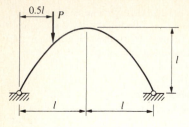

Figure I   **Figure J**

4.34 A pin-supported parabolic arch of uniform plastic moment M_0 carries a concentrated vertical load P at a quarter point as shown in Fig. I. The collapse mechanism involves a yield hinge at the point of load application, and a yield hinge at a horizontal distance ξl from the vertex. Neglecting the influence of axial forces on the yield condition, find the value of ξ that minimizes the upper bound, and hence compute the collapse load.

Answer: $\xi = 0.379$, $P = 7.85M_0/l$.

4.35 The dimensions of a parabolic arch, having a uniform cross section with plastic moment M_0, are those shown in Fig. J. The arch is built-in at both ends and is subjected to equal vertical loads P at horizontal distances l from the vertex. Neglecting the effect of the axial forces, find the value of P at collapse in terms of M_0 and l. If the cross section is rectangular with a depth equal to $0.05l$, what would be the collapse load when the normal force effect is taken into account? Assume a square yield condition, obtained by joining the vertices $N = \pm N_0$ and $M = \pm M_0$ of the interaction diagram.

Answer: $Pl = 8.0M_0$, $6.7M_0$.

4.36 A two-hinged circular arch of radius R and angular span $2\phi_0$ carries a uniformly distributed vertical load p per unit length of the span. Due to the preponderance of axial forces, the condition $|N| > \frac{1}{2}N_0$ is satisfied throughout the arch. Assuming plastic hinge rotations of amounts $2\theta_1$ and θ_2 to occur at sections $\phi = 0$ and $\phi = \phi_0 - \alpha$ respectively during collapse, show that

$$\frac{pR^2}{3M_0} = \left(1 + \frac{\theta_1}{\theta_2}\right)\left\{\left(\frac{\theta_1}{\theta_2} - 1\right)\sin^2\phi_0 + \left(1 + \frac{3h}{2R}\right)\sin^2(\phi_0 - \alpha)\right\}^{-1}$$

according to the linearized yield condition of Fig. 4.29. Compute the value of α that minimizes w when $\phi_0 = 45°$ and $R = 30h$, and determine the total vertical load P at collapse.

Answer: $\alpha \simeq 8°$, $P = 67.9M_0/R$.

4.37 The circular arch of the preceding problem carries a uniformly distributed load p per unit span over its left-hand half only. The collapse mechanism involves a sagging hinge 1 under the load, and a hogging hinge 2 in the right-hand half, at angular distances α and β respectively from the apex. If ω denotes the rotation of the leftmost segment of the arch, and (θ_1, θ_2) are the rotations at the plastic hinges, show that according to the linearized yield condition,

$$\frac{pR^2}{2M_0} = \left(1 + \frac{\theta_2}{\theta_1}\right)\left\{\frac{\omega}{\theta_1}\sin^2\phi_0 - \left(1 + \frac{h}{2R}\right)\sin^2\alpha\right\}^{-1}$$

Assuming $\phi_0 = \pi/4$, $\beta = \pi/8$, and neglecting the influence of normal forces, find the value of α that minimizes p for collapse, and estimate the corresponding upper bound.

Answer: $\alpha \simeq 11.5°$, $p = 33.1M_0/R^2$.

4.38 The frame shown in Fig. K lies in the horizontal plane and carries a concentrated vertical load W at C. All the members have a uniform thin-walled box section with fully plastic moment M_0. The yield hinges formed at the built-in ends during collapse undergo rotations ψ and θ in bending and torsion respectively. Find the ratio θ/ψ associated with the interaction relation $M^2 + \frac{3}{4}T^2 = M_0^2$ for the bending and twisting couples, and hence estimate the load W for plastic collapse.

Answer: $\theta/\psi = 0.324$, $W = 1.87M_0/l$.

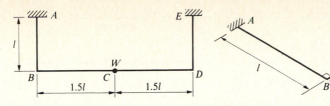

Figure K   **Figure L**

4.39 The frame of the preceding problem is subjected to a vertical load W on BD at a distance $2l$ from B, instead of at the center C. The ratios of the twisting to bending discontinuities occurring at the plastic hinges formed at A and E are denoted by x and y respectively. Considering the equilibrium of the portions EDF and ABF of the frame, where F is the point of load application, show that

$$y = \frac{2x}{1 + 3x} \qquad \frac{3 + 8y}{\sqrt{3 + 4y^2}} = \frac{6 - 4x}{\sqrt{3 + 4x^2}}$$

Compute the values of x and y satisfying these equations, and hence estimate the collapse load.
    Answer: $x = 0.228$, $y = 0.271$, $W = 1.92M_0/l$.

4.40 Figure L shows a uniform right angle bent ABD, which lies in a horizontal plane and carries a single vertical load W at C. The frame collapses by the formation of yield hinges at A, C, and D, the interaction relationship between the bending moment M and the twisting moment T being $M^2 + T^2 = M_0^2$. Denoting the ratios of the twisting to bending discontinuities at A and D by x and y respectively, show that

$$y = \frac{1}{4 - 3x} \qquad \frac{2 - y}{\sqrt{1 + y^2}} = \frac{1 + 2x}{\sqrt{1 + x^2}}$$

Compute the values of x and y, and determine the collapse load from the conditions of overall equilibrium of the frame.
    Answer: $x = y = \frac{1}{3}$, $W = 2.53M_0/l$.

4.41 A straight vertical column of length l is pin-supported at its ends, and is loaded by an axial load at the top. The column has a uniform rectangular cross section with a depth $2h$. The stress-strain curve of the material beyond the elastic limit may be represented by the equation

$$\varepsilon = \frac{\sigma}{E}\left[1 + 0.3\left(\frac{\sigma}{\sigma_0}\right)^3\right]$$

where σ_0 is a nominal yield stress equal to $E/903$. Calculate the critical values of σ/σ_0 according to the tangent and reduced modulus theories when $l/h = 28$.
    Answer: $\sigma/\sigma_0 = 1.21, 1.48$.

4.42 An initially straight column of length l is held vertical by fixing the lower end to a rigid foundation, and is subjected to a vertical load P on a line of symmetry of the upper end at a distance e from the centroid. The plastic buckling load may be approximately estimated by locating the point of intersection of the purely elastic and rigid/plastic load-deflection curves, neglecting work hardening. For a rectangular cross section of depth $2h$, show that the critical load is given by

$$\frac{h}{2e}\left(\frac{P_0}{P} - \frac{P}{P_0}\right) = \sec\left(\frac{l}{h}\sqrt{\frac{3PY}{P_0E}}\right)$$

where P_0 is the yield point load in pure compression. Compute the value of P/P_0 when $E/Y = 900$, $l/h = 30$, and $e/h = 2$.
    Answer: $P/P_0 = 0.18$.

FURTHER SOLUTIONS OF
ELASTOPLASTIC PROBLEMS

In this chapter, we shall be mainly concerned with problems where the stress distribution is either axially or spherically symmetrical. A number of important problems, such as the determination of stresses and strains in thick-walled pressure vessels and rotating discs, are of this type. The axisymmetric problem is comparatively more difficult in principle, since there are three independent stress components, even when the stresses are assumed to vary only in the radial direction. Since the directions of the principal stresses are known in advance, considerable simplifications are achieved by using Tresca's yield criterion and its associated flow rule. When the Prandtl-Reuss theory is adopted, a suitable numerical method must be used for the integration of the basic equations. For a work-hardening material, use of the Hencky theory is sometimes legitimate for a range of loading paths, and provides a great deal of mathematical simplicity. In general, the solution of the elastic/plastic problem requires a powerful numerical method, such as the finite element method described in the concluding section of this chapter.

5.1 Expansion of a Thick Spherical Shell

(i) *Elastic analysis* A thick-walled spherical shell, whose internal and external radii are a and b respectively, is subjected to uniform internal pressure p of gradually increasing magnitude. It is convenient to use spherical polar coordinates (r, θ, ϕ), where θ is the angle made by the radius vector with a fixed axis, and ϕ is the angle measured round this axis. By virtue of the spherical symmetry, $\sigma_\theta = \sigma_\phi$ everywhere in the shell, and the stresses at any given stage are functions of r only satisfying the equilibrium equation

$$\frac{\partial \sigma_r}{\partial r} = \frac{2}{r}(\sigma_\theta - \sigma_r) \tag{1}$$

For sufficiently small values of the pressure, the deformation of the shell is purely elastic. If the radial displacement is denoted by u, the stress-strain relations for the elastic shell may be written as

$$\varepsilon_r = \frac{\partial u}{\partial r} = \frac{1}{E}(\sigma_r - 2\nu\sigma_\theta)$$

$$\tag{2}$$

$$\varepsilon_\theta = \varepsilon_\phi = \frac{u}{r} = \frac{1}{E}[(1-\nu)\sigma_\theta - \nu\sigma_r]$$

Eliminating u from the above equations, and substituting for $\sigma_\theta - \sigma_r$ from the equilibrium equation, we obtain

$$\frac{\partial}{\partial r}(\sigma_r + 2\sigma_\theta) = 0$$

which is the equation of strain compatibility expressed in stresses. This shows that $\sigma_r + 2\sigma_\theta$ is independent of r. Writing $\sigma_r + 2\sigma_\theta = 3A$, the equilibrium equation (1) can be integrated by eliminating σ_θ or σ_r. This leads to the solution†

$$\sigma_r = A + \frac{B}{r^3} \qquad \sigma_\theta = A - \frac{B}{2r^3}$$

where B is also independent of r. Employing the boundary conditions $\sigma_r = 0$ at $r = b$ and $\sigma_r = -p$ at $r = a$, we get

$$A = \frac{p}{b^3/a^3 - 1} \qquad B = \frac{-pb^3}{b^3/a^3 - 1}$$

The stresses in the elastic shell, which are due to Lamé, therefore become

$$\sigma_r = -\frac{p(b^3/r^3 - 1)}{b^3/a^3 - 1} \qquad \sigma_\theta = \sigma_\phi = \frac{p(b^3/2r^3 + 1)}{b^3/a^3 - 1} \tag{3}$$

Thus σ_r is compressive and σ_θ tensile throughout the shell. The stress-strain relation (2) corresponding to ε_θ then gives

$$u = \frac{p}{E}\frac{(1-2\nu)r + (1+\nu)(b^3/2r^2)}{b^3/a^3 - 1} \tag{4}$$

If the internal pressure is increased to a critical value p_e, plastic yielding begins at the radius where the yield criterion is first satisfied. Since $\sigma_\theta = \sigma_\phi$, the Tresca and Mises criteria both reduce to

$$\sigma_\theta - \sigma_r = Y \tag{5}$$

† This solution is due to G. Lamé, *Leçons sur la théorie de l'élasticité*, Paris (1852).

It follows from (3) that $\sigma_\theta - \sigma_r$ has the greatest value at $r = a$. Hence yielding will begin at the inner radius when the pressure becomes

$$p_e = \frac{2Y}{3}\left(1 - \frac{a^3}{b^3}\right) \tag{6}$$

The radial displacement of the internal surface at this stage is

$$u_a = \frac{Ya}{3E}\left\{(1 + v) + 2(1 - 2v)\frac{a^3}{b^3}\right\}$$

If the shell is subjected to both internal and external pressures, it can be shown that yielding still begins at the internal surface when the difference of the two pressures is equal to p_e given by (6). For a very thick tube, p_e is approximately $2Y/3$.

(ii) Elastic/plastic shell With further increase in the internal pressure, the plastic zone spreads outward and the elastic/plastic boundary is a spherical surface at each stage, Fig. 5.1. In the elastic region, the stress equation is integrated as before, but the boundary condition at $r = a$ is replaced by the yield criterion (5) which must be satisfied on the plastic boundary. Denoting the radius to this boundary by c, we get

$$\sigma_r = -\frac{2Yc^3}{3b^3}\left(\frac{b^3}{r^3} - 1\right) \qquad \sigma_\theta = \frac{2Yc^3}{3b^3}\left(\frac{b^3}{2r^3} + 1\right) \qquad c \leqslant r \leqslant b \tag{7}$$

The radial displacement in the elastic region is found as

$$u = \frac{2Yc^3}{3Eb^3}\left\{(1 - 2v)r + (1 + v)\frac{b^3}{2r^2}\right\} \qquad c \leqslant r \leqslant b \tag{8}$$

In the plastic region, the stresses are required to satisfy the equilibrium equation (1), and the yield criterion (5), provided there is no strain-hardening.† Since the stresses must be continuous across the plastic boundary $r = c$, we obtain

$$\sigma_r = -\frac{2Y}{3}\left(1 - \frac{c^3}{b^3} + \ln\frac{c^3}{r^3}\right)$$
$$\qquad\qquad\qquad\qquad\qquad a \leqslant r \leqslant c \tag{9}$$
$$\sigma_\theta = \frac{2Y}{3}\left(\frac{1}{2} + \frac{c^3}{b^3} - \ln\frac{c^3}{r^3}\right)$$

The magnitude of the radial stress steadily decreases from $r = a$ to $r = b$. The circumferential stress, on the other hand, has the greatest value on the elastic/plastic boundary. Setting $r = a$ in the expression for σ_r in (9), the internal pressure is found to be†‡

$$p = \frac{2Y}{3}\left(1 - \frac{c^3}{b^3} + \ln\frac{c^3}{a^3}\right) \tag{10}$$

---

† Work-hardening can be included in the analysis in a straightforward manner assuming a linear stress-strain law. See P. Chadwick, *Int. J. Mech. Sci.*, **5**: 65 (1963).
‡ A. Reuss, *Z. angew. Math. Mech.*, **10**: 266 (1930).

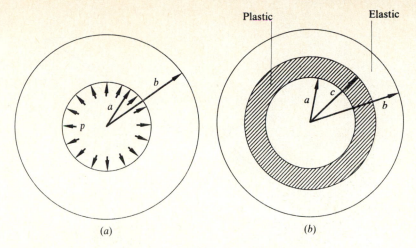

Figure 5.1 Geometry and plastic yielding in the expansion of a thick-walled spherical shell by internal pressure. (*a*) Loaded shell; (*b*) elastic and plastic domains.

If the wall ratio b/a is not too large (less than about 3), the strains in the plastic region are restricted to small values by the surrounding elastic material. The variation of the internal radius of the tube may then be neglected for calculating the pressure. Figure 5.2 shows the stress distribution in a shell of wall ratio 2.

Consider now the displacement of the elastic/plastic shell. For small strains, the equation of elastic compressibility may be expressed as

$$\frac{\partial u}{\partial r} + \frac{2u}{r} = \frac{1 - 2v}{E} (\sigma_r + 2\sigma_\theta)$$

Substituting the expressions for the stresses from (9), we obtain the differential equation for the displacement in the plastic region as

$$\frac{\partial}{\partial r} (r^2 u) = 2(1 - 2v) \frac{r^2 Y}{E} \left(\frac{c^3}{b^3} - \ln \frac{c^3}{r^3} \right)$$

Since the displacement on the plastic boundary $r = c$ is known from (8), integration of the above equation results in

$$\frac{u}{r} = \frac{Y}{E} \left\{ (1 - v) \frac{c^3}{r^3} - \frac{2}{3} (1 - 2v) \left(1 - \frac{c^3}{b^3} + \ln \frac{c^3}{r^3} \right) \right\} \qquad a \leqslant r \leqslant c \qquad (11)$$

The shell becomes completely plastic when $c = b$; the internal pressure at this stage attains the value $p_0 = 2Y \ln (b/a)$ and the corresponding displacement of the internal surface is

$$u_a = \frac{Ya}{E} \left\{ (1 - v) \frac{b^3}{a^3} - 2(1 - 2v) \ln \frac{b}{a} \right\}$$

It follows from (10) that the internal pressure attains a maximum when $c = b$. The

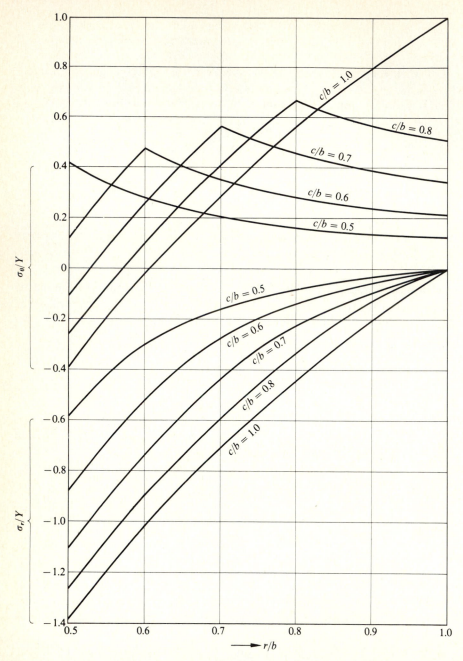

Figure 5.2 Stress distribution in a thick spherical shell of wall ratio 2.0, expanded by internal pressure in the plastic range.

pressure during the subsequent expansion steadily decreases as the change in geometry becomes significant. When the material strain-hardens to a sufficient degree, the maximum pressure occurs after some fully plastic expansion has taken place. The expansion of the shell following the pressure maximum is unstable and leads to an eventual bursting.

If the shell is unloaded from the elastic/plastic state, it will be left with residual stresses. When p/p_e is not too large, the unloading process is purely elastic, and the residual stresses are obtained by subtracting (3) from (7) and (9). The results are

$$
\sigma_r = -\frac{2Y}{3}\left(\frac{c^3}{a^3} - \frac{p}{p_e}\right)\left(\frac{a^3}{r^3} - \frac{a^3}{b^3}\right)
$$
$$
\sigma_\theta = \frac{2Y}{3}\left(\frac{c^3}{a^3} - \frac{p}{p_e}\right)\left(\frac{a^3}{2r^3} + \frac{a^3}{b^3}\right)
$$

$$c \leqslant r \leqslant b \tag{12}$$

$$
\sigma_r = -\frac{2Y}{3}\left[\frac{p}{p_e}\left(1 - \frac{a^3}{r^3}\right) - \ln\frac{r^3}{a^3}\right]
$$
$$
\sigma_\theta = Y - \frac{2Y}{3}\left[\frac{p}{p_e}\left(1 + \frac{a^3}{2r^3}\right) - \ln\frac{r^3}{a^3}\right]
$$

$$a \leqslant r \leqslant c \tag{13}$$

where p/p_e is given by (6) and (10), and is found to be less than c^3/a^3. The elastic recovery of the shell produces a circumferential compression over the inner part of the previous plastic region. The assumption of elastic unloading will be justified if the magnitude of $\sigma_r - \sigma_\theta$ nowhere exceeds the yield stress Y. The greatest magnitude of $\sigma_r - \sigma_\theta$ occurs at $r = a$, where the stress difference is $Y(p/p_e - 1)$, giving

$$
p \leqslant 2p_e
$$

as the condition for validity of (12) and (13). When this is satisfied, a reapplication of pressure less than that at the instant of unloading deforms the shell only elastically. Evidently, a secondary yielding cannot occur in an unloaded shell for which the fully plastic pressure p_0 is less than $2p_e$. The limiting wall ratio is therefore given by the equation

$$
\ln\frac{b}{a} = \frac{2}{3}\left(1 - \frac{a^3}{b^3}\right)
$$

whose solution is $b/a = 1.70$. For higher values of the wall ratio, a new plastic annulus will be formed around the inner surface due to secondary yielding when the pressure is released from a value lying between $2p_e$ and p_0.

(iii) *Solution for large strains* When the shell is very thick, large strains will occur in the plastic region. The internal pressure is still given by (10), where a represents the current internal radius of the shell. This radius must be determined for a given initial radius a_0. The total principal strains in the plastic region, whatever their magnitude, may be written as

$$\varepsilon_r = -\ln\left(\frac{\partial r_0}{\partial r}\right) \qquad \varepsilon_\theta = \varepsilon_\phi = \ln\left(\frac{r}{r_0}\right) \tag{14}$$

where r_0 is the initial radius to an element which is currently at a radius r. The equation of elastic compressibility may be written with sufficient accuracy, noting the fact that $(\sigma_r + 2\sigma_\theta)/E \ll 1$. Thus

$$\left(\frac{r_0}{r}\right)^2 \frac{\partial r_0}{\partial r} = 1 - \frac{1 - 2v}{E}(\sigma_r + 2\sigma_\theta)$$

Using the equilibrium equation (1), the above equation can be expressed in the form

$$r_0^2 \frac{\partial r_0}{\partial r} = r^2 - \frac{1 - 2v}{E} \frac{\partial}{\partial r}(r^3 \sigma_r)$$

which holds throughout the shell. Integration of this equation gives

$$\left(\frac{r_0}{r}\right)^3 = 1 - 3(1 - 2v)\frac{\sigma_r}{E} + D\frac{c^3}{r^3}$$

where D is a constant of integration. On the external surface of the shell, σ_r vanishes and the boundary condition is

$$r_0 = b\left[1 - (1 - v)\frac{Yc^3}{Eb^3}\right] \qquad r = b$$

in view of (8). This gives $D = -3(1 - v)Y/E$. Substituting for σ_r from (9), we finally obtain

$$\left(\frac{r_0}{r}\right)^3 = 1 - 3(1 - v)\frac{Yc^3}{Er^3} + 2(1 - 2v)\frac{Y}{E}\left(1 - \frac{c^3}{b^3} + \ln\frac{c^3}{r^3}\right) \tag{15}$$

The relationship between the initial and final radii of the internal surface may be written in the compact form

$$\left(\frac{a_0}{a}\right)^3 = 1 - 3(1 - v)\frac{Yc^3}{Ea^3} + 3(1 - 2v)\frac{p}{E}$$

The dimensionless pressure p/Y and the dimensionless internal displacement $a/a_0 - 1$ are plotted against c/b in Fig. 5.3 for two different b/a_0 ratios. The broken curves indicate how the pressure is overestimated if geometry changes are disregarded. When the material is incompressible ($v = 0.5$), we have

$$r^3 - r_0^3 = a^3 - a_0^3 = \frac{3Y}{2E}c^3 \tag{16}$$

This result can be directly obtained by considering the constancy of volume of the material between the internal surface and a generic surface in the elastic region, where the displacement is given by (8).

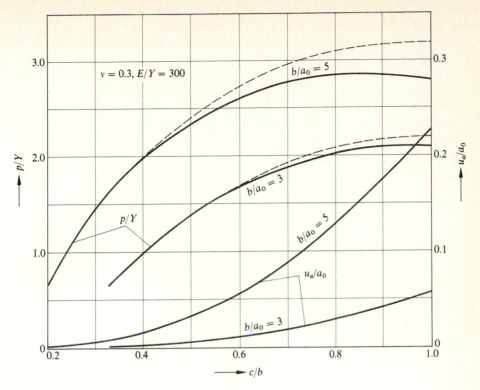

Figure 5.3 Variation of internal pressure and internal displacement with position of the plastic boundary. The broken curves are based on neglect of geometry changes.

Each element of the shell is subjected to a uniaxial compression in the radial direction, together with a superimposed hydrostatic tension equal to σ_θ. Since no strain is produced by a hydrostatic stress when the compressibility is zero, the yield criterion for an incompressible strain-hardening material may be written as

$$\sigma_\theta - \sigma_r = F(-\varepsilon_r) = F\left(2 \ln \frac{r}{r_0}\right)$$

where the function F defines the uniaxial stress-strain curve with $F(Y/E) = Y$. Inserting in the equilibrium equation (1) and integrating, we get

$$p = \frac{2}{3} Y\left(1 - \frac{c^3}{b^3}\right) + 2 \int_a^c F\left(2 \ln \frac{r}{r_0}\right) \frac{dr}{r}$$

In view of the incompressibility relation (16), the above equation can be expressed as

$$p = \frac{2}{3} Y\left(1 - \frac{c^3}{b^3}\right) + \frac{2}{3} \int_{3Y/2E}^{1 - a_0^3/a^3} F\left[-\frac{2}{3} \ln (1 - x)\right] \frac{dx}{x} \qquad (17)$$

where $x = (a^3 - a_0^3)/r^3$. For a given function F, the pressure can be calculated numerically for any assumed value of a/a_0. If the wall ratio b/a_0 is sufficiently large, the internal pressure will attain a maximum while the shell is still partly plastic. Differentiating (17) with respect to a and observing that

$$\frac{dc}{da} = \frac{2E}{3Y} \frac{a^2}{c^2}$$

the condition at the pressure maximum, which corresponds to the plastic instability, may be written as

$$\frac{2E\rho^3(\rho^3 - 1)}{3F(2 \ln \rho)} = \frac{b^3}{a_0^3} \qquad \rho = \frac{a}{a_0} \tag{18}$$

This equation is easily solved graphically for any given stress-strain curve to obtain the value of a/a_0 at instability. For a nonhardening material $(F = Y)$, the solution is

$$\left(\frac{a}{a_0}\right)^3 = \frac{1}{2}\left\{1 + \sqrt{1 + \frac{6Yb^3}{Ea_0^3}}\right\}$$

For a work-hardening material, there is a limiting wall ratio for which the pressure attains a maximum when the shell just becomes fully plastic $(c = b)$. Since the fully plastic stage corresponds to $\rho^3 - 1 = (3Y/2E)(b^3/a_0^3)$ in view of (16), the limiting wall ratio is given by

$$F\left\{\frac{2}{3}\ln\left(1 + \frac{3Y}{2E}\frac{b^3}{a_0^3}\right)\right\} = Y\left(1 + \frac{3Y}{2E}\frac{b^3}{a_0^3}\right) \tag{19}$$

It follows from (19) that the limiting wall ratio corresponds to the point on the stress-strain curve where the stress is equal to $Y \exp\left(\frac{3}{2}\varepsilon\right)$, the argument of F in (19) being then equal to ε. When the wall ratio is smaller than the limiting ratio, the internal pressure attains a maximum after the shell becomes completely plastic. In that case, the elastic strains are negligible throughout the shell, and the material may be regarded as rigid/plastic.†

(iv) Spherical cavity in an infinite medium The preceding solution for large elastic/plastic expansion of a finite shell is immediately applicable to the expansion of a spherical cavity in an infinitely extended medium. Thus, for a nonhardening material, the ratio of the initial and final cavity radii is given by (15) as

$$\left(\frac{a_0}{a}\right)^3 = 1 - 3(1 - v)\frac{Yc^3}{Ea^3} + 2(1 - 2v)\frac{Y}{E}\left(1 + 3 \ln\frac{c}{a}\right)$$

With continued cavity expansion, c/a progressively increases and a_0/a correspond-

---

† Numerical results based on the simple power law for the stress-strain curve have been presented by N. L. Svensson, *J. Appl. Mech.*, **25**: 89 (1958). For a bifurcation analysis, see H. Strifors and B. Storakers, *J. Mech. Phys. Solids*, **21**: 125 (1973). See also Y. Tomita and A. Shindo, *Int. J. Mech. Sci.*, **23**: 723 (1981).

ingly decreases. For very large expansions, c/a becomes approximately constant, the limiting value being

$$\frac{c}{a} = \left[\frac{E}{3(1-v)Y}\right]^{1/3} \tag{20a}$$

which holds even when the material work-hardens. The corresponding value of the cavity pressure is

$$p = \frac{2Y}{3}\left\{1 + \ln\frac{E}{3(1-v)Y}\right\} \tag{20b}$$

These limiting values are the same as those for the expansion of a cavity from zero radius ($a_0 = 0$). For normal prestrained metals, c/a lies between 5 and 6, while p is about $4Y$.

For an incompressible material ($v = 0.5$), the internal pressure, when the material work-hardens, is directly obtained from (17). When the cavity is expanded from zero radius, (17) reduces to

$$p = \frac{2}{3}\left\{Y + \int_{3Y/2E}^{1} F\left[-\frac{2}{3}\ln(1-x)\right]\frac{dx}{x}\right\} \tag{21}$$

where $x = a^3/r^3$. The argument of the function is infinite at the upper limit $x = 1$, the strain at the cavity surface being infinitely large. However, the integral converges because F tends to a constant value at very large strains. Integrating by parts, (21) may be written as

$$p = \frac{2}{3}Y\left(1 + \ln\frac{2E}{3Y}\right) + \frac{4}{9}\int_{3Y/2E}^{1} F'\left[-\frac{2}{3}\ln(1-x)\right]\frac{\ln x\,dx}{1-x}$$

The strain-hardening curve may be approximated by a straight line of slope $F' = T$. It is also sufficiently accurate to take the lower limit of the integral as zero. The pressure is then obtained in the closed form†

$$p = \frac{2}{3}Y\left(1 + \ln\frac{2E}{3Y}\right) + \frac{2}{27}\pi^2 T \tag{22}$$

This is also the limiting pressure for expanding a cavity from a finite internal radius. The expansion pressure (22) is quite sensitive to the rate of hardening of the material. For example, when $E/Y = 300$ and $T = Y$, the pressure is $4.93Y$, which is about 17.5 percent higher than the value $4.2Y$ for a nonhardening material having the same E/Y ratio. The theoretical value of p provides an estimate of the steady state pressure in the deep penetration of a smooth punch into a quasi-infinite medium.‡

---

† R. Hill, *The Mathematical Theory of Plasticity*, p. 104, Clarendon Press, Oxford 1950. For residual stresses due to unloading, see P. Chadwick, *Q. J. Mech. Appl. Math.*, **12**: 52 (1959).

‡ The dynamic expansion of spherical cavities has been discussed by H. G. Hopkins, *Progress in Solid Mechanics* (Eds. I. N. Sneddon and R. Hill), **1**, ch. 3, North-Holland, Groningen (1960).

5.2 Expansion of a Thick-Walled Tube

(i) *Elastic expansion and initial yielding* An important practical problem arises in the expansion of a thick-walled cylindrical tube under an internal pressure p and a longitudinal force P. The internal and external radii of the tube are a and b respectively. In the autofrettage process, the tube is either closed at both ends by rigid plugs, or provided with floating pistons which allow free axial contractions. The tube is assumed so large that plane transverse sections remain plane during the expansion. This means that the longitudinal strain ε_z is independent of the radius to the element. The stresses and strains sufficiently far from the ends do not vary along the length of the tube, and the equation of equilibrium is

$$\frac{\partial \sigma_r}{\partial r} = \frac{\sigma_\theta - \sigma_r}{r} \tag{23}$$

The z axis of the cylindrical coordinates (r, θ, z) is taken along the axis of the tube. While the tube is entirely elastic, the longitudinal stress may be written from Hooke's law as

$$\sigma_z = E\varepsilon_z + v(\sigma_r + \sigma_\theta) \tag{24}$$

where E is Young's modulus and v Poisson's ratio. Denoting the radial displacement by u, the radial strain ε_r and the circumferential strain ε_θ may be written as

$$\varepsilon_r = \frac{\partial u}{\partial r} = -v\varepsilon_z + \frac{1+v}{E}[(1-v)\sigma_r - v\sigma_\theta]$$

$$\varepsilon_\theta = \frac{u}{r} = -v\varepsilon_z + \frac{1+v}{E}[(1-v)\sigma_\theta - v\sigma_r] \tag{25}$$

Since ε_z is independent of r, the elimination of u from the above equations, and the substitution for $\sigma_\theta - \sigma_r$ from (23), leads to the compatibility equation

$$\frac{\partial}{\partial r}(\sigma_r + \sigma_\theta) = 0$$

It follows that $\sigma_r + \sigma_\theta$ and σ_z have constant values at each stage of the elastic expansion. Writing $\sigma_r + \sigma_\theta = 2A$, the equilibrium equation can be integrated to obtain Lamé's solution

$$\sigma_r = A + \frac{B}{r^2} \qquad \sigma_\theta = A - \frac{B}{r^2}$$

where A and B are constants, obtained from the boundary conditions $\sigma_r = 0$ at $r = b$ and $\sigma_r = -p$ at $r = a$. The stresses therefore become

$$\sigma_r = -p\left(\frac{b^2/r^2 - 1}{b^2/a^2 - 1}\right) \qquad \sigma_\theta = p\left(\frac{b^2/r^2 + 1}{b^2/a^2 - 1}\right) \tag{26}$$

If the resultant longitudinal load is denoted by P, the axial stress σ_z is $P/\pi(b^2 - a^2)$, since this stress is constant over the cross section. In particular, $P = 0$ for the open-end condition and $P = \pi a^2 p$ for the closed-end condition. The plane strain condition ($\varepsilon_z = 0$), sometimes considered for its simplicity, gives σ_z directly from (24) and (26). Hence

$$
\sigma_z = \begin{cases} \dfrac{p}{b^2/a^2 - 1} & \text{closed end} \\[2mm] 0 & \text{open end} \\[2mm] \dfrac{2vp}{b^2/a^2 - 1} & \text{plane strain} \end{cases} \tag{27}
$$

The radial displacement is obtained from the stress-strain equation for ε_θ. Substituting from (26) and (27), we get

$$
u = \frac{p}{E} \frac{\alpha r + (1 + v)b^2/r}{b^2/a^2 - 1} \tag{28}
$$

where α has the value $1 - 2v$ for the closed end condition, $1 - v$ for the open-end condition and $(1 + v)(1 - 2v)$ for the plane strain condition. The axial strain is obtained from (24) and (27) as

$$
E\varepsilon_z = \begin{cases} \dfrac{(1 - 2v)p}{b^2/a^2 - 1} & \text{closed end} \\[2mm] 0 & \text{plane strain} \\[2mm] -\dfrac{2vp}{b^2/a^2 - 1} & \text{open end} \end{cases} \tag{29}
$$

In all the three cases, σ_z is the intermediate principal stress. For the closed-end condition, σ_z is exactly the mean of the other two principal stresses. If Tresca's yield criterion is adopted, the onset of yielding is given by $\sigma_\theta - \sigma_r = 2k$, where k is the yield stress in pure shear. It follows from (26) that

$$
\sigma_\theta - \sigma_r = \frac{2pb^2/r^2}{b^2/a^2 - 1}
$$

which shows that the stress difference has its greatest magnitude at $r = a$. Hence yielding begins at the inner radius when the applied pressure becomes

$$
p_e = k\left(1 - \frac{a^2}{b^2}\right) \tag{30}
$$

irrespective of the end condition. If, on the other hand, the material yields according to the von Mises criterion, the end condition affects the yield pressure. It is convenient to express the Mises criterion in the form

$$
\frac{1}{2}(\sigma_\theta - \sigma_r)^2 + \frac{2}{3}\left(\sigma_z - \frac{\sigma_r + \sigma_\theta}{2}\right)^2 = 2k^2
$$

Since the second term on the left-hand side is independent of r, the yield function again has the greatest value at $r = a$, and the pressure p_e that causes yielding at $r = a$ is given by

$$\left(\frac{b^2 p_e}{b^2 - a^2}\right)^2 + \frac{1}{3}\left(\sigma_z - \frac{a^2 p_e}{b^2 - a^2}\right)^2 = k^2$$

The substitution from (27) shows that p_e is identical to (30) for the closed-end condition, the yield pressure for the other two end conditions being

$$p_e = k\left(1 - \frac{a^2}{b^2}\right)\left\{1 + (1 - 2\beta)^2 \frac{a^4}{3b^4}\right\}^{-1/2} \tag{31}$$

where $\beta = 0$ for open ends and $\beta = v$ for plane strain. The values of p_e for the three end conditions differ marginally from one another for usual values of v, the lowest pressure being that corresponding to open ends.

When a uniform pressure p is applied externally to a thick-walled tube of wall ratio b/a, the elastic distribution of σ_r and σ_θ is obtained from (26) by interchanging a and b. In this case, both the stresses are negative, σ_θ being more compressive than σ_r. Tresca's criterion requires $\sigma_\theta = -2k$ at $r = a$ for yielding to begin, and the initial yield pressure p_e is found to be the same as (30). If a thick-walled tube, which is already under an external pressure $p_2 < p_e$, is subjected to a gradually increasing internal pressure p_1, the magnitude of $\sigma_\theta - \sigma_r$ at any radius decreases when $p_1 < p_2$ and increases when $p_1 > p_2$. Yielding begins at the internal surface when p_1 exceeds p_2 by an amount equal to p_e. The elastic stress distribution in such a tube is obtained by the superposition of those due to p_1 and p_2 acting separately at $r = a$ and $r = b$ respectively.†

(ii) Elastic/plastic expansion When the internal pressure exceeds p_e, a plastic zone spreads from the inner radius, the elastic/plastic boundary at any stage being of radius c. In the elastic region ($c \leqslant r \leqslant b$), the radial and circumferential stresses are obtained from Lamé's equations, using the boundary condition $\sigma_r = 0$ at $r = b$, and the fact that the material at $r = c$ is stressed to the yield point. Adopting Tresca's yield criterion in the form $\sigma_\theta - \sigma_r = 2k$, the stress distribution in the elastic region is easily shown to be

$$\sigma_r = -\frac{kc^2}{b^2}\left(\frac{b^2}{r^2} - 1\right) \qquad \sigma_\theta = \frac{kc^2}{b^2}\left(\frac{b^2}{r^2} + 1\right)$$

$$\sigma_z = -E\varepsilon_z + 2vk\frac{c^2}{b^2} \qquad\qquad c \leqslant r \leqslant b \tag{32}$$

where ε_z depends on the end condition and cannot be determined unless the plastic region is considered, except in the case of plane strain where ε_z is zero. The radial

---

† Compound tubes have been discussed by W. R. D. Manning, *Engineering*, **163**: 349 (1947), and by B. Crossland and D. J. Burns, *Proc. Inst. Mech. Eng.*, **175**, 1083 (1961).

displacement in the elastic region is obtained from (25) and (32) as

$$u = -vr\varepsilon_z + (1 + v)\frac{kc^2}{Eb^2}\left\{(1 - 2v)r + \frac{b^2}{r}\right\} \qquad c \leqslant r \leqslant b \qquad (33)$$

It is assumed that σ_z continues to be the intermediate principal stress everywhere in the plastic region. Tresca's yield criterion then furnishes the equation

$$\sigma_\theta - \sigma_r = 2k \qquad a \leqslant r \leqslant c$$

where the material is assumed as nonhardening. Inserting in the equilibrium equation (23) and integrating, we get

$$\begin{aligned}
\sigma_r &= -k\left(1 - \frac{c^2}{b^2} + \ln\frac{c^2}{r^2}\right) \\
&\qquad\qquad\qquad\qquad a \leqslant r \leqslant c \qquad (34) \\
\sigma_\theta &= k\left(1 + \frac{c^2}{b^2} - \ln\frac{c^2}{r^2}\right)
\end{aligned}$$

on using the continuity of σ_r across $r = c$. The expressions for σ_r and σ_θ are thus independent of the end condition. The internal pressure is[†]

$$p = k\left(1 - \frac{c^2}{b^2} + \ln\frac{c^2}{a^2}\right) \qquad (35)$$

The distribution of σ_r and σ_θ is shown in Fig. 5.4 for $b/a = 2$. The magnitude of σ_r steadily decreases with the radius, while σ_θ has its greatest value at the elastic/plastic boundary. Assuming Tresca's associated flow rule, the plastic strain increments may be written as

$$d\varepsilon_\theta^p = -d\varepsilon_r^p > 0 \qquad d\varepsilon_z^p = 0$$

Since the plastic part of ε_z is zero, the axial strain is entirely elastic, which means that (24) holds in both elastic and plastic regions. Hence

$$\sigma_z = E\varepsilon_z + 2vk\left(\frac{c^2}{b^2} - \ln\frac{c^2}{r^2}\right) \qquad a \leqslant r \leqslant c$$

Solutions based on the Prandtl-Reuss theory indicate that σ_z approaches the mean of the other two principal stresses with increasing expansion when $\varepsilon_z = 0$.

To determine the axial strain ε_z for a given end condition, it is necessary to consider the equation of longitudinal equilibrium. The resultant longitudinal force acting in the tube is

$$P = 2\pi\int_a^b r\sigma_z\, dr = \pi E\varepsilon_z(b^2 - a^2) + 2\pi v\int_a^b r(\sigma_r + \sigma_\theta)\, dr$$

† L. B. Turner, *Trans. Camb. Phil. Soc.*, **21**: 377 (1909), *Engineering*, **92**: 115 (1911). The same solution applies to an annular disc for all values of $p \leqslant 2k$.

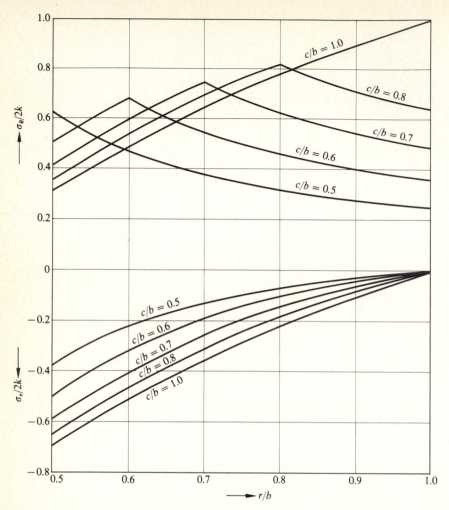

Figure 5.4 Distribution of radial and circumferential stresses in an elastic/plastic thick-walled tube subjected to internal pressure.

in view of (24). Since $r(\sigma_r + \sigma_\theta) = \partial(r^2\sigma_r)/\partial r$ by the equilibrium equation (23), we obtain

$$P = \pi[(b^2 - a^2)E\varepsilon_z + 2va^2p]$$

on using the boundary conditions for σ_r. It follows that ε_z is given by (29) even when the tube is partly plastic. The plane strain condition coincides with the closed-end condition only when the material is incompressible.

When the wall ratio is less than about 5, the strains in the plastic region are small so long as the tube remains partly plastic, and positional changes may therefore be disregarded. Since $\varepsilon_r^p + \varepsilon_\theta^p = 0$ in view of the integrated form of the associated flow rule, we have

$$\varepsilon_r + \varepsilon_\theta = \varepsilon_r^e + \varepsilon_\theta^e = \frac{1}{E}\{(1 - v)(\sigma_r + \sigma_\theta) - 2v\sigma_z\}$$

Substituting for σ_z from (24), and using the equilibrium equation (23), the above equation may be written as

$$\frac{\partial u}{\partial r} + \frac{u}{r} = -2v\varepsilon_z + \frac{1 - 2v}{2Gr}\frac{\partial}{\partial r}(r^2\sigma_r)$$

where G is the shear modulus. It is important to note that this differential equation holds in both elastic and plastic regions. The equation is readily integrated and the constant of integration determined by comparison with (33) at $r = b$, where $\sigma_r = 0$. This gives

$$\frac{u}{r} = -v\varepsilon_z + (1 - v)\frac{kc^2}{Gr^2} + (1 - 2v)\frac{\sigma_r}{2G} \tag{36}$$

which holds throughout the tube. From (29), (34), and (36), the displacement in the plastic region $(a \leqslant r \leqslant c)$ may be found for any given end condition. In particular, the displacement at the inner radius for the closed-ended tube is

$$u_a = (1 - v)\frac{kc^2}{Ga} - (1 - 2v)\frac{pa}{2G}\left\{1 + \frac{v}{1 + v}\left(\frac{b^2}{a^2} - 1\right)^{-1}\right\}$$

where p is given by (35). Similar expressions may be written down for the other two end conditions. An important feature of the plane strain condition $(\varepsilon_z = 0)$ is that the solution for any wall ratio b/a also furnishes the solutions for all lesser wall ratios. This is physically evident from the fact that the stress and strain in an outer annulus of the tube depend only on the radial pressure transmitted across the common interface, regardless of the agency through which this pressure is applied.

The analysis has been based on the assumption that σ_z is the intermediate principal stress throughout the plastic region. In view of (24), the condition $\sigma_r \leqslant \sigma_z \leqslant \sigma_\theta$ is equivalent to

$$\sigma_r \leqslant E\varepsilon_z + v(\sigma_r + \sigma_\theta) \leqslant \sigma_\theta$$

Using the yield criterion $\sigma_\theta - \sigma_r = 2k$, the inequalities may be written as

$$(1 - 2v)\sigma_r \leqslant E\varepsilon_z + 2vk \qquad E\varepsilon_z - (1 - 2v)\sigma_r \leqslant 2k(1 - v)$$

The first inequality is identically satisfied, since the right-hand side of this inequality is always positive in view of (29). The second inequality will be satisfied for all the end conditions if

$$E\varepsilon_z + (1 - 2v)p \leqslant 2k(1 - v)$$

Substituting from (29), the condition for validity of the solution may be expressed as†

$$\frac{p}{2k} \leqslant \begin{cases} \dfrac{1-v}{1-2v}\left(1-\dfrac{a^2}{b^2}\right) & \text{closed end} \\[2ex] \dfrac{1-v}{1-2v} & \text{plane strain} \\[2ex] \dfrac{(1-v)(1-a^2/b^2)}{1-2v-a^2/b^2} & \text{open end} \end{cases} \tag{37}$$

The range of wall ratios for which the solution would be valid for all elastic/plastic expansions is obtained by setting $p/2k = \ln(b/a)$. When $v = 0.3$, the limiting wall ratio is found to be 6.19 for the closed-end condition, 5.75 for the plane-strain condition, and 5.43 for the open-end condition. For still higher values of b/a, the solution will be valid so long as the internal pressure does not exceed the critical value given by (37). When the critical pressure is reached, σ_z becomes equal to σ_θ at the internal surface of the tube. An extension of the solution to higher pressures, without taking geometry changes into account, can hardly be considered as realistic.

(iii) Prandtl-Reuss theory for plane strain When the material yields according to the von Mises criterion, none of the stress components in the plastic region can be determined without recourse to the displacement. Under conditions of plane strain, the radial pressure q across the elastic/plastic boundary $r = c$ is given by the right-hand side of (31) with $\beta = v$ and a replaced by c. The stresses and displacement in the elastic region are evidently given by (26) and (28) with q and c written for p and a respectively. Consequently

$$\sigma_r = -kN\left(\frac{b^2}{r^2}-1\right) \qquad \sigma_\theta = kN\left(\frac{b^2}{r^2}+1\right) \qquad \sigma_z = 2kvN$$

$$\varepsilon_r = \frac{kN}{2G}\left(1-2v-\frac{b^2}{r^2}\right) \qquad \varepsilon_\theta = \frac{kN}{2G}\left(1-2v+\frac{b^2}{r^2}\right) \qquad c \leqslant r \leqslant b \quad (38)$$

where N is a dimensionless parameter given by

$$N = \left[\frac{1}{3}(1-2v)^2 + \frac{b^4}{c^4}\right]^{-1/2}$$

In view of the plane strain condition, the hydrostatic strain is equal to $(\varepsilon_r + \varepsilon_\theta)/3$ throughout the tube. If s_r and s_θ denote the radial and circumferential components of the deviatoric stress, then

$$\sigma_r = s_r + K(\varepsilon_r + \varepsilon_\theta) \qquad \sigma_\theta = s_\theta + K(\varepsilon_r + \varepsilon_\theta) \qquad \sigma_z = -(s_r + s_\theta) + K(\varepsilon_r + \varepsilon_\theta)$$

in both the elastic and plastic regions, where K is the bulk modulus equal to

---

† W. T. Koiter, *Biezeno Anniversary Volume*, Starn, Haarlem, p. 232 (1953).

$E/3(1-2v)$. The deviatoric stress in the plastic region must satisfy the von Mises yield criterion

$$s_r^2 + s_r s_\theta + s_\theta^2 = k^2 \qquad \text{or} \qquad s_\theta = \tfrac{1}{2}(-s_r + \sqrt{4k^2 - 3s_r^2})$$

Two of the basic equations necessary for the solution of the elastic/plastic problem is the strain compatibility equation

$$\frac{\partial \varepsilon_\theta}{\partial r} = \frac{\varepsilon_r - \varepsilon_\theta}{r} \tag{39}$$

obtained by eliminating u between the strain-displacement relations $\varepsilon_r = \partial u/\partial r$ and $\varepsilon_\theta = u/r$, and the stress equilibrium equation (23) which becomes

$$\frac{\partial s_r}{\partial r} + K \frac{\partial}{\partial r}(\varepsilon_r + \varepsilon_\theta) = \frac{1}{2r}(-3s_r + \sqrt{4k^2 - 3s_r^2}) \qquad a \leqslant r \leqslant c \tag{40}$$

Since the deviatoric principal strains are $(2\varepsilon_r - \varepsilon_\theta)/3$ and $(2\varepsilon_\theta - \varepsilon_r)/3$, the Prandtl-Reuss equations for small strains may be written as

$$\frac{2}{3} G \frac{\partial}{\partial c}(2\varepsilon_r - \varepsilon_\theta) = \frac{\partial s_r}{\partial c} + 2G\dot\lambda s_r$$

$$\frac{2}{3} G \frac{\partial}{\partial c}(2\varepsilon_\theta - \varepsilon_r) = \frac{\partial s_\theta}{\partial c} + 2G\dot\lambda s_\theta$$

for $\dot\lambda$ is a positive scalar. The elimination of $\dot\lambda$ between the above relations furnishes

$$s_\theta \frac{\partial s_r}{\partial c} - s_r \frac{\partial s_\theta}{\partial c} = \frac{2}{3} G\left[(2s_\theta + s_r) \frac{\partial \varepsilon_r}{\partial c} - (2s_r + s_\theta) \frac{\partial \varepsilon_\theta}{\partial c} \right]$$

Multiplying this relation by $2s_\theta + s_r$, and using the yield criterion, the differential equation is easily reduced to

$$\frac{\partial s_r}{\partial c} = G\left\{ \left(\frac{4}{3} - \frac{s_r^2}{k^2} \right) \frac{\partial \varepsilon_r}{\partial c} - \left(\frac{2}{3} + \frac{s_r s_\theta}{k^2} \right) \frac{\partial \varepsilon_\theta}{\partial c} \right\} \qquad a \leqslant r \leqslant c \tag{41}$$

Equations (39), (40), and (41) must be solved for ε_r, ε_θ, and s_r in the domain $a \leqslant r \leqslant c$ and $b \leqslant c \leqslant a$, the values of these quantities on the boundary $r = c$ being obtained from (38). The solution can be carried out numerically by expressing the equations in the finite difference form, and starting from the elastic/plastic boundary at each stage of the expansion.†

The radial and circumferential stresses predicted by the Prandtl-Reuss theory, for any given position of the elastic/plastic boundary, are found to be in extremely

---

† The plane strain solution has been discussed by P. G. Hodge, Jr., and G. N. White, *J. Appl. Mech.*, **17**: 180 (1950). The closed- and open-end conditions have been treated by P. V. Marcal, *Int. J. Mech. Sci.*, **7**: 229 and 841 (1965). Numerical solutions based on Tresca's yield criterion and the Prandtl-Reuss stress-strain equations are presented by R. Hill, E. H. Lee, and S. J. Tupper, *Proc. R. Soc., A,* **191**: 278 (1947), and *Proc. First U.S. Nat. Congr. Apppl. Mech.*, 561 (1951).

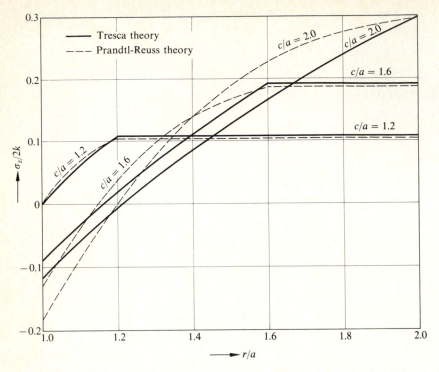

Figure 5.5 Distribution of axial stress in an elastic/plastic thick-walled tube under plane strain condition ($v = 0.3$).

close agreement with those given by the Tresca theory on the basis of the shear yield stress k. However, the distributions of the axial stress in the two solutions differ appreciably from one another, as indicated by the graphical presentation in Fig. 5.5 which corresponds to $v = 0.3$ and $b/a = 2$. Since the axial stress is relatively small compared to the other two stresses, the divergence is not important from the practical point of view. The radial displacements at the external and internal surfaces of the tube are plotted against c/a in Fig. 5.6, which shows the closeness of agreement between the two solutions.†

(iv) Residual stresses Suppose that a thick-walled tube which is rendered partially plastic by the application of an internal pressure p is completely unloaded by releasing the pressure. For sufficiently small values of p, the unloading process is fully elastic. The residual stresses are then obtained by subtracting (26) from (32)

---

† Solutions based on the Hencky equations have been given by D. N. de G. Allen and D. G. Sopwith, *Proc. R. Soc., A.*, **205**: 69 (1951) using the Tresca criterion, and by C. W. MacGregor, C. F. Coffin, and J. C. Fisher, *J. Franklin Inst.*, **26**: 245 (1948) using the von Mises yield criterion. Solutions based on $v = 0.5$ have been given by A. Nadai, *Theory of Flow and Fracture of Solids*, chaps. 30–32, McGraw-Hill Book Co., New York (1950). See also M. C. Steele, *J. Appl. Mech.*, **19**: 133 (1952).

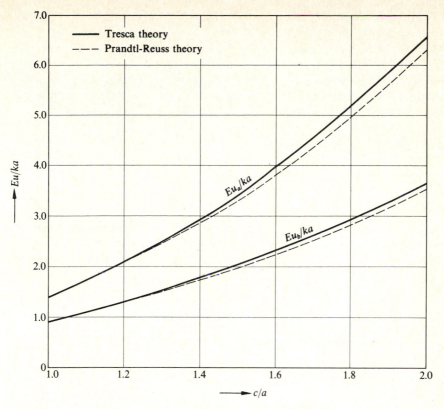

Figure 5.6 Internal and external surface displacements during the elastic/plastic expansion of a tube in plane strain ($v = 0.3$).

and (34), if Tresca's yield criterion is adopted, the result being

$$\sigma_r = -k\left(\frac{c^2}{a^2} - \frac{p}{p_e}\right)\left(\frac{a^2}{r^2} - \frac{a^2}{b^2}\right)$$

$$\sigma_\theta = k\left(\frac{c^2}{a^2} - \frac{p}{p_e}\right)\left(\frac{a^2}{r^2} + \frac{a^2}{b^2}\right) \qquad c \leqslant r \leqslant b \qquad (42)$$

$$\sigma_r = -k\left[\frac{p}{p_e}\left(1 - \frac{a^2}{r^2}\right) - \ln\frac{r^2}{a^2}\right]$$

$$\sigma_\theta = -k\left[\frac{p}{p_e}\left(1 + \frac{a^2}{r^2}\right) - \left(2 + \ln\frac{r^2}{a^2}\right)\right] \qquad a \leqslant r \leqslant c \qquad (43)$$

in view of (30) and (35). It follows that the residual σ_r is everywhere compressive with the maximum numerical value occurring at a radius $r = a\sqrt{p/p_e}$, which is less than c. The residual σ_θ is tensile in the outer part and compressive in the inner part of the tube, vanishing at a radius between $r = a$ and $r = c$. Since the axial strain is purely elastic according to Tresca's associated flow rule, it is completely removed on unloading, giving the residual axial stress

$$\sigma_z = \nu(\sigma_r + \sigma_\theta)$$

The residual σ_z is compressive within a radius $r = a \exp\left[\frac{1}{2}(p/p_e - 1)\right]$, which is also less than c. To check the validity of the assumption of elastic unloading, it is necessary to examine the magnitude of $\sigma_\theta - \sigma_r$. From (42) and (43), we get

$$\sigma_\theta - \sigma_r = 2k\frac{a^2}{r^2}\left(\frac{c^2}{a^2} - \frac{p}{p_e}\right) \qquad c \leqslant r \leqslant b$$

$$\sigma_\theta - \sigma_r = 2k\left(1 - \frac{pa^2}{p_e r^2}\right) \qquad a \leqslant r \leqslant c$$

The stress difference $\sigma_\theta - \sigma_r$ has its greatest magnitude on the internal surface, where yielding will restart if $p = 2p_e$, a possible Bauschinger effect being disregarded. The smallest wall ratio for which the secondary yielding can occur on unloading is given by

$$\ln\frac{b}{a} = 1 - \frac{a^2}{b^2}$$

which is obtained by setting $p = p_0 = 2k \ln (b/a)$. Hence, the critical wall ratio is $b/a \simeq 2.22$, and the corresponding fully plastic pressure is $p_0 \simeq 1.59k$. For $p \leqslant 2p_e$, a subsequent reloading of the tube by an internal pressure less than p results only in elastic changes in strain, whatever the wall ratio. The tube is therefore strengthened by an initial overstrain, the process of achieving this being known as *autofrettage*. It follows from above that a nonhardening tube cannot be strengthened in this way by more than a factor of two. Figure 5.7 shows the residual stresses in a thick-walled tube of wall ratio 2.0, when $c/a = 1.4$ and $\nu = 0.3$.

When $b/a > 2.22$ and $2p_e \leqslant p \leqslant p_0$, yielding occurs in the reversed sense within a radius $\rho < c$ on complete unloading from the elastic/plastic state. The residual stress distribution is a result of superposition of an additional stress system, denoted by primes, on that produced by the loading. Since $\sigma_\theta - \sigma_r$ changes from $2k$ to $-2k$ in each element of the region $a \leqslant \rho \leqslant c$ during the process of unloading, $\sigma'_\theta - \sigma'_r$ in this region must be equal to $-4k$. In the remainder of the tube, only elastic changes in stress are involved. By analogy with (32) and (34), the primed stresses may be written as

$$\sigma'_r = 2k\frac{\rho^2}{b^2}\left(\frac{b^2}{r^2} - 1\right) \qquad \sigma'_\theta = -2k\frac{\rho^2}{b^2}\left(\frac{b^2}{r^2} + 1\right) \qquad \rho \leqslant r \leqslant b$$

$$\sigma'_r = 2k\left(1 - \frac{\rho^2}{b^2} + \ln\frac{\rho^2}{r^2}\right) \qquad \sigma'_\theta = -2k\left(1 + \frac{\rho^2}{b^2} - \ln\frac{\rho^2}{r^2}\right) \qquad a \leqslant r \leqslant \rho$$

The residual σ_r and σ_θ are obtained by adding the above stresses to those given by (32) and (34). The radius ρ is determined from the boundary condition $\sigma'_r = p$ at

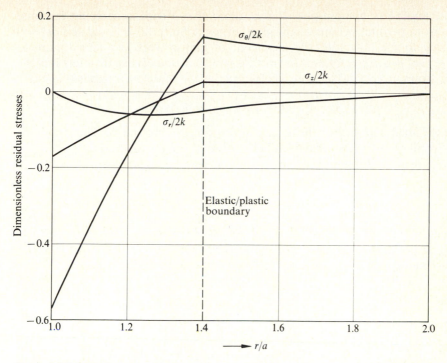

Figure 5.7 Distribution of residual stresses in a thick-walled tube ($b/a = 2$) unloaded from an elastic/plastic state.

$r = a$, the relationship between c and ρ being

$$2\left(\frac{\rho^2}{b^2} - \ln\frac{\rho^2}{ac}\right) - \frac{c^2}{b^2} = 1 \tag{44}$$

For example, when $b/a = 4$ and $c = b$, we find $\rho/a \simeq 1.28$. The residual hoop stress decreases from $-2k$ at $r = a$ to $-2.49k$ at $r = \rho$, and then steadily increases to $1.59k$ at $r = b$. The absence of a continuous cycle of plastic deformation, known as *shakedown*, is important for avoiding failure of pressure vessels by incremental collapse.† The shakedown condition for a thick-walled tube, which is repeatedly loaded and unloaded, is that the internal pressure must be less than p_0 when $b/a \leqslant 2.22$, and less than $2p_e$ when $b/a \geqslant 2.22$.

(v) *Influence of work-hardening* When the material work-hardens, Tresca's yield criterion may be written as $\sigma_\theta - \sigma_r = \sigma$, where σ is the current yield stress in

---

† The design of high-pressure containers has been discussed by T. E. Davidson and D. P. Kendall, *Mechanical Behavior of Materials under Pressure* (Ed. H. Ll. D. Pugh), p. 54, Elsevier, Amsterdam (1970).

uniaxial tension or compression. It is assumed that σ depends only on the total plastic work per unit volume of any given element. Since $d\varepsilon_r^p = -d\varepsilon_\theta^p$ and $d\varepsilon_z^p = 0$ according to the flow rule, the increment of plastic work per unit volume is $(\sigma_\theta - \sigma_r) d\varepsilon_\theta^p = \sigma d\varepsilon_\theta^p$. It follows that σ is the same function of ε_θ^p as the stress is of the plastic strain in uniaxial tension. The yield criterion therefore becomes

$$\sigma_\theta - \sigma_r = \sigma = F(\varepsilon_\theta^p) \qquad a \leqslant r \leqslant c$$

where the function F defines the uniaxial stress-plastic strain curve. The total hoop strain is still given by (36), where k is half the initial yield stress Y in simple tension. Subtracting from (36) the elastic hoop strain given by the second equation of (25), we get

$$\varepsilon_\theta^p = (1 - v^2)\left(\frac{Yc^2}{Er^2} - \frac{\sigma}{E}\right) = -\varepsilon_r^p \qquad a \leqslant r \leqslant c \tag{45}$$

in view of the yield criterion. Differentiating (45) with respect to r, and eliminating Yc^2/Er^2 by means of (45), we get

$$\frac{\partial \varepsilon_\theta^p}{\partial r} = -\frac{2}{r}\left\{\frac{\varepsilon_\theta^p + (1 - v^2)\sigma/E}{1 + (1 - v^2)H/E}\right\}$$

where $H = F'(\varepsilon_\theta^p)$. The last equation can be used to change the independent variable of (23) from r to ε_θ^p. The integration of the resulting stress equation then furnishes†

$$p = \frac{Y}{2}\left(1 - \frac{c^2}{b^2}\right) + \frac{1}{2}\int_0^{\varepsilon_0^p}\left\{\frac{1 + (1 - v^2)H/E}{\varepsilon_\theta^p + (1 - v^2)\sigma/E}\right\}\sigma \, d\varepsilon_\theta^p \tag{46}$$

where ε_0^p denotes ε_θ^p at $r = a$. The integral on the right-hand side of (46) can be evaluated numerically for any assumed value of ε_0^p. Since the corresponding value of c^2/a^2 is obtained from (45), the pressure can be calculated from (46) in a straightforward manner.

An explicit solution can be found in the case of linear work-hardening where H has a constant value. Then

$$\sigma = Y + H\varepsilon_\theta^p$$

which can be used to eliminate ε_θ^p from (45). The yield criterion therefore becomes

$$\sigma_\theta - \sigma_r = Y\left[1 + (1 - v^2)\frac{Hc^2}{Er^2}\right]\bigg/\left[1 + (1 - v^2)\frac{H}{E}\right]$$

Inserting in (23), the radial and circumferential stresses in the plastic region

---

† D. R. Bland, *J. Mech. Phys. Solids*, **4**: 209 (1956). Useful theoretical and experimental results on work-hardening tubes have been reported by B. Crossland and J. A. Bones, *Proc. Inst. Mech. Eng.*, **172**: 777 (1958). See also G. J. Franklin and J. L. M. Morrison, ibid., **174**: 947 (1960).

$(a \leqslant r \leqslant c)$ are obtained as

$$
\left[1 + (1 - v^2)\frac{H}{E}\right]\sigma_r = -\frac{Y}{2}\left\{1 - \frac{c^2}{b^2} + \ln\frac{c^2}{r^2} + (1 - v^2)\frac{H}{E}\left(\frac{c^2}{r^2} - \frac{c^2}{b^2}\right)\right\}
$$

$$
\left[1 + (1 - v^2)\frac{H}{E}\right]\sigma_\theta = \frac{Y}{2}\left\{1 + \frac{c^2}{b^2} - \ln\frac{c^2}{r^2} + (1 - v^2)\frac{H}{E}\left(\frac{c^2}{r^2} + \frac{c^2}{b^2}\right)\right\}
$$

(47)

in view of the continuity of σ_r across $r = c$. The axial stress σ_z for the appropriate end condition then follows from (24) and (29). The internal pressure for the work-hardening tube exceeds that of the nonhardening tube by the amount

$$
\Delta p = \frac{Y}{2}\left(\frac{c^2}{a^2} - \ln\frac{c^2}{a^2} - 1\right) \bigg/ \left(1 + \frac{E/H}{1 - v^2}\right)
$$

(48)

The effect of work-hardening is therefore to increase the magnitude of the stresses in the plastic region for a given radius to the elastic/plastic boundary. The internal pressure steadily increases during the elastic/plastic expansion without attaining a maximum. The assumption that σ_z is the intermediate principal stress can be shown to be valid for somewhat larger wall ratios than those for no hardening.† Yielding will not restart at the bore on unloading from an elastic/plastic state so long as the ratio p/p_e is less than $2\sigma_0/Y$, where σ_0 is the yield stress at $r = a$. The pressure-expansion curves for $H = 0$ and $H = 0.1E$ are plotted in Fig. 5.8 assuming $b/a = 2$.

(vi) *Solution for large strains* When the wall ratio is sufficiently large, it is necessary to consider geometry changes to calculate the strains in the plastic region. A drastic simplification is achieved if we assume that the material is incompressible ($v = 0.5$). The closed-end condition then coincides with the plane-strain condition with an axial stress

$$
\sigma_z = \tfrac{1}{2}(\sigma_r + \sigma_\theta)
$$

The stresses in the elastic region are still given by (32), where r is the current radius to the element, the initial radius being denoted by r_0. If a_0 and a denote the initial and current values of the internal radius of the tube, the constancy of volumes requires $r^2 - r_0^2 = a^2 - a_0^2$. For an element in the elastic region, the left-hand side of this equation is $2ru$ to the usual order of approximation, where u is expressed by (33) with $v = 0.5$ and $\varepsilon_z = 0$. Thus

$$
r^2 - r_0^2 = a^2 - a_0^2 \simeq \frac{3k}{E}c^2
$$

(49)

which gives the relationship between the initial and current radii to any element for a given position of the elastic/plastic boundary.

---

† The plastic collapse of a thick-walled tube under combined torsion and internal pressure has been investigated by B. Crossland and R. Hill, *J. Mech. Phys. Solids*, **2**: 27 (1953).

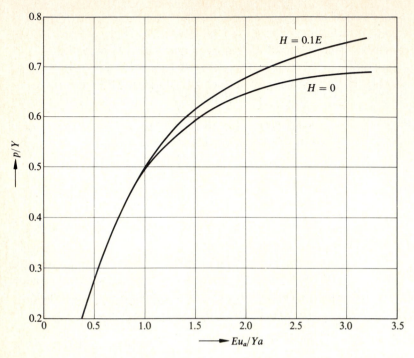

Figure 5.8 Pressure-expansion curves for a closed-ended tube with and without work-hardening, when $b/a = 2$ and $v = 0.3$.

The von Mises yield criterion coincides with the Tresca criterion with k equal to $Y/\sqrt{3}$, when the axial stress is the mean of the other two principal stresses satisfying the end condition. For a work-hardening material with $v = 0.5$, the Mises criterion may be written as

$$\sigma_\theta - \sigma_r = \frac{2}{\sqrt{3}} F\left(\frac{2}{\sqrt{3}} \ln \frac{r}{r_0}\right) \qquad a \leqslant r \leqslant c$$

where the function F defines the uniaxial stress-strain curve expressed by the equation $\sigma = F(\varepsilon)$. Evidently, $F(Y/E) = Y$. Inserting in the equilibrium equation and integrating, we get

$$p = \frac{2Y}{\sqrt{3}}\left(1 - \frac{c^2}{b^2}\right) + \frac{2Y}{\sqrt{3}} \int_a^c F\left(\frac{2}{\sqrt{3}} \ln \frac{r}{r_0}\right) \frac{dr}{r}$$

Using the incompressibility condition (49), the above equation is reduced to

$$p = \frac{2Y}{\sqrt{3}}\left(1 - \frac{c^2}{b^2}\right) + \frac{2}{\sqrt{3}} \int_{\sqrt{3} Y/E}^{1 - a_0^2/a^2} F\left[-\frac{1}{\sqrt{3}} \ln (1 - x)\right] \frac{dx}{x} \qquad (50)$$

where $x = (a^2 - a_0^2)/r^2$. The integral may be evaluated numerically for any given internal expansion specified by the ratio a_0/a. If the tube is sufficiently thick, the internal pressure attains a maximum while the tube is still partly plastic, and the subsequent expansion takes place under decreasing pressure. Differentiating (50) with respect to a and noting that $dc/da = (E/\sqrt{3}\,Y)(a/c)$ in view of (49), the condition for plastic instability marked by the pressure maximum may be written as†

$$\frac{E\rho^2(\rho^2 - 1)}{\sqrt{3}\,F(\ln \rho^2/\sqrt{3})} = \frac{b^2}{a_0^2} \qquad \rho = \frac{a}{a_0} \qquad (51)$$

If the left-hand side of the above equation is plotted against ρ^2, the onset of instability corresponds to the point on the curve where the ordinate equals the square of the wall ratio. The maximum pressure can then be found from (50) by numerical integration. For a non-hardening material, $F = Y$, and the solution of (51) becomes

$$\left(\frac{a}{a_0}\right)^2 = \frac{1}{2}\left\{1 + \left(1 + 4\sqrt{3}\,\frac{Yb^2}{Ea_0^2}\right)^{1/2}\right\}$$

For a work-hardening material, there is a limiting wall ratio for which the pressure attains a maximum when the tube is just fully plastic ($c = b$). Since at this stage $\rho^2 = 1 + \sqrt{3}\,Yb^2/Ea_0^2$, the limiting wall ratio is given by

$$F\left\{\frac{1}{\sqrt{3}}\ln\left(1 + \frac{\sqrt{3}\,Yb^2}{Ea_0^2}\right)\right\} = Y\left(1 + \frac{\sqrt{3}\,Yb^2}{Ea_0^2}\right) \qquad (52)$$

which is obtained on substitution in (51). The above equation indicates that the limiting wall ratio is directly obtained from the stress-strain curve $\sigma = F(\varepsilon)$ by finding its intersection with the curve for $Y\exp(\sqrt{3}\,\varepsilon)$ against ε. The ordinate of the point of intersection is then equal to the right-hand side of (52).

When the wall ratio is smaller than the limiting value given by (52), the pressure attains a maximum some time after the tube is fully plastic. The change in the external radius of the tube then becomes significant. Denoting the initial and final values of this radius by b_0 and b respectively, the incompressibility condition may be written as

$$r^2 - r_0^2 = a^2 - a_0^2 = b^2 - b_0^2$$

The yield criterion in this case holds throughout the tube, and the equilibrium equation gives

$$p = \frac{2}{\sqrt{3}}\int_a^b F\left(\frac{2}{\sqrt{3}}\ln\frac{r}{r_0}\right)\frac{dr}{r} = \frac{2}{\sqrt{3}}\int_{1-b_0^2/b^2}^{1-a_0^2/a^2} F\left[-\frac{1}{\sqrt{3}}\ln(1-x)\right]\frac{dx}{x} \qquad (53)$$

Differentiating (53) with respect to a, noting that $db/da = a/b$, and setting

---

† J. Chakrabarty and J. M. Alexander, *Int. J. Mech. Sci.*, **11**: 175 (1969). A bifurcation analysis has been carried out by B. Storakers, *J. Mech. Phys. Solids*, **19**: 339 (1971).

$dp/da = 0$, we obtain the condition of plastic instability in the form

$$\frac{a_0^2}{a^2} F\left(\frac{1}{\sqrt{3}} \ln \frac{a^2}{a_0^2}\right) = \frac{b_0^2}{b^2} F\left(\frac{1}{\sqrt{3}} \ln \frac{b^2}{b_0^2}\right)$$

Expressing b^2/b_0^2 in terms of a^2/a_0^2 from the incompressibility condition, the above equation may be written in the more convenient form

$$\phi(\rho^2 - I) = \phi\left\{\frac{a_0^2}{b_0^2} (\rho^2 - 1)\right\}$$

$$\phi(x) = \frac{F[\ln(1+x)/\sqrt{3}]}{1+x}$$

(54)

The function $\phi(x)$ may be plotted against x for any given function F. The curve will always have a maximum as shown in Fig. 5.9. If a pair of points A and B are located on the curve at the same vertical height such that $CB/CA = b_0^2/a_0^2$, then in view of (54),

$$CB = \rho^2 - 1 \qquad CA = \frac{a_0^2}{b_0^2} (\rho^2 - 1)$$

The quantities CA and CB directly furnish the instability strains, since the hoop strain at the inner radius is $\frac{1}{2}\ln(1 + CB)$ and that at the outer radius is $\frac{1}{2}\ln(1 + CA)$. When a/a_0 and b/b_0 are thus known, the maximum or bursting pressure may be calculated from (53). When the wall ratio has the limiting value given by (52), the point A coincides with A_0 whose abscissa is $\sqrt{3}\, Y/E$ and which corresponds to the yield point of the stress-strain curve. For wall ratios appreciably smaller than the limiting value, it would be sufficiently accurate to neglect the elastic strain, which amounts to the assumption that the material of the tube is rigid/plastic.†

(vii) Cylindrical cavity in an infinite medium An interesting situation involving large plastic strains arises in the expansion of a cylindrical cavity in an infinitely extended medium ($b = \infty$). In the elastic region, $\sigma_r + \sigma_\theta$ and σ_z are both zero, while in the plastic region, σ_z rapidly approaches the mean of the other two principal stresses. It is therefore a good approximation to take $\sigma_z = \frac{1}{2}(\sigma_r + \sigma_\theta)$ throughout the tube, and write the compressibility equation in the form

$$\frac{\partial v}{\partial r} + \frac{v}{r} = \frac{3}{2E}(1 - 2v)(\dot{\sigma}_r + \dot{\sigma}_\theta)$$

where v is the radial velocity, and the dot denotes rate of change following the particle, taking c as the time scale. If the material is nonhardening and yields

---

† A series solution for the instability of rigid/plastic tubes has been given by N. L. Svensson, *J. Appl. Mech.*, **25**: 89 (1958).

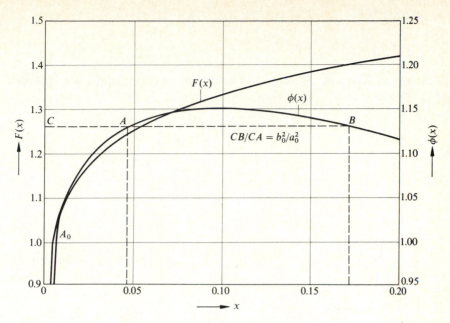

Figure 5.9 Graphical representation of the fully plastic instability condition in a thick-walled tube.

according to the von Mises criterion, σ_r and σ_θ are given by (34) with $b = \infty$ and k replaced by $Y/\sqrt{3}$. The above equation then becomes

$$\frac{\partial v}{\partial r} + \frac{v}{r} = -\sqrt{3}(1 - 2v)\frac{2Y}{Ec}$$

on neglecting a term of order Y/E in comparison with unity. Since $v = 2(1 + v)Y/\sqrt{3}\,E$ on $r = c$ in view of (33), integration of the above equation results in

$$v = -\sqrt{3}(1 - 2v)\frac{Yr}{2Ec} + (5 - 4v)\frac{Yc}{\sqrt{3}\,Er} \qquad a \leqslant r \leqslant c \qquad (55)$$

The velocity is therefore a function of the ratio r/c only. Since $v = dr/dc$ and $d(r/c) = (r/c)(dr/r - dc/c)$, Eq. (55) can be rewritten in the form

$$\frac{dr}{r} = \frac{\sqrt{3}(1 - 2v)Y/E - (5 - 4v)Yc^2/\sqrt{3}\,Er^2}{1 + \sqrt{3}(1 - 2v)Y/E - (5 - 4v)Yc^2/\sqrt{3}\,Er^2}\frac{c}{r}\,d\!\left(\frac{r}{c}\right)$$

which can be integrated under the initial conditions $r/c = 1$ and $r/r_0 = 1 + (1 + v)Y/\sqrt{3}\,E$, when an element first becomes plastic. Using the same order of approximation as introduced above, we obtain

$$\left(\frac{r_0}{r}\right)^2 = 1 + \sqrt{3}\,(1 - 2v)\frac{Y}{E} - (5 - 4v)\frac{Yc^2}{\sqrt{3}\,Er^2} \qquad a \leqslant r \leqslant c \qquad (56)$$

The ratio of the initial and final radii of the cavity is given by

$$\left(\frac{a_0}{a}\right)^2 = 1 + \sqrt{3}(1 - 2v)\frac{Y}{E} - (5 - 4v)\frac{Yc^2}{\sqrt{3}\,Ea^2}$$

As the expansion proceeds, the second term on the right-hand side soon becomes negligible. For very large expansions, a_0^2/a^2 is vanishingly small and we have

$$\frac{c}{a} \simeq \left[\frac{3E}{(5 - 4v)Y}\right]^{1/2} \tag{57}$$

This formula may be directly obtained by integrating (55) corresponding to $r = a$, with $v = da/dc$. In view of (35) and (57), the internal pressure approaches the limiting value

$$p = \frac{Y}{\sqrt{3}}\left\{1 + \ln\left[\frac{\sqrt{3}\,E}{(5 - 4v)Y}\right]\right\} \tag{58}$$

For most prestrained metals, E/Y is of order 250 to 350, while v is usually between 0.25 and 0.35. Hence c/a lies between 10 and 13, while p is normally in the range $3.3Y$ to $3.7Y$.

For a work-hardening incompressible material, the cavity pressure is given by (50) with $b = \infty$. If the cavity is expanded from zero radius, the upper limit of the integral in (50) is unity, and the integration by parts results in

$$p = \frac{Y}{\sqrt{3}}\left(1 + \ln\frac{E}{\sqrt{3}\,Y}\right) + \frac{1}{3}\int_{\sqrt{3}\,Y/E}^{1} F'\left[-\frac{1}{\sqrt{3}}\ln(1 - x)\right]\frac{\ln x\, dx}{1 - x}$$

The ratio c/a has the constant value $(E/\sqrt{3}\,Y)^{1/2}$ throughout the expansion. If a linear strain-hardening is assumed, F' has a constant value T. Taking the lower limit as zero, we then have the explicit solution[†]

$$p = \frac{Y}{\sqrt{3}}\left(1 + \ln\frac{E}{\sqrt{3}\,Y}\right) + \frac{\pi^2}{18}T \tag{59}$$

When the cavity is expanded from a finite radius, the pressure rapidly approaches this value as the expansion proceeds. As in the case of spherical cavity, the expansion pressure (59) is fairly sensitive to the rate of hardening. For example, when $E/Y = 300$ and $T = Y$, the cavity pressure is $4.10Y$, compared to $3.55Y$ for a nonhardening material.

5.3 Thermal Stresses in a Thick-Walled Tube

(i) *Elastic analysis* A long thick-walled tube made of an isotropic nonhardening material is under combined internal pressure and radial temperature gradient. It is

---

† R. Hill, *The Mathematical Theory of Plasticity*, p. 127, Clarendon Press, Oxford (1950). The work done during the expansion has been evaluated by R. D. Bhargava and C. B. Sharma, *J. Franklin Inst.*, **277**: 422 (1964).

assumed that the yield criterion and the plastic stress-strain relations are unaffected by the variation in temperature, provided the variation is not large enough to change the material properties appreciably. Thermal effects are then included in the expanding tube problem by the usual modification of the stress-strain equations. If the coefficient of linear expansion is denoted by α, the dilatation produced by a rise in temperature T is equal to αT. Subtracting this from each strain component in the generalized Hooke's law, we have the elastic stress-strain equations

$$E(\varepsilon_r - \alpha T) = \sigma_r - \nu(\sigma_\theta + \sigma_z)$$

$$E(\varepsilon_\theta - \alpha T) = \sigma_\theta - \nu(\sigma_z + \sigma_r)$$

$$E(\varepsilon_z - \alpha T) = \sigma_z - \nu(\sigma_r + \sigma_\theta)$$

Substituting for σ_z from the last equation, the first two equations may be expressed as

$$\frac{\partial u}{\partial r} = -\nu\varepsilon_z + (1 + \nu)\alpha T + \frac{1 + \nu}{E}[(1 - \nu)\sigma_r - \nu\sigma_\theta]$$

$$\frac{u}{r} = -\nu\varepsilon_z + (1 + \nu)\alpha T + \frac{1 + \nu}{E}[(1 - \nu)\sigma_\theta - \nu\sigma_r]$$

(60)

Eliminating u between these two equations, and using the equilibrium equation (23), we obtain the compatibility equation

$$\frac{\partial}{\partial r}\left(\sigma_r + \sigma_\theta + \frac{E\alpha T}{1 - \nu}\right) = 0$$

which indicates that $\sigma_r + \sigma_\theta$ varies linearly with the temperature. If we assume a steady state distribution of temperature, T satisfies the Laplace equation $\nabla^2 T = 0$, the solution of which may be written as

$$T = T_b + (T_a - T_b)\frac{\ln{(b/r)}}{\ln{(b/a)}}$$

(61)

where T_a and T_b are the temperatures at the inner and the outer surfaces respectively. It is convenient at this stage to introduce a dimensionless parameter β defined as

$$\beta = \frac{\alpha(T_a - T_b)}{2(1 - \nu)}$$

The integral of the compatibility equation is then expressed in the form

$$\sigma_r + \sigma_\theta = 2A - 2\beta E\frac{\ln{(b/r)}}{\ln{(b/a)}}$$

where A is a constant. Eliminating σ_θ from (23) by means of the above relation, and integrating, we arrive at the solution†

$$\sigma_r = A + B\frac{b^2}{r^2} - \beta E\left\{\frac{\frac{1}{2} + \ln{(b/r)}}{\ln{(b/a)}}\right\}$$

$$\sigma_\theta = A - B\frac{b^2}{r^2} + \beta E\left\{\frac{\frac{1}{2} - \ln{(b/r)}}{\ln{(b/a)}}\right\}$$

(62)

where B is also independent of r. The boundary conditions $\sigma_r = 0, r = b$ and $\sigma_r = -p, r = a$ furnish

$$A = \frac{p - \beta E}{b^2/a^2 - 1} + \frac{\beta E}{2\ln{(b/a)}} \qquad B = -\frac{p - \beta E}{b^2/a^2 - 1}$$

The radial and circumferential stresses in the purely elastic tube therefore become

$$\sigma_r = -(p - \beta E)\frac{b^2/r^2 - 1}{b^2/a^2 - 1} - \beta E\frac{\ln{(b/r)}}{\ln{(b/a)}}$$

$$\sigma_\theta = (p - \beta E)\frac{b^2/r^2 + 1}{b^2/a^2 - 1} + \beta E\frac{1 - \ln{(b/r)}}{\ln{(b/a)}}$$

(63)

The axial stress σ_z then follows from the third stress-strain relation, where ε_z must be determined from the fact that the resultant axial force is

$$P = 2\pi\int_a^b r\sigma_z \, dr = 2\pi v a^2 p + \pi E(b^2 - a^2)\varepsilon_z - 2\pi E\alpha\int_a^b rT \, dr$$

the derivation being similar to that in the previous section. When the tube has closed ends, $P = \pi a^2 p$, which gives

$$\varepsilon_z = \frac{(1 - 2v)p/E - 2(1 - v)\beta}{b^2/a^2 - 1} + \frac{(1 - v)\beta}{\ln{(b/a)}} + \alpha T_b$$

(64)

and the axial stress is then found as

$$\sigma_z = \frac{p - 2\beta E}{b^2/a^2 - 1} + \beta E\frac{1 - 2\ln{(b/r)}}{\ln{(b/a)}}$$

(65)

In view of (61), (63), and (65), the radial displacement can be calculated from the second equation of (60).

When p and β are increased to critical values, yielding may begin anywhere in the tube depending on the ratio of these parameters. We shall be concerned here with the situation where σ_z is the intermediate principal stress in the element that yields. If Tresca's yield criterion is adopted, yielding will depend on the magnitude of the stress difference

$$\sigma_\theta - \sigma_r = (p - \beta E)\frac{2b^2/r^2}{b^2/a^2 - 1} + \frac{\beta E}{\ln{(b/a)}}$$

† This solution is due to R. Lorenz, Z. Ver. deut. Ing., **51**: 743 (1907).

which has the greatest value at $r = a$, provided $p > \beta E$. Suppose that the tube is first subjected to a temperature gradient, which is followed by an application of internal pressure. Plasticity then commences at the internal surface when $\sigma_\theta - \sigma_r$ equals Y at $r = a$, and the pressure at the onset of yielding is

$$p_e = \frac{Y}{2}\left(1 - \frac{a^2}{b^2}\right) + \beta E\left\{1 - \frac{1 - a^2/b^2}{\ln(b^2/a^2)}\right\} \tag{66}$$

The effect of the thermal gradient is, therefore, to increase p_e for $\beta > 0$ and decrease p_e for $\beta < 0$. When $\beta > 0$, the necessary condition $p_e > \beta E$ leads to the restriction $\beta E/Y < \ln(b/a)$. For $\beta < 0$, the tube will yield due to the temperature difference alone unless $p_e > 0$. Hence the initial yield pressure will be correctly given by (66), only if

$$\frac{-\ln R}{R^2 \ln R^2/(R^2 - 1) - 1} < \frac{\beta E}{Y} < \ln R \tag{67}$$

where R denotes the wall ratio b/a. In view of these inequalities, it can be shown that $\sigma_r < \sigma_z < \sigma_\theta$ at $r = a$, and that $\sigma_\theta - \sigma_z$ and $\sigma_z - \sigma_r$ are each numerically less than Y throughout the tube. It may be noted in passing that yielding can occur due to the temperature gradient alone even for $\beta > 0$, when $\sigma_r - \sigma_\theta$ attains the value Y at the inner surface of the tube.†

(ii) *Elastic/plastic analysis* If the internal pressure is increased to a value greater than p_e, the tube will become plastic within a radius c. It is assumed that the axial stress continues to be the intermediate principal stress in the plastic region throughout the elastic/plastic expansion. The range of validity of this assumption will be discussed later. Tresca's yield criterion for a nonhardening material therefore becomes

$$\sigma_\theta - \sigma_r = Y \qquad a \leqslant r \leqslant c$$

In the outer elastic region, σ_r and σ_θ are still of the form (62), where A and B are to be determined from the condition $\sigma_r = 0, r = b$, and the fact that the surface $r = c$ is on the point of yielding. We thus obtain

$$A = \frac{Yc^2}{2b^2} + \frac{\beta E}{2}\frac{1 - c^2/b^2}{\ln(b/a)} \qquad B = -\frac{c^2}{2b^2}\frac{Y - \beta E}{\ln(b/a)}$$

and the expressions for σ_r and σ_θ in the elastic annulus become

$$\sigma_r = -\frac{1}{2}\left(Y - \frac{\beta E}{\ln(b/a)}\right)\left(\frac{c^2}{r^2} - \frac{c^2}{b^2}\right) - \beta E\frac{\ln(b/r)}{\ln(b/a)}$$

$$\sigma_\theta = \frac{1}{2}\left(Y - \frac{\beta E}{\ln(b/a)}\right)\left(\frac{c^2}{r^2} + \frac{c^2}{b^2}\right) + \beta E\frac{1 - \ln(b/r)}{\ln(b/a)}$$

$$c \leqslant r \leqslant b \tag{68}$$

---

† Various other cases of the initial yielding of the tube under both internal and external pressures with radial heat flow have been discussed by M. G. Derrington, *Int. J. Mech. Sci.*, **4**: 83 (1962).

It follows from (68) that $\sigma_\theta - \sigma_r$ has its greatest value at $r = c$ so long as $\beta E/Y < \ln(b/a)$. In the plastic annulus, the equilibrium equation (23) and the yield criterion readily furnish†

$$\sigma_r = -\frac{1}{2}\left(Y - \frac{\beta E}{\ln(b/a)}\right)\left(1 - \frac{c^2}{b^2} + \ln\frac{c^2}{r^2}\right) - \beta E\frac{\ln(b/r)}{\ln(b/a)}$$

$$\sigma_\theta = \frac{1}{2}\left(Y - \frac{\beta E}{\ln(b/a)}\right)\left(1 + \frac{c^2}{b^2} - \ln\frac{c^2}{r^2}\right) + \beta E\frac{1 - \ln(b/r)}{\ln(b/a)}$$

$$a \leqslant r \leqslant c \quad (69)$$

in view of the continuity of σ_r across $r = c$. The internal pressure necessary to render the tube plastic within a radius c is

$$p = \frac{1}{2}\left(Y - \frac{\beta E}{\ln(b/a)}\right)\left(1 - \frac{c^2}{b^2} + \ln\frac{c^2}{a^2}\right) + \beta E \quad (70)$$

The pressure attains a maximum when the tube just becomes fully plastic, and the fully plastic pressure is $Y \ln(b/a)$ irrespective of the temperature difference. During the elastic/plastic expansion, the internal pressure is augmented by $\beta > 0$ and reduced by $\beta < 0$. If $\beta E/Y$ is equal to $\ln(b/a)$, the whole tube becomes simultaneously plastic when the pressure attains the value $Y \ln(b/a)$.

As in the preceding section, ε_z is entirely elastic in view of Tresca's associated flow rule. The axial strain is therefore given by (64) throughout the expansion when the tube has closed ends. The elastic stress-strain relation for ε_z then furnishes

$$\sigma_z = E\varepsilon'_z + vY\frac{c^2}{b^2} + \frac{\beta E}{\ln(b/a)}\left\{v\left(1 - \frac{c^2}{b^2}\right) - \ln\frac{b^2}{r^2}\right\} \quad c \leqslant r \leqslant b$$

$$\sigma_z = E\varepsilon'_z + vY\left(\frac{c^2}{b^2} - \ln\frac{c^2}{r^2}\right) + \frac{\beta E}{\ln(b/a)}\left\{v\left(1 - \frac{c^2}{b^2} + \ln\frac{c^2}{r^2}\right) - \ln\frac{b^2}{r^2}\right\} \quad a \leqslant r \leqslant c$$

$$(71)$$

in view of (68) and (69), where ε'_z stands for $\varepsilon_z - \alpha T_b$. Thus σ_z increases with r in both the elastic and plastic regions of the tube at each stage when $\beta > 0$.

It is assumed, as before, that the strains are small so long as the tube is partly plastic. Since $\varepsilon_r^p = -\varepsilon_\theta^p$ in view of the plastic flow rule, $\varepsilon_r + \varepsilon_\theta$ is purely elastic, and hence (60) gives

$$\frac{\partial u}{\partial r} + \frac{u}{r} = -2v\varepsilon_z + 2(1 + v)\alpha T + \frac{1}{E}(1 + v)(1 - 2v)(\sigma_r + \sigma_\theta)$$

throughout the tube even when it is partly plastic. In view of (23), the above equation can be rewritten as

$$\frac{\partial}{\partial r}(ru) = -2vr\varepsilon_z + 2(1 + v)\alpha rT + \frac{1}{E}(1 + v)(1 - 2v)\frac{\partial}{\partial r}(r^2\sigma_r)$$

---

† D. R. Bland, *J. Mech. Phys. Solids*, **4**: 209 (1956). Bland also included external pressure and work-hardening in his analysis.

Substituting from (61) and integrating, we can express the solution in the form

$$\frac{u}{r} = -v\varepsilon_z + (1+v)\alpha T + (1+v)(1-2v)\frac{\sigma_r}{E} + (1-v^2)\left\{\frac{C}{r^2} + \frac{\beta}{\ln(b/a)}\right\}$$

The parameter C can be determined from the condition that the hoop strain calculated from this equation at the outer radius must be the same as that given by Hooke's law. It follows from (60) and (68) that

$$\frac{u_b}{b} = -v\varepsilon_z + (1+v)\alpha T_b + (1-v^2)\left\{\frac{Yc^2}{Eb^2} + \frac{\beta(1-c^2/b^2)}{\ln(b/a)}\right\} \tag{72}$$

Setting $r = b$ and $\sigma_r = 0$ in the above expression for u/r, and comparing with (72), we find

$$C = c^2\left(\frac{Y}{E} - \frac{\beta}{\ln(b/a)}\right)$$

Hence the radial displacement in the elastic/plastic tube is finally expressed as

$$\frac{u}{r} = -v\varepsilon_z + (1+v)\alpha T + (1+v)(1-2v)\frac{\sigma_r}{E} + (1-v^2)\left\{\frac{Yc^2}{Er^2} - \frac{\beta(c^2/r^2-1)}{\ln(b/a)}\right\} \tag{73}$$

where T is given by (61) and σ_r by (68) or (69) according as the displacement is required in the elastic or the plastic part of the tube. The variation of the internal pressure and the bore strain during the elastic/plastic expansion of a closed-ended tube of wall ratio 2 is shown in Fig. 5.10 for three different values of $\beta E/Y$. Since the uniform part of the temperature distribution causes no thermal stress, T_b has been taken as zero in calculating the results.

Equation (73) holds good even when the material work-hardens, provided Y is regarded as the initial yield stress. If the current yield stress is denoted by σ, which replaces Y in the yield criterion, the elastic part of the hoop strain is

$$\varepsilon_\theta^e = -v\varepsilon_z + (1+v)\alpha T + (1+v)(1-2v)\frac{\sigma_r}{E} + (1-v^2)\frac{\sigma}{E}$$

by (60), where σ is a given function of the plastic strain ε_θ^p. Subtracting ε_θ^e from the total strain (73), we get

$$\varepsilon_\theta^p = (1-v^2)\left\{\frac{Yc^2}{Er^2} - \frac{\sigma}{E} - \frac{\beta(c^2/r^2-1)}{\ln(b/a)}\right\} = -\varepsilon_r^p \tag{74}$$

If the stress–plastic-strain curve of the material is assumed to have a constant slope H, then the stresses in the plastic region can be determined explicitly, following the previous analysis. The internal pressure for the work-hardening tube can be shown to exceed that for the nonhardening tube by the amount

$$\Delta p = \frac{Y}{2}\left(\frac{c^2}{a^2} - \ln\frac{c^2}{a^2} - 1\right)\left(1 - \frac{\beta E/Y}{\ln(b/a)}\right)\Big/\left(1 + \frac{E/H}{1-v^2}\right) \tag{75}$$

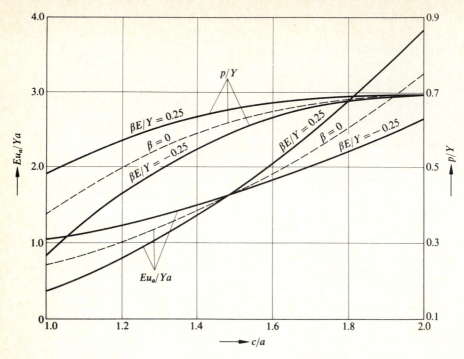

Figure 5.10 Variation of internal pressure and bore strain during the elastic/plastic expansion of a closed-ended tube ($v = 0.3$).

irrespective of the end condition. Evidently, the effect of the temperature gradient counteracts that of the work-hardening, as far as the internal pressure is concerned, for $\beta > 0$. Residual stresses on unloading from the elastic/plastic state can be calculated as in the previous section, assuming elastic changes in stress during the unloading.†

(iii) *Validity of the solution* The assumption that the axial stress is the intermediate principal stress during the elastic/plastic expansion will now be examined for a nonhardening tube with closed ends. In view of the stress-strain relation and the yield criterion, the inequalities $\sigma_\theta > \sigma_z > \sigma_r$ in the plastic region can be written as

$$(1 - 2v)\sigma_r + (1 - v)Y - E\varepsilon_z + E\alpha T > 0$$
$$-(1 - 2v)\sigma_r + vY + E\varepsilon_z - E\alpha T > 0 \tag{76}$$

where σ_r is given by (69), σ_z by (71), and T by (61). Considering the first inequality of (76), we find that the left-hand side of this inequality has the least value at $r = c$ or

---

† This has been discussed by D. R. Bland, op. cit. Elastic/plastic thermal stresses in a solid cylinder due to unsteady heat flow have been investigated by T. B. Kammash, S. A. Murch, and P. M. Naghdi, *J. Mech. Phys. Solids*, **8**: 1 (1960). See also J. H. Weiner and J. V. Huddleston, *J. Appl. Mech.*, **26**: 31 (1959).

$r = a$ according as $(\beta E/Y)\ln(b/a)$ is greater or less than $(1 - 2v)/2(1 - v)$. When $\beta E/Y \ln(b/a) > (1 - 2v)/2(1 - v)$, the inequality is found to be satisfied for all values of $c \leqslant b$. When $\beta E/Y \ln(b/a) < (1 - 2v)/2(1 - v)$, the inequality reduces to

$$-(1 - 2v)p + (1 - v)Y - E\varepsilon_z + E\alpha T_a \geqslant 0$$

If this inequality is satisfied for $c = b$, it will be satisfied for all $c \leqslant b$. The substitution for p and ε_z then furnishes

$$\frac{R^2 \ln R^2}{R^2 - 1} \leqslant \left(\ln R - \frac{\beta E}{Y}\right) \bigg/ \left\{\frac{1}{2}\left(\frac{1 - 2v}{1 - v}\right)\ln R - \frac{\beta E}{Y}\right\}$$

$$\text{for} \quad \frac{\beta E}{Y} < \frac{1}{2}\left(\frac{1 - 2v}{1 - v}\right)\ln R \quad (77)$$

where $R = b/a$. The left-hand side of the second inequality of (76) increases or decreases toward $r = c$ according as $\beta E/Y \ln R$ is greater or less than $(1 - 2v)/2(1 - v)$. Considering $\beta E/Y \ln R > (1 - 2v)/2(1 - v)$, the inequality can be written as

$$(1 - 2v)p + vY + E\varepsilon_z - E\alpha T_a \geqslant 0$$

which is satisfied for all $c \leqslant b$ provided it is satisfied for $c = a$. This gives the condition

$$\frac{R^2 \ln R^2}{R^2 - 1} \leqslant 1 + \frac{Y}{\beta E}\ln R \quad \text{for} \quad \frac{1}{2}\left(\frac{1 - 2v}{1 - v}\right)\ln R < \frac{\beta E}{Y} < \ln R \quad (78)$$

When $\beta E/Y \ln R < (1 - 2v)/2(1 - v)$, it can be shown that the second inequality of (76) is satisfied for all values of c whenever (77) is satisfied. Numerical values of the maximum permissible wall ratio, calculated from (77) and (78) for $v = 0.3$ and various values of $\beta E/Y$, are given in the following table:

$\beta E/Y$	0.798	0.707	0.654	0.448	0	-0.111	-0.220	-0.343
b/a	2.22	4.0	6.0	6.0	5.43	4.0	3.0	2.22

It remains to verify that in the elastic part of the tube ($c \leqslant r \leqslant b$), each of the stress differences $\sigma_\theta - \sigma_r$, $\sigma_\theta - \sigma_z$, and $\sigma_z - \sigma_r$ is numerically less than Y. Since $\sigma_\theta - \sigma_r$ decreases as r increases, in view of (68), becoming negative for large negative values of $\beta E/Y$, it is only necessary to ensure that $\sigma_\theta - \sigma_r$ is greater than $-Y$ at $r = b$. As the value of this stress difference is the least for $c = a$, the required inequality is

$$R^2 < \frac{-\ln R + \beta E/Y}{\ln R + \beta E/Y}$$

which is satisfied when (77) is satisfied. The other two stress differences $\sigma_\theta - \sigma_z$ and

$\sigma_z - \sigma_r$ have maximum and minimum values in the range $c \leqslant r \leqslant b$. It is necessary for the maxima to be less than Y and the minima greater than $-Y$ throughout the expansion. It turns out that these conditions are satisfied whenever (77) and (78) are satisfied.

5.4 Thermal Stresses in a Thick Spherical Shell

(i) *Elastic deformation and plastic yielding* As a further illustration of the thermal effects on stresses and strains, consider the expansion of a thick-walled spherical shell under internal pressure and steady state temperature distribution. In the completely elastic situation, the stresses and the strains can be found by the superposition of those due separately to pressure and temperature. Since the effect of internal pressure has been previously discussed (Sec. 5.1), it is necessary here to consider the temperature gradient alone. The steady state temperature in the case of spherical symmetry is given by the well-known formula

$$T = T_b + (T_a - T_b)\frac{b/r - 1}{b/a - 1} \tag{79}$$

where a and b are the internal and external radii of the sphere, and T_a and T_b the corresponding surface temperatures. Denoting the coefficient of linear expansion by α, the elastic stress-strain equations may be written as

$$\varepsilon_r = \frac{\partial u}{\partial r} = \frac{1}{E}(\sigma_r - 2v\sigma_\theta) + \alpha T$$

$$\varepsilon_\theta = \frac{u}{r} = \frac{1}{E}[(1 - v)\sigma_\theta - v\sigma_r] + \alpha T \tag{80}$$

In view of the equilibrium equation (1), the elimination of u from (80) leads to the compatibility equation

$$\frac{\partial}{\partial r}\left(\sigma_r + 2\sigma_\theta + \frac{2E\alpha T}{1 - v}\right) = 0$$

which gives, by (79),

$$\sigma_r + 2\sigma_\theta = 3A - 2\beta E\frac{b/r}{b/a - 1} \tag{81}$$

where A is independent of r, and β is a dimensionless quantity equal to $\alpha(T_a - T_b)/(1 - v)$. It may be noted that $\sigma_r + 2\sigma_\theta$ varies linearly with T. The solution of Eqs. (1) and (81) may be expressed as

$$\sigma_r = A + B\frac{b^3}{r^3} - \beta E\frac{b/r}{b/a - 1}$$

$$\sigma_\theta = A - B\frac{b^3}{2r^3} - \beta E\frac{b/2r}{b/a - 1} \tag{82}$$

Since the surfaces of the sphere are traction-free, the boundary conditions are $\sigma_r = 0$ at $r = a$ and $r = b$. Hence

$$A = \beta E \frac{b}{a}\left(\frac{b/a + 1}{b^3/a^3 - 1}\right) \qquad B = \frac{\beta E}{b^3/a^3 - 1}$$

On inserting these values in (82), the stresses in elastic sphere are obtained as

$$\sigma_r = -\beta E\left(\frac{b/r - 1}{b/a - 1} - \frac{b^3/r^3 - 1}{b^3/a^3 - 1}\right)$$

$$\sigma_\theta = -\beta E\left(\frac{b/2r - 1}{b/a - 1} + \frac{b^3/2r^3 + 1}{b^3/a^3 - 1}\right)$$

(83)

For $\beta > 0$, the inner part of the sphere tends to expand more than the outer part, giving rise to a compressive radial stress. The hoop stress is, however, compressive in the inner part and tensile in the outer part. The substitution for σ_r, σ_θ, and T into the second equation of (80) gives the radial displacement in the form

$$\frac{u}{r} = (1 - 2\nu)\frac{A}{E} - (1 + \nu)\frac{Bb^3}{2Er^3} + \frac{\beta[(1 + \nu)b/2r - (1 - \nu)]}{b/a - 1} + \alpha T_b$$

(84)

where A and B are expressed above as functions of β and b/a.

If the temperature difference is sufficiently high, yielding will occur at the radius where the stresses satisfy the yield criterion

$$|\sigma_r - \sigma_\theta| = Y$$

according to both Tresca and von Mises. It follows from (83) that

$$\sigma_r - \sigma_\theta = \beta E\left(\frac{3b^3/2r^3}{b^3/a^3 - 1} - \frac{b/2r}{b/a - 1}\right)$$

which is easily shown to have its greatest numerical value at $r = a$. Hence yielding begins at the inner surface due to hoop compression when β attains the value

$$\beta_e = \frac{Y}{E}\left(\frac{1 + a/b + a^2/b^2}{1 + a/2b}\right)$$

(85)

which indicates that the temperature difference required to cause yielding decreases as the wall ratio increases.

(ii) Yielding under combined loading If an internal pressure p is applied along with the thermal gradient, the stresses in the elastic sphere are expressed by the sum of those given by (3) and (83). Under the combined loading, therefore, we have

$$\sigma_\theta - \sigma_r = \beta E \frac{b/2r}{b/a - 1} - (\beta E - p)\frac{3b^3/2r^3}{b^3/a^3 - 1}$$

(86)

Depending on the values of $p/\beta E$ and b/a, the initial yielding may occur at the inner or the outer surface or within the wall of the sphere. Only positive values of β will be considered in what follows. The partial derivative of (86) with respect to r shows that $\sigma_\theta - \sigma_r$ steadily decreases from $r = a$ to $r = b$ so long as $p/\beta E > m_1$, where

$$m_1 = \frac{1}{9}\left(8 - \frac{a}{b} - \frac{a^2}{b^2}\right)$$

Then $\sigma_\theta - \sigma_r$ is everywhere positive, the greatest value occurring at $r = a$. Hence yielding begins at the inner surface of the sphere, and the relationship between p and β at the initial yielding is obtained from (86) as

$$3p - \beta E\left(1 - \frac{a}{b}\right)\left(2 + \frac{a}{b}\right) = 2Y\left(1 - \frac{a^3}{b^3}\right) \qquad \frac{p}{\beta E} \geqslant m_1 \qquad (87)$$

The restriction in (87) is evidently equivalent to $\beta E/Y \leqslant 3 \ (1 - a/b)$. For $p/\beta E \leqslant m_1$, the value of $\sigma_\theta - \sigma_r$ is a maximum at the radius

$$r^* = 3a\sqrt{\frac{1 - p/\beta E}{1 + a/b + a^2/b^2}} \qquad (88)$$

As the ratio $p/\beta E$ is decreased from m_1, the position of the maximum $\sigma_\theta - \sigma_r$ moves outward from $r = a$ and eventually reaches $r = b$ when $p/\beta E = m_2$, where

$$m_2 = \frac{1}{9}\left(8 - \frac{b}{a} - \frac{b^2}{a^2}\right)$$

For $m_2 \leqslant p/\beta E \leqslant m_1$, yielding will occur at radius (88) when the maximum value of $\sigma_\theta - \sigma_r$ is equal to Y. From (86) and (88), the condition for initial yielding is

$$\frac{\beta E}{Y}\sqrt{1 + \frac{b}{a} + \frac{b^2}{a^2}} = 9\left(\frac{b}{a} - 1\right)\sqrt{1 - \frac{p}{\beta E}} \qquad m_2 \leqslant \frac{p}{\beta E} \leqslant m_1 \qquad (89)$$

The above inequalities are found to be equivalent to $3(b/a - 1) \geqslant \beta E/Y \geqslant 3(1 - a/b)$. It is also necessary, for the validity of (89), that $\sigma_r - \sigma_\theta$ does not exceed Y at $r = a$. In view of (86) and (89), this amounts to the further restriction $\beta E/Y \leqslant 6(1 - a/b)$, or $p/\beta E \geqslant n_1$, where

$$n_1 = \frac{1}{9}\left(5 - 4\frac{a}{b} - 4\frac{a^2}{b^2}\right)$$

Evidently $n_1 \gtrless m_2$ for $b/a \gtrless 2$. So the inequalities in (89) must be replaced by $n_1 \leqslant p/\beta E \leqslant m_1$ when $b/a \geqslant 2$. It follows from (88) with $p = n_1 \beta E$ that yielding occurs simultaneously at $r = a$ and $r = 2a$ when

$$\frac{\beta E}{Y} = 6Y\left(1 - \frac{a}{b}\right) \qquad \frac{p}{Y} = \frac{2}{3}\left(1 - \frac{a}{b}\right)\left(5 - 4\frac{a}{b} - 4\frac{a^2}{b^2}\right) \qquad \frac{b}{a} \geqslant 2 \qquad (90a)$$

For $p/\beta E \leqslant m_2$, $\sigma_\theta - \sigma_r$ steadily increases from $r = a$ to $r = b$, the greatest numerical value occurring at $r = b$ or $r = a$ according as $p/\beta E$ is greater or less than n_2, where

$$n_2 = \frac{2(b/a - 1)^2}{3(b^2/a^2 - b/a + 1)}$$

For $n_2 \leqslant p/\beta E \leqslant m_2$, the initial yielding therefore corresponds to $\sigma_\theta - \sigma_r = Y$ at $r = b$, which gives

$$3p + \beta E \left(\frac{b}{a} - 1 \right) \left(\frac{b}{a} + 2 \right) = 2Y \left(\frac{b^3}{a^3} - 1 \right) \qquad n_2 \leqslant \frac{p}{\beta E} \leqslant m_2 \qquad (91)$$

It follows that the inequalities in (91) are equivalent to $2(b/a + a/b - 1) \geqslant \beta E/Y \geqslant 3(b/a - 1)$. Yielding occurs simultaneously at $r = a$ and $r = b$ when

$$\frac{\beta E}{Y} = 2 \left(\frac{b}{a} + \frac{a}{b} - 1 \right) \qquad \frac{p}{Y} = \frac{4a}{3b} \left(\frac{b}{a} - 1 \right)^2 \qquad \frac{b}{a} \leqslant 2 \qquad (90b)$$

For $p/\beta E$ outside the range so far considered, yielding will occur at the inner surface due to $\sigma_r - \sigma_\theta$ reaching the value Y there, the condition for the yielding being

$$\beta E \left(1 - \frac{a}{b} \right) \left(2 + \frac{a}{b} \right) - 3p = 2Y \left(1 - \frac{a^3}{b^3} \right) \qquad \frac{p}{\beta E} \leqslant n_1, n_2 \qquad (92)$$

which reduces to (85) when $p = 0$. It follows from above that the restriction $p/\beta E \leqslant n_2$ is appropriate when $b/a \leqslant 2$ and $p/\beta E \leqslant n_1$ when $b/a \geqslant 2$. Figure 5.11 shows the dependence of the location of yielding on the ratios $p/\beta E$ and b/a. The lowest curves represent simultaneous yielding at two different radii, one of which is

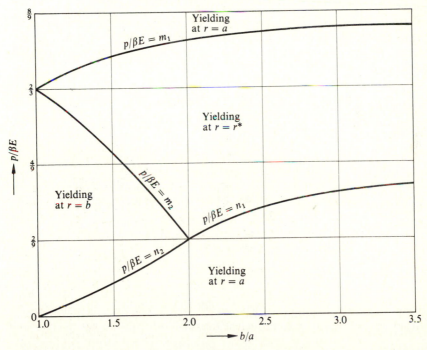

Figure 5.11 Location of the surface of initial yielding in a spherical shell under internal pressure and thermal loading.

$r = a$. It may be noted that yielding can begin at the outer surface only under the restriction $b/a \leqslant 2$.

It follows from the preceding (p, β) relations that the internal pressure required to cause yielding increases with increasing β whenever plasticity begins at the inner surface, and decreases with increasing β whenever yielding begins at the outer surface. The pressure has a maximum over the range of values of β for which initial yielding occurs within the wall. From (89), $p/\beta E$ is equal to $2/3$ at the maximum, and the corresponding value of the pressure is given by

$$\frac{p}{Y} = \frac{2\sqrt{3}\,(b/a - 1)}{\sqrt{1 + b/a + b^2/a^2}}$$

which is the greatest permissible pressure at the initial yielding. The variation of p/Y with $\beta E/Y$ at the onset of yielding is shown in Fig. 5.12 for various wall ratios.[†] The straight lines of positive slope terminating on the broken curves AB and DE correspond to yielding at the inner radius in hoop tension and hoop compression respectively. The straight lines of negative slope below the broken curve AC correspond to yielding at the outer radius. Each point within the domain ACD represents yielding at both $r = a$ and $r = b$ for two different values of b/a. The solid curves above AC correspond to yielding within the shell wall. The largest permissible values of β at the initial yielding for given wall ratios evidently correspond to the broken curve DCE.

(iii) Elastic/plastic analysis The following discussion is based on the assumption that the sphere is subjected to thermal gradient alone. For β exceeding β_e, given by (85), the sphere will become plastic within a radius c, the yield criterion being

$$\sigma_r - \sigma_\theta = Y \qquad a \leqslant r \leqslant c$$

provided the material is nonhardening. The stresses in the plastic region are immediately obtained from the equilibrium equation and the yield criterion as

$$\sigma_r = -2Y \ln \frac{r}{a} \qquad \sigma_\theta = -Y\left(1 + 2 \ln \frac{r}{a}\right) \qquad a \leqslant r \leqslant c \qquad (93)$$

satisfying the boundary condition $\sigma_r = 0$ at $r = a$. Within the elastic region, the stresses can be expressed in the form (82) where A and B are functions of c. Since $A + B = \beta E/(b/a - 1)$ in view of the boundary condition $\sigma_r = 0$ at $r = b$, we write

$$\sigma_r = B\left(\frac{b^3}{r^3} - 1\right) - \beta E\left(\frac{b/r - 1}{b/a - 1}\right)$$

$$\sigma_\theta = -B\left(\frac{b^3}{2r^3} + 1\right) - \beta E\left(\frac{b/2r - 1}{b/a - 1}\right) \qquad c \leqslant r \leqslant b \qquad (94)$$

The conditions of continuity of the stresses across $r = c$ furnish a pair of equations

---

† M. G. Derrington and W. Johnson, *Appl. Sci. Res.*, A, **7**: 408 (1958).

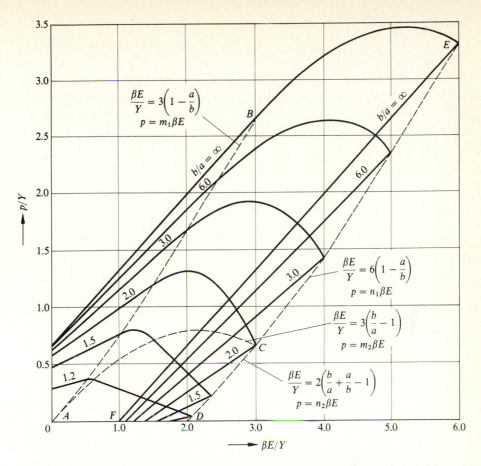

Figure 5.12 Relationship between p/Y and $\beta E/Y$ at the onset of yielding of a spherical shell for various wall ratios.

which may be solved for B and β to obtain

$$
\frac{B}{Y} = 2\frac{c^3}{b^3}\left\{\frac{1 - c/b + \ln(c/a)}{(2 + c/b)(1 - c/b)^2}\right\}
$$

$$
\frac{\beta E}{Y} = 2\left(\frac{c}{a} - \frac{c}{b}\right)\left\{\frac{1 - c^3/b^3 + \ln(c^2/a^3)}{(2 + c/b)(1 - c/b)^2}\right\}
$$

(95)

Thus all parameters affecting the stress distribution in the elastic region are expressed in terms of the single parameter c. To find the displacement u at any radius r, we consider the compressibility equation, which for small strains becomes

$$
\frac{\partial u}{\partial r} + \frac{2u}{r} = \frac{1}{E}(1 - 2v)(\sigma_r + 2\sigma_\theta) + 3\alpha T
$$

Substituting for σ_θ from (1) and T from (80), this equation can be readily integrated to obtain

$$\frac{u}{r} = (1 - 2v)\frac{\sigma_r}{E} + (1 - v)\left\{\beta\left(\frac{3b/2r - 1}{b/a - 1}\right) - C\frac{b^3}{r^3}\right\} + \alpha T_b \tag{96}$$

which is valid in both the elastic and plastic regions. Equating the displacements given by (84) and (96) at $r = b$, we find $C = 3B/2E$. Since B and β are known from (95), the displacement in the elastic or the plastic region is obtained by inserting the appropriate expression for σ_r in (96). The displacement of the internal surface is given by

$$\frac{u_a}{a} = (1 - v)\left\{\beta\left(\frac{3b/2a - 1}{b/a - 1}\right) - \frac{3B}{2E}\frac{b^3}{c^3}\right\} + \alpha T_b$$

When the plastic boundary has advanced to a radius c_1, a second plastic zone is initiated either at the outer surface or somewhere in the interior of the elastic zone.[†] The situation is that of initial yielding of a spherical shell of wall ratio b/c_1 under internal pressure $2Y \ln (c_1/a)$ and a temperature difference equal to that existing between $r = c_1$ and $r = b$. The results of the earlier analysis are directly applicable here. Since the radius $r = c_1$ is stressed to the yield point, it follows that the second plastic zone will begin at $r = b$ for $b/c_1 \leqslant 2$ and at $r = 2c_1$ for $b/c_1 \geqslant 2$. Equating to $2Y \ln (c_1/a)$ the pressure required to cause the yielding, obtained by the necessary modification of (90), we get

$$\ln\frac{c_1}{a} = \begin{cases} \dfrac{2b}{3c_1}\left(1 - \dfrac{c_1}{b}\right)^2 & \dfrac{b}{c_1} \leqslant 2 \\[3mm] \dfrac{1}{3}\left(1 - \dfrac{c_1}{b}\right)\left(5 - 4\dfrac{c_1}{b} - 4\dfrac{c_1^2}{b^2}\right) & \dfrac{b}{c_1} \geqslant 2 \end{cases} \tag{97}$$

from which b/c_1 can be determined for any given wall ratio. The value β_1 of β, corresponding to $c = c_1$, may also be found from the results in (90), where β must be replaced by $\beta_1(b/c_1 - 1)/(b/a - 1)$. Thus

$$\frac{E}{Y}\beta_1 = \begin{cases} 2\left(\dfrac{b}{a} - 1\right)\left(1 + \dfrac{c_1^2/b^2}{1 - c_1/b}\right) & \dfrac{b}{c} \leqslant 2 \\[3mm] 6\left(\dfrac{b}{a} - 1\right)\dfrac{c_1}{b} & \dfrac{b}{c_1} \geqslant 2 \end{cases} \tag{98}$$

Setting $b/c_1 = 2$ in (97) and (98), we find that the greatest value of b/a for the new yielding to start at the outer radius is $2e^{1/3} \simeq 2.79$, and the corresponding value of $\beta_1 E/Y$ is 5.57. We shall restrict ourselves to $b/a \leqslant 2.79$ in the subsequent analysis of the problem.[‡]

[†] G. R. Cowper, *J. Appl. Mech.*, **47**: 496 (1960). An analysis for the partly plastic shell simultaneously subjected to temperature and pressure loadings has been given by F. Drabble and W. Johnson, *Conf. Therm. Loading Creep*, paper No. 19, Inst. Mech. Eng. (1964).

[‡] Residual stresses on unloading from the second elastic/plastic stage have been given by W. Johnson and P. B. Mellor, *Int. J. Mech. Sci.*, **4**: 147 (1962). Transient heat flow in a solid sphere has been treated by C. Hwang, *J. Appl. Mech.*, **47**: 629 (1960).

As β is increased beyond β_1, the second plastic zone spreads inward from the outer surface, while the inner plastic zone continues to spread outward. Let ρ denote the radius to the boundary of the outer plastic region, where the stresses satisfy the yield criterion

$$\sigma_\theta - \sigma_r = Y \qquad \rho \leqslant r \leqslant b$$

The equilibrium equation, the yield criterion, and the boundary condition $\sigma_r = 0$ at $r = b$ furnish the stress distribution in the outer plastic region as

$$\sigma_r = -2Y \ln \frac{b}{r} \qquad \sigma_\theta = Y\left(1 - 2 \ln \frac{b}{r}\right) \qquad \rho \leqslant r \leqslant b \qquad (99)$$

The stresses in the inner plastic region ($a \leqslant r \leqslant c$) are still given by (93). In the elastic region, where (82) still applies, the continuity of σ_r across $r = \rho$ furnishes the stress distribution

$$\sigma_r = B\left(\frac{b^3}{r^2} - \frac{b^3}{\rho^3}\right) - \beta E\left(\frac{b/r - b/\rho}{b/a - 1}\right) - 2Y \ln \frac{b}{\rho}$$

$$c \leqslant r \leqslant \rho \qquad (100)$$

$$\sigma_\theta = -B\left(\frac{b^3}{2r^3} + \frac{b^3}{\rho^3}\right) - \beta E\left(\frac{b/2r - b/\rho}{b/a - 1}\right) - 2Y \ln \frac{b}{\rho}$$

where B depends on ρ, which is a function of c. Since the boundaries of the elastic zones are at the point of yielding, $\sigma_\theta - \sigma_r = Y$ at $r = \rho$, and $\sigma_\theta - \sigma_r = -Y$ at $r = c$. Hence

$$\frac{B}{Y} = \frac{2c^3/3b^3}{(c/\rho)(1 - c/\rho)}$$

$$\frac{\beta E}{Y} = 2\left(\frac{\rho}{a} - \frac{\rho}{b}\right)\left(1 + \frac{c^2/\rho^2}{1 - c/\rho}\right)$$

$$(101)$$

The condition of continuity of σ_r across $r = c$ then leads to the relationship between c and ρ as

$$\ln\left(\frac{c\rho}{ab}\right) = \frac{2c}{3\rho}\left(\frac{\rho}{c} - 1\right)^2 \qquad (102)$$

The radial displacement during the second elastic/plastic phase is given by (96) throughout the shell, where $C = -3B/2E$ as before, but B is now expressed by (100). The growth of the plastic zones with increasing temperature difference, and the corresponding variation of the internal and external displacements, are represented graphically in Fig. 5.13. As might be expected, the sphere becomes completely plastic only for an infinite temperature, when the elastic zone degenerates into a surface of radius $\sqrt{ab}$, across which the hoop stress becomes

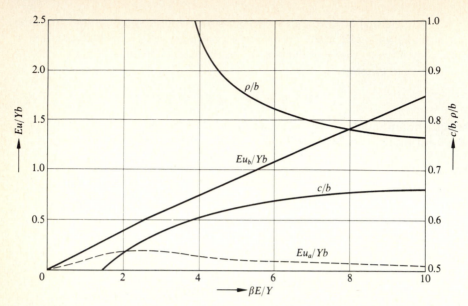

Figure 5.13 Surface displacements and positions of elastic/plastic boundary as functions of temperature in a thick spherical shell of wall ratio 2 ($v = 0.3$, $T_b = 0$).

discontinuous in the limit.[†] The theory is only valid in the temperature range for which βE and Y may be regarded as approximately constant.

5.5 Pure Bending of a Curved Bar

(i) Elastic bending A wide curved bar of uniform cross section is bent in the plane of its curvature by terminal couples M per unit width under conditions of plane strain. Since the cylindrical surfaces of the bar are free from external tractions, the stress distribution can be expected to be the same for all plane sections normal to the curved surfaces of the bar. Referring to cylindrical coordinates (r, θ, z), the bar geometry is defined by the cylindrical surfaces $r = a$ and $r = b$ and the radial planes $\theta = \pm\alpha$ (Fig. 5.14). It is also assumed that σ_r, σ_θ, and σ_z are the only nonzero stresses, which depend on r alone. Since the stress distribution is then axially symmetrical, the equation of equilibrium is

$$\frac{\partial \sigma_r}{\partial r} = \frac{\sigma_\theta - \sigma_r}{r} \tag{23}$$

So long as the bar is elastic, the stresses σ_r and σ_θ must also satisfy the compatibility equation[‡]

[†] Elastic/plastic thermal stress analysis of an annular ring has been given by J. C. Wilhoit, *Proc. 3d U.S. Nat. Congr. Appl. Mech.*, 693 (1958). Thermally stressed elastic/plastic free plates have been analyzed by J. Weiner, *J. Appl. Mech.*, **78**: 395 (1956), and H. Yuksel, *J. Appl. Mech.*, **80**: 603 (1958).

[‡] See, for example, S. Timoshenko and J. N. Goodier, *Theory of Elasticity*, p. 30, 3d ed., McGraw-Hill Book Company, New York (1970).

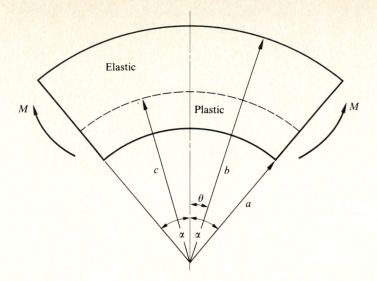

Figure 5.14 Geometry of an elastic/plastic curved bar under pure bending.

$$\left(\frac{\partial^2}{\partial r^2} + \frac{1}{r}\frac{\partial}{\partial r}\right)(\sigma_r + \sigma_\theta) = 0 \tag{103}$$

the solution of which may be written as

$$\sigma_r + \sigma_\theta = C + 2B\ln\frac{b}{r}$$

where C and B depend on the applied bending moment. The substitution in (23) then leads to the solution

$$\sigma_r = -A\left(\frac{b^2}{r^2} - 1\right) + B\ln\frac{b}{r}$$

$$\sigma_\theta = A\left(\frac{b^2}{r^2} + 1\right) - B\left(1 - \ln\frac{b}{r}\right) \tag{104}$$

satisfying the external boundary condition $\sigma_r = 0, r = b$. Here $2A$ has been written for $B + C$. The parameters A and B can be determined from the remaining boundary condition $\sigma_r = 0, r = a$, and the fact that the normal stresses σ_θ across any radial section must give rise to a couple of magnitude M per unit width. Since

$$\int_a^b (\sigma_r + \sigma_\theta)r\,dr = \int_a^b \frac{\partial}{\partial r}(r^2\sigma_r)\,dr = 0$$

in view of the equilibrium equation and the boundary conditions, we have

$$M = -\int_a^b r\sigma_\theta\,dr = \frac{1}{2}\int_a^b (\sigma_r - \sigma_\theta)r\,dr \tag{105}$$

The last expression in (105) is more convenient for calculations. The substitution from (104) then furnishes

$$A = \frac{4M}{Na^2} \ln \frac{b}{a} \qquad B = \frac{4M}{Na^2}\left(\frac{b^2}{a^2} - 1\right)$$

$$\text{(106)}$$

where
$$N = \left(\frac{b^2}{a^2} - 1\right)^2 - \frac{b^2}{a^2}\left(\ln \frac{b^2}{a^2}\right)^2$$

The radial stress is positive throughout the cross section with a maximum value at a radius $b\sqrt{2A/B}$. The hoop stress changes from positive to negative as r is increased from a to b. The neutral surface, which corresponds to $\sigma_\theta = 0$, is independent of M so long as the bar is entirely elastic.

It follows directly from the equilibrium equation and the boundary conditions that the resultant normal force across any radial section vanishes as required. Since

$$\sigma_z = v(\sigma_r + \sigma_\theta)$$

in view of the plane strain condition $\varepsilon_z = 0$, the elastic stress-strain equations in terms of the shear modulus G and Poisson's ratio v become

$$\varepsilon_r = \frac{\partial u}{\partial r} = \frac{1}{2G}[(1 - v)\sigma_r - v\sigma_\theta]$$

$$\varepsilon_\theta = \frac{u}{r} + \frac{1}{r}\frac{\partial v}{\partial \theta} = \frac{1}{2G}[(1 - v)\sigma_\theta - v\sigma_r] \qquad \text{(107)}$$

$$2\gamma_{r\theta} = \frac{\partial v}{\partial r} - \frac{v}{r} + \frac{1}{r}\frac{\partial u}{\partial \theta} = 0$$

where u and v are the radial and circumferential displacements respectively. It is assumed that a radial element at the centroid of the section $\theta = 0$ remains fixed in space during the bending. Then the conditions to be satisfied by u and v are

$$u = v = \frac{\partial v}{\partial r} = 0 \quad \text{on} \quad \theta = 0 \quad \text{and} \quad r = \frac{1}{2}(a + b) = r_0 \quad \text{(say)} \quad \text{(108)}$$

In view of (104), the solution of Eqs. (107) satisfying the end conditions (108) can be written as

$$2Gu = r\left\{A\left(1 - 2v + \frac{b^2}{r^2}\right) + B\left[(1 - v) + (1 - 2v)\ln\frac{b}{r}\right]\right\} - Kr_0 \cos\theta$$

$$\text{(109)}$$

$$2Gv = -(1 - v)Br\theta + Kr_0 \sin\theta$$

where K is a dimensionless parameter expressed by the equation

$$K = A\left(1 - 2v + \frac{b^2}{r_0^2}\right) + B\left[(1 - v) + (1 - 2v)\ln\frac{b}{r_0}\right] \qquad \text{(110)}$$

Since v is a linear function of r, each radial section remains plane during the bending, the angle of rotation of the cross section being of amount $(1 - v)B\theta/G$

in a sense that reduces the curvature of the bar. The change in the thickness $b - a$ is equal to the difference between the radial displacements at the outer and inner boundaries. From (109) and (106), the thickness change is found to be

$$\Delta(b - a) = \frac{2(1 - v)(b - a)M}{G[(b^2 - a^2) - 2ab \ln (b/a)]} \tag{111}$$

As the bending couple is increased to a critical value M_e, plastic yielding begins at the inner curved boundary. It follows from (104) that $\sigma_\theta - \sigma_r$ has its greatest value at $r = a$ where $\sigma_r = 0$ and $\sigma_z = v\sigma_\theta$. Hence according to Tresca's yield criterion, σ_θ must attain the value $2k$ at $r = a$, where k is the yield stress in shear. From (104) and (106), the initial yield couple per unit width is

$$M_e = \frac{\tfrac{1}{2}Nka^2}{1 + (b^2/a^2)[\ln (b^2/a^2) - 1]} \tag{112}$$

When $b/a = 2$, for instance, $M_e = 0.258ka^2$, the radius to the neutral surface being $1.144a$. The circumferential stress at this stage varies from $2k$ at the inner boundary to $-1.27k$ at the outer boundary. The radial stress has a maximum value of $0.276k$, occurring at a radius slightly smaller than that of the neutral surface.

(ii) The first elastic/plastic phase As the bending couple is further increased, a plastic zone is formed around the inner curved surface of the bar. In view of the symmetry of the stress distribution, the elastic/plastic boundary is a circular cylindrical surface of radius c. The stresses in the plastic region must satisfy the equilibrium equation (23) and the yield criterion

$$\sigma_\theta - \sigma_r = 2k \qquad a \leqslant r \leqslant c$$

provided σ_z continues to be the intermediate principal stress. Since $\sigma_r = 0$ at $r = 0$, we have

$$\sigma_r = 2k \ln \frac{r}{a} \qquad \sigma_\theta = 2k\left(1 + \ln \frac{r}{a}\right) \qquad a \leqslant r \leqslant c \tag{113}$$

for no work-hardening. In the elastic region $(c \leqslant r \leqslant b)$, the stresses are still expressed by (104), but the parameters A and B are now to be determined from the conditions of continuity of σ_r and σ_θ across $r = c$. Thus

$$A = \frac{2k \ln (b/a)}{1 + (b^2/c^2)[\ln (b^2/c^2) - 1]} \qquad B = 2k \frac{(b^2/c^2)[1 + \ln (c^2/a^2)] - 1}{1 + (b^2/c^2)[\ln (b^2/c^2) - 1]} \tag{114}$$

Using the expressions for $\sigma_r - \sigma_\theta$ in the elastic and plastic regions, the applied couple is obtained from (105) as†

$$M = -\frac{1}{2}k(c^2 - a^2) + \frac{1}{2}Lkc^2\left\{1 + \frac{b^2}{c^2}\left(\ln \frac{b^2}{c^2} - 1\right)\right\}^{-1} \tag{115}$$

---

† The elastic/plastic analysis for the stresses is due to B. W. Shaffer and R. N. House, *J. Appl. Mech.*, **24**: 305 (1955).

where
$$L = \left(\frac{b^2}{c^2} - 1\right)^2 - \frac{b^2}{c^2}\left\{\left(\ln\frac{b^2}{c^2}\right)\left(\ln\frac{b^2}{a^2}\right) - \left(\frac{b^2}{c^2} - 1\right)\ln\frac{c^2}{a^2}\right\}$$

The neutral surface (where σ_θ vanishes) occurs in the elastic region of the bar. The maximum radial stress also occurs in the elastic region at a radius somewhat less than that of the neutral surface.

It follows from Tresca's associated flow rule that $\varepsilon_r^p = -\varepsilon_\theta^p < 0$; so the elastic and plastic parts of ε_z individually vanish as in the case of the thick-walled tube. Hence

$$\sigma_z = v(\sigma_r + \sigma_\theta)$$

throughout the bar even when it is partly plastic. The displacements in the elastic region are expressed by (109) and (110), where A and B are now given by (114). In the plastic region, the displacements can be determined from the fact that $\varepsilon_r + \varepsilon_\theta$ is still given by Hooke's law, since the sum of the plastic parts of ε_r and ε_θ vanishes. Assuming small strains, the displacement equations become

$$\frac{\partial u}{\partial r} + \frac{u}{r} + \frac{1}{r}\frac{\partial v}{\partial \theta} = \frac{1 - 2v}{2G}(\sigma_r + \sigma_\theta)$$

$$\frac{\partial v}{\partial r} - \frac{v}{r} + \frac{1}{r}\frac{\partial u}{\partial \theta} = 0$$

$$(116)$$

where the second equation states that $\gamma_{r\theta} = 0$, since the principal axes of stress and strain rate must coincide. The solution of these equations may be taken in the form†

$$2Gu = F(r, c) - Kr_0 \cos\theta$$

$$2Gv = -(1 - v)Br\theta + Kr_0 \sin\theta$$

$$(117)$$

where K is given by (110) and (114). Since v has the same expression throughout, it is automatically continuous across $r = c$. The second equation of (116) is identically satisfied, while the first equation reduces to

$$\frac{1}{G}\frac{\partial}{\partial r}(rF) = \frac{1 - v}{G}Br + \frac{1 - 2v}{2G}\frac{\partial}{\partial r}(r^2\sigma_r)$$

in view of (23). This differential equation is readily integrated as

$$F(r, c) = r\left[(1 - v)B + (1 - 2v)\sigma_r + H\frac{b^2}{r^2}\right]$$

where H is a parameter depending on c. The above equations hold in both elastic and plastic parts of the bar. Comparison of the radial displacement given by

---

† The displacement solution has been discussed for an incompressible material by B. W. Shaffer and R. N. House, *J. Appl. Mech.*, **26**: 447 (1957), and for a compressible material by G. Eason, *Appl. Sci. Res.*, A, **9**: 53 (1960).

(117) at $r = b$ with that expressed by (109) at $r = b$ immediately shows that $H = 2(1 - v)A$. Substitution for σ_r from (113) then gives

$$F(r, c) = r\left[(1 - v)\left(B + 2A\frac{b^2}{r^2}\right) + (1 - 2v)2k \ln\frac{r}{a}\right] \qquad a \leqslant r \leqslant c \quad (118)$$

where A and B are given by (114). It may be noted that plane radial sections continue to remain plane in the plastic range. Since $2G\Delta(b - a) = F(b, c) - F(a, c)$, the change in thickness is readily obtained from (118) as

$$\Delta(b - a) = (1 - v)(b - a)\frac{k}{G}\left\{\frac{(1 - c^2/ab)\ln(b^2/a^2)}{\ln(b^2/c^2) - (1 - c^2/b^2)} - 1\right\} \quad (119)$$

When the plastic zone has spread to a sufficient extent, the outer curved boundary becomes stressed to the yield point. According to Tresca's yield criterion, σ_θ must be equal to $-2k$ at $r = b$ for yielding to start at that radius. If the radius to the elastic/plastic boundary at this stage is denoted by c_1, then (104) and (114) furnish

$$c_1^2\left(1 + \ln\frac{b}{a}\right) = b^2\left(1 - \ln\frac{ab}{c_1^2}\right) \quad (120)$$

The bending moment per unit width at the end of the first elastic/plastic phase is obtained, after some algebra, as

$$M_1 = \frac{1}{2}k\left(a^2 - b^2 + 2b^2 \ln\frac{ab}{c_1^2}\right) \quad (121)$$

Taking $b/a = 2$, we find $M_1 = 0.387ka^2$. It follows from (120) that $c_1 < b$ for all values of b/a; so yielding will always occur at $r = b$ while an outer part of the cross section is still elastic. The centroid of the section lies in the elastic region during the first elastic/plastic phase.

(iii) *The second elastic/plastic phase* For M exceeding M_1, an outer plastic zone will spread inward from the convex boundary, while the inner plastic zone continues to spread outward. The second elastic/plastic phase therefore consists of a central elastic core between a pair of plastic regions. Let the radius of the outer elastic/plastic boundary be denoted by ρ. Tresca's yield criterion requires

$$\sigma_r - \sigma_\theta = 2k \qquad \rho \leqslant r \leqslant b$$

if σ_z is the intermediate principal stress. The substitution into the equilibrium equation then furnishes

$$\sigma_r = 2k \ln\frac{b}{r} \qquad \sigma_\theta = -2k\left(1 - \ln\frac{b}{r}\right) \qquad \rho \leqslant r \leqslant b \quad (122)$$

in view of the boundary condition $\sigma_r = 0, r = b$. The stresses in the inner plastic region ($a \leqslant r \leqslant c$) are still given by (113). To obtain the stresses in the elastic region

$(c \leqslant r \leqslant \rho)$, the solution of (23) and (103) is taken in the form

$$\sigma_r = -A\left(\frac{\rho^2}{r^2} - 1\right) + B \ln \frac{\rho}{r} + 2k \ln \frac{b}{\rho}$$

$$\sigma_\theta = A\left(\frac{\rho^2}{r^2} + 1\right) - B\left(1 - \ln \frac{\rho}{r}\right) + 2k \ln \frac{b}{\rho}$$

$$c \leqslant r \leqslant \rho \qquad (123)$$

so that the radial stress is continuous across $r = \rho$. Since the radii $r = c$ and $r = \rho$ are at the point of yielding, we have

$$A = \frac{2k}{\rho^2/c^2 - 1} \qquad B = 2k\left(\frac{\rho^2/c^2 + 1}{\rho^2/c^2 - 1}\right) \qquad (124)$$

The neutral surface still occurs in the elastic region for the permissible values of b/a. The condition of continuity of σ_r across $r = c$ furnishes

$$\rho^2\left(1 - \ln \frac{ab}{c^2}\right) = c^2\left(1 + \ln \frac{\rho^2}{ab}\right) \qquad (125)$$

which is the relationship between the positions of the two elastic/plastic boundaries. Substituting for $\sigma_r - \sigma_\theta$ in (106) and integrating over the three separate regions, we have

$$M = \frac{1}{2}k\left[a^2 + b^2 - 2\rho^2\left(1 - \ln \frac{ab}{c^2}\right)\right] \qquad (126)$$

in view of (124) and (125). As the bending couple approaches the fully plastic value, the outer and inner plastic boundaries approach one another to meet at the neutral surface whose radius finally becomes $\sqrt{ab}$. The fully plastic moment per unit width is

$$M_0 = \frac{1}{2}k(b - a)^2$$

The distributions of σ_r and σ_θ during the entire elastic/plastic bending are shown in Fig. 5.15 for a series of values of M/M_0 when $b/a = 2$. The broken curve indicates the variation of the position of the neutral surface. The hoop stress is discontinuous across the neutral surface in the limiting case when the bar just becomes fully plastic.

In view of Tresca's associated flow rule, the displacement in each separate region may be expressed in the form (117), where B is given by (124) and K is yet to be determined. In the elastic region, the function F is most conveniently obtained from the stress-strain equation

$$\varepsilon_\theta = \frac{u}{r} + \frac{1}{r}\frac{\partial v}{\partial \theta} = \frac{1}{2G}[(1 - v)\sigma_\theta - v\sigma_r]$$

Substitution from (117) and (123) into the above equation immediately furnishes

$$F(r, c) = r\left\{A\left(1 - 2v + \frac{\rho^2}{r^2}\right) + B\left[(1 - v) + (1 - 2v)\ln \frac{\rho}{r}\right] + 2k(1 - 2v)\ln \frac{b}{\rho}\right\}$$

$$c \leqslant r \leqslant \rho \qquad (127)$$

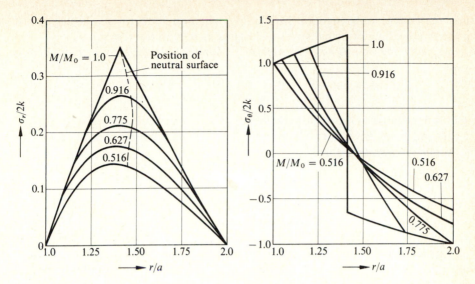

Figure 5.15 Stress distribution in a curved bar ($b/a = 2$) during elastic/plastic bending by pure couples.

In the plastic regions, the displacements must be determined from (116) as before. The integration of the resulting differential equation furnishes

$$F(r, c) = r\left[(1 - v)B + (1 - 2v)\sigma_r + H\frac{\rho^2}{r^2}\right]$$

where H has the value $2(1 - v)A$ as before, in order that this equation coincides with (127) when σ_r is expressed by (123). The substitution for σ_r from (113) and (122) then furnishes

$$F(r, c) = \begin{cases} r\left[(1 - v)\left(B + 2A\frac{\rho^2}{r^2}\right) + 2k(1 - 2v)\ln\frac{r}{a}\right] & a \leqslant r \leqslant c \\ r\left[(1 - v)\left(B + 2A\frac{\rho^2}{r^2}\right) + 2k(1 - 2v)\ln\frac{b}{r}\right] & \rho \leqslant r \leqslant b \end{cases} \quad (128)$$

where A and B are given by (124). In view of (108), the parameter r_0K is equal to $F(r_0, c)$ according to (127) when $\rho \geqslant r_0$, and according to the second equation of (128) when $\rho \leqslant r_0$. The change in thickness of the bar is readily derived from (128) and (124) as†

$$\Delta(b - a) = (1 - v)(b - a)\frac{k}{G}\left\{1 - 2\left(\frac{\rho^2/ab - 1}{\rho^2/c^2 - 1}\right)\right\} \quad (129)$$

It is easy to show that $\sigma_z < \sigma_\theta$ in the inner plastic region and $\sigma_z < \sigma_r$ in the

---

† For the plane stress bending of a curved bar according to both the Tresca and Mises criteria, see G. Eason, *Q. J. Mech. Appl. Math.*, **12**: 334 (1960).

outer plastic region throughout the bending, whatever the ratio b/a. Hence σ_z will be the intermediate principal stress so long as $\sigma_r < \sigma_z$ in the region $a \leqslant r \leqslant c$ and $\sigma_\theta < \sigma_z$ in the region $\rho \leqslant r \leqslant b$. In view of (113) and (122), these conditions become

$$(1 - 2v) \ln \frac{c}{a} - v \leqslant 0 \qquad (1 - 2v) \ln \frac{b}{\rho} - (1 - v) \leqslant 0$$

which will be satisfied throughout the bending if they are satisfied for $c = \rho = \sqrt{ab}$. The second inequality is then satisfied whenever the first one is. The validity of the preceding solution therefore requires

$$\ln \frac{b}{a} \leqslant \frac{2v}{1 - 2v}$$

For $v = 0.3$, the largest value of b/a for which the solution is valid is 4.48. For an incompressible material, the solution remains valid for all b/a ratios.

For a given b/a ratio, the elastic/plastic boundaries advance almost linearly with the bending moment as M increases to within 90 percent of the fully plastic value. The thickness change increases toward the beginning, attains a positive maximum during the first elastic/plastic phase, and then decreases rapidly during the second elastic/plastic phase. When the bar is fully plastic, the thickness change is zero. The fractional change in angle $\Delta\alpha/\alpha$, which is equal to $-(1 - v)B/G$, remains of the elastic order of magnitude so long as M is less than about $0.95M_0$,

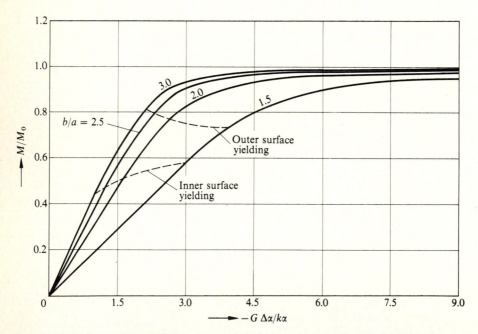

Figure 5.16 Variation of the applied moment with the fractional decrease in angle for different b/a ratios ($v = 0.3$).

but increases indefinitely as the limiting moment is approached.† Figure 5.16 shows the variation of the applied couple with the fractional change in angle for several b/a ratios.

5.6 Rotating Discs and Cylinders

(i) *Elastic discs* Consider a circular disc of uniform thickness rotating with an angular velocity of gradually increasing magnitude about an axis perpendicular to its plane and passing through the center. The thickness of the disc is assumed sufficiently small so that it is effectively in a state of plane stress ($\sigma_z = 0$). The radial equilibrium of an element of the rotating disc requires

$$r\frac{\partial \sigma_r}{\partial r} = \sigma_\theta - \sigma_r - \rho\omega^2 r^2 \tag{130}$$

where ω is the angular velocity and ρ the density of the material of the disc. Denoting the purely radial displacement by u, the relevant stress-strain equations may be written as

$$\varepsilon_r = \frac{\partial u}{\partial r} = \frac{1}{E}(\sigma_r - v\sigma_\theta)$$

$$\varepsilon_\theta = \frac{u}{r} = \frac{1}{E}(\sigma_\theta - v\sigma_r) \tag{131}$$

The elimination of u from these equations and the use of (130) lead to the compatibility equation

$$\frac{\partial}{\partial r}(\sigma_r + \sigma_\theta) = -(1 + v)\rho\omega^2 r$$

which is readily integrated to give $\sigma_r + \sigma_\theta$. The equilibrium equation then furnishes

$$\sigma_r = A + B\frac{b^2}{r^2} - \frac{3 + v}{8}\rho\omega^2 r^2$$

$$\sigma_\theta = A - B\frac{b^2}{r^2} - \frac{1 + 3v}{8}\rho\omega^2 r^2 \tag{132}$$

where b is the external radius of the disc, and the parameters A and B depend on ω only. For a solid disc, B must be zero in order that the stresses are finite at the center, while the boundary condition $\sigma_r = 0$ at $r = b$ gives $A = (3 + v)\rho\omega^2 b^2/8$. The stress distribution for the solid disc therefore becomes

$$\sigma_r = \tfrac{1}{8}\rho\omega^2(3 + v)(b^2 - r^2)$$

$$\sigma_\theta = \tfrac{1}{8}\rho\omega^2[(3 + v)b^2 - (1 + 3v)r^2] \tag{133}$$

---

† Residual stresses upon unloading the elastic/plastic moment have been discussed by D. W. Shaffer and E. E. Ungar, Office of Ordnance Research (1957).

The displacement is then obtained from (131) as

$$u = \frac{1-v}{8E} \rho\omega^2 r[(3+v)b^2 - (1+v)r^2] \tag{134}$$

Both the stresses are tensile and $\sigma_\theta \geqslant \sigma_r$, the equality holding only at $r = 0$ where the stresses have the greatest magnitude. Yielding will therefore start at the center of the disc when $\sigma_r = \sigma_\theta = Y$ at this point. If $\omega = \omega_e$ at the onset of yielding, then from (133),

$$\rho\omega_e^2 b^2 = \frac{8Y}{3+v}$$

For an annular disc with internal radius a and external radius b, the boundary conditions are $\sigma_r = 0$ at $r = a$ and $r = b$. These conditions give

$$A = \frac{3+v}{8} \rho\omega^2(a^2 + b^2) \qquad B = -\frac{3+v}{8} \rho\omega^2 a^2$$

and the elastic stress distribution becomes

$$\sigma_r = \frac{3+v}{8} \rho\omega^2\left(a^2 + b^2 - \frac{a^2 b^2}{r^2} - r^2\right)$$

$$\sigma_\theta = \frac{3+v}{8} \rho\omega^2\left(a^2 + b^2 + \frac{a^2 b^2}{r^2} - \frac{1+3v}{3+v} r^2\right) \tag{135}$$

From (131) and (135), the displacement is obtained as

$$u = \frac{(3+v)(1-v)}{8E} \rho\omega^2 r\left\{a^2 + b^2 - \left(\frac{1+v}{3+v}\right)r^2 - \left(\frac{1+v}{1-v}\right)\frac{a^2 b^2}{r^2}\right\} \tag{136}$$

The radial stress is a maximum at $r = \sqrt{ab}$, and the maximum value of σ_r is $(3+v)\rho\omega^2(b^2 - a^2)/8$. The hoop stress σ_θ has the greatest value

$$\tfrac{1}{8}\rho\omega^2[(3+v)b^2 + (1-v)a^2]$$

occurring at $r = a$. According to Tresca's yield criterion, yielding will begin at the inner radius when $\sigma_\theta = Y$. Hence the angular velocity ω_e at the initial yielding is given by

$$\rho\omega_e^2 b^2 = \frac{4Y}{(3+v) + (1-v)(a^2/b^2)} \tag{137}$$

As a tends to zero, the value of $\rho\omega_e^2 b^2$ for an annular disc approaches a limit which is half of that for a solid disc, since there is a stress concentration factor of 2 in the elastic range of deformation.†

---

† For a three-dimensional analysis of the rotating disc problem, see S. Timoshenko, and J. N. Goodier, *Theory of Elasticity*, p. 388, 3d ed., McGraw-Hill Book Company, New York (1970).

(ii) Elastic/plastic discs If the speed of rotation is further increased, the disc will consist of an inner plastic zone surrounded by an outer elastic zone. Within the plastic region, which is assumed to extend to a radius c, the stresses are required to satisfy the equilibrium equation (130) and the Tresca criterion $\sigma_\theta = Y$. For a solid disc σ_r must be finite at $r = 0$, while for an annular disc σ_r must vanish at $r = a$. The integration of (130) for a nonhardening material therefore gives

$$
\sigma_r =
\begin{cases}
Y - \dfrac{1}{3}\rho\omega^2 r^2 & \text{solid disc } (0 \leqslant r \leqslant c) \\[2ex]
Y\left(1 - \dfrac{a}{r}\right) - \dfrac{1}{3}\rho\omega^2 r^2\left(1 - \dfrac{a^3}{r^3}\right) & \text{annular disc } (a \leqslant r \leqslant c)
\end{cases}
\tag{138}
$$

The stresses in the elastic region ($c \leqslant r \leqslant b$) are of the form (132) for both solid and hollow discs. In view of the boundary condition $\sigma_r = 0$ at $r = b$, these expressions may be rewritten in the more convenient form

$$
\sigma_r = B\left(\frac{b^2}{r^2} - 1\right) + \frac{3 + v}{8}\rho\omega^2 b^2\left(1 - \frac{r^2}{b^2}\right)
$$

$$
\sigma_\theta = -B\left(\frac{b^2}{r^2} + 1\right) + \frac{1 + 3v}{8}\rho\omega^2 b^2\left(\frac{3 + v}{1 + 3v} - \frac{r^2}{b^2}\right)
\qquad c \leqslant r \leqslant b \quad (139)
$$

For a given position of the elastic/plastic boundary, B and ω can be determined from the conditions of continuity of the stresses across $r = c$. For a solid disc,

$$
B = -\frac{1}{8}(1 + 3v)\frac{\rho\omega^2 c^4}{b^2}
$$

and the elastic/plastic angular velocity is given by

$$
\frac{\rho\omega^2 b^2}{3Y} = \left\{1 + \frac{1 + 3v}{8}\left(1 - \frac{c^2}{b^2}\right)^2\right\}^{-1}
\tag{140}
$$

It is equally convenient to find c from (140) for a given ω. The complete expressions for the stresses in the elastic region of a solid disc become†

$$
\sigma_r = \frac{\rho\omega^2 b^2}{8}\left[(3 + v) - \frac{1}{3}(1 + 3v)\frac{c^4}{r^2 b^2}\right]\left(1 - \frac{r^2}{b^2}\right)
$$

$$
\sigma_\theta = \frac{\rho\omega^2 b^2}{8}\left[(3 + v) + (1 + 3v)\left\{\frac{c^4}{3b^4}\left(\frac{b^2}{r^2} + 1\right) - \frac{r^2}{b^2}\right\}\right]
\qquad c \leqslant r \leqslant b \quad (141)
$$

These stresses are the same as those at the initial yielding of a hollow disc of radius ratio b/c rotating with angular velocity ω and having a uniform radial tension $Y - \rho\omega^2 c^2/3$ at the inner raidus. Figure 5.17 shows the variation of $\rho\omega^2 b^2/Y$ in a solid elastic/plastic disc with the movement of the plastic boundary.

† See, for example, W. W. Sokolovsky, *Prinkl. Mat. Mekh.*, **12**: 87 (1948).

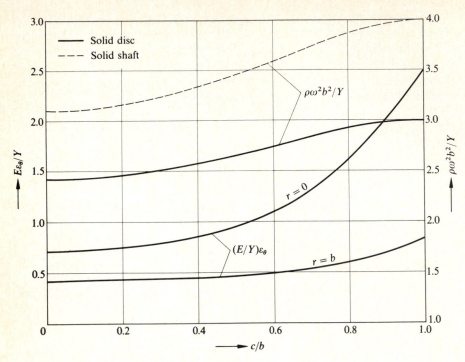

Figure 5.17 Variation of angular velocity and hoop strain with radius to the plastic boundary in rotating discs.

In the case of an annular disc, calculations similar to above result in

$$B = -\left\{\frac{ac}{2b^2}\left(Y - \frac{1}{3}\rho\omega^2 a^2\right) + \frac{1 + 3v}{24}\rho\omega^2\frac{c^4}{b^2}\right\} \tag{142}$$

$$\frac{\rho\omega^2 b^2}{3Y} = \left\{2 - \frac{a}{c}\left(1 + \frac{c^2}{b^2}\right)\right\}\Big/\left\{2 + \frac{1 + 3v}{4}\left(1 - \frac{c^2}{b^2}\right)^2 - \frac{a^3}{b^2 c}\left(1 + \frac{c^2}{b^2}\right)\right\} \tag{143}$$

Once the angular velocity has been found from (143) for an assumed position of the plastic boundary, the stresses in the elastic/plastic disc can be calculated from (138), (139), and (142). As the disc is rendered more and more plastic, ω rapidly approaches the fully plastic value ω_0, where

$$\rho\omega_0^2 b^2 = \frac{3Y}{1 + a/b + a^2/b^2} \tag{144}$$

which is independent of v. Considering $b/a = 3$, the fully plastic speed of the disc is found to be 1.324 times that at the initial yielding, assuming $v = 0.3$. The angular velocity attains a maximum when the disc is just fully plastic.

The displacements in the elastic region ($c \leqslant r \leqslant b$) are obtained by the direct substitution of the stresses into the second equation of (131). Considering the solid disc, we find

$$
\frac{Eu}{Yr} = (1 - v) + \frac{\rho\omega^2 b^2}{8Y} \left\{ (1 + 3v)\frac{c^2}{3b^2}\left[(1 - v) + (1 + v)\frac{c^2}{r^2} \right] - (1 - v^2)\frac{r^2}{b^2} \right\}
$$

$$c \leqslant r \leqslant b \quad (145)$$

The determination of the displacement in the plastic region is more involved. However, a useful approximation is achieved by adopting the total strain theory of Hencky. Assuming small strains, the plastic stress-strain equation may be written as

$$
\frac{\partial u}{\partial r} - \frac{1}{E}(\sigma_r - v\sigma_\theta) = \left(\frac{2\sigma_r - \sigma_\theta}{2\sigma_\theta - \sigma_r} \right)\left\{ \frac{u}{r} - \frac{1}{E}(\sigma_\theta - v\sigma_r) \right\}
$$

where $\sigma_\theta = Y$ in view of the yield criterion. Considering the solid disc again and substituting from (138), the above equation is reduced to

$$
\frac{\partial u}{\partial r} - \left(\frac{3Y - 2\rho\omega^2 r^2}{3Y + \rho\omega^2 r^2} \right)\frac{u}{r} = (1 - 2v)\frac{\rho\omega^2 r^2}{3E}\left(\frac{6Y - \rho\omega^2 r^2}{3Y + \rho\omega^2 r^2} \right) \quad (146)
$$

Since the displacement must be continuous across the elastic/plastic boundary, the required boundary condition is

$$
\left(\frac{u}{r} \right)_{r=c} = (1 - v)\frac{Y}{E} + \frac{v\rho\omega^2 c^2}{3E}
$$

The integration of (146) therefore leads to the displacement in the plastic region as

$$
\frac{Eu}{Yr} = \frac{1 - 2v}{5}\left(4 - \frac{\rho\omega^2 r^2}{3Y} \right) + \frac{1 + 3v}{5}\left(1 + \frac{\rho\omega^2 c^2}{3Y} \right)^{5/2}\left(I + \frac{\rho\omega^2 r^2}{3Y} \right)^{-3/2}
$$

$$0 \leqslant r \leqslant c \quad (147)$$

where ω is given by (140). The hoop strains at $r = 0$ and $r = b$ are shown in Fig. 5.17. It may be noted that the stress ratio at any given radius, once the element is overtaken by the plastic boundary, decreases slightly with the subsequent growth of the plastic zone. The displacement given by (147) cannot, therefore, be significantly different from that obtained by a numerical solution of the relevant Prandtl-Reuss equation.†

(iii) *Instability in rotating discs* When the disc is rotated in the fully plastic range, the changes in geometry become significant. For a sufficiently work-hardening material, the speed of rotation initially increases with increasing plastic strain. Since large strains are involved, it is reasonable to assume that the material is rigid/plastic. Considering an annular disc, let a_0 and b_0 denote the initial radii of

---

† The elastic/plastic stresses in a rotating ray have been analyzed by P. G. Hodge, *J. Appl. Mech.*, **24**: 311 (1955). The plastic design of rotating discs has been considered by J. Heyman, *Proc. Inst. Mech. Eng.*, **172**: 531 (1958), *Proc. 3d U.S. Nat. Congr. Appl. Mech.* (1958).

the disc, a and b being their current values. The equation of radial equilibrium is

$$\frac{\partial}{\partial r}(hr\sigma_r) = h(\sigma_\theta - \rho\omega^2 r^2) \tag{148}$$

where r is the current radius to a typical particle, and h the local thickness of the disc of uniform initial thickness h_0. If the initial radius to the particle is denoted by r_0, the radial and circumferential strains may be written as

$$\varepsilon_r = \ln\left(1 + \frac{\partial u}{\partial r_0}\right) \qquad \varepsilon_\theta = \ln\left(1 + \frac{u}{r_0}\right) \tag{149}$$

where u is the radial displacement, equal to $r - r_0$, of a typical particle.

If Tresca's yield criterion is adopted, σ_θ must be equal to the current uniaxial yield stress σ. The analysis is greatly simplified if Tresca's associated flow rule is also employed. Then ε_r vanishes and (149) gives $\partial u/\partial r_0 = 0$, which means that u is constant at any given stage. Since $hr = h_0 r_0$ in view of the incompressibility of the material, and the fact that $dr = dr_0$, equation (148) becomes

$$\frac{\partial}{\partial r_0}(r_0\sigma_r) = \frac{r_0\sigma}{u + r_0} - \rho\omega^2 r_0(u + r_0)$$

Integrating between the limits $r_0 = a_0$ and $r_0 = b_0$, and remembering that σ_r vanishes at both the limits, we have

$$\frac{1}{6}\rho\omega^2[2(b_0^3 - a_0^3) + 3u(b_0^2 - a_0^2)] = \int_{a_0}^{b_0} \frac{\sigma r_0\, dr_0}{u + r_0}$$

Adopting the work-hardening hypothesis, we write $\sigma = F(\varepsilon_\theta)$, where the function F defines the stress-strain curve in uniaxial tension. Changing the variable of the above integral to ε_θ by means of (149), and introducing dimensionless quantities $\alpha = b_0/a_0$ and $\eta = u/a_0$, the above equation may be written as

$$\frac{1}{6}\rho\omega^2 a_0^2 = \frac{\eta[f(\varepsilon_b) - f(\varepsilon_a)]}{2(\alpha^3 - 1) + 3\eta(\alpha^2 - 1)} \tag{150}$$

where

$$f(\varepsilon_\theta) = -\int_0^{\varepsilon_\theta} \frac{F(\varepsilon_\theta)}{(e^{\varepsilon_\theta} - 1)^2}\, d\varepsilon_\theta \tag{151}$$

the hoop strains at the inner and outer radii being denoted by

$$\varepsilon_a = \ln(1 + \eta) \qquad \varepsilon_b = \ln\left(1 + \frac{\eta}{\alpha}\right)$$

The fully plastic rotation of the disc will be initially stable if $d\omega/d\eta$ is positive at $\eta = 0$. Using (150) and (151), the condition for stability of the initial state is obtained as

$$\tfrac{1}{2}(\alpha^2 - 1)\rho\omega_0^2 a_0^2 < [F'(0)]\ln\alpha$$

where ω_0 is given by (144), and the prime denotes differentiation with respect to the argument. This gives

$$F'(0) > Y\left\{1 + \frac{3(\alpha^2 - 1)(\alpha - 1)}{2(\alpha^3 - 1)\ln\alpha}\right\} \tag{152}$$

If the initial rate of hardening satisfies the above inequality, the disc will become unstable at a later stage when the angular velocity attains a maximum.[†] From (150), the condition $d\omega/d\eta = 0$ gives

$$\frac{\eta[f(\varepsilon_b) - f(\varepsilon_a)]}{\alpha\phi(\varepsilon_b) - \phi(\varepsilon_a)} = 1 + \frac{3}{2}\left(\frac{\alpha^2 - 1}{\alpha^3 - 1}\right)\eta$$

where

$$\phi(\varepsilon_\theta) = F(\varepsilon_\theta)\exp(-\varepsilon_\theta) \tag{153}$$

It is convenient to use an empirical stress-strain equation expressed by $F(\varepsilon_\theta) = C\varepsilon_\theta^n$, where C and n are empirical constants. Expanding $1/(e^{\varepsilon_\theta} - 1)^2$ in powers of ε_θ, the integral in (151) is readily evaluated, giving

$$f(\varepsilon_\theta) = C\varepsilon_\theta^n\left[\frac{1}{n} + \frac{1}{(1-n)\varepsilon_\theta} - \frac{5\varepsilon_\theta}{12(1+n)}\right] \tag{154}$$

with an error of less than 1 percent over the range $0 < \varepsilon_\theta < 0.5$. When the value of η has been found from (153) and (154) by trial and error, the maximum angular velocity can be calculated either from (150), or more conveniently from the formula

$$\rho\omega^2 a_0^2 = \frac{3[\alpha\phi(\varepsilon_b) - \phi(\varepsilon_a)]}{\alpha^3 - 1} \tag{155}$$

obtained from (150) and (153). It can be shown that in the limiting case of a thin ring ($\alpha \simeq 1$), the hoop strain is $n/2$ at instability. Values of the bore strain ε_a and the speed factor $\rho\omega^2 b_0^2/C$ at the onset of plastic instability are plotted as functions of the radius ratio α in Fig. 5.18 for $n = 0.05$ and $n = 0.10$. For large values of α, the instability strain can be considerably greater than the uniaxial value n. Results based on the Hencky theory,[‡] represented by the broken curves, predict an instability strain somewhat higher than that given by the Tresca theory.[§]

(iv) *Rotating cylinders* A solid circular cylinder of radius b is rotated about its axis, the angular velocity at any instant being ω. The cylinder is assumed to be

---

[†] The present analysis is an extension of that due to H. J. Weiss and W. Prager, *J. Aero. Sci.*, **21**: 196 (1954), and a generalization of that presented by M. J. Percy and P. B. Mellor, *Int. J. Mech. Sci.*, **6**: 421 (1964). A bifurcation analysis has beeen carried out by V. Tvergaard, ibid., **20**: 109 (1978).

[‡] Solutions based on the Hencky theory have been discussed by M. Zaid, *J. Aero. Sci.*, **20**: 369 (1953), and also by M. J. Percy and P. B. Mellor, op. cit. Discs of uniform strength have been discussed by N. E. Waldren, M. J. Percy, and P. B. Mellor, *Proc. Inst. Mech. Eng.*, **180**: 111 (1966).

[§] For experimental results on the bursting speed of rotating discs, see E. N. Robinson, *Trans. ASME*, **66**: 373 (1944), and D. H. Winne and B. M. Wundt, *Trans. ASME*, **80**: 1643 (1958).

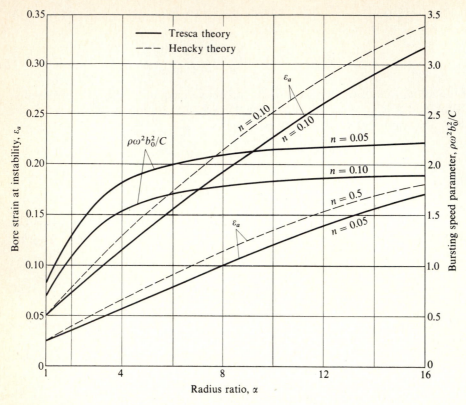

Figure 5.18 Instability strains and bursting speed parameters for hollow rotating discs of uniform thickness.

sufficiently long compared with its radius so that end effects are negligible. If the cylinder is free to contract in the axial direction, the axial strain ε_z is a negative constant at each stage. The equation of radial equilibrium is still (130), but the compatibility equation in the elastic range is obtained by replacing v with $v/(1 - v)$. It is then easy to show that

$$\sigma_r = \frac{1}{8}\left(\frac{3 - 2v}{1 - v}\right)\rho\omega^2(b^2 - r^2)$$

$$\sigma_\theta = \frac{1}{8}\rho\omega^2\left[\left(\frac{3 - 2v}{1 - v}\right)b^2 - \left(\frac{1 + 2v}{1 - v}\right)r^2\right]$$

(156)

while the cylinder is completely elastic. Since the resultant longitudinal force across the cylinder vanishes,

$$0 = \int_0^b r\sigma_z \, dr = \frac{1}{2}E\varepsilon_z b^2 + v\int_0^b r(\sigma_r + \sigma_\theta) \, dr$$

in view of the stress-strain equation. Substitution for $\sigma_r + \sigma_\theta$ from (156) results in

$$E\varepsilon_z = -\frac{1}{2} v\rho\omega^2 b^2 \qquad \sigma_z = \frac{v\rho\omega^2}{4(1-v)}(b^2 - 2r^2) \qquad (157)$$

Thus σ_z is tensile for $r < b/\sqrt{2}$ and compressive for $r > b/\sqrt{2}$. Using the stress-strain equation corresponding to ε_θ, the radial displacement is obtained from (156) and (157) as

$$u = \frac{\rho\omega^2 r}{8E(1-v)} [(3 - 5v)b^2 - (1+v)(1-2v)r^2] \qquad (158)$$

It is easy to see that $\sigma_\theta \geqslant \sigma_r > \sigma_z$ throughout the cylinder. Hence, if Tresca's yield criterion is adopted, we have to consider the stress difference

$$\sigma_\theta - \sigma_z = \frac{\rho\omega^2}{8(1-v)} [(3 - 4v)b^2 - (1-2v)r^2]$$

which has the greatest magnitude at $r = 0$. Yielding will therefore start on the axis of the cylinder for an angular velocity ω_e, where

$$\rho\omega_e^2 b^2 = 8Y\left(\frac{1-v}{3-4v}\right)$$

Since $\sigma_r = \sigma_\theta$ on the axis, this is also true for the von Mises criterion.

If the angular velocity is increased to a value somewhat larger than ω_e, there will be a plastic core of radius c, surrounded by an elastic annulus. The radial and circumferential stresses in the elastic region ($c \leqslant r \leqslant b$) may be expressed as

$$\sigma_r = B\left(\frac{b^2}{r^2} - 1\right) + \frac{1}{8}\left(\frac{3-2v}{1-v}\right)\rho\omega^2 b^2\left(1 - \frac{r^2}{b^2}\right)$$

$$c \leqslant r \leqslant b \quad (159)$$

$$\sigma_\theta = -B\left(\frac{b^2}{r^2} + 1\right) + \frac{1}{8}\rho\omega^2 b^2\left[\left(\frac{3-2v}{1-v}\right) - \left(\frac{1+2v}{1-v}\right)\frac{r^2}{b^2}\right]$$

where B is a parameter that depends on c. In the plastic region, the stresses must satisfy the yield criterion, which is taken as

$$\sigma_\theta - \sigma_z = \sigma_r - \sigma_z = Y \qquad 0 \leqslant r \leqslant c$$

The stress point is thus assumed to remain at an appropriate corner of the yield hexagon, work-hardening being neglected.† Since the stresses given by (159) must be equal to one another at $r = c$, we have

$$B = \frac{1}{8}\left(\frac{1-2v}{1-v}\right)\frac{\rho\omega^2 c^4}{b^2}$$

Within the plastic zone ($0 \leqslant r \leqslant c$), the stresses are obtained by integrating the

---

† P. G. Hodge, Jr. and M. Balban, *Int. J. Mech. Sci.*, **4**: 465 (1962).

equilibrium equation (130) with $\sigma_r = \sigma_\theta$, and using the condition of continuity across $r = c$. Thus

$$\sigma_r = \sigma_\theta = \frac{1}{8}\rho\omega^2 b^2 \left[\left(\frac{3-2v}{1-v}\right) + \left(\frac{1-2v}{1-v}\right)\frac{c^2}{b^2}\right]\left(1 - \frac{c^2}{b^2}\right) + \frac{1}{2}\rho\omega^2(c^2 - r^2)$$

$$0 \leqslant r \leqslant c \quad (160)$$

The axial stress in the plastic region is $\sigma_z = -Y + \sigma_r$ by the yield criterion, while in the elastic region, $\sigma_z = E\varepsilon_z + v(\sigma_r + \sigma_\theta)$ by Hooke's law. In view of the continuity of the stresses across $r = c$,

$$E\varepsilon_z = -Y + (1 - 2v)\sigma_c$$

where σ_c denotes the radial or hoop stress at $r = c$, and is expressed by

$$\sigma_c = \frac{\rho\omega^2 b^2}{8(1-v)}\left[(3-2v) + (1-2v)\frac{c^2}{b^2}\right]\left(1 - \frac{c^2}{b^2}\right)$$

The distribution of σ_z in the elastic and plastic regions can be conveniently expressed in terms of σ_c. Thus

$$\sigma_z = \begin{cases} -Y + \sigma_c + \dfrac{1}{2}\rho\omega^2(c^2 - r^2) & 0 \leqslant r \leqslant c \\[2mm] -Y + \sigma_c + \dfrac{v\rho\omega^2}{2(1-v)}(c^2 - r^2) & c \leqslant r \leqslant b \end{cases} \quad (161)$$

The angular velocity corresponding to a given radius to the plastic boundary can now be determined from the condition of zero resultant axial force. Using the above expressions for σ_z, and integrating, we arrive at the formula

$$\frac{\rho\omega^2 b^2}{4Y} = \left\{1 + \frac{1}{2}\left(\frac{1-2v}{1-v}\right)\left(1 - \frac{c^2}{b^2}\right)^2\right\}^{-1} \quad (162)$$

The fully plastic value of $\rho\omega^2 b^2$ is $4Y$, which may be compared with the value $3Y$ for a rotating disc without a hole. The dependence of the speed factor $\rho\omega^2 b^2/Y$ on the ratio c/b is displayed in Fig. 5.17.

The radial displacement in the elastic region can be directly obtained from (159) and (161), using Hooke's law. In particular, the external displacement is given by

$$\frac{Eu_b}{Yb} = v + \frac{\rho\omega^2 b^2}{8Y}\left(\frac{1-2v}{1-v}\right)\left[(2-3v) + (2-v)\frac{c^2}{b^2}\right]\left(1 - \frac{c^2}{b^2}\right) \quad (163)$$

In the plastic region, the displacement must satisfy the equation of elastic compressibility, and be continuous across $r = c$, while vanishing at $r = 0$. It turns out that these conditions cannot be simultaneously satisfied, and to this extent the above solution must be regarded as approximate.

The elastic compressibility soon becomes negligible, however, if the rotation is continued into the fully plastic range. Then $\varepsilon_r = \varepsilon_\theta - \frac{1}{2}\varepsilon_z$ throughout the cylinder irrespective of the rate of hardening of the material. The fully plastic stresses are easily shown to be

$$\sigma_r = \sigma_\theta = \tfrac{1}{2}\rho\omega^2(b^2 - r^2) \qquad \sigma_z = \tfrac{1}{4}\rho\omega^2(b^2 - 2r^2) \tag{164}$$

The state of stress is therefore a uniform axial compression $\sigma = \tfrac{1}{4}\rho\omega^2 b^2$, together with a varying hydrostatic tension. The rotation becomes unstable when ω attains a maximum, and this corresponds to $d\sigma/d\varepsilon = \sigma$, where ε is the axial compressive strain related to σ by the uniaxial stress-strain curve. Hence the axial strain at instability is numerically equal to the longitudinal strain at the onset of necking in simple tension.†

5.7 Infinite Plate with a Circular Hole

(i) *Plate under uniform radial tension* A large flat plate containing a small circular hole is subjected to a state of balanced biaxial stress σ. The presence of the hole will cause large additional stresses in the immediate neighborhood of the hole, but its effects will be negligible at a distance of a few diameters from its edge, theoretically vanishing at infinity. In the mathematical formulation of the problem, we therefore consider an infinitely extended plate. The nonzero stress components σ_r and σ_θ must satisfy the equilibrium equation

$$\frac{\partial \sigma_r}{\partial r} = \frac{\sigma_\theta - \sigma_r}{r} \tag{165}$$

where the origin is taken at the center of the hole. The corresponding strain components ε_r and ε_θ, assumed to be small, satisfy the compatibility equation

$$\frac{\partial \varepsilon_\theta}{\partial r} = \frac{\varepsilon_r - \varepsilon_\theta}{r} \tag{166}$$

which is readily obtained from the strain-displacement equations

$$\varepsilon_r = \frac{\partial u}{\partial r} \qquad \varepsilon_\theta = \frac{u}{r}$$

Let the radius of the hole be denoted by a. So long as the plate is completely elastic, the stress distribution is obtained from Lamé's solution as

$$\sigma_r = \sigma\left(1 - \frac{a^2}{r^2}\right) \qquad \sigma_\theta = \sigma\left(1 + \frac{a^2}{r^2}\right)$$

The hoop stress is 2σ at the edge of the hole, indicating that there is a stress concentration factor of 2 in the elastic range. Yielding occurs in hoop tension at the edge of the hole when $\sigma = Y/2$, whatever the yield criterion.

For some value of σ greater than $Y/2$, the plate will be plastic within a radius c.

---

† The fully plastic rotation of both solid and hollow cylinders has been treated by E. A. Davies and F. M. Connelly, *J. Appl. Mech.*, **26**: 25 (1959), and by E. P. J. Rimrott, *J. Appl. Mech.*, **27**: 309 (1960).

The material is assumed to be incompressible and strain-hardening according to the generalized stress-strain law†

$$\frac{\bar{\sigma}}{Y} = \left(\frac{E\bar{\varepsilon}}{Y}\right)^n \qquad \bar{\sigma} \geqslant Y \tag{167}$$

where n is a material constant. It is also assumed that the strains are small enough to justify the neglect of geometry changes. The stress distribution in the plate will be largely unaffected by the neglect of elastic compressibility.‡

Considering first the Tresca criterion, $\sigma_\theta = Y$ at the elastic/plastic boundary, and the stresses in the elastic region ($r \geqslant c$) are found to be

$$\sigma_r = \sigma - (Y - \sigma)\frac{c^2}{r^2} \qquad \sigma_\theta = \sigma + (Y - \sigma)\frac{c^2}{r^2} \qquad r \geqslant c \tag{168}$$

In the plastic region ($r \leqslant c$), the Tresca criterion may be written as $\sigma_\theta = \bar{\sigma}$. Adopting the Hencky stress-strain relation, and denoting the stress ratio σ_r/σ_θ by s, we have

$$\varepsilon_r = \bar{\varepsilon}\left(s - \frac{1}{2}\right) \qquad \varepsilon_\theta = \bar{\varepsilon}\left(1 - \frac{s}{2}\right) \qquad r \leqslant c$$

Inserting in the strain compatibility equation (166), and using the strain-hardening law (167), we get

$$\frac{1}{\bar{\sigma}}\frac{\partial \bar{\sigma}}{\partial r} = \frac{n}{2 - s}\frac{\partial s}{\partial r} - \frac{3n}{r}\left(\frac{1 - s}{2 - s}\right) \tag{169a}$$

The substitution $\sigma_\theta = \bar{\sigma}$ and $\sigma_r = s\bar{\sigma}$ in the equilibrium equation (165) leads to

$$\frac{1}{\bar{\sigma}}\frac{\partial \bar{\sigma}}{\partial r} = -\frac{1}{s}\frac{\partial s}{\partial r} + \frac{1 - s}{rs} \tag{169b}$$

The elimination of $\bar{\sigma}$ between the last two equations furnishes

$$r\frac{\partial s}{\partial r} = (1 - s)\left[\frac{2 - (1 - 3n)s}{2 - (1 - n)s}\right]$$

Since the edge of the hole is stress free, $s = 0$ at $r = a$. Integration of the above equation under this boundary condition results in

$$\frac{a}{r} = (1 - s)^{(1+n)/(1+3n)}\left[1 - (1 - 3n)\frac{s}{2}\right]^{4n/(1 - 9n^2)} \tag{170}$$

which indicates that s depends only on r for a given material. At the elastic/plastic

---

† The elastic/plastic solution for a finite hollow disc of linearly work-hardening Tresca material has been given by P. G. Hodge, Jr., *J. Appl. Mech.*, **20**: 530 (1953).

‡ B. Budiansky, *Q. Appl. Math.*, **16**: 307 (1958).

boundary $r = c$, $s = 2\sigma/Y - 1$ in view of (168). Hence

$$\frac{a}{c} = \left[2\left(1 - \frac{\sigma}{Y}\right) \right]^{(1+n)/(1+3n)} \left[\frac{3}{2}(1 - n) - (1 - 3n)\frac{\sigma}{Y} \right]^{4n/(1-9n^2)}$$

which relates the extent of the plastic zone to the magnitude of the applied stress. Evidently, the entire plate becomes plastic ($c = \infty$) when $\sigma = Y$, whatever the value of n. Eliminating r from Eqs. (169), we get

$$\frac{1}{\bar{\sigma}}\frac{\partial \bar{\sigma}}{\partial s} = -\frac{2n}{2 - (1 - 3n)s}$$

At the edge of the hole, $s = 0$ and $\bar{\sigma} = \lambda\sigma$, where λ is the elastic/plastic stress concentration factor. At the elastic/plastic boundary, $s = 2\sigma/Y - 1$ and $\bar{\sigma} = Y$. Integration of the above equation therefore furnishes

$$\bar{\sigma} = \lambda\sigma\left[1 - (1 - 3n)\frac{s}{2} \right]^{2n/(1-3n)}$$

where
$$\lambda = \frac{Y}{\sigma}\left[\frac{3}{2}(1 - n) - (1 - 3n)\frac{\sigma}{Y} \right]^{-2n/(1-3n)}$$

(171)

For a nonhardening material, the stress concentration factor is Y/σ, whatever the yield criterion. When $n = \frac{1}{3}$, the right-hand sides of the above equations become indeterminate, but direct integration of the differential equations for s and $\bar{\sigma}$ gives

$$\frac{r}{a} = e^{s/3}(1 - s)^{-2/3} \qquad \bar{\sigma} = \lambda\sigma e^{-s/3}$$

$$\lambda = \frac{Y}{\sigma}\exp\left[\frac{1}{3}\left(\frac{2\sigma}{Y} - 1\right) \right]$$

(172)

As σ is increased from $Y/2$ to Y, the stress concentration factor decreases from 2 to a limiting value $\lambda^* = 1.35$ for $n = \frac{1}{3}$, and

$$\lambda^* = \left[\frac{2}{1 + 3n} \right]^{2n/(1-3n)}$$

for other values of n. The hoop strain at the bore (equal to $\bar{\varepsilon}/2$) is still comparable to that at the initial yielding. For $\sigma = Y$, the elastic/plastic boundary disappears and the outer boundary condition becomes $s = 1$ at $r = \infty$. The stresses are then directly proportional to σ, so long as the strain remains small, and the stress concentration factor has a constant value equal to λ^*.

If the von Mises yield criterion is adopted, the stresses at the elastic/plastic boundary must satisfy the equation $\sigma_r^2 - \sigma_r\sigma_\theta + \sigma_\theta^2 = Y^2$. The stress distribution in the elastic region ($r \geqslant c$) is therefore found as

$$\sigma_r = \sigma - \frac{Yc^2}{\sqrt{3}\,r^2}\sqrt{1 - \frac{\sigma^2}{Y^2}} \qquad \sigma_\theta = \sigma + \frac{Yc^2}{\sqrt{3}\,r^2}\sqrt{1 - \frac{\sigma^2}{Y^2}} \qquad r \geqslant c \quad (173)$$

In the plastic region ($r \leqslant c$), the Mises criterion is identically satisfied by writing

$$\sigma_r = \frac{2}{\sqrt{3}} \bar{\sigma} \sin \phi \qquad \sigma_\theta = \frac{2}{\sqrt{3}} \bar{\sigma} \cos \left(\frac{\pi}{6} - \phi \right) \qquad r \leqslant c \qquad (174)$$

where ϕ is an auxiliary angle. At the edge of the hole $r = a$, the boundary condition requires $\phi = 0$, while at the plastic boundary $r = c$, the continuity of the stresses requires ϕ to have the value

$$\phi_c = \sin^{-1} \left\{ \frac{\sqrt{3}\,\sigma}{2Y} - \frac{1}{2} \sqrt{1 - \frac{\sigma^2}{Y^2}} \right\} \qquad (175)$$

It may be noted that $\phi - \pi/6$ is equal to the deviatoric angle measured from the relevant pure shear direction (Fig. 2.2). The Hencky equations for an incompressible material obeying the von Mises criterion may be expressed as

$$\varepsilon_r = \bar{\varepsilon} \sin \left(\phi - \frac{\pi}{6} \right) \qquad \varepsilon_\theta = \bar{\varepsilon} \cos \phi \qquad r \leqslant c \qquad (176)$$

where $\bar{\varepsilon}$ is related to $\bar{\sigma}$ by the strain-hardening law (167). Substitution for the stresses and strains into (165) and (166) then leads to

$$\frac{1}{\bar{\sigma}} \frac{\partial \bar{\sigma}}{\partial r} \tan \phi + \frac{\partial \phi}{\partial r} = \frac{1}{2r} (\sqrt{3} - \tan \phi)$$

$$\frac{1}{\bar{\sigma}} \frac{\partial \bar{\sigma}}{\partial r} - n \tan \phi \frac{\partial \phi}{\partial r} = -\frac{\sqrt{3}\,n}{2r} (\sqrt{3} - \tan \phi) \qquad (177)$$

from which $\bar{\sigma}$ can be eliminated to obtain the differential equation

$$r \frac{\partial \phi}{\partial r} = \frac{(\sqrt{3} - \tan \phi)(1 + \sqrt{3}\,n \tan \phi)}{2(1 + n \tan^2 \phi)}$$

In view of the boundary condition $\phi = 0$ at $r = a$, the integration of the foregoing equation gives

$$\frac{a^2}{r^2} = \frac{2}{\sqrt{3}} \cos \left(\frac{\pi}{6} + \phi \right)(\cos \phi + \sqrt{3}\,n \sin \phi)^{-4n/(1+3n^2)} \exp \left[-\sqrt{3} \left(\frac{1 - n^2}{1 + 3n^3} \right) \phi \right] \qquad (178)$$

Thus ϕ is a function of r only for a given n. The position of the elastic/plastic boundary for any given σ can be found by setting $r = c$ and $\phi = \phi_c$ in (178). The elimination of r from (177) gives

$$\frac{1}{\bar{\sigma}} \frac{\partial \bar{\sigma}}{\partial \phi} = -n \left(\frac{\sqrt{3} - \tan \phi}{1 + \sqrt{3}\,n \tan \phi} \right)$$

which is integrated under the boundary condition $\bar{\sigma} = \lambda \sigma$ at $r = a$ to obtain

$$\frac{\bar{\sigma}}{\sigma} = \lambda (\cos \phi + \sqrt{3}\,n \sin \phi)^{-n(1+3n)/(1+3n^2)} \exp \left[-\sqrt{3} \frac{n(1-n)}{1 + 3n^2} \phi \right] \qquad (179)$$

The stress concentration factor λ is finally obtained from the fact that $\bar{\sigma} = Y$ at the elastic/plastic boundary where $\phi = \phi_c$. Thus, from (175) and (179), we get†

$$\lambda = \frac{Y}{\sigma}\left\{(1 + 3n)\frac{\sigma}{2Y} + \frac{\sqrt{3}(1 - n)}{2}\sqrt{1 - \frac{\sigma^2}{Y^2}}\right\}^{n(1 + 3n)/(1 + 3n^2)}$$

$$\times \exp\left[\sqrt{3}\frac{n(1 - n)}{1 + 3n^2}\phi_c\right] \quad (180)$$

When λ has been found for a given σ, the spatial distribution of $\bar{\sigma}$ follows from (178) and (179). The limiting value of λ is

$$\lambda^* = \left(\frac{1 + 3n}{2}\right)^{n(1 + 3n)/(1 + 3n^2)} \exp\left[\frac{\pi n(1 - n)}{\sqrt{3}(1 + 3n^2)}\right]$$

corresponding to $\sigma = Y$ and $\phi_c = \pi/3$, the plate being then just fully plastic. For any further increase in σ, the stress concentration factor remains constant so long as geometry changes are negligible.

It follows from both (170) and (178) that the stress ratio remains constant at any given radius once the element has become plastic. The above solutions will therefore be identical to those based on the Prandtl-Reuss equations. The results for the stress concentration factor are shown graphically in Fig. 5.19 for various values of n. It is evident that the Tresca solution gives much simpler results without significant loss in accuracy.

(ii) *Expanding a circular hole* Consider an infinite plate of uniform thickness containing a circular hole which is subjected to gradually increasing radial pressure p around its edge. While the plate is entirely elastic, the sum of the radial and circumferential stresses is constant at each stage, in view of Lamé's solution. Since the stresses vanish at infinity, $\sigma_r + \sigma_\theta = 0$ throughout the plate. We therefore have a state of pure shear given by

$$\sigma_r = -\frac{pa^2}{r^2} \qquad \sigma_\theta = \frac{pa^2}{r^2}$$

where a is the radius of the hole. Yielding will begin at the edge of the hole when $p = k$, where k is equal to $Y/\sqrt{3}$ for the von Mises criterion. If the pressure is further increased to render the plate plastic within a radius c, the stresses in the elastic region are

$$\sigma_r = -\frac{kc^2}{r^2} \qquad \sigma_\theta = \frac{kc^2}{r^2} \qquad r \geqslant c$$

We begin with an ideally plastic material with constant yield stress k. To find

† These results have been obtained by J. Chakrabarty, unpublished work (1976). A more involved analysis, requiring the solution of a non-linear differential equation, has been given by B. Budiansky and O. L. Mangasarian, *J. Appl. Mech.*, **27**: 59 (1960). See also H. Ishilkawa, *Z. angew. Math. Mech.*, **55**: 171 (1975).

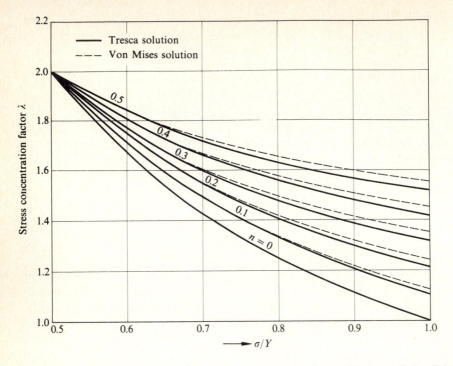

Figure 5.19 Variation of the elastic/plastic stress concentration factor with the applied radial tension.

the stresses in the plastic region ($r \leqslant c$), it is convenient to express the von Mises yield criterion in the parametric form

$$\sigma_r = -2k \sin\left(\frac{\pi}{6} + \phi\right) \qquad \sigma_\theta = 2k \sin\left(\frac{\pi}{6} - \phi\right) \qquad r \leqslant c \qquad (181)$$

Evidently, $\phi = 0$ on the elastic/plastic boundary $r = c$. The angle ϕ is the same as that which the stress vector in the deviatoric plane makes with the direction representing pure shear in the plane of the applied stresses. It may be noted that the maximum principal stress difference at any radius is $\sigma_\theta - \sigma_r = 2k \cos\phi$. Substitution in the equilibrium equation (165) gives

$$\cos\left(\frac{\pi}{6} + \phi\right)\frac{\partial\phi}{\partial r} = -\frac{\cos\phi}{r}$$

of which the integral, satisfying the condition $\phi = 0$ at $r = c$, is easily shown to be

$$\frac{c^2}{r^2} = e^{\sqrt{3}\,\phi}\cos\phi \qquad (182)$$

The internal pressure necessary to move the elastic/plastic boundary to a radius c is given parametrically by†

$$p = 2k \sin\left(\frac{\pi}{6} + \alpha\right) \qquad \frac{c^2}{a^2} = e^{\sqrt{3}\,\alpha} \cos \alpha \qquad (183)$$

where α is the value of ϕ at $r = a$. As p increases from k, α increases from zero and eventually becomes equal to $\pi/3$, giving $c/a = 1.751$, when p attains the limiting value of $2k$. The circumferential stress at the edge of the hole is then a compression of amount k. Since a nonhardening material cannot sustain a stress of magnitude greater than $2k$, the plate must start to thicken appreciably to support any further increase in the radial load.‡

Releasing the expanding pressure by an amount q is equivalent to superimposing the elastic stresses

$$\sigma'_r = \frac{qa^2}{r^2} \qquad \sigma'_\theta = -\frac{qa^2}{r^2}$$

on the above elastic/plastic stress distribution, so long as there is no secondary yielding. If q (less than p) is sufficiently large, yielding will restart at the edge of the hole when the residual value of $\sigma_r - \sigma_\theta$ is equal to $2k \cos \alpha$. It follows that secondary yielding occurs when $q = 2k \cos \alpha$. From (183), the minimum value of p for the yielding to occur on complete unloading ($q = p$) is $\sqrt{3}\,k$, corresponding to $\alpha = \pi/6$ and $c/a = 1.465$.

For $p > \sqrt{3}\,k$, the zone of secondary yielding on complete unloading will spread to a radius ρ. If $\phi = \beta$ at $r = \rho$, the value of $\sigma_r - \sigma_\theta$ superposed at this radius due to the unloading is $4k \cos \beta$. Hence, the residual stresses in the region $\rho \leqslant r \leqslant c$ are obtained by the superposition of the elastic stresses§

$$\sigma'_r = 2k\left(\frac{c^2}{r^2}\right) e^{-\sqrt{3}\beta} \qquad \sigma'_\theta = -2k\left(\frac{c^2}{r^2}\right) e^{-\sqrt{3}\beta}$$

$$\tag{184}$$

where

$$e^{\sqrt{3}\beta} \cos \beta = c^2/\rho^2 \qquad \beta \geqslant \pi/6$$

on those given by (181). In the secondary plastic zone $a \leqslant r \leqslant \rho$, the residual stresses may be expressed as

$$\sigma_r = -2k \sin\left(\psi - \frac{\pi}{6}\right) \qquad \sigma_\theta = -2k \sin\left(\psi + \frac{\pi}{6}\right) \qquad a \leqslant r \leqslant \rho \qquad (185)$$

where $\psi = \beta$ at $r = \rho$ for the stresses to be continuous across this radius. Since the hole is stress free on complete unloading, $\psi = \pi/6$ at $r = a$. Substitution from (185)

† This solution is due to A. Nadai, *Theory of Flow and Fracture of Solids*, p. 473, McGraw-Hill Book Company, New York (1950).

‡ For a finite disc of external radius b, the boundary condition at $r = c$ is modified to $\phi = \phi_c$, where $\sec \phi_c = \sqrt{1 + c^4/3b^4}$. The stresses in the elastic region are given by (32) with k replaced by $k \cos \phi_c$.

§ J. M. Alexander and H. Ford, *Proc. R. Soc.*, A, **226**: 543 (1954). These authors have also given numerical solutions for a work-hardening Prandtl-Reuss material up to large values of the strain.

into (165) results in the differential equation

$$\cos\left(\psi - \frac{\pi}{6}\right)\frac{\partial\psi}{\partial r} = \frac{\cos\psi}{r}$$

which gives, on integration, the spatial distribution of ψ as

$$\frac{r^2}{a^2} = \frac{\sqrt{3}}{2}\sec\psi\,\exp\left[\sqrt{3}\left(\psi - \frac{\pi}{6}\right)\right] \tag{186}$$

Setting $r = \rho$ and $\psi = \beta$ in (186), and using the last equation of (184), we get

$$e^{-\sqrt{3}\beta} = \left(\frac{2}{\sqrt{3}}e^{\pi/2\sqrt{3}}\right)^{-1/2}\frac{a}{c} \simeq 0.591\frac{a}{c}$$

In view of (181), (182), and (184), the residual stresses in the region $\rho \leqslant r \leqslant c$ (where $\beta \geqslant \phi \geqslant 0$) may now be written as

$$\sigma_r = -2k\left[\sin\left(\frac{\pi}{6} + \phi\right) - \left(0.591\frac{a}{c}\right)e^{\sqrt{3}\phi}\cos\phi\right]$$

$$\rho \leqslant r \leqslant c \tag{187}$$

$$\sigma_\theta = 2k\left[\sin\left(\frac{\pi}{6} - \phi\right) - \left(0.591\frac{a}{c}\right)e^{\sqrt{3}\phi}\cos\phi\right]$$

the spatial distribution of ϕ being given by (182). The residual hoop stress changes sign as ϕ varies from β at $r = \rho$ to zero at $r = c$. The residual radial stress has a maximum numerical value where $\beta - \phi$ is 0.4 radians. In the remainder of the plate, the residual stresses are

$$-\sigma_r = \sigma_\theta = k\left(1 - 1.182\frac{a}{c}\right)\frac{c^2}{r^2} \qquad r \geqslant c$$

The angle β increases from $\pi/6$ as c/a is increased from 1.465. In the limiting case of $c/a = 1.751$, $\beta = 35.93°$ and $\rho/a = 1.131$. The stress distribution in the elastic/plastic plate and the residual stresses on complete unloading are shown in Fig. 5.20 for $c/a = 1.751$. The numerically largest residual stress is of magnitude Y, occurring at the edge of the hole which is in a state of uniaxial compression.

When the material strain-hardens, an immediate simplification is achieved, without loss of generality as far as the determination of the stresses is concerned, by assuming the material to be incompressible. The von Mises yield criterion and the Hencky stress-strain relation may be written parametrically as

$$\sigma_r = -\frac{2\bar{\sigma}}{\sqrt{3}}\sin\left(\frac{\pi}{6} + \phi\right) \qquad \sigma_\theta = \frac{2\bar{\sigma}}{\sqrt{3}}\sin\left(\frac{\pi}{6} - \phi\right)$$

$$r \leqslant c \tag{188}$$

$$\varepsilon_r = -\bar{\varepsilon}\cos\left(\frac{\pi}{6} - \phi\right) \qquad \varepsilon_\theta = \bar{\varepsilon}\cos\left(\frac{\pi}{6} + \phi\right)$$

where $\bar{\sigma}$ and $\bar{\varepsilon}$ are related to one another by the strain-hardening law (167).

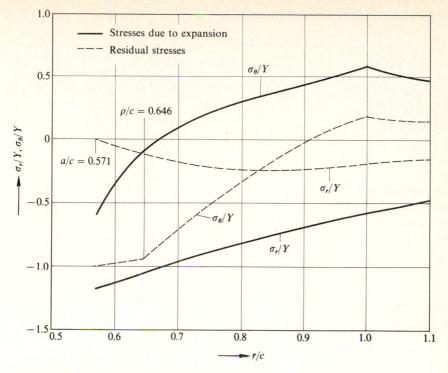

Figure 5.20 Stress distribution in an infinite plate containing a circular hole which is expanded by radial pressure.

Substitution into (165) and (166) then gives

$$(1 + \sqrt{3} \tan \phi) \frac{1}{\bar{\sigma}} \frac{\partial \bar{\sigma}}{\partial r} + (\sqrt{3} - \tan \phi) \frac{\partial \phi}{\partial r} = -\frac{2}{r}$$

$$(\sqrt{3} - \tan \phi) \frac{1}{\bar{\sigma}} \frac{\partial \bar{\sigma}}{\partial r} - n(1 + \sqrt{3} \tan \phi) \frac{\partial \phi}{\partial r} = -\frac{2\sqrt{3}\,n}{r} \qquad (189)$$

The space variable r is easily eliminated from these equations, yielding

$$\frac{1}{\bar{\sigma}} \frac{d\bar{\sigma}}{d\phi} = \frac{4n}{\sqrt{3}\,(1 - n) - (1 + 3n) \tan \phi} \qquad (190)$$

The total derivative is used in this equation because $\bar{\sigma}$ depends only on ϕ. Using the boundary condition $\bar{\sigma} = Y$ at the elastic/plastic boundary where $\phi = 0$, the above equation is integrated to

$$\frac{\bar{\sigma}}{Y} = \left[\cos \phi - \frac{1}{\sqrt{3}} \left(\frac{1 + 3n}{1 - n} \right) \sin \phi \right]^{-n(1 + 3n)/(1 + 3n^2)} \exp \left[\frac{\sqrt{3}\,n(1 - n)}{1 + 3n^2} \phi \right] \qquad (191)$$

Equations (188) and (191) give the stresses and strains as functions of ϕ which varies with both r and c. Eliminating $\bar{\sigma}$ from (189), we get

$$\left[\tan\phi + \frac{\sqrt{3}(1-n)\tan\phi - (3+n)}{\sqrt{3}(1-n) - (1+3n)\tan\phi}\right]\frac{\partial\phi}{\partial r} = \frac{2}{r}$$

which is integrated under the boundary condition $r = c$ at $\phi = 0$ to give the spatial distribution of ϕ as

$$\frac{c^2}{r^2} = \left[\cos\phi - \frac{1}{\sqrt{3}}\left(\frac{1+3n}{1-n}\right)\sin\phi\right]^{-4n/(1+3n^2)} \exp\left[\sqrt{3}\left(\frac{1-n^2}{1+3n^2}\right)\phi\right]\cos\phi$$

$$(192)$$

When $n = 0$, the above equation reduces to (182). If the right-hand sides of (191) and (192) are denoted by $f(\phi)$ and $g(\phi)\cos\phi$ respectively, the relationship between p and c can be expressed parametrically as†

$$\frac{p}{Y} = \frac{2}{\sqrt{3}}f(\alpha)\sin\left(\frac{\pi}{6}+\alpha\right) \qquad \frac{c^2}{a^2} = g(\alpha)\cos\alpha$$

where $\phi = \alpha$ corresponds to $r = a$ as before. As the expansion proceeds, $\tan\alpha$ approaches the asymptotic value $\sqrt{3}(1-n)/(1+3n)$. The validity of the infinitesimal theory requires, however, that α should be limited to values for which ε_θ at the edge of the hole is sufficiently small (less than about 5 percent).

When an element first becomes plastic, $\phi = 0$ and the stress ratio σ_r/σ_θ is -1. The subsequent variation of ϕ (and hence the stress ratio) at a given radius can be determined from (192) for various assumed values of $c/r > 1$. The stress path is therefore the same for each element. The angle which the tangent to the stress path makes with the radial path has the greatest value of $\tan^{-1}\left[\sqrt{3}(1-n)/4n\right]$ at the initial point $\phi = 0$. Since this is evidently less than $\pi/2$ for all $n > 0$, the use of the Hencky theory is justified. In Fig. 5.21, the dimensionless radial and circumferential stresses and the corresponding c/r ratio are plotted for various values of n. Each stress history curve approaches the straight line $\sigma_\theta/\sigma_r = (1 - 3n)/2$ asymptotically as the stresses are increased. The broken curve indicates points on the stress history curves where the hoop strain is ten times the uniaxial strain at the yield point.‡

(iii) *Plates of variable thickness* Consider now an infinite plate whose thickness h varies with the radial distance from the center of the hole. It is assumed that the thickness is everywhere small enough for the stresses to be averaged through the thickness. The nonzero principal stresses σ_r and σ_θ must satisfy the equilibrium

---

† More complicated solutions, based on the Ramberg-Osgood equation for the stress-strain curve, have been given by O. L. Mangasarian, *J. Appl. Mech.*, **27**: 65 (1960), using both Hencky and Prandtl-Reuss equations.

‡ Finite expansions based on the rigid/plastic assumption have been discussed by several investigators. Relevant references on the subject can be found in the paper by J. Chern and S. Nemat-Nasser, *J. Mech. Phys. Solids*, **17**: 271 (1969).

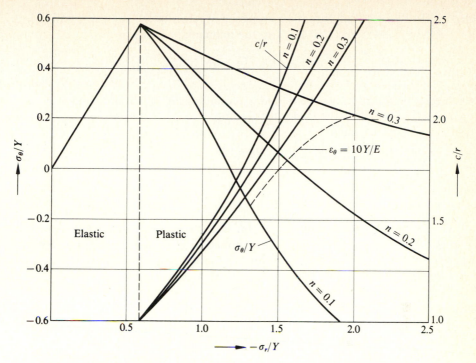

Figure 5.21 Relationship between σ_r, σ_θ and c/r in the expansion of a circular hole in a strain-hardening infinite plate.

equation, which assumes the form

$$\frac{\partial}{\partial r}(h\sigma_r) = \frac{h}{r}(\sigma_\theta - \sigma_r) \tag{193}$$

If the thickness of the plate is assumed to vary according to the equation

$$h = h_0\left(\frac{r}{a}\right)^m$$

where h_0 is the thickness at the edge of the hole and m a constant, the equilibrium equation becomes

$$r\frac{\partial\sigma_r}{\partial r} + (1+m)\sigma_r = \sigma_\theta \tag{194}$$

When m is negative, the solution will be sufficiently accurate if a is large compared to h_0. For positive values of $n < 1$, the solution will provide a good approximation for large but finite plates. While the plate is elastic, the stresses and strains obey Hooke's law. Substituting for ε_r and ε_θ in (166), and using (194), we obtain

$$\frac{\partial}{\partial r}(\sigma_r + \sigma_\theta) + (1+v)m\sigma_r = 0 \tag{195}$$

which is the compatibility equation expressed in terms of the stresses. Elimination of σ_θ between (194) and (195) leads to the differential equation

$$r^2 \frac{\partial^2 \sigma_r}{\partial r^2} + (3 + m)r \frac{\partial \sigma_r}{\partial r} + (1 + v)m\sigma_r = 0$$

which must be solved under the boundary conditions $\sigma_r = 0$, $r = \infty$ and $\sigma_r = -p$, $r = a$. We thus arrive at the stress distribution

$$\sigma_r = -p\left(\frac{a}{r}\right)^\gamma \qquad \sigma_\theta = p(\gamma - m - 1)\left(\frac{a}{r}\right)^\gamma \tag{196}$$

where
$$\gamma = 1 + \frac{m}{2} + \sqrt{1 - vm + \frac{m^2}{4}}$$

The strains are now easily found by substitution into the stress-strain relations. Since $vm < 1$ for all realistic values of m, σ_r is compressive and σ_θ tensile throughout the plate. Hence, if Tresca's yield criterion is adopted, $\sigma_\theta - \sigma_r$ must be equal to Y at $r = a$ for the yielding to begin. The pressure at the initial yielding is

$$p_e = \frac{Y}{\gamma - m}$$

When the plate is rendered plastic within a radius c by increasing the pressure to a value greater than p_e, the radial pressure transmitted across $r = c$ must be equal to p_e. The stresses in the elastic region are

$$\sigma_r = -p_e\left(\frac{c}{r}\right)^\gamma \qquad \sigma_\theta = p_e(\gamma - m - 1)\left(\frac{c}{r}\right)^\gamma \qquad r \geqslant c \tag{197}$$

If the material work-hardens, the stresses in the plastic region must be determined by using the yield criterion in the form

$$\sigma_\theta - \sigma_r = \sigma \qquad a \leqslant r \leqslant c$$

where σ is the current uniaxial yield stress of the material at any radius r. Tresca's associated flow rule furnishes $\dot{\varepsilon}_\theta^p = -\dot{\varepsilon}_r^p > 0$, $\dot{\varepsilon}_z^p = 0$, indicating that the thickness strain ε_z is purely elastic. If the amount of hardening is assumed to be a function of the plastic work per unit volume, it follows that σ is the same function of ε_θ^p as the stress is of the plastic strain in uniaxial tension. Assuming a constant slope H of the stress-strain curve, we write

$$\sigma_\theta - \sigma_r = Y + H\varepsilon_\theta^p \qquad a \leqslant r \leqslant c$$

The elastic parts of ε_r and ε_θ are expressed by Hooke's law, while the plastic parts are given by the above relation. Thus

$$\frac{H}{Y}\varepsilon_r = \left(1 + \frac{H}{E}\right)\frac{\sigma_r}{Y} - \left(1 + \frac{vH}{E}\right)\frac{\sigma_\theta}{Y} + 1$$

$$\frac{H}{Y}\varepsilon_\theta = \left(1 + \frac{H}{E}\right)\frac{\sigma_\theta}{Y} - \left(1 + \frac{vH}{E}\right)\frac{\sigma_r}{E} - 1 \tag{198}$$

If we restrict ourselves to small strains, ε_r and ε_θ must satisfy the compatibility equation (166). Substituting from (198) and using (194), we get

$$r^2 \frac{\partial^2 \sigma_r}{\partial r^2} + (3 + m)r \frac{\partial \sigma_r}{\partial r} + \left(1 + \frac{E + vH}{E + H}\right)m\sigma_r = \frac{2EY}{E + H}$$

The integration of this linear differential equation is straightforward. Excluding the case $m = 0$ (which corresponds to a flat plate), the stresses in the plastic region may be expressed as†

$$\frac{\eta \sigma_r}{Y} = 1 - A\left(\frac{c}{r}\right)^{1+\alpha} - B\left(\frac{c}{r}\right)^{1-\beta}$$

$$\frac{\eta \sigma_\theta}{Y} = (1 + m) + \beta A\left(\frac{c}{r}\right)^{1+\alpha} - \alpha B\left(\frac{c}{r}\right)^{1-\beta}$$
$$r \leqslant c \qquad (199)$$

in view of (194). Here A and B are the constants of integration, and

$$\left.\begin{array}{c}\alpha \\ \beta\end{array}\right\} = \pm \frac{m}{2} + \left[1 - m\left(\frac{E + vH}{E + H}\right) + \frac{m^2}{4}\right]^{1/2}$$

$$\eta = m\left[1 + (1 + v)\frac{H}{2E}\right]$$
$$(200)$$

It may be noted that $\alpha - \beta = m$. The conditions of continuity of σ_r and σ_θ across $r = c$ furnish

$$A = \frac{\eta + \beta - 1}{\alpha + \beta} + \frac{\eta(\alpha - 1)}{(\alpha + \beta)(\gamma - m)}$$

$$B = \frac{1 + \alpha - \eta}{\alpha + \beta} + \frac{\eta(1 + \beta)}{(\alpha + \beta)(\gamma - m)}$$
$$(201)$$

When the stresses have been found, the strains in the plastic region can be determined from (198). The plastic part of the strain is given by

$$\frac{\eta E}{Y}\varepsilon_\theta^p = -(1 + v)\frac{m}{2} + \frac{E}{H}\left[A(1 + \beta)\left(\frac{c}{r}\right)^{1+\alpha} - B(\alpha - 1)\left(\frac{c}{r}\right)^{1-\beta}\right] \qquad (202)$$

The relationship between the internal pressure and the extent of the plastic region is readily obtained by setting $\sigma_r = -p$ and $r = a$ in (199). The hoop stress vanishes at the edge of the hole when $c/a = \rho$, such that

$$\alpha B\rho^{1-\beta} - \beta A\rho^{1+\alpha} = 1 + m \qquad (203)$$

The form of the yield criterion used above is valid for all values of $c/a \leqslant \rho$. For a given rate of hardening, ρ decreases as m increases.

---

† J. Chakrabarty, *Int. J. Mech. Sci.*, **13**: 439 (1971). The particular case $m = 1$ has been analyzed by J. M. Alexander and H. Ford (op. cit.), using the Prandtl-Reuss theory.

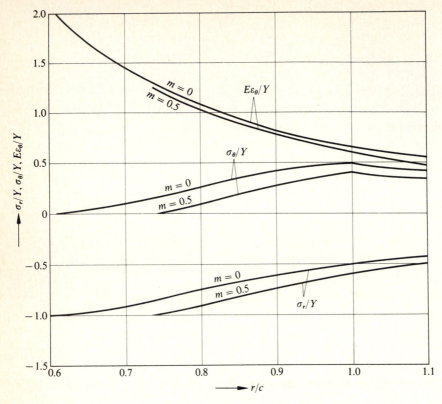

Figure 5.22 Stresses and strains in infinite plates of uniform and variable thicknesses ($v = 0.3$).

It may be noted that for a nonhardening material, $\alpha = 1$, $\beta = 1 - m$, $\eta = m$, $A = 0$ and $B = \gamma/(\gamma - m)$. Moreover, it is easily shown that

$$(\alpha - 1)\frac{E}{H} = m\left(\frac{1 - v}{2 - m}\right)$$

$$\frac{AE}{H} = \frac{m}{2 - m}\left(\frac{1 + v}{2} + \frac{1 - v}{2 - m}\frac{\gamma}{\gamma - m}\right)$$

when H is vanishingly small. The plastic strain for a nonhardening material therefore becomes†

$$\frac{E\varepsilon_\theta^p}{Y} = \frac{1 + v}{2}\left(\frac{c^2}{r^2} - 1\right) + \frac{(1 - v)\gamma}{(2 - m)(\gamma - m)}\left\{\left(\frac{c}{r}\right)^2 - \left(\frac{c}{r}\right)^m\right\} \tag{204}$$

† Large strains for the nonhardening material have been considered by T. G. Rogers, *Q. T. Mech. Appl. Math.*, **17**: 271 (1960), and for a work-hardening material by S. Namat-Nasser, *J. Mech. Phys. Solids*, **16**: 195 (1968).

The elastic part of the strain is readily obtained from Hooke's law. The distribution of stresses and strain for a nonhardening material corresponding to $v = 0.3$ and $m = 0$ and 0.5 are shown graphically in Fig. 5.22 over the range of validity of the solution.

5.8 Yielding Around a Cylindrical Cavity

(i) *Symmetrical problem* The expansion of a cylindrical cavity in an infinite medium by the application of uniform internal pressure has been previously discussed (Sec. 5.2(vii)). A more general problem of elastic/plastic deformation under combined internal pressure and independent twisting moment will now be considered. The stress distribution is axially symmetrical and the deformation occurs under plane strain condition. The relevant equations of equilibrium are

$$r \frac{\partial \sigma_r}{\partial r} = \sigma_\theta - \sigma_r \qquad \frac{\partial}{\partial r} (r^2 \tau_{r\theta}) = 0 \tag{205}$$

If the radial and circumferential displacements are denoted by u and v respectively, the components of the strain (assumed small) are given by

$$\varepsilon_r = \frac{\partial u}{\partial r} \qquad \varepsilon_\theta = \frac{u}{r} \qquad 2\gamma_{r\theta} = r \frac{\partial}{\partial r} \left(\frac{v}{r} \right) \tag{206}$$

where the polar coordinates (r, θ) are referred to the axis of the cylindrical cavity. The second of the equilibrium equations (205) is identically satisfied by taking

$$\tau_{r\theta} = -mk \frac{a^2}{r^2} \qquad 0 \leqslant m \leqslant 1$$

There is a uniformly distributed shear stress of magnitude mk round the cavity of radius a. The twisting couple per unit length of the cavity is $2\pi mka^2$ acting in the sense of increasing θ. The remaining stress components, while the material is entirely elastic, are

$$\sigma_r = -p \frac{a^2}{r^2} \qquad \sigma_\theta = p \frac{a^2}{r^2} \qquad \sigma_z = 0$$

by Lamé's solution, where p is the applied internal pressure. Plastic yielding begins on the cavity surface when p reaches the value $k\sqrt{1 - m^2}$, where k is the yield stress in pure shear.

Further radially symmetric loading, caused by suitable changes in the two loading parameters p and m, produces a plastic zone within a radius c. The nonzero stress components in the elastic region are easily found as

$$\sigma_\theta = -\sigma_r = \frac{kc^2}{r^2} \sqrt{1 - \frac{m^2 a^4}{c^4}} \qquad \tau_{r\theta} = -mk \frac{a^2}{r^2} \qquad r \geqslant c \tag{207}$$

The application of Hooke's law then gives the corresponding displacements

$$u = (1 + v)\frac{kc^2}{Er}\sqrt{1 - \frac{ma^4}{c^4}} \qquad v = (1 + v)m\frac{ka^2}{Er} \qquad r \geqslant c \qquad (208)$$

In the plastic region, it would be a good approximation to assume that $\sigma_z \simeq \frac{1}{2}(\sigma_r + \sigma_\theta)$. Then the Tresca and Mises criteria both reduce to

$$(\sigma_\theta - \sigma_r)^2 + 4\tau_{r\theta}^2 = 4k^2$$

work-hardening being neglected. The second equation of equilibrium and the yield criterion are identically satisfied by

$$\tau_{r\theta} = -mk\frac{a^2}{r^2} \qquad \sigma_\theta - \sigma_r = 2k\sqrt{1 - \frac{m^2 a^4}{r^4}} \qquad r \leqslant c \qquad (209)$$

The first equation of (205) is then integrated to obtain σ_r in the plastic region as

$$\frac{\sigma_r}{k} = -\sqrt{1 - \frac{m^2 a^4}{r^4}} - \ln\left\{\frac{r^2 - \sqrt{r^4 - m^2 a^4}}{c^2 - \sqrt{c^4 - m^2 a^4}}\right\} \qquad r \leqslant c \qquad (210)$$

which is evidently continuous across $r = c$. Hence the internal pressure is given by†

$$\frac{p}{k} = \sqrt{1 - m^2} + \ln\left\{\frac{a^2[1 - \sqrt{1 - m^2}]}{c^2 - \sqrt{c^4 - m^2 a^4}}\right\} \qquad (211)$$

For a given value of m, as the expansion proceeds, the pressure exceeds $2k \ln (c/a)$ by a quantity that rapidly approaches a constant value. When $m = 1$, for instance, $p = 0$ at the initial yielding, but the pressure soon becomes approximately equal to $k \ln (2c^2/a^2)$.

If the ratio c/a is not too large (less than 5, say), the strains in the plastic region would be small enough to justify neglect of geometry changes. Then the compressibility equation may be written as

$$\frac{\partial u}{\partial r} + \frac{u}{r} = \frac{3}{2}(1 - 2v)(\sigma_r + \sigma_\theta)$$

which is readily integrated on substitution for σ_θ from (205). Using the conditions of continuity of u and σ_r across $r = c$, we get

$$u = \frac{kr}{2E}\left\{(5 - 4v)\frac{c^2}{r^2}\sqrt{1 - \frac{m^2 a^4}{c^4}} + 3(1 - 2v)\frac{\sigma_r}{k}\right\} \qquad r \leqslant c \qquad (212)$$

where σ_r is given by (210). The expression for u furnishes $\varepsilon_\theta - \varepsilon_r$ by (206).

† These equations are due to A. Nadai, Z. Phys., **30**: 106 (1924). The solution also holds for a thin plate ($\sigma_z = 0$), yielding according to Tresca's yield criterion, provided $p/k \leqslant 1 + \sqrt{1 - m^2}$, which ensures that the algebraically greater principal stress in the plane of the plate is non-negative.

Subtracting the elastic part expressed by Hooke's law, we get

$$\varepsilon_\theta^p - \varepsilon_r^p = (5 - 4v)\frac{k}{E}\left\{\frac{c^2}{r^2}\sqrt{1 - \frac{m^2 a^4}{c^4}} - \frac{\sigma_\theta - \sigma_r}{2k}\right\} \tag{213}$$

If m is maintained constant during the expansion, the ratios of the deviatoric stresses also remain constant, and the Prandtl-Reuss equations become equivalent to the Hencky equations. We therefore write the relevant stress-strain equation in the form

$$2\gamma_{r\theta} = \frac{\tau_{r\theta}}{G} + 2\left(\frac{\varepsilon_\theta^p - \varepsilon_r^p}{\sigma_\theta - \sigma_r}\right)\tau_{r\theta}$$

where the first term represents the elastic part and the second term the plastic part. Inserting from (206), (209), and (213), we get

$$\frac{\partial}{\partial r}\left(\frac{v}{r}\right) = -\frac{mka^2}{Er^3}\left\{(5 - 4v)\left(\frac{c^4 - m^2 a^4}{r^4 - m^2 a^4}\right)^{1/2} - 3(1 - 2v)\right\}$$

In view of the continuity of v across $r = c$, the integral of the above equation becomes

$$v = \frac{kr}{2E}\left\{(5 - 4v)\frac{c^2}{ma^2}\left[1 - \sqrt{1 - \frac{m^2 a^4}{c^4}}\sqrt{1 - \frac{m^2 a^4}{r^4}}\right] - 3(1 - 2v)\frac{ma^2}{r^2}\right\}$$
$$r \leqslant c \quad (214)$$

For a given twisting moment specifying m, Eqs. (211), (212), and (214) express the displacements at any radius in the plastic region as functions of the applied internal pressure.†

(ii) Unsymmetrical problem One of the few unsymmetrical problems, for which an analytical treatment is possible, has been discussed by Galin. An infinite medium containing a cylindrical cavity is subjected to biaxial tensile stresses t_1 and t_2 under conditions of plane strain, while a uniform pressure p acts around the cavity. The three quantities t_1, t_2, and p are varied in such a way that the plastic zone, which completely surrounds the cavity, nowhere unloads during the process. It is assumed that the material is incompressible and the stresses in the plastic region are axially symmetrical. The axial stress is then equal to $\frac{1}{2}(\sigma_r + \sigma_\theta)$, and the yield criterion becomes

$$\sigma_\theta - \sigma_r = 2k$$

where k is equal to $Y/2$ for the Tresca criterion and $Y/\sqrt{3}$ for the Mises criterion. For no work-hardening, the equilibrium equation (165) and the yield criterion

---

† The combined twisting and expansion of an annular plate, for both Tresca and Mises criteria, has been discussed by R. P. Nordgen and P. M. Naghdi, *Int. J. Eng. Sci.*, **3**: 33 (1963); *J. Appl. Mech.*, **30**: 605 (1963).

furnish the stress distribution

$$\sigma_r = -p + 2k \ln \frac{r}{a} \qquad \sigma_\theta = -p + 2k\left(1 + \ln \frac{r}{a}\right) \qquad \tau_{r\theta} = 0 \qquad (215)$$

in the plastic region, in view of the stress boundary condition at $r = a$. The lines across which the shear stress has its maximum value k are given by

$$dr = \pm d\theta \qquad \text{or} \qquad r \sim \exp\left(\pm \theta\right)$$

where θ is the polar angle. The lines of shearing stress therefore form an orthogonal net of logarithmic spirals. The stresses (215) are derivable from the stress function

$$\phi = k\left\{ -\frac{1}{2} r^2\left(1 + \frac{p}{k}\right) + r^2 \ln \frac{r}{a} \right\}$$

as may be verified by the substitution into the relations

$$\sigma_r = \frac{1}{r}\frac{\partial \phi}{\partial r} \qquad \sigma_\theta = \frac{\partial^2 \phi}{\partial r^2}$$

which satisfy the equilibrium equation identically. The rectangular components of the stress are given by

$$\sigma_x + \sigma_y = \sigma_r + \sigma_\theta = -2p + 2k\left(1 + 2\ln \frac{r}{a}\right)$$

$$\sigma_y - \sigma_x = 2k \cos 2\theta \qquad \tau_{xy} = -k \sin 2\theta$$

where the polar angle θ is measured counterclockwise from the x axis, which is taken in the direction of t_1. It is convenient to introduce the complex coordinates

$$z = x + iy = re^{i\theta} \qquad \bar{z} = x - iy = re^{-i\theta}$$

where $i = \sqrt{-1}$, and write the stresses in the complex variable form

$$\sigma_x + \sigma_y = -2p + 2k\left(1 + \ln \frac{z\bar{z}}{a^2}\right)$$

$$\sigma_y - \sigma_x + 2i\tau_{xy} = 2ke^{-2i\theta} = 2k\frac{\bar{z}}{z}$$

(216)

within the plastic region. It is important to note that the plastic stress function satisfies the biharmonic equation $\nabla^4 \phi = 0$, which the elastic stress function also satisfies. Consequently, the stresses can be expressed as analytic functions of z and $\bar{z}$ in both elastic and plastic regions of the infinite medium.

We propose to map the elastic region in the z plane onto the exterior of a unit circle in a ζ plane, such that the points at infinity in the two planes correspond to one another. This is achieved by means of the conformal transformation

$$z = c\left(\zeta + \frac{m}{\zeta} + \frac{m_1}{\zeta^2} + \cdots + \right) \qquad \zeta = \rho e^{i\psi} \qquad (217)$$

where c, m, m_1, etc., are real constants to be determined from the boundary conditions. The elastic/plastic boundary, which is one of the unknowns of the problem, is transformed into the unit circle $\rho = 1$. Since the stresses must be bounded at infinity and continuous across the elastic/plastic boundary, we can immediately write

$$\sigma_x + \sigma_y = -2p + 2k\left[1 + \ln\left(\frac{1}{a^2}\frac{z\bar{z}}{\zeta\bar{\zeta}}\right)\right] = -2p + 2k\left(1 + 2\ln\frac{r}{a\rho}\right) \quad (218)$$

for $\rho \geqslant 1$, with the bar denoting complex conjugate. To find the remaining stress combination in the elastic region, consider the equilibrium equations

$$\frac{\partial\sigma_x}{\partial x} + \frac{\partial\tau_{xy}}{\partial y} = 0 \qquad \frac{\partial\tau_{xy}}{\partial x} + \frac{\partial\sigma_y}{\partial y} = 0$$

Multiplying the second of these equations by i and subtracting from the first, and observing that

$$\frac{\partial}{\partial x} = \frac{\partial}{\partial z} + \frac{\partial}{\partial\bar{z}} \qquad \frac{\partial}{\partial y} = i\left(\frac{\partial}{\partial z} - \frac{\partial}{\partial\bar{z}}\right)$$

where z and $\bar{z}$ are taken as independent variables, we obtain

$$\frac{\partial}{\partial\bar{z}}(\sigma_y - \sigma_x + 2i\tau_{xy}) = \frac{\partial}{\partial z}(\sigma_x + \sigma_y) = \frac{2k}{z} - \frac{2k/\zeta}{dz/d\zeta}$$

by (218), since z and $\bar{z}$ depend only on ζ and $\bar{\zeta}$ respectively. This is readily integrated to give

$$\sigma_y - \sigma_x + 2i\tau_{xy} = 2k\left[\frac{\bar{z}}{z} - \frac{\bar{z}/\zeta}{dz/d\zeta} + f(\zeta)\right] \quad (219)$$

In view of (216), the continuity of the stresses across the elastic/plastic boundary requires

$$f(\zeta) = \frac{\bar{z}}{\zeta(dz/d\zeta)} \qquad \text{on} \qquad \bar{\zeta} = \frac{1}{\zeta}$$

since $\zeta\bar{\zeta} = 1$ on the unit circle. Using (217), we obtain the unknown function as

$$f(\zeta) = \frac{1 + m\zeta^2 + m_1\zeta^3 + \cdots}{\zeta^2 - m - 2m_1/\zeta - \cdots}$$

Since $f(\zeta)$ must remain finite at infinity, it is evident that $m_1 = m_2 = \cdots = 0$. Hence

$$z = c\left(\zeta + \frac{m}{\zeta}\right) \qquad f(\zeta) = \frac{1 + m\zeta^2}{\zeta^2 - m} \quad (220)$$

Substitution into (219) finally gives the required stress combination

$$\sigma_y - \sigma_x + 2i\tau_{xy} = 2k\left[\frac{\bar{z}}{z} - \frac{\zeta}{\bar{\zeta}}\left(\frac{\bar{\zeta}^2 + m}{\zeta^2 - m}\right) + \frac{1 + m\zeta^2}{\zeta^2 - m}\right] \quad (221)$$

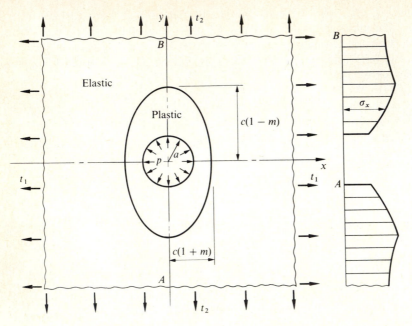

Figure 5.23 Cylindrical cavity in an infinite medium unsymmetrically loaded by biaxial tensions in plane strain ($t_1 > t_2$).

As $|\zeta|$ tends to infinity, $\bar{z}/z$ tends to $\bar{\zeta}/\zeta$, and the expression in the square bracket of the above equation tends to m. Since $\sigma_x = t_1$, $\sigma_y = t_2$, and $\tau_{xy} = 0$ at infinity, (218) and (221) furnish†

$$ m = \frac{t_2 - t_1}{2k} \qquad \ln\frac{c}{a} = \frac{1}{2}\left(\frac{p}{k} - 1\right) + \frac{t_1 + t_2}{4k} \tag{222} $$

The real and imaginary parts of the mapping function $z(\zeta)$ give

$$ x = c\left(\rho + \frac{m}{\rho}\right)\cos\psi \qquad y = c\left(\rho - \frac{m}{\rho}\right)\sin\psi $$

The elastic/plastic boundary, which corresponds to $\rho = 1$, is therefore an ellipse (Fig. 5.23) whose semiaxes are $c(1 + m)$ and $c(1 - m)$ respectively, ψ being the eccentric angle of the ellipse. Since $m \gtrless 0$ for $t_2 \gtrless t_1$ in view of (222), the major axis of the ellipse coincides with the direction of the lesser applied tension. When $t_1 = t_2$, the elastic/plastic boundary reduces to a circle of radius c. The polar coordinates (r, θ) of any point in the physical plane, corresponding to a point (ρ, ψ)

† This solution is due to L. A. Galin, *Prikl. Mat. Mekh.*, **10**: 365 (1946). For an extension of Galin's solution, based on Eqs. (209) for the stresses in the plastic region, see O. S. Parasyuk, *Prikl. Mat. Mekh.*, **13**: 367 (1948).

in the transformed plane, are given by

$$\frac{r^2}{c^2} = \rho^2 + 2\,m\rho\,\cos 2\psi + \frac{m^2}{\rho^2}$$

$$\rho \geqslant 1 \qquad (223)$$

$$\tan \theta = \left(\frac{\rho^2 - m}{\rho^2 + m}\right) \tan \psi$$

As ρ tends to infinity, r/ρ tends to c and θ tends to ψ. Returning to Eq. (221), and separating the real and imaginary parts, we get

$$\sigma_y - \sigma_x = 2k\left\{\cos 2\theta - \frac{(\rho^2 - 1)[(\rho^2 + m^2)\cos 2\psi - m(\rho^2 + 1)]}{\rho^4 - 2m\rho^2 \cos 2\psi + m^2}\right\}$$

$$\tau_{xy} = k\left\{-\sin 2\theta + \frac{(\rho^2 - 1)(\rho^2 - m^2)\sin 2\psi}{\rho^4 - 2m\rho^2 \cos 2\psi + m^2}\right\} \qquad (224)$$

The stress distribution in the elastic region ($\rho \geqslant 1$) can be calculated from (218), (223), and (224) for any suitable values of m and c. Along the y axis, the principal stress perpendicular to this axis is

$$(\sigma_x)_{\theta=\pi/2} = -p + 2k\left(1 + \ln\frac{r}{a\rho}\right) - k(1 + m)\left(\frac{\rho^2 - 1}{\rho^2 + m}\right) \qquad (225)$$

The stress rapidly decreases outward from the plastic boundary to approach its value at infinity. Taking for instance $p = 0$, $t_1 = 3k$ and $t_2 = 2.5k$, giving $m = -0.25$ and $c = 2.4a$, we find that σ_x varies from $4.2k$ at $r = 3a$ (apex of the ellipse) to $3.2k$ at $r = 7a$.

For the solution to be valid, the ellipse must completely surround the circular hole, requiring $c(1 - |m|) \geqslant a$. Hence

$$\frac{t_1 + t_2}{2k} + 2\ln\left(1 - \left|\frac{t_2 - t_1}{2k}\right|\right) \geqslant 1 - \frac{p}{k} \qquad (226)$$

in view of (222). The permissible values of t_1 and t_2 for given values of p, such that the plastic zone just surrounds the hole, are shown in Fig. 5.24. It follows that a single applied tension will not give a plastic region completely surrounding the cavity unless $p > k$, which is the pressure required to cause yielding at $r = a$ if the tensions were absent.[†]

The validity of the solution also requires that the tensions and the internal pressure are varied in such a manner that successive ellipses contain their predecessors. One way of ensuring this is to maintain a constant value of m, so that the successive ellipses are concentric with one another. Thus, the tensions are varied at the same rate $\dot{t}$ (say) once the ellipse is formed, and the solution will then hold if $\dot{c} > 0$. In view of (222), the last condition is equivalent to $\dot{t} + \dot{p} > 0$ throughout the loading.

[†] A plane strain analysis of an incompressible elastic/plastic wedge under a uniform normal pressure on one boundary has been reported by P. M. Naghdi, *J. Appl. Mech.*, **25**: 98 (1956).

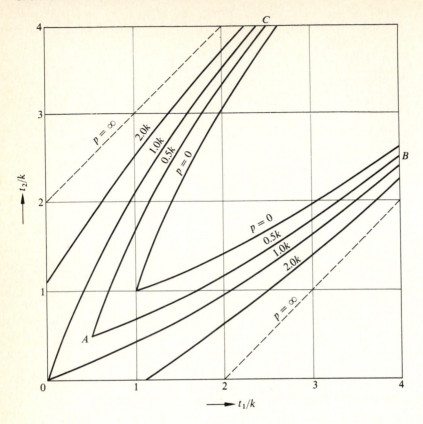

Figure 5.24 Permissible range of values of t_1, t_2, and p for the applicability of Galin's solution.

It is not known how the loads should be varied to produce the first ellipse touching the circular hole. The early part of the loading path would evidently involve a pair of discrete plastic zones adjacent to the cavity. The initial problem is not even statically determined unless the eccentricity of the ellipse is less than a certain value. For this limiting value of m, there are four diametrically opposite points on the ellipse where the tangents are along one of the maximum shear stress directions. The acute angle which each of these tangents makes with the respective radius vector is therefore equal to $\pi/4$. The included angle will always exceed $\pi/4$ if $|(dr/d\theta)| \leqslant r$. Using (223) with $\rho = 1$, the inequality may be written as

$$2m(\cos 2\theta \pm \sin 2\theta) \leqslant 1 + m^2$$

The expression on the left-hand side has its greatest value for points on the extremities of the diameters inclined at angles $\pi/8$ with respect to the major axis of the ellipse. The critical value of $|m|$ is therefore $\sqrt{2} - 1$, the critical ratio of the axes being $\sqrt{2} + 1$. For greater aspect ratios, there will be regions within the ellipse

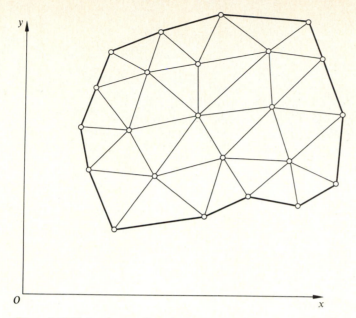

Figure 5.25 Finite element discretization of a body using triangular elements.

where the maximum shear stress trajectories cut the ellipse twice, and the stresses are no longer determined from the stress boundary conditions alone.[†]

5.9 The Finite Element Method

(i) *Deformation models* The finite element method is a powerful numerical method of solving boundary value problems in continuum mechanics. The method is based on a piecewise approximation in which the domain of interest is subdivided into a number of small regions, or finite elements (Fig. 5.25), which are assumed to deform in a prescribed manner. In selecting the velocity field, it is necessary to ensure continuity of the prescribed behavior across the boundaries of adjacent elements. The simplest analysis for plane stress and plane strain problems involve an assemblage of triangular elements in each of which the velocity varies linearly with the rectangular coordinates. Since the triangle has straight sides, the continuity of nodal velocities automatically ensures continuity of the velocity vector across the boundaries of the element.[‡]

Consider a typical triangular element whose vertices are defined by the current

---

[†] This conclusion follows from a theorem due to R. Hill, the *Mathematical Theory of Plasticity*, p. 243, Clarendon Press, Oxford (1950).

[‡] For a critical review of the finite element method in plasticity, see P. G. Hodge, Jr., *Problems of Plasticity*, p. 261, A. Sawczuk (ed.), Noordhoff International Publishing, Leyden (1974).

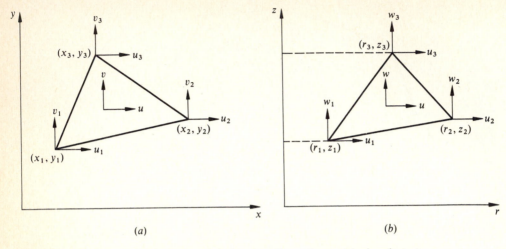

Figure 5.26 Geometry of a plane triangular element and an axisymmetric ring element.

rectangular coordinates (x_1, y_1), (x_2, y_2), and (x_3, y_3) as shown in Fig. 5.26a. The velocity components (u, v) of a typical particle in the triangle are taken in the form

$$u = \alpha_1 + \alpha_2 x + \alpha_3 y$$
$$v = \alpha_4 + \alpha_5 x + \alpha_6 y \tag{227}$$

The six constants α_i depend on the nodal velocities (u_1, v_1), (u_2, v_2), and (u_3, v_3) as shown. The substitution into the above relations furnishes

$$u_1 = \alpha_1 + \alpha_2 x_1 + \alpha_3 y_1$$
$$u_2 = \alpha_1 + \alpha_2 x_2 + \alpha_3 y_2$$
$$u_3 = \alpha_1 + \alpha_2 x_3 + \alpha_3 y_3$$

and a similar set of three equations for α_4, α_5, and α_6. The above equations are easily solved for α_1, α_2, α_3 in terms of u_1, u_2, and u_3, the result being

$$2A\alpha_1 = \sum_{i=1}^{3} a_i u_i \qquad 2A\alpha_2 = \sum_{i=1}^{3} b_i u_i \qquad 2A\alpha_3 = \sum_{i=1}^{3} c_i u_i \tag{228}$$

where

$$a_1 = x_2 y_3 - x_3 y_2 \qquad b_1 = y_2 - y_3 \qquad c_1 = x_3 - x_2$$

with the other equations obtained by cyclic permutation of the subscripts. The constant A represents the area of the triangle and is given by

$$2A = x_1(y_2 - y_3) + x_2(y_3 - y_1) + x_3(y_1 - y_2)$$

The expressions for α_4, α_5, and α_6 are obtained from those of α_1, α_2, and α_3

respectively by replacing u_i with v_i. The back substitution into (227) then furnishes the matrix equation

$$\{v\} = \begin{Bmatrix} u \\ v \end{Bmatrix} = \begin{bmatrix} N_1 & 0 & N_2 & 0 & N_3 & 0 \\ 0 & N_1 & 0 & N_2 & 0 & N_3 \end{bmatrix}\{q\} = [N]\{q\} \tag{229}$$

where

$$N_i = \frac{a_i + b_i x + c_i y}{2A} \qquad \{q\}^T = [u_1 \ v_1 \ u_2 \ v_2 \ u_3 \ v_3]$$

The vector $\{q\}$ represents the nodal degrees of freedom, and the functions N_i are termed the *shape functions* for the velocity field. The components of the strain rate corresponding to the proposed velocity field are

$$\dot{\varepsilon}_x = \frac{\partial u}{\partial x} = \alpha_2 \qquad \dot{\varepsilon}_y = \frac{\partial v}{\partial y} = \alpha_6 \qquad \dot{\gamma}_{xy} = \frac{\partial u}{\partial y} + \frac{\partial v}{\partial x} = \alpha_3 + \alpha_5$$

where $\dot{\gamma}_{xy}$ is the engineering shear strain rate. Inserting the values of $\alpha_2, \alpha_3, \alpha_5,$ and α_6 from (228), the results may be put in the matrix form

$$\{\dot{\varepsilon}\} = \begin{Bmatrix} \dot{\varepsilon}_x \\ \dot{\varepsilon}_y \\ \dot{\gamma}_{xy} \end{Bmatrix} = \frac{1}{2A} \begin{bmatrix} b_1 & 0 & b_2 & 0 & b_3 & 0 \\ 0 & c_1 & 0 & c_2 & 0 & c_3 \\ c_1 & b_1 & c_2 & b_2 & c_3 & b_3 \end{bmatrix}\{q\} = [B]\{q\} \tag{230}$$

The matrix $[B]$ involves only the nodal coordinates and is independent of x and y. The strain rates are therefore constant throughout the element. The procedure described above can be easily generalized by considering a more general form of the velocity field. Other shapes of the element may also be considered and treated in a similar manner.†

In problems of axial symmetry, the only nonzero velocities are u and w in the radial and axial directions, characterized by the coordinates r and z respectively. An axisymmetric finite element is in the form of a circular ring of constant cross section, which may be taken as triangular for simplicity, Fig. 5.26b. Assuming a linear distribution of velocity as before, we have

$$u = \alpha_1 + \alpha_2 r + \alpha_3 z$$
$$w = \alpha_4 + \alpha_5 r + \alpha_6 z \tag{231}$$

The velocity relations for the constant strain triangles are directly applicable here, with the rectangular coordinates (x, y) replaced by the cylindrical coordinates

† See, for example, C. S. Desai and J. F. Abel, *Introduction to the Finite Element Method*, Van Nostrand Reinhold, New York (1972); R. H. Gallagher, *Finite Element Analysis Fundamental*, Prentice-Hall, Englewood Cliffs, N.J. (1975); O. C. Zienkiewicz, *The Finite Element Method*, 3d ed., McGraw-Hill Book Co., London (1977); R. D. Cook, *Concepts and Applications of Finite Element Analysis*, John Wiley, New York (1981); K. J. Bathe, *Finite Element Procedures in Engineering*, Prentice-Hall Inc., Englewood Cliffs, N.J. (1982).

(r, z). The nonzero strain rates in this case are

$$\dot{\varepsilon}_r = \frac{\partial u}{\partial r} \qquad \dot{\varepsilon}_\theta = \frac{u}{r} \qquad \dot{\varepsilon}_z = \frac{\partial w}{\partial z} \qquad \dot{\gamma}_{rz} = \frac{\partial u}{\partial z} + \frac{\partial w}{\partial r}$$

Substituting from (231), and using the expressions for the relevant quantities, there results

$$\{\dot{\varepsilon}\} = \begin{Bmatrix} \dot{\varepsilon}_r \\ \dot{\varepsilon}_\theta \\ \dot{\varepsilon}_z \\ \dot{\gamma}_{rz} \end{Bmatrix} = \frac{1}{2A} \begin{bmatrix} b_1 & 0 & b_2 & 0 & b_3 & 0 \\ \lambda_1 & 0 & \lambda_2 & 0 & \lambda_3 & 0 \\ 0 & c_1 & 0 & c_2 & 0 & c_3 \\ c_1 & b_1 & c_2 & b_2 & c_3 & b_3 \end{bmatrix} \{q\} = [B]\{q\} \tag{232}$$

where $\lambda_i = b_i + (a_i + c_i z)/r$. Since the $[B]$ matrix now involves the variables r and z, the strain rate is no longer constant within the triangle. The variation is due entirely to the circumferential component $\dot{\varepsilon}_\theta$, and is negligible at sufficiently large distances from the axis of symmetry.

A plastically deforming material rapidly approaches the condition of incompressibility with increasing strain. The degrees of freedom of many common finite elements are inadequate for satisfying this constraint. Consequently, the aggregate of elements behave more stiffly than the material it is supposed to represent. A possible remedy in plane problems is to adopt patterns of constant strain triangular elements which form quadrilaterals with straight diagonals. When three of the four triangles forming a quadrilateral satisfy the incompressibility condition, the fourth triangle does so automatically.†

(ii) *Variational formulation* The nonzero components of the Jaumann stress rate in a typical finite element are represented by a column vector $(\mathring{\sigma})$, which is related to the corresponding strain rate $\{\dot{\varepsilon}\}$ by the equation

$$\{\mathring{\sigma}\} = [C]\{\dot{\varepsilon}\} \tag{233}$$

where $[C]$ is the constitutive matrix. In an elastic/plastic element obeying the Prandtl-Reuss flow rule, $[C]$ is given by Eqs. (45) and (46), Chap. 2, for plane strain and plane stress respectively. In the case of axial symmetry,

$$\{\mathring{\sigma}\}^T = [\mathring{\sigma}_r \quad \mathring{\sigma}_\theta \quad \mathring{\sigma}_z \quad \mathring{\tau}_{rz}] \qquad \{\dot{\varepsilon}\}^T = [\dot{\varepsilon}_r \quad \dot{\varepsilon}_\theta \quad \dot{\varepsilon}_z \quad \dot{\gamma}_{rz}]$$

and $[C]$ is obtained from the plane strain expression with appropriate change in subscripts, by including additional terms which involve $s_r s_\theta$, s_θ^2, $s_\theta s_z$, and $s_\theta \tau_{rz}$. For an elastic element, the constitutive matrix is that which corresponds to an infinite rate of hardening.

In general, the velocity distribution in a typical element, and the associated strain rate, can be expressed in the form

$$\{v\} = [N]\{q\} \qquad \{\dot{\varepsilon}\} = [B]\{q\} \tag{234}$$

---

† See J. C. Nagtegaal, D. M. Parks, and J. R. Rice, *Comp. Meth. Appl. Mech. Eng.*, **5**: 133 (1974).

irrespective of the choice of the element. It follows from (233) and (234) that

$$\mathring{\sigma}_{ij}\dot{\varepsilon}_{ij} = \{\dot{\varepsilon}\}^T\{\mathring{\sigma}\} = \{\dot{\varepsilon}\}^T[C]\{\dot{\varepsilon}\}$$

A consistent finite element equation is obtained on the basis of the variational principle expressed by Eq. (109), Chap. 2. When the strains are so small that positional changes are negligible, the second term in the volume integral can be omitted. The matrix form of this equation then becomes

$$\delta\left(\int \{\dot{\varepsilon}\}^T[C]\{\dot{\varepsilon}\} \, dV - 2 \int \{v\}^T\{\dot{F}\} \, dS_F \right) = 0$$

where V is the volume of the element, and S_F is that part of its boundary where traction rates are prescribed. The components $(\dot{F}_x, \dot{F}_y)$ of the traction rate are represented by the column vector $\{\dot{F}\}$. Substituting from (234), the variational principle is reduced to

$$\{\delta q\}^T\left(\int [B]^T[C][B]\{q\} \, dV - \int [N]^T\{\dot{F}\} \, dS_F \right) = 0$$

Since the variation of the nodal velocities is arbitrary, the above equation requires the expression in the parenthesis to vanish.† The rate equation of equilibrium for the element therefore becomes

$$[k]\{q\} = \{\dot{Q}\} \tag{235}$$

where $[k]$ is the *element stiffness matrix*, and $\{\dot{Q}\}$ the associated load rate vector, given by

$$[k] = \int [B]^T[C][B] \, dV \qquad \{\dot{Q}\} = \int [N]^T\{\dot{F}\} \, dS_F \tag{236}$$

For a constant strain triangular element, the volume integral in (236) is simply the matrix product times hA, where h is the thickness and A the area of the triangle. For an axisymmetric ring element, $dV = 2\pi r \, dr \, dz$ and $dS_F = 2\pi r \, ds$, where ds is a line element of the boundary, and numerical integration would be necessary for evaluating the integrals.

In problems involving large plastic strains, the change in geometry cannot be disregarded. The variational formulation then leads to an additional stiffness term in the finite element equation. To derive the second stiffness matrix, we write

$$v_i = [N_i]\{q\} \qquad \dot{\varepsilon}_{ij} = [B_{ij}]\{q\}$$

where $[N_i]$ and $[B_{ij}]$ are row vectors identical to the rows of $[N]$ and $[B]$, except for a factor of $\frac{1}{2}$ that must be associated with the shear components. The rate

---

† The incremental finite element method for small strains has been developed by P. V. Marcal and I. P. King, *Int. I. Mech. Sci.*, **9**: 143 (1967); Y. Yamada, N. Yoshimura, and T. Sakurai, *Int. J. Mech. Sci.*, **10**: 343 (1968). An initial strain approach has been discussed by J. H. Argyris, *J. R. Aeronaut. Soc.*, **69**: 633 (1965), and an initial stress approach has been used by O. C. Zienkiewicz, S. Valliappan, and I. P. King, *Int. J. Num. Meth. Eng.*, **1**: 75 (1969).

equation of equilibrium then becomes†

$$([k] + [k_s])\{q\} = \{\dot{Q}\}$$

where

$$[k_s] = \int ([N_k]_{,i}^T \sigma_{ij}[N_k]_{,j} - 2[B_{ki}]^T \sigma_{ij}[B_{kj}]) \, dV \qquad (237)$$

The comma denotes partial differentiation with respect to the coordinates, the usual summation convention being implied for the repeated indices.

Consider now the overall problem of forming the matrix equation for the assemblage from those of the individual elements. Let n denote the total number of elements, and N the total number of nodal degrees of freedom represented by the vector $\{U\}$. The load rate vector $\{\dot{Q}\}$ for a generic element e is expanded into an N vector $\{\dot{R}_e\}$ such that

$$\{\dot{R}_e\}^T = [\{0\}^T \quad \{0\}^T \quad \cdots \quad \{0\}^T \quad \{\dot{Q}\}^T \quad \{0\}^T \quad \cdots \quad \{0\}^T]$$

The position of $\{\dot{Q}\}$ in $\{\dot{R}_e\}$ corresponds to that of $\{q\}$ in $\{U\}$. The element stiffness matrix $[k]$ is similarly expanded into an $N \times N$ matrix $[K_e]$. Equation (235) is therefore equivalent to

$$[K_e]\{U\} = \{\dot{R}_e\}$$

The nodal velocity vector $\{U\}$ for the assemblage appears in each element equation when written in the expanded form. The overall equilibrium of the assemblage requires

$$\left(\sum_{e=1}^n [K_e]\right)\{U\} = \sum_{e=1}^n \{\dot{R}_e\}$$

The expression in the parenthesis is the *global stiffness matrix* $[K]$, and that on the right-hand side is the global load rate vector $\{\dot{R}\}$, giving

$$[K]\{U\} = \{\dot{R}\} \qquad (238)$$

which is the global equilibrium equation for elastic/plastic problems involving small strains. The assembly rules for $[K]$ and $\{\dot{R}\}$ are obvious from the above derivation. The global stiffness equation for large plastic strains can be established in a similar manner. For each incremental step, the stiffness matrix is always symmetric and banded. A matrix inversion is subsequently required for finding the nodal velocities, and hence the distribution of strain and stress rates throughout the deforming body.‡

---

† R. M. McMeeking and J. R. Rice, *Int. J. Solid Struct.*, **11**, 601 (1974). For an application of the large strain formulation to the extrusion of metals, see E. H. Lee, R. L. Mallett and W. H. Yang, *Comp. Meth. Appl. Mech. Eng.*, **10**: 339 (1977). See also Y. Yamada, *Comp. Struct.*, **8**: 533 (1978).

‡ Rigid/plastic finite element methods have been discussed by K. Osakada, J. Nakaro, and K. Mori, *Int. J. Mech. Sci.*, **24**: 459 (1982), and by S. I. Oh, ibid., **24**: 479 (1982). See also O. C. Zienkiewicz and P. N. Godbole, *J. Strain Anal.*, **10**: 180 (1975); J. H. Kim and S. Kobayashi, *Int. J. Mach. Tool Des. Res.*, **18**: 209 (1978); A. Chandra and S. Mukherjee, *Int. J. Mech. Sci.*, **26**: 661 (1984).

(iii) *Method of computation* In order to carry out an elastic/plastic analysis, the load necessary to initiate plastic yielding must be first estimated. The initial yield point is most conveniently obtained by solving the elastic problem corresponding to a unit load, and finding the ratio of the greatest equivalent stress $\bar{\sigma}$ to the uniaxial yield stress Y. Using the constitutive matrices appropriate for the elastic and plastic elements, the individual and global stiffnesses are constructed to obtain the increments of nodal displacement corresponding to a specified load increment ΔL. The strain and stress increments for the individual elements are then computed from (234) and (233), and are used to calculate the equivalent stress increment $\Delta\bar{\sigma}$ in each element. This enables us to identify the transition elements which become plastic during the load increment. Denoting the least value of the ratio $(Y - \bar{\sigma})/\Delta\bar{\sigma}$ in the transition region by ρ, the computation is repeated with a mean constitutive matrix for these elements. The mean value of $[C]$, obtained by adding ρ times the elastic matrix to $(1 - \rho)$ times the elastic/plastic matrix, has been found to be quite satisfactory. Since the transition region is fairly small compared to the whole structure, the mean coefficients would converge rapidly on iteration.† While continuing the solution, it is necessary at each stage to check the sign of $\Delta\bar{\sigma}$ in the plastically deforming elements. A negative sign of $\Delta\bar{\sigma}$ would imply unloading and the calculations should be modified accordingly.‡

Consider, as an example, the longitudinal tension of a strip containing a central circular hole whose diameter is one-half of the width of the strip.§ The material is aluminum having an initial yield stress $Y = 238$ MPa, Young's modulus $E = 68.7$ GPa, and Poisson's ratio $v = 0.3$. The stress-strain curve is closely approximated by a straight line with a constant plastic modulus $H = 2.21$ GPa. The finite element analysis has been carried out on the basis of a mesh of simple triangular elements as shown in Fig. 5.27a, the curved boundary of the hole being approximated by linear segments. Finer subdivisions have been used over the region where the stress and strain gradients are expected to be large. The development of the plastic enclaves during a succession of load increments, equal to 0.2 and 0.1 times the load at the initial yielding, is shown in Fig. 5.27b. These enclaves are found to be in substantial agreement with those experimentally observed. Figure 5.28 shows the variation of the maximum longitudinal strain ε with the applied tensile stress σ. The solid curve represents the computed result, and the upper broken curve represents the experimental result. Calculations based on the neglect of work-hardening furnish the lower broken curve, whose divergence

† The iterative procedure can be avoided by using the Ramberg-Osgood equation for the stress-strain curve, which allows the same constitutive matrix to be used throughout the body so long as no unloading takes place. See, for example, G. Venkateswara Rao and A. V. Krishnamurthy, *Nucl. Eng. Des.*, **17**: 297 (1971).

‡ The elastic/plastic bending of plates by the finite element method has been considered by T. Belytschko and P. G. Hodge, Jr., *J. Eng. Mech. Div. Proc. ASCE*, **98**: 227 (1972). For an analysis of elastic/plastic shells of revolution, see P. V. Marcal, *J. Amer. Inst. Aeronaut. Astronaut*, **8**: 1628 (1970).

§ This problem has been investigated by P. V. Marcal and I. P. King, op. cit., and also by O. C. Zienkiewicz, S. Valliappan, and I. P. King, op. cit. For experimental results, see P. S. Theocaris and E. Marketos, *J. Mech. Phys. Solids*, **12**: 377 (1964).

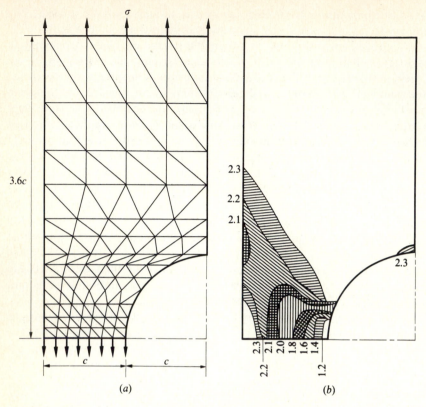

Figure 5.27 Quadrant of a perforated strip under tension. (*a*) Finite elements; (*b*) spread of plastic zone.

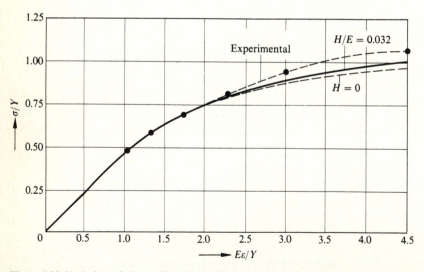

Figure 5.28 Variation of the applied stress with maximum longitudinal strain in a perforated plate. The theoretical curves are due to Marcal, and the experimental curve is due to Theocaris and Marketos.

from the work-hardening curve is apparent only at higher loads. For the nonhardening material, the plastic zone starts to grow fairly rapidly as the load is nearly twice that at the initial yielding. When $H = 0$, it is necessary at each stage to compute the increment of plastic strain that corresponds to the increment of stress, and check the sign of the equivalent plastic strain increment which must be positive for the plastically deforming elements.†

In many metal working processes, the plastic strains completely dominate the elastic strains, which are of the order of the yield stress divided by Young's modulus. Nevertheless, it is still necessary to include the elastic strains for predicting the stress distribution in regions of contained plastic flow, where the elastic and plastic strain increments are of comparable magnitudes. An analysis based on the rigid/plastic model fails to provide the necessary information, since such regions are then considered as rigid, and not enough equations are available for the solution. It is evident that a complete elastic/plastic analysis is necessary for the calculation of residual stresses, and the assessment of the possibility of forming defects such as internal or surface cracks. The changes in geometry that occur in large plastic flow problems are most conveniently allowed for by taking the current configuration at each stage as the reference configuration during a further increment of time scale. This procedure is repeated sequentially to build up the complete history of the deformation process, and the current configuration is thus available at each stage for solving the incremental stiffness equation.

Problems

5.1 A thick-walled spherical shell having an internal raidus a and an external radius b is rendered partially plastic by the application of an internal pressure p. The material of the shell hardens linearly with a constant plastic modulus H. Introducing the assumption of complete incompressibility of the material, and neglecting changes in geometry, show that the pressure necessary for the elastic/plastic boundary to have a radius c is given by

$$\left(1 + \frac{H}{E}\right)p = \frac{2}{3}Y\left\{1 - \frac{c^3}{b^3} + \ln\frac{c^3}{a^3} + \frac{H}{E}\left(\frac{c^3}{a^3} - \frac{c^3}{b^3}\right)\right\}$$

where Y denotes the uniaxial yield stress of the material.

5.2 The internal radius of a thick spherical shell increases from an initial value a_0 to a current value a when the shell is expanded into the fully plastic range by the application of internal pressure. The uniaxial stress-strain curve of the material is expressed by the equation $\sigma = F(\varepsilon)$. Neglecting elastic strains, show that the condition for plastic instability may be written as

$$\phi(\rho^3 - 1) = \phi\left\{\frac{a_0^3}{b_0^3}(\rho^3 - 1)\right\} \qquad \phi(x) = \frac{F[\frac{2}{3}\ln(1 + x)]}{1 + x}$$

where $\rho = a/a_0$ and b_0 is the initial external radius of the shell. Assuming $F(\varepsilon) = C\varepsilon^n$, where C and n are constants, compute the hoop strain on $r = a$ at instability for $b_0/a_0 = 3$ and $n = 0.2$.

   Answer: $\varepsilon_\theta = 0.154$.

---

† A finite element analysis of cracked plates in plane strain has been presented by J. L. Swedlow, *Int. J. Fract. Mech.*, **5**: 33 (1969). An analysis for a rolling disc has been given by C. Anand and H. Shaw, *Int. J. Mech. Sci.*, **19**: 37 (1977).

5.3 A multilayer cylinder having an internal radius a_0 and an external radius a_n is made up of n thick-walled tubes of the same material with a shear yield stress k. The compound cylinder is assembled by shrink-fit, and is designed in such a way that yielding occurs simultaneously in all cylinders when subjected to uniform internal pressure at $r = a_0$. Assuming Tresca's yield criterion, show that the yield pressure is a maximum when the individual cylinders have the same wall ratio, and that the maximum value of the pressure is

$$p = nk \left\{ 1 - \left(\frac{a_0}{a_n} \right)^{2/n} \right\}$$

5.4 A compound tube is made up of two thick tubes, the internal radius of the inner tube and the external radius of the outer tube being a and b respectively. The shear yield stress of the inner and outer tubes are $m^2 k$ and k respectively. If yielding occurs simultaneously in the two cylinders on the application of internal pressure, according to Tresca's criterion, show that the common radius corresponding to a maximum pressure is $\sqrt{mab}$, and that the optimum value of the pressure is

$$p = k \left(1 - 2m \frac{a}{b} + m^2 \right)$$

If the tubes are made of the same material, with $m = 1$, show that the initial radial interference required for the optimum design is

$$\delta = (b - a) \frac{2k}{E} \sqrt{\frac{a}{b}}$$

5.5 In a closed-ended thick-walled tube subjected to internal pressure, it is a good approximation to assume $\sigma_z = \frac{1}{2}(\sigma_r + \sigma_\theta)$ during the elastic/plastic expansion. Show that the maximum engineering shear strain γ in the elastic region, and the displacement u_b at the external surface $r = b$ are then given by

$$\gamma = \frac{kc^2}{Gr^2} \qquad u_b = \frac{1}{2} \left(\frac{2 - v}{1 + v} \right) \frac{kc^2}{Gb}$$

where k is the initial yield stress in pure shear, and c the radius to the elastic/plastic boundary. Assuming the same expression for γ to hold in the plastic region, and the engineering shear stress-strain curve to be given by $\tau = k(G\gamma/k)^n$, $\gamma \geq k/G$, where n is a constant, show that the pressure applied at $r = a$ is

$$p = k \left\{ 1 - \frac{c^2}{b^2} + \frac{1}{n} \left[\left(\frac{c}{a} \right)^{2n} - 1 \right] \right\}$$

5.6 A thick-walled tube with closed ends is made of a work-hardening material for which the engineering shear stress-strain curve is expressed by the equation $\tau = k(G\gamma/k)^n$, $\gamma \geq k/G$, where n is an empirical constant. Assuming $\sigma_z = \frac{1}{2}(\sigma_r + \sigma_\theta)$ throughout the tube as in the preceding problem, and estimating ε_z from the stress-strain relation in the elastic region, show that the radial displacement in the plastic region is given by

$$\frac{Eu}{kr} = (5 - 4v) \frac{c^2}{2r^2} - (1 - 2v) \left\{ \frac{3}{2n} \left[\left(\frac{c}{r} \right)^{2n} - (1 - n) \right] - \frac{c^2}{b^2} \right\}$$

where b is the external radius and c the radius to the elastic/plastic boundary. Use the equations of equilibrium and elastic compressibility.

5.7 In the case of a partially plastic open-ended tube expanded by internal pressure, the von Mises yield criterion is closely approximated by the modified Tresca criterion $\sigma_\theta - \sigma_r = 1.1Y$, where Y is the uniaxial yield stress, provided the wall ratio b/a is not too large. For the estimation of the tangential strain, it is a good approximation to assume $\sigma_z = 0$ throughout the tube satisfying the end condition. Using the value of ε_z given by the stress-strain relation in the elastic region, and neglecting work-hardening, show that the displacement in the plastic region is given by

$$\frac{Eu}{Yr} = 0.55\left\{(1 - v)\frac{c^2}{b^2} + (2 - v)\frac{c^2}{r^2} - (1 - 2v)\left(1 + \ln\frac{c^2}{r^2}\right)\right\}$$

5.8 A thick-walled tube expanded by internal pressure under conditions of plane strain has its wall ratio b/a exceeding the critical value given by inequality (37) with $p = p_0$. The material is nonhardening, and obeys Tresca's yield criterion and the associated flow rule. Show that, for sufficiently large values of the radius c to the elastic/plastic boundary, there is an inner plastic zone extending to $r = \rho$ where σ_z is equal to σ_θ, the relationship between c and ρ being

$$\ln\frac{c}{\rho} = \frac{1}{2}\left(\frac{1}{1 - 2v} + \frac{c^2}{b^2}\right)$$

Using c as the time scale, and neglecting positional changes, obtain the circumferential strain rate $\dot{\varepsilon}_\theta$ in the inner plastic region in the form

$$\frac{G\dot{\varepsilon}_\theta}{k} = 2(1 - v)\frac{c}{r^2} - \left(\frac{1 - 2v}{1 + v}\right)\frac{1}{2c}\left(1 - \frac{c^2}{b^2}\right)\left[3 - (1 - 2v)\frac{c^2\rho^2}{b^2 r^2}\right]$$

5.9 A thick-walled tube with closed ends and having a wall ratio b/a is rendered plastic within a radius c by the application of an internal pressure p. If a fraction λp of the applied pressure is then released, and a torque T is superimposed, show that the conditions for yielding not to restart are

$$\left(1 - \frac{\lambda p a^2}{p_e r^2}\right)^2 + \left(\frac{rT}{bT_e}\right)^2 < 1 \qquad a \leqslant r \leqslant c$$

$$\left(\frac{c^2}{r^2} - \frac{\lambda p a^2}{p_e r^2}\right)^2 + \left(\frac{rT}{bT_e}\right)^2 < 1 \qquad c \leqslant r \leqslant b$$

to a close approximation on the basis of the von Mises criterion, where p_e and T_e are the values of p and T required to cause yielding separately. Assuming $b/a = 1.5$, and $c/a = 1.25$, show that yielding occurs either at $r = c$ or at $r = b$, and find the range of values of λ that applies in each case.

Answer: $0 \leqslant \lambda \leqslant 0.37$, $0.37 \leqslant \lambda \leqslant 1$.

5.10 A nonhardening rigid/plastic tube of wall ratio b/a is brought to the fully plastic state by the combined action of an internal pressure p and a twisting moment T under conditions of plane strain. The ratio of the circumferential strain rate to the shear strain rate on the outer surface of the tube at the yield point is denoted by λ. Using the von Mises yield criterion and the Lévy-Mises flow rule, show that the relationship between p and T is given by

$$\frac{p}{p_0} = \frac{\sinh^{-1}(m\lambda) - \sinh^{-1}\lambda}{\ln m}$$

$$\frac{T}{T_0} = \frac{m\sqrt{1 + \lambda^2} - \sqrt{1 + m^2\lambda^2}}{m - 1}$$

where $m = b^3/a^3$, $p_0 = 2k\ln(b/a)$ and $T_0 = \frac{2}{3}\pi k(b^3 - a^3)$. Verify that the $(p/p_0, T/T_0)$ curve does not differ appreciably from a quadrant of a circle for small to moderate values of the wall ratio.

5.11 In a thick-walled spherical shell subjected to thermal loading, caused by a steady state outward flow of heat, and a relatively small internal pressure, yielding begins at the inner radius $r = a$ and spreads outward with increasing pressure p. Considering the elastic part of radius ratio b/c, which is at the onset of yielding under a combination of thermal gradient and internal pressure (at $r = c$), show that

$$\left(1 - \frac{a}{b}\right)\left[\frac{3p}{Y} + 2\left(1 - \frac{c^3}{b^3} + \ln\frac{c^3}{a^3}\right)\right] = \frac{\beta E}{Y}\left(\frac{a}{c}\right)\left(1 - \frac{c}{b}\right)^2\left(2 + \frac{c}{b}\right)$$

where β equals $\alpha/(1 - v)$ times the temperature difference between the radii $r = a$ and $r = b$. Assuming

$b/a = 3$, and $p/\beta E = \frac{1}{6}$ at the initial yielding, find the range of values c/a for which the above relationship should hold in the elastic/plastic range.

Answer: $1 \leqslant c/a \leqslant 1.082$.

5.12 A closed-ended thick-walled tube haivng a wall ratio R is subjected to an internal pressure p in the presence of a steady state temperature difference ΔT between the internal and external surfaces. The tube is completely elastic, and the flow of heat is radially outward, the coefficient of linear expansion being denoted by α. Setting $\beta = \alpha \Delta T/2(1 - v)$, show that $\sigma_\theta - \sigma_z$ has a maximum value at $r = r_1$ when $p < \beta E$, and $\sigma_z - \acute{\sigma}_r$ has a minimum value at $r = r_2$ when $p > \beta E$, such that

$$\frac{r_1^2}{a^2} = \left(1 - \frac{p}{\beta E}\right)\frac{R^2 \ln R^2}{R^2 - 1} \qquad \frac{r_2^2}{a^2} = \left(\frac{p}{\beta E} - 1\right)\frac{R^2 \ln R^2}{R^2 - 1}$$

where a is the internal radius of the tube. Assuming $R = 2$, find the range of values of $p/\beta E$ in each case for which the stationary stress difference would occur within the tube.

Answer: $0 \leqslant p/\beta E \leqslant 0.46$, $1.54 \leqslant p/\beta E \leqslant 3.16$.

5.13 An annular plate of inner radius a and outer radius b is subjected to a steady state radial temperature variation which is sufficient to render the plate partially plastic. The material is nonhardening and obeys Tresca's yield criterion with a uniaxial yield stress Y. Show that the plastic zone grows outward from the inner boundary, the radius to the elastic/plastic interface being c when the temperature at $r = a$ exceeds that at $r = b$ by ΔT, such that

$$\frac{E\alpha \Delta T}{Y} = \frac{(2 - a/c + ac/b^2) \ln (b^2/a^2)}{\ln (b^2/c^2) - (1 - c^2/b^2)}$$

where α is the coefficient of linear expansion. Show also that a second plastic zone begins to form at the outer radius when

$$\left(1 + \frac{ac}{b^2}\right) \ln \frac{b^2}{c^2} = \left(3 - \frac{a}{c} + \frac{2ac}{b^2}\right)\left(1 - \frac{c^2}{b^2}\right)$$

5.14 The temperature difference across the radial thickness of the annular plate of the preceding problem is increased beyond the point at which a second plastic zone begins to spread inward from the outer boundary. Denoting the radius to the outer elastic/plastic interface by ρ, and neglecting the effects of temperature on the material properties, show that

$$\frac{E\alpha \Delta T}{Y} = \frac{(1 + ac/\rho^2) \ln (b^2/a^2)}{1 - c^2/\rho^2}$$

where c and ρ are related by the equation

$$\left(1 + \frac{a}{c}\right) \ln \frac{\rho^2}{c^2} = \left(\frac{\rho^2}{c^2} - 1\right)\left(3 - \frac{a}{c} - \ln \frac{b^2}{c^2}\right)$$

Assuming $b/a = 2$, find the range of values of c/a for which there will be two separate plastic zones.

Answer: $1.15 < c/a < 1.46$.

5.15 An annular disc of uniform thickness is rendered fully plastic by the application of a uniform radial' pressure p at the inner boundary $r = a$. If the material yields according to von Mises criterion, show that the stress distribution may be expressed as

$$\sigma_r = -2k \sin\left(\frac{\pi}{6} + \phi\right) \qquad \sigma_\theta = 2k \sin\left(\frac{\pi}{6} - \phi\right)$$

where ϕ is a function of the radius r. Using the equilibrium equation and the boundary conditions, prove that the radius ratio b/a is given by

$$\frac{b^2}{a^2} = \frac{2}{\sqrt{3}} \cos \alpha \exp\left[\sqrt{3}\left(\frac{\pi}{6} + \alpha\right)\right]$$

where α is the value of ϕ at $r = a$. Determine the range of values of b/a for which the solution would be valid.

Answer: $b/a \leqslant 2.96$.

5.16 A uniform annular disc of internal radius a and external radius b is subjected to a uniform radial tensile stress σ along the outer edge. The material is incompressible and hardens linearly with a plastic modulus H. Assuming Tresca's yield criterion and the associated flow rule, show that the radial stress in the plastic region for a prestrained material is given by

$$\frac{\sigma_r}{Y} = 1 - \frac{a}{r} + \frac{na}{2r}\left[\left(\ln\frac{c}{a}\right)^2 - \left(\ln\frac{c}{r}\right)^2\right]$$

approximately, where c is the radius to the elastic/plastic boundary and $n = H/(E + H)$. Show also that the applied tension is given by

$$\frac{\sigma}{Y} = 1 - \frac{a}{2c}\left(1 + \frac{c^2}{b^2}\right)\left[1 - \frac{n}{2}\left(\ln\frac{c}{a}\right)^2\right]$$

5.17 A gradually increasing radial pressure p is applied round the edge of a circular hole of radius a contained in an infinite plate of uniform thickness. The uniaxial stress-strain curve of the material in the plastic range can be represented by $\sigma = Y(E\varepsilon/Y)^n$, $\varepsilon \geqslant Y/E$, where n is an empirical constant. Assuming the material to be incompressible, and adopting Tresca's yield criterion with the Hencky stress-strain relation, show that for small elastic/plastic expansions,

$$\frac{p}{Y} = s_0\left[\frac{3(1 - n)}{4 - 2(1 + 3n)s_0}\right]^{4n/(1 + 3n)}$$

where s_0 is the ratio of the applied pressure to the current yield stress at $r = a$. Show also that s_0 is related to the radius c to the elastic/plastic boundary by the equation

$$\ln\frac{c}{a} = \frac{1 - n}{1 + 3n}\left(s_0 - \frac{1}{2}\right) + \frac{8n}{(1 + 3n)^2}\ln\left[\frac{3(1 - n)}{4 - 2(1 + 3n)s_0}\right]$$

5.18 A thin infinite plate of uniform initial thickness h_0 contains a circular hole of initial radius a_0. The hole is finitely expanded by radial pressure, so that the plate is plastic within a radius c. The material is rigid work-hardening, and the uniaxial stress-strain curve is approximated by $\sigma = Y\exp(n\varepsilon)$ over the relevant range, where n is a constant. Using Tresca's yield criterion and the associated flow rule, show that the plate thickens within a radius $\rho = 0.607c$, the thickness h of an element initially at a radius r_0 being given by

$$\frac{h_0}{h} = \frac{1}{2 + n}\left\{\left(\frac{\rho}{r_0}\right)^{n/(1 + n)} + (1 + n)\left(\frac{r_0}{\rho}\right)^{(2 + n)/(1 + n)}\right\}$$

If the current radius of the hole is denoted by a, show that the relationship between a and a_0 is

$$\frac{a}{\rho} = \frac{1}{2 + n}\left\{1 + (1 + n)\left(\frac{a_0}{\rho}\right)^{(2 + n)/(1 + n)}\right\} \qquad \frac{1}{2} \leqslant \frac{a}{\rho} \leqslant 1$$

5.19 A uniform disc of external radius b, and having a concentric circular hole of radius a, is rotated about its central axis until the stress distribution is just fully plastic. The material is non-hardening, and yields according to the Tresca criterion. If the disc is subsequently brought to rest, show that the distribution of residual hoop stress for elastic unloading is given by

$$\frac{\sigma_\theta}{Y} = \left\{\frac{a}{b} - \frac{1 + 3v}{8}\left(1 + \frac{a^2}{b^2} - \frac{3r^2}{b^2}\right) - \frac{3}{8}(3 + v)\frac{a^2}{r^2}\right\}\bigg/\left(1 + \frac{a}{b} + \frac{a^2}{b^2}\right)$$

Assuming $v = 1/3$, show that a secondary yielding cannot occur on unloading if the radius ratio of the disc is less than $2 + \sqrt{2}$.

5.20 Considering the forces acting on a typical element of a disc of variable thickness h rotating with an angular velocity ω, obtain the radial equilibrium equation

$$\frac{\partial}{\partial r}(hr\sigma_r) = h(\sigma_\theta - \rho\omega^2 r^2)$$

A disc of uniform strength is defined as one in which $\sigma_r = \sigma_\theta = \sigma$ throughout the disc. Evidently, the disc must be solid, and have radial loading at the periphery $r = b$ applied through blades. Show that σ is independent of r in both elastic and plastic ranges, and that the thickness must vary from h_0 at $r = b$ according to the formula

$$h = h_0 \exp\left\{\frac{\rho\omega^2}{2\sigma}(b^2 - r^2)\right\}$$

Prove that the total mass of the disc is $2\pi(h_c - h_0)\sigma/\omega^2$, where h_c is the thickness at the center.

5.21 A nonhardening hollow disc consists of an inner part of uniform thickness h_c extending from $r = a$ to $r = c$, and an outer part of uniform strength extending from $r = c$ to $r = b$ with an external thickness h_0. The thickness is made discontinuous at $r = c$ in order to satisfy the condition of continuity of the radial force transmitted across this radius. Show that h_c is a minimum for a given fully plastic angular velocity ω if $\rho\omega^2 c^2$ is equal to the uniaxial yield stress Y, and that the minimum thickness is given by

$$\frac{h_0}{h_c} = \left\{\frac{2}{3} - \sqrt{\frac{\rho\omega^2 a^2}{Y}}\left(1 - \frac{\rho\omega^2 a^2}{3Y}\right)\right\}\exp\left\{\frac{1}{2}\left(1 - \frac{\rho\omega^2 b^2}{Y}\right)\right\}$$

5.22 A rotating hollow disc of internal radius a and external radius b is designed to have a uniform inner part of thickness h_c and a doubly conical outer part with a thickness h_0 at the periphery. The radius c at which the profile of the disc changes is so chosen that the radial force transmitted across this radius in the fully plastic state is independent of the cone angle. Show that

$$\frac{c}{b} = \frac{1}{3}\left\{\left(\frac{6}{\lambda} - 2\right)^{1/2} - 1\right\} \qquad \lambda = \frac{\rho\omega^2 b^2}{3Y} \leqslant 1$$

where Y is the constant uniaxial yield stress, the thickness ratio associated with this design being

$$\frac{h_c}{h_0} = \frac{1 + \lambda b/c - \lambda c^2/b^2}{1 - \lambda c^2/b^2 - (a/c)(1 - \lambda a^2/b^2)}$$

5.23 The thickness of a rotating hollow disc of internal radius a and external radius b varies according to the power law $h = h_0(b/r)^m$, where h_0 and m are constants. The disc is to be designed in such a way that the fully plastic state involves a uniform hoop stress $\sigma_\theta = Y$ throughout, and a radial stress $\sigma_r = Y$ at $r = b$ due to an external edge loading. Show that $mY > \rho\omega^2 b^2$ for the radial stress to decrease inward from the outer edge, where ω is the fully plastic angular velocity of the disc. Establish the relationship

$$\frac{\rho\omega^2 b^2}{Y}\left\{\left(\frac{a}{b}\right)^2 - \left(\frac{a}{b}\right)^{m-1}\right\} = \frac{3-m}{m-1}\left\{m\left(\frac{a}{b}\right)^{m-1} - 1\right\} \qquad m \neq 3$$

Assuming $b/a = 2.5$, determine the value of m that corresponds to $\rho\omega^2 b^2$ equal to Y.

Answer: $m = 2.16$.

5.24 A long hollow cylinder made of a rigid ideally plastic material has an internal radius a and an external radius b. The cylinder is rendered fully plastic by rotation about its axis with an angular velocity ω. The stress and velocity distributions at the yield point may be expressed in terms of a parameter c, such that $\sqrt{3}\,c^2/a^2$ is equal to the ratio of the maximum engineering shear strain rate at $r = a$ to the magnitude of the axial strain rate. Using the von Mises yield criterion and the Lévy-Mises flow rule, show that

$$\frac{\rho\omega^2(b^2 - a^2)}{4Y} = \frac{1}{2\sqrt{3}}\ln\left\{\frac{b^2[c^2 + \sqrt{a^4 + c^4}]}{a^2[c^2 + \sqrt{b^4 + c^4}]}\right\} = \frac{\sqrt{b^4 + c^4} - \sqrt{a^4 + c^4}}{b^2 + a^2}$$

5.25 A rigid circular disc of radius c, rotating horizontally about its central axis, is provided with several protruding rays of constant width symmetrically arranged round the periphery. Each ray is flat at the bottom and is linearly tapered from a depth h_0 at the root $r = a$ to a knife-edged tip at $r = b$. Due to the centrifugal action, a typical cross section of the ray is subjected not only to a radial force N but also to a bending moment M. Show that yielding begins at $r = a$ for $b \leqslant 7a$ and at $r = b/7$ for $b \geqslant 7a$, when the angular velocity ω satisfies the relations

$$\rho\omega^2 = \frac{12Y}{(b-a)(5b+7a)} \quad (b \leqslant 7a) \qquad \rho\omega^2 b^2 = \frac{7}{3}Y \quad (b \geqslant 7a)$$

5.26 Suppose that the speed of rotation of the tapered ray of the preceding problem is increased beyond the elastic limit, so that certain cross sections of the ray become partially plastic. For a nonhardening material, plastic collapse would occur at a critical value of the speed for which a cross section first becomes fully plastic. If the radius to the fully plastic section is denoted by ξb, and the corresponding speed factor $\rho\omega^2 b^2/Y$ is denoted by 6λ, show that

$$\lambda^2(1-\xi)^2(1+2\xi)^2 + \lambda(1-\xi^2) = 1 \qquad a/b \leqslant \xi \leqslant 1$$

Determine ξ that minimizes the speed factor, and hence estimate the collapse value of λ.

Answer: $\xi = 0.18$, $\lambda = 0.59$ ($b \geqslant 5.5a$).

5.27 A curved bar of uniform small thickness h is bounded by circular arcs of radii a and b on the concave and convex sides respectively. The plate is bent in its own plane by equal and opposite couples M, applied to the end faces $\theta = \pm\alpha$, so as to increase its curvature. If the material is ideally plastic obeying Tresca's yield criterion, and the extent of the plastic zone is sufficiently small, show that

$$M = \frac{kh}{N}\left\{b^2\left(1 - \frac{c^2}{b^2}\right)^2\left(1 - \frac{a}{2c}\right) - 2ac\left(\ln\frac{b}{c}\right)^2\right\} - kh(c^2 - a^2)$$

where k is the yield stress in pure shear, c the radius to the elastic/plastic interface, and

$$N = \ln\frac{b^2}{c^2} - \left(1 - \frac{c^2}{b^2}\right)$$

5.28 In the plane stress bending of a curved bar considered in the preceding problem, the applied moment M is assumed large enough to produce two distinct plastic zones defined by $a \leqslant r \leqslant c$ and $\rho \leqslant r \leqslant b$. Show that the radii c and ρ to the elastic/plastic boundaries are related by the equation

$$\left(1 + \frac{ac}{\rho^2}\right)\ln\frac{\rho^2}{c^2} = \left(1 - \frac{c^2}{\rho^2}\right)\left(3 - \frac{a}{c} - \ln\frac{b^2}{\rho^2}\right)$$

and that the corresponding bending moment is given by the formula

$$M = kh\left[\rho^2\ln\frac{b}{c} + \frac{a}{2c}(\rho^2 - c^2) + \frac{1}{2}(b^2 - 3\rho^2 + 2a^2)\right]$$

Compute the values of c/a and M/kha^2 in the fully plastic state when $b/a = 2$.

Answer: $c/a = 1.46$, $M/kha^2 = 0.474$.

5.29 A thin annular plate of internal radius a and external radius b is rendered partially plastic by the application of a radial pressure p and a uniform tangential stress mk ($0 < m < 1$) round the edge of the hole. If the radius to the elastic/plastic boundary is c, and the material is ideally plastic obeying the Tresca criterion, show that

$$\frac{p}{k} = \sqrt{1 - m^2} - \frac{1}{b^2}\sqrt{c^4 - m^2 a^4} + \ln\left\{\frac{c^2 + \sqrt{c^4 - m^2 a^4}}{a^2(1 + \sqrt{1 - m^2})}\right\}$$

so long as the inequality $p/k \leqslant 1 + \sqrt{1 - m^2}$ is satisfied. Find the greatest value of b/a required by the validity of the solution for all elastic/plastic stages when $m = 0$, 0.5, and 1.0.

Answer: $b/a = 2.72$, 3.23, 2.28.

5.30 Using Tresca's associated flow rule, show that the radial and circumferential plastic strains in the partially plastic plate of the preceding problem are given by

$$-\varepsilon_r^p = \varepsilon_\theta^p = \frac{2k}{Er^2}\left[\sqrt{c^4 - m^2 a^4} - \sqrt{r^4 - m^2 a^4}\right] \qquad a \leqslant r \leqslant c$$

From the condition of positive rate of plastic work, show that no plastic element will unload in an infinite plate if $k(dm/dp)$ is greater than $-f(c)$ for $\dot{p} > 0$, and less than $-f(a)$ for $\dot{p} < 0$, where

$$\frac{1}{f(r)} = \left\{\frac{1 - c^2/b^2}{c^2\sqrt{r^4 - m^2 a^4}} + \frac{1}{a^4(1 + \sqrt{1 - m^2})} - \frac{1}{c^2(c^2 + \sqrt{c^4 - m^2 a^4})}\right\}ma^4$$

Note that one of the two loading parameters p and m may decrease during continued loading of the plastic region.

THEORY OF THE SLIPLINE FIELD

6.1 Formulation of the Plane Strain Problem

In the preceding chapters, we have considered problems in which the plastic material is severely constrained by the adjacent elastic material so that the elastic and plastic strains are usually of comparable magnitudes. When the plastic zone grows to a sufficient extent, the constraint ceases to apply, and the plastic material then has freedom to flow in an unrestricted manner. In a number of important practical problems, such as drawing, extrusion, indentation, and piercing, the unrestricted plastic flow begins while the body is still partly plastic. In such cases, the elastic strain soon becomes negligible throughout the plastic zone except in a certain transition region bordering the elastic zone. It is therefore a reasonable approximation to regard the material as rigid/plastic for the determination of the stress and velocity distributions in the plastically deforming region. If the transition region is sufficiently narrow, no significant error is introduced by the neglect of elastic strains.

(i) *Basic equations* Consider the plastic flow behavior of a rigid/plastic body under conditions of plane strain. The flow is everywhere parallel to the xy plane, and the velocity field is independent of z. Evidently, the rate of extension vanishes in the z direction, which coincides with a principal axis of the strain rate. Since the material is incompressible (the elastic strains being zero), the principal strain rates in the xy plane are equal in magnitude but opposite in sign. Each incremental deformation therefore consists of a pure shear in the plane of plastic flow. For an isotropic material, the state of stress at each point is also a pure shear, together with a hydrostatic stress whose value must be equal to σ_z. It follows that

$$\sigma_z = \tfrac{1}{2}(\sigma_1 + \sigma_2) = \tfrac{1}{2}(\sigma_x + \sigma_y) \tag{1}$$

where σ_1 and σ_2 are the principal stresses in the plane of plastic flow. This expression is equally valid in nonplastic regions if we assume $v = 0.5$. The stress is then automatically continuous across the plastic boundary. Since yielding is unaffected by the hydrostatic stress, the maximum shear stress in the xy plane must be equal to the shear yield stress k. The yield criterion may therefore be written as

$$(\sigma_x - \sigma_y)^2 + 4\tau_{xy}^2 = 4k^2 \tag{2}$$

where k is equal to $Y/2$ for the Tresca criterion and $Y/\sqrt{3}$ for the von Mises criterion. If the material is assumed as nonhardening, k has a constant value throughout the plastic region. In the absence of body forces, the equations of equilibrium for quasi-static deformation reduce to

$$\frac{\partial \sigma_x}{\partial x} + \frac{\partial \tau_{xy}}{\partial y} = 0 \qquad \frac{\partial \tau_{xy}}{\partial x} + \frac{\partial \sigma_y}{\partial y} = 0 \tag{3}$$

If the nonzero components of the velocity are denoted by v_x and v_y, the condition of incompressibility of the material is expressed by the equation

$$\frac{\partial v_x}{\partial x} + \frac{\partial v_y}{\partial y} = 0 \tag{4}$$

Since in an isotropic rigid/plastic material the principal axes of stress and strain rate must coincide, there follows the stress-strain relation

$$\frac{2\dot{\gamma}_{xy}}{\dot{\varepsilon}_x - \dot{\varepsilon}_y} = \frac{2\tau_{xy}}{\sigma_x - \sigma_y}$$

Each side of this equation is equal to the tangent of twice the counterclockwise angle made by either of the common principal axes with respect to the x axis. Expressing the strain rates in terms of the velocity gradients, we get

$$(\sigma_x - \sigma_y)\left(\frac{\partial v_x}{\partial y} + \frac{\partial v_y}{\partial x}\right) = 2\tau_{xy}\left(\frac{\partial v_x}{\partial x} - \frac{\partial v_y}{\partial y}\right) \tag{5}$$

The set of five equations (2) to (5) form the basis for the calculation of the stress and velocity distributions in the plastic region. If the boundary conditions involve only the stresses, the equilibrium equations (3) and the yield criterion (2) are sufficient to determine the stresses throughout the plastic region.† Once the stresses have been found, the velocity components can be determined from equations (4) and (5). Such problems of plane plastic flow may be regarded as statically determinate. In the

---

† In the nonplastic region, the equilibrium equations are supplemented by the compatibility equation $\nabla^2(\sigma_x + \sigma_y) = 0$, which is independent of the elastic constants.

majority of practical problems, however, some of the boundary conditions are kinematical in nature, involving the velocities. The solution of such problems evidently requires consideration of the stress and velocity equations simultaneously. The plane strain problem is, therefore, not necessarily statically determined, although there are as many stress equations as the number of unknown stress components.

The significance of the nonhardening rigid/plastic theory for plane strain in relation to a compressible elastic/plastic solid may be briefly outlined. If the material obeys Tresca's yield criterion and the associated flow rule, (1) does not apply, but (2) holds in those parts of the plastic zone where σ_z is the intermediate principal stress. For a statically determinate problem, the corresponding in-plane stresses at any stage are exactly predicted by the rigid/plastic analysis.† In the remainder of the plastic zone, the yield criterion (2) breaks down, and the actual stresses differ appreciably from their rigid/plastic values. For a Mises material on the other hand, the associated Prandtl-Reuss flow rule indicates that (2) is not valid anywhere in the plastic zone unless $v = 0.5$. However, σ_z approaches the value (1) as the magnitude of the in-plane principal plastic strains progressively increases. The rigid/plastic stress distribution therefore holds increasingly well within the plastic zone, but may be appreciably in error over a region bordering the elastic/plastic interface. The applied load should converge monotonically to its theoretical yield point value as the deformation continues in the plastic range.

(ii) *Characteristics in plane strain* Through each point in the plane of plastic flow, we may consider a pair of orthogonal curves along which the shear stress has its maximum value k. These curves are called *sliplines* or shear lines. We shall denote them by α and β lines, following the convention that the line of action of the algebraically greatest principal stress makes a counterclockwise angle of $\pi/4$ with the α direction.‡ Let ϕ be the counterclockwise orientation of the α line with the x axis at a typical point P (Fig. 6.1). If the hydrostatic stress at this point is denoted by $-p$, Mohr's circle for the stress furnishes

$$\sigma_x = -p - k\sin 2\phi \qquad \sigma_y = -p + k\sin 2\phi$$

$$\tau_{xy} = k\cos 2\phi \tag{6}$$

The lines joining the pole P^* to the highest and lowest points of the circle are parallel to the tangents to the sliplines at P. The algebraically greatest and least principal stresses at P are equal to $-p + k$ and $-p - k$ respectively. In a small curvilinear element formed by two intersecting pairs of neighboring sliplines around P, the stresses acting across its faces will be directed as shown.

The quantities p and ϕ define the state of stress in each plastic element. In

---

† For a statically indeterminate problem, the stress distribution cannot be determined by a strict rigid/plastic analysis before the yield point is reached.

‡ The algebraically greatest principal stress direction therefore bisects the angle between the α and β directions taken as a right-handed pair of curvilinear axes.

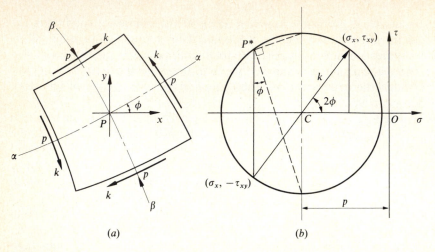

Figure 6.1 Stresses on a curvilinear element bounded by sliplines. (a) Physical plane; (b) stress plane.

view of (6), the equilibrium equations (3) for a nonhardening material become

$$\frac{\partial p}{\partial x} + 2k\left(\cos 2\phi \, \frac{\partial \phi}{\partial x} + \sin 2\phi \, \frac{\partial \phi}{\partial y}\right) = 0$$

$$\frac{\partial p}{\partial y} + 2k\left(\sin 2\phi \, \frac{\partial \phi}{\partial x} - \cos 2\phi \, \frac{\partial \phi}{\partial y}\right) = 0$$
(7)

We now proceed to enquire if the stress equations are hyperbolic, in which case there exist certain curves called characteristics across which the spatial derivatives of p and ϕ may be discontinuous. Suppose that the values of p and ϕ are given along some curve C, so that the differentials dp and $d\phi$ are known at all points on C. These quantities are expressed by the equations

$$dp = \frac{\partial p}{\partial x} dx + \frac{\partial p}{\partial y} dy \qquad d\phi = \frac{\partial \phi}{\partial x} dx + \frac{\partial \phi}{\partial y} dy$$
(8)

where dx and dy are considered along C. The set of four equations given by (7) and (8) uniquely define the first derivatives of $p/2k$ and ϕ, unless the determinant of the coefficients of these derivatives vanishes. Hence C will be a characteristic if

$$\begin{vmatrix} dx & dy & 0 & 0 \\ 0 & 0 & dx & dy \\ 1 & 0 & \cos 2\phi & \sin 2\phi \\ 0 & 1 & \sin 2\phi & -\cos 2\phi \end{vmatrix} = 0$$

When this is satisfied, the derivatives cannot be determined without further information. The derivatives may therefore be discontinuous across C. Expanding the determinant, we obtain a quadratic expression which can be written in the form

$$(dy - \tan \phi \, dx)(dy + \cot \phi \, dx) = 0$$
(9)

Hence there are two distinct characteristics through each point having slopes $\tan \phi$ and $-\cot \phi$ with respect to the x axis. The characteristics therefore coincide with the sliplines, which are inclined at angles ϕ and $\pi/2 + \phi$ with the x axis. The velocity equations may be similarly examined by considering the set of equations involving the derivatives of the components v_x and v_y. Two of these equations are provided by the differential relations

$$dv_x = \frac{\partial v_x}{\partial x} dx + \frac{\partial v_x}{\partial y} dy \qquad dv_y = \frac{\partial v_y}{\partial x} dx + \frac{\partial v_y}{\partial y} dy$$

where dv_x and dv_y are assumed known along a given curve. The remaining equations consist of the incompressibility condition (4) and the isotropy condition (5). Since $\sigma_x - \sigma_y = -2k \sin \phi$ and $\tau_{xy} = k \cos 2\phi$, the condition for nonuniqueness of the first derivatives of v_x and v_y becomes

$$\begin{vmatrix} dx & dy & 0 & 0 \\ 0 & 0 & dx & dy \\ 1 & 0 & 0 & 1 \\ \cos 2\phi & \sin 2\phi & \sin 2\phi & -\cos 2\phi \end{vmatrix} = 0$$

which again leads to Eq. (9). Hence the velocity equations are also hyperbolic, and the characteristics are the sliplines. In plane plastic flow, therefore, the characteristics of the stress and the velocity coincide, the characteristic directions at each point being those of the maximum shear stress or shear strain rate. Across these curves, the normal derivatives of the stress and velocity components may be discontinuous.

(iii) *Equations of Hencky and Geiringer* For the solution of physical problems, it is convenient to set up differential relations holding along the sliplines. Let the x and y axes be taken along the tangents to the α and β lines respectively at the considered point. Since this corresponds to $\phi = 0$, Eqs. (7) become

$$\frac{\partial p}{\partial x} + 2k \frac{\partial \phi}{\partial x} = 0 \qquad \frac{\partial p}{\partial y} - 2k \frac{\partial \phi}{\partial y} = 0$$

Thus, the tangential derivative of $p + 2k\phi$ vanishes along an α line, and that of $p - 2k\phi$ vanishes along a β line. Evidently, this result is independent of the actual orientation of the sliplines at any given point. Hence

$$p + 2k\phi = \text{const along an } \alpha \text{ line}$$
$$p - 2k\phi = \text{const along a } \beta \text{ line} \tag{10}$$

These are known as the *Hencky equations*,[†] which are simply the equilibrium equations expressed along the sliplines. It follows from (10) that the hydrostatic

---

[†] H. Hencky, *Z. angew. Math. Mech.*, **3**: 241 (1923). The Hencky equations are special cases of the more general relations for ideal soils derived earlier by F. Kötter, *Berl. Akad. Berichte*, p. 229 (1903).

pressure varies along a slipline by an amount equal to $2k$ times the angle turned through along it, an increase in pressure corresponding to a clockwise turning of an α line or a counterclockwise turning of a β line.

The velocity equations can be reduced in a similar manner by introducing the components u and v along the α and β lines respectively. The rectangular components v_x and v_y are expressed in terms of u and v by the equations

$$v_x = u \cos \phi - v \sin \phi$$
$$v_y = u \sin \phi + v \cos \phi$$

(11)

When the x and y axes are taken along the tangents to the α and β lines, $\sigma_x = \sigma_y = - p$ and $\tau_{xy} = k$. Equations (4) and (5) then indicate that

$$\left(\frac{\partial v_x}{\partial x} \right)_{\phi = 0} = \left(\frac{\partial v_y}{\partial y} \right)_{\phi = 0} = 0$$

These results imply that the rate of extension vanishes along the sliplines. Substituting for v_x and v_y from (11) into the above derivatives, and then setting $\phi = 0$, we get

$$\frac{\partial u}{\partial x} - v \frac{\partial \phi}{\partial x} = 0 \qquad \frac{\partial v}{\partial y} + u \frac{\partial \phi}{\partial y} = 0$$

Since the derivatives in the above equations are along the tangents to the sliplines when $\phi = 0$, the differential relations along them are obtained as

$$du - v \, d\phi = 0 \qquad \text{along an } \alpha \text{ line}$$
$$dv + u \, d\phi = 0 \qquad \text{along a } \beta \text{ line}$$

(12)

which are due to Geiringer.† It follows from (10) and (12) that the hydrostatic pressure and the tangential velocity remain constant along a straight slipline $\phi = $ const. When both families of sliplines are straight, the stress is uniform in that region, but the velocity is not necessarily uniform.‡

Since the material is incompressible, the conservation of mass requires the normal component of velocity to be continuous across any curve. The tangential component may, however, be discontinuous. The line of discontinuity must be regarded as the limit of a narrow region in which the rate of shearing in the tangential direction is infinitely large. The tangent to the curve therefore coincides with a direction of the maximum shear stress, which means that the discontinuity is a slipline. If the velocity is discontinuous across an α line, v is continuous while u changes by a finite amount Δu on crossing the discontinuity. It follows from (12), applied to each side of the discontinuity, that $d(\Delta u) = 0$ along the discontinuity. The jump in the tangential velocity therefore remains constant along a slipline.

† H. Geiringer, *Proc. 3d Int. Cong. Appl. Mech., Stockholm,* **2**: 185 (1930).

‡ The lines of principal stress under plane strain condition have been discussed by M. A. Sadowsky, *J. Appl. Mech.,* **8**: 74 (1941).

(iv) *Influence of work-hardening* We shall now examine how the slipline field equations are modified when the work-hardening of the material is taken into account. The shear yield stress k at any given particle is a function of the maximum shear strain following the motion of the particle. Let the magnitude of the maximum engineering shear strain rate be denoted by $\dot{\gamma}$. Then the equivalent stress and strain rates are $\sqrt{3}\,k$ and $\dot{\gamma}/\sqrt{3}$ respectively, their ratio being the current slope H of the uniaxial stress-strain curve. This gives $\dot{\gamma} = 3k/H$. If ψ is the counterclockwise angle made by the direction of motion of the particle with the positive α direction, then the rate of extension in the direction of flow is

$$\dot{\varepsilon} = \frac{1}{2}\dot{\gamma}\sin 2\psi = \frac{3\dot{k}}{2H}\sin 2\psi$$

If the resultant velocity of the particle is denoted by w, then $\dot{\varepsilon} = \partial w/\partial s$, which is the space derivative of w along the instantaneous flowline. Since $\dot{k}$ is the rate of change of k following the particle, the above equation becomes

$$\frac{\partial w}{\partial s} = \frac{3}{2H}\left(\frac{\partial k}{\partial t} + w\frac{\partial k}{\partial s}\right)\sin 2\psi \tag{13}$$

where t denotes the time scale. The quantities w and ψ are related to u and v by the equations

$$w = \sqrt{u^2 + v^2} \qquad \psi = \tan^{-1}\frac{v}{u} \tag{14}$$

When the problem is one of steady state, the field quantities at any given point are independent of t. The work-hardening relation (13) may then be written as

$$\frac{dw}{w} = \frac{3dk}{2H}\sin 2\psi \qquad \text{along a flowline} \tag{15}$$

In the case of steady motion, the flowlines are identical to the paths of the particles. These curves are in fact the characteristics across which the normal derivative of k may become discontinuous.

The Geiringer equations (12) are unaffected by the work-hardening, since they aee based on relations that are independent of the yield stress. Hence, the sliplines are still the characteristics of the velocity field. It can be shown that the characteristics of stress and velocity coincide even when the material work-hardens. The Hencky equations must of course be modified to allow for the variation of the yield stress. Substituting from (6) into the equilibrium equations (3), and setting $\phi = 0$ on differentiation as before, we get

$$\frac{\partial p}{\partial x} + 2k\frac{\partial \phi}{\partial x} = \frac{\partial k}{\partial y}$$

$$\frac{\partial p}{\partial y} - 2k\frac{\partial \phi}{\partial y} = \frac{\partial k}{\partial x}$$

where the rectangular axes are now along the local slipline directions. The above equations will be independent of the orientation of the axes if $\partial/\partial x$ and $\partial/\partial y$ are replaced by the tangential derivatives $\partial/\partial s_\alpha$ and $\partial/\partial s_\beta$ along the α and β lines respectively. Hence†

$$dp + 2k\,d\phi = \frac{\partial k}{\partial s_\beta}\,ds_\alpha \qquad \text{along an } \alpha \text{ line}$$

$$\tag{16}$$

$$dp - 2k\,d\phi = \frac{\partial k}{\partial s_\alpha}\,ds_\beta \qquad \text{along a } \beta \text{ line}$$

These relations are complicated because of the terms involving the derivatives of k. Consequently, the hydrostatic pressure generally varies even along a straight slipline. A process of successive approximation will usually be necessary to determine the stress and velocity distributions under given boundary conditions.

Strictly speaking, a velocity discontinuity cannot occur in a work-hardening material for which the strain rate must always remain finite. However, there can be certain narrow transition regions through which the velocity changes rapidly and continuously. If the thickness of such a transition region is neglected for an approximation, the yield stress will be discontinuous across the resulting velocity discontinuity. The line of discontinuity cannot be considered as a slipline for the material on both sides of it, since the unequal values of the shear stress acting across this line will violate the condition of tangential equilibrium. The maximum shear directions are therefore discontinuous across a line of assumed velocity discontinuity. The subsequent discussion of slipline fields will be based on the assumption that the material is ideally plastic.‡

6.2 Properties of Slipline Fields and Hodographs

(i) Geometry of the slipline field A field of orthogonal families of sliplines has certain interesting geometrical properties which are frequently used in the solution of physical problems. We begin our discussion by considering a curvilinear quadrilateral $ABCD$ bounded by the α lines AB and CD and the β lines AC and BD (Fig. 6.2a). In view of (10), the pressure difference between C and B may be written as

$$p_C - p_B = (p_C - p_A) + (p_A - p_B) = 2k(\phi_B + \phi_C - 2\phi_A)$$

$$p_C - p_B = (p_C - p_D) + (p_D - p_B) = 2k(2\phi_D - \phi_B - \phi_C)$$

Since the right-hand sides of the above equations must be equal to one another, we

---

† These equations have been essentially given by D. G. Christopherson, P. L. B. Oxley, and W. B. Palmer, *Engineering*, **186**: 113 (1958). See also H. G. Hopkins, *Problems of Plasticity*, A. Sawczuk (ed.), p. 253, Noordhoff International Publishing, Leyden (1974).

‡ A semi-graphical method of finding the slipline field from an experimentally determined flow field has been discussed by J. Chakrabarty, *Proc. Int. Conf. Prod. Eng.*, Tokyo (1980). See Prob. 6.4 at the end of this chapter.

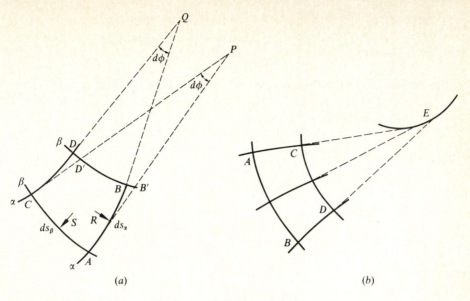

(a)   (b)

Figure 6.2 Slipline field geometries for establishing Hencky's theorems.

obtain the following relationship between the nodal values of ϕ and p:

$$\phi_C - \phi_D = \phi_A - \phi_B \qquad p_C - p_D = p_A - p_B \qquad (17)$$

This is known as *Hencky's first theorem*,[†] which states that the angle between the tangents to a pair of sliplines of one family at the points of intersection with a slipline of the other family is constant along their lengths. In other words, if we pass from one slipline to another of the same family, the angle turned through and the change in hydrostatic pressure are the same along each intersecting slipline.

It follows that if a segment AC of a slipline is straight, the corresponding segment BD of any slipline of the same family will also be straight (Fig. 6.2b). The straight segments must be of equal lengths, since the intersecting curved sliplines are their orthogonal trajectories. Indeed, the curved sliplines have the same evolute E, which is the locus of the centers of curvature along either of them. Both AB and CD may therefore be described by unwinding a taut string from the evolute. The ends of the string in the two cases will be separated from one another by a distance equal to the length of each straight segment. It follows from (12) that the normal component of velocity changes by a constant amount in passing from one straight slipline to another along any intersecting curved slipline.

Let the radii of curvature of the α and β lines be denoted by R and S respectively. These radii will be taken as positive if the α and β lines, regarded as a right-handed pair of curvilinear axes, rotate in the counterclockwise and clockwise

† H. Hencky, *Z. angew. Math. Mech.*, **3**: 241 (1923).

senses respectively. Hence R and S are defined by the equations

$$\frac{1}{R} = \frac{\partial \phi}{\partial s_\alpha} \qquad \frac{1}{S} = -\frac{\partial \phi}{\partial s_\beta} \tag{18}$$

Let the sliplines of Fig. 6.2a be now regarded as infinitesimally close to one another. The lengths of the arcs AB and AC are denoted by ds_α and ds_β, their radii of curvature being denoted by R and S respectively. The angular spans of the arcs AC and BD are each equal to $d\phi$ by Hencky's first theorem. Since the angles turned through by AB and CD are also equal, it follows from geometry that

$$BD \simeq B'D' \simeq (S - ds_\alpha)\, d\phi$$

which is correct to second order. This shows that if $S + dS$ is the radius of curvature of BD, then $dS = -ds_\alpha$. Similarly, if $R + dR$ is the radius of curvature of CD, then $dR = -ds_\beta$. These results may be expressed as

$$\frac{\partial R}{\partial s_\beta} = -1 \qquad \frac{\partial S}{\partial s_\alpha} = -1 \tag{19}$$

These equations state that as we move along a slipline, the radii of curvature of the sliplines of the other family change by the distance traveled. This is known as *Hencky's second theorem*. An alternative form of (19), more suitable for practical calculations, is obtained by combining these equations with (18). Thus

$$\begin{aligned} dS + R\, d\phi &= 0 \qquad \text{along an } \alpha \text{ line} \\ dR - S\, d\phi &= 0 \qquad \text{along a } \beta \text{ line} \end{aligned} \tag{20}$$

If the normal derivative of ϕ is discontinuous across a slipline, (18) indicates that the curvature of the sliplines of the other family is also discontinuous. This means that R may be discontinuous across a β line, and S may be discontinuous across an α line. Since (20) must hold on each side of the discontinuity, the jump in the radius of curvature is constant along the slipline. It follows from (10) that the normal derivative of p is discontinuous across a slipline involving a discontinuity in curvature of the sliplines of the other family.

Since the radius of curvature of AB exceeds that of CD by an amount equal to the arc length AC, the centers of curvature P and Q must lie on the involute of AC or BD. This leads to *Prandtl's theorem*,† which states that as we proceed along a slipline of one family, the centers of curvature of the sliplines of the other family form an involute of the given slipline. The radius of curvature decreases in magnitude as we move toward the concave side of a given family of sliplines. If the plastic zone extends sufficiently far, the radius of curvature eventually vanishes to form a cusp. In that case, the neighboring sliplines of the other family run together, their envelope being a limiting line‡ across which the slipline of the previous

---

† L. Prandtl, *Z. angew. Math. Mech.*, **3**: 401 (1925).

‡ Some general properties of limiting lines have been examined by W. Prager and P. C. Hodge, *Theory of Perfectly Plastic Solids*, p. 150, Wiley and Sons (1951).

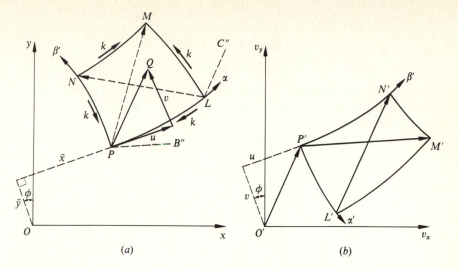

Figure 6.3 Correspondence between slipline field and hodograph. (a) Physical plane; (b) hodograph plane.

family cannot be continued (see Fig. 6.17). The envelope of one family of sliplines is evidently the locus of the cusps of the sliplines of the other family. When the envelope degenerates into a point, the result is a centered fan which occurs frequently in practical applications. The radii of curvature of all sliplines passing through the center of the fan have the same value at this point.†

For the calculation of the rectangular coordinates of the nodal points of a slipline field network, it is often convenient to introduce the Mikhlin‡ variables $(\bar{x}, \bar{y})$, which are related to (x, y) by the equations

$$\bar{x} = x \cos \phi + y \sin \phi$$
$$\bar{y} = -x \sin \phi + y \cos \phi \tag{21}$$

These new variables are the coordinates of a typical point P with respect to axes passing through a fixed origin O and parallel to the slipline directions at P (Fig. 6.3a). The inversion of (21) gives

$$x = \bar{x} \cos \phi - \bar{y} \sin \phi$$
$$y = \bar{x} \sin \phi + \bar{y} \cos \phi \tag{22}$$

from which the rectangular coordinates can be found when $\bar{x}$ and $\bar{y}$ are known

---

† Several properties of slipline fields and certain special solutions have been discussed by A. M. Freudental and H. Geiringer, *Handbuch der Physik*, **6**, Springer Verlag, Berlin (1958).
‡ S. G. Mikhlin, *Akad. Nauk. USSR* (1934); S. A. Khristianovich, *Mat. Sb. Nov. Ser.*, **1**: 511 (1936).

along the sliplines. The variations of $\bar{x}$ and $\bar{y}$ can be immediately written down as

$$d\bar{x} = (\cos \phi \, dx + \sin \phi \, dy) + \bar{y} \, d\phi$$
$$d\bar{y} = (-\sin \phi \, dx + \cos \phi \, dy) - \bar{x} \, d\phi \tag{23}$$

The expression in the parenthesis of the first equation vanishes along a β line, and that of the second equation vanishes along an α line, in view of (9). The above expressions therefore become

$$d\bar{y} + \bar{x} \, d\phi = 0 \qquad \text{along an } \alpha \text{ line}$$
$$d\bar{x} - \bar{y} \, d\phi = 0 \qquad \text{along a } \beta \text{ line} \tag{24}$$

These relations are analogous to (12) and (18). Consider, now, the infinitesimal variation of $\bar{x}$ along an α line and of $\bar{y}$ along a β line. Then the expressions in the parentheses of (23) are equal to the respective arc elements ds_α and ds_β. Hence

$$\frac{\partial \bar{x}}{\partial s_\alpha} = 1 + \frac{\bar{y}}{R} \qquad \frac{\partial \bar{y}}{\partial s_\beta} = 1 + \frac{\bar{x}}{S} \tag{25}$$

These are important relations connecting $\bar{x}$, $\bar{y}$, and the radii of curvature of the sliplines. It may be noted that no special property of the slipline field has been used in the derivation of (24) and (25), which are therefore valid for any two orthogonal families of curves.

(ii) *The hodograph and its significance* The velocity distribution in plane plastic flow is represented graphically by a plane diagram known as the *hodograph*. The resultant velocity at a typical point P in the physical plane is represented in magnitude and direction by the vector $O'P'$ in the hodograph, where O' is a fixed origin known as the pole of the hodograph (Fig. 6.3b). Evidently, the rectangular coordinates of P' in the hodograph plane are the rectangular components of the velocity at P. As P describes the slipline field, P' traces out the hodograph net. To obtain the correspondence between the slipline field and the hodograph, consider the variations dv_x and dv_y in the neighborhood of P. Equations (11) furnish

$$dv_x = (du - v \, d\phi) \cos \phi - (dv + u \, d\phi) \sin \phi$$
$$dv_y = (du - v \, d\phi) \sin \phi + (dv + u \, d\phi) \cos \phi$$

In each of the above equations, the expression in the first parenthesis vanishes along an α line and that in the second parenthesis vanishes along a β line, as may be seen from (12). Hence

$$dv_y/dv_x = \begin{cases} -\cot \phi & \text{along an } \alpha \text{ line} \\ \tan \phi & \text{along a } \beta \text{ line} \end{cases}$$

giving the slopes of the hodograph traces at P'. On the other hand, the slopes of the corresponding sliplines at P are $\tan \phi$ and $-\cot \phi$ respectively. It follows that the tangent to a slipline at any point is orthogonal to the corresponding tangent to its image in the hodograph.† This geometrical property is a consequence of the fact that the change in velocity between the neighboring points on a slipline is directed along its normal, because the rate of extension is zero along the slipline. The angle turned through along a slipline PL is identical to that along its image $P'L'$ in the hodograph. It follows that the geometrical properties of Hencky and Prandtl are equally applicable to the hodograph.

By means of hodographs, tentative slipline fields can be rapidly checked for positive plastic work. The rate of plastic work per unit volume is equal to $k\dot{\gamma}$, where $\dot{\gamma}$ is the engineering shear rate associated with the local α and β directions forming a right-handed pair. According to our convention, the algebraically greater principal stress at P acts along the internal bisector of the angle between the sliplines at P. It follows that the right angle at P must instantaneously decrease, giving a positive value of $\dot{\gamma}$. The expression for the shear rate may be obtained from the relation

$$\dot{\gamma} = \left(\frac{\partial v_x}{\partial y} + \frac{\partial v_y}{\partial x} \right)_{\phi = 0}$$

Substituting from (11), and adopting the usual operators to denote the tangential derivatives, we have

$$\dot{\gamma} = \frac{\partial u}{\partial s_\beta} + \frac{\partial v}{\partial s_\alpha} + \frac{u}{R} + \frac{v}{S} \qquad (26)$$

Let α' and β' denote a right-handed pair of orthogonal directions in the hodograph plane, corresponding to the α and β directions in the physical plane. Figure 6.3 shows that the α' and β' directions are obtained by a 90° clockwise rotation from the α and β directions. The radii of curvature of the α' and β' lines are denoted by R' and S', defined in a similar manner to R and S. The corresponding line elements along the α and α' curves are in the ratio $R:R'$, and those along the β and β' curves are in the ratio $S:S'$. Hence the tangential derivatives along the hodograph lines are given by

$$\frac{\partial}{\partial s_{\alpha'}} = \frac{R}{R'} \frac{\partial}{\partial s_\alpha} \qquad \frac{\partial}{\partial s_{\beta'}} = \frac{S}{S'} \frac{\partial}{\partial s_\beta}$$

It follows from Fig. 6.3 that $(-v, u)$ have the same significance in the hodograph plane as $(\bar{x}, \bar{y})$ have in the physical plane. The hodograph equations corresponding to (25) may therefore be written down as

$$\frac{\partial v}{\partial s_\alpha} + \frac{u}{R} = -\frac{R'}{R} \qquad \frac{\partial u}{\partial s_\beta} + \frac{v}{S} = \frac{S'}{S} \qquad (27)$$

It is interesting to note that the sum of left-hand sides of (27) is identical to the

---

† W. Prager, *Trans. Inst. Tech., Stockholm*, no. 65 (1953). See also A. P. Green, *J. Mech. Phys. Solids*, **2**: 73 (1953).

right-hand side of (26). We therefore arrive at the result[†]

$$\dot{\gamma} = \frac{S'}{S} - \frac{R'}{R} > 0 \tag{28}$$

for the rate of plastic work to be positive. If S and S' have the same sign while R and R' have opposite signs, then $\dot{\gamma}$ is known to be positive from an inspection of the general forms of the two networks. When S'/S and R'/R are both positive, the following graphical method would be useful for checking the slipline field for positive work rate.[‡]

Let $P'L'M'N'$ be a curvilinear mesh of the hodograph corresponding to a slipline mesh $PLMN$. The velocity of M relative to P is equal to the vector $P'M'$, and the velocity of N relative to L is equal to the vector $L'N'$. The rate of plastic work will be positive if the component of $P'M'$ parallel to PM is in the direction of PM, and the component of $L'N'$ parallel to LN is directed opposite to LN. In other words, the scalar product of the vectors PM and $P'M'$ must be positive and that of LN and $L'N'$ must be negative. Thus, if PM'' and LN'' are drawn parallel to $P'M'$ and $L'N'$ respectively in the direction of these vectors, then the angle MPM'' must be acute and the angle NLN'' obtuse in order that the rate of plastic work is positive. If the velocity is discontinuous across a slipline, the jump in the tangential velocity must correspond to a relative sliding that is consistent with the direction of the shear stress across the slipline. The requirement of positive plastic work rate imposes a restriction on the shape of the hodograph in relation to that of the slipline field.

In the solution of special problems, a velocity discontinuity often occurs across a slipline that separates a rigid region from a deforming one. A nonrotating rigid region is mapped into a single point in the hodograph. The particles on the deforming side of the discontinuity are mapped into a circular arc, with center representing the rigid region, and with radius equal to the magnitude of the velocity discontinuity (see Fig. 6.9b). If, on the other hand, the rigid region is rotating with an angular velocity ω, any curve in this region, including the bounding slipline, is mapped as a geometrically similar curve[§] rotated through $90°$ in the same sense as that of ω. This conclusion follows from the fact that the radius vector from the center of rotation to a typical point in the physical plane is orthogonal to the radius vector from the pole to the corresponding point in the hodograph, while the lengths of these radius vectors are in the constant ratio $1:\omega$. If the velocity is discontinuous across the bounding slipline, the deforming side of this slipline will be mapped into a parallel curve at a normal distance equal to the velocity discontinuity.

(iii) *Completeness of solutions* In order to solve any particular problem in plane plastic flow, a slipline field must be found in the region where the deformation is

---

† A. P. Green, *Q. J. Mech. Appl. Math.*, **6**: 223 (1953).

‡ H. Ford, *Advanced Mechanics of Materials*, p. 519, Longman Green and Co. (1963).

§ A. P. Green, *J. Mech. Phys. Solids*, **2**: 73 (1953).

assumed to occur. The proposed field must satisfy all the stress and velocity boundary conditions, as well as the condition of positive rate of plastic work. It is also necessary to ensure that the yield criterion is nowhere violated in the rigid region. A partial solution in which the rigid region has not been examined will be regarded as incomplete. It is strictly an upper bound solution, since the associated velocity field is kinematically admissible. Suppose that the stress distribution of the incomplete solution is now extended to the entire body in a statically admissible manner, so that it is acceptable also as a lower bound solution. Such an extended solution will be regarded as a complete solution.† The yield point load corresponding to a complete solution is the actual one, although the associated deformation mode is not necessarily actual (Sec. 2.6). If a complete solution does not exist, the yield point load for a partly plastic body will naturally be overestimated in accordance with the upper bound theorem.

There may be more than one complete solution to a problem under given boundary conditions. The associated deformation mode in each case is compatible with the stress distribution, which is uniquely defined in the region where deformation can occur in any mode. It follows that a deformation mode associated with the stress field in one complete solution is also compatible with the stress field in any other complete solution. This means that if a region is necessarily rigid in a known complete solution, it must be rigid in all other complete solutions. If it were not so, the deformation mode in another solution would not be compatible with the stress field of the known solution. To determine the extent of the deformable region occupied by the complete set of modes, it is therefore necessary to consider only one complete solution of the problem.‡ The stress distribution in the rigid region of a complete solution cannot be regarded as actual for a solid whose elastic modulus has become infinitely large.

6.3 Stress Discontinuities in Plane Strain

(i) *Conditions at a discontinuity* We have seen that the normal derivative of the stress may be discontinuous across a slipline. The possibility of the stress itself being discontinuous has already been established in connection with the fully plastic bending and torsion discussed in Chap. 3. The general properties of stress discontinuities will now be examined under conditions of plane strain. Equilibrium demands that the normal stress σ_n and the shear stress τ must be continuous across any curve (Fig. 6.4a). If the curve is a line of stress discontinuity, the components σ_t and σ_t' acting parallel to the curve will be different from one another. Since the material on both sides of the discontinuity is assumed as plastic, σ_t and σ_t' are the two possible values of σ satisfying the yield criterion

$$(\sigma - \sigma_n)^2 + 4\tau^2 = 4k^2$$

† J. F. W. Bishop, *J. Mech. Phys. Solids*, **2**: 44 (1953).
‡ J. F. W. Bishop, A. P. Green, and R. Hill, *J. Mech. Phys. Solids*, **4**: 256 (1956).

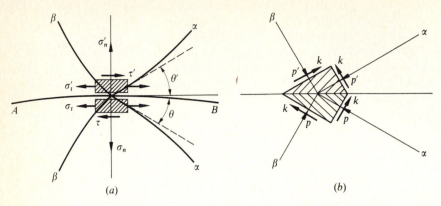

(a)   (b)

Figure 6.4 Stress discontinuity across a surface under conditions of plane strain.

If σ_n and τ are regarded as given, the roots of the above equation may be written as

$$\sigma_t, \sigma_t' = \sigma_n \pm 2\sqrt{k^2 - \tau^2}$$

The difference $\sigma_t - \sigma_t'$ representing the stress discontinuity is therefore of magnitude $4\sqrt{k^2 - \tau^2}$. In Mohr's representation of the stress, two circles of radius k can be drawn through a given point (σ_n, τ), the distance between the centers of the circles being $2\sqrt{k^2 - \tau^2}$, which is the magnitude of the jump in the hydrostatic pressure.†

It is instructive to consider the line of stress discontinuity in relation to the sliplines.‡ Let θ and θ' denote the acute angles made by the α lines with the discontinuity on the lower and upper sides respectively. The mean compressive stress on the two sides of the discontinuity are denoted by p and p'. Equations (6) may be applied to each side of the discontinuity with the x and y axes taken along the tangent and the normal to this curve. It immediately follows that the shear stress will be continuous if $\theta' = \theta$, which means that the sliplines are reflected in the line of stress discontinuity. The condition of continuity of the normal stress then becomes

$$-p - k \sin 2\theta = -p' + k \sin 2\theta$$

The pressure jump $p' - p$ is therefore equal to $2k \sin 2\theta$. If the α lines point to the left instead of to the right, $p' - p$ would be equal to $-2k \sin 2\theta$. The magnitude of the pressure jump in each case is

$$|p' - p| = 2k \sin 2\theta \tag{29}$$

---

† The maximum stress discontinuity of amount $4k$ occurs in the familiar example of the plastic bending of beams.

‡ W. Prager, *Courant Anniversary Volume*, p. 289, Interscience Publishers, New York (1948). For a general discussion of stress discontinuities, see W. Prager, *Proc. 2d U.S. Nat. Cong. Appl. Mech.*, pp. 21–32 (1954).

The algebraically greater value of the pressure occurs on the upper side of the discontinuity when the positions of the sliplines are as shown in Fig. 6.4. If these positions are interchanged, the pressure would decrease on crossing the discontinuity from the lower side to the upper side. Since $\sin 2\theta = \sin(\pi - 2\theta)$, the acute angle θ in (29) may be regarded as the inclination of the α lines or the β lines with the line of discontinuity. Equation (29) may be obtained directly from the condition of equilibrium of a quadrilateral element with sides perpendicular to the sliplines (Fig. 6.4b).

The curvatures of the sliplines become discontinuous across a line of stress discontinuity. The amount of jump in the curvatures can be established by considering the tangential derivatives of p and ϕ along the discontinuity. For the lower side of the discontinuity, the tangenital derivative is given by the operator

$$\frac{\partial}{\partial s} = \cos\theta\,\frac{\partial}{\partial s_\alpha} + \sin\theta\,\frac{\partial}{\partial s_\beta}$$

where the arc lengths are assumed to increase in the sense from A to B. Applying this operator to ϕ and p, and using (18) and (10), we obtain the relations

$$\frac{\partial\phi}{\partial s} = \frac{\cos\theta}{R} - \frac{\sin\theta}{S}$$

$$\frac{\partial p}{\partial s} = -2k\left(\frac{\cos\theta}{R} + \frac{\sin\theta}{S}\right)$$

(30)

The corresponding expressions for the upper side of the discontinuity are obtained by replacing ϕ, p, R, and S by the corresponding primed variables, and changing the sign of θ. Since $\phi' - \phi = 2\theta$ and $p' - p = 2k\sin 2\theta$, we have

$$\left(\frac{1}{R'} - \frac{1}{R}\right)\cos\theta + \left(\frac{1}{S'} + \frac{1}{S}\right)\sin\theta = 2\,\frac{d\theta}{ds}$$

$$-\left(\frac{1}{R'} - \frac{1}{R}\right)\cos\theta + \left(\frac{1}{S'} + \frac{1}{S}\right)\sin\theta = 2\cos 2\theta\,\frac{d\theta}{ds}$$

This shows that the changes in curvatures depend not only on the angle θ but also on its derivative along the discontinuity. It follows from above that[†]

$$\frac{1}{R'} - \frac{1}{R} = 2\sin\theta\tan\theta\,\frac{d\theta}{ds}$$

$$\frac{1}{S'} + \frac{1}{S} = 2\cos\theta\cot\theta\,\frac{d\theta}{dS}$$

(31)

These relations furnish the jumps in the curvatures of the sliplines at any point of the line of discontinuity. If S and S' are both positive, θ increases along AB, and the

---

[†] These conditions have been established by W. Prager, *Courant Anniversary Volume*, 189 (1948).

curvature of the α line is then algebraically greater on the upper side of the discontinuity than on the lower side.

A line of stress discontinuity may be regarded as the limit of a narrow transition region through which the tangential stress component varies rapidly in a continuous manner. For equilibrium, the normal and shear stress components must be very nearly constant through this region. There are in fact two possible values of the tangential component producing plastic states for given normal and shear components. It follows that the change in stress must occur through a central elastic strip where the stress components do not satisfy the yield criterion. Within the framework of the rigid/plastic theory, a line of stress discontinuity should be regarded as an inextensible but perfectly flexible filament separating two plastic regions. A velocity discontinuity, on the other hand, occurs across a slipline, which may be considered as the limit of a plastic strip across which the tangential component of velocity varies rapidly. Hence, a velocity discontinuity cannot be permitted across a line of stress discontinuity.† Slipline fields involving simultaneous stress and velocity discontinuities are, however, sometimes useful for deriving approximate solutions.

In the approximate solution of physical problems, it is sometimes necessary to consider stress discontinuities separating plastic regions from nonplastic ones. Referring to Fig. 5.4a, suppose that the region above the discontinuity is nonplastic, while that below the discontinuity is stressed to the yield point. The expressions for σ_n and τ are obtained as before, but those for the other two stresses are modified to

$$\sigma'_n = -p' + \tau_0 \sin 2\theta' \qquad \tau' = \tau_0 \cos 2\theta'$$

where $\tau_0 < k$ is the maximum shear stress in the nonplastic region. The continuity conditions $\sigma_n = \sigma'_n$ and $\tau_0 = \tau'_0$ then lead to the relations

$$\frac{\tau_0}{k} = \frac{\cos 2\theta}{\cos 2\theta'} \qquad \frac{p' - p}{k} = \frac{\sin 2(\theta + \theta')}{\cos 2\theta'} \tag{32}$$

The sign of $p' - p$ in (32) must be reversed, as before, with the reversal of the positions of the sliplines. It is evident that $\theta' \leqslant \theta$, where the equality applies when either the upper region is also plastic or the discontinuity is along a principal stress direction. In the latter case, the magnitude of the hydrostatic pressure jump can have any value between k and $2k$.

Consider, now, the possibility of several straight discontinuities meeting at a point and separating regions of constant stress. Assuming these regions to be plastic, it can be shown that at least four such discontinuities must meet at a common point.‡ In particular, when the discontinuous stress field admits an axis

† See, for example, E. H. Lee, *Proc. 3d Symp. Appl. Math.*, p. 213, McGraw-Hill Book Co., New York (1950).

‡ A. Winzer and G. F. Carrier, *J. Appl. Mech.*, **15**: 261 (1948).

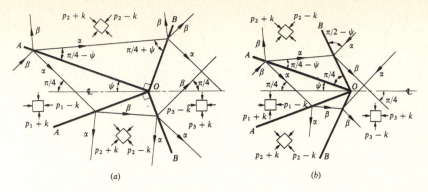

Figure 6.5 Stress discontinuities meeting on an axis of symmetry. (a) 90° reflection; (b) 45° reflection.

of symmetry, the lines of discontinuity must meet on this axis,† the included angle between the pair of discontinuities on either side of the axis being either $\pi/2$ or $\pi/4$. The two possible situations are illustrated in Fig. 6.5, where OA and OB are the stress discontinuities with OA inclined at an acute angle ψ to the axis of symmetry. The sliplines are reflected at each discontinuity and are inclined at 45° to the axis of symmetry outside the angles AOB.

In the case of the 90° reflection, the acute angles made by the α line with OA and OB are $\pi/4 - \psi$ and $\pi/4 + \psi$ respectively. By (29), the hydrostatic pressure increases by the amount of $2k \cos 2\psi$ on crossing each discontinuity from left to right, the net pressure jump being

$$p_3 - p_1 = 4k \cos 2\psi$$

The 45° reflection, on the other hand, involves angles $\pi/4 - \psi$ and $\pi/2 - \psi$ which the α line makes with OA and OB respectively. The hydrostatic pressure jump is of amount $2k \cos 2\psi$ across OA and $2k \sin 2\psi$ across OB, so that

$$p_3 - p_1 = 2k(\cos 2\psi - \sin 2\psi)$$

It may be noted that the directions of the algebraically greater and smaller principal stresses to the right of OB for the 45° reflection are opposite to those for the 90° reflection, these directions to the left of OA being assumed identical in the two cases. Intersecting discontinuities, though unlikely to occur in actual physical situations, are useful for obtaining lower bound approximations to practical problems.‡

(ii) Conditions at a stress singularity Consider a two-dimensional wedge whose

---

† F. Ellis, *J. Strain Anal.*, **2**: 52 (1967). An angle of reflection lying between $\pi/4$ and $\pi/2$ is possible when one of the domains is nonplastic.

‡ Stress discontinuities of variable intensity have been discussed by A. Winzer and G. F. Carrier, *J. Appl. Mech.*, **16**: 346 (1949).

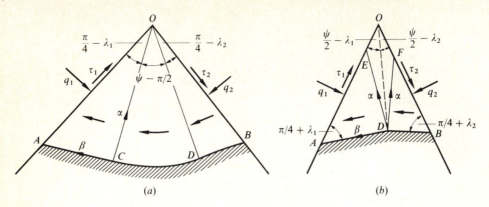

Figure 6.6 Plastic yielding of a wedge of included angle α, equal to $\psi - \lambda_1 - \lambda_2$. (a) $\psi \geqslant \pi/2$; (b) $\psi \leqslant \pi/2$.

inclined plane faces are subjected to normal pressures q_1, q_2 and tangential tractions τ_1 and τ_2 near the vertex. For arbitrary values of these tractions, the vertex O is a singularity of stress. Depending on the magnitudes and directions of the applied shear stresses, in relation to the vertex angle α, there are two possible slipline fields for the state of stress in the wedge at the yield point.† In Fig. 6.6a, the plastic material consists of two uniformly stressed zones OAC and OBD separated by a centered fan field OCD in which the sliplines are radial lines and circular arcs. Suppose that the plastic boundary $ACDB$ is an α line inclined to AO and BO at angles $\pi/4 + \lambda_1$ and $\pi/4 + \lambda_2$ respectively. Then the angles made by the α lines with AO and BO are $\pi/4 - \lambda_1$ and $\pi/4 - \lambda_2$ respectively. It follows from (6) that the shear stresses across OA and OB are given by

$$\tau_1 = k \sin 2\lambda_1 \qquad \tau_2 = k \sin 2\lambda_2 \tag{33}$$

Since these stresses cannot exceed k in magnitude, both λ_1 and λ_2 must lie between $-\pi/4$ and $\pi/4$. The mean compressive stresses along OA and OB are

$$p_1 = q_1 + k \cos 2\lambda_1 \qquad p_2 = q_2 - k \cos 2\lambda_2 \tag{34}$$

The hydrostatic pressure is constant along AC and BD, but there is an increase in pressure along CD by an amount equal to $2k$ times the fan angle $\psi - \pi/2$, where $\psi = \alpha + \lambda_1 + \lambda_2$. Thus $p_2 - p_1 = k(2\psi - \pi)$, which on substitution from (34) becomes

$$q_2 - q_1 = k(\cos 2\lambda_1 + \cos 2\lambda_2 + 2\psi - \pi) \tag{35a}$$

$$\psi = \alpha + \lambda_1 + \lambda_2 \geqslant \frac{\pi}{2}$$

---

† The results of this section are essentially due to R. Hill, *J. Mech. Phys. Solids*, **2**: 278 (1954), who investigated the problem directly from the equilibrium equations and the yield criterion.

Since the normal component of velocity must be zero along the plastic boundary *ACDB*, it follows from the Geiringer equations (12) that $u = 0$ and $v = \text{const}$ throughout the plastic region. The velocity is discontinuous across *ACDB* above which the material moves with a constant speed parallel to the β lines. The motion must take place from right to left in order to have a positive rate of plastic work.

The slipline field for $\psi \leqslant \pi/2$ is shown in Fig. 6.6*b*, which contains a stress discontinuity *OD* separating two regions of constant stress. The angles *OAD* and *OBD* are equal to $\pi/4 + \lambda_1$ and $\pi/4 + \lambda_2$ respectively, as before. From geometry, the angle *EDF* is $\pi/2 - \psi$, where ψ is again equal to $\alpha + \lambda_1 + \lambda_2$. Hence, the acute angle made by the α lines with the discontinuity is $\pi/4 - \psi/2$. By (29), the jump in the hydrostatic pressure is $p_1 - p_2 = 2k \cos \psi$, which gives

$$q_2 - q_1 = k(\cos 2\lambda_1 + \cos 2\lambda_2 - 2 \cos \psi)$$

$$\psi = \alpha + \lambda_1 + \lambda_2 \leqslant \frac{\pi}{2}$$

(35*b*)

in view of (34). A typical point *P* in the field *EDF* may be specified by the distances (ξ, η) from *D* of the points where the α and β lines through *P* meet the discontinuity *OD*. Excluding a translation of the entire wedge, the velocity of any particle on *OD* must be perpendicular to *OD*, where $\xi = \eta$. Since *u* is constant along $\xi = \text{const}$, and *v* is constant along $\eta = \text{const}$, the velocity distribution in *EDF* may be written as

$$u = \pm f(\xi)(1 - \sin \psi) \qquad v = f(\eta) \cos \psi$$

where the upper sign applies to the region *ODE*, and the lower sign to the region *ODF*. The velocity is evidently continuous across *OD*. In the regions *ADE* and *BDF*, the velocity field is given by $u = 0$ and $v = f(\eta) \cos \psi$, where $f(0) = 0$, so that the velocity is continuous across *DE* and *DF*, and also across *AD* and *BD*. The function *f* is otherwise arbitrary, except that its first derivative must be positive for the rate of plastic work to be positive.

An overriding restriction follows from the geometrical fact that the angles *ODE* and *ODF* cannot exceed the angles *DEA* and *DFB* respectively. This means that both λ_1 and λ_2 must be less than or equal to $\psi/2$. Inserting the expression for ψ, we obtain the inequality

$$|\lambda_1 - \lambda_2| \leqslant \alpha$$

(36)

which imposes a limitation on the permissible values of λ_1 and λ_2 in relation to the vertex angle α.

Consider now the situation where the plastic boundary is an α line. The angles which this slipline makes with the boundaries are $\pi/4 - \lambda_1$ and $\pi/4 - \lambda_2$ respectively in order that τ_1 and τ_2 can still be expressed in the form (33). Equations (34) are, however, modified to

$$p_1 = q_1 - k \cos 2\lambda_1 \qquad p_2 = q_2 + k \cos 2\lambda_2$$

Proceeding as before, the pressure difference $p_1 - p_2$ is found to be equal to $k(2\psi^* - \pi)$ for the continuous field and $-2k \cos \psi^*$ for the discontinuous field, where $\psi^* = \alpha - \lambda_1 - \lambda_2$. Inserting the value of p_1 and p_2, we get

$$q_2 - q_1 = k(\pi - 2\psi^* - \cos 2\lambda_1 - \cos 2\lambda_2) \qquad \psi^* \geqslant \frac{\pi}{2}$$

$$q_2 - q_1 = k(2\cos \psi^* - \cos 2\lambda_1 - \cos 2\lambda_2) \qquad \psi^* \leqslant \frac{\pi}{2} \tag{37}$$

For given values of λ_1, λ_2, and α, Eqs. (35) give the maximum value, and Eqs. (37) the minimum value of $q_2 - q_1$ for the yield point state. The vertex of a wedge (with arbitrary curved boundaries) will not be overstressed so long as the difference of the normal pressures near the corner lies between these limits. Some particular cases will now be examined.

(a) Let $\tau_1 = \tau_2 = 0$, implying that the vertex is under normal pressures alone. Setting $\lambda_1 = \lambda_2 = 0$ in (35) and (37), we obtain

$$|q_2 - q_1| \leqslant \begin{cases} 2k(1 + \alpha - \pi/2) & \alpha \geqslant \pi/2 \\ 2k(1 - \cos \alpha) & \alpha \leqslant \pi/2 \end{cases} \tag{38}$$

The elastic limit is reached, at a smaller pressure difference, on the sides of the vertex for $\alpha < \pi/2$, and on the bisector of the wedge angle for $\alpha > \pi/2$. When $\alpha = \pi/2$, the singularity disappears, and the vertex becomes plastic when $|q_2 - q_1| = 2k$.

(b) The shearing stresses have the largest magnitude k and are both directed either toward or away from the vertex. Thus $\tau_1 = -\tau_2 = \pm k$ and $\lambda_1 = -\lambda_2 = \pm \pi/4$. In view of (36), it is first necessary that $\alpha \geqslant \pi/2$. Equations (35) and (37) then give

$$|q_2 - q_1| \leqslant k(2\alpha - \pi) \tag{39}$$

(c) Both the shear stresses are of magnitude k, but one of them is directed toward the vertex and the other one away from the vertex (as in Fig. 6.6). Setting $\tau_1 = \tau_2 = k$, which means $\lambda_1 = \lambda_2 = \pi/4$, the permissible range is found as

$$2k\alpha \geqslant q_2 - q_1 \geqslant \begin{cases} -2k(\alpha - \pi) & \alpha \geqslant \pi \\ 2k\sin \alpha & \alpha \leqslant \pi \end{cases} \tag{40}$$

(d) On one side the shear stress is zero, while on the other side the shear stress has the greatest value k. Thus $\tau_1 = 0$ and $\tau_2 = k$, giving $\lambda_1 = 0$ and $\lambda_2 = \pi/4$; then (36) requires $\alpha \geqslant \pi/4$. From (35) and (37), the permissible values of the pressure difference are given by

$$1 + 2\left(\alpha - \frac{\pi}{4}\right) \geqslant \frac{q_2 - q_1}{k} \geqslant \begin{cases} -1 - 2\left(\alpha - \frac{3\pi}{4}\right) & \alpha \geqslant \frac{3\pi}{4} \\ -1 + 2\cos\left(\alpha - \frac{\pi}{4}\right) & \alpha \leqslant \frac{3\pi}{4} \end{cases} \tag{41}$$

Let AO and its continuation represent a stress-free boundary ($q_1 = \tau_1 = 0$), while OB is a slipline ($\tau_2 = k$) intersecting the boundary at angles α and $\pi - \alpha$. Applying (41) to each vertex angle in turn, we find that the only values of α for which there exists a pressure on OB common to both ranges are $\pi/4$ and $3\pi/4$. Hence, a slipline cannot intersection a free surface with continuous slope at an acute angle other than $\pi/4$.

6.4 Construction of Slipline Fields and Hodographs

The plastic region generally consists of a number of subsidiary domains with the curvature of the sliplines changing discontinuously across their boundaries. The construction of the slipline field depends on the boundary conditions. When the problem is statically determined, the slipline field is uniquely defined by the stress boundary conditions, and will represent a plastically deforming zone if the stress field can be associated with a nonzero velocity field. When the problem is not statically determined, as is generally the case, the velocity boundary conditions impose restrictions on the slipline field itself. Consequently, the construction of the slipline field will involve a simultaneous consideration of the velocity field or, equivalently, the hodograph. In the simplest class of problems of this kind, one or more starting sliplines can be guessed in advance, and the network of sliplines constructed from them using the stress boundary conditions. The solution can be subsequently tested to see if the velocity boundary conditions are also satisfied. In general, a process of trial and error would be necessary if the standard numerical and graphical methods discussed below are adopted. A direct method of treating the statically indeterminate problem is provided by the matrix formulation discussed in the next section. Here we shall consider the types of boundary conditions that arise in the construction of slipline fields and hodographs, leaving aside the question of finding the initial slipline.

(i) *First boundary-value problem* Suppose that the positions of two intersecting sliplines AB and AC are given (Fig. 6.7a). They may be considered as the boundary of a previously calculated slipline field in some area to the left. It is required to

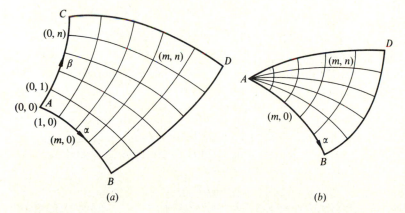

Figure 6.7 Slipline fields for the first boundary-value problem. (a) Regular domain; (b) singular domain.

construct the slipline field to the right of AB and AC, assuming the region to be plastic. It is known from the theory of characteristics that the field is uniquely determined within the curvilinear rectangle $ABDC$ formed by the given sliplines AB and AC. Let AB be an α line and AC a β line. These sliplines are divided into a number of small arcs by the points $(1, 0), (2, 0), (3, 0) \ldots$ along AB, and the points $(0, 1), (0, 2), (0, 3) \ldots$ along AC. A typical nodal point (m, n) within the field is the intersection of the sliplines through the base points $(m, 0)$ and $(0, n)$. Since the values of ϕ are known at all points on AB and AC, the hydrostatic pressures at these points are readily found from the Hencky equations (10), provided p is given at one point, say A. The values of p and ϕ at any nodal point (m, n) are then obtained from the Hencky relations as

$$p(m, n) = p(m, 0) + p(0, n) - p(0, 0)$$

$$\phi(m, n) = \phi(m, 0) + \phi(0, n) - \phi(0, 0)$$

It is often convenient to choose a constant angular distance $\Delta\phi$ between the successive base points. Then the angular span of any slipline between two adjacent nodal points has the constant value $\Delta\phi$ throughout the field. In an equiangular net such as this, one family of diagonal curves passing through opposite nodal points are contours of constant p, while the other family of diagonal curves are contours of constant ϕ. Along the first family of curves the principal stresses have constant magnitudes, and along the second family of curves the principal axes have fixed directions.[†]

The simplest approximate method of calculating the coordinates of the modal points consists in replacing each slipline arc by a chord with slope equal to the mean of the terminal slopes. The actual slope dy/dx at any point is $\tan\phi$ for an α line and $-\cot\phi$ for a β line. The approximation applied to a pair of intersecting arcs leads to a pair of simultaneous equations for the coordinates of each nodal point. An alternative and more accurate method is based on Eqs. (24), the values of $\bar{x}$ and $\bar{y}$ at the base points being found from (21). The finite difference equations for an equiangular net become

$$\bar{x}(m, n) - \bar{x}(m, n - 1) = \tfrac{1}{2}[\bar{y}(m, n) + \bar{y}(m, n - 1)]\mu\,\Delta\phi$$

$$\bar{y}(m, n) - \bar{y}(m - 1, n) = -\tfrac{1}{2}[\bar{x}(m, n) + \bar{x}(m - 1, n)]\lambda\,\Delta\phi \tag{42}$$

where λ and μ are equal to 1 or -1 depending on whether ϕ increases or decreases toward the point (m, n) from the two neighboring points along the respective sliplines. When the values of $\bar{x}$ and $\bar{y}$ have been found at $(m, n - 1)$ and $(m - 1, n)$, those at (m, n) can be calculated from (42). The coordinates of each nodal point can be finally determined by using (22). A third method consists in evaluating the radii of curvature of the slipline from (20), using a finite difference form[‡] analogous to (42). The values of S along AB and R along AC are obtained by a straightforward

---

[†] This was pointed out by R. von Mises, Z. angew. Math. Mech., **5**: 147 (1925).

[‡] For details, see R. Hill, The Mathematical Theory of Plasticity, p. 145, Clarendon Press, Oxford (1950).

numerical integration of (20). The coordinates of the nodal points are obtained from the fact that

$$dx = R \cos \phi \, d\phi \qquad dy = R \sin \phi \, d\phi$$

along an α line. These equations may be integrated by using mean values of $R \cos \phi$ and $R \sin \phi$ between two successive nodal points.

The first boundary-value problem is associated with a velocity distribution in which the normal component of the velocity is prescribed along the given intersecting sliplines. Thus, v is given along AB, and u is given along AC. The application of (12) readily furnishes u along AB and v along AC. The velocity components of a typical nodal point (m, n) are expressed in terms of those at the neighboring points $(m - 1, n)$ and $(m, n - 1)$ by the finite difference equations

$$u(m, n) - u(m - 1, n) = \tfrac{1}{2}[v(m, n) + v(m - 1, n)]\lambda \, \Delta\phi$$
$$v(m, n) - v(m, n - 1) = -\tfrac{1}{2}[u(m, n) + u(m, n - 1)]\mu \, \Delta\phi \tag{43}$$

When the mean values of u and v have been found by solving (43) for each nodal point (m, n), the cartesian components of the velocity can be determined from (11). The prescribed normal velocities may be such that the tangential velocities are discontinuous across the given sliplines, the jump in the velocity component being constant along each slipline.

If the radius of curvature of one of the given sliplines, say AC, is allowed to vanish, while the change in angle between A and C is held constant (Fig. 6.7b), we obtain a centered fan ABD defined by the slipline AB and the angular difference at A. All α lines pass through A, which is a singularity of stress since the hydrostatic pressure at this point has a different value for each slipline. It follows from (20) that all α lines have the same radius of curvature at A, where the β line is of zero radius. The values of p and ϕ at a typical nodal point (m, n) are found as before, with $p(0, n)$ and $\phi(0, n)$ referring to the appropriate α line at A. The velocity field within the fan ABD can be uniquely determined if the normal component of velocity is specified along AB and BD. The calculations can be carried out as before, starting from the point B. The values of u and v at the singular point A will depend on the particular slipline considered through A.

(ii) *Second boundary-value problem* Consider the situation where the normal and tangential tractions are prescribed along a given curve AC. We are required to construct the slipline field below AC on the assumption that the material is plastic. This field is uniquely defined within the curvilinear triangle ABC bounded by the sliplines through A and C (Fig. 6.8a). At each point on AC, the normal stress component acting parallel to the boundary can have two different values satisfying the yield criterion. Physical considerations will indicate the correct value of the stress, and hence specify the values of p and ϕ along the boundary. It is not generally possible to choose an equiangular net unless AC is a contour of constant p or of constant ϕ. The former occurs when the boundary is acted upon by a uniform normal traction and zero tangential traction. The latter occurs when a constant frictional stress acts along the boundary with an arbitrary distribution of normal stress.

Let the given curve AC be divided into an arbitrary number of small segments by the points $(1, 1)$, $(2, 2)$, $(3, 3)$, etc. A typical nodal point (n, m) in the field is defined by the intersection of the sliplines passing through (m, m) and (n, n) on the boundary. Assuming AB to be an α line, we have

$$p(n, m) - p(m, m) = 2k[\phi(m, m) - \phi(n, m)]$$

$$p(n, m) - p(n, n) = 2k[\phi(n, m) - \phi(n, n)]$$

(44)

in view of the Hencky equations (10). Since the values of p and ϕ are known at all points on AC, these equations furnish p and ϕ at any point within the field. The rectangular coordinates of the nodal points can be found by any one of the methods described for the first boundary-value problem. It is necessary, in this case, to start from the given curve AC along which the values of $\bar{x}$ and $\bar{y}$ can be found from (21), using the equation of the curve in the parameteric form $x = x(\phi)$ and $y = y(\phi)$. The radii of curvature of the sliplines along AC can be determined from (30) using the known variation of p and ϕ along this curve. The set of equations of type (42) must be modified to allow for any variation of $\Delta\phi$ between the successive nodal points.

The velocity solution associated with the second boundary-value problem involves both u and v being prescribed along the given curve AC. The velocity components at the nodal points of the field ABC can be uniquely determined by the successive application of (43), taking due account of any variation of the angular distance $\Delta\phi$. Suppose, now, that AC is a boundary separating the plastic material from the nonplastic one. Since the material is rigid/plastic, the nonplastic material above AC moves as a rigid body. The velocity distribution in the plastic region ABC can be obtained by superposing a uniform velocity on that based on the boundary conditions $u = v = 0$ along AC. The plastic material therefore moves as a rigid whole attached to the nonplastic material. It follows that, in general, there will be two types of rigid regions in a partly plastic body. In one, the material is stressed below the yield point, and in the other, the material is plastic but the deformation is prevented by the nonplastic material. The rigid and the deforming parts of the plastic zone are separated from one another by one or more sliplines across which the tangential velocity can be discontinuous. In a real material, the rigid part of the plastic zone would correspond to a region where the elastic and plastic strains are of comparable magnitudes.

(iii) *Third boundary-value problem* In the third or mixed boundary-value problem, we are given a slipline AB and a curve AC along which ϕ is known at each point. The value of ϕ on AB at A is generally different from that on AC at A. The slipline field in the region between AB and AC then involves a centered fan ABD having an angular span equal to the difference of the two values of ϕ (Fig. 6.8b). The fan is uniquely determined by the given slipline AB as in the first boundary-value problem. The remaining field ADC is uniquely defined by the known slipline AD and the given condition along AC. Let the equation of this curve be expressed in the parameteric form $x = x(\phi), y = y(\phi)$,

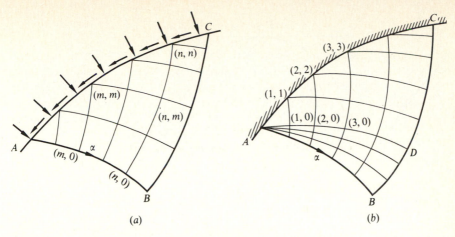

Figure 6.8 Slipline domains in the neighborhood of a curved boundary. (a) Second boundary-value problem; (b) third boundary-value problem with a singularity.

where x and y are known functions of ϕ along AC. Let AD be an α line, which is divided into a number of small arcs by the points (1, 0), (2, 0), (3, 0), etc. To locate the point (1, 1) on AC, the corresponding value $\phi(1, 1)$ is determined by trial and error so that the first equation of (42) is satisfied with $m = n = 1$, and $\Delta\phi = \phi(1, 1) - \phi(1, 0)$. The values of ϕ at all other nodal points on the α line through (1, 1) are then obtained from Hencky's theorem. The calculation of $(\bar{x}, \bar{y})$ at these nodal points is similar to that for the first boundary-value problem. The procedure is repeated until the whole field is covered, a new nodal point being located on AC at the beginning of each stage.

It is evident that the slipline net cannot be equiangular unless ϕ is constant along the curve AC. An example of constant ϕ occurs in the situation where AC is a straight boundary having a constant frictional stress along its length. Then ϕ is obtained for all the nodal points directly from Hencky's theorem, and the coordinates of the nodal points can be separately calculated. When the frictional stress is zero along a straight boundary, so that all sliplines meet it at $45°$, the field is identical to that defined by AD and its reflection in the given boundary (first boundary-value problem).

The third boundary-value problem is usually coupled with a velocity solution in which the normal component of velocity is prescribed along both AB and AC. To calculate the velocity distribution in the field ABD, it is also necessary to know the normal component of velocity along BD, as in the first boundary-value problem. The solution for the velocity field in ABD furnishes the normal component of velocity along AD. Thus v is known along the slipline AD, while a linear relationship between u and v is given along the curve AC. Considering the point (1, 1) on this curve, one of the equations for the velocity components $u(1, 1)$ and $v(1, 1)$ is obtained by setting $m = n = 1$ in the second equation of (43), with $\Delta\phi = \phi(1, 1) - \phi(1, 0)$. The relationship between $u(1, 1)$ and $v(1, 1)$ then furnishes the velocity at (1, 1). Using these values of u and v, the velocity components at (2, 1), (3, 1) ... are calculated successively as in the first boundary-value problem. The process is repeated to cover the whole field.

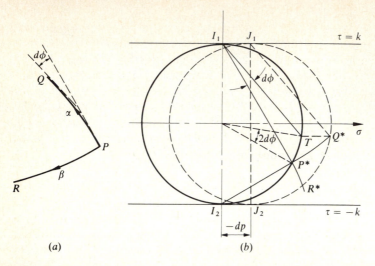

(a)   (b)

Figure 6.9 Geometrical representation of the Hencky equations. (a) Physical plan; (b) stress plane.

In the solution of special problems, the boundary AC frequently corresponds to a rigid die, so that the velocity component normal to the boundary vanishes at each point. The normal component of velocity is a given function of ϕ along the bounding sliplines AB and BC whose shapes are unknown. It is assumed, for simplicity, that the value of ϕ is given on AB at A. Then the velocities at the nodal points of the fan ABD can be determined without having to know the actual shape of the slipline field. The velocity solution can be subsequently built up in the region ACD starting from the point C on the boundary. Since the shape of this boundary is a known function of ϕ, the nodal points on the slipline CD can be chosen at such intervals of ϕ that the values of u and v at $(1, 1)$, $(2, 2)$, etc., satisfy the required boundary condition. This fixes the positions of the nodal points on the given boundary AC. Starting from these points on AC, the values of $\bar{x}$ and $\bar{y}$ can be calculated at all the nodal points of the field ACD, as in the second boundary-value problem. The slipline AD having been found, the construction of the remaining field ADB is identical to that for the first boundary-value problem.

(iv) *Prager's geometrical method* An interesting geometrical interpretation of Hencky's pressure equations with reference to Mohr's stress plane has been given by Prager.† In Fig. 6.9, P^* is the pole of the Mohr circle, with center C, representing the state of stress at a generic point P in the plastic material. The directions of the α and β lines at P are parallel to P^*I_1 and P^*I_2 respectively, where I_1 and I_2 are the highest and lowest points of the Mohr circle. The broken circle with center D corresponds to the state of stress at a neighboring point Q on the α line, the new pole being at Q^*. The

---

† W. Prager, *Trans. R. Inst. Tech.*, Stockholm, no. 65 (1953). See also W. Prager, *Introduction to Plasticity*, chap. 4, Addison-Wesley Pub. Co., New York (1959).

change in position of the pole from P^* to Q^* can be produced by a rotation of the solid circle about its center, bringing P^* to T, followed by a translation of the Mohr circle, moving T to Q^*. Since TI_1 is parallel to the new α direction Q^*J_1, the angle P^*I_1T is equal to the angle $d\phi$ turned through along PQ. The angle subtended by the arc P^*T at the center C is therefore $2d\phi$, and the length of this arc is equal to $2k\,d\phi$. The length TQ^*, on the other hand, is equal to the distance $-dp$ between the centers of the two circles. Since $-dp = 2k\,d\phi$ along an α line by (10), it follows that the pole travels by equal amounts during rotation and translation of the Mohr circle. In the stress plane, the α line through P is therefore mapped into a curve generated by rolling the Mohr circle without sliding on the straight line $\tau = k$. The curve is evidently a cycloid. Similarly, the image of the β line through P is a cycloid described by the pole when the Mohr circle is rolled without sliding on the line $\tau = -k$. The instantaneous centers of rotation at each stage are the highest and lowest points of the circle. Hence I_1P^* and I_2P^* are normal to the cycloidal elements P^*Q^* and P^*R^* respectively. It follows that the tangents to a slipline and its cycloidal image at the corresponding points are mutually orthogonal. This means that the angle turned through along a slipline segment is equal to that along its cycloidal image.

Consider first the situation where a pair of intersecting sliplines AM and AN are given (Fig. 6.10). The state of stress at the point of intersection A is represented by the pole A^* of the Mohr circle whose position along the σ axis depends on the hydrostatic pressure at A. The point A^* is such that I_1A^* and I_2A^* are parallel to the α and β directions at A. The cycloid images A^*B^* and A^*C^* corresponding to the small slipline segments AB and AC are drawn through A^* such that the angles turned through along the sliplines and their respective images are identical. After B^* and C^* have been located, the cycloid images of the remaining sliplines through B and C can be drawn to obtain the point of intersection D^* in the stress plane. The slipline segments BD and CD can be constructed graphically by assuming each one of them to be a circular arc whose chord has a slope equal to the mean of the terminal slopes of the segment. Once D has been located by the intersection of these chords, the arcs BC and CD can be drawn as smooth curves tangential to the known slipline directions at B, C, and D.

We now discuss the hodograph for the particular case in which the material on either side of the pair of intersecting sliplines undergoes rigid body translation in a specified direction. The velocity to the left of AC is represented by the vector PQ, and that to the right of AB is represented by the vector PR, where P is the pole of the hodograph (Fig. 6.10c). The velocity at A, considered in the deforming zone, is represented by the vector PA', where A' is the intersection of the lines through Q and R drawn parallel to the slipline directions at A. Hence QA' and RA' denote the velocity discontinuities across AC and AB respectively. Since the magnitude of the velocity discontinuity must remain constant along a slipline, the segments AB and AC are mapped into the circular arcs $A'B'$ and $A'C'$ with centers R and Q respectively. The angles subtended by these arcs at the centers are equal to the angles turned through along the respective sliplines. The hodograph field $A'B'D'C'$ can be constructed from $A'B'$ and $A'C'$ in a manner identical to that used for the slipline field. It is evident that velocity boundary conditions impose restrictions on the hodograph in the same way

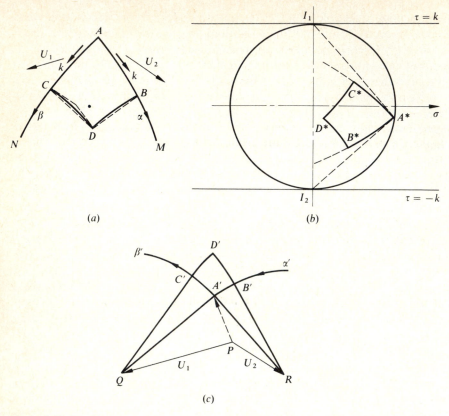

Figure 6.10 Graphical treatment of the first boundary-value problem. (*a*) Physical plane; (*b*) stress plane; (*c*) hodograph.

as stress boundary conditions do on the slipline field. The shape of the hodograph in relation to the slipline field is further restricted by the requirement of positive rate of plastic work as explained in Sec. 6.3.

When the surface tractions across a boundary curve *AC* are given (Fig. 6.8*a*), two circles of radius k can be drawn in the stress plane through the stress point corresponding to a typical point on the boundary. The appropriate Mohr circle in this case may be identified from the consideration of the nature of the problem. Since the angular distances between the successive nodal points are again known in advance, the slipline field can be constructed by the approximate small arc process as before, starting from the boundary *AC*. The corresponding cycloidal net can be drawn separately from the known angular spans of the various segments. In the solution of the mixed boundary-value problem, the construction of the slipline field and its cycloidal net should generally be carried out simultaneously by a trial-and-error process, using the correspondence between the physical plane and the stress plane. When both components of the velocity are known along a given boundary of a plastic zone, for which the slipline field has been found, the hodograph can be constructed by the small arc process using the orthogonality of the corresponding tangents to the sliplines and their hodograph images. Conversely, when the velocity field in a region adjacent to a boundary curve has been determined, the corresponding slipline field can be constructed in a similar manner provided the slipline directions are known at each point of the boundary.

6.5 Analytical and Matrix Methods of Solution

(i) *The analytical theory* Consider the first boundary-value problems in which a pair of intersecting sliplines CA and CB are given. Let CA and CB be taken as a right-handed pair of curvilinear axes, the angles turned through along them being denoted by α and β respectively (Fig. 6.11). By Hencky's first theorem, these are the angles turned through to reach a generic point P along either pair of orthogonal sliplines. The coordinate curves $\alpha = $ const and $\beta = $ const are therefore the sliplines with the base curves CA and CB corresponding to $\beta = 0$ and $\alpha = 0$ respectively. We shall regard α and β as algebraically increasing along CA and CB irrespective of their sense of rotation. Referring to Fig. 6.11a, we have

$$\phi = \phi_0 + \beta - \alpha \qquad p = p_0 + 2k(\alpha + \beta)$$

where p_0 and ϕ_0 are the values of p and ϕ at C. Since $d\phi = -d\alpha$ along an α line and $d\phi = d\beta$ along a β line, Eqs. (12) may be written as

$$\frac{\partial u}{\partial \alpha} = -v \qquad \frac{\partial v}{\partial \beta} = -u \tag{45}$$

except when one family of sliplines is straight. Similarly, Eqs. (20) and (24) become

$$\frac{\partial S}{\partial \alpha} = R \qquad \frac{\partial R}{\partial \beta} = S \tag{46}$$

$$\frac{\partial \bar{y}}{\partial \alpha} = \bar{x} \qquad \frac{\partial \bar{x}}{\partial \beta} = \bar{y} \tag{47}$$

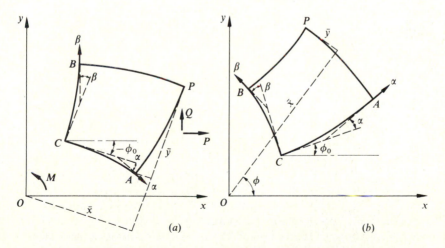

Figure 6.11 Slipline nets with canonical variables and moving coordinates.

The elimination of either of the dependent variables from each pair of the above equations shows that u, v, R, S, $\bar{x}$, and $\bar{y}$ separately satisfy the same differential equation[†]

$$\frac{\partial^2 f}{\partial \alpha \, \partial \beta} = f \tag{48}$$

which is known as the equation of telegraphy. It applies to the plain strain problem only when both families of sliplines are curved. Setting $ds_\alpha = -R \, d\alpha$ and $ds_\beta = -S \, d\beta$ in (25), we get

$$\frac{\partial \bar{x}}{\partial \alpha} + \bar{y} = -R \qquad \frac{\partial \bar{y}}{\partial \beta} + \bar{x} = -S \tag{49}$$

which are useful relations between the Mikhlin variables and the radii of curvature of the sliplines. From (47) and (49), it is easy to show that

$$\frac{\partial}{\partial \alpha} (\bar{x} + \bar{y}) = \bar{x} - \bar{y} - R \qquad \frac{\partial}{\partial \beta} (\bar{x} - \bar{y}) = \bar{x} + \bar{y} + S \tag{50}$$

These relations are sometimes useful in the solution of special problems. Returning to the governing differential equation (48), we observe that it has the following particular solutions:[‡]

$$f(\alpha, \beta) = \begin{cases} \cos(\alpha - \beta), \sin(\alpha - \beta), \exp[\pm(\alpha + \beta)] \\ (\alpha + \beta)[\cos(\alpha - \beta), \sin(\alpha - \beta)], (\alpha - \beta)\exp[\pm(\alpha + \beta)] \end{cases} \tag{51}$$

The boundary conditions in physical problems cannot be generally satisfied by these elementary functions alone. To obtain a general solution of (48), we consider the function

$$f_n(\alpha, \beta) = \left(\frac{\alpha}{\beta}\right)^{n/2} I_n(2\sqrt{\alpha\beta}) \tag{52}$$

where I_n is the modified Bessel function of the first kind and of an integer order $n \geqslant 0$. It is defined as

$$I_n(z) = \sum_{s=0}^{\infty} \frac{(z/2)^{n+2s}}{s!\,(n+s)!}$$

It is evident that $I_0(0) = 1$, $I_n(0) = 0$ for $n \geqslant 1$, and $I_0'(z) = I_1(z)$, where the prime denotes differentiation with respect to z. The recurrence relations satisfied by $I_n(z)$ are

$$I_{n-1}(z) - I_{n+1}(z) = \frac{2n}{z} I_n(z)$$

$$I_{n-1}(z) + I_{n+1}(z) = 2I_n'(z) \tag{53}$$

[†] C. Carathéodory and E. Schmidt, *Z. angew. Math. Mech.*, **3**: 468 (1923).

[‡] These solutions may be generalized by writing $c\alpha$ for α and β/c for β in the trigonometric and exponential functions, and $c^2\alpha$ for α in the algebraic coefficients, where c is an arbitrary constant.

Substituting the expression for $I_n(2\sqrt{\alpha\beta})$ in (52), we immediately obtain the series expansion

$$f_n(\alpha, \beta) = \sum_{s=0}^{\infty} \frac{\alpha^{n+s}\beta^s}{(n+s)!\,s!} \tag{54}$$

which is frequently useful in the solution of special problems. Differentiating (54) partially with respect to α and β, we get

$$\frac{\partial f_n}{\partial \alpha} = f_{n-1} \qquad \frac{\partial f_n}{\partial \beta} = f_{n+1} \tag{55}$$

The derivatives of $f_n(\beta, \alpha)$ are obtained by merely interchanging α and β in (55). It follows that both $f_n(\alpha, \beta)$ and $f_n(\beta, \alpha)$ satisfy equation (48). Hence the general solution of (48) can be written as†

$$f(\alpha, \beta) = \sum_{n=0}^{\infty} \left[a_n f_n(\alpha, \beta) + c_n f_{n+1}(\beta, \alpha) \right] \tag{56}$$

where a_n and c_n are arbitrary constants to be determined from the boundary conditions. From (54) and (56), it is easily shown that

$$f(\alpha, 0) = \sum_{n=0}^{\infty} a_n \frac{\alpha^n}{n!} \qquad f(0, \beta) = a_0 + \sum_{n=1}^{\infty} c_{n-1} \frac{\beta^n}{n!} \tag{57}$$

These equations express the power series expansion of f along the base sliplines. If the value of the function is given along both these curves, the coefficients of the expansions are readily found and the required field variable then follows from (56).

When the curvature of one of the base sliplines, say CA, is reversed as in Fig. 6.11b, it is necessary to change the sign of α in the preceding theory. Since the argument of I_n in (52) then becomes complex, it is convenient to replace $f_n(\alpha, \beta)$ by

$$g_n(\alpha, \beta) = \left(\frac{\alpha}{\beta}\right)^{n/2} J_n(2\sqrt{\alpha\beta}) \tag{58}$$

where n is zero or a positive integer, and J_n is the Bessel function of the first kind defined as

$$J_n(z) = \sum_{s=0}^{\infty} (-1)^s \frac{(z/2)^{n+2s}}{s!\,(n+s)!}$$

It has the properties $J_0(0) = 1$, $J_n(0) = 0$ for $n \geqslant 1$, $J'_0(z) = -J_1(z)$. Further, $I_n(iz) = i^n J_n(z)$, where i represents $\sqrt{-1}$. The series expansion form of (58) is

$$g_n(\alpha, \beta) = \sum_{s=0}^{\infty} \frac{\alpha^{n+s}(-\beta)^s}{(n+s)!\,s!} \tag{59}$$

---

† H. Geiringer, *Memorial des Sciences Mathematiques*, 86 Gauthier Villars, Paris (1937); J. Chakrabarty, *Int. J. Mech. Sci.*, **21**: 477 (1979).

and the partial derivatives of $g_n(\alpha, \beta)$ are found as

$$\frac{\partial g_n}{\partial \alpha} = g_{n-1} \qquad \frac{\partial g_n}{\partial \beta} = -g_{n+1} \tag{60}$$

Equation (48), with the appropriate sign change, is therefore satisfied by both $g_n(\alpha, \beta)$ and $g_n(\beta, \alpha)$, and the general solution becomes

$$f(\alpha, \beta) = \sum_{n=0}^{\infty} [a_n g_n(\alpha, \beta) + c_n g_{n+1}(\beta, \alpha)] \tag{61}$$

which again reduces to (57) on the base sliplines $\beta = 0$ and $\alpha = 0$. The solution is of the form (56) whenever the radii of curvature of the base sliplines have the same sign, and of the form (61) whenever they have opposite signs.

For given shapes of the base sliplines, the differential equation for f remains unchanged in form if the independent variables are changed to ξ and η defined as

$$\xi = \alpha_0 \pm \alpha \qquad \eta = \beta_0 \pm \beta$$

where α_0 and β_0 are constants, provided both the upper signs or both the lower signs are taken. Then (α, β) may be replaced by (ξ, η) in the appropriate solution (56) or (61). When other signs are considered in the above expressions, the solution is obtained by writing (ξ, η) for (α, β) in (61) if R and S are of the same sign, and in (56) if they are of opposite signs. It follows from (55) and (60) that

$$f'_n(\alpha, \xi) = f_{n-1}(\alpha, \xi) \pm f_{n+1}(\alpha, \xi)$$
$$g'_n(\alpha, \xi) = g_{n-1}(\alpha, \xi) \mp g_{n+1}(\alpha, \xi) \tag{62}$$

where the prime denotes differentiation with respect to the independent variable α. The upper and lower signs in (62) correspond with the upper and lower signs in the expression for ξ.

(ii) *A series method of solution* A general method of treating slipline field problems will now be discussed in relation to the first boundary-value problem in which the radii of curvature of the base sliplines have the same sign. Referring to Fig. 6.11a, the base-line radii of curvature are expanded in the convergent power series[†]

$$-R(\alpha, 0) = \sum_{n=0}^{\infty} a_n \frac{\alpha^n}{n!} \qquad -S(0, \beta) = \sum_{n=0}^{\infty} b_n \frac{\beta^n}{n!} \tag{63}$$

where a_n and b_n are given constants. The first of these conditions is satisfied by assuming the solution for $-R(\alpha, \beta)$ in the form (56). In view of (46), the remaining boundary condition then gives $c_n = b_n$. The expression for R and S therefore become

$$-R(\alpha, \beta) = \sum_{m,n=0}^{\infty} \left[a_n \frac{\alpha^{m+n}}{(m+n)!} \frac{\beta^m}{m!} + b_n \frac{\alpha^m}{m!} \frac{\beta^{m+n+1}}{(m+n+1)!} \right]$$

$$-S(\alpha, \beta) = \sum_{m,n=0}^{\infty} \left[a_n \frac{\alpha^{m+n+1}}{(m+n+1)!} \frac{\beta^m}{m!} + b_n \frac{\alpha^m}{m!} \frac{\beta^{m+n}}{(m+n)!} \right]$$

† D. F. J. Ewing, *J. Mech. Phys. Solids*, **15**: 105 (1967).

This solution can be verified directly by inspection and is readily shown to be uniformly and absolutely convergent. Rearranging the subscripts, the above expressions can be written in the form

$$-R(\alpha, \beta) = \sum_{n=0}^{\infty} r_n(\beta) \frac{\alpha^n}{n!} \qquad -S(\alpha, \beta) = \sum_{n=0}^{\infty} s_n(\alpha) \frac{\beta^n}{n!} \qquad (64)$$

where r_n and s_n are constant along α and β lines respectively, and are given by

$$r_n(\beta) = \sum_{m=0}^{n} a_{n-m} \frac{\beta^m}{m!} + \sum_{m=n+1}^{\infty} b_{m-n-1} \frac{\beta^m}{m!}$$

$$s_n(\alpha) = \sum_{m=0}^{n} b_{n-m} \frac{\alpha^m}{m!} + \sum_{m=n+1}^{\infty} a_{m-n-1} \frac{\alpha^m}{m!} \qquad (65)$$

We thus have the power series expansions for R along an α line and S along a β line. If the curvature of both the base sliplines is reversed, we may use the series expansions (63) and (64) with the negative signs omitted. Equations (46), with due changes in sign of α and β, then indicate that the resulting modifications of (65) are such as to have the sign of the second series for both r_n and s_n reversed.

The series expansions for $\bar{x}$ and $\bar{y}$ in terms of α and β would be analogous to those for R and S. Alternatively, the expression for either R or S may be used to determine $\bar{x}$ and $\bar{y}$. Considering the second equations of (47), (49), and (64), it can be easily shown that

$$\bar{x}(\alpha, \beta) = \sum_{n=0}^{\infty} t_n(\alpha) \frac{\beta^n}{n!} \qquad \bar{y}(\alpha, \beta) = \sum_{n=0}^{\infty} t_{n+1}(\alpha) \frac{\beta^n}{n!}$$

where $$t_{n+2}(\alpha) + t_n(\alpha) = s_n(\alpha) \qquad (66)$$

It is important to note that $t_0(\alpha)$ and $t_1(\alpha)$ are the values of $\bar{x}$ and $\bar{y}$ respectively along the α baseline. Hence $t_n(\alpha)$ can be readily found from the above recurrence relationship for any given β line $\alpha = \alpha_0$. When $\bar{x}$ and $\bar{y}$ have been found, the rectangular coordinates follow from (23). Consider now the convolution integrals

$$U(\alpha, \beta) = - \int_0^{\beta} t S(\alpha, t) \sin(\beta - t)\, dt$$

$$V(\alpha, \beta) = - \int_0^{\beta} t S(\alpha, t) \cos(\beta - t)\, dt \qquad (67)$$

which are taken along a given β line from its intersection with the α baseline. It follows from direct differentiation that

$$\frac{\partial U}{\partial \beta} = V \qquad \frac{\partial V}{\partial \beta} + U = -\beta S$$

Hence, the solution for U and V, vanishing on $\beta = 0$, can be written as

$$U(\alpha, \beta) = \sum_{n=1}^{\infty} u_n(\alpha) \frac{\beta^n}{n!} \qquad V(\alpha, \beta) = \sum_{n=1}^{\infty} u_{n+1}(\alpha) \frac{\beta^n}{n!}$$

where $$u_{n+2}(\alpha) + u_n(\alpha) = n s_{n-1}(\alpha) \qquad \text{with} \qquad u_1 = 0 \qquad (68)$$

Let (P, Q) be the rectangular components of the resultant traction acting across the line segment AP (Fig. 6.11a) due to the material on the right of this slipline. The counterclockwise angle which is made by the tangent to AP with the y axis at a generic point (α, t) is $\phi_0 - \alpha + t$. It follows from geometry that

$$P = kX - (p_0 + 2k\alpha)Y + 2k \int_0^\beta tS(\alpha, t) \cos(\phi_0 - \alpha + t) \, dt$$

$$Q = kY + (p_0 + 2k\alpha)X + 2k \int_0^\beta tS(\alpha, t) \sin(\phi_0 - \alpha + t) \, dt$$

where p_0 is the hydrostatic pressure at C, and (X, Y) denote the coordinate differences $(x_P - x_A, y_P - y_A)$. Using (67), the above equations can be expressed as

$$\frac{P}{k} = X - \left(\frac{p_0}{k} + 2\alpha\right)Y - 2(V \cos\phi + U \sin\phi)$$

$$\frac{Q}{k} = Y + \left(\frac{p_0}{k} + 2\alpha\right)X + 2(U \cos\phi - V \sin\phi)$$

(69)

where $\phi = \phi_0 + \beta - \alpha$. Similar expressions may be written down for the components of the resultant force across an α line.† The resultant counterclockwise moment about the origin, of the distribution of tractions on AP, is easily shown to be given by

$$\frac{M}{k} = -\int_0^\beta \left\{\left[\frac{p_0}{k} + 2(\alpha + t)\right]\bar{y}(\alpha, t) + \bar{x}(\alpha, t)\right\} S(\alpha, t) \, dt$$

(70)

When $\bar{x}$, $\bar{y}$, and S have been found by the summing of the respective power series, M/k can be evaluated by a straightforward numerical integration.

(iii) *Matrix formulation* The method outlined above is particularly suitable for evaluating the slipline field quantities in problems of the direct type where a pair of initial sliplines are either given or can be reasonably guessed. There are however many important problems of the indirect type in which the shape of none of the sliplines, or their hodograph images, can be deduced in advance. In order to deal with such cases, it is convenient to recast the foregoing theory in matrix language.‡ We shall be concerned here with the radii of curvature of the sliplines from which the other relevant quantities can be determined by using the series method. We begin by writing (64) in the matrix form

$$-R = [1, \alpha_1, \alpha_2, \ldots] \begin{Bmatrix} r_1 \\ r_2 \\ r_3 \\ \vdots \end{Bmatrix} \qquad -S = [1, \beta_1, \beta_2, \ldots] \begin{Bmatrix} s_1 \\ s_2 \\ s_3 \\ \vdots \end{Bmatrix}$$

(71)

---

† A graphical method of finding the resultant force acting across a slipline has been proposed by W. Johnson and A. G. Mamalis, *Int. J. Mech. Sci.*, **20**: 47 (1978).

‡ I. F. Collins, *Proc. R. Soc., A*, **303**: 317 (1968); P. Dewhurst and I. F. Collins, *Int. J. Num. Methods Eng.*, **7**: 357 (1973). The sign convention employed by these authors was somewhat different.

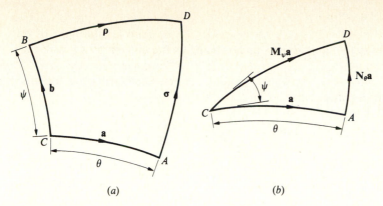

Figure 6.12 Vector representation of sliplines. (*a*) Regular domain; (*b*) singular domain. The arrow indicates the intrinsic direction.

where $\alpha_n = \alpha^n/n!$ and $\beta_n = \beta^n/n!$ $(n = 1, 2, 3, \ldots)$. The column vectors $\{r_n\}$ and $\{s_n\}$ may be considerd as representing the radii of curvature of typical sliplines $\beta = $ const and $\alpha = $ const respectively. Let $\boldsymbol{\rho}$ and $\boldsymbol{\sigma}$ denote these vectors for the sliplines BD and AD, considered through a typical point D with angular coordinates θ and ψ (Fig. 6.12*a*). These vectors are defined in such a way that their sign is always positive, irrespective of the sense of the curvature. Eqs. (65) may now be written in the matrix form†

$$\boldsymbol{\rho} = \mathbf{M}_\psi \mathbf{a} + \mathbf{N}_\psi \mathbf{b} \qquad \boldsymbol{\sigma} = \mathbf{M}_\theta \mathbf{b} + \mathbf{N}_\theta \mathbf{a} \tag{72}$$

where $\mathbf{a}$ and $\mathbf{b}$ are column vectors having elements a_n and b_n respectively (with $n = 1, 2, 3, \ldots$), and representing the radii of curvature of the base sliplines CA and CB. The operators M and N are square matrices expressed as

$$\mathbf{M}_\theta = \begin{bmatrix} 1 & 0 & 0 & \cdots \\ \theta_1 & 1 & 0 & \cdots \\ \theta_2 & \theta_1 & 1 & \cdots \\ \cdots & \cdots & \cdots & \cdots \end{bmatrix} \qquad \mathbf{N}_\theta = \begin{bmatrix} \theta_1 & \theta_2 & \theta_3 & \cdots \\ \theta_2 & \theta_3 & \theta_4 & \cdots \\ \theta_3 & \theta_4 & \theta_5 & \cdots \\ \cdots & \cdots & \cdots & \cdots \end{bmatrix} \tag{73}$$

where $\theta_n = \theta^n/n!$. If one of the base sliplines, CB (say), degenerates into a point ($\mathbf{b} = 0$), the radii of curvature of CD and AD of the fan CAD are given by $\mathbf{M}_\psi \mathbf{a}$ and $\mathbf{N}_\theta \mathbf{a}$ respectively. It follows that $\mathbf{M}$ and $\mathbf{N}$ are matrix operators which generate the singular field on the convex side of a given slipline (Fig. 6.12*b*). Thus $\mathbf{M}$ is associated with sliplines of the same family and $\mathbf{N}$ with sliplines of the opposite family.‡ It may be noted that $\mathbf{M}_\theta \mathbf{M}_\psi = \mathbf{M}_{\theta + \psi}$.

Consider now the power series expansion of the radii of curvature of the sliplines DB and DA with respect to the new base point D. Taking due account of the

---

† This is an analytical statement of the superposition principle discussed in Sec. 6.8.
‡ Evidently, the singular field on the concave side of a given slipline would be generated by the inverse operators $\mathbf{M}^{-1}$ and $\mathbf{N}^{-1}$ in a similar way.

sign convention, we write

$$R(\alpha, \psi) = \sum_{n=0}^{\infty} \bar{r}_n(\theta, \psi) \frac{\xi^n}{n!} \qquad S(\theta, \beta) = \sum_{n=0}^{\infty} \bar{s}_n(\theta, \psi) \frac{\eta^n}{n!}$$

where $\xi = \theta - \alpha$ and $\eta = \psi - \beta$, while $\bar{r}_n$ and $\bar{s}_n$ are new coefficients. On the other hand, (64) directly furnishes

$$R(\alpha, \psi) = \sum_{n=0}^{\infty} r_n(\psi) \frac{\alpha^n}{n!} \qquad S(\theta, \beta) = \sum_{n=0}^{\infty} s_n(\theta) \frac{\beta^n}{n!}$$

where $\alpha = \theta - \xi$ and $\beta = \psi - \eta$. Using the binomial expansions of $(\theta - \xi)^n$ and $(\psi - \eta)^n$, and comparing the coefficients of ξ^n and η^n on the right-hand sides of the above equations, we obtain

$$\bar{r}_n(\theta, \psi) = (-1)^n \sum_{m=n}^{\infty} r_m(\psi) \frac{\theta^{m-n}}{(m-n)!}$$

and a similar expression for $\bar{s}_n(\theta, \psi)$. Denoting the column vectors $\{\bar{r}_n\}$ and $\{\bar{s}_n\}$ by ρ' and σ' respectively, these equations may be written in the matrix form

$$\rho' = R_\theta \rho \qquad \sigma' = R_\psi \sigma \tag{74}$$

where

$$R_\psi = \begin{bmatrix} 1 & \psi_1 & \psi_2 & \cdots \\ 0 & -1 & -\psi_1 & \cdots \\ 0 & 0 & 1 & \cdots \\ \cdots & \cdots & \cdots & \cdots \end{bmatrix} \tag{75}$$

The direction indicated by arrows in Figs. 6.12 and 6.13 represents the *intrinsic*

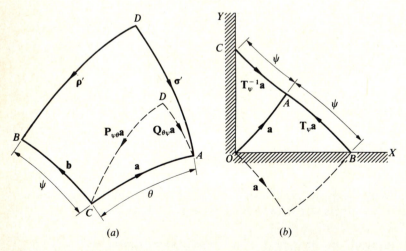

(a)   (b)

Figure 6.13 Vector representation of sliplines with intrinsic directions opposite to those of the base curves.

direction in which the characteristic angle, whose powers are involved in the respective series expansion, progressively increases. Since ξ and η increases along DB and DA, the intrinsic directions for ρ' and σ' are opposite to those for ρ and σ. The matrix $\mathbf{R}_\psi$ is therefore a *reversion operator* which reverses the intrinsic direction of a given slipline of angular span ψ. It has the property $\mathbf{R}_\psi^2 = \mathbf{I}$, where $\mathbf{I}$ is the unit matrix operator. We now introduce operators $\mathbf{P}$ and $\mathbf{Q}$ defined as

$$\mathbf{P}_{\theta\psi} = \mathbf{R}_\psi\mathbf{M}_\theta \qquad \mathbf{Q}_{\theta\psi} = \mathbf{R}_\psi\mathbf{N}_\theta \tag{76}$$

The geometrical interpretation for $\mathbf{P}$ and $\mathbf{Q}$ is analogous to that for $\mathbf{M}$ and $\mathbf{N}$, the only difference being in the associated intrinsic direction (Fig. 6.13a). Thus $\mathbf{P}_{\psi\theta}\mathbf{a}$ and $\mathbf{Q}_{\theta\psi}\mathbf{b}$ represent the radii of curvature of the sliplines DC and DA of the centered fan CAD with the intrinsic directions pointing toward C and A respectively. Premultiplying the two equations of (72) by $\mathbf{R}_\theta$ and $\mathbf{R}_\psi$ respectively, and using (74) and (76), we obtain

$$\rho' = \mathbf{P}_{\psi\theta}\mathbf{a} + \mathbf{Q}_{\psi\theta}\mathbf{b} \qquad \sigma' = \mathbf{P}_{\theta\psi}\mathbf{b} + \mathbf{Q}_{\theta\psi}\mathbf{a} \tag{77}$$

These equations express the vectors representing the radii of curvature of DB and DA with respect to D in terms of those of CA and CB with respect to C. Alternatively, we may regard DB and DA as base sliplines, and express $\mathbf{a}$ and $\mathbf{b}$ as linear functions of ρ' and σ'. Taking due account of the modifications resulting from the reversed curvature of the new baselines, it can be shown that

$$\mathbf{a} = \mathbf{P}_{\psi\theta}\rho' - \mathbf{Q}_{\psi\theta}\sigma' \qquad \mathbf{b} = \mathbf{P}_{\theta\psi}\sigma' - \mathbf{Q}_{\theta\psi}\rho' \tag{78}$$

Eliminating ρ' and σ' between (77) and (78), we obtain a pair of matrix equations which can be simultaneously satisfied if

$$\mathbf{P}_{\theta\psi}\mathbf{Q}_{\theta\psi} = \mathbf{Q}_{\theta\psi}\mathbf{P}_{\psi\theta} \qquad \mathbf{P}_{\theta\psi}^2 - \mathbf{Q}_{\theta\psi}\mathbf{Q}_{\psi\theta} = \mathbf{I} \tag{79}$$

These identities are sometimes useful for establishing the validity of a proposed slipline field. It is interesting to note that $\mathbf{P}_{\psi\psi}$ and $\mathbf{Q}_{\psi\psi}$ are mutually commutative, namely $\mathbf{P}_{\psi\psi}\mathbf{Q}_{\psi\psi} = \mathbf{Q}_{\psi\psi}\mathbf{P}_{\psi\psi}$.

If the origin of the curvilinear coordinates (α, β) is moved from C to D, it is necessary to replace α by $\theta + \alpha$ and β by $\psi + \beta$ in the series expansions (64). Proceeding as before, it can be shown that the vectors representing the sliplines through D with the intrinsic directions unchanged, are obtained by pre-multiplying ρ and σ by the matrices $\mathbf{S}_\theta$ and $\mathbf{S}_\psi$ respectively, where $\mathbf{S}_\psi$ is a shift operator given by (75) with all the negative signs suppressed.

An important special case arises in the construction of the slipline field between a given curve OA, represented by the vector $\mathbf{a}$, and two frictionless planes OX and OY (Fig. 6.13b). The field may be regarded as being generated by OA and its reflection in the planes. The vector $\mathbf{T}_\psi\mathbf{a}$, representing the bounding curve BA on the convex side of OA, is obtained by setting $\mathbf{b} = \mathbf{a}$ and $\theta = \psi$ in (77). Thus

$$\mathbf{T}_\psi = \mathbf{P}_{\psi\psi} + \mathbf{Q}_{\psi\psi} \qquad \mathbf{T}_\psi^{-1} = \mathbf{P}_{\psi\psi} - \mathbf{Q}_{\psi\psi} \tag{80}$$

where the second result follows from (79) with $\theta = \psi$. The vector $\mathbf{T}_\psi^{-1}\mathbf{a}$ evidently represents the curve CA, on the concave side of OA, so that the operator $\mathbf{T}_\psi$ can generate the original curve OA from CA.

For computational purposes, the infinite dimensional vectors and matrices introduced in the above formulation must be truncated to a suitable size of dimension N (say). The error involved in this approximation can be estimated by considering the next higher dimension. Using this formulation, the problem of finding an unknown initial slipline may be reduced to one of matrix inversion, after a simultaneous consideration of the slipline field and the associated hodograph. The matrix method of determining the initial slipline fails, however, when the boundary conditions are such that the basic problem is mathematically nonlinear.

(iv) *Riemann's method of integration* The solution of (48) may also be expressed in an integral form using Riemann's method, which consists in finding a particular analytical solution known as Green's function. For the problem under consideration, the Green's function is required to have a constant value along the characteristics through a typical point D (Fig. 6.12a). The angular coordinates of D are denoted by (θ, ψ) to distinguish it from an arbitrary point (α, β) within the field. Then Green's function for the problem may be written as

$$F(\alpha, \beta) = I_0[2\sqrt{(\theta - \alpha)(\psi - \beta)}]$$

which corresponds to (52) with $n = 0$ and $(\theta - \alpha, \psi - \beta)$ written for (α, β) respectively. The above function evidently satisfies the differential equation (48) and assumes a value of unity along the curves $\alpha = \theta$ and $\beta = \psi$. Since

$$F \frac{\partial^2 f}{\partial \alpha \, \partial \beta} = Ff = f \frac{\partial^2 F}{\partial \alpha \, \partial \beta}$$

in view of (48), it is readily verified by direct differentiation that

$$\frac{\partial}{\partial \beta} \left(F \frac{\partial f}{\partial \alpha} - f \frac{\partial F}{\partial \alpha} \right) = \frac{\partial}{\partial \alpha} \left(f \frac{\partial F}{\partial \beta} - F \frac{\partial f}{\partial \beta} \right)$$

which is the necessary and sufficient condition for the expression

$$\left(F \frac{\partial f}{\partial \alpha} - f \frac{\partial F}{\partial \alpha} \right) d\alpha + \left(f \frac{\partial F}{\partial \beta} - F \frac{\partial f}{\partial \beta} \right) d\beta$$

to be a perfect differential. Hence the line integral of this expression round any closed curve is zero. We therefore write

$$\oint \left\{ \left(F \frac{\partial f}{\partial \alpha} - f \frac{\partial F}{\partial \alpha} \right) d\alpha + \left(f \frac{\partial F}{\partial \beta} - F \frac{\partial f}{\partial \beta} \right) d\beta \right\} = 0$$

where the integral is taken round $CADBC$ in a counterclockwise sense. Since $F = 1$ along $AD(\alpha = \theta)$ and $DB(\beta = \psi)$, and consequently $\partial F/\partial \alpha = 0$ on AD and $\partial F/\partial \beta = 0$ on DB, the integral is readily evaluated along these curves, and the above equation is reduced to

$$2f_D = f_A + f_B + \int_C^A \left(F \frac{\partial f}{\partial \alpha} - f \frac{\partial F}{\partial \alpha} \right) d\alpha + \int_C^B \left(F \frac{\partial f}{\partial \beta} - f \frac{\partial F}{\partial \beta} \right) d\beta$$

where the subscripts denote the values of f at the specified points, and the integrals are taken along the base curves CA and CB. Integrating by parts, and then substituting for $F(\alpha, \beta)$, we finally obtain†

$$f(\theta, \psi) = I_0(2\sqrt{\theta\psi})f(0, 0) + \int_0^\theta I_0(2\sqrt{\xi\psi}) \frac{\partial f}{\partial \alpha} \, d\alpha + \int_0^\psi I_0(2\sqrt{\theta\eta}) \frac{\partial f}{\partial \beta} \, d\beta \quad (81)$$

where $\xi = \theta - \alpha$ and $\eta = \psi - \beta$. Equation (81) expresses the value of the function f at any point of the field in terms of its boundary values along the base curves. Taking $f = R$, for instance, we have $\partial f/\partial \beta = S$. The integrands are therefore known when R and S are given along CA and CB respectively. When $R(\alpha, \beta)$ has been found from (81), the rectangular coordinates of any point can be determined from the relations

$$\frac{\partial x}{\partial \alpha} = -R \cos (\phi_0 + \beta - \alpha) \qquad \frac{\partial y}{\partial \alpha} = R \sin (\phi_0 + \beta - \alpha) \quad (82)$$

on integrating along the α lines. Alternatively, these coordinates may be found through $\bar{x}$ and $\bar{y}$ calculated from (81) and (47). If one of the base sliplines is curved in the opposite sense, it is only necessary to replace I_0 by J_0 in Eq. (81). Riemann's method is also applicable when the function f and one of its derivatives, say $\partial f/\partial \alpha$, are given along any curve passing through A and B. Since $\partial f/\partial \beta$ can be found from $\partial f/\partial \alpha$ and the known variation of f along the given curve AB, the value of f at any point in the field may be determined as before by considering CAB as the relevant integration path.

6.6 Explicit Solutions for Direct Problems

(i) *Field defined by circular arcs of equal radii—I* As a first example to illustrate the analytical theory, consider the situation where each of the base sliplines CA and CB is a circular arc of radius $\sqrt{2}\,a$ (Fig. 6.14). Then $R = -\sqrt{2}\,a$, $S = -\sqrt{2}\,a(1 + \alpha)$ on $\beta = 0$ and $S = -\sqrt{2}\,a$, $R = -\sqrt{2}\,a(1 + \beta)$ on $\alpha = 0$, in view of (46). A comparison with (57), where f is taken to stand for R, indicates that $a_0 = c_0 = -\sqrt{2}\,a$ and $a_n = c_n = 0$ for $n \geqslant 1$. These values immediately furnish R, while S follows from it on using (46). The radii of curvature of the sliplines at any point (α, β) therefore become

$$R = -\sqrt{2}\,a\left[I_0(2\sqrt{\alpha\beta}) + \sqrt{\frac{\beta}{\alpha}} I_1(2\sqrt{\alpha\beta}) \right]$$

$$S = -\sqrt{2}\,a\left[\sqrt{\frac{\alpha}{\beta}} I_1(2\sqrt{\alpha\beta}) + I_0(2\sqrt{\alpha\beta}) \right] \quad (83)$$

It follows that $R(\alpha, \beta) = S(\beta, \alpha)$ as expected in view of the symmetry of the field.

---

† C. Carathéodory and E. Schmidt, *Z. angew. Math. Mech.*, **3**: 468 (1923); R. Hill, *The Mathematical Theory of Plasticity*, p. 153, Clarendon Press, Oxford (1950).

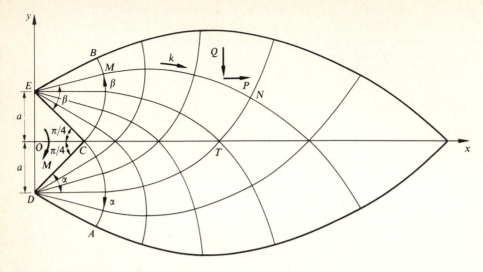

Figure 6.14 Symmetrical field defined by circular arcs of equal radii.

The leading terms in the brackets of (83) are the numerical values of R and S for a centered fan defined by an α baseline of unit radius. Similarly, the last terms in the brackets correspond to a centered fan defined by a β baseline of unit radius.

To obtain the rectangular coordinates, let the x axis be taken along the axis of symmetry and the y axis through the centers D and E of the circular arcs. The counterclockwise angle made by the α line with the x axis at a generic point N is $\phi = -\pi/4 + \beta - \alpha$. By (22), the coordinates of the slipline field may be written as

$$\sqrt{2}\,x = (\bar{x} + \bar{y})\cos(\beta - \alpha) + (\bar{x} - \bar{y})\sin(\beta - \alpha)$$
$$\sqrt{2}\,y = (\bar{x} + \bar{y})\sin(\beta - \alpha) - (\bar{x} - \bar{y})\cos(\beta - \alpha)$$

(84)

It is convenient here to work with $\bar{x} + \bar{y}$ and $\bar{x} - \bar{y}$ instead of $\bar{x}$ and $\bar{y}$ separately. From geometry, the boundary conditions are easily derived as

$$\bar{x} + \bar{y} = \sqrt{2}\,a(1 + \sin\alpha) \qquad \bar{x} - \bar{y} = -\sqrt{2}\,a(1 - \cos\alpha) \qquad \text{on } \beta = 0$$
$$\bar{x} + \bar{y} = \sqrt{2}\,a(1 + \sin\beta) \qquad \bar{x} - \bar{y} = -\sqrt{2}\,a(1 - \cos\beta) \qquad \text{on } \alpha = 0$$

The solution for $\bar{x} + \bar{y}$ and $\bar{x} - \bar{y}$ at a generic point in the field may therefore be written in the form

$$\frac{\bar{x} + \bar{y}}{\sqrt{2}\,a} = I_0(2\sqrt{\alpha\beta}) + 2F_1(\alpha, \beta) + \sin(\beta - \alpha)$$

(85)

$$\frac{\bar{x} - \bar{y}}{\sqrt{2}\,a} = I_0(2\sqrt{\alpha\beta}) - 2F_2(\alpha, \beta) - \cos(\beta - \alpha)$$

where the functions F_1 and F_2 must be of the form (56) and satisfy the conditions $F_1(0, \beta) = F_2(0, \beta) = 0$ and $F_1(\alpha, 0) = \sin \alpha$, $F_2(\alpha, 0) = 1 - \cos \alpha$. The coefficients c_n are therefore zero, while a_n are readily found from (57) using the series expansions of $\sin \alpha$ and $\cos \alpha$. Thus

$$F_1(\alpha, \beta) = \sum_{m=0}^{\infty} (-1)^m \left(\frac{\alpha}{\beta}\right)^{m+1/2} I_{2m+1}(2\sqrt{\alpha\beta})$$

$$F_2(\alpha, \beta) = \sum_{m=0}^{\infty} (-1)^m \left(\frac{\alpha}{\beta}\right)^{m+1} I_{2m+2}(2\sqrt{\alpha\beta}) \tag{86}$$

These functions are easily calculated† from a standard table for $\alpha \leqslant \beta$, which corresponds to $y \geqslant 0$. Because of symmetry, the slipline field for $\alpha \geqslant \beta$ is the mirror-image of that for $\alpha \leqslant \beta$. The functions $F_1(\alpha, \beta)$ and $F_2(\alpha, \beta)$ are related to one another by the equations

$$\frac{\partial F_1}{\partial \alpha} = I_0(2\sqrt{\alpha\beta}) - F_2 \qquad \frac{\partial F_2}{\partial \alpha} = F_1 \tag{87}$$

Similar expressions may be written down for the derivatives with respect to β, noting the fact that both F_1 and F_2 satisfy (48). Employing the series expansion for the modified Bessel functions it can be shown that

$$F_1(\beta, \alpha) - F_1(\alpha, \beta) = \sin(\beta - \alpha)$$

$$F_2(\beta, \alpha) + F_2(\alpha, \beta) = I_0(2\sqrt{\alpha\beta}) - \cos(\beta - \alpha) \tag{88}$$

which are useful properties of the functions F_1 and F_2. From (84) and (85), the rectangular coordinates of any point (α, β) are finally obtained as‡

$$\frac{x}{a} = [I_0(2\sqrt{\alpha\beta}) + 2F_1(\alpha, \beta)] \cos(\beta - \alpha)$$

$$+ [I_0(2\sqrt{\alpha\beta}) - 2F_2(\alpha, \beta)] \sin(\beta - \alpha)$$

$$\frac{y}{a} = 1 + [I_0(2\sqrt{\alpha\beta}) + 2F_1(\alpha, \beta)] \sin(\beta - \alpha) \tag{89}$$

$$- [I_0(2\sqrt{\alpha\beta}) - 2F_2(\alpha, \beta)] \cos(\beta - \alpha)$$

In view of (88), the above equations satisfy the conditions $x(\alpha, \beta) = x(\beta, \alpha)$ and $y(\alpha, \beta) = -y(\beta, \alpha)$, required by the symmetry of the field about the x axis. The last condition also ensures that $y = 0$ on the axis of symmetry. The distance of a generic point on the axis from the origin O is

$$x(\alpha, \alpha) = a[I_0(2\alpha) + A_0(2\alpha)]$$

where

$$A_0(2\alpha) = 2F_1(\alpha, \alpha) = \int_0^{2\alpha} I_0(z) \, dz \tag{90}$$

---

† The functions F_1, F_2, as well as similar ones occurring subsequently, may be evaluated directly from the double power series obtained by using (54) or (59), whichever is appropriate.

‡ J. Chakrabarty, *Int. J. Mech. Sci.*, **21**: 477 (1979).

The last equality follows from the repeated use of the second relation of (53) for successive values of n.

It may be easily verified by direct differentiation that Eqs. (87) are satisfied by the integral representation

$$F_1(\alpha, \beta) = \int_0^\alpha I_0(2\sqrt{t\beta}) \cos(\alpha - t)\, dt$$

$$F_2(\alpha, \beta) = \int_0^\alpha I_0(2\sqrt{t\beta}) \sin(\alpha - t)\, dt \tag{91}$$

The integrals are taken along a given α line, starting from the point where it is intersected by the β baseline. We now introduce two additional functions L and N defined as

$$L(\alpha, \beta) = \int_0^\alpha F_1(t, \beta) \cos(\alpha - t)\, dt$$

$$N(\alpha, \beta) = \int_0^\alpha F_1(t, \beta) \sin(\alpha - t)\, dt \tag{92}$$

Similar integrals involving F_2 may be obtained by writing $F_1 = \partial F_2 / \partial t$ in (92) and then integrating by parts. Differentiating (92) partially with respect to α and β, and remembering that $F_1(0, \beta) = 0$, we get

$$\frac{\partial L}{\partial \alpha} + N = F_1 \qquad \frac{\partial N}{\partial \alpha} = L \qquad \frac{\partial L}{\partial \beta} = N \tag{93}$$

The last two equations indicate that both L and N satisfy (48). The solution to these differential equations satisfying the conditions $L(0, \beta) = N(0, \beta) = 0$ may be written as

$$L(\alpha, \beta) = \sum_{m=1}^\infty (-1)^{m+1} m \left(\frac{\alpha}{\beta}\right)^m I_{2m}(2\sqrt{\alpha\beta})$$

$$N(\alpha, \beta) = \sum_{m=1}^\infty (-1)^{m+1} m \left(\frac{\alpha}{\beta}\right)^{m+1/2} I_{2m+1}(2\sqrt{\alpha\beta}) \tag{94}$$

which may be verified by direct substitution in (93). The new functions L and N are related to F_1 and F_2 by the equations

$$2L(\alpha, \beta) = (\alpha + \beta) F_1(\alpha, \beta) - \sqrt{\alpha\beta}\, I_1(2\sqrt{\alpha\beta})$$

$$2N(\alpha, \beta) = (\alpha + \beta) F_2(\alpha, \beta) + F_1(\alpha, \beta) - \alpha I_0(2\sqrt{\alpha\beta}) \tag{95}$$

which are easily shown to satisfy the differential equations (93) and the boundary conditions $L(0, \beta) = N(0, \beta) = 0$. Both (94) and (95) give $2L(\alpha, 0) = \alpha \sin\alpha$, $2N(\alpha, 0) = \sin\alpha - \alpha \cos\alpha$. It follows from (88) and (95) that

$$2L(\beta, \alpha) - 2L(\alpha, \beta) = (\alpha + \beta) \sin(\beta - \alpha)$$

$$2N(\beta, \alpha) + 2N(\alpha, \beta) = 2F_1(\alpha, \beta) + \sin(\beta - \alpha) - (\alpha + \beta) \cos(\beta - \alpha) \tag{96}$$

In the solution of physical problems we are often required to calculate integrals of the type $\int x\, d\alpha$ and $\int y\, d\alpha$ to find the components of the resultant traction across a slipline. Setting $\alpha = t$ in (89) and using (91) and (92), it can be shown that

$$\int_0^\alpha x(t, \beta)\, dt = a[(F_1 - F_2 + 4L) \cos (\beta - \alpha) + (F_1 - F_2 - 4N) \sin (\beta - \alpha)]$$

$$\int_0^\alpha y(t, \beta)\, dt = a\alpha + a[(F_1 - F_2 + 4L) \sin (\beta - \alpha) - (F_1 - F_2 - 4N) \cos (\beta - \alpha)]$$

(97)

where all the functions are considered at (α, β). The integrals involving F_2 have been reduced to those in terms of F_1 by using equations (87). Interchanging α and β in (97), and using (88) and (96), it is easily shown that

$$\int_0^\beta x(t, \alpha)\, dt = y(\alpha, \beta) + \int_0^\alpha x(t, \beta)\, dt$$

$$\int_0^\beta y(t, \alpha)\, dt = x(\alpha, \beta) - \int_0^\alpha y(t, \beta)\, dt - a(1 + \alpha + \beta)$$

(98)

These relations may be used to calculate the values of the integrals (97) for $\alpha \geqslant \beta$ when those for $\alpha \leqslant \beta$ are known. Accurate numerical values of x/a and y/a, as well as their associated integrals, are presented in Table A-1 for a $10°$ equiangular net and in Table A-2 for a $15°$ equiangular net (Appendix). Numerical values of the relevant mathematical functions used in the analysis are given in Tables A-3 and A-4 (Appendix).

If p_0 denotes the hydrostatic pressure at C, the normal compressive stress acting at a generic point (α, α) on the axis of symmetry is of magnitude $p_0 + k(1 + 4\alpha)$. Assuming CD and CE to be sliplines, the resultant normal force transmitted across the axis between the origin O and a typical point (α, α) may be written as

$$F = (p_0 + k)x(\alpha, \alpha) + 4k \int_0^\alpha \alpha \left(\frac{dx}{d\alpha} \right) d\alpha$$

on the basis of a unit width of the material. Since

$$\alpha \frac{dx}{d\alpha} = 2\alpha[I_1(2\alpha) + I_0(2\alpha)] = \frac{d}{d\alpha}[\alpha I_0(2\alpha) + \alpha I_1(2\alpha)] - I_0(2\alpha)$$

in view of (90) and (53), the integration can be carried out explicitly. Using (90) again, the result may be put in the form

$$F = (p_0 - k)x(\alpha, \alpha) + 2ka[(1 + 2\alpha)I_0(2\alpha) + 2\alpha I_1(2\alpha)]$$

(99)

If the α and β lines are interchanged, it is only necessary to change the sign of k in the above expression.

Consider now the resultant force acting across a typical slipline EMN, the material in the region ECB being regarded as plastic (Fig. 6.14). It is convenient to assume $p_0 = 0$ for the present purpose, so that interchanging the α and β lines of the field would result in a mere sign reversal of the force components.† Let (P, Q) denote the forces exerted on EN per unit width by the material above this slipline in the directions of the positive x axis and the negative y axis respectively. Since EN is an α line, it follows from Hencky's equations that

$$P = kx_N + 2k \int_A^N (t + \beta) \, dy \qquad Q = k(a - y_N) + 2k \int_A^N (t + \beta) \, dx$$

where t denotes the angle turned through along the slipline, and (x_N, y_N) are the coordinates of N. Integrating by parts, the above relations may be expressed as

$$P = k[x_N + 2(\alpha + \beta)y_N - 2\beta a - 2 \int_0^\alpha y(t, \beta) \, dt]$$

$$Q = k[(a - y_N) + 2(\alpha + \beta)x_N - 2 \int_0^\alpha x(t, \beta) \, dt]$$

(100)

The forces P and Q are functions of the angular coordinates (α, β) of the generic point N. In view of the symmetry of the field, (100) also gives the magnitude of the forces across the β line extending from D to the field point (β, α). It follows from (100) and (98) that

$$P(\beta, \alpha) + P(\alpha, \beta) = 2ka \qquad Q(\beta, \alpha) = Q(\alpha, \beta)$$

relating the force components for $\alpha \geqslant \beta$ to those for $\alpha \leqslant \beta$. These results are consistent with the fact that the resultant force across DE is a horizontal tension of magnitude $2ka$. Numerical values of $P/2ka$ and $Q/2ka$ are included in Tables A-1 and A-2 (Appendix). The dimensionless clockwise moment M/ka^2 of the tractions about O, calculated numerically using (70) and the symmetry condition, appears in the last column of Table A-1 over a range of $90°$ for α and β. It is important to note that $M(\beta, \alpha) = M(\alpha, \beta)$ in view of the symmetry of the field about the x axis.

(ii) *Field defined by equal logarithmic spirals* As another example, consider the slipline field defined by a pair of identical logarithmic spirals OA and OB with poles at D and E, each at a distance b from the axis of symmetry (Fig. 6.15). The angle turned through along each spiral is equal to the angle swept over by the radius vector to the curve from the corresponding pole. Since the radius vector makes an angle of $45°$ with the tangent to the spiral at each point, the polar equations of the base sliplines OA and OB are $r = be^{-\alpha}$ and $r = be^{-\beta}$ respectively. The radius of curvature at any point of either baseline is numerically equal to $\sqrt{2}\,r$, which gives

$$R = -\sqrt{2}\,be^{-\alpha} \qquad S = -\sqrt{2}\,b(2 - e^{-\alpha}) \qquad \text{on } \beta = 0$$

$$R = -\sqrt{2}\,b(2 - e^{-\beta}) \qquad S = -\sqrt{2}\,be^{-\beta} \qquad \text{on } \alpha = 0$$

in view of (46). The radii of curvature at any point within the field may therefore be

---

† It is easy to allow for any nonzero value of p_0 whose effect is to augment the hydrostatic pressure at each point of the field by the same amount.

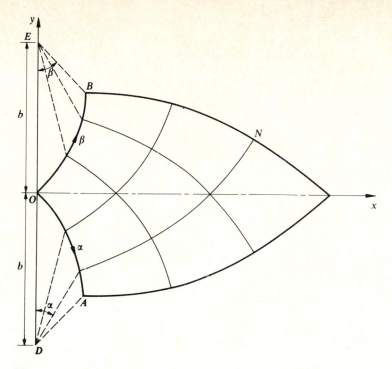

Figure 6.15 Slipline field defined by identical logarithmic spirals.

written as

$$R = -\sqrt{2}\,b[2I_0(2\sqrt{\alpha\beta}) - 2G(\alpha, \beta) - \exp(-\alpha - \beta)]$$

$$S = -\sqrt{2}\,b[2G(\alpha, \beta) + \exp(-\alpha - \beta)]$$

(101)

the exponential terms being suggested by (51) and the nature of the boundary conditions. Evidently, the function G is required to satisfy the conditions $G(0, \beta) = 0$, $G(\alpha, 0) = 1 - e^{-\alpha}$. From (57), it is easily found that

$$G(\alpha, \beta) = \sum_{n=0}^{\infty} (-1)^n \left(\frac{\alpha}{\beta}\right)^{(n+1)/2} I_{n+1}(2\sqrt{\alpha\beta})$$

(102)

Which can be calculated for $\alpha \leqslant \beta$ using a table. It is interesting to note that $R + S$ has the magnitude $2\sqrt{2}\,bI_0(2\sqrt{\alpha\beta})$ everywhere. The symmetry requirement $R(\alpha, \beta) = S(\beta, \alpha)$ is fulfilled in view of the relation

$$G(\alpha, \beta) + G(\beta, \alpha) = I_0(2\sqrt{\alpha\beta}) - \exp(-\alpha - \beta)$$

(103)

which may be verified by using the series expansion of the modified Bessel functions. The coordinates of the field may be obtained by determining $\bar{x} + \bar{y}$ and $\bar{x} - \bar{y}$ as before. It follows from geometry that $x = be^{-\alpha}\sin\alpha$, $y = -b(1 - e^{-\alpha}\cos\alpha)$ on $\beta = 0$ and $x = be^{-\beta}\sin\beta$, $y = b(1 - e^{-\beta}\cos\beta)$ on $\alpha = 0$. Using (84), we obtain the boundary conditions

$$\bar{x} + \bar{y} = \sqrt{2}\,b \sin \alpha \qquad \bar{x} - \bar{y} = \sqrt{2}\,b\,(\cos \alpha - e^{-\alpha}) \qquad \text{on } \beta = 0$$

$$\bar{x} + \bar{y} = \sqrt{2}\,b \sin \beta \qquad \bar{x} - \bar{y} = -\sqrt{2}\,b\,(\cos \beta - e^{-\beta}) \qquad \text{on } \alpha = 0$$

which are very similar to those of the previous problem. The solution for $\bar{x} + \bar{y}$ and $\bar{x} - \bar{y}$ satisfying these conditions are similarly obtained as

$$\frac{\bar{x} + \bar{y}}{\sqrt{2}\,b} = 2F_1(\alpha, \beta) + \sin(\beta - \alpha)$$

$$\frac{\bar{x} - \bar{y}}{\sqrt{2}\,b} = -2F_2(\alpha, \beta) + 2G(\alpha, \beta) - \cos(\beta - \alpha) + \exp(-\alpha - \beta)$$

(104)

where F_1 and F_2 are given by (86) and G by (102). Substitution in (84) now furnishes the rectangular coordinates of the nodal points of the field as[†]

$$\frac{x}{b} = 2F_1 \cos(\beta - \alpha) + [2G - 2F_2 + \exp(-\alpha - \beta)] \sin(\beta - \alpha)$$

$$\frac{y}{b} = 1 + 2F_1 \sin(\beta - \alpha) - [2G - 2F_2 + \exp(-\alpha - \beta)] \cos(\beta - \alpha)$$

(105)

It follows from Eqs. (88) and (103) that the conditions $x(\alpha, \beta) = x(\beta, \alpha)$ and $y(\alpha, \beta) = -y(\beta, \alpha)$ are automatically satisfied, as required by the symmetry of the field with respect to the x axis. The distance of a generic point on the axis ($\beta = \alpha$) from the origin is $bA_0(2\alpha)$ where $A_0(2\alpha)$ is defined in (90). Numerical values of $G(\alpha, \beta)$, $R/\sqrt{2}\,b$, $S/\sqrt{2}\,b$, x/b and y/b for a $10°$ equiangular net are given in Table A-5 (Appendix).

(iii) *Field defined by circular arcs of unequal radii* Suppose that the base sliplines CA and CB are circular arcs of unequal radii, the distance DE between the centers being $2a$ (Fig. 6.16). Let λ be the counterclockwise angle made with ED by the tangent to the α line at C. Since the radii of curvature of the α and β baselines are $2a \sin \lambda$ and $2a \cos \lambda$ respectively, the expressions for R and S at any point (α, β) within the field become

$$R = -2a\left\{ I_0(2\sqrt{\alpha\beta}) \sin \lambda + \sqrt{\frac{\beta}{\alpha}}\, I_1(2\sqrt{\alpha\beta}) \cos \lambda \right\}$$

$$S = -2a\left\{ I_0(2\sqrt{\alpha\beta}) \cos \lambda + \sqrt{\frac{\alpha}{\beta}}\, I_1(2\sqrt{\alpha\beta}) \sin \lambda \right\}$$

(106)

To find the rectangular coordinates of the field defined by CA and CB, we choose our x axis along OC which is perpendicular to the line joining the centers D and E.

[†] J. Chakrabarty, *Int. J. Mech. Sci.*, **21**: 477 (1979).

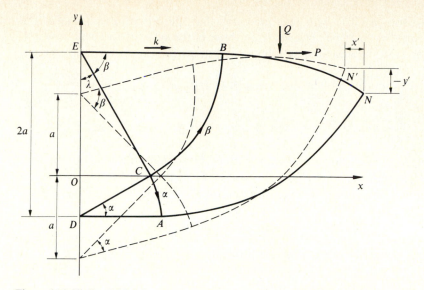

Figure 6.16 Slipline field defined by circular arcs of unequal radii. The associated symmetrical field is shown broken.

If the origin is taken temporarily at C, the boundary conditions may be written as

$$\bar{x} = 2a \sin \lambda \sin \alpha \qquad \bar{y} = 2a \sin \lambda(1 - \cos \alpha) \qquad \beta = 0$$

$$\bar{x} = 2a \cos \lambda(1 - \cos \beta) \qquad \bar{y} = 2a \cos \lambda \sin \beta \qquad \alpha = 0$$

The solution for $\bar{x}$ and $\bar{y}$, satisfying these conditions, can be immediately written down in terms of the functions F_1 and F_2 as

$$\bar{x} = 2a\{F_1(\alpha, \beta) \sin \lambda + F_2(\beta, \alpha) \cos \lambda\}$$

$$\bar{y} = 2a\{F_2(\alpha, \beta) \sin \lambda + F_1(\beta, \alpha) \cos \lambda\}$$

The coordinates of the field, when the y axis is taken through C, are obtained by substituting for $(\bar{x}, \bar{y})$ in (22) with $\phi = -\pi/2 + (\lambda + \beta - \alpha)$. If the y axis is now taken along OE as shown, it is only necessary to add $OC = a \sin 2\lambda$ to the expression for x. Using (88), we finally obtain the rectangular coordinates at any point (α, β) as

$$\frac{x}{a} = 2I_0(2\sqrt{\alpha\beta}) \cos \lambda \sin (\lambda + \beta - \alpha) + 2F_1(\alpha, \beta) \cos (\beta - \alpha)$$

$$- 2F_2(\alpha, \beta) \sin (\beta - \alpha)$$

$$\frac{y}{a} = 2 \cos^2 \lambda - 2I_0(2\sqrt{\alpha\beta}) \cos \lambda \cos (\lambda + \beta - \alpha)$$

$$+ 2F_1(\alpha, \beta) \sin (\beta - \alpha) + 2F_2(\alpha, \beta) \cos (\beta - \alpha)$$

$$(107)$$

The comparison of (107) with (89) indicates that the coordinates (x, y) of the unsymmetrical field may be obtained from those of the symmetrical field $(\lambda = \pi/4)$ for the same (α, β) by adding quantities (x', y') given by

$$\frac{x'}{a} = [\sin (2\lambda + \beta - \alpha) - \cos (\beta - \alpha)]I_0(2\sqrt{\alpha\beta})$$

$$\frac{y'}{a} = \cos 2\lambda - [\cos (2\lambda + \beta - \alpha) + \sin (\beta - \alpha)]I_0(2\sqrt{\alpha\beta})$$

(108)

The symmetrical field (shown broken) is defined by circular arcs having centers at $(0, \pm a)$. It follows from (108) that x' is positive or negative according as $\beta - \alpha$ is greater or less than $\pi/4 - \lambda$. From (108) and (91), it is easy to show that

$$\int_0^\alpha x'(t, \beta) \, dt = a[\sin (2\lambda + \beta - \alpha) - \cos (\beta - \alpha)]F_1(\alpha, \beta)$$

$$+ a[\cos (2\lambda + \beta - \alpha) + \sin (\beta - \alpha)]F_2(\alpha, \beta)$$

$$\int_0^\alpha y'(t, \beta) \, dt = a\alpha \cos 2\lambda - a[\cos (2\lambda + \beta - \alpha) + \sin (\beta - \alpha)]F_1(\alpha, \beta)$$

$$+ a[\sin (2\lambda + \beta - \alpha) - \cos (\beta - \alpha)]F_2(\alpha, \beta)$$

(109)

For $\alpha \leqslant \beta$, these integrals may be readily evaluated by using the tabulated values of F_1 and F_2. When $\alpha \geqslant \beta$, it is necessary to use (88) to obtain F_1 and F_2 from the tabulated data. Assuming the region ECB to be plastic, the components of the resultant force across a typical α line EN, reckoned positive as shown, are found as

$$P = k\left[x_N + 2(\alpha + \beta)y_N - 4a\beta \cos^2 \lambda - 2\int_0^\alpha y(t, \beta) \, dt \right]$$

$$Q = k\left[2a \cos^2 \lambda - y_N + 2(\alpha + \beta)x_N - 2\int_0^\alpha x(t, \beta) \, dt \right]$$

(110)

The derivation of these expressions is identical to that of (100). Numerical values of the relevant parameters for a 15° equiangular net are given in Table A-6 (Appendix), corresponding to $\lambda = 0, 15,$ and 30°. When $\lambda = 0$, the points C and O coincide with D, and the field reduces to a centered fan defined by a circular arc of radius $2a$, the x axis being tangential to this arc at the origin.

(iv) Field defined by circular arcs of equal radii—II Consider the field in which each of the base sliplines is again a circular arc of radius $\sqrt{2}\,a$, but the curvature of the α baseline is now opposite to that of the previous problem.† In Fig. 6.17, the field defined by the circular arcs CA and CB is therefore without an axis of symmetry. The radii of curvature of the sliplines, satisfying Eqs. (46) with $-\alpha$ written for α, and the boundary conditions $R = \sqrt{2}\,a$ on $\beta = 0$ and $S = -\sqrt{2}\,a$ on $\alpha = 0$, can be

† Unpublished work of the author (1976).

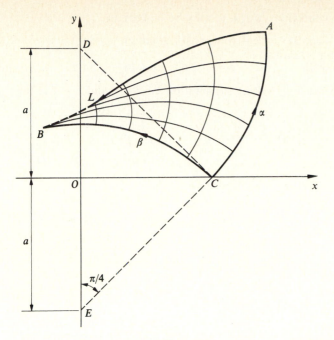

Figure 6.17 Unsymmetrical field defined by circular arcs of equal radii. The broken curve *BL* is a limiting line.

written down in terms of the Bessel functions J_0 and J_1, the result being

$$R = \sqrt{2}\,a\left[J_0(2\sqrt{\alpha\beta}) - \sqrt{\frac{\beta}{\alpha}}\,J_1(2\sqrt{\alpha\beta})\right]$$

$$S = -\sqrt{2}\,a\left[J_0(2\sqrt{\alpha\beta}) + \sqrt{\frac{\alpha}{\beta}}\,J_1(2\sqrt{\alpha\beta})\right]$$

(111)

For a given α, R decreases as β is increased. Hence there exists a limiting line *BL* along which R vanishes, and which forms an envelope of the intersecting β lines. Values of α and β defining this limiting line are given in the following table:

$\alpha°$	0	10	20	30	40	50	60
$\beta°$	57.30	52.60	48.39	44.70	41.43	38.54	36.0
x/a	−0.301	−0.189	−0.090	−0.004	0.067	0.137	0.176
y/a	0.382	0.412	0.448	0.489	0.536	0.596	0.643

Choosing the origin of rectangular coordinates at the midpoint of the line of centers DE, the boundary conditions for $(\bar{x}, \bar{y})$ may be written as

$$\bar{x} + \bar{y} = -\sqrt{2}\,a(1 - \cos\alpha) \qquad \bar{x} - \bar{y} = \sqrt{2}\,a(1 + \sin\alpha) \qquad \text{on } \beta = 0$$

$$\bar{x} + \bar{y} = \sqrt{2}\,a(1 - \cos\beta) \qquad \bar{x} - \bar{y} = \sqrt{2}\,a(1 - \sin\beta) \qquad \text{on } \alpha = 0$$

These conditions can be satisfied by writing the solution for $\bar{x} + \bar{y}$ and $\bar{x} - \bar{y}$ in the form

$$\frac{\bar{x} + \bar{y}}{\sqrt{2}\,a} = J_0(2\sqrt{\alpha\beta}) - 2F_2^*(\alpha, \beta) - \cos(\alpha + \beta)$$

$$\frac{\bar{x} - \bar{y}}{\sqrt{2}\,a} = J_0(2\sqrt{\alpha\beta}) + 2F_1^*(\alpha, \beta) - \sin(\alpha + \beta)$$

(112)

where F_1^* and F_2^* must be of the form (61), and assume the values $F_1^*(0, \beta) = F_2^*(0, \beta) = 0$, and $F_1^*(\alpha, 0) = \sin\alpha$, $F_2^*(\alpha, 0) = 1 - \cos\alpha$. All these conditions are satisfied by taking

$$F_1^*(\alpha, \beta) = \sum_{m=0}^{\infty} (-1)^m \left(\frac{\alpha}{\beta}\right)^{m+1/2} J_{2m+1}(2\sqrt{\alpha\beta})$$

$$F_2^*(\alpha, \beta) = \sum_{m=0}^{\infty} (-1)^m \left(\frac{\alpha}{\beta}\right)^{m+1} J_{2m+2}(2\sqrt{\alpha\beta})$$

(113)

These functions can be calculated for $\alpha \leqslant \beta$ using Table A-8 for the Bessel functions. Evidently, both F_1^* and F_2^* are solutions of (48) with the appropriate sign change in α. Using the series formula (59), it can be shown that

$$F_1^*(\alpha, \beta) + F_1^*(\beta, \alpha) = \sin(\alpha + \beta)$$

$$F_2^*(\alpha, \beta) + F_2^*(\beta, \alpha) = J_0(2\sqrt{\alpha\beta}) - \cos(\alpha + \beta)$$

(114)

The rectangular coordinates of a generic point (α, β) are found by setting $\phi = \pi/4 + \alpha + \beta$ in (22), and then substituting from (101), resulting in

$$\frac{x}{a} = [J_0(2\sqrt{\alpha\beta}) + 2F_1^*(\alpha, \beta)]\cos(\alpha + \beta)$$

$$- \{J_0(2\sqrt{\alpha\beta}) - 2F_2^*(\alpha, \beta)]\sin(\alpha + \beta)$$

(115)

$$\frac{y}{a} = -1 + [J_0(2\sqrt{\alpha\beta}) + 2F_1^*(\alpha, \beta)]\sin(\alpha + \beta)$$

$$+ [J_0(2\sqrt{\alpha\beta}) - 2F_2^*(\alpha, \beta)]\cos(\alpha + \beta)$$

In view of (114) and (115), the coordinates at any point (β, α) are related to those at (α, β) by the equations

$$x(\alpha, \beta) + x(\beta, \alpha) = 2aJ_0(2\sqrt{\alpha\beta})\cos(\alpha + \beta)$$

$$y(\alpha, \beta) + y(\beta, \alpha) = 2aJ_0(2\sqrt{\alpha\beta})\sin(\alpha + \beta)$$

(116)

Values of x and y along the envelope BL are included in the above table. It follows from (116) that the radius vector from the origin O to any point (α, α) is of length $aJ_0(2\alpha)$, and is inclined at an angle 2α to the x axis. The radial direction is therefore a principal stress direction at each point of the curve $\beta = \alpha$. Accurate numerical values of x/a, y/a, $R/\sqrt{2}\,a$, $-S/\sqrt{2}\,a$, for a $10°$ equiangular net are given in Table A-7 (Appendix).

6.7 Some Mixed Boundary-Value Problems

(i) *Slipline field near a straight limiting line* Consider the mixed boundary-value problem in which we are required to determine the slipline field between a circular base slipline OA of radius a (Fig. 6.18), and a plane boundary OB along which the shear stress has its maximum value k. Thus OB is the envelope of one family of sliplines, and is at right angles to the sliplines of the other family. If OA is an α line, the radius of curvature of the β lines must vanish along OB, where $\alpha = \beta$. Hence the boundary conditions for R and S are

$$R(\alpha, 0) = -a \qquad S(\alpha, \alpha) = 0$$

These conditions, together with the equation $S = \partial R/\partial \beta$, furnish the solution

$$R(\alpha, \beta) = -a\left[I_0(2\sqrt{\alpha\beta}) - \frac{\beta}{\alpha} I_2(2\sqrt{\alpha\beta}) \right]$$

$$S(\alpha, \beta) = -a\left(\sqrt{\frac{\alpha}{\beta}} - \sqrt{\frac{\beta}{\alpha}} \right) I_1(2\sqrt{\alpha\beta})$$

(117)

The β base curve, which lies above the limiting line, has a positive curvature.

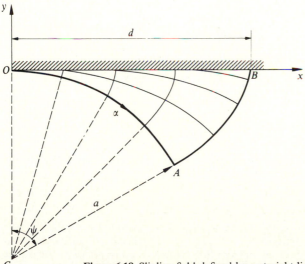

Figure 6.18 Slipline field defined by a straight limiting line and a given circular arc.

Let the x and y axes be taken along and perpendicular to OB with the origin at O. Then the boundary conditions for $\bar{x}$ and $\bar{y}$ can be written as

$$\bar{x}(\alpha, 0) = a \sin \alpha \qquad \bar{y}(\alpha, 0) = a(1 - \cos \alpha) \qquad \bar{y}(\alpha, \alpha) = 0$$

The last two conditions are sufficient to find $\bar{y}$, from which $\bar{x}$ follows in view of the equation $\bar{x} = \partial \bar{y}/\partial \alpha$. The solution for $\bar{x}$ and $\bar{y}$ therefore becomes

$$\bar{y}(\alpha, \beta) = a[I_0(2\sqrt{\alpha\beta}) - 2F_2(\beta, \alpha) - \cos(\alpha - \beta)]$$

$$\bar{x}(\alpha, \beta) = a\left[-\sqrt{\frac{\beta}{\alpha}} I_1(2\sqrt{\alpha\beta}) + 2F_1(\beta, \alpha) + \sin(\alpha - \beta) \right] \tag{118}$$

where F_1 and F_2 are given by (86). Substituting in (22) with $\phi = \beta - \alpha$, the rectangular coordinates of a generic point (α, β) are obtained as†

$$\frac{x}{a} = \left[2F_1(\beta, \alpha) - \sqrt{\frac{\beta}{\alpha}} I_1(2\sqrt{\alpha\beta}) \right] \cos(\alpha - \beta)$$

$$+ [I_0(2\sqrt{\alpha\beta}) - 2F_2(\beta, \alpha)] \sin(\alpha - \beta) \tag{119}$$

$$-\frac{y}{a} = 1 + \left[2F_1(\beta, \alpha) - \sqrt{\frac{\beta}{\alpha}} I_1(2\sqrt{\alpha\beta}) \right] \sin(\alpha - \beta)$$

$$- [I_0(2\sqrt{\alpha\beta}) - 2F_2(\beta, \alpha)] \cos(\alpha - \beta)$$

It is interesting to note that the right-hand sides of (119) differ from those of (89), with α and β interchanged, by terms that depend on $R(\alpha, \beta)$ for the field defined by opposed circular arcs of unit radius.‡ The distance of a generic point on the boundary from the origin O is

$$x(\alpha, \alpha) = a[A_0(2\alpha) - I_1(2\alpha)] \tag{120}$$

where $A_0(z)$ is defined in (90). Values of x/a and y/a at the nodal points of a $15°$ equiangular net are given in Table A-9 (Appendix).

Let p_0 denote the hydrostatic pressure at O. Then the normal pressure at any point on the plane boundary is $p_0 + 4k\alpha$. If the angular span of OA is denoted by ψ, the resultant normal force acting on OB per unit width, in view of (120), may be written as

$$Q = p_0 d + 4ka \int_0^\psi 2\alpha[I_0(2\alpha) - I_1'(2\alpha)] \, d\alpha$$

where $d = x(\psi, \psi)$. Since the expression within the integral is equal to $I_1(2\alpha) \, d\alpha$ by

---

† J. Chakrabarty, *Int. J. Mech. Sci.*, **21**: 477 (1979).

‡ If we construct the field between a pair of circular arcs of radius $\sqrt{2}\,a$ with centers at $(0, \pm a)$, and drop a perpendicular from the center of curvature of the α line at a typical point (α, β) of the field on to the line of action of the major principal stress at (α, β), then the foot of this perpendicular defines the corresponding point (α, β) of the field shown in Fig. 6.18.

(53), we finally obtain

$$Q = p_0 d + 2ka[I_0(2\psi) - 1] \tag{121}$$

The components of the resultant force per unit width across a typical β line through C, when the region COA is plastic with $p_0 = 0$, are given by expressions similar to (100). From (119), (91) and (92), it is easily shown that

$$\int_0^\beta x(\alpha, t)\, dt = a[(4L - 2F_2)\cos(\alpha - \beta) + (I - 4N)\sin(\alpha - \beta)]$$
$$\tag{122}$$
$$-\int_0^\beta y(\alpha, t)\, dt = a\beta + a[(4L - 2F_2)\sin(\alpha - \beta) - (I - 4N)\cos(\alpha - \beta)]$$

where $I = \sqrt{\beta/\alpha}\, I_1(2\sqrt{\alpha\beta})$, and all other functions correspond to the angular coordinates (β, α). Numerical values of these integrals are included in Table A-9.

An explicit solution may also be obtained in the important situation where the base slipline BE belongs to the field defined by a pair of opposed circular arcs of equal radii (Fig. 6.19). The plane boundary AF passes through the center A of the circular arc BC of radius $\sqrt{2}\, a$ and angular span χ. The radius of curvature of BE is evidently given by the first equation of (83) with $\beta = \chi$. Thus

$$R = -\sqrt{2}\, a\left[I_0(2\sqrt{\alpha\chi}) + \sqrt{\frac{\chi}{\alpha}}\, I_1(2\sqrt{\alpha\chi}) \right] \qquad \text{along } BE$$

which corresponds to $\beta = 0$ for the field BEF. Introducing new variables ξ and η through the transformation

$$\xi = \chi + \alpha \qquad \eta = \chi + \beta$$

the radii of curvature of the sliplines in BEF, satisfying Eqs. (46) and the boundary

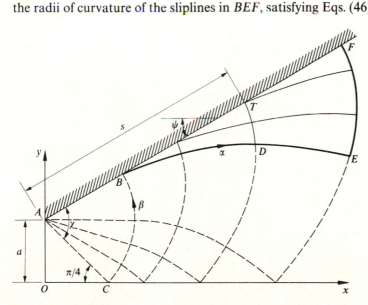

Figure 6.19 Slipline field between a limiting line and an axis of symmetry.

conditions on $\beta = 0$ and $\beta = \alpha$, can be written as

$$-\frac{R}{\sqrt{2}\,a} = I_0(2\sqrt{\alpha\eta}) + \sqrt{\frac{\eta}{\alpha}}\,I_1(2\sqrt{\alpha\eta}) - \sqrt{\frac{\beta}{\xi}}\,I_1(2\sqrt{\xi\beta}) - \frac{\beta}{\xi}\,I_2(2\sqrt{\xi\beta})$$

$$-\frac{S}{\sqrt{2}\,a} = I_0(2\sqrt{\alpha\eta}) + \sqrt{\frac{\alpha}{\eta}}\,I_1(2\sqrt{\alpha\eta}) - I_0(2\sqrt{\xi\beta}) - \sqrt{\frac{\beta}{\xi}}\,I_1(2\sqrt{\xi\beta})$$

(123)

It follows that S is discontinuous across BE by an amount $\sqrt{2}\,a$. Consider now the solution for $\bar{x}$ and $\bar{y}$, when the origin of coordinates is as shown. Setting $\beta = \chi$ in (85) furnishes the boundary condition

$$\bar{y} = \sqrt{2}\,a[F_1(\alpha, \chi) + F_2(\alpha, \chi)] + a\cos\left(\chi - \alpha - \frac{\pi}{4}\right) \qquad \text{on } BE$$

The solution for $\bar{y}$ which reduces to the above expression when $\beta = 0$, and assumes the value $a\cos(\chi - \pi/4)$ on $\beta = \alpha$, can be immediately written as

$$\frac{\bar{y}}{\sqrt{2}\,a} = F_1(\alpha, \eta) + F_2(\alpha, \eta) - F_1(\beta, \xi) - F_2(\beta, \xi) + \frac{1}{\sqrt{2}}\cos\left(\eta - \alpha - \frac{\pi}{4}\right)$$

the partial derivative of which with respect to α gives $\bar{x}$. The results may be conveniently put in the form

$$\frac{\bar{x} + \bar{y}}{\sqrt{2}\,a} = I_0(2\sqrt{\alpha\eta}) + 2F_1(\alpha, \eta) - 2F_2(\beta, \xi) - \sqrt{\frac{\beta}{\xi}}\,I_1(2\sqrt{\beta\xi}) + \sin(\eta - \alpha)$$

$$\frac{\bar{x} - \bar{y}}{\sqrt{2}\,a} = I_0(2\sqrt{\alpha\eta}) - 2F_2(\alpha, \eta) + 2F_1(\beta, \xi) - \sqrt{\frac{\beta}{\xi}}\,I_1(2\sqrt{\beta\xi}) - \cos(\eta - \alpha)$$

(124)

Substituting from (124) into (84), with β replaced by η, furnishes the rectangular coordinates for the field BEF. Comparison with (89) then shows that these coordinates may be expressed as

$$x(\alpha, \beta) = x^\circ(\alpha, \eta) + x'(\alpha, \beta)$$

$$y(\alpha, \beta) = y^\circ(\alpha, \eta) + y'(\alpha, \beta)$$

where (x°, y°) denote the rectangular coordinates for the symmetrical field defined by a pair of circular arcs through C, while (x', y') are additional coordinates given by†

$$\frac{x'}{a} = \left[2F_2(\beta, \xi) + \sqrt{\frac{\beta}{\xi}}\,I_1(2\sqrt{\beta\xi})\right]\cos(\eta - \alpha)$$

$$+ \left[2F_1(\beta, \xi) - \sqrt{\frac{\beta}{\xi}}\,I_1(2\sqrt{\beta\xi})\right]\sin(\eta - \alpha) \quad (125a)$$

---

† These results have been obtained by J. Chakrabarty in an unpublished work (1976).

$$\frac{y'}{a} = \left[2F_2(\beta, \xi) + \sqrt{\frac{\beta}{\xi}} I_1(2\sqrt{\beta\xi}) \right] \sin(\eta - \alpha)$$

$$- \left[2F_1(\beta, \xi) - \sqrt{\frac{\beta}{\xi}} I_1(2\sqrt{\beta\xi}) \right] \cos(\eta - \alpha) \quad (125b)$$

For an equiangular net with $\beta \leqslant \alpha \leqslant \chi$, these coordinates may be calculated from the tabulated values of the relevant functions. It follows from the condition of symmetry of the $(x°, y°)$ field that

$$\int_0^\beta x(\alpha, t)\, dt = \int_\chi^\eta x°(t, \alpha)\, dt + \int_0^\beta x'(\alpha, t)\, dt$$

$$\int_0^\beta y(\alpha, t)\, dt = -\int_\chi^\eta y°(t, \alpha)\, dt + \int_0^\beta y'(\alpha, t)\, dt$$

The second integrals on the right-hand side can be expressed, as before, in terms of known functions. After some algebraic manipulation, using (98), we finally obtain

$$\int_0^\beta x(\alpha, t)\, dt = \int_0^\alpha x°(t, \eta)\, dt - \int_0^\alpha x°(t, \chi)\, dt - y°(\alpha, \chi) + y(\alpha, \beta) + aE$$

$$\int_0^\beta y(\alpha, t)\, dt = \int_0^\alpha y°(t, \eta)\, dt - \int_0^\alpha y°(t, \chi)\, dt + x°(\alpha, \chi) - x(\alpha, \beta) + a(\beta - F)$$

(126)

where E and F denote the expressions in the upper and lower square brackets respectively of (97) with $\cos(\beta - \alpha)$ and $\sin(\beta - \alpha)$ replaced by $\sin(\eta - \alpha)$ and $\cos(\eta - \alpha)$ respectively, the functions F_1, F_2, L, and N being considered at (β, ξ). Table A-10 (Appendix) gives numerical values of the relevant parameters of the field for different values of $\psi = \chi - \pi/4$.

Let s denote the distance from A of a generic point T on the plane boundary. Since $\bar{x} + \bar{y} = s + \sqrt{2}\, a \sin \chi$ along BE by simple geometry, it follows from (124) that

$$s = \sqrt{2}\, a \left[I_0(2\sqrt{\alpha\xi}) - \sqrt{\frac{\alpha}{\xi}} I_1(2\sqrt{\alpha\xi}) + 2F_1(\alpha, \xi) - 2F_2(\alpha, \xi) \right] \quad (127)$$

In view of (62), the differentiation of the above expression with respect to α furnishes

$$\alpha \frac{ds}{d\alpha} = \sqrt{2}\, a \left[(1 + \chi) \sqrt{\frac{\alpha}{\xi}} I_1(2\sqrt{\alpha\xi}) + \chi \left(\frac{\alpha}{\xi} \right) I_2(\sqrt{\alpha\xi}) \right]$$

The normal pressure acting on the plane boundary at (α, α) is equal to $p_0 + 4k\alpha$ by Hencky's equations. The resultant normal force per unit width transmitted across AT is

$$Q = p_0 s + 4k \int_0^s \alpha\, ds = p_0 s + 4k \int_0^\alpha \alpha \left(\frac{ds}{d\alpha} \right) d\alpha$$

where p_0 is the hydrostatic pressure on AB. Inserting from the preceding equation, and noting the fact that

$$\frac{d}{d\alpha}[F_2(\alpha, \xi)] = \sqrt{\frac{\alpha}{\xi}} I_1(2\sqrt{\alpha\xi}) \qquad \frac{d}{d\alpha}[F_1(\alpha, \xi)] = I_0(2\sqrt{\alpha\xi})$$

in view of (62), we finally obtain

$$Q = p_0 s + 4\sqrt{2}\,ka\left\{(1 + \chi)F_2(\alpha, \xi) + \chi\left[\sqrt{\frac{\alpha}{\xi}} I_1(2\sqrt{\alpha\xi}) - F_1(\alpha, \xi)\right]\right\} \quad (128)$$

where s is given by (127). When p_0/k is specified, the value of Q/ka can be readily calculated from (128) as a function of α.

When the shape of the given base slipline is arbitrary, its radius of curvature $R(\alpha, 0)$ may be expressed in the form of the power series (63) with known values of a_n. The boundary condition $S(\alpha, \alpha) = 0$ then furnishes the relation $b_n = -a_{n-1}$ with $b_0 = 0$. Once the coefficients have been found, the series method of solution may be employed to derive the slipline field.

(ii) *Generating a stress-free boundary* A base slipline LM is given (Fig. 6.20). We propose to determine the slipline field LMN such that the normal and tangential tractions along the unknown boundary LN are zero. Let LM be an α line, the other baseline being unknown. The vanishing of the normal and shear stresses across LN requires $p = -k$ along the boundary, and all sliplines meet the boundary at $45°$. It follows from the Hencky equations that $\beta = \alpha$ along the boundary, 2α being the

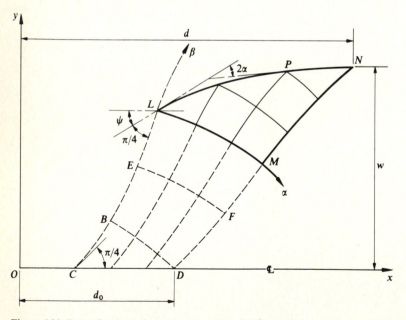

Figure 6.20 Stress-free boundary associated with slipline field containing an axis of symmetry.

clockwise angle made by the tangent to the boundary at a generic point P with that at L. Considering a triangular element at P formed by a pair of intersecting sliplines, we find $-R\,d\alpha = S\,d\beta = \sqrt{2}\,\rho\,d\alpha$, where ρ is the radius of curvature of LN at P. This leads to the boundary condition

$$-R(\alpha, \alpha) = S(\alpha, \alpha) = \sqrt{2}\,\rho$$

where R and S are related to one another by the equations

$$\frac{\partial R}{\partial \beta} = -S, \qquad \frac{\partial S}{\partial \alpha} = R \tag{129}$$

Let the quantities $-R(\alpha, 0)$ and $S(0, \beta)$ be given by the right-hand sides of (63), where a_n are known constants. Then the solution for $-R(\alpha, \beta)$ may be expressed in the form (61) with $c_n = b_n$. In view of (129), the boundary condition $R = -S$ leads to the recurrence relationship†

$$b_{n+1} - b_n = a_n + a_{n+1} \qquad \text{(with } b_0 = a_0\text{)}$$

for the unknown coefficients b_n. The radius of curvature at a generic point of LN is given by

$$\sqrt{2}\,\rho = a_0 J_0(2\alpha) + \sum_{n=1}^{\infty} (a_n + b_{n-1}) J_n(2\alpha) \tag{130}$$

Since ρ is equal to $\frac{1}{2}(ds/d\alpha)$, where s is the arc length measured from L, the shape of the stress-free boundary can be determined from (130). Changing the sign of β in (47) and (49), and combining these equations, the boundary condition $R = -S$ can be written in terms of $\bar{x}$ and $\bar{y}$ as

$$\frac{\partial}{\partial \alpha}(\bar{x} + \bar{y}) = \frac{\partial}{\partial \beta}(\bar{x} + \bar{y}) \qquad \text{on } \alpha = \beta$$

If $\bar{x} + \bar{y}$ is expressed in the form (61), where a_n and c_n are new constants, it follows from above that $c_n = a_{n+1}$. Thus, $\bar{x} + \bar{y}$ is symmetrical with respect to α and β everywhere in the field. The quantities $\bar{x}$ and $\bar{y}$ are directly obtainable from (66), using the series method.

Suppose that the known slipline field to the left of LM contains an axis of symmetry which is taken as the x axis. The β line $CBEL$ emanating from the point C on the axis turns through an angle ψ to reach L. Then the angle which the α direction at L makes with x axis is $\phi_0 = -(\pi/4 - \psi)$. Substituting $\phi = -\pi/4 + \psi - \alpha - \beta$ in (22) gives

$$\sqrt{2}\,x = (\bar{x} + \bar{y})\cos(\psi - \alpha - \beta) + (\bar{x} - \bar{y})\sin(\psi - \alpha - \beta)$$
$$\sqrt{2}\,y = (\bar{x} + \bar{y})\sin(\psi - \alpha - \beta) - (\bar{x} - \bar{y})\cos(\psi - \alpha - \beta) \tag{131}$$

An explicit solution can be found if CE is assumed to be a circular arc of radius

---

† D. F. J. Ewing, *J. Mech. Phys. Solids*, **15**: 105 (1967).

$\sqrt{2}\,a$. It is also assumed that a portion EL of the bounding slipline is straight. Then the radius of curvature at any point of LM numerically exceeds that at the corresponding point of EF by the length of each straight segment. Denoting this length by $\sqrt{2}\,c_0$, we have

$$-\frac{R}{\sqrt{2}} = a\left[I_0(2\sqrt{\alpha\psi}) + \sqrt{\frac{\psi}{\alpha}}\, I_1(2\sqrt{\alpha\psi}) \right] + c_0 \qquad \text{along } LM$$

in view of (83). This is the boundary condition for R on $\beta = 0$ for the required field LMN. The solution for $R(\alpha, \beta)$ may therefore be assumed in the form

$$-\frac{R}{\sqrt{2}} = a\left[I_0(2\sqrt{\alpha\eta}) + \sqrt{\frac{\eta}{\alpha}}\, I_1(2\sqrt{\alpha\eta}) \right] + \sum_{n=0}^{\infty} c_n \left(\frac{\beta}{\xi}\right)^{n/2} I_n(2\sqrt{\xi\beta}) \qquad (132)$$

where c_n are unknown coefficients, and the additional variables ξ and η are defined as

$$\xi = \psi - \alpha \qquad \eta = \psi - \beta$$

Each term of (132) satisfies (48). Employing the boundary condition $S = -R$, and remembering that $\xi = \eta$ when $\alpha = \beta$, we find that $c_0 = a$, $c_1 = 3a$, and $c_n = 4a$ for all $n \geqslant 2$. It follows that the straight segments are of length $\sqrt{2}\,a$. It is convenient at this stage to introduce the function

$$H(\alpha, \beta) = \sum_{n=0}^{\infty} \left(\frac{\alpha}{\beta}\right)^{(n+1)/2} I_{n+1}(2\sqrt{\alpha\beta}) \qquad (133)$$

numerical values of which are included in Table A-5. Evidently $H(0, \beta) = 0$ and $H(\alpha, 0) = e^{\alpha} - 1$. It is not difficult to show that

$$H(\alpha, \beta) + H(\beta, \alpha) = -I_0(2\sqrt{\alpha\beta}) + \exp(\alpha + \beta) \qquad (134)$$

From (132) and (129), the radii of curvature of the sliplines in the region of LMN may now be written as

$$-\frac{R}{\sqrt{2}\,a} = I_0(2\sqrt{\alpha\eta}) + \sqrt{\frac{\eta}{\alpha}}\, I_1(2\sqrt{\alpha\eta})$$

$$+ I_0(2\sqrt{\xi\beta}) - \sqrt{\frac{\beta}{\xi}}\, I_1(2\sqrt{\xi\beta}) + 4H(\beta, \xi)$$

$$(135)$$

$$\frac{S}{\sqrt{2}\,a} = -I_0(2\sqrt{\alpha\eta}) - \sqrt{\frac{\alpha}{\eta}}\, I_1(2\sqrt{\alpha\eta})$$

$$+ 3I_0(2\sqrt{\xi\beta}) + \sqrt{\frac{\xi}{\beta}}\, I_1(2\sqrt{\xi\beta}) + 4H(\beta, \xi)$$

Setting $\beta = \alpha$ in either of these equations, we obtain the radius of curvature at a generic point of LN as

$$\frac{\rho}{a} = 2\left\{I_0(2\sqrt{\alpha\xi}) + \frac{\psi - 2\alpha}{2\sqrt{\alpha\xi}} I_1(2\sqrt{\alpha\xi}) + 2H(\alpha, \xi)\right\} \tag{136}$$

As the generic angle α increases from zero, ρ steadily increases from $a(2 + \psi)$, reaching the value $2ae^{\psi}$ when $\alpha = \psi/2$ in view of the identity

$$I_0(\psi) + 2H\left(\frac{\psi}{2}, \frac{\psi}{2}\right) = e^{\psi}$$

obtained by setting $\alpha = \beta = \psi/2$ in (134). It may be noted that $\alpha = \beta = \psi/2$ corresponds to the point N where the tangent to the boundary is parallel to the axis of symmetry.

As ψ approaches $\pi/2$, LM is intersected by sliplines of its own family before M is reached. The minimum value of ψ for this to happen corresponds to $S = 0$ at $\alpha = \psi/2$ and $\beta = 0$. The second equation of (135) therefore gives

$$I_0(\sqrt{2}\,\psi) + \frac{1}{\sqrt{2}} I_1(\sqrt{2}\,\psi) = 3$$

leading to $\psi \simeq 86.1°$. For higher values of ψ, the solution can be extended by introducing stress discontinuities. For $\psi < 86.1°$, S vanishes on $\beta = 0$ at some point beyond M for which α exceeds $\psi/2$.

Along the straight sliplines, $\bar{x}$ remains constant, while $\bar{y}$ increases by the amount $\sqrt{2}\,a$ on passing from EF to LM. If the origin O is taken at a distance a from C, it follows from (85) that

$$\frac{\bar{x} + \bar{y}}{\sqrt{2}\,a} = 1 + I_0(2\sqrt{\alpha\psi}) + 2F_1(\alpha, \psi) + \sin(\psi - \alpha) \qquad \text{on } LM$$

The solution for $\bar{x} + \bar{y}$, which is symmetrical with respect to α and β, and which reduces to the above expression when $\beta = 0$ may be written as

$$\frac{\bar{x} + \bar{y}}{\sqrt{2}\,a} = I_0(2\sqrt{\alpha\eta}) + I_0(2\sqrt{\beta\xi}) + 2F_1(\alpha, \eta) + 2F_1(\beta, \xi) + \sin(\psi - \alpha - \beta) \tag{137a}$$

Substitution into the first equation of (50) then furnishes

$$\frac{\bar{x} - \bar{y}}{\sqrt{2}\,a} = I_0(2\sqrt{\alpha\eta}) - I_0(2\sqrt{\beta\xi}) - F_2(\alpha, \eta) - 4H(\beta, \xi) - \cos(\psi - \alpha - \beta) \tag{137b}$$

For a given ψ, the right-hand sides of (137) can be calculated from Tables A-3 and A-5 when $\alpha + \beta \leqslant \psi$, and all angles are integral multiples of $10°$. The rectangular coordinates of the slipline net are then obtained from (131). Setting $\beta = \alpha$, we obtain the parametric equation of the boundary LN in the form

$$\frac{x}{a} = 2\{I_0(2\sqrt{\alpha\xi}) + 2F_1(\alpha, \xi)\} \cos(\psi - 2\alpha)$$

$$- 4\{F_2(\alpha, \xi) + H(\alpha, \xi)\} \sin(\psi - 2\alpha) \tag{138}$$

$$\frac{y}{a} = 1 + 2\{I_0(2\sqrt{\alpha\xi}) + 2F_1(\alpha, \xi)\} \sin(\psi - 2\alpha)$$

$$+ 4\{F_2(\alpha, \xi) + H(\alpha, \xi)\} \cos(\psi - 2\alpha)$$

where $\xi = \psi - \alpha$ as before. If the tangent to the boundary at N is parallel to the x axis, the angle turned through by each of the segments LM, MN, BD, and DF is equal to $\psi/2$. Since

$$I_0(\psi) - F_2\left(\frac{\psi}{2}, \frac{\psi}{2}\right) = 1$$

in view of (88), the rectangular coordinates of the point N are obtained as†

$$d = 2a[I_0(\psi) + A_0(\psi)] = 2d_0 \qquad w = a(2e^\psi - 1) \tag{139}$$

where d_0 represents the distance OD, and $A_0(\psi)$ is defined by (90). It follows that the center of curvature of LN at N is situated at a distance a below the axis of symmetry.

Consider next the stress-free boundary emanating from the slipline field defined by equal logarithmic spirals through C, which is taken to coincide with the origin O. The field is assumed to extend right up to LM without any intervening region of straight sliplines. From (129), and the condition of continuity of R across LM, the radii of curvature of the sliplines within LMN are obtained as

$$-\frac{R}{\sqrt{2b}} = 2I_0(2\sqrt{\alpha\eta}) - 2G(\alpha, \eta) + 2H(\beta, \xi) - \exp(\beta - \alpha - \psi)$$

$$\tag{140}$$

$$\frac{S}{\sqrt{2b}} = 2I_0(2\sqrt{\beta\xi}) - 2G(\alpha, \eta) + 2H(\beta, \xi) - \exp(\beta - \alpha - \psi)$$

It is readily verified that the expression for R in (140) for $\beta = 0$ is identical to that in (101) for $\beta = \psi$, implying the continuity of R. Moreover, S changes sign on crossing the slipline LM and the jump in the radius of curvature has the constant magnitude $2\sqrt{2}b$. The radius of curvature of the boundary LN is given by

$$\frac{\rho}{b} = 2\left\{I_0(2\sqrt{\alpha\xi}) + 2\sum_{m=1}^{\infty}\left(\frac{\alpha}{\xi}\right)^m I_{2m}(2\sqrt{\alpha\xi})\right\} - e^{-\psi} \tag{141}$$

---

† These results have been derived by D. J. F. Ewing, *J. Mech. Phys. Solids*, **16**: 81 (1968), using a lengthy algebraic method, and by M. Sayir, *Z. angew. Math. Phys.*, **20**: 298 (1969), using Riemann's method of integration. A mass flux method has also been employed by D. J. F. Ewing, *J. Mech. Phys. Solids*, **16**: 267 (1968). The present analysis is due to J. Chakrabarty, *Int. J. Mech. Sci.*, **21**: 477 (1979).

where $\xi = \psi - \alpha$, the second term in the curled bracket being equal to $H(\alpha, \xi) - G(\alpha, \xi)$. The radius of curvature increases from the value $b(2 - e^{-\psi})$ at $\alpha = 0$ to be^{ψ} at $\alpha = \psi/2$, in view of the identity

$$I_0(\psi) + 2 \sum_{m=1}^{\infty} I_{2m}(\psi) = \cosh \psi$$

obtained from (103) and (134). The condition of continuity of $\bar{x}$ and $\bar{y}$ across LM and the symmetry of $\bar{x} + \bar{y}$ with respect to α and β lead to

$$\frac{\bar{x} + \bar{y}}{\sqrt{2}\,b} = 2F_1(\alpha, \eta) + 2F_1(\beta, \xi) + \sin(\psi - \alpha - \beta) \tag{142}$$

in view of (104). The expression for $\bar{x} - \bar{y}$ then follows from (50) as before, and the rectangular coordinates are finally obtained from (131). In particular, the coordinates of the point N, where the tangent to the boundary is parallel to the x axis, are found to be†

$$d = 2bA_0(\psi) = 2d_0 \qquad w = b(e^{\psi} - 1) \tag{143}$$

The center of curvature of the boundary at N lies on the straight line parallel to the x axis drawn through the opposite pole. It is interesting to note that the relationship $d = 2d_0$ holds for both the slipline fields considered here.‡

6.8 Superposition of Slipline Fields

(i) *The basic principle* An important consequence of the linearity of the governing differential equations is that any two slipline fields can be suitably combined to generate a third field. If the two fields are positioned in such a way that a chosen pair of points and their associated slipline directions are coincident, then the third field can be obtained by the addition or subtraction of the position vectors to points with identical slipline directions. It follows that the line segment between any pair of points in the generated field is the vector resultant of the line segments between the corresponding pairs of points in the given fields. The radii of curvature of the sliplines at the corresponding points combine algebraically, and so do the arc lengths of any finite segments of the corresponding curves. Since the tangent to a generated curve is parallel to the corresponding tangents to the given curves, the generated net is orthogonal and possesses the geometrical properties of Hencky and Prandtl.§

Any regular domain can be generated from a pair of centered fans by using the principle of vectorial superposition. Situations where the centered fans are either both outside or both inside the regular domain are shown in Fig. 6.21. Let OA and

---

† D. J. F. Ewing, *J. Mech. Phys. Solids*, **16**: 81 (1968).

‡ Since the superposition of a uniform hydrostatic pressure leaves the slipline field unchanged, the stress-free boundary discussed here also defines a curve along which the normal pressure has a constant value.

§ R. Hill, *J. Mech. Phys. Solids*, **15**: 255 (1967).

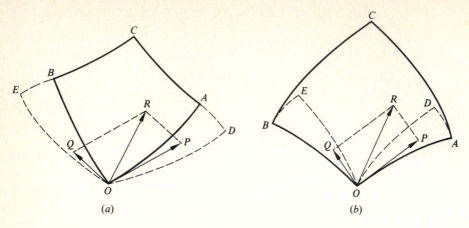

Figure 6.21 Vectorial superposition of singular fields to generate regular domains.

OB be a pair of base curves which are regarded as sliplines of opposite families. In order to generate the domain $OACB$, the centered fans OAD and OBE must be defined over the same range of values of (α, β) as that of the regular field. It is therefore necessary to continue the fans round O as far as the curves OD and OE, so that the tangents to sliplines at D and E are parallel to those at C. According to the principle of superposition, if P, Q, R are three corresponding points in the centered fans and the regular domain, then OR is the vector resultant of OP and OQ. In

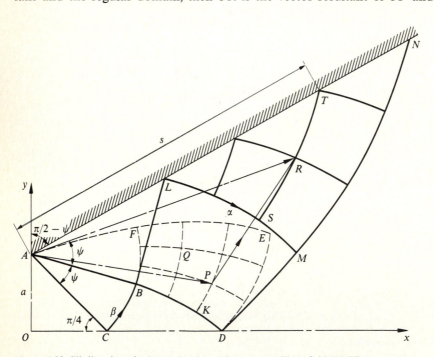

Figure 6.22 Slipline domain LMN developed from an auxiliary field $BDEF$.

particular, as the generating point R traces the slipline BC, the corresponding points P and Q in the singular fields move from O to D and from B to E respectively. The rectangular coordinates of the regular field may be expressed as

$$x = f_1(\alpha, \beta) + f_2(\alpha, \beta)$$

$$y = g_1(\alpha, \beta) + g_2(\alpha, \beta)$$

(144)

where (f_1, g_1) and (f_2, g_2) denote the values of (x, y) for the two singular fields referred to the same pair of rectangular axes.

To illustrate a variant of the graphical technique, consider the slipline field in the neighborhood of a frictionless plane boundary AN (Fig. 6.22). The point A is the center of the fan ACB in which the sliplines are radii and circular arcs. AB and AC make an angle of $45°$ with AN and OD respectively, where OD is an axis of symmetry. From geometry, the angle of inclination of AN with CD is equal to the fan angle ψ. The slipline domain CBD is one-half of the field defined by the circular arc CB and its reflection in CD. Since all sliplines meet CD and AN at $45°$, the angle turned through by each of the segments BD, LM, and MN is equal to ψ. The field LMN is one-half of the network defined by LM and its reflection in AN. The radius of curvature at a generic point on LM numerically exceeds that at the corresponding point on BD by a constant amount equal to AB. The field LMN and its reflection in LN are therefore obtained by the superposition of the field $BDEF$, which is an extension of the field CBD over an angular range ψ, and its reflection in AN. Choosing an equiangular net, a nodal point in the field LMN is located by the vector AR obtained by adding the vectors AP and PR. The position of P relative to B is similar to that of R relative to L. The vector PR is equal to the image of the associated vector AQ in AN. Thus PR is equal in length to AQ and is inclined to AN at angle equal to NAQ. It follows that each of the diagonal points from B to E is situated at the same distance from A as from the corresponding nodal points along AN. In other words, the distance of each nodal point on LN from A is bisected at the foot of the perpendicular on AN from the corresponding diagonal point of the extended field.

To express the graphical construction in mathematical terms, let $f(\alpha, \beta)$ and $g(\alpha, \beta)$ denote the values of x/a and $1 - y/a$ at any point (α, β) of the field defined by the circular arc CF and the axis of symmetry CD. The rectangular components of the vectors AP and AQ can be immediately written down in terms of f and g, while those of PR are obtained from the fact that the counterclockwise orientation of PR with respect to the x axis exceeds the clockwise orientation of AQ by an angle 2ψ. The rectangular coordinates of a generic point R resulting from the vectorial superposition may therefore be written as[†]

$$x = a[f(\alpha, \psi + \beta) + f(\beta, \psi + \alpha) \cos 2\psi - g(\beta, \psi + \alpha) \sin 2\psi]$$

$$y = a[1 - g(\alpha, \psi + \beta) + f(\beta, \psi + \alpha) \sin 2\psi + g(\beta, \psi + \alpha) \cos 2\psi]$$

(145)

[†] The radii of curvature at any point (α, β) in the field LMN are obtained by the addition of the corresponding values at the points $(\alpha, \psi + \beta)$ and $(\beta, \psi + \alpha)$ in the field $BDEF$. For the special case $\psi = \pi/2$, the expressions for R and S have been derived from the corresponding Riemann integrals by J. Grimm, *Ing.-Archiv.*, **44**: 79 (1975).

where (α, β) are referred to the new base point L. It follows from (145), or directly from the geometry of the construction, that the distance of a typical boundary point T from the corner A is

$$s = 2a[f(\alpha, \xi) \cos \psi - g(\alpha, \xi) \sin \psi]$$

where $\xi = \psi + \alpha$. Substituting the expressions for $f(\alpha, \xi)$ and $g(\alpha, \xi)$, which are given by (89), we obtain

$$s = 2a[I_0(2\sqrt{\alpha\xi}) + 2F_1(\alpha, \xi)] \tag{146}$$

Since the hydrostatic pressure at T is equal to $p_0 + 2k(\psi + 2\alpha)$ by Hencky's equations, where p_0 is the pressure at C, the resultant normal force across AT is

$$Q = [p_0 + k(1 + 2\psi)]s + 4k \int_{2a}^{s} \alpha \, ds$$

$$= [p_0 + k(1 + 2\psi)]s + 4k\left(\alpha s - \int_0^\alpha s \, d\alpha\right)$$

Substituting from (146), and using the relations $dF_1/d\alpha = I_0$ and $dL/d\alpha = F_1$, where F_1 and L are considered at (α, ξ), we finally obtain

$$Q = [(p_0 - k) + 2k(\psi + 2\alpha)]s + 4ka[I_0(2\sqrt{\alpha\xi}) - 4L(\alpha, \xi)] \tag{147}$$

For given values of ψ and α, the functions in the square bracket as well as the ratio s/a are readily obtained from Table A-3 or A-4 (Appendix).

(ii) *A mixed boundary-value problem* Consider a frictionless plane boundary AD passing through an arbitrary stress singularity A, the slipline AC being concave to the plane making an angle of $45°$ (Fig. 6.23). The singular field on the convex side is ABC in which the same angle ψ is turned through by the segments AB, AC, and BC. The regular field ACD is formed by the vectorial superposition† of the singular field ACB and its reflection in AD. We observe that for an equiangular mesh, the vectors AE_1, AE_2, etc., are the resultants of vectors AF_1, AF_2, etc., and their images in the plane AD. The distances $AE_1, AE_2, \ldots, AD$ are therefore bisected by the perpendiculars from $F_1, F_2, \ldots, B$ to the plane AD. Since equal angles are turned through along BC and CD, the tangents to the sliplines at B are inclined at $45°$ with AD.

Suppose that the initial slipline AB, taken as the α baseline, is a circular arc of radius $\sqrt{2}\,\rho$. Let the overall longitudinal and transverse dimensions of the field be denoted by $2a$ and b respectively. Then the distance of B from the vertical through A is equal to a. It follows from geometry that

$$a = \rho(\cos \psi + \sin \psi - 1) \qquad b = \rho(1 - \cos \psi + \sin \psi) \tag{148}$$

The angle which the chord AB makes with AD is evidently equal to $\pi/4 + \psi/2$. If

---

† R. Hill, *J. Mech. Phys. Solids*, **15**: 255 (1967).

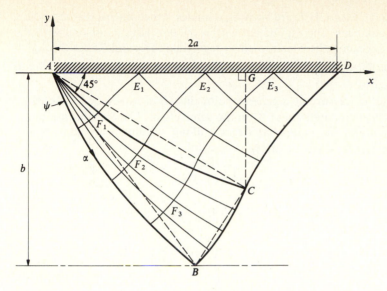

Figure 6.23 Slipline field with a singularity on a smooth plane, illustrating the principle of vectorial superposition.

the rectangular axes are taken through A as shown, we have $\bar{x} = \sqrt{2}\rho \sin \alpha$ and $\bar{y} = -\sqrt{2}\rho(1 - \cos \alpha)$ along the slipline AB. The formal solution for $\bar{x}$ and $\bar{y}$ within the fan ABC may therefore be written as

$$\bar{x}(\alpha, \beta) = \sqrt{2}\rho F_1^*(\alpha, \beta) \qquad \bar{y}(\alpha, \beta) = -\sqrt{2}\rho F_2^*(\alpha, \beta)$$

where F_1^* and F_2^* are defined by (113). Setting $\alpha = \beta = \psi$ in the above expressions, and using (114), we have

$$\sqrt{2}\bar{x} = \rho \sin 2\psi \qquad \sqrt{2}\bar{y} = -\rho[J_0(2\psi) - \cos 2\psi]$$

The rectangular coordinates of the point C are obtained by inserting from above into (22), with $\phi = -(\pi/4 - \psi)$. When ψ is small (less than 20° say), we may write $J_0(2\psi) \simeq 2 \cos \psi - 1$, with a maximum error of about 0.2 percent. This leads to the approximate results

$$AG = x(\psi, \psi) = b \cos \psi = a(1 + \sin \psi)$$
$$CG = -y(\psi, \psi) = a \cos \psi = b(1 - \sin \psi) \tag{149}$$

in view of (148). The above approximation is equivalent to replacing the curve OC by a circular arc of radius $\sqrt{2}\rho \cos \psi$. It follows from above that the angle CAG is equal to $\pi/4 - \psi/2$, and consequently the angle BAC is $\theta \simeq \psi$ to the same order of approximation. It also follows from above that

$$AC = AB \cos \psi = AD \cos \left(\frac{\pi}{4} - \frac{\psi}{2}\right)$$

Hence the angles ACB and ACD are both right angles, which means that the points B, C, and D are collinear. The slipline net within the fan ABC, consistent with the above approximation, may be constructed by regarding the α and β lines through a generic point (α, β) as circular arcs of radii $\sqrt{2}\rho \cos \beta$ (except near O) and $\sqrt{2}\rho \sin \alpha$ respectively.

The hydrostatic pressure at a generic point on AB is $p_0 - 2k\psi$ by Hencky's equation, where p_0 is the pressure at A on the same curve. The distribution of the normal pressure p and the tangential stress k along AB has a resultant whose horizontal component per unit width is

$$P = p_0 b - 2k \int \alpha\, dy - ka$$

where the integral is taken along the slipline BA, on which the expression for y may be written approximately as

$$y(\alpha, 0) = -b + \rho[1 - \cos (\psi - \alpha) + \sin (\psi - \alpha)]$$

Substituting in the above integral, and integrating between the appropriate limits, we get

$$P = p_0 b + k(a - 2\rho\psi) \tag{150}$$

in view of (148). The normal pressure on the plane boundary AD is $p_0 + k + 2k(\psi - 2\alpha)$ at a point which is situated at a distance $2x(\alpha, \psi - \alpha)$ from A, where $x(\alpha, \beta)$ is the abscissa to a generic point of the fan ABC. From geometry,

$$x(\alpha, \psi - \alpha) \simeq \rho(\cos \alpha + \sin \alpha - 1) \cos (\psi - \alpha)$$

Integrating along AD, the resultant compressive force per unit width on the plane boundary is obtained as

$$Q = 2a(p_0 + k) + 4k\rho(\psi - \sin \psi) \tag{151}$$

As ψ tends to zero, ρ tends to infinity, such that $\rho\psi \simeq a \simeq b$. The slipline field then reduces to a net of orthogonal straight lines inclined at $45°$ with AD. The normal pressure along AD then has a constant value equal to $p_0 + k$.

(iii) *Geometrical similarity* We begin with the configuration of Fig. 6.24, where AD and DF are two mutually perpendicular frictionless planes. The slipline field domains DCA and DCF can be generated as before from the same centered fan DCE; it is evident that the slipline segments AC, CD, CF, and CE all have the same angular span. Then DF is the resultant of the vector DE and its image in DF, while DA is the resultant of the vector DE and its image in DA. The triangles AED and DEF are therefore isosceles. It follows that A, E, and F are collinear, and E is the midpoint of the line AF. For every point P on CE, there are corresponding points Q and R on AC and CF respectively, such that the tangents at these three points are parallel to one another. Considering the parallelograms formed by the vector DP and its reflected counterparts, it is easily shown that the points P, Q, and R are collinear, and P is the midpoint of QR. The domain ACD may also be generated

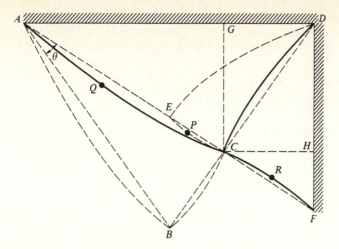

Figure 6.24 Geometrical similarity of slipline field domains associated with frictionless plane boundaries.

from the fan ACB in a similar manner, where equal angles are turned through along AC and BC.

Suppose, now, that the curves AC and DC are similar in the ratio $\lambda : 1$. This means that the radii of curvature of AC and DC at points situated at equal angular distances from A and D are in the ratio $\lambda : 1$. The singular fields ACB and DCE are then also similar in the same ratio, and so are the regular fields ACD and DCF generated by the singular ones on their concave sides. The curves DC and FC are therefore similar in the ratio $\lambda : 1$, and the curves AC and FC are similar in the ratio $\lambda^2 : 1$. Thus, C is the center of similitude of the curves CA, CD, and CF. Since the tangents at A and F to the respective sliplines are parallel to one another, the straight line joining A and F must pass through C. In fact, the line joining any trio of corresponding points P, Q, and R will necessarily pass through C. Since the two fields $ABCD$ and $DECF$ are identical except in scale, we have

$$\frac{AB}{DC} = \frac{DE}{FC} = \frac{1}{2}\left(\frac{AC}{FC} + 1\right) = \frac{1}{2}(\lambda^2 + 1)$$

$$\frac{BC}{DC} = \frac{EC}{FC} = \frac{1}{2}\left(\frac{AC}{FC} - 1\right) = \frac{1}{2}(\lambda^2 - 1)$$

It follows that the curves CA, CD, CF, CB, and BA are all similar to one another in the ratios $\lambda : 1 : 1/\lambda : \frac{1}{2}(\lambda^2 - 1) : \frac{1}{2}(\lambda^2 + 1)$, and the points B, C, and D are collinear.[†] The curves CB and CE are evidently similar in the ratio $\lambda : 1$. The similarity of the domains ACD and DCF furnishes

$$\frac{AG}{CG} = \frac{DH}{CH} = \frac{CG}{CH} = \lambda$$

[†] I. F. Collins, *Proc. R. Soc.*, A, **303**: 317 (1968).

where CG and CH are perpendiculars to AD and DF respectively. It follows that the angles CAD and CDF are each equal to $\cot^{-1} \lambda$, and the straight lines AF and BD are mutually orthogonal. The angle θ between the line segments AB and AC is given by

$$\cot \theta = \frac{AC}{BC} = \frac{2\lambda}{\lambda^2 - 1}$$

Let the angular span of each slipline be such that B and F are situated at the same distance from AD. The domains DCF and FCB are then similar in the ratio $\lambda : 1$, and the curves CF and CB are also similar in the same ratio. Hence $(\lambda^2 - 1)/2\lambda = \lambda$, giving $\lambda = AD/DF = \sqrt{2}$, and $\theta = \cot^{-1} 2\sqrt{2} \simeq 19.47°$. Since this angle is sufficiently small, the similarity conditions would be satisfied to a close approximation by adopting the solution given in 6.8(ii) with $\psi \simeq \theta$.

Problems

6.1 Figure A shows a quadrilateral element $ABCD$, of which the sides AB and AD are tangential to the sliplines through A. The upper face DC is parallel to AB, while BC is normal to the slipline AE at a point whose angular distance from A is $d\phi$. Considering AE as an α line and a β line in turn, obtain Hencky's pressure equations from the condition of equilibrium of the element in the direction parallel to AB. Show that the derivatives of the stresses are discontinuous across a slipline when the curvature of the other family of sliplines changes abruptly across it.

6.2 Considering the velocity components in the characteristic directions at the extremities of a typical slipline element, and using the fact that the rate of extension vanishes along the sliplines, obtain Geiringer's equations for the velocity field. Show that the radial and tangential velocities at any point in a centered fan, where the sliplines are radial lines and circular arcs, must be of the form

$$v_r = -f'(\theta) \qquad v_\theta = f(\theta) + g(r)$$

where r is the distance measured from the center of the fan and θ the angle measured from an arbitrary radial line through the center.

6.3 One of the plane faces of a wedge, made of a rigid/plastic material, is acted upon by normal and tangential tractions near its vertex as shown in Fig. B. Draw the possible slipline fields that correspond to the yield point of the wedge tip, and hence determine the normal pressure q in each case as a function

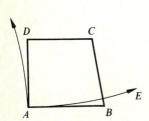

Figure A

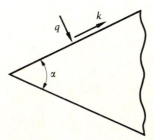

Figure B

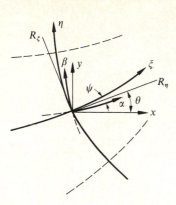

Figure C

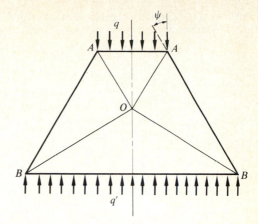

Figure D

of the wedge angle $\alpha \leqslant \pi/2$. Find also the condition for the wedge tip not to be overstressed, when the direction of the shear stress is reversed.

6.4 Figure C shows the flowlines and one of their orthogonal trajectories within a plastically deforming region. Prove that the counterclockwise angle ψ made by the local flow direction with the positive α direction is given by

$$\cot 2\psi = \frac{1}{2}\left(\frac{1}{w}\frac{\partial w}{\partial s_\eta} + \frac{R_\eta}{R_\xi}\right)$$

where R_ξ and R_η are the local radii of curvature of the flowline and the trajectory respectively, reckoned positive as shown. The above relation may be used for obtaining the slipline field from an experimentally determined flow field.

6.5 A truncated wedge of semiangle ψ is uniformly loaded along its top and bottom faces as shown in Fig. D. Assuming stress discontinuities OA and OB, which bisect the angles at the corners A and B, show that the upper and lower face pressures at the yield point are

$$q = 2k(1 + \sin \psi) \qquad q' = 2k(1 - \sin \psi)$$

Does the discontinuous stress field represent the actual state of stress in the wedge at the yield point?

6.6 A nonhardening elastic/plastic material, obeying Tresca's yield criterion and the associated flow rule, is plastically deformed under conditions of plane strain. The strains are assumed small, so that positional changes and rotations of the element may be disregarded. Show that the Hencky equations are unaffected by the inclusion of elastic strains, so long as σ_z is the intermediate principal stress, while the Geiringer equations are modified to

$$G(du - v\,d\phi) = -[k\dot\phi + (\tfrac{1}{2} - v)\dot p]\,ds_a \qquad \text{along an } \alpha \text{ line}$$

$$G(dv + u\,d\phi) = [k\dot\phi - (\tfrac{1}{2} - v)\dot p]\,ds_\beta \qquad \text{along a } \beta \text{ line}$$

where G is shear modulus and v is Poisson's ratio for the material, the superposed dot denoting the time derivative.

6.7 Consider a slipline field defined by a pair of identical base curves of negative curvature. Prove that (a) every slipline in the field is a logarithmic spiral when the baselines are logarithmic spirals having a common pole on the axis of symmetry; and (b) every slipline in the field is a cycloid when the baselines are cycloids generated by identical circles rolling in opposite directions.

6.8 Referring to Fig. 6.16, show that the components of the resultant force acting across the slipline *ECA*, when the hydrostatic pressure vanishes at *C*, are given by

$$\frac{P}{2ka} = \sin 2\lambda - [\cos (\alpha - \lambda) + 2\alpha \sin (\alpha - \lambda)] \sin \lambda$$

$$\frac{Q}{2ka} = \cos 2\lambda + [-\sin (\alpha - \lambda) + 2\alpha \cos (\alpha - \lambda)] \sin \lambda$$

Show also that the resultant clockwise moment of the tractions on *ECA* about the origin *O* is given by

$$\frac{M}{2ka^2} = \sin 2\lambda + (1 - \cos 2\lambda)\left(\alpha - \frac{P}{2ka} \right)$$

6.9 Show that the radius of curvature of a typical α line at a given point (θ, ψ) of the field of Fig. 6.17, where the base sliplines are circular arcs each of radius $\sqrt{2}\,a$, is given by

$$R = \sqrt{2}\,a\left\{ J_0(2\sqrt{\theta\psi}) - \int_0^\psi J_0[2\sqrt{\theta(\psi - \beta)}]d\beta \right\}$$

Derive the final expressions for *R* and *S*, and compute the values of ψ along the envelope *BL* corresponding to $\theta = 0, 10, 20,$ and $30°$, using interpolations based on Table A-8.

6.10 Employ the Riemann method of integration to derive Eqs. (83) for the radii of curvature of the α and β lines of the field defined by a pair of opposed circular arcs of equal radii. Using the expression for R/a, find the values of x/a and y/a by numerical integration with a $10°$ equiangular net considered over the range $\alpha \leqslant \beta \leqslant 30°$.

6.11 Using a dimension $N = 4$ for truncating the matrices **M** and **N** introduced in Sec. 6.5(iii), find the column vector $\boldsymbol{\sigma}$ representing the slipline *AB* of Fig. 6.18, when $\psi = 15, 30,$ and $45°$. Hence calculate the values of x/a and y/a for the slipline field defined by the circular arc and the limiting line, considering a $15°$ equiangular net with $\beta \leqslant \alpha \leqslant 45°$, and employing the series method of solution.

6.12 Show that the rectangular coordinates for the slipline field defined by a pair of circular arcs *CA* and *CB* (Fig. 6.16), having radii *a* and *b* respectively, are given by

$$x = af(\alpha, \beta) + bg(\beta, \alpha) \qquad y = ag(\alpha, \beta) + bf(\beta, \alpha)$$

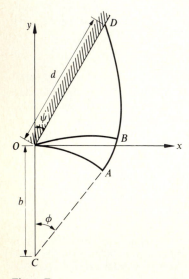

Figure E

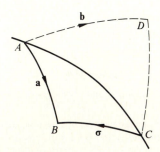

Figure F

when the x and y axes are taken along the tangents to the α and β lines at C. The functions f and g at any point (α, β) denote x/a and y/a for the singular field defined by CA on its convex side. Taking $b/a = 2$, and using Table A-6, calculate the values of x/a and y/a for a $15°$ equiangular net with $\alpha \leqslant \beta \leqslant 45°$.

6.13 The field $OABD$ of Fig. E consists of a singular domain OAB and a regular domain OBD adjacent to a frictionless plane OD. The initial slipline OA is a circular arc of radius b with center on the negative y axis. Using the principle of vectorial superposition, show that the distance of D from the origin O is

$$d = \sqrt{2}\, b[F_1(\phi, \xi) + F_2(\phi, \xi)]$$

where $\xi = \pi/4 - \psi + \phi$, and that the normal compressive force transmitted across the inclined plane OD is

$$Q = [p_0 + k(1 + 4\phi)]d - 4\sqrt{2}\, k[L(\phi, \xi) + N(\phi, \xi)]b$$

where p_0 denotes the hydrostatic pressure at O considered on the plane boundary.

6.14 It is required to construct the slipline field ABC between a given slipline AB, turning through an angle θ and a stress-free boundary AC whose position is not known in advance (Fig. F). Representing the sliplines AB and AD by vectors $\mathbf{a}$ and $\mathbf{b}$ respectively, obtain the relationship $(\mathbf{D} - \mathbf{I})\mathbf{b} = (\mathbf{D} + \mathbf{I})\mathbf{a}$, where $\mathbf{I}$ is the identity matrix and

$$\mathbf{D} = \begin{bmatrix} 0 & 1 & 0 & 0 & \cdots \\ 0 & 0 & 1 & 0 & \cdots \\ 0 & 0 & 0 & 1 & \cdots \\ \cdots & \cdots & \cdots & \cdots & \cdots \end{bmatrix}$$

Show that the vector $\boldsymbol{\sigma}$ representing the slipline CB is given by $\boldsymbol{\sigma} = \mathbf{F}\mathbf{a}$, where

$$\mathbf{F} = \mathbf{R}_\theta \mathbf{M}_\theta^{-1}[(\mathbf{D} - \mathbf{I})^{-1}(\mathbf{D} + \mathbf{I}) - \mathbf{N}_\theta \mathbf{R}_\theta]$$

where $\mathbf{M}_\theta$ and $\mathbf{R}_\theta$ are matrix operators given by Eqs. (73) and (75) respectively.

SEVEN

STEADY PROBLEMS IN PLANE STRAIN

We shall be concerned here with problems of steady motion in which the stress and velocity distributions do not vary with time. The rigid material enters the deforming zone on one side, and after passing through this region becomes rigid again on the other side. Steady state problems are not statically determined, since the shape and position of the plastic boundary are restricted by velocity boundary conditions required to maintain the flow. The steady state condition is satisfied, to a close approximation, in continuous metal-forming processes such as rolling, drawing, and extrusion. Since large strains are involved in such cases, the assumption of rigid/plastic material would be sufficiently good. The results for a nonhardening material should be a good approximation for a real material having a sharp yield point and a low subsequent rate of hardening. It is often possible, as we shall see, to include work-hardening in an approximate manner by introducing a simple correction factor. Although the majority of the solutions given in this chapter and the next are incomplete in the sense that the stress state in the rigid regions has not been examined, they are otherwise satisfactory. In each case, the proposed slipline field represents only that part of the plastic region which is undergoing deformation.

7.1 Symmetrical Extrusion Through Square Dies

Extrusion is the process of forming a metal billet, held in a container, by forcing it through a shaped die with a moving ram.† The process is said to be direct when the

---

† For practical aspects of the extrusion process, see C. E. Pearson and R. N. Parkins, *The Extrusion of Metals*, Chapman and Hall, London (1960). A critical review of the early theoretical literature on extrusion is due to J. F. W. Bishop, *Met. Rev., Inst. Met.*, **2**: 361 (1957).

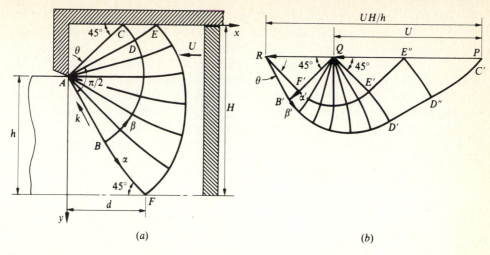

Figure 7.1 Extrusion through a smooth square die with reductions less than 50 percent. (*a*) Slipline field; (*b*) hodograph.

ram and the billet move together against the stationary die, and the product also moves in the same direction. In the inverted extrusion, the ram is held fixed and the die is forced into the billet causing an opposite movement of the product. The pressure builds up rapidly as the billet is initially compressed, following which the extrusion proceeds steadily. When the die is square-faced, some material in the corner between the die and the container wall is held back and not extruded with the rest of the billet. This material, known as dead metal, effectively converts the square die into a curved one with a perfectly rough surface.

(i) *Frictionless extrusion for small reductions* Figure 7.1 shows the slipline field† for the plane strain direct extrusion of a billet when the ratio of the initial to the final cross-sectional areas, known as the *extrusion ratio*, is less than 2. The container wall and the die are well lubricated so that frictional forces are negligible. The dead metal boundary AC, which is assumed straight, must be a slipline inclined at 45° with the container wall. The field ABC is a centered fan in which the sliplines are radii and circular arcs. $BCEF$ is one-half of the field defined by the circular arc BC and its reflection in the wall. The angle CAD is so chosen that the exit slipline ABF intersects the axis of symmetry at 45°. The angle BAD is 90°, being equal to the angle turned through by the slipline EF. If the angle CAD is denoted by θ, the *fractional reduction* corresponding to an initial thickness $2H$ and a final thickness $2h$ may be expressed as

$$r = 1 - \frac{h}{H} = \left\{ 1 + I_0 \left[2\sqrt{\theta\left(\frac{\pi}{2} + \theta\right)} \right] + 2F_1\left(\theta, \frac{\pi}{2} + \theta\right) \right\}^{-1}$$

† R. Hill, *J. Iron Steel Inst.*, **158**: 177 (1948).

in view of the second equation of (89), Chap. 6. When θ is an integral multiple of 10 or 15°, $1/r$ may be directly obtained from Table A-1 or A-2, being equal to the value of y/a corresponding to the angular coordinates $(\theta, \pi/2 + \theta)$.

The stress distribution in the plastic region can be determined by using the Hencky equations which give the mean compressive stress at any point in terms of the value p_0 on AB. The pressure p_0 is obtained from the condition that the resultant axial force acting across the exit slipline ABF must vanish. It follows from physical considerations that the algebraically greater principal stress at F is along the axis, the α and β lines being identified as shown. The normal pressure at any point on BF is $p_0 - 2k\alpha$, where α is the angle turned through by the tangent to the curve. Hence the resultant longitudinal force across ABF is

$$p_0 h - 2k \int_B^F \alpha \, dy - kd = 0$$

where d is the distance of F from the exit plane of the die, the ratio $d/(H - h)$ being obtained from Table A-1 or A-2. The pressure p_0 is therefore given by

$$\frac{p_0}{2k} = \frac{d}{2h} + \frac{\theta H}{h} - \int_0^\theta \left(\frac{y}{h}\right) d\alpha \tag{1}$$

The average pressure q on the die is equal to the magnitude of the principal stress on AC in the direction normal to the die. Using Hencky's theorem, we get

$$q = p_0 + k(1 + \pi + 2\theta)$$

The die pressure therefore increases as the reduction decreases. Let the mean pressure exerted by the ram, known as the *extrusion pressure*, be denoted by p_e. For equilibrium, the total ram load $p_e H$ per unit width must be equal to the horizontal thrust $q(H - h)$ on the die. This gives $p_e = rq$, leading to[†]

$$\frac{p_e}{2k} = r\left\{\frac{1 + \pi}{2} + \left(\frac{2 - r}{1 - r}\right)\theta + \frac{r}{1 - r}\left[\frac{d}{2a} - \int_0^\theta \left(\frac{y}{a}\right) d\alpha\right]\right\} \tag{2}$$

where $a = H - h$. The integral on the right-hand side is obtained from the Table, with $\alpha = \theta$ and $\beta = \pi/2 + \theta$. Since r and d/h are known functions of θ, the extrusion pressure is obtained as a function of the reduction r parametrically through θ. In the special case of 50 percent reduction, $\theta = 0$ and $d = h$, yielding $p_e = k(1 + \pi/2)$. The results for $r \leqslant 0.5$ are summarized in Table 7.1. The extrusion pressure may be conveniently expressed by the empirical formula

$$p_e/2k = 0.1 + 0.3r(8.4 - r) \qquad r < 0.5 \tag{3}$$

which is correct to within 0.5 percent. The solution does not hold for very small reductions (less than about 5.4 percent), when the die pressure reaches the value $2k(1 + \pi)$ causing the plastic zone to spread round the corner A to the free surface of the extruded billet.

[†] Although the problem can be solved more directly by using the values of the resultant force given in Table A-1 or A-2, this procedure is adopted here to illustrate a general method of solution.

Table 7.1 Extrusion through a smooth container with $r \leqslant 0.5$

θ	0°	5°	10°	20°	30°	45°	60°
r	0.500	0.428	0.336	0.268	0.196	0.122	0.076
$q/2k$	2.571	2.595	2.705	2.817	3.035	3.410	3.842
$p_e/2k$	1.285	1.110	0.909	0.755	0.595	0.416	0.289

The associated velocity field is represented by the hodograph (Fig. 7.1b). The velocity of the rigid material entering the deforming zone is represented by the vector PQ of magnitude U. Since the material is incompressible, the velocity of the extruded billet, represented by the vector PR, is of magnitude UH/h. The tangential components of the velocity are evidently discontinuous across the sliplines through F, whose image F' is the intersection of the straight lines through Q and R parallel to the slipline directions at F. The velocity discontinuity propagates along the sliplines AE, EF, and FA with a constant magnitude equal to $(U/\sqrt{2})r/(1-r)$. The segments $F'B'$ and $F'F'$ are therefore circular arcs defining the region $F'B'D'E'$ of the hodograph. The points immediately above the slipline DE are mapped into $D''E''$ separated from $D'E'$ by a constant normal distance equal to the velocity discontinuity. The velocity is continuous across the dead metal boundary AC which is mapped into the pole P of the hodograph. The container wall CE is mapped into PE'' where all characteristics meet at an angle of $45°$. It is apparent from the figure that the radii of curvature of the β and β' curves have the same sign, while those of the α and α' curves are of opposite signs. Moreover, the velocity discontinuities across AE, EF, and FA are in the directions of the shear stress across these curves. The rate of plastic work is therefore everywhere positive.

(ii) *Complete solution for $r = 0.5$* The preceding solution is incomplete in the sense that the stress distribution in the rigid regions has not been examined. This will now be carried out for the special case of 50 percent reduction (Fig. 7.2). A statically admissible stress field beyond the exit slipline AB consists of a stress discontinuity AO separating a stress-free region on the left from the uniformly stressed plastic region AOB on the right. The stress in the dead metal zone ANC is also assumed uniform with the principal stresses as shown. To obtain a statically admissible stress field in the region ahead of the die, we extend the slipline field† up to the principal stress trajectory CF through C. The principal stresses at each point of the trajectory are directed along the tangent and the normal to the curve, which is inclined at $45°$ with the sliplines. It follows from the curvilinear triangle formed by an infinitesimal arc of the trajectory and a pair of intersecting sliplines that $R\,d\alpha = -S\,d\beta$. In view of Eqs. (83) of Chap. 6, this condition becomes

$$d[I_0(2\sqrt{\alpha\beta})] = -I_0(2\sqrt{\alpha\beta})\,d(\alpha + \beta)$$

† J. M. Alexander, *Q. Appl. Math.*, **19**: 31 (1961).

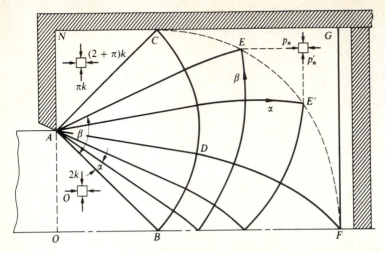

Figure 7.2 Statically admissible stress field for extrusion with 50 percent reduction.

along the trajectory, the base point for (α, β) being taken at B. Since $\alpha = 0$ when $\beta = \pi/2$, the above equation is readily integrated to

$$\exp\left(\alpha + \beta - \frac{\pi}{2}\right) I_0(2\sqrt{\alpha\beta}) = 1 \tag{4}$$

which is the equation of the trajectory CF. The point D, where the trajectory intersects the axis of symmetry, corresponds to $\alpha = \beta = \psi$. This gives $\psi \simeq 35.1°$, while OF is found to be approximately 2.8 times OA.

Since the mean compressive stress at B is equal to k, the normal pressure transmitted across the trajectory at E is $p_n = 2k(\alpha + \beta)$ by Hencky's equations. Let CF be a line of stress discontinuity. The conditions of symmetry and zero wall friction are satisfied by assuming that the principal stresses in the material to the right of CF are everywhere parallel and perpendicular to the axis. The normal and shear stresses will be continuous across CF if the state of stress immediately to the right of CF is a hydrostatic compression of magnitude equal to the normal pressure acting across CF. It is also assumed that within the region CFG, the horizontal principal stress is constant along a horizontal line and the vertical principal stress is constant along a vertical line.

The principal compressive stresses at the point defined by the horizontal and vertical lines through E and E' are p_n and p'_n, where p'_n is the normal pressure across CF at E'. The stress distribution will be statically admissible if the maximum value of $p_n - p'_n$, which occurs at G, is less than $2k$. This is indeed satisfied since the pressure difference at G is $(\pi - 4\psi)k \simeq 0.69k$. The statically admissible stress field is extended further to the right of FG by assuming that the stress distribution along any vertical line is the same as that along FG. The extrusion pressure $k(1 + \pi/2)$ is

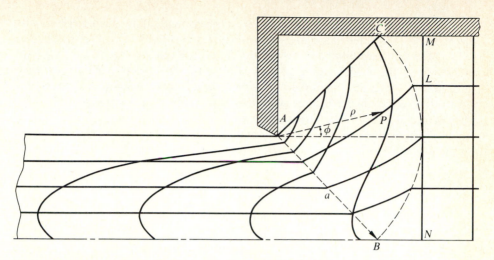

Figure 7.3 Distortion of a square grid in extrusion with 50 percent reduction.

therefore the actual extrusion pressure since it is both an upper and a lower bound.†

(iii) *Distortion of a square grid* In general, the trajectories of the particles in the deforming zone can be found by a graphical procedure from the known direction of the velocity vector throughout the field. A simple analytical solution exists, however, for the special case of 50 percent reduction (Fig. 7.3). Denoting the radial and tangential velocities by u and v respectively, the boundary conditions may be written as $u = -U \cos \phi$ along BC and $v = -\sqrt{2}\, U$ along AB. In view of the Geiringer equations, the velocity field in the sector ABC becomes

$$u = -U \cos \phi \qquad v = -U\left(\frac{1}{\sqrt{2}} - \sin \phi\right) \qquad (5)$$

The velocity therefore depends only on ϕ. The trajectories of the particles in the fan ABC are given by the differential equation

$$\frac{1}{\rho}\frac{d\rho}{d\phi} = \frac{u}{v} = \frac{\sqrt{2}\cos \phi}{1 - \sqrt{2}\sin \phi}$$

Let ϕ_0 be the angular position of a given particle when it enters the deforming zone. Since $\rho = a$ when $\phi = \phi_0$, the integration of the above equation results in

$$\rho = a\left(\frac{1 - \sqrt{2}\sin \phi_0}{1 - \sqrt{2}\sin \phi}\right) \qquad (6)$$

---

† A statically admissible extension of the slipline field for $r < 0.5$ under modified boundary conditions has been treated by J. Grimm, *Ing.-Arch.*, **44**: 209 (1975).

In order to compare the theory with experiment, it is often necessary to determine the distorted shape of a uniform square grid marked on the undeformed billet.† The gridlines originally parallel to the central axis assume the shape of the trajectories of the particles lying on them. To obtain the distortion of the transverse gridlines, consider the time taken by a particle to move along its trajectory. Let $t = 0$ correspond to the instant when a given gridline MN is tangential to the circular arc BC (Fig. 7.3). The particle which is initially at the point L on MN reaches the circular arc at $t = a (1 - \cos \phi_0)/U$. Since $v = \rho\, d\phi/dt$, the time taken by the particle to reach a generic point P is

$$t = \frac{a}{U} \left\{ (1 - \cos \phi_0) + \int_{\phi_0}^{\phi} \frac{\rho}{v}\, d\phi \right.$$

Substituting from (5) and (6), and integrating, we obtain

$$t = \frac{a}{U} \left\{ (1 - \cos \phi_0) + 2(1 - \sqrt{2} \sin \phi_0)[f(\phi) - f(\phi_0)] \right\} \tag{7}$$

where

$$f(\phi) = \sqrt{2} \coth^{-1}\left(\sqrt{2} - \tan \frac{\phi}{2} \right) - \cos \phi\, (1 - \sqrt{2} \sin \phi)^{-1} \tag{8}$$

The time t^* required by a given particle to reach the exit slipline AB corresponds to $\phi = \pi/4$. Note that $f(-\pi/4) = 0.5$. The locus of the positions of the particles at a given time $t < t^*$ can be determined from (7) and (8) by calculating ϕ for various values of ϕ_0. For $t > t^*$, we have $t - t^* = x/2U$, where x is the axial distance of a particle from AB.

If the length of each side of the squares of the approaching grid is denoted by c, successive transverse gridlines come in contact with the arc BC at a regular interval of time c/U. The deformed shape of these lines is given by the locus of the positions of the particles, originally at the nodal points of MN, corresponding to $t = mc/U$ ($m = 1, 2, 3, \ldots$). Several curves determined in this way are shown in Fig. 7.3 for $c = a/2\sqrt{2}$. It gives the distorted shape of the grid when a transverse line, originally five meshes to the right of M, has reached the position MN. The longitudinal grid lines in the extruded billet are straight and equally spaced at a constant distance $c/2$ from one another. The transverse lines are also equally spaced after extrusion, the axial distance between two successive lines being twice that before extrusion. Apart from a cusp on the axis, the distorted grid is similar to that observed in cylindrical billets.

The angle of the cusp for $r \leqslant 0.5$ can be found as follows. Let a particle S leave the deforming zone at an infinitesimal distance δy from the central axis. The particle enters the zone at a distance $(H/h)\, \delta y$ from the axis, and traverses an axial distance of $(H/h + 1)\, \delta y$ to reach the exit slipline AF. Since the axial component of

† R. Hill, *J. Iron Steel Inst.*, **158**: 177 (1948). A graphical method based on the hodograph has been discussed by J. M. Alexander, *Proc. Conf. Technol. Eng. Manuf., Instn. Mech. Eng.*, 155 (1958).

the velocity immediately above F is $(U/2)(H/h + 1)$, given by the hodograph, the time of traverse is $2 \, \delta y/U$. During this time, a particle T on the axis, originally at a distance $(H/h) \, \delta y$ to the right of F, moves through a distance $(3 - H/h)(H/h) \, \delta y$. Hence the axial distance between the two particles S and T at the end of the time interval is $(H/h - 1)^2 \, \delta y$. It follows, therefore, that an element of any transverse gridline on the axis makes an acute angle of $\cot^{-1} (H/h - 1)^2$ with the axis after deformation. The cusp is evidently less pronounced for relatively small reductions. The semiangle of the cusp is equal to $45°$ when the reduction is 50 percent.

(iv) Frictionless extrusion for large reductions An obvious slipline field for reductions greater than 50 percent consists in interchanging the positions of the central axis and the container wall in Fig. 7.1. In that case, it is not difficult to show that the mean extrusion pressure for any reduction r is equal to $r/(1 - r)$ times that for a reduction $1 - r$. The extrusion pressure for $r \geqslant 0.5$ can therefore be obtained from that calculated for $r \leqslant 0.5$. The solution requires the die face to be partially rough. An empirical formula for $r > 0.5$, correct to within 1.5 percent, is

$$\frac{p_e}{2k} = 0.41 + 1.26 \ln \left(\frac{1}{1 - r} \right) \qquad 0.5 < r < 0.9 \qquad (9)$$

When the die face and the container wall are both perfectly smooth, the exit slipline for $2 < H/h < 3$ must be partly curved.[†] The solution consists of two types of field, one applying to extrusion ratios between 2 and $\sqrt{2} + 1$, and the other between $\sqrt{2} + 1$ and 3. The former involves a dead metal region adjacent to the die face, while the latter is without any dead metal. The mean extrusion pressure only marginally exceeds that given by the approximate formula

$$p_e = 2k\left(1 + \frac{\pi}{2}\right)r \qquad \frac{1}{2} \leqslant r \leqslant \frac{2}{3} \qquad (10)$$

which is exact for $r = \frac{1}{2}$ and $r = \frac{2}{3}$, the maximum error being less than 1 percent, corresponding to $H/h = \sqrt{2} + 1$. The mean die pressure differs only slightly from the constant value $2k(1 + \pi/2)$. By an extension of the statically admissible stress field of Fig. 7.2, it is easily shown that (10) corresponds to a strict lower bound to the extrusion pressure for all reductions.

For $r \geqslant \frac{2}{3}$, the slipline field and the hodograph are those shown in Fig. 7.4, the dead metal zone being absent.[‡] The exit slipline AC is straight as before, inclined at $45°$ to the axis of symmetry. ABC is a $90°$ centered fan, and $BCEF$ is defined by the circular arc BC and its reflection in the axis. The slipline segments BF and GJ are parallel curves at a distance $\sqrt{2} \, h$ from one another. The angle turned through by each of the segments DE, BF, GJ, and JK is equal to the angle CAD, denoted by θ. This angle is such that N coincides with the meeting point of the die and the

---

† W. A. Green, *J. Mech. Phys. Solids*, **10**: 225 (1962); I. F. Collins, ibid, **16**: 137 (1968).

‡ Slipline field solutions for large reductions, based on various frictional conditions along the die and the container wall, have been given by W. Johnson, *J. Mech. Phys. Solids*, **4**: 191 (1956).

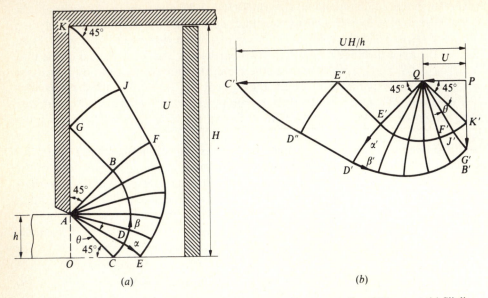

Figure 7.4 Extrusion through a smooth square die with reductions greater than 66.7 percent. (a) Slipline field; (b) hodograph.

container wall. The field GJK is a part of that discussed in Sec. 6.8 with $\psi = \pi/2$, the extrusion ratio being given by

$$\frac{H}{h} = 1 + 2\left\{ I_0\left[2\sqrt{\theta\left(\frac{\pi}{2} + \theta\right)} \right] + 2F_1\left(\theta, \frac{\pi}{2} + \theta \right) \right\}$$

which follows from (146), Chap 6, with $p_0 = k$, $s = H - h$, $a = h$ and $\alpha = \theta$. The extrusion ratio is also obtainable directly from Table A-1 or A-2 (Appendix) by subtracting unity from twice the value of y/a corresponding to the angular coordinates $(\theta, \pi/2 + \theta)$.

The mean compressive stress along the exit slipline AC must be equal to k in order to have zero longitudinal force across this line. The Hencky equations then furnish the hydrostatic pressure at any other point in the field. By (147), Chap. 6, the mean die pressure q may be written as

$$\frac{q}{2k} = \frac{\pi}{2} + 2\theta + \frac{2h}{H - h}[I_0(2\sqrt{\theta\xi}) - 4L(\theta, \xi)]$$

where $\xi = \pi/2 + \theta$. Since the extrusion pressure p_e is equal to rq by the condition of overall equilibrium, we have

$$\frac{p_e}{2k} = r\left(\frac{\pi}{2} + 2\theta\right) + 2(1 - r)[I_0(2\sqrt{\theta\xi}) - 4L(\theta, \xi)] \tag{11}$$

The expression in the square bracket of (11) can be evaluated using Table A-3 or A-4. The results of the calculation are presented in Table 7.2, an empirical formula for

Table 7.2 Results for frictionless extrusion with $r \geqslant \frac{2}{3}$

θ	0°	15°	30°	45°	60°	75°	90°
r	0.667	0.814	0.892	0.935	0.961	0.976	0.985
$q/2k$	2.571	2.725	3.028	3.404	3.823	4.273	4.746
$p_e/2k$	1.714	2.218	2.701	3.183	3.672	4.171	4.677

the extrusion pressure being

$$\frac{p_e}{2k} = 0.63 + 0.96 \ln \left(\frac{1}{1 - r} \right) \qquad r > \frac{2}{3} \tag{12}$$

which is correct to within 2 percent for $r < 0.95$. The greatest reduction for which the proposed slipline field applies is 98.5 percent, corresponding to $\theta = \pi/2$.

The uniform velocities of the material before entering and after leaving the deforming zone are represented by the vectors PQ and PC' in the hodograph (Fig. 7.4b). Since the particles on the die face AN must move vertically downward, a velocity discontinuity of magnitude $\sqrt{2} \, U$ propagates along the slipline $KJFE$. At the point E on the axis, the velocity suffers a second discontinuity of the same magnitude across the slipline EDA. The velocity increases along EC from the magnitude $3U$ at E to reach the exit value UH/h at C, there being no discontinuity across the line AC. The velocity is constant in magnitude and direction along each straight slipline in the regions ABC and $BFJG$. The triangular region ABG therefore moves downward as a rigid body with a velocity represented by the vector PG'. The hodograph is identical in shape to the slipline field, and the rate of plastic work is seen to be everywhere positive.†

The work done by the ram per unit time per unit width of the billet is $p_e HU$. Since the volume of material passing through the die, in unit time, is UH per unit width, p_e is equal to the work done per unit volume of the material. If the sheet were reduced in thickness by compressing it between smooth plates under plane strain condition, the work done per unit volume would have been $2k \ln (H/h)$. This is represented by the broken curve in Fig. 7.5, where the actual extrusion pressure calculated from the slipline fields is also plotted as a function of the fractional reduction. The relative expenditure of the redundant work, not directly contributing to the reduction, is seen to be quite considerable. The mean die pressure q, having a minimum value of $2k(1 + \pi/2)$, is represented by the uppermost solid curve.

---

† The solution may be shown to be complete by considering a statically admissible extension of the slipline field into the rigid region. See J. M. Alexander, *Q. Appl. Math.*, **19**: 31 (1961); J. Grimm, *Ing.-Arch.*, **44**: 79 (1975).

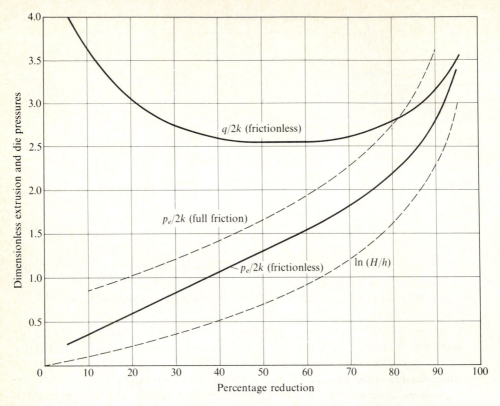

Figure 7.5 Extrusion pressure and die pressure for plane strain extrusion through a square die.

(v) *Inverted extrusion and piercing* In the case of inverted extrusion, the frictional resistance between the billet and the container is absent, since there is no relative movement between the two. The slipline fields for direct and inverted extrusions are obviously identical. The velocity field for the inverted extrusion is obtained from that for the direct extrusion by superposing a uniform velocity U from left to right. The billet is thereby brought to rest and the die is pushed into it with a velocity U. The pole of the hodograph is shifted from P to Q, the vector QP then representing the velocity of the die.

The process of piercing, in which the billet is held in a container and hollowed out by forcing a punch in the middle, is a variant of the inverted extrusion. If the problem is considered as one of plane strain and the punch is flat-ended, the slipline field for piercing is similar to that for steady state extrusion. When the container wall is smooth,† the slipline field and hodograph for piercing with reductions greater than 50 percent are as shown in Fig. 7.6. The initial and final thicknesses of

---

† R. Hill, *J. Iron Steel Inst.*, **158**: 177 (1948). A possible slipline field for piercing in a rough container has been suggested by W. Johnson and H. Kudo, *The Mechanics of Metal Extrusion*, p. 55, Manchester University Press (1962).

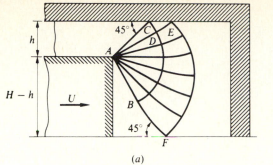
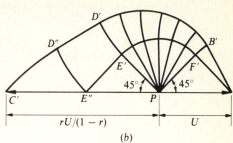

(a)   (b)

Figure 7.6 Slipline field and hodograph for steady state piercing in a smooth container with reductions greater than 50 percent.

the billet are $2H$ and $2h$ respectively giving a fractional reduction $r = (H - h)/H$. The punch and the emerging billet move in opposite directions with speeds U and $U(H - h)/h$ respectively. The hodograph of the process is similar to that for extrusion (Fig. 7.1), but the pole P now takes the position of the singularity Q, while PR is now equal to the punch speed U.

The mean punch pressure in piercing with reduction r is equal to the mean die pressure in steady state extrusion with reduction $1 - r$. This can be easily shown by superposing a uniform hydrostatic tension of amount equal to the mean die pressure on the stress system of Fig. 7.1, and then changing the sign of all the stresses. The slipline field for piercing with reductions less than 50 percent is therefore identical to that for extrusion process, except that the positions of the container wall and the central axis are interchanged. The curve for the steady state punch pressure against reduction is therefore symmetrical about $r = 0.5$, and is directly given by the mean die pressure in extrusion (Fig. 7.5) for reductions less than 50 percent. The pressure therefore has a minimum value of $2k(1 + \pi/2)$ corresponding to a reduction of 50 percent.

A false head of dead metal remains attached to the punch during the steady state operation, a certain amount of friction on the punch probably being necessary for the assumed dead metal. When the reduction is less than 50 percent, the punch pressure needed to begin piercing is $2k(1 + \pi/2)$, which is the yield pressure for indentation by a flat punch. For reductions greater than 50 percent, piercing would not begin until the punch pressure attains the steady state value. As the reduction is deceased to very small values, the punch pressure approaches a limiting value which should be approximately equal to the cavity pressure required for expanding in an infinite medium (Sec. 5.2(vii)). These results are in broad agreement with experiments† on the piercing of cylindrical billets, except for the rounding of the pressure-penetration curve due to elastic strains and work-hardening.

† E. Siebel and E. Fangmeier, *Mitt. Kaiser-Wilhelm-Inst. Eisenforsch.*, **13**: 28 (1931); W. Johnson and J. B. Haddow, *Int. J. Mach. Tool Des. Res.*, **2**: 1 (1962).

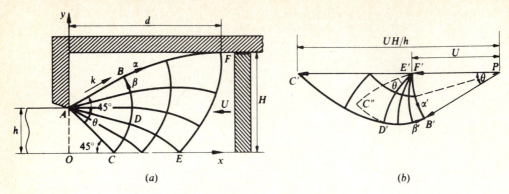

Figure 7.7 Extrusion through a square die when the container wall is perfectly rough. (*a*) Slipline field; (*b*) hodograph.

(vi) *Perfectly rough container wall* Some estimation of the influence of friction can be obtained by assuming that the container wall is perfectly rough so that the frictional stress reaches the value k where there is relative sliding.† The exit slipline AC is inclined at 45° with the axis of symmetry, and the angle of the centered fan CAB is so chosen that the slipline ABF is tangential to the wall (Fig. 7.7*a*). The angle ABD is 45°, being equal to the angle turned through by EF. The slipline field outside the fan is defined by the circular arc BC and the axis of symmetry. The existence of the dead metal zone to the left of AF requires certain frictional conditions on the die face. If the angle θ is assumed to be an integral multiple of 15°, the extrusion ratio is obtained directly from Table A-2 as the value of y/a corresponding to $(\theta, \pi/4 + \theta)$. The extrusion ratio may be expressed mathematically as

$$\frac{H}{h} = 1 + \sqrt{2}\left[F_1\left(\theta, \frac{\pi}{4} + \theta\right) + F_2\left(\theta, \frac{\pi}{4} + \theta\right) \right]$$

The mean compressive stress at a generic point of BF is $k(1 + \pi/2 + 2\theta + 2\alpha)$ by Hencky's equations. The resultant longitudinal thrust acting across the dead metal boundary ABF furnishes the mean die pressure

$$q = k\left(1 + \frac{\pi}{2} + 2\theta\right) + \frac{k}{H - h}\left(d + 2\int_B^F \alpha \, dy\right)$$

Integrating by parts, the extrusion pressure $p_e = rq$ may be expressed as

$$\frac{p_e}{2k} = \frac{1}{2}\left(1 + \frac{\pi}{2}\right)r + (1 + r)\theta + (1 - r)\left\{ \frac{d}{2h} - \int_0^\theta \left(\frac{y}{h}\right) d\alpha \right\} \tag{13}$$

The expression in the curly bracket can be readily evaluated by using the appropriate table (Appendix). The results of the calculation are given in Table 7.3,

---

† A slipline field solution for partially rough containers has been given by W. Johnson, *J. Mech. Phys. Solids*, **3**: 218 (1955).

Table 7.3 Results for symmetrical extrusion with a perfectly rough container

θ	5°	10°	15°	20°	30°	45°	60°	75°	90°
r	0.118	0.225	0.320	0.406	0.550	0.708	0.812	0.881	0.925
d/h	1.646	1.905	2.195	2.522	3.311	4.962	7.465	11.321	17.376
$p_e/2k$	0.890	1.069	1.248	1.428	1.784	2.316	2.846	3.395	3.901
μ		0.337	0.288	0.241	0.196	0.112	0	0.090	0.152	0.175

and a graphical plot of the extrusion pressure is shown in Fig. 7.5. This value of the pressure corresponds to the position of the ram at F. When the ram is farther back at a distance D from the die face, an additional term equal to $(D - d)/2H$ must be added to the right-hand side of (13) to allow for the frictional drag between the billet and the container. An empirical formula for the nominal extrusion pressure (13), correct to within 1 percent over the range $0.15 < r < 0.8$, may be written as

$$\frac{p_e}{2k} = 0.75 + 1.26 \ln\left(\frac{1}{1-r}\right) \tag{14}$$

The minimum coefficient of friction on the die face required for the validity of the slipline field may be estimated on the assumption that the rigid corner at A is stressed to the yield limit. It follows from Eqs. (33) and (35), Chap. 6, with $\tau_1 = k$, $\tau_2 = -k\cos 2\theta$, $q_1/k = 1 + \pi/2 + 2\theta$, $|\tau_2| = \mu q_2$ and $\alpha = \pi/2 - \theta$, that the minimum coefficient of friction is

$$\mu = \frac{|\cos 2\theta|}{1 + \pi/2 + 2\theta + \sin 2\theta} \tag{15}$$

Thus, μ increases as θ increases or decreases from $\pi/4$. The maximum reduction for which the slipline field is applicable is 92.5 percent corresponding to $\theta = \pi/2$, when AB coincides with the die face. The frictional stress on the die then attains the shear yield stress k, requiring a coefficient of friction not less than $2/(2 + 3\pi) \simeq 0.175$. The minimum values of μ for different reductions are included in Table 7.3.

The slipline field is associated with a velocity distribution which is everywhere continuous except across the dead metal boundary. The magnitude of the velocity discontinuity across this boundary is equal to the uniform speed U of the billet approaching the die. The entry and exit sliplines are mapped into the points F' and C' lying on the horizontal through the pole P of the hodograph (Fig. 7.7b). The slipline domain $FBDE$ is mapped into the singular field $F'B'D'$, where $F'B'$ is a circular arc of radius U. The remaining field $E'D'C'$ of the hodograph is the vector resultant of the fan $E'D'C''$, which is an extension of the original fan at E' through an angle θ, and its reflection in the line $E'C'$. An inspection of the slipline field and its hodograph net indicates that the rate of plastic work is nowhere negative.†

---

† Possible slipline fields for work-hardening materials have been discussed by L. E. Farmer and P. L. B. Oxley, *J. Mech. Phys. Solids*, **19**: 369 (1971).

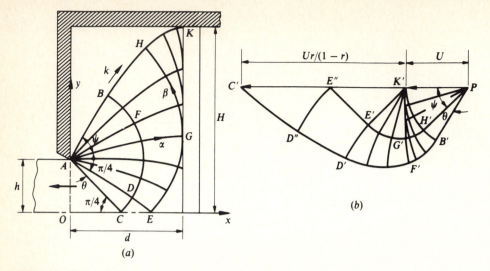

Figure 7.8 Slipline field and hodograph for extrusion of short slugs.

The load on the ram steadily decreases during the extrusion until the ram comes nearly in contact with the entry slipline. The process then ceases to be one of steady state, and the solution becomes strongly dependent on the roughness of the punch. The extrusion pressure rapidly varies as the residual length of the billet continues to decrease. The measured value of the extrusion pressure at the transition between the steady and unsteady states is found to be in close agreement† with the theoretical pressure (14). Due mainly to the breakdown of lubricants, the steady state extrusion pressure from a lubricated container frequently approaches the value attained in the unlubricated extrusion.

(vii) *Extrusion of short slugs* In the steady state solutions of the extrusion problem, it has been tacitly assumed that the slug is long enough for the slipline field to remain fixed in space during the entire process. When the slug length becomes less than a certain minimum, the deforming zone changes continuously as the extrusion proceeds, and the slipline field increasingly spreads over the face of the punch.‡ Figure 7.8a shows the instantaneous slipline field for extrusion of a short slug when the container wall and the punch face are assumed to be perfectly rough. The slipline *EGH* is tangential to the punch face whose instantaneous distance from the exit plane is denoted by *d*. Above the point *G*, all *β* lines meet the punch face tangentially, a condition that defines the region *GHK*. The slipline field is

† W. Johnson, *J. Mech. Phys. Solids*, **4**: 269 (1956). See also E. G. Thomsen, *J. Appl. Mech.*, **23**: 225 (1956). A semiexperimental method of solving the extrusion problem has been discussed by A. H. Shabaik and E. G. Thomsen, *J. Eng. Indust., Trans. ASME*, **91B**: 543 (1969).

‡ The final unsteady process in extrusion has been investigated by W. Johnson, *J. Mech. Phys. Solids*, **5**: 193 (1957). See also T. Murakami and H. Takahashi, *Int. J. Mech. Sci.*, **23**: 77 (1981).

completely specified by the angles θ and ψ, defining the two physical parameters d/h and H/h of the extrusion process.

The velocity is continuous across the exit slipline AC, while a discontinuity of amount U propagates along the dead metal boundary $KHBA$. The velocity is also discontinuous across the sliplines ADE and EG, the magnitude of this discontinuity being determined by the velocity of the plastic element at G relative to the punch. Since the rigid material above ABK is at rest, this slipline is mapped into the circular arc $B'K'$ of radius U and angular span $\theta + \psi$, as shown in Fig. 7.8b. The region $K'B'F'G'$ of the hodograph is then obtained from the fact that the characteristics are normal and tangential to the limiting line $K'G'$. The curve $G'F'$ and the circular arc $G'E'$ define the region $G'E'D'F'$, and the hodograph is completed in the usual manner to reach the point C' which maps the exit slipline AC. The plastic work rate is easily shown to be nowhere negative.

For computational purposes, it is convenient to assign a value for θ, thereby fixing the ratio d/h, and vary H/h by considering a number of values of ψ lying between 0 and $\pi/2 - \theta$. A continuous extrusion of a given slug is not, therefore, directly examined. Assuming a value of ψ, for a selected value of θ, the field GHK may be constructed by using the condition $\bar{x} = d$ along GK, and the continuity of $(\bar{x}, \bar{y})$ across GH. It may be noted that R is discontinuous across GH by an amount equal to its value at G immediately to the left of GH. The values of $\bar{x}$ and $\bar{y}$ along GH are obtained from (85), Chap. 6, while those within the field GHK are calculated by using (42), Chap. 6. The rectangular coordinates of the field then follow from the transformation relations. The extrusion pressure p_e is finally obtained from the equation

$$\frac{p_e}{2k} = \frac{r}{2}\left(1 + \frac{\pi}{2}\right) + (1 + r)(\psi + \theta) + (1 - r)\left\{\frac{d}{2h} - \int_0^\theta \left(\frac{y}{h}\right) d\alpha - \int_0^\psi \left(\frac{y}{h}\right) d\alpha\right\}$$

(16)

the derivation of which is similar to that of (13). The first integral of (16) is taken over the slipline BH and is directly obtained from Table A-1, while the second integral taken over HK is found by a numerical procedure.

When $\theta = 0$, the punch face is tangential to the circular arc BC, giving $d = \sqrt{2}\,h$. The ratio $(H - h)/h$ is then $\sqrt{2}$ times the value of x/a given by Table A-9 with $\alpha = \beta = \psi$. The total extrusion force $p_e H$ is equal to $\sqrt{2}\,kh + Q$, where Q is given by (121), Chap. 6, with $a = \sqrt{2}\,h$, $d = H - h$ and $p_0 = k(1 + \pi/2)$. The extrusion pressure is therefore given by

$$\frac{p_e}{2k} = \frac{r}{2}\left(1 + \frac{\pi}{2}\right) + \frac{1 - r}{\sqrt{2}}\,[2I_0(2\psi) - 1]$$

(17)

If r exceeds the value 0.773, which corresponds to $\psi = \pi/2$, shearing would occur along the die face over a distance greater than $\sqrt{2}\,h$, and the slipline field should be modified accordingly. No analytical solution seems possible in that case, although a numerical solution would be straightforward.

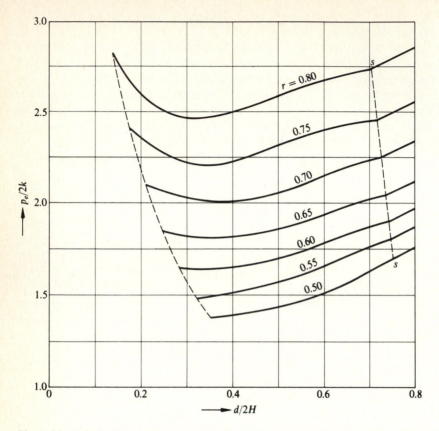

Figure 7.9 Variation of extrusion pressure with slug length during unsteady extrusion in plane strain (after *W. Johnson*).

For any given value of ψ, the ratio H/h and d/H can be calculated for each assumed value of θ. A plot of $p_e/2k$ and d/H against r is then made for each particular value of θ. These graphical plots readily furnish the variation of $p_e/2k$ with d/H corresponding to any given reduction in thickness. The results are displayed in Fig. 7.9 for a number of reductions, the points representing $\theta = 0$ being indicated by the broken line on the left. The dotted line ss represents the steady state solution with d/H just sufficient to contain the appropriate slipline field. Beyond this line, $p_e/2k$ increases linearly with d/H due to the frictional drag on the slug. The extrusion pressure first decreases and then increases as the extrusion is continued into the unsteady range. It is interesting to note that a fairly rapid reduction in pressure occurs over an appreciable range of reductions in the billet length.

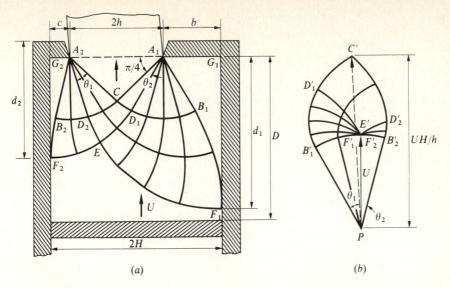

Figure 7.10 Unsymmetrical extrusion with perfectly rough container walls. (*a*) Slipline field; (*b*) hodograph.

7.2 Unsymmetrical and Multihole Extrusion

(i) *End extrusion* Consider the situation where the aperture of the die is not central with respect to the container.† When the container wall is perfectly rough, the slipline field and the hodograph are modified to those shown in Fig. 7.10. The exit sliplines A_1C and A_2C are inclined at 45° to the die opening A_1A_2, but the fans centered at A_1 and A_2 are of unequal angular spans. The field defined by the circular arcs CB_1 and CB_2 is extended as far as F_1 and F_2 where the sliplines are normal and tangential to the container walls. If the angles turned through by B_1F_1 and B_2F_2 are denoted by θ_1 and θ_2 respectively, the fan angles CA_1B_1 and CA_2B_2 are equal to $\pi/4 + \theta_1$ and $\pi/4 + \theta_2$ respectively, by Hencky's theorem.

The velocity discontinuity occurring across the dead metal boundaries $A_1B_1F_1$ and $A_2B_2F_2$ is of amount U, equal to the speed of the approaching billet. The centered fans $E'B_1'D_1'$ and $E'B_2'D_2'$ in the hodograph are defined by the circular arcs $E'B_1'$ and $E'B_2'$ of radius U, with fan angles $\pi/4 \pm (\theta_1 - \theta_2)$. The remainder of the hodograph is defined by the base curves $E'D_1'$ and $E'D_2'$ (first boundary-value problem). The billet is extruded obliquely with velocity PC', making an acute angle η with the shorter die face. Evidently, the thickness of the extruded billet is somewhat smaller than the width of the die aperture.

The eccentricity of the die may be defined as $e = (b - c)/(b + c)$, where b and c denote the widths of the die on either side of the aperture. If the initial and final thicknesses of the billet are $2H$ and $2h$ respectively, then $b + c = 2(H - h) = 2rH$,

---

† A. P. Green, *J. Mech. Phys. Solids*, **3**: 189 (1954).

where r is the nominal reduction in thickness.† Hence

$$\frac{b}{H} = (1 + e)r \qquad \frac{c}{H} = (1 - e)r$$

where e ranges from zero to one. The lengths d_1 and d_2 covered by the dead metal zones are identical to those for symmetrical extrusion with extrusion ratios $(b + h)/h$ and $(c + h)/h$ respectively, the corresponding reductions being

$$r_1 = \frac{(1 + e)r}{1 + er} \qquad r_2 = \frac{(1 - e)r}{1 - er} \tag{18}$$

The mean normal pressures on the dies A_1G_1 and A_2G_2 are p_1/r_1 and p_2/r_2 respectively, where p_1 and p_2 are the extrusion pressures for symmetrical extrusion with reductions r_1 and r_2 respectively. The overall vertical equilibrium requires $2p_eH = (p_1/r_1)b + (p_2/r_2)c$, which gives

$$p_e = \tfrac{1}{2}\{(1 + er)p_1 + (1 - er)p_2\} \tag{19}$$

in view of the last two equations. Adopting the empirical formula (14), and using (18), we have

$$\frac{p_1}{2k} = 0.75 + 1.26 \ln\left(\frac{1 + er}{1 - r}\right) \qquad \frac{p_2}{2k} = 0.75 + 1.26 \ln\left(\frac{1 - er}{1 - r}\right)$$

Substitution in (19) shows that the extrusion pressure in unsymmetrical extrusion exceeds that in symmetrical extrusion for the same reduction r by an amount Δp_e, where

$$\frac{\Delta p_e}{2k} = 0.63\left\{er \ln\left(\frac{1 + er}{1 - er}\right) + \ln\left(1 - e^2r^2\right)\right\} \tag{20}$$

The pressure Δp_e increases with e from a minimum value of zero at $e = 0$. Equation (20) gives the value of the pressure when the ram has moved to F_1. In general, we have to add the quantity $k(2D - d_1 - d_2)/2H$ to the right-hand side of (19) to allow for the frictional drag between the billet and the walls. The corresponding modification of (20) involves the additional term $(2d - d_1 - d_2)/4H$ for $\Delta p_e/2k$, where d is the dead metal length in the symmetrical extrusion for the same reduction.

When the container wall is perfectly smooth and the reduction is sufficiently large, a simple modification of the slipline field for symmetrical extrusion is possible if b and c are both greater than h. The proposed slipline field and the associated hodograph are shown in Fig. 7.11, their construction being similar to that for the rough container. The extruded material is seen to emerge straight but oblique to the axis of the orifice. The extrusion pressure is still given by (19), where p_1 and p_2 are related to r_1 and r_2 by the empirical formula (9). The amount by which the pressure in unsymmetrical extrusion exceeds that in symmetrical

---

† The true reduction in thickness is equal to $1 - (1 - r)\sin\eta$ where η is defined above.

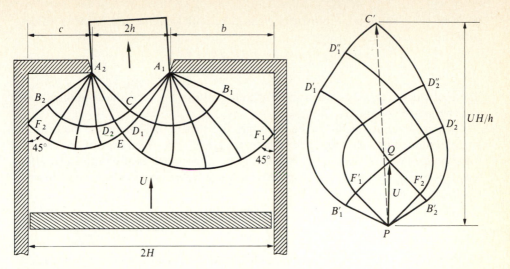

Figure 7.11 Slipline field and hodograph for unsymmetrical extrusion with smooth container walls ($b > h$, $c > h$).

extrusion is again given by (20), where $e \leqslant 2 - 1/r$ in view of the restriction b, $c \geqslant h$. The effect of eccentricity over this range is therefore identical for smooth and rough containers within the accuracy of the empirical equations.

The solution for smooth containers outside the above range is extremely complicated. Green† has proposed two types of slipline field, one applying to the range $b, c < h$, and the other applying to the range $b > h > c$. In both cases, the exit sliplines are curved, and the extruded material rotates to form a curved product with constant curvature. The center of curvature lies on the same side of the orifice as the larger dead metal zone. However, recent computations have revealed that the second type of field holds not only for $b > h$, but also over a part of the range $b < h$. Moreover, there is an upper limit on the eccentricity for both fields, depending on the reduction, beyond which the solution ceases to hold. Plasticine experiments by Green have shown that a curved product always results when $e > 2 - 1/r$, while a straight product is invariably obtained when $e < 2 - 1/r$.

The results for unsymmetrical extrusion can be used to analyze the problem of extrusion through two equal holes symmetrically placed in a square die.‡ When the container is perfectly smooth, one-half of the slipline field for symmetrical two-hole extrusion is identical to Fig. 7.11 with one of the container walls replaced by the axis of symmetry. The difference between the extrusion pressures in symmetrical single-hole and two-hole extrusions from a smooth container is therefore

---

† A. P. Green, op. cit.; N. S. Das, N. R. Chitkara, and I. F. Collins, *Int. J. Num. Meth. Eng.*, **11**: 1379 (1977).

‡ Symmetrical two-hole and three-hole extrusions have been examined by L. C. Dodeja and W. Johnson, *J. Mech. Phys. Solids*, **5**: 267 (1957).

given by (20), where the eccentricity e can take both positive and negative values with the restriction

$$-\left(2 - \frac{1}{r}\right) \leqslant e \leqslant \left(2 - \frac{1}{r}\right)$$

The eccentricity vanishes when the orifice is midway between the container wall and the axis of symmetry.

When the container is perfectly rough, the slipline field between the central axis and the orifice axis is the same as before, while that between the orifice axis and the container wall is identical to Fig. 7.10. The extrusion pressure is given by (19), where p_1 and p_2 correspond to symmetrical single-hole extrusion from smooth and rough containers respectively. Thus

$$p_1 = 0.41 + 1.26 \ln\left(\frac{1 + er}{1 - r}\right) \qquad p_2 = 0.75 + 1.26 \ln\left(\frac{1 - er}{1 - r}\right)$$

Substituting in (19), we find that the amount Δp by which the nominal extrusion pressure in two-hole extrusion exceeds that in single-hole extrusion for a given reduction may be expressed by the empirical formula

$$\frac{\Delta p_e}{2k} = 0.63\left\{er \ln\left(\frac{1 + er}{1 - er}\right) + \ln(1 - e^2r^2)\right\} - 0.17(1 - er) \qquad (21)$$

This solution is valid so long as the width of the central die is not less than the width of the orifice, the restriction on e being identical to that for the smooth container. For a given reduction, the extrusion pressure is a minimum when $e = 0$ for the smooth container, and when $e \simeq -0.134/r$ for the rough container. The minimum value of $\Delta p_e 2k$ for the rough container is -0.18 approximately, irrespective of the reduction. The variation of $\Delta p_e/2k$ with e at constant reductions is shown in Fig. 7.12 for both smooth and rough containers.

(ii) *Side extrusion* A billet of initial thickness H is extruded through an orifice of width $2h$ on one side of a container. The end of the container is assumed to be sufficiently far away from the orifice so as to have no effect on the zone of deformation. The slipline field is shown in Fig. 7.13, where the exit sliplines AC and BC are equally inclined to the container wall. The field is defined by the circular arcs CD and CE of equal radii but having unequal angular spans θ and ϕ which depend on the wall friction. The point F moves up the wall as the friction between the material and the container is increased.[†] If λ denotes the acute angle which the tangent to the slipline DF at F makes with the wall, it follows from Hencky's first theorem that

$$\phi - \theta = \frac{\pi}{4} - \lambda$$

[†] A. P. Green, *J. Mech. Phys. Solids*, **3**: 189 (1954); W. Johnson, P. B. Mellor and D. M. Woo, ibid., **6**: 203 (1958).

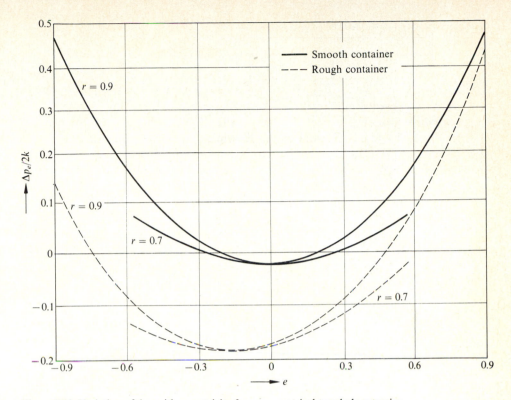

Figure 7.12 Variation of Δp_e with eccentricity for a symmetrical two-hole extrusion.

The hydrostatic pressure on AC and BC must be equal to k in order that the resultant horizontal force across the orifice becomes zero. Since the hydrostatic pressure at F has the value $k(1 + 2\theta + 2\phi)$ by Hencky's equations, the ratio of the tangential traction and the normal pressure at F is

$$\mu = \frac{\cos 2\lambda}{1 + 2\theta + 2\phi - \sin 2\lambda} \tag{22}$$

The coefficient of friction μ therefore depends on λ and one of the fan angles. For given values of θ and ϕ, the ratio H/h is obtained from the appropriate table (Appendix). The minimum value of H/h for which the slipline field holds is $\sqrt{2} \cos \lambda$, when θ vanishes. The upper limit of H/h corresponds to the situation where AE is inclined at an angle λ to the container wall. For higher H/h ratios, the yield criterion would be violated in the rigid corner at A.

    If the speed of the ram is denoted by U, the condition of incompressibility of the material requires the component of velocity of the extruded billet along the axis of the orifice to be equal to UH/h. The hodograph is defined by the circular arcs $F'D'$ and $F'E'$ of radii $U \cos \lambda$ and $U \sin \lambda$, which are the magnitudes of the velocity discontinuity across the sliplines FDB and FEA respectively. The material below

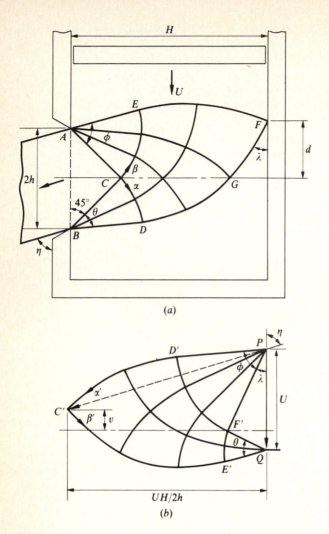

Figure 7.13 Sideways extrusion through a rough container. (*a*) Slipline field; (*b*) hodograph.

BDF is at rest. The velocity of the extruded billet is represented by the vector *PC'* which is inclined at an angle η to the vertical. It follows from simple geometry that

$$\tan \eta = \frac{H/2h}{\cos^2 \lambda - (v/U)} \tag{23}$$

The ratio v/U is obtained from Table A-6 as the value of $y/2a$ corresponding to the angular coordinates (θ, ϕ).

Since the mean compressive stress at *C* is equal to *k* by the condition of zero horizontal pull across the orifice, the vertically downward force exerted by the ram is $kH + Q$, where *Q* denotes the vertical force across *AEF* with zero hydrostatic pressure at C. The extrusion pressure p_e, when the ram touches EF, is given by

$$\frac{p_e}{2k} = \frac{1}{2}\left(1 + \frac{Q}{kH}\right) \tag{24}$$

The ratios $Q/2kh$ and H/h, which correspond to the angular coordinates (θ, ϕ), may be directly obtained from Table A-1 or A-2. The results for a perfectly rough container wall ($\lambda = 0$) are given in Table 7.4. The velocity discontinuity across the slipline AEF disappears in this case, and the points E' and F' coincide with Q. If the material is extruded by a pair of oppositely moving rams, the slipline field would consists of the net $ACGF$ and its reflection in GC, a triangular zone of dead metal being formed between FG and its reflected image. The extrusion pressure required by each ram is identical to (24), but the direction of motion of the extruded billet depends on the ratio of the speeds of the two rams. The billet is extruded along the axis of the orifice when the rams have equal speeds.

When the container wall is perfectly smooth ($\lambda = \pi/4$), the fan angles θ and ϕ are equal to one another, and the slipline field becomes symmetrical about the axis of the orifice. The extrusion pressure may be represented by the empirical formula

$$\frac{p_e}{2k} = 0.85\left(\frac{H}{h}\right)^{0.52} \qquad 2.0 < \frac{H}{h} < 8.5 \tag{25}$$

which is correct to within 1.5 percent of the exact solution. The hodograph in this case is identical to the slipline field, and the angle η assumes the value $\tan^{-1}(H/h)$ when the extrusion is carried out by a single ram as shown. The variation of the dimensionless extrusion pressure with the ratio H/h is shown in Fig. 7.14 for the limiting cases of perfectly smooth and perfectly rough containers. The extrusion pressure and the mode of deformation predicted by the theory have been confirmed by experiments.[†]

The solution for the single-hole extrusion from a smooth container also applies to the situation where the material is extruded through two collateral holes of width $2h$ on opposite sides of a container of width $2H$ and having an arbitrary state of roughness of the wall. When the two holes are of unequal widths $2h_1$ and $2h_2$

Table 7.4 Data for side extrusion with a perfectly rough container

θ	$0°$	$15°$	$30°$	$45°$	$60°$	$75°$
H/h	1.414	2.195	3.311	4.961	7.465	11.330
d/h	1.000	1.472	2.222	3.420	5.336	8.412
$p_e/2k$	1.285	1.492	1.806	2.195	2.639	3.123
$\eta°$	45.00	58.26	67.96	74.70	79.32	82.45

[†] A comparison with experiments based on the extrusion of lead billets has been made by T. F. Jordan and E. G. Thomsen, *J. Mech. Phys. Solids*, **4**: 184 (1955).

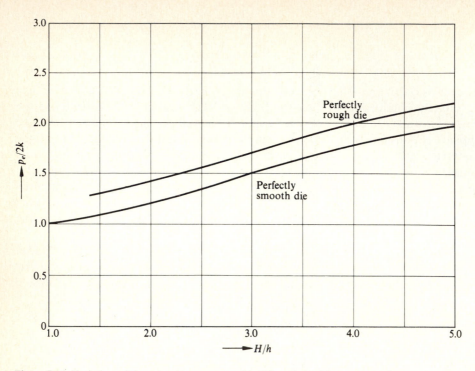

Figure 7.14 Variation of the extrusion pressure with H/h ratio in sideways extrusion.

such that $h_1 + h_2 < 2H$, the slipline field is a combination of two geometrically similar ones for one-sided extrusion from a smooth container. The extrusion pressure is therefore identical to that for the one-sided extrusion with a container width H and an orifice width $2h = h_1 + h_2$. The direction of motion of the extruded billet on either side of the container makes an angle of $\tan^{-1}(H/h)$ with that of the ram, irrespective of the frictional condition on the wall.[†]

(iii) *Combined end and side extrusion* Suppose that the material is extruded simultaneously through an end orifice and a pair of side orifices.[‡] The particular case of symmetrical extrusion through holes of equal sizes will be examined here, but the solution can be easily extended for holes of unequal widths. In Fig. 7.15a is shown one-half of the slipline field for the general case where the meeting point F of the two kinds of field is situated at some distance away from the center line of the side orifice. The field for the side extrusion involves unequal fan angles θ_2 and θ_4 at the corners A and B respectively. The field corresponding to the end extrusion is

---

[†] A problem of partial side extrusion has been treated by W. Johnson, *J. Mech. Phys. Solids,* **5**: 193 (1957).

[‡] W. Johnson, P. B. Mellor, and D. M. Woo, *J. Mech. Phys. Solids,* **6**: 203 (1958).

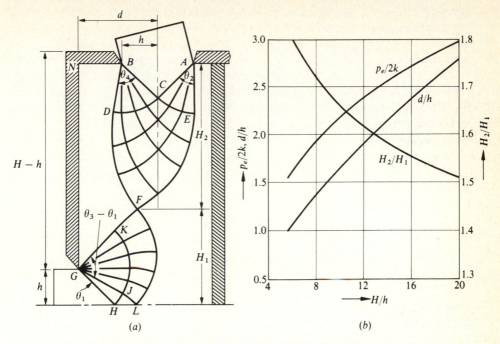

Figure 7.15 Simultaneous end and side extrusion. (*a*) Slipline field; (*b*) results for $\theta_2 = \theta_4$.

defined by the angles θ_1 and θ_3 subtended by the circular arcs HJ and HK respectively. Since the mean compressive stress is equal to k along the exit sliplines, the conditions of continuity of the hydrostatic pressure and the slipline directions at F furnish

$$\theta_3 - \theta_4 = \theta_2 - \theta_1 \qquad \theta_3 + \theta_4 = \frac{\pi}{2} + (\theta_2 + \theta_1)$$

These equations may be solved for θ_3 and θ_4 to obtain

$$\theta_3 = \frac{\pi}{4} + \theta_2 \qquad \theta_4 = \frac{\pi}{4} + \theta_1$$

It follows that $\theta_2 - \theta_1 \leqslant \pi/4$ for $\theta_2 \leqslant \theta_4$, which constitutes the practical range. The angles θ_1 and θ_2 may be taken as independent parameters defining the slipline field. For given values of H/h and d/h, these angles may be determined from the conditions (*a*) the sum of the horizontal distances of F from G and C is equal to d, and (*b*) the sum of the vertical distances of F from H and A is equal to the semithickness H. The extrusion pressure may be calculated from the relation

$$\frac{p_e}{2k} = \frac{1}{2}\left(1 + \frac{P + Q}{kH}\right) \tag{26}$$

where P and Q are the magnitudes of the horizontal thrust exerted by the dead metal across the boundaries GKF and BDF respectively, when the hydrostatic pressure at H and C are taken to be zero. The ratios $P/2kh$ and $Q/2kh$ are obtainable from Table A-1 or A-2, the appropriate angular coordinates being (θ_1, θ_3) and (θ_2, θ_4) respectively. The solution is independent of the wall friction so long as BD is inclined at an angle greater than $45°$ to the container wall. When $\theta_1 = \theta_2 = 0$, the slipline field degenerates into a pair of $45°$ centered fans, giving a value $\sqrt{2} + 1$ for both H/h and d/h ratios, the corresponding value of $p_e/2k$ being $(2 - \sqrt{2})(1 + \pi/4) \simeq 1.046$.

Consider now the special case when the common point F is on the vertical center line. The slipline field for the side extrusion is then symmetrical about the orifice axis, requiring

$$\theta_2 = \theta_4 = \frac{\pi}{4} + \theta_1 \qquad \theta_3 = \frac{\pi}{2} + \theta_1$$

The entire slipline field may be specified by the angle θ_1, the ratios d/h, H_1/h and H_2/h being obtainable from the appropriate table. The force exerted by the ram to cause simultaneous extrusion through the three orifices is the sum of those for two side extrusions, each over a depth H_2, and an end extrusion over a depth $2H_1$. The extrusion pressure is therefore given by

$$Hp_e = H_1 p_1 + H_2 p_2$$

where p_1 and p_2 are the extrusion pressures for the separate end and side extrusion processes respectively. Adopting the empirical formulas (9) and (25) we get

$$\frac{p_e}{2k} = \frac{H_1}{H} \left(0.41 + 1.26 \ln \frac{H_1}{h} \right) + 0.85 \frac{H_2}{H} \left(\frac{H_2}{h} \right)^{0.52} \tag{27}$$

where $2H$ is the initial sheet thickness. The calculated values of $p_e/2k$, H_2/H_1 and d/h are plotted against H/h in Fig. 7.15b. The minimum value of H/h for this solution to apply is 5.644, which corresponds to $\theta_1 = 0$.

The region $GKFDBN$ is a dead metal zone where the material remains attached to the container during the extrusion. The sheet extruded through the end orifice moves along the axis with a speed equal to UH_1/h, where U is the speed of the rigid material approaching the deforming zone. The material leaving the side orifice moves in the direction FB, when $\theta_2 = \theta_4$, with a speed approximately equal to $UH_2/2h$. The ratio of the two speeds of the extruded billet, having the value $2H_1/H_2$ to a close approximation, increases as the ratio H/h is increased.

7.3 Sheet Drawing Through Tapered Dies

(i) Frictionless drawing with small reduction Consider the steady state process in which a metal sheet of initial thickness $2H$ is reduced to a final thickness $2h$ by pulling it symmetrically through a wedge-shaped die having an included angle 2ψ. The die is assumed to be perfectly smooth so that all sliplines meet the die face at

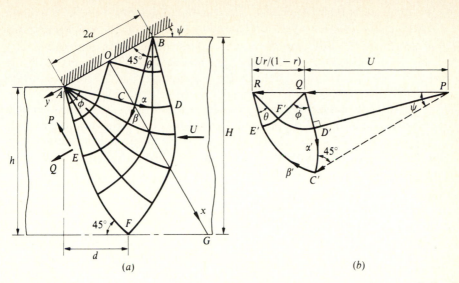

Figure 7.16 Sheet drawing through a smooth tapered die with $H/h \leqslant 1 + 2 \sin \psi$. (a) Slipline field; (b) hodograph.

45°. When the reduction in thickness is sufficiently small, the slipline field involves a single point F of the axis of symmetry (Fig. 7.16a). The triangle ABC is a region of constant stress, while ACE and BCD are centered fans continued round the corners A and B. The circular arcs CD and CE define the remaining part $DCEF$ of the slipline field.[†] The fan angles θ and ϕ are such that the point F lies on the axis of symmetry. Since the counterclockwise angle made by AE with the axis is $\pi/4 + \phi - \psi$, it follows from Hencky's first theorem that

$$\phi - \theta = \psi$$

Let (x_0, y_0) be the coordinates of F referred to the rectangular axes as shown, with the origin at O, the midpoint of AB. If G is the point where the x axis is intersected by the axis of symmetry, it follows from geometry that

$$OG = \tfrac{1}{2}(H + h) \sec \psi = x_0 + y_0 \tan \psi$$

If the die contact length is denoted by $2a$, which is equal to $(H - h) \operatorname{cosec} \psi$, the fractional reduction $r = (H - h)/H$ may be expressed, in view of the above relation, as

$$\frac{2}{r} = 1 + \frac{x_0}{a} \cot \psi + \frac{y_0}{a} \tag{28}$$

from which r may be calculated as a function of either ϕ or θ for a given die angle,

[†] R. Hill and S. J. Tupper, *J. Iron Steel Inst.*, **159**: 353 (1948).

using the appropriate table (Appendix). The maximum drawing ratio for which the present field is valid is $1 + 2 \sin \psi$, corresponding to $\theta = 0$ and $\phi = \psi$, the greatest value of the reduction being

$$r = \frac{2 \sin \psi}{1 + 2 \sin \psi} \tag{29}$$

The hydrostatic pressure in the region ABC is equal to $q - k$, where q denotes the magnitude of the uniform die pressure.† If the sheet is drawn without any back pull, the force required for the drawing must equal the longitudinal component of the thrust on the die. If the mean drawing stress is denoted by t, then $ht = (H - h)q$ for longitudinal equilibrium, giving

$$t = \left(\frac{H}{h} - 1\right)q = \left(\frac{r}{1 - r}\right)q$$

Let (P, Q) denote the components of the resultant force per unit width across the slipline AEF due to the rigid material on its left, when the hydrostatic pressure of C is assumed to be zero. Then the forces acting on BDF in the negative x and y directions due to the rigid material approaching the die are of magnitudes $2ka - P$ and Q respectively. Since the actual hydrostatic pressure at C is equal to $q - k$, the condition of zero longitudinal force across BDF may be written as

$$(q - k)H + (2ka - P) \sin \psi - Q \cos \psi = 0$$

In view of the relationship between t and q, and the identity $rH = 2a \sin \psi$, the above equation reduces to

$$\frac{t}{2k} = \frac{r}{2}\left\{1 + \frac{r}{1 - r}\left(\frac{P}{2ka} + \frac{Q}{2ka} \cot \psi\right)\right\} \tag{30}$$

The ratios $P/2ka$ and $Q/2ka$ are directly obtained from Table A-1 or A-2, the corresponding angular coordinates being (θ, ϕ). In the limiting case $\theta = 0$, the mean drawing stress t is equal to $2k(1 + \psi) - q$ by the Hencky relation. Since $rq = (1 - r)t$, the dimensionless drawing stress in this particular case becomes

$$\frac{t}{2k} = \frac{2(1 + \psi) \sin \psi}{1 + 2 \sin \psi}$$

in view of (29). For a given die angle, as the reduction is decreased, the drawing stress decreases but the die pressure increases with increasing gradient. When the reduction is sufficiently small, the die pressure becomes large enough to cause the plastic zone to spread round the corner B to the surface of the undrawn sheet. By (38), Chap. 6, with $q_2 - q_1 = q$ and $\alpha = \pi - \psi$, this die pressure is $q = 2k(1 + \pi/2 - \psi)$, which sets a limit of validity of the solution. The limit is known

---

† The hydrostatic stress at F is predominantly tensile. It becomes compressive for $\psi < 18°$ when the reduction approaches (29). See B. Dodd and D. A. Scivier, *Int. J. Mech. Sci.*, **17**: 663 (1975).

as the *bulge limit*, and the reduction for which it occurs increases with increasing die angle.

Since the material to the right of *BDF* moves as a rigid body, the normal component of velocity is constant along the straight slipline *BD*. It follows from Geiringer's equations that the normal component is constant on each slipline through *B*, and in particular on *BC*. Similarly, the normal component of velocity is constant on each slipline through *A*, and in particular on *AC*. The region *ABC* therefore moves with a uniform velocity which must be parallel to the die face. If the speed of the sheet ahead of the die is denoted by *U*, the speed of the drawn sheet is UH/h in view of the incompressibility of the material. In the hodograph of the process (Fig. 7.16*b*), the velocities of the material entering and leaving the die are represented by the vectors *PQ* and *PR* respectively. A velocity discontinuity of amount $rU/(1 - r)$ occurs across the bounding sliplines *AEF* and *BDF*, while the uniform velocity of the triangle *ABC* is given by the vector *PC′*. An inspection of the slipline field and the hodograph indicates that the plastic work rate is everywhere positive.

The ratios $t/2k$ and $q/2k$ are plotted in Figs. 7.17 and 7.18 as functions of the reduction *r* for various semiangles of the die. Each solid curve terminates at the bulge limit represented by the broken curve on the left. The relationship between the die angle and the reduction corresponding to this limit may be expressed empirically as

$$r = \psi\left(0.23 + \frac{\psi}{9}\right) \tag{31}$$

For reductions smaller than this, a standing wave of plastic material is formed in front of the die, as has actually been observed in the drawing of sheets and round bars.†

(ii) Frictionless drawing with large reduction The slipline field and the hodograph for reductions greater than (29) are shown in Fig. 7.19, where the deformation extends over a finite portion of the axis of symmetry. From geometry, the angle of the fan *ABC* is equal to the die semiangle ψ. The angle *ACD*, denoted by θ, must be such that the entry slipline *EFJK* passes through the corner *K*. Since the sliplines meet the die at 45°, the angles turned through by *GJ* and *JK* are both equal to θ. The field *GJK* may be constructed by the method of vectorial superposition discussed in Sec. 6.8(i). Using Eq. (146), Chap. 6, where $s = (H - h) \operatorname{cosec} \psi$ and $\alpha = \theta$, the extrusion ratio may be written as

$$\frac{H}{h} = 1 + 2[I_0(2\sqrt{\theta\xi}) + 2F_1(\theta, \xi)] \sin \psi \tag{32}$$

where $\xi = \psi + \theta$. Since the hydrostatic pressure at *C* is $p_0 = k - t$, where *t* is the

---

† J. G. Wistreich, *Proc. Inst. Mech. Eng.*, **169**: 123 (1955); R. W. Johnson and G. W. Rowe, ibid., **182**: 521 (1968).

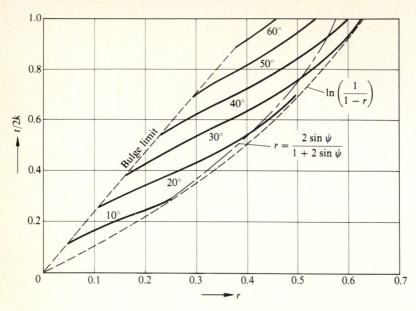

Figure 7.17 Variation of the mean die pressure with the fractional reduction for various semiangles of a tapered die.

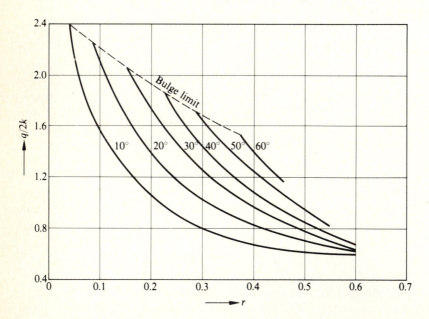

Figure 7.18 Variation of the mean drawing stress with reduction in sheet drawing without friction.

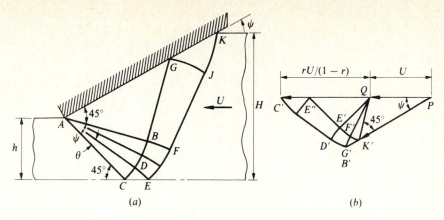

Figure 7.19 Sheet drawing through a smooth tapered die with $H/h \geqslant 1 + 2 \sin \psi$. (a) Slipline field; (b) hodograph.

mean drawing stress, it follows from (147), Chap. 6, that

$$\frac{q+t}{2k} = \psi + 2\theta + \frac{2h}{H-h} [I_0(2\sqrt{\theta\xi}) - 4L(\theta, \xi)] \sin \psi$$

where q is the mean die pressure. The actual pressure on the die is constant over AG, but increases steadily along GK. Since $rq = (1 - r)t$ for equilibrium, the mean drawing stress is obtained as

$$\frac{t}{2k} = r(\theta + \xi) + 2(1 - r)[I_0(2\sqrt{\theta\xi}) - 4L(\theta, \xi)] \sin \psi \tag{33}$$

The values of I_0, F_1, and L are readily obtained from the appropriate table (Appendix) for any assumed value of θ. Equations (32) and (33) therefore give the relationship between t and r parametrically through θ. The maximum reduction for which the slipline field is applicable corresponds to $\theta = \psi$. No drawing is possible when $t \geqslant 2k$, for which there is necking of the drawn sheet outside the die.

The associated velocity field is represented by the hodograph, which is explained in the same way as that for extrusion with large reductions. There is a velocity discontinuity of amount $\sqrt{2} U \sin \psi$ propagating along $AEFJK$, where U is the uniform speed of the material entering the die. The velocity is continuous across the exit slipline, and the rate of plastic work is everywhere positive. The part of the hodograph to the left of QK' is identical in shape to the slipline field.

The solid curves in Fig. 7.17, representing the drawing stress, do not extend as far as the reduction that marks the limit of the present field. The limiting value of the reduction for $\psi \leqslant \pi/6$ is approximately equal to $\psi(2 - \psi)$. The broken curve represents the ideal drawing stress which is equal to $2k \ln (H/h)$ corresponding to homogeneous compression. The relative amount of redundant work, represented by the difference between the solid and broken curves, decreases as the reduction is increased. The ratio of the actual drawing stress to the ideal drawing stress is known as the redundant work factor. For all die angles and reductions, this factor is found to depend only on the ratio b/a, where $2b$ denotes the length of the circular arc with center at the virtual apex of the wedge and joining the midpoints of the two areas of die contact. The drawing stress may therefore be written in the form[†]

$$\frac{t}{2k} = f\left(\frac{b}{a}\right) \ln\left(\frac{1}{1-r}\right) \qquad \frac{b}{a} = \left(\frac{2}{r} - 1\right)\psi \tag{34}$$

It is also found that $2kf(b/a)$ is very nearly equal to the mean die pressure in the compression of a strip between a pair of smooth flat dies, the ratio of the strip thickness to the die width being b/a (sec. 8.1). The redundant work factor f for various values of b/a is given in Table 7.5, an empirical formula correct to within 1 percent being

$$f\left(\frac{b}{a}\right) = \begin{cases} 0.93 + 0.07\left(\frac{b}{a}\right)^2 & 1.0 \leqslant b/a \leqslant 2.5 \\[3mm] 0.85\left(\frac{b}{a}\right)^{0.52} & 2.5 \leqslant b/a < 8.0 \end{cases} \tag{35}$$

The work done per unit volume of the drawn sheet is equal to the mean drawing stress t. Since the equivalent stress for a nonhardening Mises material is $\sqrt{3}\,k$, the mean equivalent strain of the drawn sheet is equal to $t/\sqrt{3}\,k$. It is reasonable to suppose that the same mean effective strain occurs in the process whatever the strain-hardening characteristic of the material. Now, the work done per unit volume of a work-hardening material is $\int \bar{\sigma}\, d\bar{\varepsilon}$, where the integral represents the area under the equivalent stress-strain curve. It follows that the mean drawing stress t^* for a work-hardening sheet is equal to the area under the true stress-strain curve up to a total strain of $t/\sqrt{3}\,k$, the value of which is obtained from the slipline field theory. If the stress-strain curve of the material is expressed by the simple power law $\sigma = C\varepsilon^n$, where C and n are empirical constants, the mean drawing stress becomes

$$t^* = C \int_0^{t/\sqrt{3}\,k} \varepsilon^n \, d\varepsilon = \frac{C}{1+n}\left(\frac{t}{\sqrt{3}\,k}\right)^{1+n}$$

Table 7.5 Redundant work factor in sheet drawing without friction

b/a	$f(b/a)$	b/a	$f(b/a)$	b/a	$f(b/a)$	b/a	$f(b/a)$
8.716	2.571	4.567	1.881	2.014	1.218	0.800	1.021
8.046	2.481	3.791	1.703	1.605	1.103	0.746	1.033
7.117	2.345	3.190	1.550	1.383	1.050	0.707	1.040
6.216	2.199	2.792	1.442	1.183	1.014	0.670	1.033
5.434	2.057	2.440	1.342	1.000	1.000	0.500	1.000

[†] A. P. Green and R. Hill, *J. Mech. Phys. Solids*, **1**: 31 (1953); A. P. Green, *Proc. Inst. Mech. Eng.*, **174**: 847 (1960).

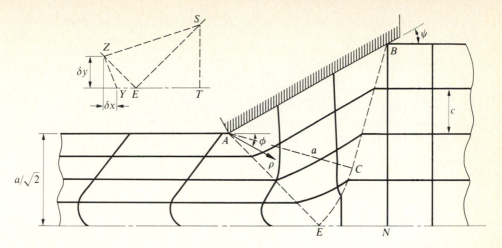

Figure 7.20 Distorted square grid in a sheet drawn through a smooth tapered die of 30° semiangle.

It may be noted that C is $\sqrt{3}$ times the shear yield stress of the material when it is prestrained by $\bar{\varepsilon} = 1$. Substitution from (34) into the above expression gives

$$t^* = \frac{C}{1 + n}\left\{\frac{2}{\sqrt{3}}f\left(\frac{b}{a}\right)\ln\left(\frac{1}{1 - r}\right)\right\}^{1 + n}$$

(36)

The redundant work factor for the work-hardening material is therefore equal to $(1 + n)$th power of that for the nonhardening material. To avoid yielding of the drawn sheet, t^* must be less than the mean yield stress at the exit, which requires $f(b/a) \ln(H/h) < 1 + n$, setting an upper limit on the reduction in thickness.

(iii) *Distortion of a square grid* The deformed shape of a uniform square grid† scribed on the undrawn sheet can be determined analytically when the reduction is given by (29). Then the relevant range of angles for sheet drawing is $0 < \psi < \pi/4$, which corresponds to $t < 2k$. Let (ρ, ϕ) denote the polar coordinates of any point in the centered fan ACE, where ϕ is measured in the clockwise sense from the horizontal through A (Fig. 7.20). The radial and circumferential velocities u and v, satisfying the Geiringer equations and the boundary conditions, may be written as

$$u = -U\cos\phi \qquad v = U(\sqrt{2}\sin\psi + \sin\phi) \tag{37}$$

Thus, u is continuous across CE and v is continuous across AE, the speed of the material leaving the fan being $U(1 + 2\sin\psi)$. The uniform speed of the material in ABC moving along the die face is $U(\cos\psi + \sin\psi)$.

The gridlines originally parallel to the axis of symmetry are deformed into the streamlines while passing through the die. The streamlines in ABC are evidently all

---

† The influence of die friction has been discussed by W. Johnson, R. Sowerby, and R. D. Venter, *Plane Strain Slipline Fields for Metal Deformation Processes*, Pergamon Press, Oxford (1982).

parallel to the die face. Within the sector ACE, the streamlines are given by the differential equation

$$\frac{1}{\rho}\frac{d\rho}{d\phi} = \frac{u}{v} = -\frac{\cos\phi}{\sqrt{2}\sin\psi + \sin\phi}$$

A typical streamline lying below the one passing through C may be identified by the angular position ϕ_0 of the point where the streamline meets the circular arc CE of radius a. The integral of the above equation then becomes

$$\frac{\rho}{a} = \frac{\sqrt{2}\sin\psi + \sin\phi_0}{\sqrt{2}\sin\psi + \sin\phi} \qquad \phi_0 \geqslant \frac{\pi}{4} - \psi \tag{38}$$

The path of a particle entering the plastic sector across AC may be identified by the radius ρ_0 to the point of intersection of its trajectory with AC. The equation for this streamline is

$$\frac{\rho}{\rho_0} = \frac{\cos(\pi/4 - \psi)}{\sqrt{2}\sin\psi + \sin\phi} \qquad \rho_0 \leqslant a$$

Any transverse gridline is deformed into a curve which is the locus of the particles on this line after a given interval of time. Consider a transverse line that coincides with BN at the initial instant $t = 0$. The time taken by a particle on BN to move to a generic point on its path across CE is

$$t = \frac{a}{U}(\sqrt{2}\cos\psi - \cos\phi_0) + \int_{\phi_0}^{\phi}(\rho/v)\,d\phi$$

in view of the relation $v = \rho(d\phi/dt)$. The first term on the right-hand side denotes the time when the particle reaches the boundary CE. The integral can be exactly evaluated on substitution from (37) and (38), resulting in the expression

$$t = \frac{a}{U}\left\{(\sqrt{2}\cos\psi - \cos\phi_0) + (\sqrt{2}\sin\psi + \sin\phi_0)[f(\phi) - f(\phi_0)]\right\} \tag{39a}$$

where

$$f(\phi) = \left\{-\frac{\cos\phi}{\lambda + \sin\phi} + \frac{2\lambda}{\sqrt{\cos 2\psi}}\tanh^{-1}\left[\sqrt{\frac{1-\lambda}{1+\lambda}}\tan\left(\frac{\pi}{4} - \frac{\phi}{2}\right)\right]\right\}\sec 2\psi \tag{40}$$

with $\lambda = \sqrt{2}\sin\psi$. If the particle enters the deforming zone across BC, its position in the sector, given by a typical angle ϕ, would correspond to the time

$$t = \frac{\rho_0}{U}\left\{\sin\left(\frac{\pi}{4} - \psi\right) + \cos\left(\frac{\pi}{4} - \psi\right)\left[f(\phi) - f\left(\frac{\pi}{4} - \psi\right)\right]\right\}$$

$$+ \frac{a - \rho_0}{U}\sec\left(\frac{\pi}{4} - \psi\right) \tag{39b}$$

Since the material in ABC is in a state of uniform simple shear, the transverse gridlines in this region are inclined at a contant angle to the vertical. The acute angle made by these lines with the die face is easily shown to be equal to $\cot^{-1} (2 \sin^2 \psi)$. The time t^* at which a given particle reaches the exit slipline AE can be found from (39) and (40) by setting $\phi = \pi/4$. For $t > t^*$, the particle moves horizontally through a distance $U(t - t^*)(1 + 2 \sin \psi)$ measured from its point of exit on AE.

The vertical gridlines of the undrawn sheet assume the position BN at a regular time interval c/U, which c is the length of each side of the original square mesh. The deformed shape of the grid is obtained by finding the points of intersection of the appropriate streamlines with the loci of the positions of the nodal particles at instants $t = mc/U$ ($m = 1, 2, 3, \ldots$). Owing to the velocity discontinuities occurring across the sliplines through E, a cusp is formed on the axis of symmetry with a semiangle that depends on ψ. The shape of the distorted grid for $\psi = 30°$ is shown in Fig. 7.20, assuming $c = a/2\sqrt{2}$. Apart from a cusp on the axis, the calculated distortion closely resembles that experimentally observed in wire drawing.

The semiangle of the cusp may be determined analytically by considering the velocity distribution in the neighborhood of the point E (Fig. 7.20). A particle leaving the deforming zone at Z, situated at a distance δy from the axis of symmetry, must have entered the zone at S whose distance from the axis is $(1 + 2 \sin \psi) \delta y$. Since the axial component of the velocity of the particle during this motion is $(1 + \sin \psi)U$ in view of the hodograph, while the axial distance between S and T is $2(1 + \sin \psi) \delta y$ by geometry, the time taken by the particle to move from S to T is $2\delta y/U$. During this time interval a particle on the axis, originally situated vertically below the previous one, moves from T to E with a speed U and then from E to Y with a speed $(1 + 2 \sin \psi)U$. Hence

$$EY = (\delta y/U)[2 - (1 + 2 \sin \psi)](1 + 2 \sin \psi)U = (1 - 4 \sin^2 \psi) \delta y$$

and the axial distance between the two particles finally becomes $\delta x = (4 \sin^2 \psi) \delta y$. It follows that an element of transverse gridline makes an angle $\cot^{-1} (\delta x/\delta y) = \cot^{-1} (4 \sin^2 \psi)$ with the axis of symmetry as it moves out of the deforming zone. The cusp is evidently more pronounced for larger die angles.

(iv) *Influence of friction and back pull* The most convenient way of taking die friction into account is to introduce a suitable correction factor into the frictionless theory, assuming a constant coefficient of friction μ between the material and the die. If t' and q' are the mean drawing stress and die pressure respectively when the friction is present, the condition of longitudinal equilibrium requires

$$T = 2(1 + \mu \cot \psi)(H - h)q'$$

where $T = 2ht'$. The component of the resultant force per unit width acting on each half of the die perpendicular to the axis of drawing is

$$Q = (\cot \psi - \mu)(H - h)q'$$

The elimination of $(H - h)q'$ between the above relations furnishes the coefficient of friction

$$\mu = \frac{1 - (2Q/T)\tan\psi}{\tan\psi + (2Q/T)} \tag{41}$$

Thus, μ can be experimentally determined from the measurement of the loads T and Q. The equation of longitudinal equilibrium may be rearranged to yield

$$t' = (1 + \mu\cot\psi)\frac{rq'}{1-r} = t(1 + \mu\cot\psi)\frac{q'}{q} \tag{42}$$

which is an important relationship between t'/t and q'/q. The assumption $q' \simeq q$ immediately gives the often quoted friction factor $1 + \mu\cot\psi$, which increasingly overestimates the drawing stress as the reduction is increased.

A fairly accurate estimate of the ratio t'/t is obtained by considering the longitudinal equilibrium of a thin slice of the sheet having vertical faces of semiheights y and $y + dy$, the material being assumed plastic between the planes of entry and exit.† If the mean longitudinal stresses across these faces are denoted by σ and $\sigma + d\sigma$, and the local die pressure is denoted by s, the equation of longitudinal equilibrium of the slice may be written as

$$\frac{d}{dy}(y\sigma) + (1 + \mu\cot\psi)s = 0$$

It is further assumed that the yield criterion may be written in the form $s + \sigma \simeq 2k$. The elimination of s from the equilibrium equation furnishes the first-order linear differential equation

$$y\frac{d\sigma}{dy} - \mu\sigma\cot\psi = -2k(1 + \mu\cot\psi) \tag{43}$$

Introducing the boundary conditions $\sigma = 0$ when $y = H$, and $\sigma = t'$ when $y = h$, we obtain the solution

$$\frac{t'}{2k} = \left(1 + \frac{\tan\psi}{\mu}\right)\left\{1 - \left(\frac{h}{H}\right)^{\mu\cot\psi}\right\}$$

Since the redundant shearing at entry and exit is neglected in this analysis, the above formula underestimates the drawing stress except at large reductions. Indeed, when $\mu = 0$, the predicted drawing stress has the ideal value $2k\ln(H/h)$. The ratio of the drawing stresses with and without friction therefore becomes

$$\frac{t'}{t} = \left(1 + \frac{\tan\psi}{\mu}\right)\frac{1 - (1-r)^{\mu\cot\psi}}{\ln[1/(1-r)]}$$

† G. Sachs and L. J. Klingler, *J. Appl. Mech.*, **14**: A-88 (1947). These authors treated the more general case where die profile is arbitrary. The angle ψ in (43) then represents the inclination of the local tangent to the profile.

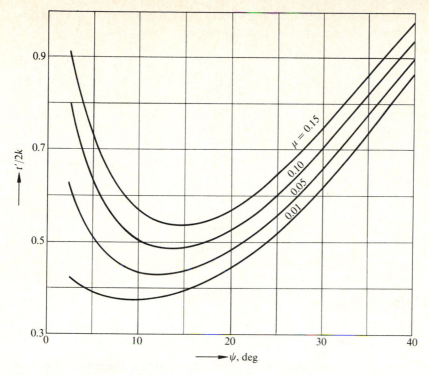

Figure 7.21 Variation of the drawing stress with die semiangle for $r = 0.3$ and various values of μ.

which is found to agree with the slipline field solution,[†] including the friction, to within 1.5 percent over most of the range of values of μ, r, and ψ. Inserting from (34), the above equation may be written as[‡]

$$\frac{t'}{2k} = \left(1 + \frac{\tan \psi}{\mu}\right)\{1 - (1 - r)^{\mu \cot \psi}\} f\left\{\left(\frac{2}{r} - 1\right)\psi\right\} \tag{44}$$

For a given reduction and coefficient of friction, the ratio t'/t decreases with increasing ψ. Since t increases as the die angle is increased, there is an optimum value of ψ for which t' is a minimum. This is illustrated in Fig. 7.21, where $t'/2k$ is plotted against ψ for various values of μ at a reduction of 30 percent. The curves follow the same trend as that experimentally observed in wire-drawing.[§]

[†] The results of the slipline field solution for drawing through rough dies have been presented by H. C. Rogers and L. F. Coffin, *Int. J. Mech. Sci.*, **13**: 141 (1971). A slipline field predicting a midplane cracking has been discussed by B. Dodd and H. Kudo, *Int. J. Mech. Sci.*, **22**: 67 (1980).

[‡] When the material work-hardens, a good approximation to the mean drawing stress is obtained by multiplying the right-hand side of (36) by the same friction factor as that given by (44).

[§] J. G. Wistreich, *Proc. Inst. Mech. Eng.*, **169**: 123 (1955); Useful experimental results have been reported by R. M. Caddell and A. G. Atkins, *J. Eng. Indust., Trans. ASME*, **91B**: 664 (1969).

When μ is sufficiently small, a close approximation is achieved by expanding $(1 - r)^{\mu \cot \psi}$ in ascending powers of $\mu \cot \psi$ and neglecting terms of order higher than μ^2, the result being

$$\frac{t'}{t} \simeq 1 + \mu\left(1 - \frac{1}{2}\ln\frac{1}{1-r}\right)\cot\psi \tag{45}$$

Since the optimum die angle is sufficiently small for usual values of r and μ, the redundant work factor for the associated drawing stress is given by the first equation of (35). Substituting for $t/2k$ into (45), and minimizing $t'/2k$ with respect to ψ, the optimum die angle may be shown to be very closely approximated by the formula

$$\psi_0 \simeq 1.49\left(\frac{\mu r^2}{2-r}\right)^{1/3} - 0.08\mu(2 - r) \tag{46}$$

Suppose, now, that a tensile force F per unit width is applied to the sheet on the entry side. The effect of this back pull is to increase the drawing force T and decrease the horizontal die load P by amounts that are proportional to F. Denoting the drawing force without back pull by T_0, we write

$$T = T_0 + (1 - \eta)F \qquad P = T_0 - \eta F$$

where η is a nondimensional quantity known as the *back-pull factor*. Since $T - P = F$ by the above relations, the condition of longitudinal equilibrium is identically satisfied. The permissible amount of back tension is limited by the fact that the corresponding drawing force T cannot exceed the value $4kh$ for which the drawn sheet would neck outside the die.

When the friction is absent, the back pull F causes a uniform hydrostatic tension $F/2H$ at every point of the deforming zone. All boundary conditions are satisfied and the slipline field remains unchanged. It is assumed that the rigid material is able to support the additional boundary stresses. The drawing stress is then augmented by $F/2H$ and the die pressure is reduced by the same amount. Since the reduction in the die load is $(H - h)F/H = rF$, the back-pull factor η is equal to the fractional reduction r. When the die is rough, a useful approximation is achieved by assuming the same reduction in die pressure as that in the frictionless situation.† Since the longitudinal die load is then reduced by the amount $(1 + \mu \cot \psi)rF$, we get

$$\eta \simeq r(1 + \mu \cot \psi)$$

This formula is found to overestimate the back-pull factor by a few percent for typical values of r, ψ, and μ.

A more accurate formula for the back-pull factor is obtained by considering the solution of (43) under the modified boundary conditions $\sigma = F/2H$ at $y = H$ and $\sigma = T/2h$ at $y = h$. Denoting the drawing force by T_0 when $F = 0$ the solution

---

† R. Hill, *The Mathematical Theory of Plasticity*, p. 176, Clarendon Press, Oxford (1950).

may be expressed as†

$$T = T_0 + F\left(\frac{h}{H}\right)^{1 + \mu \cot \psi}$$

This relation holds even when work-hardening is taken into account. It follows from the definition of the back-pull factor that

$$\eta = 1 - (1 - r)^{1 + \mu \cot \psi} \tag{47}$$

Calculations based on the slipline field indicate that (47) slightly underestimates the back-pull factor‡ over the relevant range of values of r, ψ, and μ. The application of back pull permits much higher drawing speeds while reducing the margin of safety against the possibility of necking of the drawn sheet.§

7.4 Extrusion Through Tapered Dies

(i) *Extrusion without friction* The extrusion process is similar to sheet drawing, the main difference being that the material in this case is pushed through the die instead of being pulled. When there is no friction, the stress in the deforming part of the plastic region is obtained from that in sheet drawing by the addition of a uniform hydrostatic pressure of magnitude equal to the mean drawing stress. The slipline field in extrusion is identical to that in drawing for the same die angle and reduction, and the extrusion pressure is therefore equal to the mean drawing stress for the same reduction, provided the container wall is also smooth. The mean die pressure in extrusion is, however, H/h times that in drawing. The slipline fields and hodographs of Figs. 7.16 and 7.19 are directly applicable to the extrusion process without the limitations of bulging and necking.¶

The smallest reduction for which the field of Fig. 7.16 is valid depends on two possibilities. Firstly, under frictionless conditions, the slipline BD cannot make an angle less than $\pi/4$ with the container wall, which means that the angle θ cannot exceed $\pi/2 - \psi$. Secondly, the die pressure for very small reductions may rise to the value $2k(1 + \pi/2 + \psi)$, causing the plastic zone to spread round the corner A to the free surface of the extruded billet. The plastic deformation is then locally confined near the free surface while most of the billet passes through undeformed. The greatest reduction for which the field of Fig. 7.19 is applicable is $\theta = \psi$ for reasons

---

† This is the plane strain analog of the result for a wire-drawing analysis by R. W. Lunt and G. D. S. MacLellan, *J. Inst. Met.*, **72**: 65 (1946). For an approximate analysis of wire drawing, including redundant work and work-hardening, see A. G. Atkins and R. M. Caddell, *Int. J. Mech. Sci.*, **10**: 15 (1968). See also, W. A. Backofen, *Deformation Processing*, Addison-Wesley, Reading, Mass. (1972).

‡ R. Hill, *J. Mech. Phys. Solids*, **1**: 142 (1953); J. F. W. Bishop, ibid., **2**: 39 (1953).

§ J. G. Wistreich, *J. Iron Steel Inst.*, **157**: 417 (1947); *Met. Rev.*, *Inst. Met.*, **3**: 97 (1958).

¶ W. Johnson, *J. Mech. Phys. Solids*, **3**: 218 (1955). Unsymmetrical extrusion through wedge-shaped dies has been considered by W. Johnson, P. B. Mellor, and D. M. Woo, ibid., **6**: 203 (1958). The initial nonsteady state in the extrusion process has been examined by L. I. Kronsjo, *Int. J. Mech. Sci.*, **11**: 281 (1969).

of geometry. This limiting value of r is tabulated below as a function of the die semiangle.

ψ	10°	20°	30°	40°	50°	60°	90°
r	0.330	0.581	0.746	0.847	0.906	0.943	0.985

For still higher reductions, with a given die angle, the stress and velocity boundary conditions can be simultaneously satisfied only if the exit slipline is taken as curved.† The situation is analogous to that encountered in the compression of a block between smooth flat dies with nonintegral width/height ratios greater than unity (Sec. 8.5). The extrusion pressure in this range differs only marginally from the ideal value $2k \ln (H/h)$ which is approached in an oscillatory manner as the reduction in thickness is increased. The problem is most conveniently solved by the matrix method which directly furnishes the shape of the initial slipline.

The dimensionless extrusion pressure based on the slipline fields of Figs. 7.16 and 7.18 is presented graphically in Fig. 7.22. For relatively large reductions, the extrusion pressure may be conveniently expressed by the empirical equation

$$\frac{p_e}{2k} = a + b \ln \left(\frac{1}{1 - r} \right)$$

where a and b are constants for a given die angle. Values for a and b for different values of ψ are given in Table 7.6 which includes the lowest and highest values of r for which these constants furnish sufficiently accurate values of p_e.

Frictionless conditions over the die are realized to a large extent in hydrostatic extrusion where the billet is surrounded by a fluid within the container. The pressure of the surrounding fluid is increased to a sufficiently high value so that a slight forward pull is able to initiate the extrusion. Sometimes, the billet is extruded into a pressurized chamber to inhibit the formation of tensile cracks in the extruded product. Since friction between the billet and the container is absent, long billets

Table 7.6 Empirical constants for extrusion through smooth dies

ψ	15°	30°	45°	60°	75°	90°
a	0.06	0.13	0.22	0.33	0.46	0.63
b	0.92	0.92	0.93	0.94	0.95	0.96
r_1	0.35	0.50	0.58	0.63	0.65	0.67
r_2	0.60	0.85	0.85	0.90	0.90	0.95

† A. P. Green, *Phil. Mag.*, **42**: 900 (1951); P. Dewhurst and I. F. Collins, *Int. J. Num. Meth. Eng.*, **7**: 357 (1973).

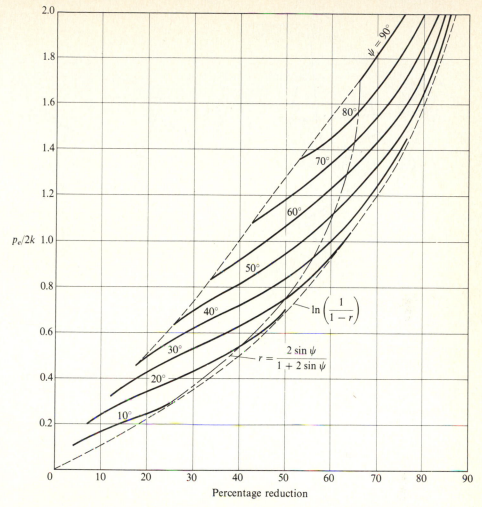

Figure 7.22 Variation of the frictionless extrusion pressure with the reduction for various semiangles of the die.

can be used in hydrostatic extrusion. The improved lubrication of the die permits the use of low die angles, thereby minimizing the redundant work.†

(ii) *Extrusion with Coulomb friction* Consider the plane strain extrusion through a wedge-shaped die which is partially rough, the frictional stress along the die being

---

† Hydrostatic extrusion has been described in detail by H. Ll. D. Pugh (ed.), *Mechanical Behaviour of Materials under Pressure*, Elsevier, Amsterdam (1970). See also J. M. Alexander and B. Lengyel, *Hydrostatic Extrusion*, Mills and Boon, London (1971). Some limitations of the process have been discussed by R. Hill and D. W. Kim, *J. Mech. Phys. Solids*, **22**: 73 (1974).

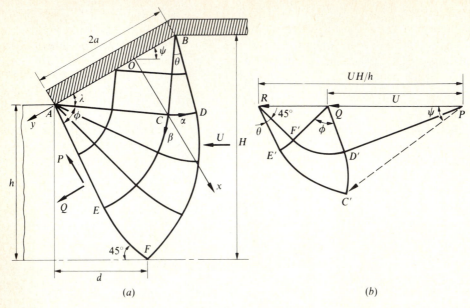

Figure 7.23 Extrusion through a rough tapered die for small reductions. (a) Slipline field; (b) hodograph.

assumed to be a constant fraction of the normal pressure.† When the reduction in thickness is sufficiently small, the acute angle made by the α lines with the die face has a constant value $\lambda < \pi/4$ as shown in Fig. 7.23. The normal pressure on the die is uniformly distributed, its magnitude being denoted by q. Since the shear stress on the die has the magnitude $k \cos 2\lambda$, the coefficient of friction along the die face is given by

$$\mu = \frac{k}{q} \cos 2\lambda \tag{48}$$

Since the horizontal component of the normal and tangential forces on the die must be equal to the total ram load, the extrusion pressure p_e may be written as

$$\frac{p_e}{k} = r\left(\frac{q}{k} + \frac{\cos 2\lambda}{\tan \psi}\right) \tag{49}$$

which reduces to $p_e = rq$ when $\lambda = \pi/4$. The slipline field is built up from the centered fans BCD and ACE of angular spans θ and ϕ respectively. The application of Hencky's first theorem furnishes the relationship.

$$\phi - \theta = \frac{\pi}{4} + \psi - \lambda$$

† W. Johnson, *J. Mech. Phys. Solids*, **3**: 224 (1955).

Let OC be drawn perpendicular to the die face AB from C. Since $OB = AB \sin^2 \lambda$, the height of B above O is equal to $(H - h) \sin^2 \lambda$. If (x_0, y_0) are the coordinates of F with respect to the rectangular axes through O as shown, it can be shown from simple geometry that the reduction r is given by

$$\frac{1}{r} = \sin^2 \lambda + \frac{1}{2}\left(\frac{x_0}{a}\cot\psi + \frac{y_0}{a}\right) \tag{50}$$

where $2a$ denotes the length AB. The values of (x_0, y_0) correspond to the angular coordinates (θ, ϕ) of the field defined by the circular arcs CD and CE having unequal radii.

The resultant longitudinal tension per unit width across the slipline AEF corresponding to zero hydrostatic pressure at C is $P \sin\psi + Q \cos\psi$, where P and Q are the component forces directed as shown. Since the mean compressive stress at C is equal to $q - k \sin 2\lambda$, the condition of zero longitudinal pull on the extruded billet becomes

$$(q - k \sin 2\lambda)h = P \sin\psi + Q \cos\psi$$

Using the geometrical relationship

$$\sin\psi = \frac{H - h}{2a} = \frac{r}{1 - r}\left(\frac{h}{2a}\right)$$

in the preceding equation, the die pressure may be expressed in the form

$$\frac{q}{k} = \sin 2\lambda + \frac{r}{1 - r}\left(\frac{P}{2ka} + \frac{Q}{2ka}\cot\psi\right) \tag{51}$$

When ψ, λ, and θ are all integral multiples of $15°$, the fractional reduction and the die pressure can be readily calculated from (50) and (51), using Table A-6. The coefficient of friction and the extrusion pressure then follow from (48) and (49) respectively. The determination of the extrusion pressure for a number of reductions with a given value of μ would evidently involve a great deal of interpolation. A limit of applicability of the field corresponds to $\theta = 0$, and the limiting reduction depends on the coefficient of friction according to the parametric relationship

$$r = \frac{\sqrt{2}\sin\psi}{\cos\lambda + \sqrt{2}\sin\psi}$$

$$\mu = \frac{\cos 2\lambda}{1 + \sin 2\lambda + \pi/2 + 2\psi - 2\lambda}$$

The limiting reduction decreases as the coefficient of friction increases. Another limit occurs when $\lambda = 0$, and the corresponding value of μ is equal to k/q, which increases with increasing reduction. The shear stress along the die face is then equal to k, and the solution for any given reduction is unchanged by still higher values of the coefficient of friction.

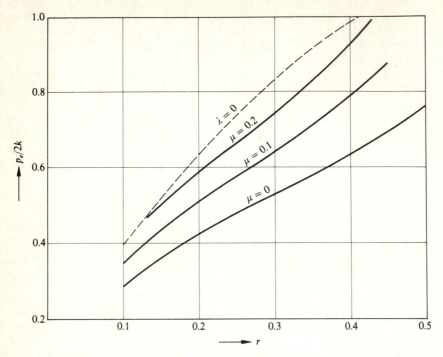

Figure 7.24 Variation of the extrusion pressure with fractional reduction and coefficient of friction ($\psi = 30°$).

Let the mean extrusion pressure for a die with a coefficient of friction μ be identical to the mean drawing stress with a coefficient of friction μ^*, corresponding to the same die angle and reduction.† The slipline fields in the two cases will be identical if the associated die pressures q and q^* are such that the frictional stresses μq and $\mu^* q^*$ are equal. Since $q(1 + \mu \cot \psi) = q^*(1 + \mu^* \cot \psi)$ by hypothesis, the relationship between μ and μ^* is

$$\mu^* = \frac{\mu}{1 - r(1 + \mu \cot \psi)} \tag{52}$$

For given values of r and ψ, the extrusion through a tapered die therefore corresponds to the drawing with somewhat higher coefficient of friction. This correspondence between μ and μ^* holds so long as the coefficient of friction is less than or equal to that for which shearing would occur along the die face.

The general effect of friction is to decrease the die pressure and increase the extrusion pressure for given die angle and reduction. The dimensionless extrusion pressure is plotted in Fig. 7.24 as a function of the reduction for $\psi = 30°$ and

---

† The extrusion pressure is identical to the drawing stress for the same values of r, ψ, and λ, but the required coefficient of friction is different due to the difference in die pressure.

various values of μ, the limit $\lambda = 0$ being represented by the broken curve. The extrusion pressure based on (34) and (44), with μ^* written for μ, is found to be remarkably close to that given by the slipline field solution. As in the drawing process, there is an optimum die angle for which the extrusion pressure is a minimum with specified values of r and μ. The solution is not valid for very small reductions when a surface distortion would occur in an attempted extrusion as a result of the die pressure becoming exceedingly high.

(iii) *Extrusion with sticking friction* When the container wall is perfectly rough, the slipline field of Fig. 7.7 holds for all semiangles greater than θ, provided the die has a sufficient degree of roughness. The extrusion pressure for a wedge-shaped die is therefore the same as that for the square die up to the reduction for which $\theta = \psi$. In this limiting state, the slipline AB coincides with the die face, requiring a coefficient of friction not less than $(1 + \pi/2 + 2\psi)^{-1}$. The maximum coefficient of die friction needed for smaller reductions is obtained from (15) with $\psi - \theta$ written for θ in the trigonometric functions. The limiting value of the extrusion ratio is given by $\theta = \psi$, the corresponding reduction r being given in the following table:

ψ	15°	30°	45°	60°	75°	90°
r	0.320	0.550	0.708	0.812	0.881	0.925

For still higher reductions, when the die and the container are both perfectly rough, the slipline field and the hodograph are as shown in Fig. 7.25. The field

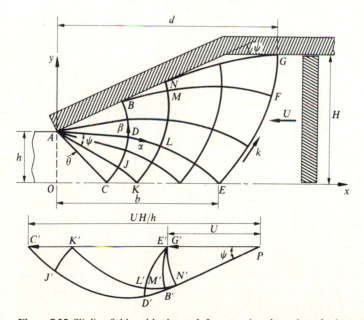

Figure 7.25 Slipline field and hodograph for extrusion through perfectly rough die and container wall.

involves a zone of dead metal whose outer boundary NG is tangential to the die and the container wall. By Hencky's first theorem, the angle turned through by each of the segments NG, MF, LE, and DJ is equal to the die semiangle ψ. The field $CBFE$ is the usual extension of the centered fan ABC where the sliplines are straight lines and circular arcs. The remaining field $BNGF$ is part of the slipline field defined by BF and a straight limiting line coinciding with the die face (Sec. 6.7). For a given die angle, the extrusion ratio depends on the angle CAJ, denoted by θ, which can have any value between zero and $\pi/4$. The angles turned through by the segments BM, MN, and FG are each equal to θ. When ψ and θ are both integral multiples of $15°$, the extrusion ratio H/h may be obtained directly from Table A-10, the angular coordinates of G being $(\psi + \theta, \theta)$ referred to the base point B.

Since the normal pressure on the exit slipline AC must be equal to k, the hydrostatic pressure at any point on EFG is $k(1 + 4\psi + 4\theta + 2\phi)$, where ϕ is the angle turned through along the slipline. The resultant horizontal force exerted by the ram per unit width across EG is

$$P = k(1 + 4\psi + 4\theta)H + 2k \int_E^G \phi \, dy - k(d - b)$$

where d and b are the distances of G and E from the exit plane. Integrating by parts, and noting that $d\phi = d\beta$ along EFG, we get

$$P = k\left(1 + \frac{\pi}{2} + 4\psi + 4\theta\right)H - k(d - b) - 2k \int_{\theta+\psi}^{\pi/4+\psi} y \, d\beta - 2k \int_0^\theta y \, d\beta \quad (53)$$

The first integral on the right-hand side is taken along the segment EF and the second integral along the segment FG. Using (98), Chap. 6, the first integral may be expressed as

$$\int_{\theta+\psi}^{\pi/4+\psi} y(\psi + \theta, \beta) \, d\beta = \int_0^{\theta+\psi} y(t, \theta + \psi) \, dt + \int_0^{\theta+\psi} y\left(t, \frac{\pi}{4} + \psi\right) dt$$

$$+ \left(1 + \frac{\pi}{4} + 2\psi + \theta\right)h - x\left(\theta + \psi, \frac{\pi}{4} + \psi\right) \quad (54)$$

The angular coordinates in the above equation are referred to the point C on the axis. Combining (53) and (54), the extrusion pressure $p_e = P/H$ may be written as

$$\frac{p_e}{2k} = r\left(1 + \frac{\pi}{4} + 2\psi\right) + (1 + r)\theta - \frac{1}{2} + (1 - r)\left(\frac{x_F}{h} - \frac{d - b}{2h} - \frac{c}{h}\right)$$

$$\quad (55)$$

$$c = \int_0^{\theta+\psi} y\left(t, \frac{\pi}{4} + \psi\right) dt + \int_0^{\theta+\psi} y(t, \theta + \psi) \, dt + \int_0^\theta y(\theta + \psi, t) \, dt$$

When θ and ψ are integral multiples of $15°$, the ratios x_F/h, b/h and the first two integrals for c/h are obtained from Table A-2, while d/h and the last integral for c/h are obtained from Table A-10.

The dimensionless extrusion pressure given by (55) is plotted against the

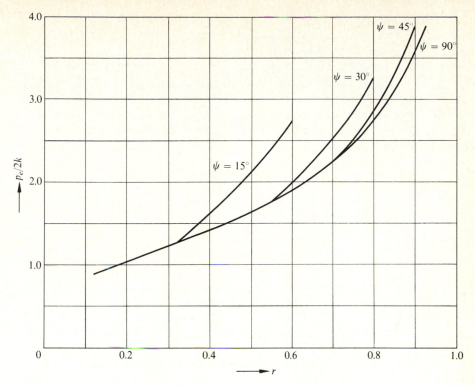

Figure 7.26 Variation of the extrusion pressure with fractional reduction under conditions of sticking friction.

fractional reduction in Figure 7.26 for three different die angles.† The curve for $\psi = 90°$ is based on the field of Fig. 7.7. Over the range of applicability of the present slipline field, the extrusion pressure increases as the die angle decreases with a given reduction. This is due to the fact that the deforming zone spreads out to a greater extent for smaller die angles, producing a greater amount of frictional work. For $\psi > 60°$, the extrusion pressure is practically independent of the die angle.

The dead metal boundary NG, which is a line of velocity discontinuity, is mapped into the circular arc $N'G'$, whose center P is the pole of the hodograph. The velocity is continuous across EG which is therefore mapped into a single point G'. The singularity at G' and the condition of symmetry about $E'K'$ define the region of $G'N'L'K'$, which maps the slipline domain $GNKE$. The construction of the region $N'B'D'L'$ is similar to that of $BNGF$. The remainder of the hodograph is built up from the known characteristics $L'D'$ and $L'K'$, and the condition of symmetry about $K'C'$. The curvatures of the sliplines and their hodograph images indicate that the plastic work rate is nowhere negative.

---

† J. Chakrabarty, unpublished work (1982).

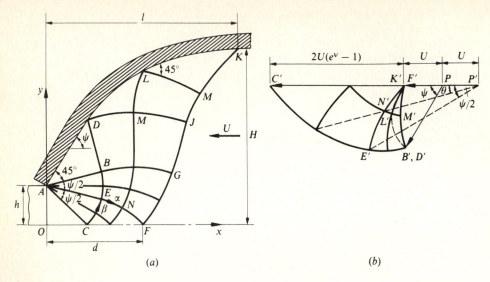

Figure 7.27 Extrusion through a constant pressure die without friction. (a) Slipline field; (b) hodograph.

7.5 Extrusion Through Curved Dies

(i) *Constant pressure dies* We begin with the extrusion through a smooth concave die with zero entry angle (Fig. 7.27). The shape of the die is such that the normal pressure is constant along the profile.† The length of the straight lower part AD, inclined at an angle ψ to the central axis, is twice the thickness of the extruded billet. The centered fan ABC, of angle ψ, is extended in the usual manner to obtain the region $CFGJDB$, where DJ and BG are parallel curves at a distance $\sqrt{2}\,h$ from one another. The curved profile DK and the associated field DJK are identical to those discussed in Sec. 6.7, in relation to the stress-free boundary. Thus AE is the bisector of the fan angle ABC, and the sliplines DJ and JK turn through the same angle $\psi/2$. The normal pressure on the die has the constant value $q = 2k(1 + \psi)$, and the mean extrusion pressure is

$$p_e = rq = 2k(1 + \psi)r \tag{56}$$

where r is the fractional reduction depending on ψ. Since the extrusion ratio H/h is equal to $2e^{\psi} - 1$, we have

$$\psi = \ln\left(\frac{2 - r}{2 - 2r}\right) \tag{57}$$

The largest possible reduction is 88.4 percent corresponding to $\psi = \pi/2$, although the slipline field is strictly valid for ψ less than about 86°. The theoretical efficiency of the die, expressed as the ratio of the ideal extrusion pressure to the actual

† R. Sowerby, W. Johnson, and S. K. Samanta, *Int. J. Mech. Sci.*, **10**: 231 (1968).

extrusion pressure (56), is extremely high. For example, when $\psi = 60°$, the reduction is 78.7 percent and the efficiency is 96 percent. For small reductions, the redundant work is only about $r^2/24$ times the homogeneous work. The total axial length l of the die is twice the distance $OF = d$. If the same die is used for drawing, the mean drawing stress t is the same as the extrusion pressure, but the reduction is limited to 62.5 percent corresponding to $t = 2k$.

In the hodograph of this process, the slipline $FGJK$ is mapped into a single point F' at a distance U from the pole P. The region DJK of the slipline field is mapped into part of a singular field bounded by the broken line $D'K'$, representing the image of the curved external boundary DK. The singular field $F'B'E'$ maps the slipline domain $GBDJ$. A typical point L' on the broken curve, corresponding to a point L on DK, is such that PL' is parallel to the tangent to the boundary at L. The triangular region moves as a rigid body with a velocity represented by the vector PD'. The velocity field in the domain CEF can be constructed entirely in terms of the elementary functions contained in the first row of (51), Chap. 6. Since the velocity changes in magnitude from U at F to $(2e^\psi - 1)U$ at C, the velocity components (u, v) along the sliplines, satisfying the equation $v = -\partial u/\partial \alpha$ and the condition $u = v$ on $\alpha = \beta$, may be written as

$$u = -U\left\{\sqrt{2} \exp{(\psi - \alpha - \beta)} - \cos\left(\frac{\pi}{4} + \alpha - \beta\right)\right\}$$

$$v = -U\left\{\sqrt{2} \exp{(\psi - \alpha - \beta)} - \sin\left(\frac{\pi}{4} + \alpha - \beta\right)\right\}$$

(58)

The rectangular components of the velocity, defining the shape of the hodograph net with pole P, are readily obtained from (11), Chap. 6, with $\phi = -(\pi/4 + \alpha - \beta)$. It follows that the field $C'E'F'$ consists of logarithmic spirals with pole at P' where $PP' = U$. In particular, the polar equations of $C'E'$ and $E'F'$ are found to be $w = 2U \exp{(\psi - \theta)}$ and $w = 2U \exp{(\theta)}$ respectively, where θ is measured from $P'C'$. The singular field $F'E'B'$, defined by the spiral $F'E'$, can be determined in terms of Bessel functions following the method of Sec. 6.6. It can be shown that

$$PB' = U[1 + 2J_1(\psi) + 2J_3(\psi) + \cdots]$$

Within the centered fan ABC, the velocity is constant along any radial line through A. There is no velocity discontinuity, and the rate of plastic work is nowhere negative.

The slipline field and the hodograph can be modified in a straightforward manner when the die is partially rough. The angular span of CE is still equal to $\psi/2$, but angle BAD is now equal to λ (less than $\pi/4$). All α lines in the region DJK are inclined at the same angle λ to die profile. The normal pressure on the die has the constant value

$$q = k\left(1 + \sin 2\lambda + \frac{\pi}{2} + 2\psi - 2\lambda\right)$$

(59)

The coefficient of friction is $\mu = (k/q) \cos 2\lambda$, and the mean extrusion pressure is given by

$$p_e = rq + k(1 - r)\frac{l}{h}\cos 2\lambda \tag{60}$$

where l is the axial length of the die. The slipline field in the region DJK can be determined from the condition $S = -R \tan \lambda$ along DK. It follows that $\bar{x} \tan \lambda + \bar{y}$ is a symmetrical function of α and β, and can be written down from the boundary condition along DJ. There is no simple geometrical relation for the extrusion ratio, but a sufficiently accurate empirical formula for small to moderate coefficients of friction is

$$\frac{H}{h} = 1 + (e^\psi - 1)\sqrt{2}\sec \lambda \tag{61}$$

When the friction is small, the ratio l/h is affected only marginally by the value of μ for a given angle ψ. As the reduction in thickness is increased to large values depending on λ, the slipline DJ is intersected by sliplines of its own family. The minimum value of ψ for this to happen is given by

$$I_0(\sqrt{2}\,\psi) + \frac{1}{\sqrt{2}}I_1(\sqrt{2}\,\psi) = \sec^2 \lambda + \tan \lambda \tag{62}$$

When $\lambda = \pi/6$, for instance, the slipline field is valid for $\psi \leqslant 49°$. For larger values of ψ, the field can be continued only by introducing stress discontinuities. It is important to note that the solution for the partially rough die does not contain the limiting case of full friction when the die profile coincides with a slipline. The entire slipline field is then an extension of the centered fan ABC. The normal pressure varies along the die profile, and the extrusion pressure is given by the upper broken curve of Fig. 7.5.

(ii) Graphical solution for arbitrary profiles In general, the slipline field for a curved die is more complicated than those hitherto considered. The construction must be carried out by a trial-and-error process to propose a field that satisfies the stress and velocity boundary conditions. An inverse semi-graphical method is sometimes convenient for the solution of problems involving curved dies.† A hodograph of suitable shape is first assumed so that the velocity boundary conditions are satisfied everywhere. The hodograph image of the curved boundary is then determined such that it is compatible with the frictional condition on the die face. The slipline field is finally constructed from the orthogonality of the slipline elements and their hodograph images.

Consider the extrusion through a smooth convex die of arbitrary contour and

---

† This is an adaptation of a graphical method employed by W. Johnson, *Int. J. Mech. Sci.*, **4**: 323 (1962); M. J. Hillier, ibid., **4**: 529 (1962); M. J. Hillier and W. Johnson, ibid., **5**: 191 (1963). For a numerical approach to the problem, see W. W. Sokolovsky, *J. Mech. Phys. Solids*, **10**: 353 (1962). The extrusion of initially curved sheets has been investigated by W. Johnson, *Int. J. Mech. Sci.*, **8**: 163 (1966).

having semiangles χ and ψ at the entry and exit respectively (Fig. 7.28). The reduction is assumed to be sufficiently small so that the slipline field contains a single point F of the axis of symmetry, and there are stress singularities at the entry and exit points of the die. The constant velocity discontinuity propagating along the entry and exit sliplines is taken to be of $\sqrt{2}$ units. Let θ and ϕ denote the fan angles at B and A respectively. Since the sliplines meet the die at 45°, the angles turned through along AE and BD are of amounts $\psi + \theta - \phi$ and $\chi + \theta - \phi$ respectively. The hodograph is begun with the centered fans $RF'E'$ and $QF'B'$, each of radius $\sqrt{2}$ and having angular spans θ and $\theta + \chi$. The point D' is located on $F'B'$ such that angle $QF'D'$ is equal to ϕ. If the reduction is not too small, ϕ is less than ψ, and the point A' lies on the arc $E'F'$ produced, the angle subtended by $F'A'$ being $\psi - \phi$. The hodograph field $F'E'B''B'$, defined by circular arcs of equal radii, is easily constructed using Table A-1. The field defined by the circular arcs $F'A'$ and $F'D'$ can be constructed graphically or numerically, but the rectangular coordinates of the various nodal points of this field for a 10° equiangular net are directly obtained from Table A-7.

The pole P of the hodograph is located on RQ produced, making $PR = 2/r$, where r is the fractional reduction. Since the ratio PQ/PR is then equal to h/H, the vectors PQ and PR represent the entry and exit velocities of the billet. For given die angles and reduction, the fan angles θ and ϕ can be so chosen that PA'' and PB'' intersect the hodograph curves at 45°. The angles which these lines make with PR are then equal to the entry and exit semiangles of the die. The rectangular coordinates (ξ, η) of the nodal points of the field $F'B'B''E'$ are known functions of the curvilinear coordinates (α', β') referred to $F'E'$ and $F'B'$ as the baselines. If these

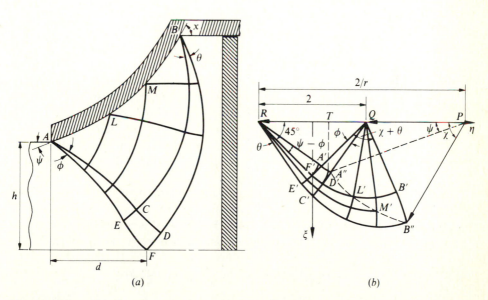

(a) (b)

Figure 7.28 Extrusion through a smooth convex die. (a) Slipline field; (b) hodograph.

functions are denoted by $f(\alpha', \beta')$ and $g(\alpha', \beta')$ respectively, it follows from geometry that

$$\frac{2}{r} = 1 + f(\theta, \chi + \theta) \cot \chi + g(\theta, \chi + \theta) \tag{63}$$

which can be solved for θ using Table A-1 or A-2, when r and χ are given. The angle ϕ is similarly determined from Table A-7, using the condition $PT/TA'' = \cot \psi$. Having located the points A'' and B'', it is possible to draw lines from P to intersect each characteristic through R at $45°$. The locus $A''L'M'N'B''$ of the points of intersection is the image of the die profile, whatever its shape. When the shape AB is given, the points L, M, N can be located on it by drawing tangents to the profile parallel to the corresponding velocity vectors. The triangular field ABC adjacent to the die is mapped into the region $A''B''C'$. The maximum reduction for which the construction is valid is

$$r^* = \frac{2 \sin \chi}{1 + 2 \sin \chi} \tag{64}$$

corresponding to $\theta = 0$ (B'' coinciding with B'). This is the same as that in the case of a wedge-shaped die of semiangle χ for the type of slipline field considered. It also follows from geometry that ϕ vanishes (A'' coinciding with A') when $r = 2 \sin \psi$.

The slipline field is now completely defined by the hodograph. Starting from the profile AB, the field can be constructed in small segments of circular arc with the tangents at the extremities perpendicular to the corresponding tangents to the hodograph elements. The distribution of the hydrostatic pressure is determined from the Hencky equations, the pressure p_0 at A on the exit side being obtained from the condition of zero longitudinal force across AEF. Thus

$$\frac{p_0}{2k} = \frac{d}{2h} - \int \frac{y}{h} d\alpha \tag{65}$$

where the integral is taken along AF; y is the height of a generic point on AF above the axis and α the angle turned through along AF in the clockwise sense. Note that $d\alpha$ is positive along AE and negative along EF. The normal pressure on the die decreases from entry to exit, and the resultant longitudinal component of the die load furnishes the mean extrusion pressure. Calculations based on $\chi = 90°$, $\psi = 30°$, and $r = \frac{2}{3} = r^*$ for a circular profile give an extrusion pressure of $2.82k$. This is about 2 percent higher than that for a smooth wedge-shaped die of semiangle $60°$, which is the mean semiangle of the convex die.

As the reduction is decreased, ϕ increases and becomes equal to ψ when $H = h(1 + 2 \sin \psi)$, the points A' and A'' coinciding with F' and D' respectively. For still smaller reductions, A' lies on $E'F'$, and the fan ACE is mapped into a part of the region $F'E'C'D'$. In that case, ϕ can be determined from the equation

$$\frac{2}{r} = 1 + f(\phi - \psi, \phi) \cot \psi + g(\phi - \psi, \phi) \tag{66}$$

using the table (Appendix). In all cases, the rate of plastic work is easily shown to be

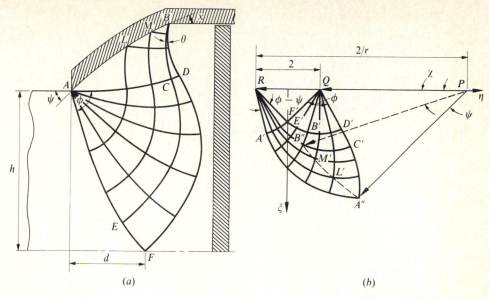

Figure 7.29 Extrusion through a smooth concave die. (*a*) Slipline field; (*b*) hodograph.

everywhere positive, justifying the shape of the proposed slipline field. By a slight modification of the method, a partially rough die with a constant frictional stress along it can also be treated.†

The slipline field solution for extrusion through a concave die is very similar to that for a convex die. When the reduction is small, the slipline field involves centered fans of angles θ and ϕ at the entry and exit respectively (Fig. 7.29). Under frictionless conditions, the angles turned through along AE and BD are $\phi - \theta - \psi$ and $\phi - \theta - \chi$ respectively. The entire slipline field is mapped into the region $F'A'A''D'$ defined by the circular arcs $F'A'$ and $F'D'$, subtending angles $\phi - \psi$ and ϕ respectively at the centers. For given values of r, χ, and ψ, the fan angles θ and ϕ can be obtained from (63) and (66) as before. The image $A''B''$ of the curved profile passes through the points where the characteristics through R are intersected at 45° by lines drawn from the pole P. The angles which PB'' and PA'' make with PR are equal to the entry and exit semiangles of the die. The slipline field is drawn as before from the nodal points thus established on the boundary AB. The construction is valid for all reductions less than r^*, given by (64), when the entry fan disappears. The die pressure increases from entry to exit, and is determined when the hydrostatic pressure p_0 on the exit side of A is calculated from (65), noting the fact that $d\alpha$ is negative throughout the exit slipline. Considering a concave die with

---

† A slipline field solution for drawing through smooth cylindrical dies with zero exit angle has been given by T. C. Firbank and P. R. Lancaster, *Int. J. Mech. Sci.*, **6**: 415 (1964). See also T. C. Firbank, P. R. Lancaster, and G. McArthur, ibid., **8**: 541 (1966).

$\chi = 20°$ and $\psi = 40°$, the extrusion pressure[†] for the limiting reduction of 41 percent is found to be $1.61k$. This may be compared with the pressure $1.32k$ for a wedge-shaped die of $30°$ semiangle.

For certain die angles and reductions, it is likely that sliplines of the same family would cross one another near the steepest part of the die. The construction of the slipline field in such cases would naturally involve stress discontinuities. The present method cannot be applied, without modification, when the exit angle vanishes for the convex die or the entry angle vanishes for the concave die, since the velocity discontinuity then disappears.[‡]

7.6 Ideal Die Profiles in Drawing and Extrusion

(i) *Basic principles* The plastic flow is known as streamlined when the velocity vector is everywhere along a principal stress direction. The streamlines therefore coincide with one family of trajectories of principal stress. Since the velocity vector bisects the slipline directions at each point, $u = \pm v$ throughout the field, where the upper or lower sign is taken according to whether the streamline is in the direction of the algebraically greater or lesser principal stress. The Geiringer equations are then immediately integrated, and it follows that the magnitude w of the resultant velocity is proportional to $\exp(\pm\phi)$ along an α line and $\exp(\mp\phi)$ along a β line. Consequently, the traces of the sliplines in the hodograph plane are logarithmic spirals irrespective of the form of the slipline field. The application of the Hencky equations then reveals that

$$w = c \exp(\mp p/2k) \tag{67}$$

throughout the field, where c is a constant. The deformation mode would be physically possible if the rate of plastic work is everywhere positive. Hence a material line element along a streamline must increase in length when the direction of flow is along the major principal stress and decrease in length when it is along the minor principal stress. In other words, the flow must proceed in the direction of increasing or decreasing speed according as the principal stress in this direction is major or minor. It follows from (67) that in either case the flow must occur in the direction of decreasing hydrostatic compression or algebraically increasing mean principal stress.[§]

We now consider the steady state drawing or extrusion of an ideally plastic material through a frictionless die of suitable shape. To eliminate redundant work, it is necessary to ensure that material elements are not sheared longitudinally while passing through any infinitesimal streamtube. In addition, the material should enter and leave the deforming zone smoothly, without any velocity discontinuity. It

---

[†] L. I. Kronsjo and P. B. Mellor, *Int. J. Mech. Sci.*, **8**: 515 (1966).

[‡] Upper bound solution for plane strain extrusion through curved dies have been obtained by W. Johnson, *Appl. Sci. Res.*, **7A**: 437 (1958).

[§] R. Hill, *J. Mech. Phys. Solids*, **14**: 245 (1966). For a general three-dimensional discussion, see R. Hill, ibid., **15**: 223 (1967).

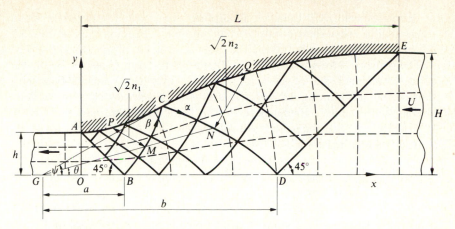

Figure 7.30 Slipline field and distorted square grid for drawing through an ideal sigmoidal die.

follows that the die must have zero entry angle and its profile must be such that the flow within the die is streamlined with the major principal stress occurring along the streamlines. Since the material enters and leaves the die with uniform speeds, it follows from (67) that the entry and exit sliplines must be straight. Moreover, the difference between the major principal stresses across these sliplines is $2k$ times the logarithm of the ratio of the exit and entry speeds. The mean drawing stress or the extrusion pressure therefore has the ideal value $2k \ln (H/h)$. In the case of drawing, the reduction must be less than 63.2 percent to avoid necking of the drawn sheet outside the die.

Let us consider a curvilinear segment momentarily orthogonal to the streamlines in a steady state process. Since the principal axes of stress and strain rate coincide, the orthogonality will persist while the segment moves in the field. This means that transverse lines originally perpendicular to the direction of flow will deform into orthogonal trajectories to the streamlines while passing through the deforming zone. It follows that a uniform square grid marked in the rigid material entering the die will become uniformly rectangular after leaving the die. The emerging material is, therefore, uniformly strained, although the deformation within the die is not homogeneous.

(ii) Solution for a sigmoidal die Let the streamlined die be sigmoidal in shape, so that the entry and exit angles both vanish, and there is no singularity anywhere in the field.† The design of the ideal die profile for symmetrical drawing or extrusion then involves a slipline field of the type shown in Fig. 7.30. The starting slipline BC is concave to the exit, intersecting the axis of symmetry BD at 45°. This will

---

† O. Richmond and M. L. Davenpeck, *Proc. 4th U.S. Nat. Cong. Appl. Mech.*, 1053 (1962); M. L. Davenpeck and O. Richmond, *J. Eng. Ind.*, **6**: 425 (1965). See also H. Takahasi, *Proc. Jpn. Soc. Mech. Eng.*, 9 (1967).

uniquely define the field BCD in which all sliplines are concave in the same sense, and the flow will be everywhere convergent as required. The remaining fields ABC and CDE can be constructed by taking the entry and exit sliplines DE and AB as straight. The principal stress trajectories AC and CE through C define the profile ACE of a streamlined die of sigmoidal shape. Since the velocity boundary conditions require $u = v$ along AB, BD, DE, and AE, the velocity equations will be satisfied only if u equals v everywhere in the field. The streamlines therefore coincide with the major principal stress trajectories.

The magnitude of the velocity at any point in the field CDE is Ue^θ, where θ is the angle turned through along an α line from the entry slipline DE. Similarly, the velocity at any point in the field ABC, defined by the angle θ turned through along a β line from AB, is of magnitude $U(H/h)e^{-\theta}$. Since the velocity remains constant in magnitude and direction along each straight slipline, the regions ABC and CDE are mapped into a pair of orthogonal logarithmic spirals in the hodograph. The continuity of the velocity at C furnishes $H = he^{2\psi}$, where ψ is the angle turned through by either of the sliplines BC and CD. Hence the fractional reduction is

$$r = 1 - \exp(-2\psi) \tag{68}$$

The distribution of the hydrostatic pressure in the deforming region can be calculated by Hencky's equations, using the boundary condition $p = -k$ along DE for drawing and $p = k$ on AB for extrusion. The normal pressure on the die decreases from $2k$ at E to $2k(1 - 2\psi)$ at A in the case of drawing, and increases from $2k$ at A to $2k(1 + 2\psi)$ at E in the case of extrusion. The mean drawing stress or extrusion pressure is $4k\psi$ or $2k \ln(H/h)$ as explained earlier.

Suppose that BC is a logarithmic spiral with pole G on the central axis. Then all sliplines of the field BCD are logarithmic spiral having the same pole. The polar equations of BC and CD are $\xi = ae^\theta$ and $\xi = be^{-\theta}$ respectively, where $BG = a$, $DG = b$, and $b = ae^{2\psi}$. It follows that the streamlines in this region are radially converging to the pole. The magnitude of the velocity at a radial distance ξ from G is $(b/\xi)U$ in view of the incompressibility and continuity conditions. It is to be noted that GC is tangential to the die profile and is of length $\sqrt{ab}$.

Let GMN be any radial line inclined at angle θ to the axis, intersecting BC and CD at M and N respectively. The straight sliplines MP and MN, drawn through M and N, meet the die profile at P and Q respectively. The angles turned through by AP and EQ are each equal to θ, and the tangents to the profile at P and Q are both parallel to MN. Let $PM = \sqrt{2}\,n_1$ and $QN = \sqrt{2}\,n_2$. Since the radii of curvature of the spirals BC and CD are numerically equal to $\sqrt{2}\,ae^\theta$ and $\sqrt{2}\,be^{-\theta}$ respectively, $S = -\sqrt{2}(ae^\theta - n_1)$ along AC and $R = -\sqrt{2}(be^{-\theta} + n_2)$ along CE. Considering a triangular element formed by the sliplines through the extremities of a small arc of the profile of angular span $d\theta$, we get $dn_1 = S\,d\theta$ along AC and $dn_2 = R\,d\theta$ along CE. Hence the differential equations for n_1 and n_2 are

$$\frac{dn_1}{d\theta} - n_1 = -\sqrt{2}\,ae^\theta \qquad \frac{dn_2}{d\theta} + n_2 = -\sqrt{2}\,be^{-\theta}$$

and the boundary conditions are $n_1 = n_2 = 0$ when $\theta = \psi$. The integration of the above equations results in

$$n_1 = \sqrt{2}\, a(\psi - \theta)e^\theta \qquad n_2 = \sqrt{2}\, b(\psi - \theta)e^{-\theta} \tag{69}$$

Evidently, n_2 is greater than n_1 for all values of θ less than ψ. Since $n_1 = \sqrt{2}\, h$ and $n_2 = \sqrt{2}\, H$ when $\theta = 0$, we have

$$a = \frac{h}{\psi} \qquad b = \frac{H}{\psi} \qquad \frac{b}{a} = \frac{H}{h} = e^{2\psi} \tag{70}$$

The length of the axis covered by the slipline field is $BD = b - a = rH/\psi$, and it follows that the total axial length of the die is

$$L = H\left\{(2 - r) + \frac{2r}{\ln\left[1/(1 - r)\right]}\right\} \tag{71}$$

The radius of curvature of the die profile is $\rho = ds/d\theta$, where the arc length s is assumed to increase from entry to exit. Since $ds = \sqrt{2}\, dn_1$ along the curve AC and $ds = -\sqrt{2}\, dn_2$ along CE, we get

$$\rho = \begin{cases} -2a(1 - \psi + \theta)e^\theta & \text{along } AC \\ 2b(1 + \psi - \theta)e^{-\theta} & \text{along } CE \end{cases}$$

in view of (69). The radius of curvature numerically increases from $2a(1 - \psi)$ at A to $2b(1 + \psi)$ at E, while there is a discontinuity of amount $4ae^\psi$ in the value of ρ at C. The rectangular coordinates of points on the die profile can be found from geometry, using (69). Choosing the x and y axes as shown, we obtain

$$\frac{x}{a} = [\cos\theta - (\psi - \theta)(\cos\theta + \sin\theta)]e^\theta$$
$$\frac{y}{a} = [\sin\theta + (\psi - \theta)(\cos\theta - \sin\theta)]e^\theta \tag{72}$$

on the exit side of the die (along AC), and

$$\frac{x}{b} = [\cos\theta + (\psi - \theta)(\cos\theta - \sin\theta)]e^{-\theta}$$
$$\frac{y}{b} = [\sin\theta + (\psi - \theta)(\cos\theta + \sin\theta)]e^{-\theta} \tag{73}$$

on the entry side (along CE). These equations define the die profile parametrically through θ. When ψ is increased to 1 rad, a becomes equal to h, and A becomes the center of curvature of the spiral BC at B. For still higher values of ψ, some of the straight sliplines will intersect one another near A. Hence the maximum reduction for which the field is valid is $r = 1 - e^{-2} \simeq 0.865$.

Suppose, now, that the slipline field in the central region BCD is an equiangular net. The streamlines passing through the diagonal points of this network meet the entry and exit sliplines at points that are equidistant from one another. A typical streamline inclined at an angle θ_0 to the axis in the radial flow region can be extended on either side by using (72) and (73) with ψ replaced by $\theta_0 > \theta$. The diagonal streamlines may be regarded as the distorted longitudinal lines of a uniform square grid marked on the undeformed sheet. When a transverse gridline passes through the zone BCD, the part of the line lying inside this region is deformed into a circular arc with center at G. Let t denote the time taken by a typical particle on the axis to move from D to a generic point at a distance ξ from G. Since $d\xi/dt = -(b/\xi)U$ while the particle remains on BD, we obtain

$$\xi = b\sqrt{1 - \frac{tU}{H}\ln\frac{H}{h}} \tag{74}$$

in view of (70). Denoting the length of each side of the original mesh by c, the deformed transverse gridlines in BCD may be obtained from (74) by setting $tU/c = 1, 2, \ldots$ The time t^* taken by the particle to move from D to B is given by

$$\frac{t^*U}{H} = \frac{r(2-r)}{\ln[1/(1-r)]}$$

To complete the construction of the distorted grid within the die, it is only necessary to extend the transverse lines into the regions ABC and CDE as trajectories intersecting the sliplines at $45°$. The time required by each particle to travel between the planes of entry and exit is

$$T = \frac{H}{U}\left\{1 + (1-r)^2 + \frac{r(2-r)}{\ln[1/(1-r)]}\right\} \tag{75}$$

The shape of the distorted grid as it passes through the die is shown by broken lines in Fig. 7.30, assuming $c = H/3$. The square grid becomes rectangular on leaving the die with the longitudinal spacing multiplied by H/h and the transverse spacing divided by the same ratio.†

(iii) *The minimum length die* For practical purposes, it is desirable to minimize the length of the streamlined die required to produce a given reduction. This is achieved by considering a nonzero exit angle with a singularity at the exit point of the die (Fig. 7.31). Assuming a straight end AD of the die, the slipline field is started from the centered fan ABC of angle ψ, equal to the semiangle of the die at the exit. The circular arc CB defines the field CBF, which is extended above by taking the β lines as straight. If AD is equal to the thickness $2h$ of the outgoing sheet, the straight slipline through B will meet the die at D. The concave part DE of the die is defined by the principal stress trajectory through D, intersecting the sliplines at $45°$. The entry angle of the die evidently vanishes, and the velocity is everywhere continuous. The Hencky equations furnish $4k\psi$ as the drawing stress or the extrusion pressure,

---

† A slipline field solution for extrusion through a sigmoidal die in the form of a cosine curve has been presented by S. K. Samanta, *J. Mech. Phys. Solids*, **19**: 300 (1971).

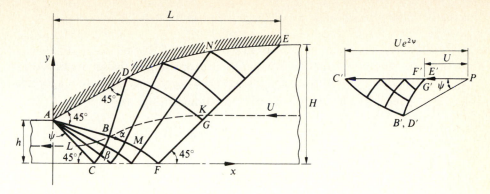

Figure 7.31 Slipline field and hodograph for drawing through a streamlined die of minimum length.

indicating that $2\psi = \ln(H/h)$ as before. In view of (90), Chap. 6, the axial length of the die is

$$L = H\{1 + [I_0(2\psi) + A_0(2\psi)]e^{-2\psi}\} \tag{76}$$

Equations (68) and (76) give the relationship between L/H and r parametrically through ψ. When $r = 0.5$, for instance, $L/H = 1.92$ compared to the value 2.94 required for the sigmoidal die.

Let (α, β) denote the curvilinear coordinates of the field CBF with respect to CB and its reflection in the axis as the baselines. The resultant velocity at any point on the straight slipline through a generic point on BF has the magnitude $U \exp(\psi - \alpha)$. Within the fan ABC, the magnitude of the velocity along any radial line is $U \exp(2\psi - \beta)$. In the hodograph plane, the regions ABC and $BDEF$ are therefore mapped into the logarithmic spirals $C'B'$ and $B'F'$ with pole P. Each characteristic in the field $C'B'F'$ is also a logarithmic spiral, the velocity at a generic point in CBF being of magnitude $U \exp(2\psi - \alpha - \beta)$ and directed along the major principal stress.

Let MN be any straight slipline of length n, intersecting BF and DE at M and N respectively. The angles turned through by BM and DN are each equal to α. The radius of curvature of the α line at N is numerically equal to $n - R_0$, where R_0 is the radius of curvature of the slipline BF at M. By (83), Chap. 6,

$$R_0 = -\sqrt{2}\,h\left\{I_0(2\sqrt{\alpha\psi}) + \sqrt{\frac{\psi}{\alpha}}\,I_1(2\sqrt{\alpha\psi})\right\}$$

Consider a neighboring point on the profile at an angular distance $d\alpha$ from N. The length of the straight slipline changes by an amount $dn = (n - R_0)\,d\alpha$ as we move along an infinitesimal arc of the profile. Hence

$$\frac{dn}{d\alpha} - n = -R_0$$

It may be verified by direct substitution that the above differential equation, subject to the boundary condition $n = \sqrt{2}\,h$ at $\alpha = 0$, has the exact solution[†]

$$n = \sqrt{2}\,h[I_0(2\sqrt{\alpha\psi}) + 2H(\alpha, \psi)] \tag{77}$$

where $H(\alpha, \psi)$ is defined by (133), Chap. 6. The expression in the square bracket is equal to $e^{2\psi}$ when $\alpha = \psi$. Since $n = \sqrt{2}\,H$ at E, we recover the result $H = he^{2\psi}$. The equation of the curve DE may be expressed parametrically as

$$x = x_0(\alpha) + n\cos\left(\frac{\pi}{4} + \psi - \alpha\right)$$

$$y = y_0(\alpha) + n\sin\left(\frac{\pi}{4} + \psi - \alpha\right) \tag{78}$$

where (x_0, y_0) are the coordinates of the typical point M from which n is measured. The die profile and the associated slipline field can be readily established by using Tables A-1, A-3, and A-5.

Since the velocity varies as $e^{-\alpha}$ in the region $BDGF$, and the material is incompressible, the length of the straight characteristic between the die and a given streamline must vary as e^{α} along the streamline. This property enables us to draw the streamlines in the region $BDEF$, starting from selected points on the exit slipline EF. It follows that the streamline passing through the point B intersects EF at K, such that $EK/DB = e^{\psi}$, the depth of K below E being $\sqrt{Hh}$ or $H\sqrt{1-r}$. All streamlines lying below KBL pass through the central field CBF, intersecting the sliplines at $45°$. In the region ABD, which moves with a uniform speed $U\sqrt{H/h}$, the streamlines are obviously parallel to AD. The streamlines in the centered fan ABC are logarithmic spirals with pole at A, and the vertical distance between A and L is $h\sqrt{1-r}$, where L is the exit point of the streamline.

The time taken by a typical particle on the axis to move from F to a generic point on CF is obtained from the equation $dx/dt = -U\exp(2\psi - 2\alpha)$. Substituting from (90), Chap. 6, and rearranging the terms, the equation of motion may be expressed as

$$\frac{d}{dt}[e^{2\alpha}I_0(2\alpha)] = -\frac{U}{h}e^{2\psi}$$

Employing the initial condition $\alpha = \psi$ at $t = 0$, and using the fact that $e^{2\psi}$ is equal to H/h, the solution is obtained as

$$\frac{tU}{h} = I_0(2\psi) - (1 - r)e^{2\alpha}I_0(2\alpha) \tag{79}$$

which is valid for $0 \leqslant \alpha \leqslant \psi$. Equation (79) enables us to locate the positions on the axis of the transverse lines of an originally square grid. The deformed pattern of these gridlines may now be obtained by constructing the principal stress

---

[†] J. Chakrabarty, unpublished work (1977).

trajectories through the appropriate points on the axis. The total time taken by any particle to travel from the plane of entry to the plane of exit is

$$T = \frac{H}{U}\left[1 + (1 - r)I_0\left(\ln \frac{1}{1 - r} \right) \right] \tag{80}$$

The stress distribution in the deforming region can be readily found by using Hencky's equations. The normal pressure along the curved portion of the die decreases from $2k$ at E to $2k(1 - \psi)$ at D in the case of drawing, and increases from $2k(1 + \psi)$ at D to $2k(1 + 2\psi)$ at E in the case of extrusion.

7.7 Limit Analysis of Plane Strain Extrusion

The slipline field solutions discussed in the preceding sections are incomplete in the sense that no attempt has been made (except in a special case) to extend the stress field into the assumed rigid zones in a statically admissible manner. It is not shown, therefore, that the proposed slipline fields meet the requirements of the lower bound theorem of limit analysis. On the other hand, the velocity field associated with the slipline field in each case is kinematically admissible as required by the upper bound theorem. Consequently, the incomplete solutions are strictly upper bound solutions, although the estimated value of the external load is not likely to exceed the actual value significantly. Lower and upper bound solutions for the symmetrical extrusion through straight dies using approximate stress and velocity distributions will be discussed in what follows.

(i) *Lower bound solutions* An intuitively obvious lower bound for the extrusion pressure p_e is $2k \ln (H/h)$ corresponding to homogeneous compression. Under frictionless conditions, this may be rigorously demonstrated by considering stress discontinuities across a pair of concentric circular arcs drawn through the extremities of the die with the center at the virtual apex. The stress distribution in the region between the considered circular arcs is taken to be that corresponding to a thick cylinder under a uniform external pressure equal to p_e. A stress-free state on the exit side and a uniform hydrostatic compression of intensity p_e on the entry side are also assumed to complete the stress field that is statically admissible.

To obtain an improved lower bound for frictionless extrusion through a square-faced die, consider the discontinuous stress fields of Fig. 7.32 which hold for two different ranges of reduction.† The proposed discontinuity patterns are such that all the stress boundary conditions are satisfied, the material to the right of the die and the exit plane being assumed plastic everywhere except region $LFBDN$ in (a) and region $LBEDN$ in (b). The boundary conditions on the exit side are accommodated by assuming the material to the left of the exit plane to be entirely stress-free. Using (29), Chap. 6, the stresses in the plastic regions are readily

---

† The discontinuous stress field for $R \geqslant 3$ is due to J. M. Alexander, *Q. Appl. Math.*, **19**: 31 (1961). The remaining lower bound solutions presented here have been obtained by J. Chakrabarty in an unpublished work (1971).

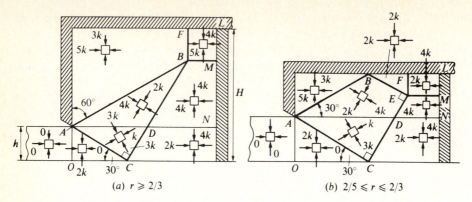

(a) $r \geqslant 2/3$ (b) $2/5 \leqslant r \leqslant 2/3$

Figure 7.32 Discontinuous stress fields for frictionless extrusion through a square die.

evaluated from the known hydrostatic pressure in AOC, and the fact that the α lines are inclined at 15° to each of the discontinuities AC, AD, AB, and at 75° to the discontinuity CD. The magnitude and direction of the principal compressive stresses throughout the field are as indicated in the figure.

Since BD, BE, and DE coincide with principal stress directions in the plastic regions to their left, the boundary and continuity conditions require a state of hydrostatic compression to exist on the other side of each of these inclined discontinuities. The continuity of the stresses in the nonplastic zones is maintained by introducing additional discontinuities through B and E in fields (a) and (b) respectively. Both the stress fields are statically admissible, and involve a uniform die pressure equal to $5k$. Equating the punch load p_eH to the die load $5k(H - h)$, we obtain the lower bound

$$p_e = 5k\left(1 - \frac{h}{H}\right) = 5kr \tag{81}$$

The two discontinuity patterns become identical when points B and E coincide, the corresponding extrusion ratio being $H/h = \cot^2 30° = 3$. Thus, field (a) applies to $r \geqslant 2/3$ and field (b) applies to $r \leqslant 2/3$. The smallest reduction for which field (b) holds is that for which E coincides with D, resulting in

$$h(\cot 30° + \tan 30°) = 2(H - h) \cot 30°$$

which gives $r = 2/5$. For still smaller reductions, BE intersects AD at some point J while DE becomes nonexistent. The stress distribution on the entry side would be statically admissible if we replace EF by a new discontinuity EG, where G is on the axis vertically below E. The material to the right of EG is under a hydrostatic compression of intensity $2k$, while that between EM and JN must be under a vertical compression of magnitude $2k$. The extrusion pressure is still given by (81), since the die pressure is unaffected by the modified stress distribution.

A lower bound solution for wedge-shaped dies may be derived in a similar manner, assuming frictionless conditions as before. Depending on the reduction in relation to die angle, two types of discontinuity pattern must be considered as in Fig. 7.33. The condition of zero tangential traction along the die face and the container wall can be satisfied by introducing a pair of discontinuities through A inclined at an angle $\psi/2$ with one another, where ψ is the die angle, and a discontinuity through B drawn perpendicular to the die face. The discontinuities through C are mutually perpendicular, and the proposed angle $\phi = (\pi - \psi)/4$ between AC and the axis of symmetry furnishes the most suitable lower bound.

In field (a), the discontinuities CD and AD meet at a point D, and DE is parallel to the die face, the remaining discontinuities through D and E being horizontal and vertical. Assuming the region $OABEDG$ to be stressed to the yield point, the hydrostatic pressure jump across the sides of the triangle ACD is obtained from the fact that the α lines make an angle $\psi/4$ with AC and AD, and an angle $\pi/2 - \psi/4$ with CD. The magnitudes of the hydrostatic pressure are easily found by following the successive reflection of the sliplines (see Sec. 6.3(i)). These pressures are

$$p_1 = k\left(1 + 2\sin\frac{\psi}{2}\right) \qquad p_2 = p_3 = k\left(1 + 4\sin\frac{\psi}{2}\right)$$

In both the fields, the triangle BEF is under a hydrostatic compression $p_3 - k$. The principal stresses shown in the different regions of the field constitute a statically admissible state of stress. Since the extrusion pressure is r times the die pressure, the lower bound is

$$p_e = 2kr\left(1 + 2\sin\frac{\psi}{2}\right) \tag{82}$$

Field (a) holds for all reductions greater than that for which DE vanishes. Since the projection of AD on the die face is equal to the die length for the limiting

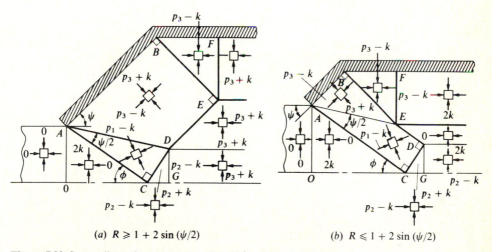

(a) $R \geqslant 1 + 2\sin(\psi/2)$ (b) $R \leqslant 1 + 2\sin(\psi/2)$

Figure 7.33 Stress discontinuity patterns for frictionless extrusion through a tapered die, where $\phi = \pi/4 - \psi/4$.

reduction, it follows from simple geometry that

$$h \sec \frac{\psi}{2} = (H - h) \operatorname{cosec} \psi$$

which reduces to $H/h = 1 + 2\sin(\psi/2)$. For smaller extrusion ratios, field (b) becomes appropriate. The discontinuities through A and B then meet at E, and ED is drawn perpendicular to CD. The hydrostatic pressure jump across ED is of the same amount as that across AC but occurs in the opposite sense. The stress distribution on the entry side is therefore modified as shown. Since $p_2 - k = p_3 - k \leqslant 2\sqrt{2}k$, the stress field is still statically admissible, giving the same lower bound for the extrusion pressure. For $\psi = \pi/2$, the lower bound pressure given by (82) is $2(\sqrt{2} + 1)kr$, which is slightly inferior to (81).

When the die is perfectly rough, but the container wall is still smooth, Fig. 7.34, a statically admissible stress field is easily constructed for a wedge-shaped die with $\psi \leqslant \pi/4$. The state of stress in the plastic material adjacent to the die involves a discontinuity through A inclined at $3\pi/8 + \psi/2$ to the exit plane and a discontinuity through B inclined at $3\pi/8 - \psi/2$ to the container wall. In field (a), these discontinuities do not meet, and are joined by a third discontinuity CD

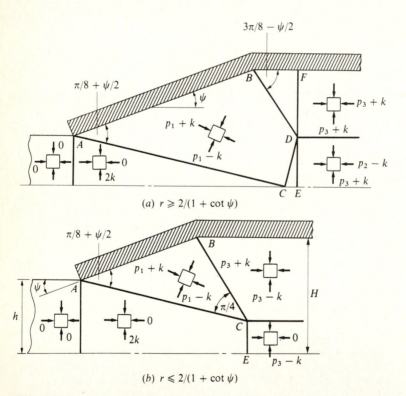

(a) $r \geqslant 2/(1 + \cot \psi)$

(b) $r \leqslant 2/(1 + \cot \psi)$

Figure 7.34 Stress discontinuity patterns for extrusion with smooth container and rough die. The vertical compressive stresses in CDE and BDF are $p_2 + k$ and $p_3 - k$ respectively.

perpendicular to AC. By geometry, the angles of inclination of the α lines with AC, CD, and BD are $\pi/8 + \psi/2$, $3\pi/8 - \psi/2$, and $3\pi/8 + \psi/2$ respectively. Using (29), Chap. 6, the hydrostatic pressures in the associated plastic regions $ABDC$, CDE, and BDF are found to be

$$p_1 = k\left[1 + 2\sin\left(\frac{\pi}{4} + \psi\right)\right] \quad p_2 = k\left[1 + 4\sin\left(\frac{\pi}{4} + \psi\right)\right]$$

$$p_3 = k(1 + 2\sqrt{2}\sin\psi)$$

Since $p_2 - p_3 < 2\sqrt{2}\,k$, the material to the right of ED is stressed below the yield limit. For equilibrium, the distribution of uniform normal pressure p_1 and shear stress k along the die face must produce a horizontal resultant equal to $p_e H$. The extrusion pressure is therefore obtained as

$$p_e = kr(1 + \cot\psi)(1 + \sqrt{2}\sin\psi) \tag{83}$$

When the reduction is decreased to a limiting value, points D and C coincide on the axis of symmetry. The discontinuity pattern of field (b) is applicable for all smaller reductions, giving the same extrusion pressure. Since $p_3 \leqslant 3k$ for $\psi \leqslant \pi/4$, the yield limit is not exceeded in the rigid zone to the right of CE. At the limiting reduction, the length BC becomes equal to $H \sec(\pi/8 + \psi/2)$, and the application of the sine rule to triangle ABC furnishes

$$(H - h)(\sin\psi + \cos\psi) = 2H\sin\psi$$

Hence the limiting reduction is $r = 2/(1 + \cot\psi)$. When $\psi = \pi/4$, field (b) holds for all reductions, the corresponding lower bound being $4kr$ with AC and BC reduced to horizontal and vertical discontinuities through A and B respectively.

(ii) Upper bound solutions An effective method of constructing kinematically admissible velocity fields for plane strain problems is to divide the material into a number of zones, each one of which is assumed to move as a rigid block with shearing along the interfaces.[†] A simple velocity pattern for extrusion through a wedge-shaped die is shown in Fig. 7.35a, where AC and BC are lines of velocity discontinuity emanating from the die and meeting on the axis of symmetry. As the punch advances toward the die with a unit speed, a typical particle on the entry side reaches BC and instantaneously changes its path to proceed parallel to the die face. On crossing AC, the particle suffers a second change in velocity and moves parallel to the axis again with a speed equal to the extrusion ratio R. The acute angles

---

[†] Various upper bound solutions for plane strain extrusion have been discussed by W. Johnson, *Proc. Inst. Mech. Eng.*, **173**: 61 (1959); H. Kudo, *Int. J. Mech. Sci.*, **1**: 57 and 229 (1960); J. Halling and L. A. Mitchell, *J. Mech. Eng. Sci.*, **6**: 240 (1964); and R. G. Fenton, *J. Basic Eng.*, **9**: 45 (1968). See also W. Johnson and P. B. Mellor, *Engineering Plasticity*, chap. 13, Van Nostrand-Reinhold Company, London (1973); B. Avitzur, *Metal Forming: Processes and Analysis*, McGraw-Hill Book Company, New York (1969).

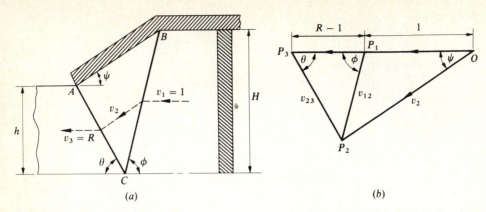

Figure 7.35 Single triangular velocity discontinuity pattern and the associated hodograph for plane strain extrusion.

made by the discontinuities with the axis are denoted by θ and ϕ, one of which can be independently chosen for given die angle and reduction.

The associated hodograph shown in Fig. 7.35b is started from the vector OP_1 representing the velocity of the rigid region adjacent to the punch face. The lines drawn through O and P_1 parallel to AB and BC respectively locate the point P_2, defining the velocity v_2 of the rigid triangle ABC and the velocity discontinuity v_{12} across BC. A line from P_2 drawn parallel to AC intersects OP_1 produced at P_3, where OP_3 is the speed R of the extruded sheet, and P_2P_3 the velocity discontinuity v_{23} across AC. It follows from the geometry of the triangles OP_1P_2 and OP_2P_3 that

$$v_2 = \frac{\sin \phi}{\sin (\phi - \psi)} = \frac{R \sin \theta}{\sin (\theta + \psi)}$$

which gives

$$\cot \theta + \cot \psi = R(\cot \psi - \cot \phi) \tag{84}$$

In view of this relationship, the velocity discontinuities may be expressed as

$$v_{12} = \frac{\sin \psi}{\sin (\phi - \psi)} = \frac{\operatorname{cosec} \phi}{\cot \psi - \cot \phi}$$

$$v_{23} = \frac{R \sin \psi}{\sin (\phi + \psi)} = \frac{\operatorname{cosec} \theta}{\cot \psi - \cot \phi}$$

The container wall is assumed to be perfectly smooth as in previous solutions. When the die is perfectly smooth, the energy is dissipated only by shearing along the discontinuities AC and BC. Equating the rate of external work done by the punch to the rate of internal energy disssipation readily gives

$$Hp_e = k(BC \cdot v_{12} + AC \cdot v_{23})$$

Substituting for v_{12} and v_{23}, and noting that $BC = H \operatorname{cosec} \phi$ and $AC = h \operatorname{cosec} \theta$, the extrusion pressure may be written in the dimensionless form

$$\frac{p_e}{k} = \frac{\cosec^2 \theta + R \cosec^2 \phi}{R(\cot \psi - \cot \phi)}$$

The best upper bound corresponds to the value of θ or ϕ for which the extrusion pressure is a minimum. Differentiating the above equation with respect to ϕ, and using the fact that $d\theta/d\phi = -R(\cosec^2 \phi/\cosec^2 \theta)$, the minimum extrusion pressure is found to be

$$p_e = 2k[(R - 1) \cot \psi - (R + 1) \cot \phi] \qquad (85)$$

the condition for the minimum being

$$\cosec^2 \theta + R \cosec^2 \phi = 2R(\cot \theta - \cot \phi)(\cot \psi - \cot \phi)$$

Using (84), the above equation may be expressed as a quadratic in $\cot \phi$, and the solution is obtained as

$$\cot \phi = \cot \psi - \frac{1}{\sqrt{R}} \cosec \psi$$

Substitution into (85) now furnishes the best upper bound† for the extrusion pressure as a function of the die angle and the reduction. Thus

$$\frac{p_e}{2k} = \frac{2 - r}{\sqrt{1 - r}} \cosec \psi - 2 \cot \psi \qquad (86)$$

When the die is perfectly rough, the material is sheared along the die face, and the energy dissipation increases by the amount $k \cdot AB \cdot v_2$. The extrusion pressure is then given by the modified expression

$$\frac{p_e}{k} = \frac{\cosec^2 \theta + R \cosec^2 \phi + (R - 1) \cosec^2 \psi}{R(\cot \psi - \cot \phi)}$$

The condition for the best upper bound leads to a second relationship between θ and ϕ as before. The minimum value of the extrusion pressure is still expressed in the form (85), but the discontinuity angle ϕ is now given by

$$\cot \phi = \cot \psi - \sqrt{\frac{2}{R + 1}} \cosec \psi$$

For given values of R and ψ, θ is smaller and ϕ is greater for the rough die than for the smooth die. The best value of the upper bound for the rough die becomes

$$\frac{p_e}{2k} = \sqrt{\frac{2(2 - r)}{1 - r}} \cosec \psi - 2 \cot \psi \qquad (87)$$

The preceding solutions provide good approximations to the extrusion

---

† A. P. Green, *British Iron and Steel Res. Assoc. Report*, MW/B/7 (1952).

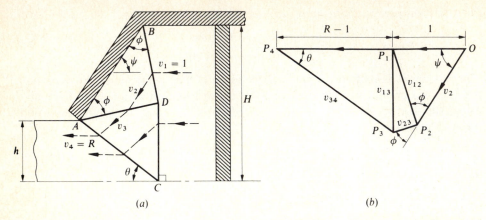

Figure 7.36 Double triangular velocity discontinuity pattern and the associated hodograph for plane strain extrusion.

pressure when the reduction is sufficiently large. For moderate to small reductions, the upper bound must be improved by introducing additional discontinuities. A convenient expression for the extrusion pressure is again possible if we consider the discontinuity pattern of Fig. 7.36, where identical angles are made by AD and BD with the die face, while CD is perpendicular to the axis. The velocity of the material triangle ACD is represented by OP_3 in the hodograph, where P_3 is the meeting point of lines from P_1 and P_2 drawn parallel to CD and AD respectively. The velocity triangle OP_1P_2 is constructed as before, and the hodograph is completed by the line P_3P_4 which is parallel to AC. It follows from the geometry of the hodograph that

$$v_{12} = \frac{\sin \psi}{\sin \phi} \qquad v_{13} = (R - 1) \tan \theta \qquad v_{34} = (R - 1) \sec \theta$$

$$v_{23} = -\frac{\cos (\psi + \phi)}{\cos (\psi - \phi)} \qquad v_{12} = \left(\frac{\tan \phi - \cot \psi}{\tan \phi + \cot \psi} \right) \frac{\sin \psi}{\sin \phi}$$

Equating the expression for v_{12}/v_{13} given by above to that furnished by the velocity triangle $P_1P_2P_3$, we obtain

$$\cot \theta = \frac{R - 1}{2} (\tan \phi + \cot \psi) \qquad (88)$$

as the relationship between θ and ϕ for a given geometry of the process. The upper bound for the extrusion pressure for a perfectly smooth die is given by

$$Hp_e = k(BD \cdot v_{12} + CD \cdot v_{13} + AD \cdot v_{23} + AC \cdot v_{34})$$

where

$$BD = AD = \frac{(H - h) \sec \phi}{2 \sin \psi} \qquad CD = \frac{H + h}{2} - \frac{(H - h) \tan \phi}{2 \tan \psi} \qquad AC = \frac{h}{\sin \theta}$$

Substituting into the above upper bound relation, inserting the expressions for the velocity discontinuities, and eliminating ϕ by means (88), the extrusion pressure may be written in the compact form

$$\frac{p_e}{k} = (2 - r) \cot \theta + \frac{r}{1 - r} (2 - r + r \cot^2 \psi) \tan \theta - 3r \cot \psi$$

Setting to zero the derivative of p_e with respect to $\tan \theta$, the minimum extrusion pressure is found to correspond with

$$\cot^2 \theta = \frac{r}{1 - r} \left(1 + \frac{r}{2 - r} \cot^2 \psi \right)$$

and the best value of the upper bound is obtained as[†]

$$\frac{p_e}{2k} = \sqrt{r(2 - r)\left(2 + \frac{r}{1 - r} \csc^2 \psi \right)} - \frac{3}{2} r \cot \psi \tag{89}$$

When $\psi = 90°$, Eq. (89) gives a lower value of the extrusion pressure than (86) does for all reductions. When $\psi = 30°$, Eq. (89) gives a lower extrusion pressure for $r < 0.31$. The range of reductions for which (89) provides a better bound increases as the die angle is increased.

An upper bound solution for a perfectly rough die face may be obtained by including the term $k \cdot AB \cdot v_2 = k(H - h)(\cot \phi + \cot \psi)$ in the expression for Hp_e. A closed form solution for the minimum extrusion pressure is no longer possible, except for a square die ($\psi = 90°$). The best upper bound in this case is easily shown to be

$$\frac{p_e}{2k} = (2 - r) \cot \theta = \sqrt{\frac{r(2 - r)(4 - r)}{2(1 - r)}} \qquad \psi = \frac{\pi}{2} \tag{90}$$

When the reduction is sufficiently small in relation to the die angle, a better upper bound is obtained on the assumption that a zone of dead metal is attached to the die face. Then the extrusion pressure is identical to that for a perfectly rough die with a smaller die angle α. Using (87) with α written for ψ, and minimizing the extrusion pressure with respect to α, we obtain the upper bound formula

$$\frac{p_e}{2k} = 2 \tan \alpha = \sqrt{\frac{2r}{1 - r}} \qquad \alpha < \psi \tag{91}$$

which is independent of the frictional condition on the die face. In the case of a perfectly smooth die, the extrusion pressure (91) is lower than (89) for $r < 0.586$ when $\psi = 90°$, and for $r < 0.06$ when $\psi = 30°$. For a perfectly rough square die, the upper bound estimate (91) is lower than (90) in the range $r < 3 - \sqrt{5} \simeq 0.764$. The upper and lower bound solutions are compared with one another in Fig. 7.37 for

---

[†] J. Chakrabarty, unpublished work (1982).

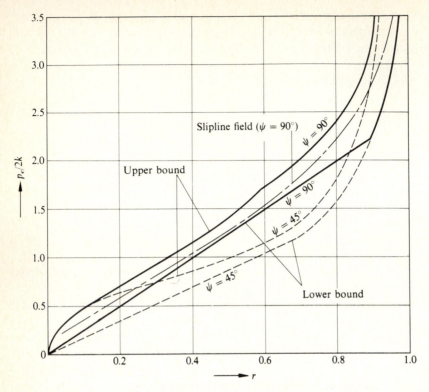

Figure 7.37 Comparison of upper and lower bounds for frictionless extrusion in plane strain.

the perfectly smooth die with 45 and 90° semiangles. The slipline field solution for the square die is also included in the figure for comparison.

The block sliding modes of deformation used in the preceding solutions can be applied without modification for the estimation of upper bounds with coulomb friction along the die face. Since the resultant traction on the die face makes an angle $\lambda = \tan^{-1} \mu$ with the inward normal, the magnitude of the resultant die reaction is $p_e H / \sin(\psi + \lambda)$, and the rate of work done by its tangential component is $-p_e H v_2 \sin \lambda / \sin \beta$, where $\beta = \psi + \lambda$. This must be added to the punch work to obtain $p_e H (\cot \beta - \cot \phi)/(\cot \psi - \cot \phi)$ as the net work done by the external forces per unit time, in view of the velocity field of Fig. 6.35. The best upper bound is still given by (85), but the expression for $\cot \phi$ becomes much more complicated. When μ is sufficiently small, terms of order μ^2 and higher may be neglected, and the upper bound then reduces to[†]

$$\frac{p_e}{2k} \simeq \left(1 + \frac{\mu \operatorname{cosec} \psi}{\sqrt{1-r}}\right)\left(\frac{2-r}{\sqrt{1-r}} \operatorname{cosec} \psi - 2 \cot \psi\right) \tag{92}$$

[†] I. F. Collins, *J. Mech. Phys. Solids*, **17**: 323 (1969).

The range of reductions for which this formula provides a reasonable bound is about the same as that for the frictionless situation. However, the friction factor represented by the expression in the first parenthesis predicts with sufficient accuracy the ratio of the extrusion pressures with and without friction over a wide range of die angles and reductions.

7.8 Cold Rolling of Strips

(i) General considerations We shall investigate the process in which a metal sheet or strip is reduced in thickness by passing it between a pair of cylindrical rolls having their axes parallel to one another. In cold rolling, the radius R of the rolls is usually more than 50 times the initial strip thickness. If the width of the strip is at least five times the length of the arc of contact, the nonplastic material prevents the lateral spread, and the deformation takes place effectively under plane strain condition. Due to the pressure of the rolled stock, the rolls are themselves flattened so as to increase the arc of contact by as much as 20 to 25 percent or even more. It will be assumed, for simplicity, that the part of the rolls in contact with the strip is deformed into a cylindrical surface of a larger radius R'. Since the volume of the material passing through each vertical plane per unit time is the same, the speed of the strip steadily increases as it moves through the roll gap. On the entry side, the peripheral speed of the rolls is higher than that of the strip, and consequently the frictional forces draw the strip into the roll gap. On the exit side, the strip moves faster than the rolls, and the frictional forces therefore oppose the delivery of the strip. It follows that there is a neutral point N somewhere on the arc of contact where the strip moves at the same speed as that of the rolls.

Let O be the center of the upper roll and O' the center of curvature of the arc of contact AB (Fig. 7.38). The deformation of the roll is assumed to be such tht O' lies on the straight line joining O and the midpoint C of the arc of contact. If we ignore the small part of the arc of contact that arises from the elastic recovery of the rolled strip, O' must be vertically above the exit point A. The angle of contact $AO'B$, denoted by α, is defined by the radius R' of the arc of contact, and the difference $h_1 - h_2$ between the initial and final strip thicknesses. Let ϕ be the angular distance of a generic point on the arc of contact measured from the plane of exit. If the normal pressure at this point is denoted by q, the local frictional stress is equal to μq, where μ is a constant coefficient of friction.† Taking due account of the direction of the frictional stress on either side of the neutral point, the resultant vertical force per unit width of the roll is obtained as

$$P = R' \int_0^\alpha q \cos \phi \, d\phi + \mu R' \left(\int_{\phi_n}^\alpha q \sin \phi \, d\phi - \int_0^{\phi_n} q \sin \phi \, d\phi \right)$$

---

† For the measurement of the distribution of frictional stress in cold rolling, see G. T. Van Rooyen and W. A. Backofen, *J. Iron Steel Inst.*, **186**: 235 (1957); the measurement of the coefficient of friction along the arc of contact has been carried out by L. Lai-Seng and J. G. Lenard, *J. Eng. Mat. Technol., Trans. ASME,* **106**: 139 (1984).

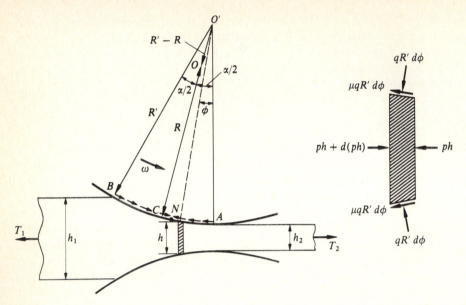

Figure 7.38 Geometry of strip rolling, showing the forces acting on a slice considered on the exit side.

where ϕ_n is the angular position of the neutral point N. Since the angle of contact is generally less than 0.15 rad, the expression in the bracket is negligible. It is therefore sufficiently accurate to write

$$P = R' \int_0^\alpha q \cos \phi \, d\phi \simeq R' \int_0^\alpha q \, d\phi \tag{93}$$

Let T_2 and T_1 be the front and back tensions respectively applied per unit width of the strip. Denoting the difference $T_2 - T_1$ by T, the condition of equilibrium of the horizontal forces acting on the strip may be written as

$$T = 2R' \int_0^\alpha q \sin \phi \, d\phi + 2\mu R' \left(\int_0^{\phi_n} q \cos \phi \, d\phi - \int_{\phi_n}^\alpha q \cos \phi \, d\phi \right)$$

Replacing $\sin \phi$ by ϕ and $\cos \phi$ by unity, with an error not exceeding 1 percent, we obtain the simplified expression

$$T = 2R' \int_0^\alpha q\phi \, d\phi + 2\mu R' \left(\int_0^{\phi_n} q \, d\phi - \int_{\phi_n}^\alpha q \, d\phi \right) \tag{94}$$

The torque G per unit width acting on each roll is the resultant moment about the spindle axis O due to the forces along the arc of contact. The frictional forces produce a positive torque on the entry side and a negative torque on the exit side. The lever arm of an infinitesimal force $\mu q R' \, d\phi$ is equal to R to a close approximation. Since the lines of action of the normal forces nearly pass through O, their contribution to the moment is negligible. Hence

$$G = \mu R R' \left(\int_{\phi_n}^{\alpha} q \, d\phi - \int_0^{\phi_n} q \, d\phi \right) \tag{95}$$

Since this expression involves the difference of two quantities of the same order of magnitude, a small error in either of them could lead to a large error in the torque. If the moment of the normal pressure is taken into account, the necessary modification of (95) follows from the fact that the lever arm of the force $qR' \, d\phi$ is approximately equal to $\pm (R' - R)(\alpha/2 - \phi)$, depending on which side of OO' the normal lies. We therefore have to add the quantity

$$- R'(R' - R) \int_0^{\alpha} q \left(\frac{\alpha}{2} - \phi \right) d\phi$$

to the right-hand side of (95). This additional term is usually negative, and its magnitude is generally negligible. Eliminating the terms in μ between (94) and (95), we obtain an alternative expression for the torque, more suitable for numerical evaluation. This is

$$G = RR' \int_0^{\alpha} q\phi \, d\phi - \frac{1}{2} RT \tag{96}$$

The first term on the right-hand side is approximately equal to the moment of the vertical component of the roll pressure distribution with respect to the roll axis. The moment of the horizontal components of the normal and tangential forces along the arc of contact, which is equal to that of their resultant of magnitude $\frac{1}{2}T$, is expressed by the last term. When the moment of the normal pressure is included, (96) is modified to

$$G = R'^2 \int_0^{\alpha} q\phi \, d\phi - \frac{1}{2} RT - \frac{1}{2}(R' - R)\alpha P \tag{96a}$$

If a strip of metal is rolled under gradually increasing back tension, a stage is eventually reached when the rolls are just at the point of slipping. Since the neutral point is then forced to the exit, Eqs. (93) and (95) furnish

$$\mu = \frac{G}{RP}$$

as the condition for slipping. The coefficient of friction can be determined experimentally from the above formula by plotting the measured value of the ratio G/RP against the applied back tension for any suitable pass reduction. In practical cold rolling, the coefficient of friction is less than 0.1, and so the frictional stress μq is almost invariably less than the shear yield stress k.

If the exit speed of the strip is denoted by U, the mean speed over the plane of entry is $U(h_2/h_1)$. Hence the rate of work done by the applied front and back tensions per unit width are $T_2 U$ and $-T_1 U \, (h_2/h_1)$ respectively. The rolls on the other hand do work at the rate $2G\omega$ per unit width, where ω is the angular velocity of the rolls. Since the volume of material rolled per unit time is Uh_2, the total work done per unit volume is

$$W = \frac{2G\omega}{Uh_2} + \frac{T_2}{h_2} - \frac{T_1}{h_1}$$

The mean speed of the material at the neutral plane is $U(h_2/h_n)$, where h_n is the strip thickness at the neutral point. Since this speed is approximately equal to the peripheral speed $R\omega$ of the rolls, we have $Uh_2 = \omega Rh_n$. Substituting in the above relation, we get

$$W = \frac{2G}{Rh_n} + (t_2 - t_1) \tag{97}$$

where t_2 and t_1 are the mean front and back tension stresses. A part of this total energy is expended for the plastic compression of the strip, the remaining part being dissipated by the friction between the strip and the rolls. The ratio of the homogeneous work of compression to that of the total work W is called the *efficiency* of the rolling process. The actual efficiency of rolling is somewhat less than this due to the frictional losses in the roll neck bearings.

The relative difference between the speeds of the strip and the rolls at the point of exit is known as the *forward slip* denoted by s. It follows from the preceding argument that

$$s = \frac{U}{R\omega} - 1 \simeq \frac{h_n}{h_2} - 1 = \frac{2R'}{h_2}(1 - \cos\phi_n) \simeq \frac{R'}{h_2}\phi_n^2 \tag{98}$$

This relation is sometimes used to determine the position of the neutral point from the measurement of the forward slip.[†] If a fine line is scribed on the rolls parallel to the roll axis, then s is obtained from the fact that the ratio of the distance between the successive imprints on the rolled strip to the roll circumference is equal to the speed ratio $U/R\omega$. Unless the front tension is too high, the forward slip is less than 0.1 in cold rolling.

(ii) *Elastic deformation of the rolls* The pressure distribution over the arc of contact produces elastic bending as well as flattening of the cylindrical rolls. The effect of bending is usually compensated in practice by introducing a slight convexity or camber in the roll profile.[‡] The roll-flattening, however, is an important factor in the theoretical estimation of the roll force and torque. The simplest method of taking the roll-flattening into account is to assume, as we have done, that the deformed arc of contact is circular in shape. This approximation involves the actual distribution of roll pressure to be replaced by an elliptical one

[†] J. Puppe, *Stahl Eisen*, **29**: 161 (1909). See also L. Underwood, *The Rolling of Metals*, Chapman and Hall, London (1950).

[‡] The design of roll cambers has been discussed at length by E. C. Larke, *The Rolling of Strip, Sheet and Plate*, chap. 3, 2d ed., Chapman and Hall, London (1963).

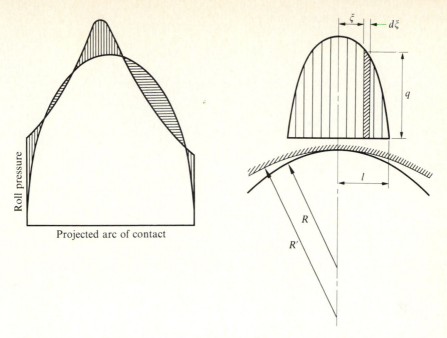

Roll pressure

Projected arc of contact

Figure 7.39 The equivalent elliptic distribution of roll pressure for the estimation of roll flattening.

giving the same total load.† The effect of the frictional forces is probably small and is therefore neglected. The deformation of the roll is then identical to that of an elastic cylinder of radius R pressed against a rigid concave surface of radius R' by an external force P per unit length in the axial plane (Fig. 7.39). Since the length of the arc of contact is small compared to its radius of curvature, the cylinder may be regarded as a semi-infinite medium under an elliptical distribution of loading.

It is reasonable to suppose that the deformation of the cylinder takes place under conditions of plane strain. Then the vertical displacement of its surface at any distance x from the central axis, produced by an elemental load $q\,d\xi$ per unit length acting at a distance ξ, is‡

$$dw = \frac{(1 - v_r^2)q}{\pi E_r}[C + \ln(x - \xi)^2]\,d\xi$$

where E_r is Young's modulus and v_r Poisson's ratio for the rolls, while C is a constant. If the length of the arc of contact is denoted by $2l$, the resultant normal

---

† The accuracy of this approximation has been examined mathematically by D. R. Bland, *Proc. Inst. Mech. Eng.*, **163**: 141 (1950).

‡ See, for example, S. Timoshenko and J. N. Goodier, *Theory of Elasticity*, p. 109, 3d ed., McGraw-Hill Book Co., New York (1970).

displacement at the considered point is

$$w = \frac{1 - v_r^2}{\pi E_r} \int_{-l}^{l} q[C + \ln(x - \xi)^2] \, d\xi$$

Since q is a function of ξ only, the differentiation of this expression with respect to x gives

$$\frac{dw}{dx} = \frac{2(1 - v_r^2)}{\pi E_r} \int_{-l}^{l} \frac{q \, d\xi}{x - \xi} \tag{99}$$

In view of the constant change in curvature of the cylindrical surface over the region of contact, d^2w/dx^2 has a constant value, giving

$$\frac{dw}{dx} = \left(\frac{1}{R} - \frac{1}{R'} \right) x$$

The last two equations will be compatible only if the distribution of q is elliptical. Since the area under the pressure distribution curve must be equal to P, the intensity of the maximum pressure is $2P/\pi l$, and the pressure distribution is given by

$$q = \frac{2P}{\pi l} \sqrt{1 - \frac{\xi^2}{l^2}}$$

In view of (99), and the last two equations, the relationship between the applied load and the change in curvature becomes

$$\frac{4(1 - v_r^2)P}{\pi^2 E_r l^2} \int_{-l}^{l} \frac{\sqrt{l^2 - \xi^2}}{x - \xi} \, d\xi = \left(\frac{1}{R} - \frac{1}{R'} \right) x$$

The integral on the left-hand side is easily shown to be equal to πx. Hence

$$\frac{1}{R} - \frac{1}{R'} = \frac{4(1 - v_r^2)P}{\pi E_r l^2}$$

From geometry, the length of the arc of contact is $2l \simeq R'\alpha \simeq \sqrt{R'\delta}$, where δ is the thickness change $h_1 - h_2$, known as the *draft*. The radius of the deformed arc of contact is therefore given by the formula (due to Hitchcock[†])

$$R' = R\left(1 + \frac{P}{c\delta} \right) \qquad c = \frac{\pi E_r}{16(1 - v_r^2)} \tag{100}$$

The constant c depends on the material of the rolls, its value for steel rolls being 2.99 (10^3) tons/in^2 or 4.62 (10^4) MN/m^2. When δ is vanishingly small, the force required to bring the strip to the yield point is sufficient to produce a finite arc of contact between the strip and the rolls. In this case, the elastic deformation of the strip must be considered for a realistic estimation of R'.

† J. H. Hitchcock, *Roll Neck Bearings*, app. I, ASME Publication (1935).

(iii) *Von Karman's theory of rolling* It is assumed at the outset that each element of the strip is uniformly compressed between the rolls while passing through the roll gap. The vertical compression of the strip is accompanied by a horizontal force which is increasingly compressive as the neutral point is approached from either side. Let p denote the mean horizontal pressure over a vertical section specified by its angular coordinate ϕ (Fig. 6.38). The resultant compressive force acting over this section per unit width is hp, where h is local strip thickness. Consider a thin slice of the strip between two vertical sections corresponding to the angles ϕ and $\phi + d\phi$. Since the horizontal components of the normal and tangential forces acting on the ends of the slice are $qR' \sin \phi \, d\phi$ and $\mu q R' \cos \phi \, d\phi$ respectively, the equation of equilibrium may be written as

$$\frac{d}{d\phi}(hp) = 2qR'(\sin \phi \pm \mu \cos \phi) \tag{101}$$

where the upper sign applies to the exit side and the lower sign to the entry side of the neutral point. We now assume that the material is everywhere plastic between the planes of entry and exit, and the principal compressive stresses at each point on a vertical section are approximately equal to p and q. Then the yield criterion may be written in the form

$$q - p = 2k$$

where the shear yield stress k generally varies along the arc of contact. The value of $2k$ at a generic point on the arc of contact is approximately equal to the ordinate of the compressive stress-strain curve, obtained under plane strain condition, corresponding to an abscissa equal to $\ln (h_1/h)$. Alternatively, the variation of the yield stress can be estimated by rolling a length of the strip in a succession of passes and carrying out a tensile test at the end of each pass.[†] This gives the tensile yield stress $\sqrt{3} \, k$ as a function of the thickness ratio h/h_1.

The theory of rolling expressed by the differential equation (101) and the yield criterion is due to von Karman.[‡] The equilibrium equation can be reduced further by eliminating p and using the relation $dh/d\phi = 2R' \sin \phi$. Adopting the usual approximation $\sin \phi \simeq \phi$ and $\cos \phi \simeq 1$, since the angle of contact is small, we arrive at the governing equation

$$h\frac{d}{d\phi}(q - 2k) \mp 2\mu R'q = 4kR'\phi \tag{102}$$

where

$$h = h_2 + 2R'(1 - \cos \phi) \simeq h_2 + R'\phi^2$$

The boundary conditions are $p = -t_2$, $q = 2k_2 - t_2$ at $\phi = 0$, and $p = -t_1$, $q = 2k_1 - t_1$ at $\phi = \alpha$, where k_1 and k_2 are the values of k at the entry and exit planes

---

[†] H. Ford, *Proc. Inst. Mech. Eng.*, **159**: 115 (1948).
[‡] Th. von Karman, *Z. angew. Math. Mech.*, **5**: 139 (1925).

respectively.† From geometry,

$$\alpha \simeq \sqrt{\frac{h_1 - h_2}{R'}} = \sqrt{\frac{rh_1}{R'}}$$

where r is the fractional reduction. The experimentally derived relationship between the yield stress $2k$ and the thickness ratio h/h_1 may be expressed in the form

$$2k = 2k_0\left\{1 - m\left(\frac{h}{h_1}\right)^n\right\} \tag{103}$$

where k_0, m, and n are empirical constants. The nonhardening material corresponds to $m = 0$. It is convenient at this stage to change the independent variable from ϕ to ψ, where

$$\psi = \tan^{-1}\sqrt{\frac{R'}{h_2}}\phi = \tan^{-1}\left\{\sqrt{\frac{r}{1-r}}\frac{\phi}{\alpha}\right\} \tag{104}$$

Inserting from (103) into (102), and noting that $h = h_2 \sec^2 \psi$ and $\sqrt{R'/h_2}\,d\phi = \sec^2\psi\,d\psi$ in view of (104), the equilibrium equation is reduced to

$$\frac{dq}{d\psi} \mp 2aq = 4k_0[1 - m(1 + n)(1 - r)^n \sec^{2n}\psi]\tan\psi \tag{105}$$

where

$$a = \mu\sqrt{\frac{R'}{h_2}} = \frac{\mu}{\alpha}\sqrt{\frac{r}{1-r}}$$

The solution of the first-order linear differential equation (105) is uniquely defined by the boundary condition $q = 2k_2 - t_2$ at $\psi = 0$ and $q = 2k_1 - t_1$ at $\psi = \psi_0$, where

$$\psi_0 = \tan^{-1}\sqrt{\frac{r}{1-r}} = \sin^{-1}\sqrt{r} \tag{106}$$

Using the standard method of solving linear first-order equations, the distribution of roll pressure along the arc of contact can be expressed in the nondimensional form

$$\frac{q}{2k_0} = \left[1 - \frac{t_2}{2k_0} - m(1 - r)^n + f(\psi)\right]e^{2a\psi} \qquad \text{exit side}$$

$$\tag{107}$$

$$\frac{q}{2k_0} = \left[\left(1 - \frac{t_1}{2k_0} - m\right)e^{2a\psi_0} + g(\psi) - g(\psi_0)\right]e^{-2a\psi} \qquad \text{entry side}$$

on inserting the values of k_1 and k_2 from (103). The functions $f(\psi)$ and $g(\psi)$ in the

---

† Numerical results, for the average pressure and the peak pressure, based on a graphical solution of (102), have been presented by W. Trinks, *Blast Furn. Steel Plant*, **25**: 617 (1933).

above expressions are defined by

$$f(\psi) = 2 \int_0^\psi [1 - m(1 + n)(1 - r)^n \sec^{2n} \psi] e^{-2a\psi} \tan \psi \, d\psi$$

$$g(\psi) = 2 \int_0^\psi [1 - m(1 + n)(1 - r)^n \sec^{2n} \psi] e^{2a\psi} \tan \psi \, d\psi$$

(108)

The roll pressure steadily increases as the neutral point $\psi = \psi_n$ is approached from the entry and exit points of the arc of contact. The neutral point is determined from the fact that the two values of q given by (107) are identical at $\psi = \psi_n$. Since rolling is possible only for $\psi_n > 0$, the inequality

$$\left(1 - \frac{t_1}{2k_0} - m\right)e^{2a\psi_0} - \left(1 - \frac{t_2}{2k_0}\right) + m(1 - r)^n > g(\psi_0)$$

must always be satisfied. When $m = 0$, the integrals (108) can be evaluated explicitly by writing $\tan \psi \simeq \psi(1 + \psi^2/3)$ with sufficient accuracy over the relevant range, resulting in

$$f(\psi) \simeq \frac{1}{2a^2}\left(1 + \frac{1}{2a^2}\right)(1 - e^{-2a\psi}) - \frac{\psi}{a}\left(1 + \frac{1}{2a^2} + \frac{\psi}{2a} + \frac{\psi^2}{3}\right)e^{-2a\psi}$$

$$g(\psi) \simeq -\frac{1}{2a^2}\left(1 + \frac{1}{2a^2}\right)(e^{2a\psi} - 1) + \frac{\psi}{a}\left(1 + \frac{1}{2a^2} - \frac{\psi}{2a} + \frac{\psi^2}{3}\right)e^{2a\psi}$$

(109)

The error in the actual roll pressure at any point due to this approximation will be less than 0.5 percent.[†] Typical pressure distribution curves for $m = 0$, $a = 1.5$, and $r = 0.3$ are shown in Fig. 7.40, which indicates how the neutral point is moved backward and forward by the application of the front and back tensions respectively.[‡]

When the distribution of roll pressure has been found, the roll force and torque can be calculated from the expressions (93) and (96), which become

$$\frac{P}{2k_0 h_2} = \frac{R'}{h_2} \int_0^\alpha \left(\frac{q}{2k_0}\right) d\phi = \sqrt{\frac{R'}{h_2}} \int_0^{\tan\psi_0} \left(\frac{q}{2k_0}\right) d\xi$$

$$\frac{G}{2k_0 R h_2} = \frac{R'}{h_2} \int_0^\alpha \left(\frac{q}{2k_0}\right)\phi \, d\phi - \frac{T}{4k_0 h_2} = \int_0^{\tan\psi_0} \left(\frac{q}{2k_0}\right)\xi \, d\xi - \frac{T}{4k_0 h_2}$$

(110)

where $\xi = \sqrt{R'/h_2}\,\phi = \tan\psi$ is a convenient new variable. For given values of a, r, m, n, $t_1/2k_0$, and $t_2/2k_0$, the integrals appearing in (110) can be evaluated by a numerical procedure, using (107) and the calculated value of ψ_n defining the

---

† A solution for the pressure distribution, based on the approximation $\tan\psi \simeq \psi$, has been discussed by A. Nadai, *J. Appl. Mech.*, **6**, A-55 (1939).

‡ The elastic property of the strip extends the arc of contact to include a region of elastic compression at the entry and a region of elastic recovery at the exit. The former is usually of little significance, but the latter can have an appreciable effect on the roll torque (see Prob. 7.33).

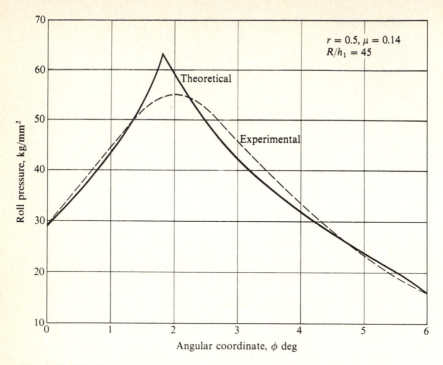

Figure 7.40 Distribution of normal roll pressure in the cold rolling of annealed aluminum.

position of the neutral point. Since P and R' are interdependent, they have to be found simultaneously before the torque can be calculated. A trial value of R' used in (110) for the calculation of P must be repeatedly altered until (100) is satisfied to a sufficient accuracy.

Figure 7.41 shows the theoretical and experimental pressure distributions for rolling without tension of a strip of annealed aluminum with $R/h_1 = 45$, $r = 0.5$, and $\mu = 0.14$, the roll-flattening being neglected. The theoretical curve has been derived by Orowan† with the help of graphical integration of an equation substantially the same as (102), using the actual stress-strain curve of the material. The experimental curve has been obtained by Siebel and Lueg,‡ who measured the pressure distribution by the piezoelectric method using a pressure-transmitting pin embedded in the roll surface. The agreement between theory and experiment is fairly good, except that the theoretical pressure peak at the neutral point is absent in the measured curve. The rounding of the peak is due mainly to the fact that a

† E. Orowan, *Proc. Inst. Mech. Eng.*, **150**: 140 (1943).

‡ E. Siebel and W. Lueg, *Mitt. K.W. Inst. Eisenf.*, **15**: 1 (1933). See also W. Lueg, *Stahl Eisen*, **53**: 346 (1933); L. Underwood, *The Rolling of Metals*, Chapman and Hall (1950). For the measurement of roll pressure distribution, see also F. A. R. Al-Salehi, T. C. Firbank, and P. R. Lancaster, *Int. J. Mech. Sci.*, **15**: 693 (1973).

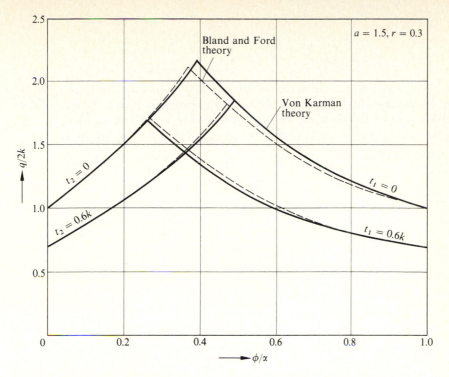

Figure 7.41 Roll pressure distributions in cold rolling with and without strip tensions.

zone of material near the neutral point actually remains nonplastic during the rolling process.†

(iv) An approximate analysis An exact solution of the von Karman equation requires numerical integration even when work-hardening is neglected. For practical purposes, an approximate solution due to Bland and Ford has been found to be generally adequate.‡ The approximation is arrived at by starting with Eq. (101) from which p is eliminated by the yield criterion. Setting $\sin \phi \simeq \phi$ and $\cos \phi \simeq 1$, we express the equilibrium equation as

$$\frac{d}{d\phi}\left\{2kh\left(\frac{q}{2k}-1\right)\right\} = 2R'q(\phi \pm \mu)$$

† E. Orowan (op. cit.) developed a theory of rolling in which the inhomogeneity of deformation is taken into account by an assumed stress distribution across each vertical section. Numerical computations based on Orowan's theory have been carried out by M. Cook and E. C. Larke, *J. Inst. Met.*, **71**: 557 (1945). See also J. B. Hockett, *Trans. ASME*, **52**: 675 (1960); V. Venter and A. Abdo-Rabbo, *Int. J. Mech. Sci.*, **22**: 83 and 93 (1980).

‡ D. R. Bland and H. Ford, *Proc. Inst. Mech. Eng.*, **159**: 189 (1948). See also H. Ford, *Metall. Rev.*, **2**: 5 (1957).

Carrying out the differentiation gives

$$kh \frac{d}{d\phi}\left(\frac{q}{2k}\right) + \left(\frac{q}{2k} - 1\right)\frac{d}{d\phi}(kh) = R'q(\phi \pm \mu)$$

If the rate of work-hardening is sufficiently low, $d(kh)$ would be small compared to kh since k increases in the direction of decreasing thickness. If, in addition, the strip tensions are small or moderate, $(q/2k - 1)$ would be small compared to $d(q/2k)$. Then the second term on the left-hand side of the above equation may be neglected, resulting in the simplification

$$\frac{d}{d\phi}\left(\frac{q}{2k}\right) = \frac{2R'}{h}(\phi \pm \mu)\frac{q}{2k} \tag{111}$$

where $h \simeq h_2 + R'\phi^2$. The approximation is also good for an annealed material with no back tension, since q is then very nearly equal to $2k$ over the part of the entry side where the strain-hardening is most pronounced. In view of the boundary conditions $q/2k = 1 - t_2/2k_2$ at $\phi = 0$ and $q/2k = 1 - t_1/2k$ at $\phi = \alpha$, the solution of (111) is

$$\frac{q}{2k} = \frac{h}{h_2}\left(1 - \frac{t_2}{2k_2}\right)e^{2a\psi} \qquad \text{exit side}$$

$$\frac{q}{2k} = \frac{h}{h_1}\left(1 - \frac{t_1}{2k_1}\right)e^{2a(\psi_0 - \psi)} \qquad \text{entry side} \tag{112}$$

where a, ψ, and ψ_0 are the same as before. For given ratios $t_1/2k_1$ and $t_2/2k_2$, the roll pressure at each point is proportional to the local yield stress. The condition of continuity of $q/2k$ at the neutral point $\phi = \phi_n$ furnishes

$$\tan^{-1}\sqrt{\frac{R'}{h_2}}\,\phi_n = \frac{1}{2}\sin^{-1}\sqrt{r} - \frac{1}{4a}\ln\left\{\frac{h_1(1 - t_2/2k_2)}{h_2(1 - t_1/2k_1)}\right\} \tag{113}$$

It is interesting to note that the neutral point is independent of the manner in which the yield stress varies along the arc of contact. The minimum coefficient of friction for which rolling is possible under given tensions is

$$\mu^* = \sqrt{\frac{h_2}{R'}}\ln\left\{\frac{1}{1 - r}\left(\frac{1 - t_2/2k_2}{1 - t_1/2k_1}\right)\right\}\Big/(2\sin^{-1}\sqrt{r}) \tag{114}$$

For this critical value of μ the neutral point falls at the point of exit. The angle of contact α must be less than the angle of friction $\tan^{-1}\mu \simeq \mu$, in order that the rolls can draw the strip into the roll gap. It follows that unless the strip is forced into the roll gap, the reduction cannot exceed the value $\mu^2 R'/h_1$. The roll pressure distribution for a nonhardening material is compared with that obtained from the von Karman theory in Fig. 7.40, which illustrates the accuracy of the above approximation.†

To find the roll force and torque, we have to insert the expressions for q into (110) and integrate over the arc of contact, using the fact that $h = h_2(1 + \xi^2)$. If the material is nonhardening ($k_1 = k_2 = k$), and the strip tension are absent ($t_1 = t_2 = 0$), we have

$$\frac{P}{P^*} = \sqrt{\frac{1-r}{r}} \left\{ \int_0^{\xi_n} (1 + \xi^2)e^{2a\psi}\, d\xi + (1 - r)e^{2a\psi_0} \int_{\xi_n}^{\xi_0} (1 + \xi^2)e^{-2a\psi}\, d\xi \right\}$$

$$\frac{G}{G^*} = 2\left(\frac{1-r}{r}\right)\left\{ \int_0^{\xi_n} \xi(1 + \xi^2)e^{2a\psi}\, d\xi + (1 - r)e^{2a\psi_0} \int_{\xi_n}^{\xi_0} \xi(1 + \xi^2)e^{-2a\psi}\, d\xi \right\} \tag{115}$$

where $P^* = 2k\sqrt{R'rh_1}$ and $G^* = kRrh_1$. These are the values of P and G for a uniform distribution of normal pressure equal to $2k$ along the arc of contact. In the above integrals, $\psi = \tan^{-1} \xi$ and

$$\xi_n = \tan\left(\frac{1}{2}\sin^{-1}\sqrt{r} - \frac{1}{4a}\ln\frac{1}{1-r}\right) \qquad \xi_0 = \sqrt{\frac{r}{1-r}} \tag{116}$$

The dimensionless values of roll force and torque, calculated numerically from (115), are plotted in Fig. 7.42 as functions of the reduction r and the parameter $\mu\sqrt{R'/h_1}$. In view of (97), the work done per unit volume of the strip for no applied tension may be written as

$$W = \frac{2G/Rh_2}{1 + (R'/h_2)\phi_n^2} = 2k\left(\frac{r}{1-r}\right)\frac{G/G^*}{1 + \xi_n^2}$$

where ξ_n^2 is equal to the forward slip in view of (98), its value being generally small compared to unity. Since the work of homogeneous compression amounts to $2k \ln (h_1/h_2)$, the efficiency of rolling for any given reduction is approximately equal to the ratio of the quantity $[(1 - r)/r] \ln [1/(1 - r)]$ to the appropriate value of G/G^* obtained directly from the graph. The efficiency decreases as the parameter $\mu\sqrt{R'/h_1}$ increases.

The deformed roll radius R', corresponding to a given radius R, must be determined from Hitchcock's formula (100) by successive approximation, using a suitable starting value of R'. For this purpose, it is convenient to express the roll force by the empirical formula

$$\frac{P}{P^*} = 1.02 + r(1.5 + 1.6r^2)\mu\sqrt{\frac{R'}{h_1}} - 1.9r^2 \tag{117}$$

† An approximate theory in which the arc of contact is replaced by a pair of chords has been put forward by A. T. Tselikov, *Mettalurg* (Russian), **6**: 61 (1936). See also his book, *Stresses and Strains in Metal Rolling*, Moscow (1967). For an approximate solution, obtained by setting $q \simeq 2k$, $\sin \phi \simeq \phi$ and $\cos \phi \simeq 1$ on the right-hand side of (101), see E. Siebel, *Stahl Eisen*, **45**: 1563 (1925). An approximate theory based on purely kinematical considerations has been discussed by B. Avitzur, *J. Eng. Ind., Trans. ASME*, **86**: 31 (1964).

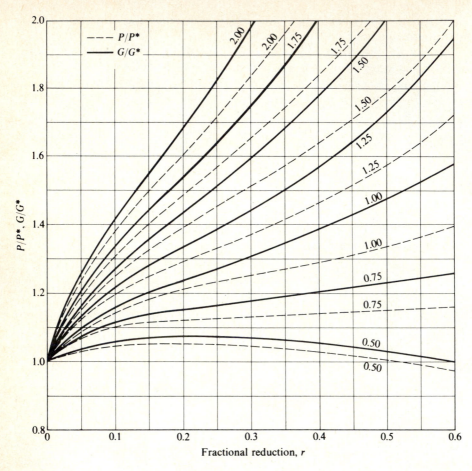

Figure 7.42 Variation of roll force and torque with reduction in cold rolling for various values of $\mu\sqrt{R'/h_1}$.

which is correct to within ± 1.5 percent over the ranges $0.1 \leqslant r \leqslant 0.5$ and $0.5 \leqslant \mu\sqrt{R'/h_1} \leqslant 1.5$, in relation to the results presented in Fig. 7.42. Substitution into (100) then furnishes the quadratic equation

$$\left\{\frac{h_1}{\mu^2 R} - \lambda\sqrt{r}(1.5 + 1.6r^2)\right\}\left(\mu\sqrt{\frac{R'}{h_1}}\right)^2 - \lambda\left(\frac{1.02}{\sqrt{r}} - 1.9r\sqrt{r}\right)\mu\sqrt{\frac{R'}{h_1}} - 1 = 0$$

$$(118)$$

where $\lambda = 2k/\mu c$. It may be noted that $\mu\sqrt{R'/h_1}$ depends only on the parameters $\mu\sqrt{R/h_1}$, λ and r. The variation of R'/R with $2k/\mu c$ for various reductions is displayed in Fig. 7.43, using two different values of $\mu\sqrt{R/h_1}$. For a given value of $2k/\mu c$, the ratio R'/R increases as the reduction is decreased, since there is a greater

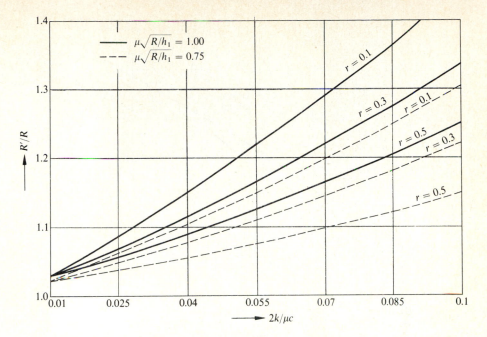

Figure 7.43 Curves for roll flattening in the cold rolling of metal strips.

concentration of the roll force at smaller reductions producing an increased local distortion of the roll surface. When the reduction in thickness is extremely small, the effect of the elastic distortion of the strip becomes significant, particularly for thin hard materials.†

When the material work-hardens, it is convenient to define the quantities P^* and G^* in such a way that they represent the values of P and G for a distribution of normal pressure equal to the local yield stress at each point of the arc of contact. Since

$$\sqrt{\frac{R'}{h_1}}\,\phi = \sqrt{\frac{h - h_2}{h_1}} = \sqrt{r - e}$$

where $e = (h_1 - h)/h_1$, denoting the fractional reduction at a generic point of the arc of contact, we have

$$P^* = R'\int_0^\alpha 2k\,d\phi = \sqrt{R'h_1}\left[2k_1\sqrt{r} + 2\int_{k_1}^{k_2}\sqrt{r - e}\,dk\right]$$

$$G^* = RR'\int_0^\alpha 2k\phi\,d\phi = R\int_{h_2}^{h_1} k\,dh = Rh_1\int_0^r k\,de$$

(119)

† The elastic compression of the strip has been approximately treated by D. R. Bland and H. Ford, *J. Iron Steel Inst.*, **171**: 245 (1952). A finite element solution for the elastic/plastic problem has been discussed by T. Tamano, *J. Japan Soc. Tech. Plast.*, **14**: 766 (1973).

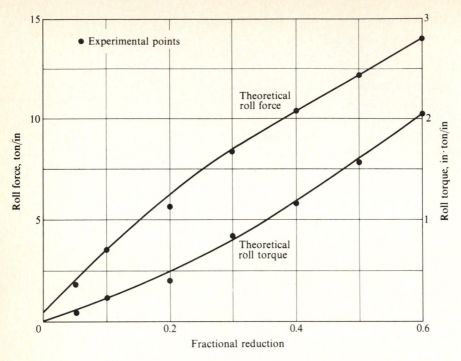

Figure 7.44 Comparison of theoretical and experimental roll force and torque in cold rolling.

The first relation of (119) follows on integrating by parts and using the fact that $\alpha = \sqrt{rh_1/R'}$. For any given reduction in thickness, the integrals in (119) can be readily evaluated from an experimentally determined $(2k, e)$ curve. To determine the roll force and torque for a work-hardening material, it is assumed that the ratios P/P^* and G/G^* are independent of work-hardening. This means that P and G can be obtained by multiplying the appropriate ordinates of Fig. 7.42 by the corresponding values of P^* and G^* given by (119). The approximation is obviously equivalent to using suitable mean values of the yield stress, separately defined for the roll force and the torque. The radius R' is furnished by (118) with $\lambda = 2\bar{k}/\mu c$, where

$$2\bar{k} = 2k_1 + 2 \int_{k_1}^{k_2} \sqrt{1 - \frac{e}{r}}\, dk$$

In Fig. 7.44, the theoretical values of the roll force and the torque are compared with experimental results for high conductivity copper,† using 5-in roll radius and

---

† The experimental results have been obtained by H. Ford, *Proc. Inst. Mech. Eng.*, **159**: 115 (1948). For recent experimental results on ring rolling, see A. G. Mamalis, W. Johnson, and J. B. Hawkyard, *J. Mech. Eng. Sci.*, **10**: 196 (1976).

0.05-in initial strip thickness, the coefficient of friction being equal to 0.086. In the theoretical estimation of roll force and torque, work-hardening has been allowed for in the manner described above, using the actual stress-strain curve of the material. Apart from the experimental scatter, the agreement between the calculated and the measured values is quite satisfactory.†

(v) *Influence of strip tensions* Tensions are often applied in cold rolling in order to ensure adequate control of the dimensional accuracy of the rolled strip. For a prestrained material, the effect of tensions is to reduce the roll pressure at each point approximately by the factor $t_1/2k_1$ on the entry side and by the factor $t_2/2k_2$ on the exit side. The position of the neutral point is moved forward by the application of a back tension and backward by the application of a front tension. The permissible amounts of front and back tensions are limited by the fact that the neutral point must lie between the planes of entry and exit. Each of the ratios $t_1/2k_1$ and $t_2/2k_2$ must of course be less than unity so that the strip does not neck before entering or after leaving the roll gap. Using (110) and (112), the roll force and torque can be expressed in terms of the functions

$$f_1(\xi), f_2(\xi) = \int_0^\xi (1 + \xi^2) \exp(\mp 2a \tan^{-1} \xi)\, d\xi$$

$$g_1(\xi), g_2(\xi) = \int_0^\xi \xi(1 + \xi^2) \exp(\mp 2a \tan^{-1} \xi)\, d\xi \tag{120}$$

where the upper sign applies to the first function and the lower sign to the second function in each case. Assuming average values of the yield stress over the arc of contact, the roll force and torque can be written as

$$\frac{P}{P^*} = \sqrt{\frac{1-r}{r}} \left\{ \left(1 - \frac{t_2}{2k_2}\right) f_2(\xi_n) + (1-r)\left(1 - \frac{t_1}{2k_1}\right) e^{2a\psi_0}[f_1(\xi_0) - f_1(\xi_n)] \right\} \tag{121}$$

$$\frac{G'}{G^*} = 2\left(\frac{1-r}{r}\right)\left\{ \left(1 - \frac{t_2}{2k_2}\right) g_2(\xi_n) + (1-r)\left(1 - \frac{t_1}{2k_1}\right) e^{2a\psi_0}[g_1(\xi_0) - g_1(\xi_n)] \right\}$$

where P^* and G^* are given by (119), $G' = G + \frac{1}{2}RT$, and $\xi_n = \sqrt{R'/h_2}\, \phi_n$, the values of ψ_0 and ξ_0 being $\sin^{-1}\sqrt{r}$ and $\sqrt{r/(1-r)}$ respectively. The various functions occurring in the above expressions can be evaluated numerically for any given values of a, r, $t_1/2k_1$, and $t_2/2k_2$. The results may be conveniently put in the form

$$P = P^*\left(1 - \frac{t_1}{2k_1}\right) F(a, r, b) \qquad G = G^*\left(1 - \frac{t_1}{2k_1}\right) H(a, r, b) - \frac{1}{2} RT \tag{122}$$

† Useful experimental works on cold rolling have been reported by I. Y. Tarnovskii, A. A. Pozdeyev, and V. B. Lyashkov, *Deformation of Metal During Rolling* (trans. from Russian), Pergamon Press, Oxford (1965).

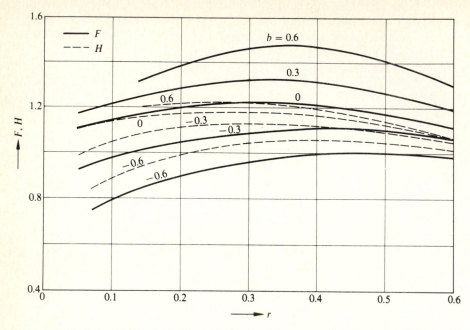

Figure 7.45 Variation of F and H with r and b for $a = 1$ in relation to rolling with tension.

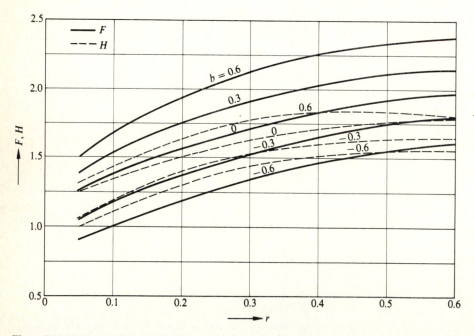

Figure 7.46 Variation of F and H with r and b for $a = 2$ in relation to rolling with tension.

where F and H are dimensionless functions of a, r, and the parameter

$$b = \ln \left(\frac{1 - t_2/2k_2}{1 - t_1/2k_1} \right)$$

The strain-hardening characteristic of the material enters only through P^* and G^*. The quantities F and H are shown graphically[†] in Figs. 7.45 and 7.46 as functions of r, corresponding to $a = 1$ and $a = 2$, and for a range of values of b. The results are subject to the restriction $0 < \phi_n < \alpha$, which is equivalent to

$$-\left(2a \sin^{-1} \sqrt{r} + \ln \frac{1}{1 - r} \right) < b < \left(2a \sin^{-1} \sqrt{r} - \ln \frac{1}{1 - r} \right) \tag{123}$$

Since the neutral point is independent of applied tensions when $t_1 = t_2$, the functions F and H corresponding to $b = 0$ are identical to P/P^* and G/G^* for rolling without tensions.[‡]

The approximation leading to (112) breaks down in the presence of high back tensions in an annealed strip, and consequently the roll pressure is underestimated on the entry side of the arc of contact. A modification of the theory is, however, possible if we consider the solution of (102) for two identical passes, one with tensions t_1 and t_2 applied to the strip and the other without tensions. It is assumed that R' is the same in the two passes, so that the values of R would be different. If q denotes the pressure with tensions and q_0 the pressure without tensions, then (102) gives

$$h \frac{d}{d\phi} (q_0 - q) \mp 2\mu R'(q_0 - q) = 0$$

The distribution of q_0 corresponding to either the upper or the lower sign is assumed to extend up to the neutral point defined by the distribution of q. A straightforward integration gives

$$q_0 - q = t_2 \, e^{2a\psi} \qquad \text{exit side}$$

$$q_0 - q = t_1 \, e^{2a(\psi_0 - \psi)} \qquad \text{entry side}$$

So far the analysis has been exact. We now introduce the assumption that the expression for q_0 is the same as that given by (112) with $t_1 = t_2 = 0$. Then the roll pressure distribution becomes[§]

$$q = \left(\frac{2kh}{h_2} - t_2 \right) e^{2a\psi} \qquad \text{exit side}$$

$$q = \left(\frac{2kh}{h_1} - t_1 \right) e^{2a(\psi_0 - \psi)} \qquad \text{entry side} \tag{124}$$

[†] H. Ford, F. Ellis, and D. R. Bland, *J. Iron Steel Inst.*, **168**: 57 (1951); **171**: 239 (1952). See also W. C. F. Hessenberg and R. B. Sims, *J. Iron Steel Inst.*, **168**: 155 (1951).

[‡] For a numerical comparison of various rolling theories, including the effects of work-hardening and strip tensions, see J. M. Alexander, *Proc. R. Soc., A*, **326**: 525 (1972). A finite element solution has been given by G. Li and S. Kobayashi, *J. Eng. Ind., Trans. ASME*, **104**: 55 (1982).

[§] D. R. Bland and R. B. Sims, *Proc. Inst. Mech. Eng.*, **167**: 371 (1953).

This solution is in close agreement with that obtained by the numerical integration of (102). The neutral point corresponds to $\psi = \psi_n$, given by

$$\psi_n = \frac{1}{2}\sin^{-1}\sqrt{r} - \frac{1}{4a}\ln\left\{\frac{\sec^2\psi_n - t_2/2k_n}{(1-r)\sec^2\psi_n - t_1/2k_n}\right\} \tag{125}$$

where k_n is the shear yield stress at the neutral section $\phi = \phi_n$. The above equation may be solved for ψ_n by trial and error, noting the fact that

$$k_n = k_0[1 - m(1-r)^n\sec^{2n}\psi_n]$$

in view of (103). The roll force and the torque can be found numerically from Eqs. (110) and (124), using an assumed value of R'.

Approximate expressions for the roll force and torque may be developed on the assumption that the change in roll pressure produced by the applied tensions varies linearly along the arc of contact according to the relationship

$$q_0 - q = t_2\left\{1 + (\sqrt{1-r}\,e^{a\psi_0} - 1)\frac{\phi}{\phi_n}\right\} \qquad 0 \leqslant \phi < \phi_n$$

$$q_0 - q = t_1\left\{1 + \left(\frac{e^{a\psi_0}}{\sqrt{1-r}} - 1\right)\left(\frac{\alpha-\phi}{\alpha-\phi_n}\right)\right\} \qquad \phi_n < \phi \leqslant \alpha$$

where ϕ_n corresponds to rolling of the strip without any tension. A discontinuity in roll pressure is allowed at $\phi = \phi_n$, the terminal values of $q_0 - q$ at this point being indentical to those given by the preceding theory. Inserting in (93) and (96), and denoting the no-tension values of roll force and torque by P_0 and G_0 respectively, we obtain†

$$\frac{P_0 - P}{\sqrt{R'h_2}} = \frac{1}{2}\left\{t_2(1 + \sqrt{1-r}\,e^{a\psi_0})\xi_n + t_1\left(1 + \frac{e^{a\psi_0}}{\sqrt{1-r}}\right)\left(\sqrt{\frac{r}{1-r}} - \xi_n\right)\right\} \tag{126}$$

$$\frac{G_0 - G}{Rh_2} \simeq \frac{1}{2}\left\{(t_2 - t_1)(1 + \xi_n^2) + \frac{2}{3}t_1\left(\frac{e^{a\psi_0}}{\sqrt{1-r}} - 1\right)\left(\sqrt{\frac{r}{1-r}} - \xi_n\right)^2\right\} \tag{127}$$

where ξ_n is given by (116). Equation (126) is consistent with the fact that the roll force is decreased more by back tension than by an equal stress applied as a front tension. The effect of the tensions on roll flattening may be disregarded in using (126) and (127).

As a very rough approximation, it may be assumed that the effect of tensions is to decrease the roll pressure by t_1 on the entry side and by t_2 on the exit side of the neutral plane. Using Eq. (96) and the fact that the forward slip is usually small, it is easily shown that

$$G \simeq G_0 + \tfrac{1}{2}Rh_2(t_1 - t_2)$$

---

† There seems to be an approximately linear relationship between G/PR and T/P. See R. Hill, *Proc. Inst. Mech. Eng.*, **163**: 135 (1950).

which agrees reasonably with experiment. The above relationship indicates that the work expended per unit volume of the material is practically unaffected by the applied tensions. Consequently, no significant advantage could be gained by rolling with tensions higher than those necessary to produce a properly coiled strip.†

(vi) *Minimum thickness in cold rolling* For a given reduction in pass, there is a limiting entry thickness below which it is impossible to achieve the reduction by cold rolling.‡ At this critical value of the strip thickness, an increased roll force merely flattens the roll surface still further, thereby increasing the arc of contact between the strip and the rolls. The longitudinal compressive stresses induced by the additional frictional forces are sufficient to suppress plastic yielding of the material in the roll gap. A limit is therefore set to the reduction that can be allowed in a single pass for very thin materials. From the theoretical point of view, if P/h_1 and R'/R are plotted against R/h_1 for a given material and specified values of r, μ, t_1, and t_2, both the curves will eventually have an infinite slope at a definite value of R/h_1, beyond which there is no solution to the rolling problem. The maximum value of R/h_1 obtained for a given set of rolling conditions gives the minimum entry thickness of the material for the given rolling schedule.§

We begin our discussion by considering the situation where no tensions are applied to the strip. The roll force calculated from (115) may be conveniently expressed by the empirical formula

$$\frac{P}{P^*} = \frac{1}{2}\left\{1 + \exp\left[(0.9 + r^2)\mu\sqrt{\frac{R'r}{h_1}}\right]\right\}$$

which is correct to within 2.5 percent over the relevant range. Inserting into Hitchcock's formula (100) furnishes

$$r\left(\frac{R'}{R} - 1\right) = \frac{\bar{k}}{c}\sqrt{\frac{R'r}{h_1}}\left\{1 + \exp\left[(0.9 + r^2)\mu\sqrt{\frac{R'r}{h_1}}\right]\right\} \qquad (128)$$

where $\bar{k}$ is the mean shear yield stress for the given reduction. When the strip thickness is a minimum, the derivative of R/h_1 with respect to the ratio R'/R must vanish. Using (128), the minimum condition may be written as

$$\frac{h_1}{R} = \frac{\bar{k}}{2c}\sqrt{\frac{h_1}{R'r}}\left\{1 + \left[1 + (0.9 + r^2)\mu\sqrt{\frac{R'r}{h_1}}\right]\exp\left[(0.9 + r^2)\mu\sqrt{\frac{R'r}{h_1}}\right]\right\} \qquad (129)$$

---

† Methods of controlling the thickness of the rolled strip have been discussed by W. C. F. Hessenberg and R. B. Sims, *Proc. Inst. Mech. Eng.*, **166**: 75 (1952). See also R. B. Sims and P. R. A. Briggs, *Sheet Met. Ind.*, **31**: 181 (1954); G. W. Alderton and W. C. F. Hessenberg, *Met. Rev.*, **1**: 239 (1956).

‡ This question has been discussed by R. Hill and I. M. Longman, *Sheet Met. Ind.*, **28**: 705 (1951), using an approximate expression for the roll force. The limiting case $r = 0$ has been treated by M. D. Stone, *Iron Steel Eng.*, **30**: 61 (1953).

§ The problem has been solved graphically by R. B. Sims, *J. Iron Steel Inst.*, **178**: 19 (1954). The present solution is due to J. Chakrabarty, unpublished work (1978).

Multiplying this equation by R'/h_1 and then eliminating R'/R by means of (128), we obtain

$$\sqrt{\frac{R'r}{h_1}} \left\{ \left[(0.9 + r^2)\mu \sqrt{\frac{R'r}{h_1}} - 1 \right] \exp\left[(0.9 + r^2)\mu \sqrt{\frac{R'r}{h_1}} \right] - 1 \right\} = \frac{2cr}{k} \quad (130)$$

This transcendental equation can be solved for $\mu\sqrt{R'r/h_1}$ for any assumed values of r and $2k/\mu c$. The corresponding value of $h_1/\mu^2 R$ then follows from (129). These calculations immediately furnish the ratio R'/R, and the associated roll force is then obtained from (100). The results are shown graphically in Fig. 7.47. When r is vanishingly small, $\mu\sqrt{R'r/h_1}$ approaches a limiting value equal to 1.421, obtained by equating the expression in the curly bracket of (130) to zero. This is independent of the yield stress. The minimum strip thickness and the corresponding roll force

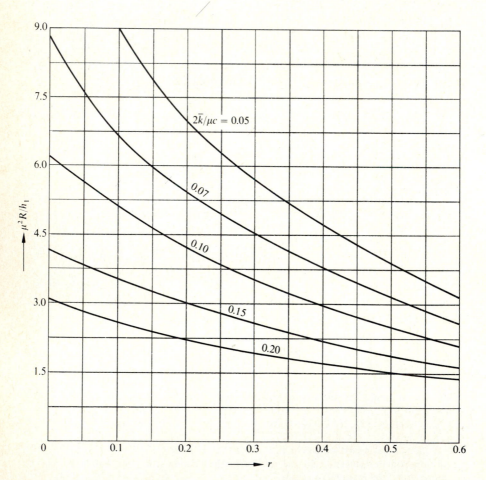

Figure 7.47 Minimum strip thickness in cold rolling as a function of the reduction, the coefficient of friction, and the mean yield strength.

for a vanishingly small reduction are

$$h_1 = 1.616\mu R \frac{2k}{c} \qquad \frac{P}{2k} = 5.273 R \frac{2k}{c}$$

A rigorous analysis of the problem when tensions are applied to the strip will be very complicated. As a rough approximation, it may be assumed that the effect of tensions is to reduce the roll force by the factor $\bar{t}/2\bar{k}$, where $\bar{t}$ is the mean of the front and back tension stresses. It is then only necessary to write $2\bar{k} - \bar{t}$ for $2\bar{k}$ in the above equations for the calculation of the minimum strip thickness. Evidently, it is possible to roll thinner gauges by applying tensions to the strip. The minimum thickness in cold rolling is found to be very sensitive to the value of the coefficient of friction, an accurate estimation of which is always a difficult problem.

The elastic deformation of the strip has a significant effect on the minimum thickness for extremely small reductions. The limiting thickness below which it is impossible to roll the strip corresponds to a completely elastic arc of contact symmetrical about the line joining the roll centers. The central section of the strip is just at the point of yielding under the action of a vertical roll pressure q_0 and a horizontal compressive stress p_0. The distribution of roll pressure is symmetrical about the center of the arc of contact, and may be approximately represented by the equation†

$$q = q_0 \cos \frac{\pi x}{2l}$$

where l is the semilength of contact, and x is measured from the center of contact. The roll force per unit width is

$$P = 2 \int_0^l q \, dx = 2q_0 \int_0^l \cos \frac{\pi x}{2l} \, dx = \frac{4q_0 l}{\pi}$$

When no tension is applied to the strip, the longitudinal equilibrium of the material on either side of the central section requires $\mu P = p_0 h$, where h is the strip thickness. The compressive normal stresses at the central section therefore become

$$p_0 = \frac{\mu P}{h} \qquad q_0 = \frac{\pi P}{4l}$$

The substitution into the yield criterion $q_0 - p_0 = 2k$ gives

$$\frac{P}{h} \left(\frac{\pi h}{4l} - \mu \right) = 2k$$

The thickness of the strip is reduced by the amount l^2/R' at the central section, giving a compressive thickness strain equal to $l^2/R'h$. In view of the plane strain

---

† A solution for the limiting thickness based on a parabolic distribution of pressure has been given by H. Ford and J. M. Alexander, *J. Inst. Met.*, **88**, 193 (1960).

condition, the elastic stress-strain relation furnishes

$$\frac{l^2}{R'h} = \frac{1 - v^2}{E}\left(q_0 - \frac{v}{1-v}p_0\right) = \frac{1-v^2}{E}\left(\frac{\pi P}{4l} - \frac{v}{1-v}\frac{\mu P}{h}\right)$$

where E and v are the usual elastic constants for the strip material. Substituting for l^2/R' given by Hitchcock's equation, and then eliminating P/h by means of the preceding equation, we get

$$\frac{l}{R}\left(\frac{\pi}{4} - \frac{\mu l}{h}\right) = (1 - v^2)\frac{8k}{\pi E_r} + \left\{(1 - v^2)\frac{\pi h}{4l} - \mu v(1 + v)\right\}\frac{2k}{E} \qquad (131)$$

Differentiating this expression with respect to l, and setting $dh/dl = 0$, the condition for the thickness to be a minimum may be written as

$$\frac{l}{R}\left(\frac{\pi}{4} - \frac{2\mu l}{h}\right) + (1 - v^2)\frac{2k}{E}\left(\frac{\pi h}{4l}\right) = 0$$

This is a cubic equation in l/h, and the solution is easily shown to be

$$\frac{l}{h} \simeq \frac{\pi}{8\mu} + \frac{8\mu R}{h}(1 - v^2)\frac{2k}{E}$$

to a close approximation. The minimum value of the strip thickness is then given by (131), the result being

$$\frac{h}{\mu R} \simeq \frac{64}{\pi^2}\left\{(1 - v_r^2)\frac{8k}{\pi E_r} + \mu(1 + v)(2 - 3v)\frac{2k}{E}\right\} \qquad (132)$$

By letting E tend to infinity, we obtain the rigid/plastic solution $h \simeq 1.62\mu R(2k/c)$, which may be compared with the result previously obtained by an entirely different approach. The neglect of the elastic distortion of the strip appreciably underestimates the minimum thickness for which rolling is possible. The above analysis is essentially unchanged if equal front and back tensions are applied to the strip, the only necessary modification being the yield stress $2k$ replaced by $2k - t$, where t is the applied tensile stress.†

7.9 Analysis of Hot Rolling

(i) The technological theory In the hot rolling process, the material is initially heated to a sufficiently high temperature at which there is a marked decrease of the yield stress. The speed of rolling has a significant effect at such elevated temperatures, the yield stress being increased by increasing the rate of deformation. The frictional resistance in hot rolling is generally so high that the tangential stress at each point of the roll surface attains the local yield stress in shear, representing a state of *sticking friction* over the entire arc of contact. In view of the incompressi-

† A sandwich technique for rolling thin hard materials has been discussed by R. R. Arnold and P. W. Whitton, *Proc. Inst. Mech. Eng.*, **173**: 241 (1959).

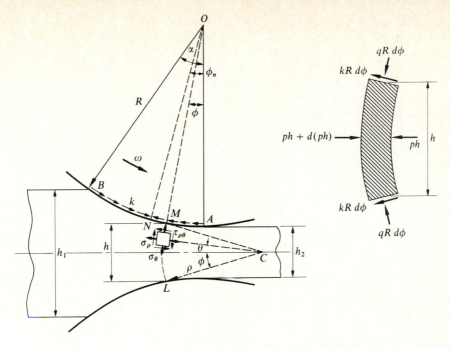

Figure 7.48 Stresses in an element in the roll gap, and the resultant forces acting on a curved slice.

bility of the material, the average longitudinal speed over a vertical section steadily increases from the plane of entry to the plane of exit. The tangential stress tends to pull the material into the roll gap on the entry side, and oppose the material being extruded out on the exit side.

Let LM be a circular arc orthogonal to the rolls and having its center C on the axis of symmetry of the rolled stock (Fig. 7.48). The radius of the arc is $\rho = \frac{1}{2}h \csc \phi$, where ϕ is the angular distance of L and M with respect to the plane of exit, and h the local thickness of the slab or strip. Introducing polar coordinates (ρ, θ) of a generic point on the arc LM, the shear stress component $\tau_{\rho\theta}$ is assumed to be proportional to the distance of the point from the axis of symmetry. Since the shear stress has the magnitude k on the roll surface, we have $\tau_{\rho\theta} = k(\sin \theta / \sin \phi)$, which furnishes

$$\sigma_\rho - \sigma_\theta = 2\sqrt{k^2 - \tau_{\rho\theta}^2} = 2k\sqrt{1 - \frac{\sin^2 \theta}{\sin^2 \phi}} \tag{133}$$

in view of the yield criterion under conditions of plane strain. If the mean horizontal pressure across LM is denoted by p, the resultant horizontal thrust transmitted across the circular arc is

$$ph = -2\int_0^\phi \rho\sigma_\rho \cos \theta \, d\theta = -h \int_0^\phi \sigma_\rho \left(\frac{\cos \theta}{\sin \phi}\right) d\theta$$

It is further assumed that σ_θ has a constant value equal to $-q$ along the arc LM, where q is the roll pressure at L and M. Substitution from (133) into the above integral then furnishes the yield condition[†]

$$q - p = \frac{\pi}{2} k$$

For small values of ϕ, the above relation is also obtained on the assumption that the stress distribution along LM is approximately the same as that in a plastic material compressed between a pair of rough platens inclined to one another at an angle 2ϕ (Sec. 8.6(vi)).

Consider now the equilibrium of a thin curved slice of the material contained between a pair of cylindrical surfaces orthogonal to the rolls at angular distances ϕ and $\phi + d\phi$ from the exit plane. Since the resultant horizontal component of the forces acting on the slice must vanish for equilibrium, we have

$$\frac{d}{d\phi}(ph) = 2R(q \sin \phi \pm k \cos \phi)$$

where R is the roll radius. The upper sign holds on the exit side and the lower sign on the entry side. Since the material is sufficiently soft in hot rolling, the elastic distortion of the rolls is generally small and is therefore neglected. Using the geometrical relationship $dh = 2R \sin \phi \, d\phi$, and substituting from the yield criterion, the above equation may be expressed as

$$\frac{dq}{2k} = \frac{\pi}{4}\left(\frac{dk}{k} + \frac{dh}{h}\right) \pm \frac{R \cos \phi \, d\phi}{h_2 + 2R(1 - \cos \phi)} \tag{134}$$

The boundary conditions are $p = 0$ and $q = \pi k/2$ at both $\phi = 0$ and $\phi = \alpha$, where α is the angle of contact. The above equation can be integrated exactly if k is assumed to have a mean value $\bar{k}$ over the arc of contact. The solution is greatly simplified with negligible error by using a slight approximation in the final result. Denoting the entry and exit thicknesses of the strip by h_1 and h_2 respectively, the distribution of roll pressure may be expressed as

$$\frac{q^+}{2\bar{k}} = \frac{\pi}{4}\left(1 + \ln \frac{h}{h_2}\right) + c\sqrt{\frac{R}{h_2}} \tan^{-1} \sqrt{\frac{R}{h_2}} \phi - \frac{1}{2}\phi$$

$$\frac{q^-}{2\bar{k}} = \frac{\pi}{4}\left(1 + \ln \frac{h}{h_1}\right) + c\sqrt{\frac{R}{h_2}}\left(\tan^{-1} \sqrt{\frac{R}{h_2}} \alpha - \tan^{-1} \sqrt{\frac{R}{h_2}} \phi\right) - \frac{1}{2}(\alpha - \phi) \tag{135}$$

where q^+ and q^- denote the pressures on the exit and entry sides respectively, and $c = 1 + (h_2/2R)$. The angular distribution of the dimensionless roll pressure given by (135) is shown by the solid lines in Fig. 7.49. Since the pressure must be

---

[†] E. Orowan, *Proc. Inst. Mech. Eng.*, **150**: 140 (1943). The derivation given here is an extension of that of Orowan.

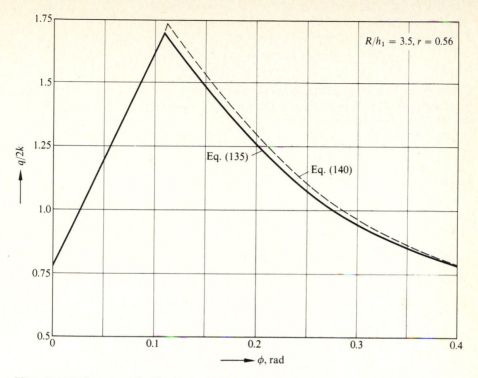

Figure 7.49 Roll pressure distribution in the hot rolling process.

continuous at the neutral point $\phi = \phi_n$, we have

$$\sqrt{\frac{R}{h_2}}\,\phi_n = \tan\left\{\frac{1}{2}\tan^{-1}\sqrt{\frac{R}{h_2}}\,\alpha - \frac{1}{2c}\sqrt{\frac{h_2}{R}}\left(\frac{\pi}{4}\ln\frac{1}{1-r} + \frac{\alpha}{2} - \phi_n\right)\right\} \quad (136)$$

This equation may be solved by successive approximation, starting with a suitable trial value of ϕ_n on the right-hand side. Accurate values of ϕ_n/α and h_n/h_2, covering a wide range of values of R/h_2 and r, are given in Table 7.7. The torque per unit width can be calculated from the formula

$$G = \bar{k}R^2\left(\int_{\phi_n}^{\alpha} d\phi - \int_0^{\phi_n} d\phi\right) = \bar{k}R^2(\alpha - 2\phi_n) \quad (137)$$

obtained by setting $R' \simeq R$ and $\mu q = \bar{k}$ in (95). The roll force per unit width may be written as

$$P = R\int_0^{\alpha} q\cos\phi\,d\phi + R\left(\bar{k}\int_{\phi_n}^{\alpha}\sin\phi\,d\phi - \bar{k}\int_0^{\phi_n}\sin\phi\,d\phi\right)$$

The expression in the parenthesis represents the contribution from the tangential stresses. Integrating the first integral by parts, and using (134) and the fact that

$h + 2R \cos \phi = 2cR$, we get

$$\frac{P}{2\bar{k}R} = \frac{\pi}{4}\left(\sin \alpha - \int_{h_2}^{h_1} \sin \phi \, \frac{dh}{h}\right) + \frac{c}{2}\left(\int_{h_n}^{h_1} \frac{dh}{h} - \int_{h_2}^{h_n} \frac{dh}{h}\right)$$

Although the first integral can be evaluated exactly, it is convenient to introduce the same approximation as that leading to (135). The result is

$$\frac{P}{2\bar{k}R} = \frac{\pi}{4}(1 + c)\sqrt{\frac{h_2}{R}}\tan^{-1}\sqrt{\frac{R}{h_2}}\alpha + c\left(\frac{1}{2}\ln\frac{1}{1-r} - \ln\frac{h_n}{h_2} - \frac{\pi}{4}\alpha\right) \quad (138)$$

where r is the fractional reduction in thickness, while

$$\alpha = \cos^{-1}\left(1 - \frac{rh_1}{2R}\right) \qquad \frac{h_n}{h_2} = 1 + \frac{2R}{h_2}(1 - \cos \phi_n) \quad (139)$$

For α less than about 30°, the roll force predicted by (138) cannot differ by more than 1 percent from that based on the exact solution.

When the angle of contact is fairly small (less than about 10°), it is sufficiently accurate to solve (134) with the approximations $\cos \phi \simeq 1$ and $2(1 - \cos \phi) \simeq \phi^2$, giving the pressure distribution†

$$\frac{q^+}{2k} = \frac{\pi}{4}\left(1 + \ln\frac{h}{h_2}\right) + \sqrt{\frac{R}{h_2}}\tan^{-1}\sqrt{\frac{R}{h_2}}\phi$$

$$\frac{q^-}{2k} = \frac{\pi}{4}\left(1 + \ln\frac{h}{h_1}\right) + \sqrt{\frac{R}{h_2}}\left(\tan^{-1}\sqrt{\frac{R}{h_2}}\alpha - \tan^{-1}\sqrt{\frac{R}{h_2}}\phi\right)$$

$$(140)$$

The broken curve of Fig. 7.49 indicates that (140) is a good approximation for the roll pressure distribution even for moderate angles of contact.‡ Since $\alpha \simeq \sqrt{rh_1/R}$, the neutral angle corresponding to (140) is given by

$$\sqrt{\frac{R}{h_2}}\phi_n = \tan\left(\frac{1}{2}\sin^{-1}\sqrt{r} - \frac{\pi}{8}\sqrt{\frac{h_2}{R}}\ln\frac{1}{1-r}\right) \quad (141)$$

This expression is also obtained from (136) by setting $c \simeq 1$, and taking $\phi_n \simeq \alpha/2$ as a first approximation. The torque is given by (137) and (141), while the formula for

---

† R. B. Sims, *Proc. Inst. Mech. Eng.*, **168**: 191 (1954). Accurate values of roll force and torque based on Sims' theory have been tabulated by E. C. Larke, *The Rolling of Strip and Sheet and Plate*, p. 363, Chapman and Hall (1963).

‡ An approximate theory based on a triangular friction hill has been put forward by E. Orowan and K. J. Pascoe, *Iron Steel Inst.*, Special Report, no. 34, p. 124 (1946). A modification of the theory has been discussed by J. M. Alexander and H. Ford, *Prager Anniversary Volume*, p. 191, Collier-Macmillan, London (1963). An approximate solution based on equilibrium across assumed shear planes has been discussed by J. W. Green and J. F. Wallace, *J. Mech. Eng. Sci.*, **4**: 136 (1962). See also J. W. Green, L. G. M. Sparling, and J. F. Wallace, ibid., **6**: 219 (1964).

Table 7.7 Neutral angle and neutral thickness ratios in hot rolling

R/h_2	r	ϕ_n/α	h_n/h_2	R/h_2	r	ϕ_n/α	h_n/h_2
5	0.15	0.4532	1.0108	50	0.05	0.4812	1.0122
	0.10	0.4285	1.0204		0.10	0.4688	1.0244
	0.20	0.3869	1.0376		0.20	0.4461	1.0498
	0.30	0.3478	1.0522		0.30	0.4222	1.0764
	0.40	0.3083	1.0640		0.40	0.3966	1.1050
	0.50	0.2668	1.0723		0.50	0.3687	1.1361
	0.60	0.2218	1.0756		0.60	0.3362	1.1699
10	0.05	0.4652	1.0114	100	0.05	0.4848	1.0124
	0.10	0.4460	1.0221		0.10	0.4740	1.0250
	0.20	0.4126	1.0426		0.20	0.4538	1.0515
	0.30	0.3803	1.0622		0.30	0.4326	1.0800
	0.40	0.3468	1.0806		0.40	0.4086	1.1113
	0.50	0.3110	1.0975		0.50	0.3830	1.1468
	0.60	0.2714	1.1118		0.60	0.3513	1.1853
25	0.05	0.4757	1.0119	150	0.05	0.4864	1.0125
	0.10	0.4613	1.0236		0.10	0.4771	1.0253
	0.20	0.4351	1.0475		0.20	0.4571	1.0522
	0.30	0.4083	1.0715		0.30	0.4367	1.0815
	0.40	0.3803	1.0966		0.40	0.4140	1.1143
	0.50	0.3496	1.1224		0.50	0.3884	1.1509
	0.60	0.3146	1.1492		0.60	0.3578	1.1922

the roll force becomes

$$\frac{P}{2\bar{k}R} = \frac{\pi}{2}\sqrt{\frac{h_2}{R}}\sin^{-1}\sqrt{r} + \frac{1}{2}\ln\frac{1}{1-r} - \ln\frac{h_n}{h_2} - \frac{\pi}{4}\sqrt{\frac{rh_1}{R}} \tag{142}$$

on substituting from (140) into (93). This result is identical to (138) with the approximations $c \simeq 1$ and $\alpha \simeq \sqrt{rh_1/R}$. It may be noted that Eq. (137) also follows from (96) with $R' \simeq R$, and the roll pressure distribution (140). The identity is due to the fact that (140) is an exact solution of (134) on the assumption $\phi \ll 1$, for which (95) and (96) are completely equivalent.

The roll force and torque calculated from (138) and (137) are shown non-dimensionally in Fig. 7.50, in terms of the parameters $P^* = 2\bar{k}R\alpha$ and $G^* = \bar{k}R\alpha^2$. When α is sufficiently small, P^* and G^* are equal to the values P and G corresponding to a normal pressure equal to $2\bar{k}$ at each point of the arc of contact. The work done per unit volume, which is equal to $2G/Rh_n$ by (97), can be readily estimated. The assumption of sticking friction throughout the arc of contact is strictly valid if the coefficient of friction μ exceeds $2/\pi \simeq 0.637$. Frictional coefficients of this order are normally expected in the hot working of metals.†

† The spread in hot rolling has been studied by A. Helmi and J. M. Alexander, *J. Iron Steel Inst.*, **206**: 1110 (1968); N. R. Chitkara and W. Johnson, *J. Basic Eng., Trans. ASME*, **88**: 489 (1966). For an application of the plane strain theory to the planetary hot rolling process, see L. G. M. Sparling, *J. Mech. Eng. Sci.*, **4**: 257 (1962). Section rolling has been discussed by Z. Wusatowski, *Fundamentals of rolling*, Pergamon Press, Oxford (1969).

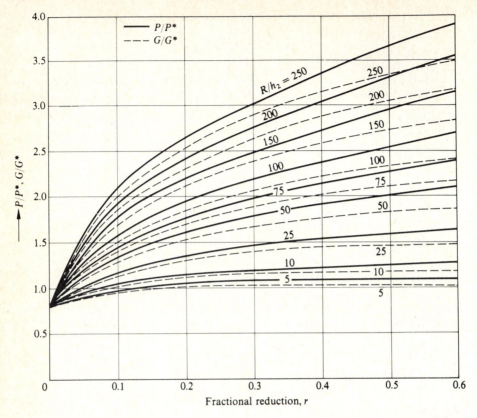

Figure 7.50 Variation of roll force and torque with reduction in hot rolling for various radius/thickness ratios.

(ii) *Influence of strain rate and temperature* The hot working of metals is characterized by the circumstance where the rate of thermal softening is sufficient to annihilate the rate of work-hardening. Except for a brief initial rise, the yield stress of the material is almost independent of the total strain. The temperature at which hot working begins is of the order of the recrystallization temperature, which depends on the rate of deformation. The absolute recrystallization temperature of different metals, under similar speeds of deformation, is roughly proportional to their absolute melting temperature. The ratio of the absolute working temperature to the absolute melting temperature is known as the homologous temperature. For usual speeds of rolling, hot working begins when the homologous temperature is somewhat greater than 0.6. If the speed of rolling is small, relatively low temperatures should produce hot working.

In general, the yield stress of the material at a given temperature depends on the strain and the strain rate, both of which vary from entry to exit. The strain rate or the rate of deformation λ is defined as the rate at which the thickness of the rolled stock decreases in relation to the current thickness in the roll gap. Thus

$$\lambda = -\frac{\dot{h}}{h} = -\frac{2R}{h}\dot{\phi}\sin\phi$$

where the dot denotes the rate of change. It is assumed for the present purpose that vertical plane sections remain plane during the motion. In view of the constancy of volume, $\dot{\phi} = -\omega(h_n \cos\phi_n)/(h\cos\phi)$, where ω is the angular velocity of the rolls. This gives

$$\lambda = 2\omega\left(\frac{Rh_n}{h^2}\right)\cos\phi_n \tan\phi \qquad (143)$$

The strain rate vanishes at the plane of exit, and its value gradually increases toward the plane of entry, attaining a maximum at an angle $\phi \simeq (3R/h_2 - 1)^{-1/2}$ to a close approximation. The maximum value of the strain rate is approximately given by

$$\lambda_0 \simeq \frac{9}{8}\omega\frac{h_n}{h_2}\sqrt{\frac{R/h_2}{3 - h_2/R}}$$

The variation of the yield stress along the arc of contact follows a similar pattern. Typical variations of λ and k through the pass are shown graphically in Fig. 7.51. In

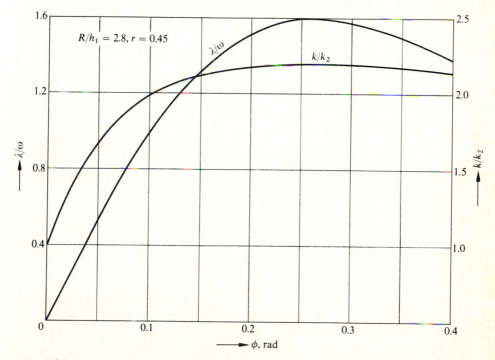

Figure 7.51 Variation of strain rate and yield stress with angular position. The curve for k/k_2 is based on a typical $(\sigma, \dot{\varepsilon})$ relation.

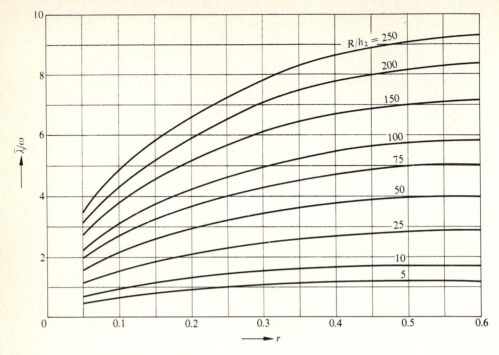

Figure 7.52 Dependence of the mean strain rate on the reduction r and the ratio R/h_2.

many cases, a useful approximation is achieved by considering a mean value of the rate of deformation. The mean strain rate $\bar{\lambda}$ in hot rolling is usually defined as

$$\bar{\lambda} = \frac{1}{\alpha} \int_0^\alpha \lambda \, d\phi = \left(\frac{\omega h_n}{\alpha} \right) \cos \phi_n \int_{h_2}^{h_1} \sec \phi \, \frac{dh}{h^2}$$

in view of (143). For practical purposes, it is sufficiently accurate to write $h_n - h_2 \simeq R\alpha^2/4$, and use an approximate mean value of $\sec \phi$ equal to $\sec \phi_n$. Then, a straightforward integration results in

$$\bar{\lambda} \simeq \omega \left(1 - \frac{3}{4} r \right) \sqrt{\frac{r}{1-r} \frac{R}{h_2}} \tag{144}$$

to a close approximation. The ratio $\bar{\lambda}/\omega$ is plotted as a function of r in Fig. 7.52 for various values of R/h_2. For given r and ω, the mean strain rate increases as the ratio R/h_2 is increased.[†]

The variation of the yield stress in hot rolling is usually determined from the stress-strain behavior of the material in simple compression. The specimen is

---

[†] The temperature distribution in hot rolling, using an upper bound approach, has been examined by W. Johnson and H. Kudo, *Int. J. Mech. Sci.*, **1**: 175 (1960). A simplified upper bound solution has been presented by R. Piispanen, R. Eriksson, and O. Piispanen, *Bleche Röhre*, **8**: 819 (1967).

compressed in a special testing machine, known as the *cam plastometer*, where the lower platen is moved upward by a logarithmic cam rotating with a constant angular velocity. The rate of straining therefore remains constant during the compression. The relationship between the yield stress Y and the strain rate $\dot{\varepsilon}$, derived from a series of stress-strain curves obtained at different strain rates, may be expressed by the empirical equation

$$Y = C\dot{\varepsilon}^n$$

where C and n are constants depending on the strain and the temperature.† Since the effective strain rate in plane strain compression is $\dot{\varepsilon} = 2\lambda/\sqrt{3}$, the plane strain yield stress $2k$ in hot rolling may be written as

$$2k = \frac{2C}{\sqrt{3}}\left(\frac{2\lambda}{\sqrt{3}}\right)^n \tag{145}$$

Introducing the local fractional reduction $e = (h_1 - h)/h_1$, and using the approximation $h - h_2 \simeq R\phi^2$ with $\phi_n \simeq \alpha/2$, the local strain rate λ may be expressed in the more convenient form

$$\lambda = \frac{\omega R(4 - 3r)}{2h_1(1 - e)^2} \tan \sqrt{\frac{(r - e)h_1}{R}}$$

Let the mean values of k required for the calculation of the roll force and the torque be denoted by $\bar{k}_P$ and $\bar{k}_G$ respectively. Following the analysis of the cold rolling process, these stresses may be defined as

$$\bar{k}_P = \frac{1}{\alpha}\int_0^\alpha k\,d\phi \simeq \frac{1}{\sqrt{r}}\int_0^{\sqrt{r}} k\,df$$

$$\bar{k}_G = \frac{1}{\delta}\int_{h_2}^{h_1} k\,dh = \frac{1}{r}\int_0^r k\,de \tag{146}$$

where δ is the draft $h_1 - h_2$, and f denotes the quantity $\sqrt{r - e}$. For practical purposes, a mean yield stress may be calculated directly from (145), using the mean strain rate $\bar{\lambda}$, and the mean coefficients

$$\bar{C} = \frac{1}{r}\int_0^r C(e)\,de \qquad \bar{n} = \frac{1}{r}\int_0^r n(e)\,de$$

These integrals can be evaluated explicitly if suitable empirical expressions are used for $C(e)$ and $n(e)$ at a mean working temperature. When the effects of temperature and strain rate are taken into account, reasonable agreements may be obtained between theory and experiment.‡

---

† The values of C and n for a number of metals at various strains and temperatures have been given by J. F. Alder and V. A. Phillips, *J. Inst. Met.*, **83**: 80 (1954).

‡ For experimental results on hot rolling, see O. Emicke and K. H. Lucas, *Sheet Met. Ind.*, **17**: 611 (1943); G. Wallquist, *J. Iron Steel Inst.*, **177**: 142 (1954); R. Stewartson, *Proc. Inst. Mech. Eng.*, **168**: 201 (1954).

(iii) *A slipline field solution* The construction of slipline fields for hot rolling is a problem of the indirect type where no initial slipline is known at the outset. The problem can be conveniently treated by the matrix method if we consider the special case when a rigid zone of material covers the entire arc of contact. The following analysis is based on the existence of such a rigid zone, together with the assumption of sticking friction over the arc of contact. The solution corresponds to roll geometries that are encountered in the first few stands of a typical hot rolling schedule. For this range of roll geometries, the technological theory of rolling is generally regarded as insufficiently accurate.

In Fig. 7.53, the rigid zone $AEFG$ rotates with the rolls with an angular velocity ω. There is a velocity discontinuity of constant amount propagating along the sliplines ABD and DFG, where FG is a circular arc of radius ρ. The velocity is therefore continuous across the entry slipline but discontinuous across the exit slipline. The material enters and leaves the roll gap with velocities represented by the vectors PC' and PQ respectively in the hodograph, where P is the pole (not shown). Since the material $AEFG$ rotates as a rigid body, the boundaries AEF and FG are mapped into geometrically similar curves $A^*E'F'$ and $F'G'$, rotated through $90°$ in the counterclockwise sense (Sec. 6.2(ii)). Thus $F'G'$ is a circular arc of radius $\omega\rho$, which is equal to the magnitude of the velocity discontinuity. The particles immediately above and below the discontinuity ABD are mapped into the curves $A'B'D'$ and $A''B''D''$ respectively, separated by a distance $\omega\rho$. The image of the roll surface AG is the circular arc A^*G' having an angular span equal to the angle of contact. Since the peripheral speed of the roll is ωR, it follows that the exit speed of the rolled stock is equal to $\omega(R + \rho)$. The fact that the rigid zone $AEFG$ is not overstressed is demonstrated by the continuation (shown broken) of the stress field into the rigid zone.† The rate of plastic work is found to be nowhere negative.

The curve AE is taken as the base slipline of the field, and is represented by the column vector $\mathbf{a}$ formed by the coefficients in the power series expansion of the radius of curvature in terms of the angular distance from A (Sec. 6.5(iii)). The angles turned through along AE, CB, and BE are denoted by θ, ψ, and ϕ respectively. Since AEC is a centered fan generated on the convex side of AE, the curve CE is represented by the vector $\mathbf{Q}_{\theta\eta}\mathbf{a}$, where $\mathbf{Q}$ is the appropriate matrix operator and $\eta = \psi + \phi$. If EF is continued to meet the axis of symmetry at J, then the curve JFE is given by $\mathbf{T}_\eta\mathbf{Q}_{\theta\eta}\mathbf{a}$, and consequently the curve FE is represented by

$$\sigma = \mathbf{S}_\phi\mathbf{T}_\eta\mathbf{Q}_{\theta\eta}\mathbf{a} \tag{147}$$

where $\mathbf{S}$ is the shift operator and $\mathbf{T}$ the smooth boundary operator. The geometrically similar curves A^*E' and $F'E'$ are evidently denoted by $\omega\mathbf{a}$ and $\omega\sigma$ respectively, while the circular arc $F'D'$ is denoted by $\omega\rho\mathbf{c}$, where $\mathbf{c}$ is the column

---

† The slipline field has been proposed by B. A. Druyanov, *Plastic Flow of Metals* (trans. from Russian), vol. I, p. 80, Plenum Pub. Corp., New York (1971). The present analysis is due to P. Dewhurst, I. F. Collins, and W. Johnson, *J. Mech. Eng. Sci.*, **15**: 439 (1973). The special case $\phi = 0$ had been considered earlier by N. R. Chitkara and W. Johnson, *Proc. 5th MTDR Conf.*, p. 391 (1965). Druyanov analyzed the same field using Riemann's method of integration.

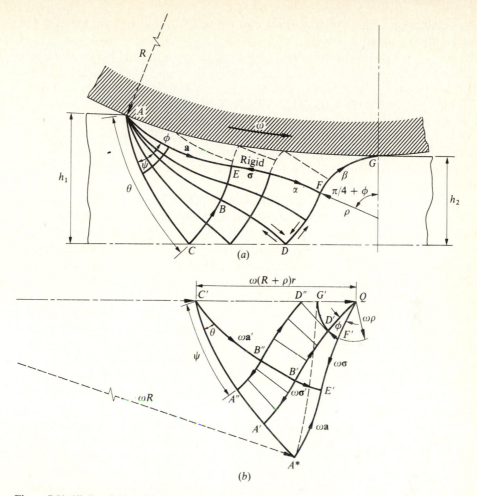

Figure 7.53 Slipline field and hodograph for hot rolling of a flat sheet.

vector $[1, 0, 0, \ldots]^T$. It follows that the vector representing the curve $B'D'$ is $\omega(\mathbf{P}_{\phi\psi}\boldsymbol{\sigma} + \rho\mathbf{Q}_{\phi\psi}\mathbf{c})$, which added to the vector $\omega\rho\mathbf{c}$ gives the curve $B''D''$. Denoting the curve $C'B''$ by the vector $\omega\mathbf{a}'$, we have

$$\mathbf{a}' = \mathbf{T}_\psi\mathbf{R}_\psi[\mathbf{P}_{\phi\psi}\boldsymbol{\sigma} + \rho(\mathbf{I} + \mathbf{Q}_{\phi\psi})\mathbf{c}] \qquad (148)$$

where $\mathbf{I}$ is the unit matrix and $\mathbf{R}$ the reversion operator. We shall subsequently use the relations $\mathbf{R}_\psi\mathbf{P}_{\phi\psi} = \mathbf{M}_\phi$ and $\mathbf{R}_\psi\mathbf{Q}_{\phi\psi} = \mathbf{N}_\phi$. If $\omega\boldsymbol{\sigma}'$ represents the curve $B'E'$, then

$$\boldsymbol{\sigma}' = \rho\mathbf{P}_{\psi\phi}\mathbf{c} + \mathbf{Q}_{\psi\phi}\boldsymbol{\sigma} \qquad (149)$$

The curve $B''A''$ is given by $\omega\mathbf{N}_\psi\mathbf{a}'$, and the curve $B'A'$ by $\omega(\mathbf{N}_\psi\mathbf{a}' + \rho\mathbf{c})$. Expressing the vector representation of A^*E' in terms of those of $B'E'$ and $B'A'$, we have

$$\mathbf{a} = \mathbf{P}_{\phi\theta}(\mathbf{N}_\psi\mathbf{a}' + \rho\mathbf{c}) + \mathbf{Q}_{\phi\theta}\boldsymbol{\sigma}'$$

The substitution from (148) and (149) into the last relation, and the use of (147), lead to the matrix equation

$$(\mathbf{I} - \mathbf{D})\mathbf{a} = \rho\mathbf{Ec} \tag{150}$$

where

$$\mathbf{D} \equiv (\mathbf{Q}_{\phi\theta}\mathbf{Q}_{\psi\phi} + \mathbf{P}_{\phi\theta}\mathbf{N}_{\psi}\mathbf{T}_{\psi}\mathbf{M}_{\phi})\mathbf{S}_{\phi}\mathbf{T}_{\eta}\mathbf{Q}_{\theta\eta}$$

$$\mathbf{E} = \mathbf{P}_{\phi\theta}[\mathbf{I} + \mathbf{N}_{\psi}\mathbf{T}_{\psi}(\mathbf{R}_{\psi} + \mathbf{N}_{\phi})] + \mathbf{Q}_{\phi\theta}\mathbf{P}_{\psi\phi} \tag{151}$$

The problem of finding the initial slipline, denoted by **a**, is therefore reduced to a simple matrix inversion. Calculations indicate that the initial slipline does not differ significantly from a circular arc.

For any assumed value of ρ/h_1, the slipline field is completely defined by the three angles θ, ψ, and ϕ, which furnish the three fundamental parameters R/h_1, h_2/h_1 and the back tension. The mean compressive stress p_0 at G is determined from the condition of zero resultant front tension across DFG. For a given angle ψ, the values of θ and ϕ corresponding to a zero back tension must be found before the field can be considered as valid for the hot rolling process. Only those values of the field angles for which h_2/h_1 is less than unity are relevant to the process. Once the distribution of hydrostatic pressure has been found by using the Hencky equations, the roll force and torque can be calculated by numerical integration along the boundary $AEFG$. Due to the presence of the rigid zone, it is not possible to ascertain the precise distribution of roll pressure over the arc of contact.

The largest value of R/h_1 corresponding to a given reduction is attained when $\phi = 0$. The matrix equation for the fundamental vector **a** in this limiting situation is easily shown to be

$$(\mathbf{I} - \mathbf{Q}_{\psi\theta}\mathbf{T}_{\psi}^2\mathbf{Q}_{\theta\psi})\mathbf{a} = \rho(\mathbf{I} + \mathbf{Q}_{\psi\theta}\mathbf{T}_{\psi})\mathbf{c} \tag{152}$$

As ϕ increases from zero, θ also increases but ψ decreases. The ratios R/h_1 and ρ/h_1 also decrease until the circular arc FG degenerates into a single point and the velocity discontinuity disappears. The matrix equation (150) then becomes homogeneous, and the vector **a** reduces to the *eigenvector* of the matrix **D** corresponding to a unit eigenvalue. This extreme situation occurs, however, for values of R/h_1 that are too small to be of any practical interest.

For a given reduction, the normal roll pressure p_0 at G decreases with decreasing roll radius and becomes zero before the eigencase $\rho = 0$ is reached. The solution cannot be valid for values of R/h_1 less than that corresponding to $p_0 = 0$. The computed values of the roll force are plotted in Fig. 7.54 as solid lines extending between the limits $\phi = 0$ and $p_0 = 0$. The broken curves are based on (138) and (139) and are included here for comparison. The difference between the two solutions, as far as the roll force is concerned, is about 10 percent over the range of validity of the field. The torque values given by the slipline field are, however, found to be in close agreement with those obtained from the formula (137). The ratio ρ/R increases appreciably with the reduction, but decreases only slightly with the roll radius for a given reduction. Since the roll pressure is quite small near the exit point, the tangential stress near the exit will not satisfy the

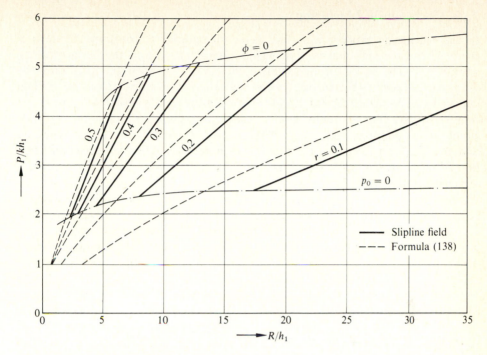

Figure 7.54 Comparison of slipline field and technological theories of hot rolling in relation to the roll separating force.

condition of sticking friction. Nevertheless, the estimated roll force and torque should not be significantly in error.†

When the ratio R/h_1 exceeds that given by the limit $\phi = 0$, the slipline field modifies in such a way that there is a plastically deforming region bordering on the roll surface. The problem then becomes nonlinear in nature, and the construction of the slipline field requires a prohibitive amount of trial and error. The shape of the initial slipline cannot be determined by the process of matrix inversion. Each half of the slipline field involves a central rigid zone having its apex on the axis of symmetry, and a narrow rigid zone extending between the point of entry and an intermediate point on the roll surface. The field on the entry side consists of a singular domain defined by a curved entry slipline, and several regular domains in one of which the sliplines are normal and tangential to the roll surface. The exit slipline is also curved and defines a regular field in which the arc of contact is again a limiting line.‡

---

† A slipline solution for asymmetrical hot rolling has been discussed by I. F. Collins and P. Dewhurst, *Int. J. Mech. Sci.*, **17**: 643 (1975).

‡ Such a solution has been discussed by J. M. Alexander, *Proc. Inst. Mech. Eng.*, **169**: 1021 (1955). Approximate constructions of the field have been considered by F. A. A. Crane and J. M. Alexander, *J. Inst. Met.*, **96**: 289 (1968).

7.10 Mechanics of Machining

In the machining process, a surface layer of material of constant thickness is removed by a wedge-shaped tool which travels parallel to the surface of the workpiece.† The speed of the relative movement between the tool and the workpiece, known as the cutting speed, normally ranges from 1 to 10 m/s. When cutting in unlubricated conditions and at low cutting speeds (less than 0.5 m/s), the removed metal chip is usually discontinuous, being made up of small segments produced by periodic fracture.‡ Extremely brittle materials, such as magnesium and gray cast iron, form discontinuous chips under all practical cutting conditions. Sometimes, a cap of dead metal builds up around the cutting edge from which it separates out at intervals. The formation of built-up edges is apparently restricted to metals containing two or more phases.§ When the material is ductile and the tool face lubricated, the chip forms a continuous coil, and the process may be considered as one of steady state. We shall be concerned here with the steady orthogonal machining in which the cutting edge of the tool is perpendicular to the direction of its relative motion. This condition is satisfied in some planing and broaching operations, and is approximately realized in others, such as lathe turning. Many of the results for the two-dimensional theory may be applied qualitatively to three-dimensional machining processes.¶

(i) *Basic concepts and early theories* It is assumed that the depth of cut is small compared to its width so that the deformation occurs essentially in plane strain. There is sufficient experimental evidence to indicate that at normal cutting speeds the plastic deformation is confined in a narrow region springing from the cutting edge of the tool. It is therefore reasonable to assume that the deformation consists of simple shear across a plane AB inclined at an angle ϕ to the direction of cutting (Fig. 7.55). In a real material, the *shear plane* is the limit of a narrow zone in which the velocity changes rapidly but continuously. The angle α which the tool face

---

† The basic physical principles of the machining process have been discussed by M. C. Shaw, *Metal Cutting Principles*, Clarendon Press, Oxford (1984). For relevant metallurgical details, see E. M. Trent, *Metal Cutting*, Butterworth Publishing Company, London (1977). A great deal of useful theoretical and experimental work on machining has been reported by N. N. Zorev, *Metal Cutting Mechanics* (trans. from Russian), Pergamon Press, Oxford (1966). For a comprehensive review on the subject, see T. H. C. Childs and G. W. Rowe, *Report on the Progress in Physics*, The Institute of Physics, London, **36**: 223–288 (1973). See also, A. Bhattacharya and I. Ham, *Design of Cutting Tools*, Society of Manufacturing Engineers, Dearborn, Mich. (1969).

‡ The discontinuous process has been studied theoretically by E. H. Lee, *J. Appl. Mech.*, **76**: 189 (1954), and experimentally by N. H. Cook, I. Finnie, and M. C. Shaw, *Trans. ASME*, **76**: 153 (1954). See also J. Bannerjee and W. B. Palmer, *Proc 6th MTDR Conf.*, Pergammon Press, Oxford (1966).

§ The formation of built-up edges has been extensively studied by E. M. Trent, *J. Inst. Prod. Eng.*, **38**: 105 (1959); and by W. B. Heginbotham and S. L. Gogia, *Proc. Inst. Mech. Eng.*, **175**: 892 (1961). See also H. Takayama and T. Ono, *J. Eng. Ind.*, **90**: 335 (1968).

¶ The mechanics of three-dimensional machining has been considered by G. V. Stabler, *Proc. Inst. Mech. Eng.*, **165**: 14 (1951); M. C. Shaw, N. H. Cook, and P. A. Smith, *Trans. ASME*, **74**: 1055 (1952); E. Usui, M. Masuko, and A. Hirota, *J. Eng. Ind., Trans. ASME*, **100**: 222 (1978); E. Usui and A. Hirota, ibid., **100**: 229 (1978).

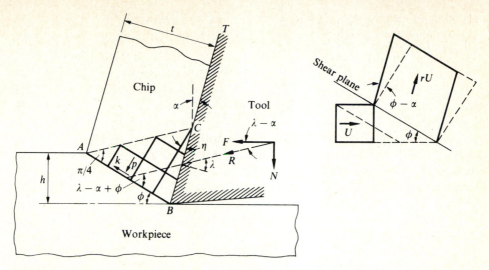

Figure 7.55 Geometry of orthogonal machining with a possible slipline field.

makes with the normal to the machined surface is known as the *rake angle*, reckoned positive when the tool face is inclined as shown. The angle between the rake face BT and the shear plane AB is evidently $\pi/2 - \phi + \alpha$. The shear angle ϕ can be found experimentally from the measurement of the chip thickness ratio r, which is the ratio of the depth of cut h to the chip thickness t. Thus

$$r = \frac{h}{t} = \frac{\sin \phi}{\cos (\phi - \alpha)}$$

$$(153)$$

or

$$\tan \phi = \frac{r \cos \alpha}{1 - r \sin \alpha}$$

Let the workpiece be considered as moving relative to the cutting tool with a uniform velocity U. Continuity requires that the velocity of the chip sliding up the tool face is rU. It follows from the geometry of the shearing process (inset of Fig. 7.55) that the engineering shear strain experienced by the material is

$$\gamma = \cot \phi + \tan (\phi - \alpha) = \frac{\cos \alpha}{\sin \phi \cos (\phi - \alpha)} \tag{154}$$

For a given rake angle, the shear strain has a minimum value of $2 \tan (\pi/4 - \alpha/2)$ corresponding to $\phi = \pi/4 + \alpha/2$, for which AB bisects the angle between the rake face and the machined surface. Machining involves very large amounts of shear strain, values of γ greater than 3.0 being fairly common.

It is assumed that the frictional condition at the chip-tool interface can be represented by a mean coefficient of friction $\mu = \tan \lambda$, where λ is the corresponding angle of friction. The direction of the resultant tool force R therefore makes an angle λ with the normal to the rake face. The component of the resultant force in the direction of cutting is known as the *cutting force*, denoted by F per unit width. If the component normal to the cutting direction per unit width is N, then from geometry,

$$N = F \tan (\lambda - \alpha)$$

or
$$\tan \lambda = \frac{N + F \tan \alpha}{F - N \tan \alpha} \tag{155}$$

Thus λ can be found from experimentally measured values of F and N. For a given material, λ decreases as the cutting speed is increased. Since μ generally varies along the rake face, only an average value of λ can be found. For an isotropic material, the shear stress across AB is equal to k, and the normal stress across it is uniformly distributed.† Since the resultant force transmitted across the shear plane makes an angle $\lambda - \alpha + \phi$ with the direction of the shear stress on it, the normal pressure across the shear plane is

$$p = k \tan (\lambda - \alpha + \phi) \tag{156}$$

The cutting force is most conveniently obtained by considering the normal and shear forces acting on the shear plane. Resolving in the cutting direction, we have

$$F = h(p + k \cot \phi) \tag{157}$$

Since the cutting speed is denoted by U, the external work done per unit time per unit width of cut is FU, and the corresponding volume of metal removed is hU. Hence the work done per volume is F/h. In order to relate ϕ to α and λ, Merchant assumed that ϕ is such that the work done per unit volume is a minimum.‡ Equations (156) and (157) then furnish

$$\cos (\lambda - \alpha + \phi) = \sin \phi$$

or
$$\phi = \frac{\pi}{4} - \frac{1}{2} (\lambda - \alpha) \tag{158}$$

This optimum value of ϕ gives $F = 2kh \cot \phi$, the normal pressure p on the shear plane being $k \cot \phi$. Experimental results indicate that although ϕ is approximately a linear function of $\lambda - \alpha$, Eq. (158) tends to overestimate the shear angle quite appreciably,§ except for small values of $\lambda - \alpha$.

---

† The distribution of normal and shear stresses across the shear plane in a real material has been tentatively discussed by M. C. Shaw, *Int. J. Mech. Sci.*, **22**: 673 (1980).

‡ See H. Earnst and M. E. Merchant, *Trans. Am. Soc. Met.*, **29**: 299 (1941); and M. E. Merchant, *J. Appl. Phys.*, **16**: 267 and 318 (1945). A minimum energy criterion, based on a constant frictional force (rather than an angle of friction) has been discussed by G. W. Rowe and P. T. Spick, *J. Eng. Ind., Trans. ASME*, **89**: 530 (1967).

§ Merchant (op. cit.) attempted to improve the agreement by assuming k to be a function of p. This, however, is inconsistent with the observed plastic behavior of ductile metals. See, for example, J. M. Alexander and R. C. Brewer, *Manufacturing Properties of Materials*, chap. 7, Van Nostrand, London (1963).

Lee and Shaffer applied† the theory of slipline fields to derive the required expression for the shear angle. They assumed a triangular plastic region ABC above the shear plane with sliplines parallel and perpendicular to AB (Fig. 7.55). If the nonplastic material above AC is assumed stress-free, no stress is transmitted across AC which is therefore inclined at an angle $\pi/4$ with the sliplines. There is a hydrostatic pressure jump of amount k across AC. Since the stress is uniform in the plastic region, the normal pressure on the shear plane is $p = k$. It immediately follows from (156) that

$$\phi = \frac{\pi}{4} - (\lambda - \alpha) \tag{159}$$

which means that AC is parallel to the line of action of the resultant tool force. Hence the α lines are inclined to the tool face at an angle

$$\eta = \frac{\pi}{4} - \lambda$$

This result may be otherwise obtained from the fact that the ratio of the shear stress to the normal pressure on the tool face is equal to $\tan \lambda$. The cutting force becomes

$$F = kh(1 + \cot \phi)$$

Since the velocity on either side of the plastic zone corresponds to rigid body motion, the Geiringer equations indicate that the plastic field ABC also moves as a rigid body. The deformation occurs, therefore, entirely across the shear plane AB. When $\lambda = \pi/4$, the frictional stress on the tool face is equal to k, the maximum which the chip material can transmit. For larger values of λ, shear flow will occur at the tool-chip interface, requiring $\phi = \alpha$. A serious limitation of (159) is that the predicted shear angle can vanish under practical conditions, implying an infinite cutting force. The solution is not valid for $\lambda < \alpha$, since the material in the corner A below the shear plane is then overstressed.‡

(ii) *Solution considering work-hardening* A realistic solution for the machining problem should take into account the ability of the material to work-harden. If the shear plane approximation is still retained for simplicity, it must be supposed that the yield stress is discontinuous across the shear plane. Since the chip material must be considerably harder than the workpiece, it is reasonable to envisage a state of stress in which a part of the workpiece adjacent to the shear plane is brought to the yield point, the material above the shear plane being nonplastic. The strain-

---

† E. H. Lee and B. W. Shaffer, *J. Appl. Mech.*, **18**: 405 (1951). These authors have also presented a solution for a built-up nose involving an unknown coefficient of friction between the bottom of the nose and the machined surface.

‡ R. Hill, *J. Mech. Phys. Solids*, **3**: 47 (1954), considering the possibility of stress singularities occurring at A and B, has presented a range of permissible values of ϕ. Merchant's solution lies outside this range, and Lee and Shaffer's solution forms one boundary of it. Unfortunately, there are many experimental values of ϕ outside Hills' permitted range.

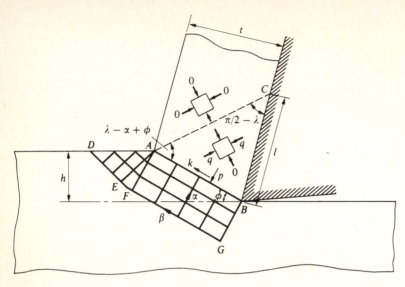

Figure 7.56 An alternative field, applicable to work-hardening materials ($\phi \leqslant \pi/4$).

hardening of the material and the subsequent elastic unloading would occur in a narrow transition zone, the limit of which is assumed to coincide with the shear plane.† Across this plane, not only the yield stress but also the maximum shear directions become discontinuous (Sec. 6.1(iv)).

The slipline field $BADEFG$ shown in Fig. 7.56 holds for $\phi \leqslant \pi/4$, and consists of two regions of constant stress separated by a fan of angular span $\pi/4 - \phi$. In the triangular region ADE, the sliplines meet the stress-free surface AD at $\pi/4$, while in the region $ABGF$ the sliplines are parallel and perpendicular to AB. It follows from the Geiringer equations and the velocity boundary conditions that no deformation occurs in the assumed plastic region, and consequently there is no strain-hardening here either. Since the hydrostatic pressure along the stress-free boundary must be equal to the initial shear yield stress k, the normal pressure acting on the shear plane AB is

$$p = k\left(1 + \frac{\pi}{2} - 2\phi\right) \qquad \phi \leqslant \frac{\pi}{4} \tag{160a}$$

When ϕ exceeds $\pi/4$, the centred fan AEF must be replaced by a stress discontinuity inclined at an angle $3\pi/8 - \phi/2$ with AD. The angle made by the α lines with the discontinuity is $\phi/2 - \pi/8$, giving the pressure

$$p = k\left[1 - 2\sin\left(\phi - \frac{\pi}{4}\right)\right] \qquad \phi \geqslant \frac{\pi}{4} \tag{160b}$$

† There is usually a narrow zone of secondary shear next to the rake face where large frictional forces cause local plastic deformation of the chip. This is a boundary phenomenon observed in several metal working process, and cannot be regarded as peculiar to metal cutting.

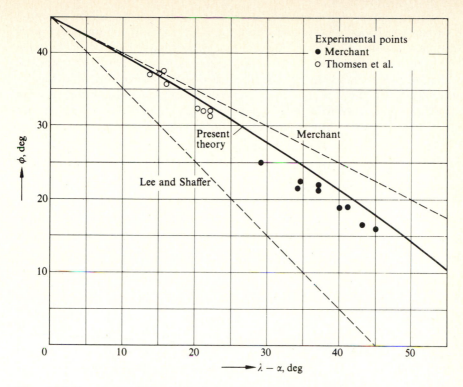

Figure 7.57 Shear-angle relationship in the orthogonal machining of metals.

The extent of the assumed plastic zone is indeterminate. It is easy to show that the nonplastic material in the corner B is able to support the tractions along the slipline BG. Equations (156) and (157) continue to hold with k denoting the initial yield stress. It follows from (156) that†

$$\tan^{-1}(p/k) - \phi = \lambda - \alpha \tag{161}$$

where p/k is now given by (160) as a function of the shear plane angle. The variation of ϕ with $\lambda - \alpha$ for $\lambda \geqslant \alpha$ is represented by the solid curve in Fig. 7.57. This curve lies between the broken straight lines given by (158) and (159), the former being tangential to the curve at $\phi = 45°$, which corresponds to $\lambda = \alpha$ in all the three theories. The experimental points‡ included for comparison indicate that the shear angle relationship given by (160) and (161) is a definite improvement over the other two for a range of values of λ and α.

† J. Chakrabarty, *Proc. Int. Conf. Prod. Eng.*, New Delhi (1977).

‡ M. E. Merchant, *J. Appl. Phys*, **16**: 267 (1945); and E. G. Thomsen, J. T. Lapsley, and R. C. Grassie, *Trans. ASME*, **75**: 591 (1953). Useful experimental data on machining have also been reported by D. Kececioglu, *Trans. ASME*, **80**: 158 (1958). See also E. G. Thomsen, C. Y. Yang, and S. Kobayashi, *Mechanics of Plastic Deformation in Metal Processing*, Macmillan, New York (1965)

A statically admissible stress field in the chip can be constructed by assuming that a triangular region ABC is under a uniform compression q in the direction of the resultant tool force. AC is a line of stress discontinuity inclined in the same direction, the material above AC being assumed stress-free. The normal pressure and the shear stress acting on the tool face are $q \cos^2 \lambda$ and $q \sin \lambda \cos \lambda$ respectively, satisfying the required frictional condition. The stress is also discontinuous across AB (except for $\phi = \pi/4$), which is not a slipline for the chip material. The condition of continuity of the normal and shear stresses across AB furnishes

$$q = 2k \operatorname{cosec} 2(\lambda - \alpha + \phi)$$

The yield criterion will not be violated in ABC if $q \leqslant 2k'$, where k' is the shear yield stress of the work-hardened material above the shear plane, its magnitude depending on the shear strain as well as on the mean strain rate and temperature. Using (156), the condition for validity of the solution may be expressed as

$$\frac{k'}{k} \geqslant \frac{1}{2}\left(\frac{p}{k} + \frac{k}{p}\right) \tag{162}$$

As ϕ decreases from $\pi/4$, the right-hand side of (162) increases from unity, approaching the value 1.48 when ϕ tends to zero. The right-hand side also increases as ϕ increases from $\pi/4$, but its value is less than 1.5 for $\phi < 63°$ or $\lambda - \alpha > -42°$. In view of the large amount of shear strain involved in the process, the inequality (162) will be satisfied in a wide range of practical situations. The minimum yield stress ratio k'/k, determined from (160) and (162) as a function of ϕ, is plotted against $\lambda - \alpha$ in Fig. 7.58. The specific cutting force F/kh, obtained from (157) and (160), is also displayed in the same figure as a function of $\lambda - \alpha$.

When the inequality (162) is violated by (160), the region ABC must be stressed to the yield point with the α lines inclined at an angle $\pi/4 - \lambda$ to the rake face. The workpiece is then completely nonplastic. The shear stress across AB is still equal to k, but the normal pressure on AB is modified so as to satisfy the yield criterion $q = 2k'$. The result is the equality in (162), giving

$$\frac{p}{k} = \frac{k'}{k} \pm \sqrt{\left(\frac{k'}{k}\right)^2 - 1} \tag{163}$$

where the upper sign holds for $\lambda > \alpha$ and the lower sign for $\lambda < \alpha$. The shear plane angle is now given by (161) and (163), and the cutting force by (157) and (163). Since k'/k cannot be determined before ϕ is known, a process of trial and error will be necessary to obtain consistent values of these quantities. For a given value of k'/k, the shear angle relationship in this range is represented by a straight line drawn parallel to the lower broken line of Fig. 7.57, and terminating at a point on the solid curve.†

---

† Work-hardening has been considered in a semiempirical manner, taking a finite width of the shear zone, by W. B. Palmer and P. L. B. Oxley, *Proc. Inst. Mech. Eng.*, **173**: 623 (1959). See also R. N. Roth and P. L. B. Oxley, *J. Mech. Eng. Sci.*, **14**: 85 (1972).

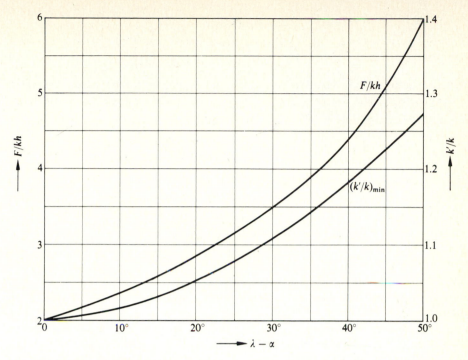

Figure 7.58 Dimensionless cutting force and minimum yield stress ratio for machining of work-hardening materials.

In the machining process, there is so little constraint on the flow of metal that the initial conditions may be expected to influence the final steady state. There can be many theoretically possible steady state solutions, each one involving a kinematically admissible mode of deformation for the cutting operation.† A kinematically valid solution giving a lower cutting resistance cannot, however, be regarded as more acceptable for the prediction of the shear plane angle, even when work-hardening is absent, since the limit theorems do not apply to such quantities.

(iii) *Temperature and strain rate* The energy expended in the machining process, for producing plastic deformation and overcoming frictional resistance, is mostly converted into heat. A part of the heat generated in the shear zone flows into the workpiece and a part is carried away into the chip. Since the rate of total energy consumption per unit width of cut is equal to FU, the rate at which energy is dissipated in the shear zone is $FU - F'(h/t)U$, where F' is the frictional force along the chip-tool interface. The fact that the line of action of the resultant tool force makes an angle $\lambda - \alpha$ with the direction of the cutting force F (Fig. 7.55), immediately gives $F' = F \sin \lambda / \cos (\lambda - \alpha)$. The rate of heat generation per unit

---

† See, for example, P. Dewhurst, *Proc. R. Soc.* (*London*), Ser. A, **360**: 587 (1978).

width due to the shearing therefore becomes

$$\dot{Q} = FU \left\{ 1 - \frac{\sin \lambda \sin \phi}{\cos (\lambda - \alpha) \cos (\phi - \alpha)} \right\}$$

on inserting from (153). To obtain the chip temperature, let β denote the proportion of the generated heat that is convected into the chip. The average temperature rise of the material leaving the shear zone may then be written as

$$\Delta T = \frac{\beta \dot{Q}}{hU\rho c} = \frac{\beta F}{h\rho c} \left\{ 1 - \frac{\sin \lambda \sin \phi}{\cos (\lambda - \alpha) \cos (\phi - \alpha)} \right\} \tag{164}$$

where ρ is the density and c the specific heat of the workpiece material. Available theoretical and experimental results[†] on this problem suggest that β is given with sufficient accuracy by the empirical equation

$$\beta = 0.5 + 0.13 \ln \left(\frac{hU\rho c}{\kappa} \tan \phi \right) \tag{165}$$

where κ is the thermal conductivity of the workpiece. For any given values of λ and α, the temperature rise can be computed with an appropriate value of the shear plane angle ϕ.

The temperature distribution in the chip has been obtained analytically on the simplifying assumption[‡] that there is a uniform heat source along the chip-tool interface with no heat loss through the remaining chip surface. The maximum temperature rise $\Delta T'$ in the chip then occurs at the point where the chip loses contact with the tool, and is given by

$$\Delta T' = 1.13 \frac{F'}{t} \sqrt{\frac{hUt}{\kappa \rho c l}}$$

where l is the length of the tool-chip contact. The total temperature rise of the material as it leaves the tool is $\Delta T + \Delta T'$, whose value is usually between 600 and 1000°C. The substitution for F'/t furnishes

$$\Delta T' = \frac{1.13 F \sin \lambda \sin \phi}{h \cos (\lambda - \alpha) \cos (\phi - \alpha)} \sqrt{\frac{hUt}{\kappa \rho c l}} \tag{166}$$

The ratio l/t may be approximately estimated from the configuration of Fig. 7.56, in which the contact pressure between the chip and the tool vanishes at all points beyond C. It is easily shown that

$$l/t = \tan \lambda + \tan (\phi - \alpha)$$

[†] An analytical solution of the problem of partition of heat between the chip and the workpiece has been presented by J. H. Weiner, *Trans. ASME*, **77**: 1331 (1955). The problem has been studied experimentally by G. Boothroyd, *Proc. Inst. Mech. Eng.*, **177**: 789 (1963).

[‡] A. C. Rapier, *Br. J. Appl. Phys.*, **5**: 400 (1954). The temperature distribution in the tool due to frictional heating along the rake surface has been calculated numerically by B. T. Chao and K. J. Trigger, *Trans. ASME*, **77**: 1107 (1955), and **80**: 311 (1958).

The strain rate occurring in the shear zone cannot be estimated on the basis of the shear plane model, since the thickness of the shear zone is assumed to vanish. Detailed measurement of plastic flow in this region has revealed that the maximum shear strain rate varies markedly across the shear zone with a peak value attained at the position of the shear plane. The mean value of the shear strain rate is found to be given with reasonable accuracy by the empirical relation[†]

$$\dot{\gamma} = C \, \Delta v / b$$

where Δv is the magnitude of the overall change in velocity in the shear zone, b is the length of the shear zone, and C a constant whose value generally lies between 6 and 10. Since Δv must be very closely equal to the velocity discontinuity $U \cos \alpha / \cos (\phi - \alpha)$ across the shear plane, we obtain the approximate formula

$$\dot{\gamma} = \frac{8U \cos \alpha \sin \phi}{h \cos (\phi - \alpha)} \tag{167}$$

For normal cutting speeds and depths of cut, the mean strain rate in machining lies between 10^4 to $10^5/\text{s}$. When the mean values of the strain rate and temperature have been calculated, the mean yield stress k' of the chip material next to the shear plane can be found from an experimentally determined variation of the flow stress $\sqrt{3} \, k'$ of the fully hardened material with the velocity modified temperature

$$T_m = T\left(1 - m \ln \frac{\dot{\gamma}}{\sqrt{3}} \right)$$

where m is an empirical constant and T the absolute temperature. For practical purposes, it is reasonable to assume a linear variation of the yield stress with T_m over the relevant range of values of $\dot{\gamma}$ and T.

(iv) *Machining with sticking friction* A possible slipline field[‡] for the machining of an ideally plastic material, when the frictional stress over the tool face is equal to the yield stress in shear, is shown in Fig. 7.59. The field begins with the centered fan ABC, where AC is perpendicular to the tool face, and AB is inclined at some angle ψ to AC. The circular arc BC and the frictional condition along CE uniquely define the remaining domain BCE. The broken line AD is a line of stress discontinuity, above which the material is stress free. The equilibrium of the isosceles rigid triangle ACD requires that the normal pressures on AC and CD are each equal to k. By Eq. (121), Chap 6, the normal component of the force exerted on CE per unit width for a nonhardening material is

$$Q = k\{l + 2t[I_0(2\psi) - 1]\} \tag{168}$$

[†] M. G. Stevenson and P. L. B. Oxley, *Proc. Inst. Mech. Eng.*, **184**: 561 (1970). For an approximate analysis based on this relation, see W. F. Hastings, P. L. B. Oxley, and M. G. Stevenson, ibid., **188**: 245 (1974).

[‡] H. Kudo, *Int. J. Mech. Sci.*, **7**: 43 (1965), J. Chakrabarty, ibid., **21**: 477 (1979).

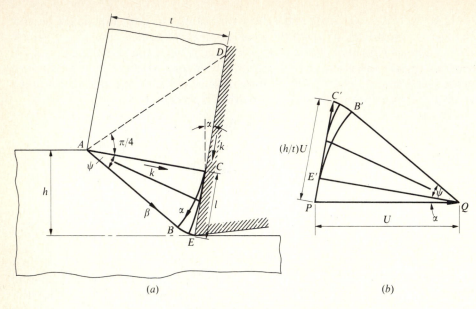

Figure 7.59 Machining with sticking friction. (a) Slipline field; (b) hodograph.

where l denotes the dimension CE, and is given by the relation

$$l = t[A_0(2\psi) - I_1(2\psi)]$$

From geometry, the depth of cut h is equal to $l \cos \alpha + t \sin \alpha$, and the substitution from above gives

$$h = t\{\sin \alpha + [A_0(2\psi) - I_1(2\psi)] \cos \alpha\} \qquad (169)$$

The normal and tangential components of the resultant tool force per unit width are $Q + kt$ and $k(l + t)$ respectively. Hence the horizontal cutting force is

$$F = kl(\cos \alpha + \sin \alpha) + kt\{[2I_0(2\psi) - 1] \cos \alpha + \sin \alpha\}$$

in view of (168). Substituting for l and using (169), we obtain (after some algebraic manipulation) the result

$$\frac{F}{kh} = 1 + \tan \alpha + \frac{2I_0(2\psi) - \sec^2 \alpha}{A_0(2\psi) - I_1(2\psi) + \tan \alpha} \qquad (170)$$

The solution is kinematically admissible, since the slipline field is associated with a consistent hodograph (Fig. 7.59). Assuming the tool to be at rest, the velocity of the workpiece is represented by the vector PQ having a magnitude U. There is a velocity discontinuity of amount $U \cos \alpha$ across the slipline ABE, which is therefore mapped into the circular arc $B'E'$ in the hodograph. The field $E'B'C'$ is similar to the field CBE, the vector PC' representing the chip velocity with magnitude $(h/t)U$. The velocity is constant along each straight slipline of the centered fan ABC, which is mapped into the curve $B'C'$.

Table 7.8 Numerical data for matching with a perfectly rough tool

α	$0°$	$10°$	$20°$	$30°$	$40°$	$45°$
ψ	$39.8°$	$30.4°$	$22.2°$	$13.8°$	$5.2°$	$0°$
t/h	1.331	1.373	1.397	1.409	1.413	1.414
F/kh	3.77	3.31	2.91	2.54	2.17	2.00

The angle ψ may be determined by minimizing the cutting force correspond-ing to given rake angle and depth of cut. In view of the relations $I_0'(z) = I_1(z)$, $A_0'(z) = I_0(z)$, and $zI_0(z) = d(zI_1)/dz$, the condition $dF/d\psi = 0$ furnishes the result

$$I_0(2\psi) - 2\psi[A_0(2\psi) - I_1(2\psi) + \tan\alpha] = \tfrac{1}{2}\sec^2\alpha \qquad (171)$$

When ψ has been calculated from (171) for any given α, the cutting ratio t/h and the specific cutting force F/kh can be calculated from (169) and (170) respectively. The results for various rake angles are given in Table 7.8. As α tends to $\pi/4$, the angle ψ tends to zero and the ratio t/h tends to $\sqrt{2}$, while the plastic zone degenerates into a single shear plane perpendicular to the tool face.

(v) *Solution for curled chips* In the continuous chip formation, the chip is almost invariably found to be curled. The sheared material slides over the rake face of the tool and finally curls away, thereby breaking contact with the tool at a distance l from its tip. A possible slipline field and the corresponding hodograph for the curly chip formation in an ideally plastic material are shown in Fig. 7.60. The defor-mation is assumed to occur in the region BCD adjacent to the tool face. The

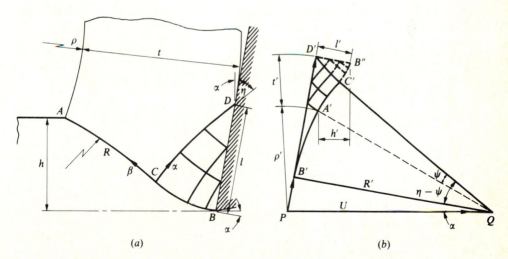

(a)   (b)

Figure 7.60 Formation of curled chips in machining. (a) Slipline field; (b) hodograph.

work-chip interface ACB is a curve across which the velocity is discontinuous. The material above ACD undergoes rigid body rotation, relative to the tool, about the center of curvature of the curled chip. Then AC must be a circular arc in order that the motion of the rigid chip is compatible with its relative sliding over the workpiece.

The conditions of equilibrium of the chip can be satisfied by taking the plastic/rigid boundary CD as a circular arc identical to CA. Since the distribution of normal and shear stresses along CA and CD are then also identical, their resultant must be a single force acting along the bisector of the right angle at C. Using Hencky's pressure equation along either curve, the resultant force can be expressed in terms of the hydrostatic pressure p_0 at D (or A), the angular span ψ, and the radius R of each of the circular arcs. Equating this resultant force to zero, the relationship between p_0 and ψ is obtained as[†]

$$\frac{p_0}{k} = 1 + \frac{2(\psi - \sin \psi)}{\cos \psi + \sin \psi - 1} \tag{172}$$

Since ψ is small over the relevant range, the second term on the right-hand side is equal to $\psi^2/3$ to a close approximation.

The slipline field BCD is associated with frictional stresses that decrease along the tool face from B to D. In view of the practical difficulty in lubricating the rake face near the tool tip, it is reasonable to suppose that a state of sticking friction exists at B. The slipline BC then meets the tool face at $90°$. The frictional stress at D is defined by the angle η which the tangent to CD at D makes with the tool face. The restrictions imposed on the slipline field by the velocity boundary conditions are such that the frictional stress distribution along the entire chip-tool interface is determined when η is given.

The velocity of the workpiece relative to the tool is represented by the vector PQ in the hodograph. The magnitude of the velocity discontinuity QB' remains unchanged along the slipline BCA, whose image is the circular arc $B'A'C'$ of radius $R' = U \cos \alpha$. The angles subtended by $B'C'$ and $C'A'$ at Q are $\eta - \psi$ and ψ respectively. Since the velocity distribution along CD corresponds to a rigid body rotation, its image $C'D'$ is also a circular arc of radius R' and angular span ψ. The image of the chip-tool interface is $B'D'$ parallel to BD. The hodograph field $B'C'D'$ can be easily completed since it is part of the standard field defined by circular arcs of equal radii.

Let ρ denote the radius of curvature of the concave side of the chip, of thickness t, rotating about its geometrical center of curvature with angular velocity ω. The peripheral speeds of the concave and convex sides of the chip are equal to the magnitudes of the vectors PA' and PD' respectively in the hodograph. If these magnitudes are denoted by ρ' and $\rho' + t'$ respectively, then $\rho' = \omega\rho$ and $t' = \omega t$. The

[†] H. Kudo, *Int. J. Mech. Sci.*, **7**: 43 (1965). Chip curling has been studied experimentally by R. S. Hahn, *Trans. ASME*, **75**: 581 (1953), and by T. H. C. Childs, *Int. J. Mech. Sci.*, **13**: 373 (1971), ibid., **14**: 359 (1972). Interesting models have been proposed by J. G. Horne, *Int. J. Mech. Sci.*, **20**: 739 (1978). See also S. Ramalingam, E. D. Doyle, and D. M. Turley, *J. Eng. Ind.*, *Trans. ASME*, **102**: 177 (1980).

continuity of the material during the chip formation requires

$$Uh = \omega\left(\rho + \frac{t}{2}\right)t \quad \text{or} \quad \rho' = \frac{Uh/t}{1 + t/2\rho} \tag{173}$$

which must be satisfied if ψ is correctly chosen for given α and η. However, since h/t is yet unknown, this cannot be verified without considering the slipline field.

The construction of the slipline BCD is simplified by drawing a similar field $B''C'D'$ directly on the hodograph, where $D'B''$ is perpendicular to the actual tool face BD. This is done graphically by using the fact that the tangent to a slipline at any point of the network $D'C'B''$ is parallel to the corresponding tangent of the hodograph net $D'C'B'$. Let the horizontal distance between B'' and A' be denoted by h'. In view of the similarity of $A'C'B''D'C'$ with $ACBDC$, the dimensions R', l', h', t', and ρ' are in the same ratios with one another as are R, l, h, t, and ρ. The problem is thus completely solved once ψ is adjusted to satisfy (173) with reasonable accuracy.

The distributions of normal and shear stresses along BD are obtained from the inclinations of the sliplines to the tool face. The frictional stress decreases from k at B to $k \cos 2\eta$ at D; the normal pressure slightly increases from B to D. If the rake angle is too large, the workpiece material in corner A may be overstressed. To avoid yielding of the chip material in corner A, the normal pressure p_0 must be less than $k(1 - \pi/2 + 2\delta)$, where δ is the angle between the chip surface and the slipline at A. Since p_0 exceeds k only by a few percent, this condition would be satisfied.

The results of the calculation based on $\eta = 45$, 30, and 15°, and for various values of α, are shown graphically in Fig. 7.61. The frictional stress at D corresponding to these values of η are 0, $0.5k$, and $0.866k$ respectively. For a given rake angle and depth of cut, the effect of increasing the friction is to increase not only the cutting force but also the chip thickness, the chip radius of curvature, and the contact length.[†] When there is sticking friction over the entire tool face ($\eta = 0$), the plastic zone BCD disappears and the slipline ACB reduces to a straight line inclined at an angle α to the horizontal.

(vi) *Restricted chip-tool contact* In the conventional machining process considered so far, actual measurements indicate that the frictional stress is nearly constant over a lower part of the contact region, but the stress decreases rapidly over the remaining upper part.[‡] If, however, the length of chip-tool contact is limited by using a properly shaped tool, the frictional stress is constant over a major part of

---

[†] A more complicated slipline field, in which the curve ACB is replaced by a singular domain, has been proposed by H. Kudo, op. cit., and discussed in detail by E. Usui and K. Takada, *J. Jpn Soc. Precis. Eng.*, **33**: 23 (1967). A finite element solution has been presented by K. Iwata, K. Osakada, and Y. Terasaka, *J. Eng. Mater. Technol., Trans. ASME*, **106**: 132 (1984).

[‡] See, for example, P. W. Wallace and G. Boothroyd, *J. Mech. Eng. Sci.*, **6**: 74 (1964), and A. Bhattacharya, *Proc. 6th MTDR Conf.*, p. 491, Pergamon Press (1966). The stress distribution in the tool has been studied photoelastically by H. Chandrasekharan and D. V. Kapoor, *J. Eng. Ind., Trans. ASME*, **87**: 495 (1965), and also by E. Amini, *J. Strain Anal.*, **3**: 206 (1968).

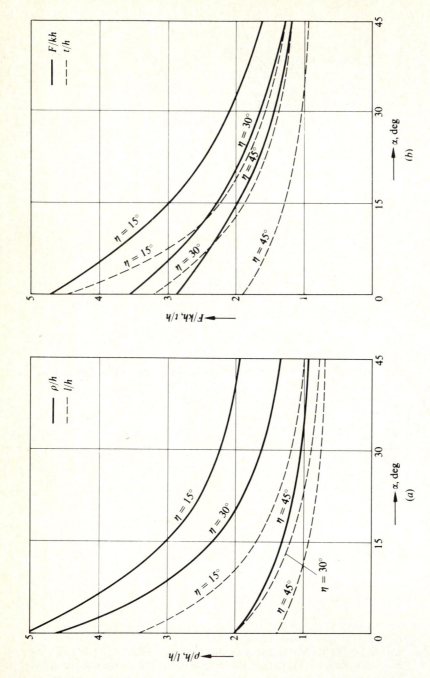

Figure 7.61 Computed results based on Fig. 7.60. (*a*) Chip radius and contact length; (*b*) cutting force and chip thickness.

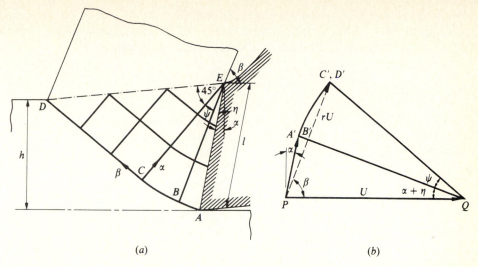

Figure 7.62 Machining with restricted contact tool. (*a*) Slipline field; (*b*) hodograph.

the contact region. In this case, the cutting force increases with the length of chip-tool contact, so long as this is smaller than the natural contact length for the unrestricted cutting tool.

The simplest slipline field† and the corresponding hodograph for machining with a restricted contact tool are shown in Fig. 7.62. The right-angled triangle ABE adjacent to the tool face AE is a field of constant stress, the angle of inclination η depending on the frictional condition. BCE is a centered fan of angular span ψ, while CD is a straight work-chip interface of length equal to $l \cos \eta$. The material beyond ED is assumed stress free. The statical equilibrium of the rigid triangle ECD requires the hydrostatic pressure to have the value k along DC and CE. The normal pressure acting on AE is equal to $k(1 + 2\psi + \sin 2\eta)$, and the frictional stress is of magnitude $k \cos 2\eta$. The resultant horizontal component of the normal and tangential forces on the tool face is the cutting force F, given by

$$\frac{F}{kl} = (1 + 2\psi) \cos \alpha + \sin (2\eta + \alpha) \tag{174}$$

where l is the length of the restricted tool face. From geometry, the ratio h/l can be expressed as

$$\frac{h}{l} = \cos \alpha - \sqrt{2} \cos \eta \sin \left(\frac{\pi}{4} - \psi - \eta - \alpha \right) \tag{175}$$

† W. Johnson, *Int. J. Mech. Sci.*, **4**: 323 (1962); E. Usui and K. Kikachi, *J. Jpn Soc. Precis. Eng.*, **29**: 436 (1963); E. Usui, K. Kikachi, and K. Hoshi, *J. Eng. Ind.*, *Trans. ASME*, **86**: 95 (1964). More complex slipline fields, giving smaller cutting force, have been proposed by H. Kudo, *Int. J. Mech. Sci.*, **7**: 43 (1965).

For a given frictional stress on the tool face, Eqs. (174) and (175) give the relationship between the specific cutting force F/kh and the dimensionless contact length l/h parametrically through ψ. The coefficient of friction is

$$\mu = \frac{\cos 2\eta}{1 + 2\psi + \sin 2\eta} \tag{176}$$

When μ is given, η can be found from the above equation for any assumed ψ. For sticking friction ($\eta = 0$), this equation gives the minimum coefficient of friction required for its existence.

The material approaching the tool with a uniform velocity U, suffers a velocity discontinuity of amount $U \cos \alpha / \cos \eta$ across the slipline $ABCD$. The region ABE moves with a uniform velocity represented by the vector PA', parallel to AE. The velocity of the chip is represented by the vector PC' inclined at an angle β to the direction of the workpiece velocity. It follows from the geometry of the hodograph that

$$\tan \beta = \frac{\sin (\psi + \eta + \alpha)}{\sec \alpha \cos \eta - \cos (\psi + \eta + \alpha)} \tag{177}$$

Evidently, the chip is more inclined to the workpiece than the rake face is, in agreement with what is actually observed. The fact that the vector PC' is of magnitude $(h/t)U$ leads to the relation

$$\left(\frac{h}{t}\right)^2 = 1 - 2 \cos \alpha \sec \eta \cos (\psi + \eta + \alpha) + \cos^2 \alpha \sec^2 \eta \tag{178}$$

The solution cannot be valid for large values of l/h, since F/kh must be independent of l/h when l exceeds the length of chip-tool contact for the unrestricted tool face. An approximate estimate of the limit of validity of the solution may be obtained by assuming the greatest specific cutting force F/kh to be approximately $\cos \eta$ times the value given in Table 7.8 for the same rake angle. For given α and η, the greatest value of ψ for which the solution holds is then obtainable from (174), and the corresponding value of l/h follows from (175). The variation of the specific cutting force with the length ratio is shown in Fig. 7.63 for $\eta = 0$ and different values of α. The cutting force increases with the contact length until the limit is reached.

Consider, now, the distortion of a square grid during the deformation, in the special case of an overall sticking friction existing along the rake face. Let (ρ, θ) denote the polar coordinates of any point in the fan BEC with respect to E and the downward vertical. Since $\eta = 0$, the normal component of velocity along BE must be $U \cos \alpha$. If the radial and circumferential velocities are denoted by u and v respectively, it follows from Geiringer's equations and the continuity conditions that the velocity field in BEC is given by

$$u = -U \sin \theta \qquad v = U(\cos \alpha - \cos \theta)$$

The slipline components of the velocity in DCE are obtained by setting $\theta = \psi + \alpha$

Figure 7.63 Specific cutting resistance for orthogonal machining with a restricted contact tool ($\eta = 0$).

in the above expressions. The trajectories of the particles in BEC are given by

$$\frac{1}{\rho}\frac{d\rho}{d\theta} = \frac{u}{v} = -\frac{\sin\theta}{\cos\alpha - \cos\theta}$$

or

$$\rho = l\left(\frac{\cos\alpha - \cos\theta_0}{\cos\alpha - \cos\theta}\right)$$

(179)

where θ_0 denotes the angular position at which a given particle enters the fan across BC. The time taken by the particle to move along its trajectory on crossing the boundary BC is

$$\tau = \int_{\theta_0}^{\theta} \frac{\rho\,d\theta}{v} = \frac{l}{U}(\cos\alpha - \cos\theta_0)\int_{\theta_0}^{\theta} \frac{d\theta}{(\cos\alpha - \cos\theta)^2}$$

since $v = \rho(d\theta/d\tau)$ along the trajectory of the particle. Performing the integration, the solution may be written as

$$\frac{\tau U}{l} = (\cos\alpha - \cos\theta_0)[f(\theta_0) - f(\theta)]$$

(180)

where

$$f(\theta) = \operatorname{cosec}^2\alpha\left\{\frac{\sin\theta}{\cos\alpha - \cos\theta} + \cot\alpha \ln\left(\frac{\tan\theta/2 - \tan\alpha/2}{\tan\theta/2 + \tan\alpha/2}\right)\right\}$$

(181)

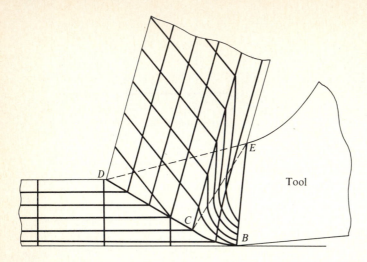

Figure 7.64 Distortion of grids in machining with limited contact tools and sticking friction.

for all values of $\alpha > 0$. When $\alpha = 0$, the above expression for $f(\theta)$ becomes indeterminate, but an independent calculation furnishes

$$f(\theta) = \frac{1}{2} \cot \frac{\theta}{2} \left(1 + \frac{1}{3} \cot^2 \frac{\theta}{2} \right) \qquad \alpha = 0$$

Using the above relations, it is possible to determine the position of selected points on an initially vertical line after given intervals of time. The deformed shape of an original grid is given by the intersection of the trajectories of the particles and the loci of their positions at equal intervals. Figure 7.64 shows the pattern of grid distortion for $\alpha = 5°$, $\psi = 25°$, and $\eta = 0$, when a vertical gridline originally four spaces to the left of B coincides with the vertical through B. The actual deformation observed in machining tests† strikingly resembles the theoretically predicted pattern, which involves a zone of intense shear near the cutting edge.‡

Problems

7.1 In Fig. 7.1, let P denote the resultant force per unit width exerted on the exit slipline by the adjacent rigid material in the direction of its motion, when the mean compressive stress at C is assumed to vanish.

---

† E. Usui, K. Kikuchi, and K. Hoshi, *J. Eng. Ind., Trans. ASME*, **86**: 95 (1964). These authors have also studied the variation of coefficient of friction along the rake face.
‡ The influence of elastic deformations on the chip formation, when cutting with an unrestricted tool, has been investigated by T. H. C. Childs, *Int. J. Mech. Sci.*, **22**: 457 (1980).

Show that

$$\frac{p_e}{2k} = r\left\{\frac{1}{2} + \left(\frac{r}{1-r}\right)\frac{P}{2ka}\right\}$$

where $a = H - h$. Using the appropriate table (Appendix) verify the numerical results of Table 7.1. Extend this table for $r > 0.5$ assuming the die face to be covered by a dead metal.

7.2 Draw the slipline field and the hodograph for extrusion through a perfectly rough square die under conditions of Coulomb friction along the container wall. Show that the extrusion pressure is given by the expression

$$\frac{p_e}{2k} = \frac{r}{2} + (1-r)\frac{P}{2kh}$$

where P is the longitudinal thrust per unit width exerted by the dead metal on its boundary when the hydrostatic pressure vanishes along the exit slipline. Assuming the angle between the container wall and the dead metal boundary to be $\lambda = 30°$, calculate the coefficient of friction, the fractional reduction, and the extrusion pressure when the angle turned through by the dead metal boundary is $\theta = 15, 30$ and 45.

Answer: $\mu = 0.132, 0.103,$ and $0.085; r = 0.624, 0.761$ and $0.850; p_e/2k = 1.676, 2.229,$ and $2.774.$

7.3 Figure A shows the slipline field for incipient extrusion through a square die when the length of the slug is $\sqrt{2}$ times that of the die. Frictionless conditions along the die and the container wall, and a state of sticking friction along the punch face, are assumed to exist. Show that the extrusion pressure may be written as

$$\frac{p_e}{2k} = r\left\{\frac{1+\pi}{2} + \frac{a}{h}\left[\frac{1}{2}\left(1 - \frac{d}{a}\right) + \int_0^\theta \left(\frac{x}{a}\right)d\alpha\right]\right\}$$

where $a = \sqrt{2(H-h)}$, and the integral is taken along the slipline segment BE. Use Table A-9 to evaluate the extrusion pressure and the fractional reduction when $\theta = 15, 30,$ and $45°$. Construct the associated hodograph.

Answer: $r = 0.398, 0.318,$ and $0.252; p_e/2k = 1.041, 0.857,$ and $0.698.$

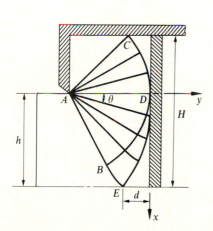

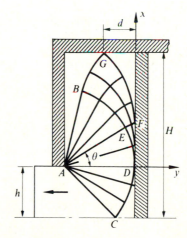

Figure A   **Figure B**

7.4 The slipline field for incipient extrusion of a short slug is shown in Fig. B, where the length of the slug is $\sqrt{2}$ times the emerging sheet thickness. The face of the punch is assumed to be perfectly rough, and the container wall perfectly smooth. Considering the tractions across the dead metal boundary

ABG, show that

$$\frac{p_e}{2k} = r\left(\frac{1+\pi}{2} + 2\theta\right) + \frac{1-r}{\sqrt{2}}\left\{1 - \frac{d}{a} - 2\int_0^\theta \left(\frac{x}{a}\right)d\alpha\right\}$$

where $a = \sqrt{2}\, h$. Using Table A-9, calculate the values of $p_e/2k$ and r for $\theta = 15, 30,$ and $45°$. Sketch the hodograph associated with the slipline field.

Answer: $r = 0.602, 0.682,$ and 0.748; $p_e/2k = 1.603, 1.957,$ and 2.359.

7.5 The slipline field for side extrusion from a perfectly rough container caused by a pair of opposed rams is shown in Fig. C. The speeds of the right-hand and left-hand rams are U and nU respectively, where $0 < n < 1$. Draw the hodograph of the process and hence show that the angle η^* which the extruded sheet makes with the container wall is given by

$$\tan \eta^* = \left(\frac{1+n}{1-n}\right)\tan \eta$$

where η is the angle for one-sided extrusion with identical wall friction. When $\lambda = 15°$, $\theta = 30°$, and $n = 0.3$, find the extrusion pressure and the angles η and η^*.

Answer: $p_e/2k = 1.654$, $\eta = 68.0°$, $\eta^* = 77.7°$.

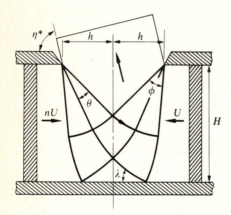

Figure C

7.6 In the combined end and side extrusion process discussed in Sec. 7.2(iii) for arbitrary d/h and H/h ratios, prove that the angle η between the direction of the side extrusion and the plane of the side orifice is given by

$$\tan \eta = \frac{H/2h}{\sin^2 \lambda + v/U}$$

where λ is the acute angle made by the α direction at F with the axis of symmetry, and v the horizontal distance of the image of C from the pole of the hodograph. Assuming $d/h = 2.74$, and a total fan angle of $90°$ at the exit point, find the values of H/h, $p_e/2k$, and η.

Answer: $H/h = 6.97$, $p_e/2k = 1.855$, $\eta = 83.2°$.

7.7 Using the notation of Fig. 7.16, representing the frictionless sheet drawing through a tapered die, show that the hydrostatic pressure p_0 at the axial point F is given by

$$\frac{p_0}{k} = r\left(\frac{P}{2ka} + \frac{Q}{2ka}\cot \psi - 1\right) - 2(\theta + \phi)$$

where P and Q depend on the angular coordinates (θ, ϕ) of F. Considering the case $\psi = 20°$, obtain the variation of p_0/k with the reduction r over the range covered by the slipline field. Find the range of values of ψ for which p_0 becomes negative for all reductions.

Answer: $\psi > 17.9°$.

7.8 In the plane strain drawing of a sheet through a rough tapered die, the bulge limit depends not only on the die angle but also on the coefficient of friction. Show that the die pressure in this limiting state becomes

$$q = k\left[1 + \sin 2\lambda + \frac{\pi}{2} - 2(\psi - \lambda)\right]$$

where λ is the angle of inclination of the β lines with the die face. Assuming $\mu = 0.1$, calculate the values of λ, r, $q/2k$, and $t/2k$ at the bulge limit corresponding to $\psi = 15°$.
  Answer: $\lambda = 32.9°$; $r = 0.079$, $t/2k = 0.241$.

7.9 Over the practical range of reductions and coefficients of friction, show that the mean drawing stress for a rough wedge-shaped die may be written as

$$t' \simeq t\left\{1 + \left(1 - \frac{r}{2}\right)\mu \cot \psi\right\}$$

Using the appropriate empirical expression for the redundant work factor, prove that the optimum die angle is given by the equation

$$\psi^3 + \left(\frac{2 - r}{4}\right)\mu\psi^2 - 3.32\left(\frac{\mu r^2}{2 - r}\right) = 0$$

in which the second term may be neglected as a first approximation. Show that the second approximation then coincides with Eq. (46).

7.10 Show that for small coefficients of friction, and small-to-moderate reductions in thickness, the ratio of the extrusion pressures with and without friction over a wedge-shaped die may be written on the basis of equations (45) and (52) as

$$\frac{p'_e}{p_e} \simeq 1 + \frac{1}{2}\left(\frac{2 - r}{1 - r}\right)\mu \cot \psi$$

Assuming $\mu = 0.1$ and $\psi = 30°$, calculate the dimensionless extrusion pressure for $r = 0.2$, 0.3, and 0.4. Determine the optimum value of ψ when $\mu = 0.1$ and the reduction is 30 percent.
  Answer: $p'_e/2k = 0.507$, 0.646, and 0.785, $\psi = 15.6°$.

7.11 The combined effect of friction and work-hardening may be estimated by approximating the relevant part of the stress-strain curve by a straight line with an initial yield stress k_1 and a final yield stress k_2. Considering the solution of Eq. (43), where k is now a variable quantity, prove that the ratio of the drawing stresses with and without friction becomes

$$\frac{t'}{t} = (1 + m)[1 - (1 - r)^m]\left\{\frac{1}{m\varepsilon_2} - \frac{1}{6}\left(\frac{k_2 - k_1}{k_2 + k_1}\right)\right\}$$

to a close approximation, where $\varepsilon_2 = \ln (H/h)$ and m denotes the quantity $\mu \cot \psi$. Note that the second term in the curly bracket is negligible except for annealed materials together with large values of $m\varepsilon_2$.

7.12 Draw the slipline field and the hodograph for symmetrical extrusion through a perfectly rough tapered die when the deforming zone involves a single point on the axis of symmetry. Assuming a 15° die semiangle, calculate the fractional reduction, the mean die pressure, and the extrusion pressure when the entry fan angle is 0, 15, and 30°. Determine in each case the minimum coefficient of friction required by the given frictional condition.

7.13 Figure D shows a possible slipline field that predicts the formation of a surface strip in an attempted extrusion through a smooth wedge-shaped die when the reduction is very small. Show that the extrusion pressure and the direction of chip travel are given by

$$p_e = 2kr\left(1 + \frac{\pi}{2} + \psi\right) \qquad \tan \eta = \frac{\sin \psi}{1 + \sin \psi}$$

Rederive the expression for p_e using the velocity distribution and the upper bound theorem.

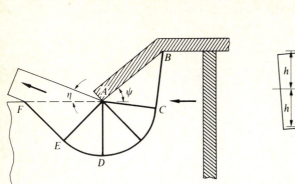

Figure D

Figure E

7.14 Figure E shows the slipline field for unsymmetrical extrusion through a smooth wedge-shaped die with unequal angles of inclination ψ_1 and ψ_2. The field applies to a particular value of the nominal extrusion ratio $(H_1 + H_2)/2h$. Prove that the extrusion pressure is given by

$$\frac{p_e}{2k} = \frac{(1 + \psi_1)\sin\psi_1 + (1 + \psi_2)\sin\psi_2}{1 + \sin\psi_1 + \sin\psi_2}$$

Draw the associated hodograph and hence show that the direction of motion of the extruded billet makes an acute angle η with the plane of the orifice, where $\cot\eta = (H_1 - H_2)/(H_1 + H_2)$.

7.15 Suppose that the nominal reduction for the unsymmetrical extrusion is increased from the preceding value such that the ratio H_1/H_2 remains constant. Assuming $\psi_1 = 45°$ and $\psi_2 = 30°$, show that the extrusion pressure for any reduction r may be expressed by the empirical equation

$$\frac{p_e}{2k} = 0.184 + 0.925 \ln\left(\frac{1}{1 - r}\right)$$

Find the range of reductions for which this formula is expected to be sufficiently accurate.
Answer: $0.55 < r < 0.84$.

7.16 Figure F shows a particular slipline field for unsymmetrical extrusion through a perfectly rough wedge-shaped die when the container wall is also perfectly rough. Assuming $\psi_1 = 45°$ and $\psi_2 = 30°$, find the eccentricity, the nominal reduction, and the extrusion pressure when the position of the ram is at F. Excluding the effect of the frictional drag between the billet and the wall, show that the extrusion

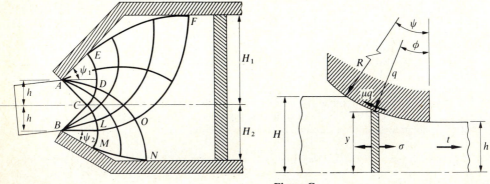

Figure F

Figure G

pressure for smaller reductions, with the ratio H_1/H_2 held constant, is given by the empirical formula

$$\frac{p_e}{2k} = 9.78 + 1.26 \ln\left(\frac{1}{1-r}\right)$$

Answer: $e = 0.329$, $r = 0.646$, $p_e/2k = 2.398$ for the geometry of Fig. F.

7.17 Figure G shows the drawing of a sheet through a convex die of circular profile with zero exit angle. Using the yield criterion in the form $\sigma + q = 2k$, obtain the equilibrium equation

$$y \, d\sigma + 2k \, dy + \mu q R \cos \phi \, d\phi = 0$$

Assuming a mean value of q, equal to that for frictionless drawing, show that the drawing stress t for a sufficiently small angle of contact ψ is given by

$$\frac{t}{2k} = \ln\frac{1}{1-r} + \mu\left(1 - \frac{1}{2}\ln\frac{1}{1-r}\right)\sqrt{\frac{2R}{h}} \tan^{-1}\sqrt{\frac{r}{1-r}}$$

Adopting a redundant work factor, equal to that for a wedge-shaped die of semiangle $\psi/2$, determine $t/2k$ when $r = 0.25$, $\mu = 0.1$, and $R/h = 5$.

Answer: $t/2k = 0.448$.

7.18 Draw the slipline field for symmetrical extrusion through a partially rough constant pressure die with a wedge-shaped end of semiangle ψ. Using the geometry of the slipline field, prove that the length of the axis of symmetry covered by the field for a given final thickness is independent of the coefficient of friction. If the modification of the field of Fig. 7.27 is such that

$$l \simeq \left[1 + \cos\left(\frac{\pi}{4} - \lambda\right)\right] d$$

as an approximation for low die friction, calculate the reduction in thickness and the extrusion pressure when $\psi = 30°$ and $\mu = 0.1$. Compare this extrusion pressure with that in frictionless extrusion for the same reduction.

Answer: $r = 0.456$, $p_e/2k = 1.151$.

7.19 A lower bound solution for frictionless extrusion through a square die may be obtained by assuming stress discontinuity across the exit plane AD and the sides of an isosceles right-angled triangle ABC as shown in Fig. H. The angle ϕ which AC makes with the axis of symmetry depends on the extrusion ratio R. Show that the stress field is statically admissible, giving the lower bound

$$\frac{p_e}{2k} = 2\left(\frac{R^2 - 1}{R^2 + 1}\right)$$

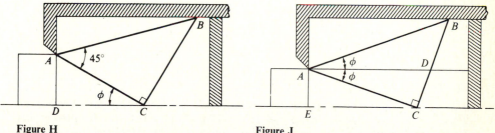

Figure H   **Figure J**

7.20 An improved lower bound for medium reductions is furnished by the discontinuity pattern of Fig. J, where the material in the top right-hand corner is assumed nonplastic. The stress field coincides with that of Fig. 7.32 when $R = 3$, and degenerates into that of Fig. H when $R = \sqrt{2} + 1$. Show that a permissible state of stress exists between these limiting cases, and that the extrusion pressure is

$$p_e = 4k\left(3 - r - \frac{1}{r}\right)$$

7.21 An upper bound solution for extrusion through a square die may be obtained by assuming the single triangular velocity discontinuity pattern of Fig. K, which involves a dead metal region adjacent to the die face. Considering the associated hodograph, and minimizing the extrusion pressure with respect to the dead metal boundary angle α, derive Eq. (91) for the best upper bound estimate.

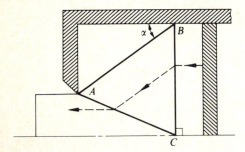

Figure K

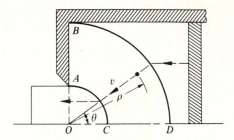

Figure L

7.22 For a perfectly rough square die with large reductions, a reasonable upper bound approximation is achieved by assuming velocity discontinuity across a pair of concentric circular arcs AC and BD, and a purely radial flow of material in the region between them (Fig. L). Prove that the radial velocity v is of magnitude $\cos\theta/\rho$ per unit speed of the ram, and that the extrusion pressure is given by

$$\frac{p_e}{2k} = 1 + E\left(\frac{\sqrt{3}}{2}, \frac{\pi}{2}\right)\ln\frac{1}{1-r}$$

where $E(\sqrt{3}/2, \pi/2)$ is the complete elliptic integral of the second kind. Note that this solution is valid for all frictional conditions along the die face.

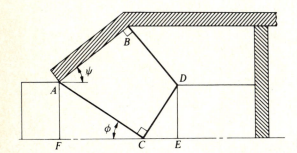

Figure M

7.23 A statically admissible stress field for frictionless extrusion through a wedge-shaped die may be constructed by using the stress discontinuity pattern of Fig. M, where $\phi = \pi/4 - \psi/2$ and BD is normal to the die face. Determine the principal stresses in each separate region of the field, and hence obtain the lower bound

$$p_e = 2kr(1 + \sin\psi)$$

Show that the proposed field is geometrically possible for $\sin\psi/(1 + \sin\psi) \leqslant r \leqslant 2\sin\psi/(1 + \sin\psi)$. Discuss the lower bound solution outside this range of reductions.

7.24 For $\psi > 45°$, an improved lower bound may be found by considering the discontinuity pattern shown in Fig. N, where AD is parallel to the axis of symmetry and DE is parallel to the die face. Show that the field is valid for $R \geqslant 1 + 2 \cos \psi$, and is statically admissible, giving the extrusion pressure

$$p_e = 2kr(1 + \cos \psi - \cos 2\psi)$$

Indicate how the stress field may be modified for lower extrusion ratios without affecting the lower bound value.

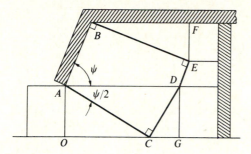

Figure N   **Figure O**

7.25 For moderate die angles and reductions, a reasonable upper bound estimate for frictionless extrusion through a wedge-shaped die is furnished by the velocity discontinuity pattern of Fig. O. Considering the associated hodograph, derive the upper bound expression

$$\frac{p_e}{k} = (2 - r) \tan \psi + \left(\frac{r^2}{1 - r} \right) \cot \psi$$

Verify that the same upper bound follows from the velocity field of Fig. K with the die face taking the place of the dead metal boundary.

7.26 An upper bound solution for extrusion through a perfectly rough wedge-shaped die may be found by using the double triangular velocity discontinuity pattern of Fig. 7.36. Assuming $\phi = \pi/4$, obtain the extrusion pressure in the form

$$\frac{p_e}{2k} = \frac{2}{1 + \cot \psi} + \frac{r(2 - r) \cot \psi + r^2}{4(1 - r)}$$

Compare this upper bound with (87) for $\psi = 45°$ over the range $0.1 \leqslant r \leqslant 0.9$, and present your results in tabular form.

7.27 The stress discontinuity pattern of Fig. P may be used to obtain a lower bound solution for frictionless extrusion through a convex die whose profile is a quarter circle. The stress distribution is assumed radially symmetrical in the region $CABD$, where CD is a circular arc of radius H, the radial compressive stress at any radius being denoted by q. Show that a statically admissible stress field exists for $r > 0.2$ approximately, and that the extrusion pressure is given by

$$\frac{p_e}{2k} = \frac{r}{1 - r} \ln \frac{1}{r}$$

7.28 An upper bound solution for extrusion through a smooth convex die of circular profile may be obtained by using the velocity discontinuity pattern of Fig. Q. The material in the curvilinear triangle ABC is assumed to rotate as a rigid body about the center O of the circular arc AB. The discontinuities AC and BC are also circular arcs whose radii and angular spans can be found from the hodograph. Show that the upper bound is

$$\frac{p_e}{2k} = r \left\{ \frac{\pi}{4} + \left[1 + \left(\frac{r}{1 - r} \right)^2 \right] \sin^{-1} \left(\frac{1 - r}{\sqrt{2} \, r} \right) \right\}$$

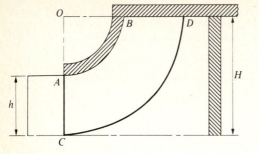

Figure P

Figure Q

This formula can be expected to provide a reasonable approximation to the extrusion pressure for $r > 0.5$.

7.29 The velocity discontinuity pattern of Fig. R represents the simultaneous forward and backward extrusion of a slug of thickness $2a$, resulting in two emerging sheets of equal thickness h moving in opposite directions. The discontinuity angle θ depends on the reduction r and the thickness ratio a/H. Show that the assumed velocity field permits a simultaneous operation for $r > 0.5$, and that the predicted upper bound is

$$\frac{P_e}{2k} = \frac{1}{2} \left\{ \frac{rH}{a} + \frac{a}{H(1-r)} \right\}$$

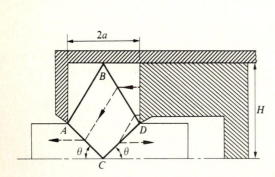

Figure R

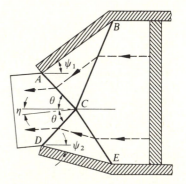

Figure S

7.30 Figure S shows the velocity discontinuity pattern for unsymmetrical extrusion through a smooth wedge-shaped die. The eccentricity of the die and the inclination of the die faces are such that the corners B and E are on the same vertical line. Assuming $\theta = 45°$, and a nominal reduction of 50 percent, calculate the upper bound on the extrusion pressure and the corresponding direction of extrusion when the die semiangles are $\psi_1 = 40°$ and $\psi_2 = 20°$.

Answer: $p_e/2k = 0.84$, $\eta = 12.9°$.

7.31 An approximate solution for the cold rolling of a strip without applied tensions may be obtained by setting $q \simeq 2k$ in the second term of Eq. (102). Derive the corresponding roll pressure distribution neglecting work-hardening, and show that the roll force per unit width is given by

$$\frac{P}{P^*} = 2\sqrt{\frac{1-r}{r}} \sin^{-1}\sqrt{r} + \frac{\mu}{\alpha}\left\{ \ln\frac{1}{1-r} - 2\ln(1 + \xi_n^2) \right\} - 1$$

where $\xi_n = \sqrt{R'/h_2}\,\phi_n$, and $P^* = 2k\sqrt{R'rh_1}$, the neutral angle being identical to that given by (116). Is the above equation expected to underestimate or overestimate the actual roll force?

7.32 When the neutral point and the roll force are known, a useful approximation to the torque may be obtained by assuming a triangular distribution of the roll pressure giving the same total force. If the position of the pressure peak is taken as that of the preceding theory, show that

$$\frac{G}{G^*} = \frac{2P}{3P^*}\left\{1 + \sqrt{\frac{1-r}{r}}\tan\left[\frac{1}{2}\sin^{-1}\sqrt{r} - \frac{1}{4\mu}\sqrt{\frac{h_2}{R'}}\ln\frac{1}{1-r}\right]\right\}$$

Plot the variation of P/P^* and G/G^* with r for $\mu\sqrt{R'/h_1} = 1$, using the above expressions, and compare it with that of Fig. 7.42.

7.33 In cold rolling, the elastic recovery of the strip on the exit side may be approximately taken into account by assuming the longitudinal stress in the elastic region to have a constant value. Using the stress-strain relation, show that the elastic arc of contact increases the roll force by ΔP and decreases the roll torque by $\mu R\Delta P$, where

$$\frac{\Delta P}{P^*} \simeq \frac{2k_2 - t_2}{3\bar{k}}\sqrt{\frac{(1-r)\lambda}{r + (1-r)\lambda}} \qquad \lambda = (1 - v^2)\left(\frac{2k_2 - t_2}{E}\right)$$

E and v are the elastic constants for the strip, and P^* is the mean yield stress $2\bar{k}$ times the plastic arc of contact. Prove that the radius of the deformed arc of contact is given by

$$\frac{R'}{R} = 1 + \frac{P + \Delta P}{c h_1[\sqrt{r + (1-r)\lambda} + \sqrt{(1-r)\lambda}\,]^2}$$

7.34 The relationship between the yield stress $2k$ and the fractional reduction e in the plane strain compression of a copper strip can be expressed in the empirical form

$$2k = 540\{1 - 0.5(1-e)^{3.5}\}\text{ MN/m}^2$$

Assuming an overall reduction of 30 percent produced by cold rolling, determine the mean values of the yield stress appropriate for the evaluation of the roll force and the torque. Using Fig. 7.42 for P/P^* and G/G^*, calculate the values of P and G when $R = 180$ mm, $h_1 = 2.5$ mm, and $\mu = 0.08$.
   Answer: $2\bar{k}_p = 407$ MN/m , $2\bar{k}_G = 380$ MN/m^2, $P = 6.01$ MN/m, and $G = 34.15$ kN·m/m.

7.35 A strip of annealed copper having an initial thickness 1.8 mm is cold rolled in a single pass to a reduction of 26.5 percent using 180-mm diameter steel rolls. A gradually increasing back tension is applied to the strip until the rolls are just at the point of slipping. The yield stress curve of the material may be represented by the equation

$$2k = 650\{1 - 0.8(1-e)^{5.0}\}\text{ MN/m}^2$$

If the coefficient of friction between the strip and the rolls is 0.07, determine the applied back tension, and the corresponding roll force and torque, when the rolls are about to slip.
   Answer: $t_1 = 107.15$ MN/m^2, $P = 2.597$ MN/m, $G = 16.36$ kN·m/m.

7.36 A mild steel sheet 50 mm wide and 10 mm thick is to be rolled in a hot rolling mill to a thickness of 5 mm at 1000°C, using 400-mm diameter steel rolls driven at 60 rpm. The results of the simple compression test carried out at the same temperature may be expressed by the equation $Y = C\varepsilon^n$. In terms of the fractional reduction e in plane strain compression, the constants C and n may be expressed by the empirical relations

$$C = 65 + 320e - 450e^2$$

$$n = 0.126 - 0.24e + 0.46e^2$$

which furnish the yield stress in meganewtons per square meter. Calculate the roll force and the torque, and estimate the power required to drive the mill if the mechanical efficiency is 90 percent.
   Answer: $P = 10.57$ MN/m, $G = 131.56$ kN·m/m, $W = 918.5$ kW.

7.37 An approximate expression for the roll separating force in the hot rolling process may be obtained by assuming a linear distribution of roll pressure with a gradient corresponding to the mean sheet thickness on either side of the neutral point. Considering a mean yield stress $2\bar{k}$, and assuming $\phi_n \simeq \alpha/2$ as a first approximation, show that

$$\frac{P}{P^*} = \frac{\pi}{4} + \frac{2}{8 - 5r}\sqrt{\frac{rR}{h_1}}$$

The mean yield stress may be approximately estimated on the basis of the mean strain rate $\bar{\lambda}$, and the mean empirical constants $\bar{C}$ and $\bar{n}$. Using the data of the preceding problem, compute the roll force per unit width from the above formula.

Answer: 10.88 MN/m.

7.38 An upper bound solution for the hot rolling problem may be obtained by assuming velocity discontinuities across a pair of circular arcs AD and DB with centers of curvature at C and E on the exit plane (Fig. T). The discontinuity BD is tangential to the roll surface at the exit point and meets the axis of strip at 45°. Sketch the hodograph and derive the upper bound expression

$$\frac{G}{2kh_12} = 1.457(1 - r)^2\left\{\frac{\pi}{4} + \text{cosec}^2\,\beta \cot^{-1}\left(\frac{1-r}{\sqrt{2}-1} + \frac{2R}{h_1}\sin\alpha\right)\right\}$$

where α is the angle of contact, r the fractional reduction, and

$$2\cot\beta = \frac{\sqrt{2}-1}{1-r}\left\{1 + \left(\frac{2R}{h_1}\right)^2\sin^2\alpha\right\} - \frac{1-r}{\sqrt{2}-1}$$

Assume that there is no relative motion between the material and the roll. Calculate the value of $G/2kh_1^2$ when $r = 0.4$ and $R/h_1 = 10^2$, using the above relations.

Answer: 2.98.

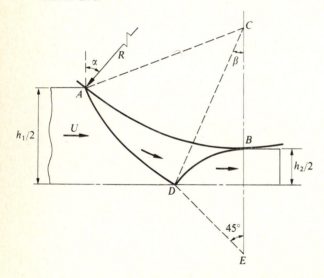

Figure T

7.39 In the orthogonal machining of nonhardening materials, a permissible state of stress in the workpiece may be constructed by assuming stress discontinuities across the straight lines shown in Fig. U. The stress state in the triangle ABC is the same as that on the shear plane itself, while stress-free conditions are assumed to exist in the regions DCE, GAM, and BFN. Show that the configuration is geometrically possible if the shear plane angle ϕ does not exceed $3\pi/4 - \psi$, where

$$\psi = \frac{\pi}{2} - \sin^{-1}\left(\frac{1}{2} - \frac{p}{2k}\right)$$

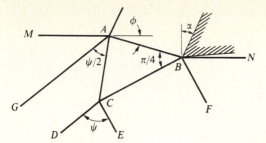

Figure U

Assuming an angle of friction $\lambda = \pi/8$, find the range of values of α for which the chip material would yield at the corner A without overstressing the workpiece.

Answer: $\alpha \leqslant 62°$.

7.40 A possible slipline field for machining with a built-up nose BCE is shown in Fig. V, where the material is assumed stress free above the discontinuity AD. Show that the relationship between the angles ϕ and ψ is

$$\phi - \psi = \frac{\pi}{4} - (\lambda - \alpha)$$

Show also that the workpiece material in the corner A will not be overstressed if $\psi < (\lambda - \alpha)/2$. Find the minimum coefficient of friction between the nose and the workpiece for which the solution would be statically admissible when $\lambda - \alpha = 0$, 30, and 45°. Assuming the frictional stress to be uniform along the base BE, calculate the value of F/kh corresponding to the limiting state of stress in the corner when $\alpha = 0$ and $\lambda = 30°$.

Answer: $\mu = 1$, 0.760, and 0.656; $F/kh = 3.356$.

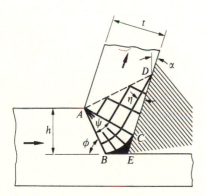

Figure V

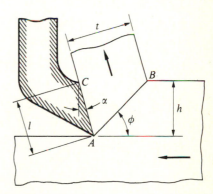

Figure W

7.41 Figure W shows the shear plane model of chip formation in orthogonal machining with a restricted contact tool. The workpiece is assumed to move horizontally with a unit speed toward a stationary tool whose rake face is of length l. Show that under frictionless conditions, the rate of energy dissipation is a minimum when the shear plane angle ϕ has the Merchant value $\pi/4 + \alpha/2$. For a perfectly rough tool, show that the minimum energy criterion leads to the shear angle relationship

$$\cot \phi = \sec \alpha \sqrt{1 + m \cos \alpha} - \tan \alpha$$

where $m = l/h$. Calculate the shear plane angle and the specific chip thickness when $\alpha = 0$ and $30°$, assuming $m = 1$.

Answer: $\phi = 35.6°$, $45°$; $t/h = 1.414$, 1.366.

7.42 A two-dimensional machining operation is carried out with a cutting speed of 1.5 m/s and a depth of cut of 0.25 mm, the material of the workpiece being plain carbon steel having an initial temperature 25°C and a density 7.86 (10^3) kg/m^3. The specific heat c and the thermal conductivity κ of the material vary with the absolute temperature T according to the linear laws

$$c = 416 + 0.104T(\text{J/kg} \cdot \text{K})$$

$$\kappa = 60.5 - 0.035T(\text{W/m} \cdot \text{K})$$

If the rake angle of the tool is 5°, and the mean coefficient of friction along the tool-chip interface is 0.32, compute the temperature of the chip leaving the tool face. Use Eqs. (160) and (161) for the shear angle ϕ, and assume $k = 245$ MPa for the workpiece.

EIGHT

NONSTEADY PROBLEMS IN PLANE STRAIN

We now turn to two-dimensional problems in which the stress and velocity at any given point vary from one instant to another. It is generally necessary, for simplicity, to restrict the discussion to the estimation of the yield point load, under which the deformation just begins in a rigid/plastic body. This load is effectively the same as that for which the overall distortion in a nonhardening elastic/plastic body, under identical boundary conditions, increases rapidly in relation to the change in load. The yield point load is generally approached in an asymptotic manner, and is very nearly attained while the deformation is still of the elastic order of magnitude. In a number of interesting situations, the plastic zone develops in such a way that the configuration remains geometrically similar throughout the deformation. If the configuration at each stage is scaled down by a certain factor so as to obtain the same geometrical figure, the stress and velocity at any given point remain constant. The problem is therefore analogous to that of steady motion, except that it is necessary to satisfy the velocity boundary condition required for the maintenance of the geometrical similarity.

8.1 Indentation by a Flat Punch

(i) *Semi-infinite medium* As a typical example of the problem of incipient plastic flow, consider the indentation of a semi-infinite body by a flat rigid punch under conditions of plane strain (Fig. 8.1). As soon as a load is applied by the punch, plastic zones begin to form at the corners A, where the punch pressure has its greatest value. Since the material is rigid/plastic, no deformation is possible until the applied load is sufficient to cause the plastic zones to spread across the bottom

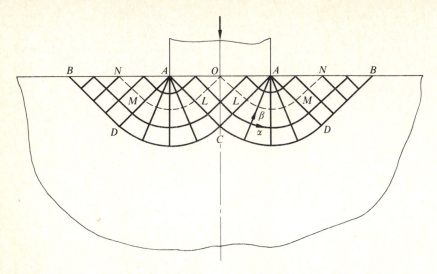

Figure 8.1 Slipline fields for the indentation of a semi-infinite medium by a flat punch.

of the punch. Following the fusion of the plastic zones spreading from both sides, the indentation proceeds as the material is pushed upward by the sides of the punch. The slipline field, due to Prandtl,[†] consists of a pair of 90° centered fans with adjacent regions of constant stress. Since the mean compressive stress has the value k along the free boundaries AB, it follows from Hencky's equation that the punch pressure at the yield point has the value

$$q = 2k\left(1 + \frac{\pi}{2}\right) \tag{1}$$

This pressure is uniformly distributed over the width of the punch. The yield point state would of course be entirely different if a gradually increasing uniform pressure were applied over the width AA.

The material triangle ACA moves vertically downward as a rigid body with a speed equal to the punch speed U. Since the material below the sliplines CDB is rigid, the normal component of velocity must be zero along these boundaries. It follows from the Geiringer equations that the particles outside the triangle ACA move outward along the sliplines parallel to the boundary CDB with a constant speed $U/\sqrt{2}$, while velocity discontinuities of the same amount occur across $ACDB$. The field is independent of the frictional condition over the punch face, and it has actually been verified by experiment.[‡] Beneath the field $ACDB$, there exists a

---

† L. Prandtl, *Göttinger Nachr., Math. Phys. Kl.*, 74 (1920). This work preceded Hencky's general theory by three years.

‡ See, for example, A. Nadai, *Theory of Flow and Fracture of Solids*, vol. 2, p. 470, McGraw-Hill Book Co., New York (1963).

further plastic region where the material is stressed to the yield point but the deformation is prevented by the constraint of the nonplastic matrix.†

An alternative solution, proposed by Hill,‡ consists of a deformation mode in which the material below the sliplines $OLMN$ remains rigid, the velocity discontinuity propagating along these sliplines being of amount $\sqrt{2}\,U$. Evidently, the particles in each half of the deforming zone move with a speed $\sqrt{2}\,U$ along the sliplines parallel to the boundary $OLMN$. In view of the relative sliding between the material and the punch face, the solution is valid only for a perfectly smooth punch, while the punch pressure is still given by (1). From the thoretical point of view, any slipline field having its boundary lying between $OLMN$ and $ACDB$ represents a possible solution to the problem, giving the same punch pressure at the yield point of the semi-infinite block.§

An extension of the Prandtl field into the rigid region is shown in Fig. 8.2, only one half of the field being considered.¶ The circular arc CD and the condition of symmetry uniquely define the field to the left of DPQ. The slipline BJ, which is parallel to DP, defines the stress-free boundary BN, where the tangent at N is parallel to the axis of symmetry of the field. Since the sliplines of one family begin to intersect one another†† near J, it is necessary to introduce the stress discontinuity JM to continue the slipline field beyond JKN. The discontinuity bisects at each point the angles between the sliplines of the same family. The field immediately to the right of the discontinuity must be compatible with the slipline JN, and the hydrostatic pressure jump across JM must be compatible with its inclination to the sliplines at each point. These requirements are sufficient to construct the fields JKM and KMN. It turns out that the angles turned through along JK, KM, and HM are approximately equal to one another. The discontinuity line NMR is then constructed through M and N, using relations (32), Chap. 6, such that the material below this line is in a state of uniaxial stress (of varying magnitude) parallel to the axis of symmetry. Evidently, the hydrostatic tension must be equal to the maximum shear stress everywhere in this region. It is found that the material in this region is stressed below the yield limit. An additional discontinuity NS is considered through N parallel to the axis, and the material on the right of BNS is assumed unstressed. The extended field therefore constitutes a complete solution of the indentation problem. The perpendicular distances of N from OR and OB are given by

$$\frac{w}{a} = 2e^{\pi/2} - 1 \simeq 8.621 \qquad \frac{d}{a} \simeq 7.288$$

---

† If the material is annealed, the familiar sinking-in mode of deformation would occur at a somewhat smaller load, the displaced material being accommodated by the elastic resilience of the block.

‡ R. Hill, *Q. J. Mech. Appl. Math.*, **2**: 40 (1949). A finite element solution has been given by C. H. Lee and S. Kobayashi, *Int. J. Mech. Sci.*, **12**: 349 (1970).

§ The indeterminacy of the deformation mode may be removed by specifying a traction rate, and considering the nonhardening material as the limit of a work-hardening material (Sec. 2.7(iii)).

¶ J. F. W. Bishop, *J. Mech. Phys. Solids*, **2**: 43 (1953). For a different treatment, see M. Sayir and H. Ziegler, *Ing. Arch.* **36**: 294 (1968).

†† The intersection first occurs on BJ at an angular distance of 2.4° from J, but this has a negligible effect on further extension of the slipline field into the rigid region.

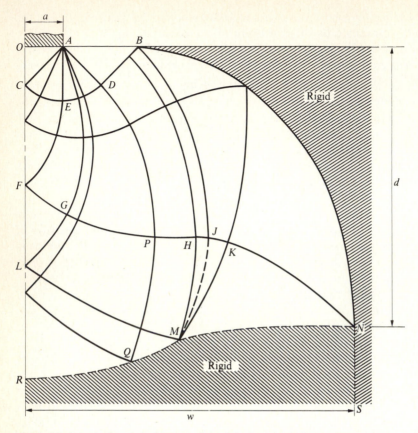

Figure 8.2 A statically admissible extension of the Prandtl field into the rigid region.

in view of (139), Chap. 6. The depth of R below O is found to be about $8.74a$, where a denotes the semiwidth of the punch.

The material lying outside $BNMR$ is nonplastic and therefore rigid. Since the velocity must be continuous across the curves BN, MN, and MR, which are not characteristics, the regions BNJ, MNK, and MRL must also be rigid. The normal component of velocity therefore vanishes along the sliplines JK and MK, which means that the regions JKM and JHM are rigid, the velocity being continuous across the stress discontinuity JM. Following the same argument, the regions $LMHG$, LFG, $BJFE$, and CEF are successively shown to be rigid, bearing in mind that the normal component of velocity is zero along the axis of symmetry. It follows that the Prandtl field $OCDB$ covers the whole of the deformable region, outside which the material is necessarily rigid.†

---

† The indentation of a surface by a pair of punches in close proximity has been discussed by S. A. Meguid, I. F. Collins, and W. Johnson, *Int. J. Mech. Sci.*, **19**: 1 (1977).

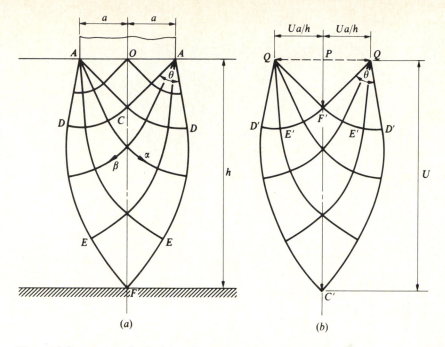

Figure 8.3 Indentation of a block by a flat punch when $1 \leqslant h/a \leqslant 4.77$. (a) Slipline field; (b) hodograph.

(ii) *Medium of finite depth* A block of material of thickness h, resting on a rigid foundation, is indented by a flat punch of width $2a$. When the ratio h/a is less than a certain critical value, plastic zones originating from the corners of the punch spread through the thickness of the block to bring it to the yield point. If the foundation is smooth, the slipline field consists of a pair of identical centered fans ACD and the associated field $CDFD$, defined by the equal circular arcs CD (Fig. 8.3). The triangular region ACA is uniformly stressed,† and the sliplines at F meet the foundation at 45°. Since the material is incompressible, the rigid material on either side of the field moves horizontally with a speed Ua/h, where U denotes the speed of the punch. The velocities of the rigid ends are represented by the vectors PQ in the hodograph. The sliplines DF are mapped into the circular arcs $D'F'$ having a radius equal to the magnitude of the velocity discontinuity occurring across the pair of sliplines $ADEF$. These circular arcs define the remainder $F'D'C'D'$ of the hodograph, with PC' denoting the velocity of the punch. The material immediately above F abruptly changes its direction of motion without change in speed on crossing the discontinuities.

The uniform pressure q on the punch is equal to $p_0 + k$, where p_0 is the mean

---

† The slipline field is due to L. Prandtl, *Z. angew. Math. Mech.* **3**: 401 (1923). See also R. Hill, *J. Iron Steel Inst.*, **156**: 513 (1947). For experimental confirmation, see F. Körber and E. Siebel, *Mitt. Kaiser-Wilhelm-Inst. Eisenforsch.*, **10**: 97 (1928).

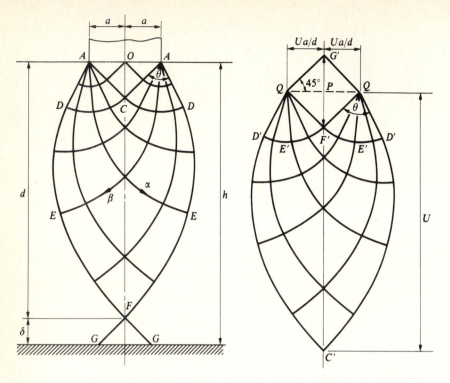

Figure 8.4 Slipline field and hodograph for punch indentation when $4.77 \leqslant h/a \leqslant 8.77$.

compressive stress at C. This is determined from the condition that the resultant horizontal force acting across the vertical section OF must be zero. Using Eq. (99), Chap. 6, with the necessary sign change for the terms in k, there results

$$\frac{q}{2k} = \frac{a}{h}\{I_0(2\theta) + 2\theta[I_0(2\theta) + I_1(2\theta)]\} \tag{2}$$

where θ is the angular span of the centered fans ACD. The ratio h/a is directly obtained as a function of θ from the appropriate table in the Appendix. As h/a increases from unity, the principal compressive stress at F normal to the foundation progressively decreases. When $h/a \simeq 4.77$ ($\theta \simeq 55.1°$), the normal stress vanishes at F, and the transverse stress is equal to the tensile yield stress $2k$. The ratio $q/2k$ at this stage attains the value 1.923. For higher values of h/a, the stress normal to the foundation becomes tensile, and the field is therefore no longer valid.

When $h/a > 4.77$, the slipline field is modified by adding a right-angled triangle GFG having a uniform tensile stress $2k$ parallel to the foundation (Fig. 8.4). The material in the triangle moves upward with a velocity represented by the vector PG' in the corresponding hodograph. The magnitude of this velocity is equal to the outward speed of the rigid material on either side of the slipline field. Let δ denote the height of the triangle GFG, and d the depth of the point F below the punch face,

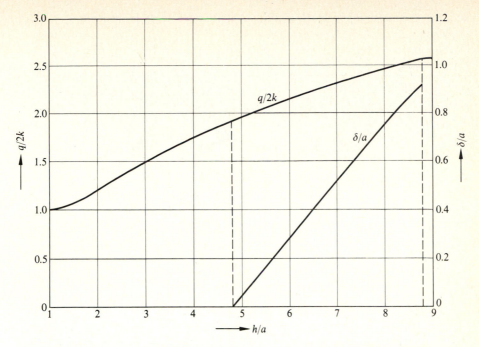

Figure 8.5 Relationship between $q/2k$, h/a, and δ/a at the yield point of a block indented by a flat punch.

so that $h = d + \delta$. The condition of zero resultant thrust across the vertical section through O can be written as

$$qd - 2ka\{I_0(2\theta) + 2\theta[I_0(2\theta) + I_1(2\theta)]\} = 2k\delta$$

Since the mean compressive stress is $-k$ at F, it follows from the Hencky equations that $q = 4k\theta$. Substituting in the above equation, and using (90), Chap. 6, we get

$$\frac{\delta}{a} = 2\theta[2F_1(\theta, \theta) - I_1(2\theta)] - I_0(2\theta) \tag{3}$$

from which δ/a, and hence h/a, can be calculated as functions of θ. The relationship between $q/2k$, h/a, and δ/a is shown graphically in Fig. 8.5, the numerical values of these quantities being given in Table 8.1. The punch pressure attains the value $2k(1 + \pi/2)$ when $h/a \simeq 8.77$ ($\theta \simeq 73.65°$), the ratio δ/a at this stage being 0.91. For $h/a > 8.77$, the indentation will always begin at the pressure given by (1), and the deformation mode will then consist of a displacement of the material to the free surface of the block.[†]

When the foundation is not smooth, the punch pressure necessary to begin the indentation increases to a value q'. The frictional resistance on either side of the

[†] Slipline fields predicting rotation of the ends of the strip toward the indenter have been discussed by P. Dewhurst, *Int. J. Mech. Sci.*, **16**: 923 (1974).

Table 8.1 Indentation of a block resting on a smooth foundation using a flat punch

$\theta°$	h/a	$q/2k$	$\theta°$	h/a	δ/a	$q/2k$
0	1.000	1.000	55	4.754	0	1.921
15	1.605	1.104	60	5.639	0.205	2.094
30	2.440	1.342	65	6.654	0.438	2.269
45	3.644	1.667	70	7.815	0.698	2.443
50	4.162	1.790	73.65	8.772	0.910	2.571

foundation plane is $\mu a q'$ per unit width, where μ is the coefficient of friction. It is reasonable to assume that the effect of friction is to increase the mean compressive stress at each point by a constant amount. Then the punch pressure with friction exceeds that without friction by the same amount, which is $(\mu a/h)q'$, giving†

$$q' = \frac{q}{1 - \mu a/h} \qquad \frac{q}{2k} \leqslant 2.571 \qquad (4)$$

The slipline field for the symmetrical indentation of a rectangular block of height $2h$ between a pair of opposed dies of which $2a$ is identical to that shown in Fig. 8.3, except that the supporting plane is replaced by the horizontal plane of symmetry.‡ The solution holds for all values of h/a lying between 1.0 and 8.72, the tensile stress acting across the horizontal axis at the center of the block being permissible so long as fracture is not produced by the triaxial tension. The punch pressure becomes equal to $2k(1 + \pi/2)$ when $h/a \simeq 8.72(\theta \simeq 77.4°)$; the vertical tensile stress at F equals $0.26k$ at this stage. The punch pressure for $h/a \leqslant 8.72$ may be written as $q = 2kf(h/a)$, where f is obtainable from Table 7.5. The punch pressure remains constant for $h/a \geqslant 8.72$, since the field of Fig. 8.1 then becomes operative.§

(iii) *The critical width* If the width of the block is reduced to a critical value, the plastic zone spreads out to the sides of the block when the load attains the yield point.¶ A single slipline $FGHJK$ then extends from the axis of symmetry to the

---

† R. Hill, *The Mathematical Theory of Plasticity*, p. 257, Clarendon Press, Oxford (1950). A slipline field solution for friction on the foundation has been given by W. Johnson and D. M. Woo, *J. Appl. Mech.*, **25**: 64 (1958). The slipline field is identical to that for side extrusion through a rough container produced by a pair of opposed rams.

‡ If a finite block of width/thickness ratio less than unity is compressed by a pair of overlapping dies, the die pressure is $2k$ for all conditions of friction. A triangular zone of material moves rigidly with each die, while velocity discontinuities occur along straight sliplines passing through the corners of the block and inclined at 45° to the die face.

§ The theoretical predictions have been experimentally confirmed by A. B. Watts and H. Ford, *Proc. Inst. Mech. Eng.*, **1B**: 448 (1952). See also B. B. Murdi and K. N. Tong, *J. Mech. Phys. Solids*, **4**: 121 (1956).

¶ This is a special case of a more general problem discussed by R. Hill, *Phil. Mag.*, **41**: 745 (1950).

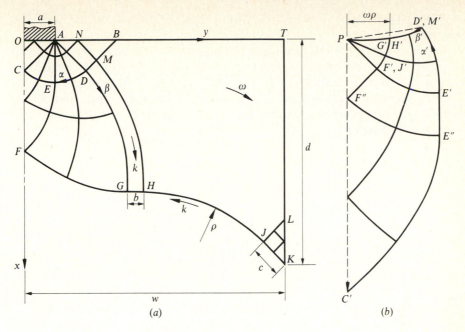

Figure 8.6 Estimation of the critcial width of a block indented by a flat punch. (*a*) Slipline field; (*b*) hodograph.

lateral surface, meeting it at an angle of 45° (Fig. 8.6). The rigid shoulder $BMHJLT$ is assumed to be displaced sideways by sliding over the slipline HJK. Since any rigid body motion in a plane is necessarily a rotation about an instantaneous center, HJ must be a circular arc of some radius ρ. The segment JK, of length c, is assumed straight, so that the triangle JKL is under a uniform compression $2k$ parallel to the surface. Let b denote the length of the straight segment GH, which forms a part of the continuation $DMHG$ adjacent to the standard field $CDGF$, defined by the circular arc CD and the axis of symmetry. Since the net angle turned through in moving from K to F along the slipline is zero, the hydrostatic pressure at F is equal to k. The radius AE must bisect the fan ACD in order that the hydrostatic pressure in the triangle ABD is compatible with that at F. It follows that the angles subtended by CE, FG, and HJ are each equal to $\pi/4$.

The three unknown parameters ρ, b, and c must be determined from the conditions of equilibrium of the rigid shoulder $BMHJLT$. The equilibrium is established by setting to zero the resultant force components and the resultant moment of the distribution of tractions along the boundary $ADGHJK$. The hydrostatic pressure at any point of HJ, situated at an angular distance ϕ from H, is $p = k(1 - \pi/2 + 2\phi)$ by Hencky's equations. The distribution of normal pressure p and tangential stress k along HJ produces a resultant force whose components per unit width of the interface are $-k\rho(\sqrt{2} - 1)$ and $-k\rho(\pi/2 - 1)$ referred to the axes of x and y respectively. The vertical and horizontal components of the force acting on GH are $kb(\pi/2 - 1)$ and $-kb$, while those on JK are $-\sqrt{2}\,kc$ and zero respectively.

Let $(P, -Q)$ denote the rectangular components of the force per unit width exerted on ADG by the plastic material on its left when the hydrostatic pressure is assumed to vanish at C. Since $p = (1 + \pi)k$ at C, the actual force acting on ADG has components $P - k(1 + \pi)(y_0 - a)$ and $-Q + k(1 + \pi)x_0$ per unit width, where (x_0, y_0) are the rectangular coordinates of G. The condition for overall force equilibrium is therefore expressed as

$$(\sqrt{2} - 1)\frac{\rho}{a} - \left(\frac{\pi}{2} - 1\right)\frac{b}{a} + \frac{\sqrt{2}\,c}{a} = \frac{P}{ka} - (1 + \pi)\left(\frac{y_0}{a} - 1\right)$$

$$\left(\frac{\pi}{2} - 1\right)\frac{\rho}{a} + \frac{b}{a} = (1 + \pi)\frac{x_0}{a} - \frac{Q}{ka}$$

(5)

The normal and tangential tractions along the circular arc HJ are statically equivalent to their resultant acting at the center of curvature of JH, together with a counterclockwise couple of magnitude $\frac{1}{4}\pi k\rho^2$ per unit width. The clockwise moment about G due to the tractions along ADG, corresponding to zero hydrostatic pressure at C, is equal to $M - Py_0 - Qx_0$ per unit width, where M is the clockwise moment about the origin O. The actual hydrostatic pressure at C is responsible for an additional clockwise moment of amount $\frac{1}{2}k(1 + \pi)[x_0^2 + (y_0 - a)^2]$ about the point G. Considering all the moments about G, the equation for overall couple equilibrium is obtained as

$$\frac{1}{2}\left(\frac{\rho + c}{a}\right)^2 - \left(\frac{\pi}{2} - 1\right)\frac{\rho^2}{2a^2} + \frac{b}{a}\left[(\sqrt{2} - 1)\frac{\rho}{a} - \left(\frac{\pi}{2} - 1\right)\frac{b}{2a} + \frac{\sqrt{2}\,c}{a}\right]$$

$$= \frac{M}{ka^2} - \frac{Py_0 + Qx_0}{ka^2} + \frac{1 + \pi}{2}\left[\frac{x_0^2}{a^2} + \left(\frac{y_0}{a} - 1\right)^2\right] \quad (6)$$

The ratios x_0/a, y_0/a, P/ka, and Q/ka are directly obtained from Table A-2, the appropriate angular coordinates being $(45°, 90°)$. The value of M/ka^2 can be found by interpolation using Table A-1. Substituting for b/a and c/a, obtained from (5) as linear functions of ρ/a, Eq. (6) is reduced to a quadratic in ρ/a. The final results are

$$\frac{\rho}{a} \simeq 5.686 \qquad \frac{b}{a} \simeq 0.490 \qquad \frac{c}{a} \simeq 0.937$$

These results furnish the width ratio $w/a \simeq 8.593$. The depth of K below the plane surface is given by $d/a \simeq 7.290$, which is very nearly equal to twice the ratio of OF to OC.

Let ω be the angular velocity of the instantaneous rotation of the rigid shoulder at the yield point. The resulting velocity discontinuity propagating along $KJHGFEA$ is then of amount $\omega\rho$. Since the material beneath $FHJK$ is at rest, this slipline is mapped into the circular arc $H'J'$ whose center P is the pole of the hodograph (Fig. 8.6). The image of the boundary HM must be a geometrically similar curve, so that the radius of curvature at any point of $H'M'$ is ω times that at the corresponding point of H. The two curves $G'F'$ and $G'D'$ uniquely define the image field $G'F'E'D'$ (first boundary-value problem). The points immediately to the left of the slipline FEA are mapped into the curve $F''E''$, which is parallel to $F'E'$, the distance between them being equal to the velocity discontinuity $\omega\rho$. The remaining field $F''E''C'$ of the hodograph is defined by $F''E''$ and its reflection in $F''C'$. The angular velocity ω is determined from the fact that the vector PC' represents the velocity of the punch. The material in the triangle ADB moves sideways with a velocity represented by the vector PD'. It follows from the sense of rotation of the sliplines and their images, and the direction of shearing along the velocity discontinuities, that the rate of plastic work is everywhere positive.

If the stress field could be continued throughout the block without overloading, the solution would be complete, and the width of the block would then represent the actual critical width. Otherwise the width is less than critical, since the actual yield point load associated with this width is lower than the Prandtl value. On the other hand, the width corresponding to the statically admissible field of Fig. 8.2 must exceed the critical value, since the yield point load in this case is higher than the Prandtl value. It follows that the critical value of the ratio w/a lies between the bounds 8.593 and 8.621, which are remarkably close to one another.†

8.2 Indentation by a Rigid Wedge

(i) *Frictionless indentation* Consider the penetration of a smooth rigid wedge into a semi-infinite mass of rigid/perfectly-plastic material, so that the bisector of the wedge angle 2ψ is perpendicular to the plane surface of the medium.‡ The slipline field maintains its shape during the indentation, but increases in size by a scale factor equal to the depth of penetration c. As the wedge is pressed into the medium, a raised coronet is formed on either side of the wedge (Fig. 8.7). The slipline field on each side consists of a centered fan BCD separating a pair of uniformly stressed triangular regions ABC and BDE, the displaced surface BE being assumed straight. All sliplines meet AB and BE at $45°$ in order to satisfy the stress boundary conditions. If the fan angle BCD is denoted by θ, the angle of inclination of BE to the original plane surface is $\psi - \theta$. Since the height of E above A is equal to c, it follows from geometry that

$$b[\cos\psi - \sin(\psi - \theta)] = c \qquad (7)$$

where b denotes the length $AB = BE$. In view of the incompressibility of the material, the triangles AOE and ABE must be equal in area, the triangle AFE being common to both. Since the perpendicular distance of E from AB is equal to $b\cos\theta$, we obtain

$$c[\sin\psi + \cos(\psi - \theta)] = b\cos\theta \qquad (8)$$

---

† When the width of the block is less than critical, $q/2k$ is given to a close approximation by the right-hand side of (90) with h replaced by the semiwidth w.

‡ R. Hill, E. H. Lee, and S. J. Tupper, *Proc. R. Soc.*, A, **188**: 273 (1947).

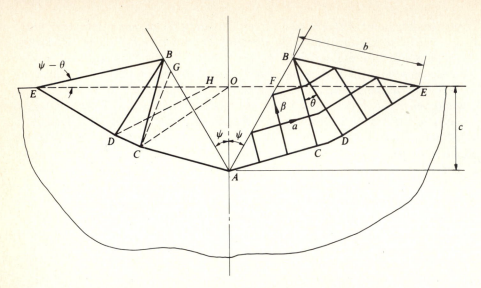

Figure 8.7 Indentation of a semi-infinite block by a rigid lubricated wedge.

The elimination of b/c between (7) and (8) furnishes the relationship between θ and ψ as

$$\cos(2\psi - \theta) = \frac{\cos\theta}{1 + \sin\theta} = \tan\left(\frac{\pi}{4} - \frac{\theta}{2}\right)$$

which is readily solved for ψ in terms of θ. The result is

$$\psi = \frac{1}{2}\left\{\theta + \cos^{-1}\tan\left(\frac{\pi}{4} - \frac{\theta}{2}\right)\right\} \qquad (9)$$

When θ has been found for a given ψ, the ratio h/c can be calculated from (7) or (8). It may be noted that $\theta < \psi$ for all nonzero values of $\psi < \pi/2$. When $\psi = \pi/2$, θ also has the same value, and the slipline field represents only the incipient mode at the yield point.

Since the normal component of velocity vanishes on the plastic/rigid boundary $ACDE$, the velocity component along the straight sliplines normal to the boundary must be zero by Geiringer's equation. It follows that the steamlines coincide with the sliplines parallel to the boundary, and the velocity along them has a constant magnitude equal to $\sqrt{2}\sin\psi$ throughout the field, the downward speed of the wedge being taken as unity with c as the time scale. The velocities of the lip BE and the vertex A have components $\sin\psi$ and $\cos(\psi - \theta)$ respectively along the normal to BE. The perpendicular distance of A from BE therefore increases at the rate $\sin\psi + \cos(\psi - \theta)$, which remains constant during the indentation. It follows that the left-hand side of (8) represents the distance of A from BE when the depth of

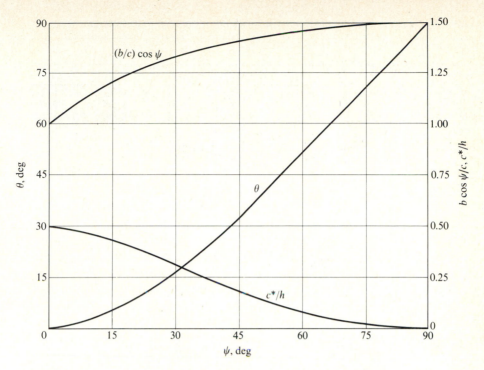

Figure 8.8 Variation of geometrical parameters with wedge semiangle in the indentation of a block.

penetration is c. Since this distance is also equal to $b \cos \theta$, Eq. (8) is recovered from the kinematical standpoint.†

The mean compressive stress increases from k on the free surface BE to a value $k(1 + 2\theta)$ on the wedge face AB by Hencky's equation. The uniformly distributed normal pressure on the wedge is therefore given by

$$q = 2k(1 + \theta) \tag{10}$$

Equations (9) and (10) express the relationship between q and ψ parametrically through θ. The vertical load per unit width of the wedge is $2qb \sin \psi$, and the work done per unit width is $qcb \sin \psi$. Table 8.2 gives the values of θ, $q/2k$, and c/b for various semiangles of the wedge. The results have been found to be in good agreement with experiments using a lubricated wedge.‡ The variations of θ and $(b/c) \cos \psi$ with the semiangle ψ are represented by the solid curves of Fig. 8.8. It follows from Eqs. (8) and (7) that $(b/c) \cos \theta$ and $(b/c) \cos \psi$ tend to their limiting values of 2 and 3/2 respectively as ψ tends to $\pi/2$.

A consideration of the trajectories of the particles (see below) indicates that the

---

† The effect of work-hardening is to decrease the slope of the raised lip while increasing its length for a given depth of penetration.

‡ D. S. Dugdale, *J. Mech. Phys. Solids*, **2**: 14 (1953).

Table 8.2 Results for indentation of a block by a smooth symmetrical wedge

$\psi°$	0	15	30	45	60	75	90
$\theta°$	0	5.43	17.34	32.94	50.84	70.11	90
c/b	1.0	0.800	0.647	0.498	0.341	0.174	0
$q/2k$	1.0	1.095	1.302	1.575	1.887	2.224	2.571
$h/b*$	2.0	1.858	2.089	2.932	4.295	6.230	8.772
$c*/h$	0.5	0.430	0.310	0.170	0.079	0.028	0
V/U	0	0.268	0.216	0.225	0.197	0.163	0.127
$\phi°$	0	0	7.51	21.99	39.07	56.21	73.65

material initially occupying the region OAC has consistently moved parallel to AC after being plastic. Consequently, this material currently remains in the region GAC, where OG is in the direction of the velocity. Similarly, the material which is finally in the triangle BDE has consistently moved parallel to DE after becoming plastic. The initial position of this material is therefore HDE, where H is the intersection of OE with the parallel to DE through B. While the deformation in these two regions consists of simple shear in the respective directions of motion, the material originally in $OCDH$ is finally displaced to $GCDB$ through a complicated mode of deformation. The particles initially lying on the surface OH are drawn along the wedge face so as to remain in contact with it over the length GB in the final state.

(ii) Distortion of a square grid The geometrical configuration in which the characteristic length c appears as the unit of length is called the *unit diagram*. A particle P whose position vector is $\mathbf{r}$ in the actual physical plane is represented in the unit diagram by a point P' whose position vector is $\boldsymbol{\rho} = \mathbf{r}/c$. In view of the geometrical similarity of the process, the stress and velocity at each stage are functions of the single variable $\boldsymbol{\rho}$. If the velocity $d\mathbf{r}/dc$ of the particle P is denoted by $\mathbf{v}$, the velocity $d\boldsymbol{\rho}/dc$ of the corresponding point P' in the unit diagram is given by†

$$c(d\boldsymbol{\rho}/dc) = \mathbf{v} - \boldsymbol{\rho} \tag{11}$$

which is readily obtained on differentiating the relation $\mathbf{r} = c\boldsymbol{\rho}$. The above equation indicates that P' moves toward a focus S whose position vector is $\mathbf{v}$ in the unit diagram. The path of the particle P in the physical plane is represented by the trajectory of the point P' in the unit diagram.

Since the velocity $\mathbf{v}$ has the constant magnitude $\sqrt{2} \sin \psi$ throughout the deforming region, the focus for the entire field must lie on a circular arc of radius $\sqrt{2} \sin \psi$ with centre O', which is the image of the fixed origin O in the unit diagram (Fig. 8.9). The arc NM of angular span θ is defined by the radii $O'N$ and $O'M$ drawn parallel to AC and DE respectively. Let $A'C'D'E'B'$ be the image of the slipline field

---

† R. Hill, E. H. Lee, and S. J. Tupper, op. cit. An approximation to the distortion, based on assumed stress discontinuities, has been presented by P. G. Hodge, *J. Appl. Mech.*, **17**: 257 (1950).

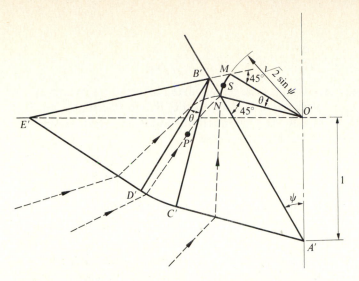

Figure 8.9 Unit diagram showing trajectories of particles in wedge indentation.

ACDEB. Since $O'N$ is inclined to $A'B'$ at an angle of 45°, it follows from the geometry of the triangle $A'O'N$ that N lies on $A'B'$. Further, M lies on $E'B'$ produced, since the sum of the projections of $O'M$ and $O'A'$ perpendicular to $B'E'$ is equal to the projection of $A'B'$ in view of Eq. (8). It is important to note that $O'A'NM$ is actually the hodograph of the process, with $O'A'$ representing the velocity of the wedge and $A'N$ the velocity discontinuity along the wedge face.

Until a particle is overtaken by the expanding slipline field, its velocity is zero, and consequently the trajectory of its image in the unit diagram is a straight line directed toward O'. When the image point crosses the boundary $A'C'D'E'$, the particle has been engulfed by the plastic flow field. There are three kinds of trajectory in the unit diagram, depending on whether the point enters the image field by crossing $A'C'$, $C'D'$, or $D'E'$. In the domains $A'C'B'$ and $D'E'B'$, the trajectories are straight lines directed toward the foci N and M respectively. In the remaining domain $C'D'B'$, the trajectory is curved, and the focus moves from N to M as the point moves from $D'B'$ to $C'B'$. The focus S at any instant is the intersection of the arc NM with the perpendicular from O' to $B'P'$. It may be noted that all particles originally situated on the same radius through O have the same trajectory in the unit diagram.

To find the distortion of an originally square grid, it is necessary to calculate the final position of the corner of a square whose initial position is $\mathbf{r}_0$. Let c_0 (less than c) be the depth of penetration of the wedge when the corner is first overtaken by the expanding slipline field. A line drawn through O' in the direction of $\mathbf{r}_0$ meets the boundary $A'C'D'E$ at a point whose position vector is $\boldsymbol{\rho}_0 = \mathbf{r}_0/c_0$. This point subsequently moves along the trajectory through the tip of the vector $\boldsymbol{\rho}_0$ as the depth of penetration is increased from c_0 to c. If the distance traversed by the point during this interval, and the final focal distance, are denoted by s and $f(s)$

respectively, then by (11),

$$\frac{dc}{c} = \frac{ds}{f(s)} \qquad \text{or} \qquad \ln\frac{c}{c_0} = \int_0^s \frac{ds}{f(s)} \tag{12}$$

When the trajectory has been found, s can be calculated in terms of c by using (12), and the final position $\mathbf{r} = c\mathbf{\rho}$ therefore follows. In $A'C'B'$ and $D'E'B'$, the result is simplified by the fact that $f(s) = d - s$, where d is the distance of the point $\mathbf{\rho}_0$ from the appropriate focus N or M. Then the solution is

$$\frac{c}{c_0} = \frac{d}{d-s}$$

The integral in (12) must be evaluated numerically in the domain $C'D'E'$. The calculated distortion of the grid for a 60° wedge is shown in Fig. 8.10.

(iii) *Influence of friction* When the wedge is rough, and the wedge angle not too large, the slipline field is modified as shown in Fig. 8.11a, only one half of the field being considered. The raised lip remains straight, but is reduced in both height and slope with BE exceeding AB. The straight α lines now meet the wedge face at an angle $\lambda < \pi/4$, so that the normal pressure on the contact surface is $p + k \sin 2\lambda$ and the frictional stress is $k \cos 2\lambda$, where p is the hydrostatic pressure in the region ABC. Since $p = k(1 + 2\theta)$ by Hencky's equation, where θ is the fan angle CBD, the coefficient of friction is

$$\mu = \frac{\cos 2\lambda}{1 + 2\theta + \sin 2\lambda} \tag{13}$$

in view of the relative motion between the wedge face and the material. The normal pressure q acting on the wedge face is

$$q = k(1 + 2\theta + \sin 2\lambda) \tag{14}$$

For a given value of μ, λ and θ depend on the wedge semiangle ψ. From geometry, the angle made by the raised lip BE with the horizontal is $\psi - \phi$, where

$$\phi = \theta + \lambda - \pi/4$$

If the contact length AB is denoted by b, then the lip BE is of length $\sqrt{2}\,b \cos \lambda$. The fact that E lies on the original surface is expressed by the relation

$$b[\cos \psi - \sqrt{2} \cos \lambda \sin (\psi - \phi)] = c \tag{15}$$

The condition of continuing geometrical similarity during the penetration furnishes the relation

$$c[\sin \psi + \sqrt{2} \cos \lambda \cos (\psi - \phi)] = \sqrt{2}\,b \cos \lambda \cos \phi \tag{16}$$

which is easily derived from the constancy of volume. For selected values of λ and ψ, the corresponding values of ϕ and b/c can be calculated from (15) and (16) by a

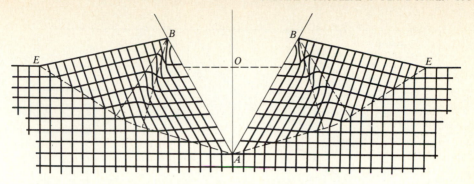

Figure 8.10 Deformed square grid after indentation by a lubricated rigid wedge.

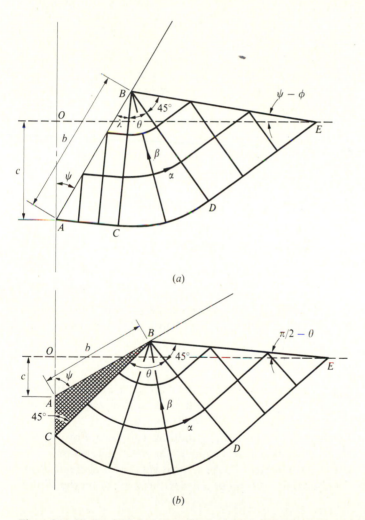

(a)

(b)

Figure 8.11 Slipline fields for indentation by a rough wedge. (a) Without false nose; (b) with false nose.

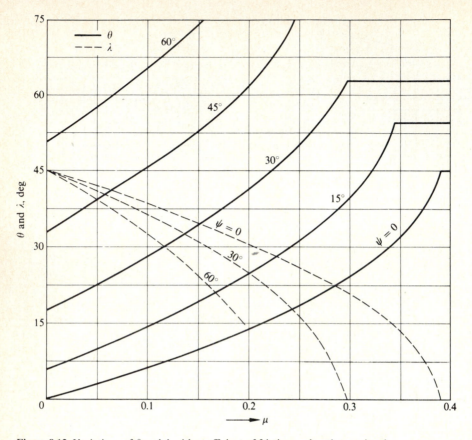

Figure 8.12 Variations of θ and λ with coefficient of friction and wedge semiangle.

process of trial and error.† The variations of θ and λ with μ for different values of ψ are shown in Fig. 8.12. When θ and λ have been found, the contact pressure q can be calculated from (14). The vertical load per unit width of the wedge is easily shown to be

$$P = 2qb(\sin \psi + \mu \cos \psi)$$

Computations indicate that the effect of friction is to decrease b/c by a fraction $\frac{1}{4}\mu \sin 2\psi$, and increase $q/2k$ by a fraction $\mu(1 + \frac{1}{4} \sin 2\psi)$ to a close approximation. When ψ tends to zero, b/c tends to unity and θ tends to $\pi/4 - \lambda$.

As μ increases, one or other of two limiting situations can arise. For $\psi \leqslant \pi/4$, the angle λ vanishes, and the frictional stress is equal to the shear yield stress k, with $\mu = 1/(1 + 2\theta)$. As ψ increases from 0 to 45°, the critical value of μ decreases from 0.389 to 0.257 approximately. Higher values of μ are without effect on the slipline

† J. Grunzweig, I. M. Longman, and N. J. Petch, *J. Mech. Phys. Solids,* **2**: 81 (1954).

field. When $\psi \geqslant \pi/4$, the solution is valid for those values of μ for which AC is inclined to the downward vertical at an angle less than $\pi/4$, in order that the rigid corner at A is not overstressed. The limiting value of λ in this case is $\psi - \pi/4$. Numerical values of θ, b/c, μ, and P/kc corresponding to the limiting state are given in Table 8.3.

For obtuse-angled wedges, when the friction is sufficiently high, the slipline field involves a 90°-cap of dead metal ABC attached to the wedge face† as shown in Fig. 8.11b. There is a velocity discontinuity of amount $U/\sqrt{2}$ propagating along $BCDE$, where U is the downward speed of the wedge. If AB is taken to be of length b, then

$$BE = 2b \sin \psi \qquad AC = b(\sin \psi - \cos \psi)$$

The acute angle between the lip BE and the wedge axis is equal to the fan angle θ, which means that the lip angle is $\pi/2 - \theta$. Hence the depth of penetration is

$$c = b(\cos \psi - 2 \sin \psi \cos \theta) \tag{17}$$

Since the triangles ABE and OAE must be equal in area due to the incompressibility of the material, we have

$$c(1 + 2 \sin \theta) = 2b \sin (\psi + \theta)$$

Eliminating b/c between the last two relations, we obtain after some algebra,

$$\cot \psi = 4 \cos \theta(1 + \sin \theta) \tag{18}$$

Since the material in the dead metal cap is not necessarily plastic, the actual distribution of the contact pressure cannot be estimated. The vertical load per unit width of the indenter is

$$P = 4kb(1 + \theta) \sin \psi \tag{19}$$

in view of the uniform normal pressure $k(1 + 2\theta)$ and the tangential stress k acting along BC, which is of length $\sqrt{2} \, h \sin \psi$. Numerical values of θ, b/c, and P/kc for various semiangles of the wedge are given in Table 8.4.

Table 8.3 Results for the limiting field corresponding to Fig. 8.11a

ψ°	0	15	30	45	60	75	90
θ°	45.00	54.74	67.67	82.79	84.12	86.29	90.00
b/c	1.000	1.196	1.459	1.888	2.777	5.568	∞
P/kc	2.00	4.11	7.43	10.10	23.74	53.91	∞
μ	0.389	0.344	0.297	0.257	0.195	0.102	0

† W. Johnson, F. U. Mahtab, and J. B. Haddow, *Int. J. Mech. Sci.*, **6**: 329 (1964); J. B. Haddow, *Int. J. Mech. Sci.*, **9**: 159 (1967).

Table 8.4 Indentation by a rough obtuse-angled wedge, Fig. 8.11b

ψ	45°	50°	60°	70°	80°	90°
θ	82.79	83.95°	85.85°	87.38°	88.73°	90°
b/c	1.888	2.078	2.669	3.905	6.587	∞
P/kc	10.10	15.70	23.10	37.04	76.92	∞
μ	0.257	0.240	0.193	0.027	0.013	0

The value of P/kc in the false nose solution are found to be slightly lower than those corresponding to the limiting frictional condition in the solution without the false nose. For $77° < \psi < 90°$, the false nose solution gives a lower value of P/kc than for the frictionless wedge. Since the limit theorems are not directly applicable to this problem, in which the change in geometry is unspecified, it is not possible to say that the lower load gives a better bound. The minimum coefficient of friction required by the false nose solution may be estimated by assuming a uniform plastic state of stress existing in the cap. The minimum value of μ is then obtained by setting $\lambda = \psi - \pi/4$ in (13), and is included in Table 8.4. This value is slightly smaller than the limiting value of μ in the preceding solution.

(iv) *Critical penetration for finite thickness* Suppose that a block of finite thickness h rests on a smooth horizontal base and is penetrated normally by a smooth symmetrical wedge of semiangle ψ. For sufficiently small values of the depth of penetration, the deformation mode is identical to that for a semi-infinite block. At some stage of the indentation, the element of the rear surface directly below the wedge tip is stressed to the yield limit, and a second plastic zone begins to spread upward from the foundation. Eventually, the upper plastic zone extends continuously downward to join the lower plastic zone. At this stage, the constraint fails and the rigid ends begin to separate by sliding over the foundation. In the proposed slipline field, Fig. 8.13, the two bounding sliplines $BFGHS$ intersect one another at a point T which is situated at a height δ above the foundation.[†] The defining angle CBF, denoted by ϕ, depends on the wedge semiangle ψ. The critical depth of penetration c^* and the associated length of contact b^* are determined from the condition that the resultant horizontal thrust across $BATS$ is zero.

Since the sliplines meet AT at 45°, the angular span of the fan ACJ is equal to $\pi/2 - \psi$. The material triangle STS is under a uniform tensile stress $2k$ acting parallel to SS. The stress is also uniform in the square $AJIJ$, where the horizontal principal stress is less than $2k$. Since the hydrostatic pressure has the value k at E and $-k$ at T, the contact pressure along AB may be written as

$$q = 2k\left(\frac{\pi}{2} - \psi + 2\phi\right) = 2k(1 + \theta) \tag{20}$$

[†] The slipline field is due to R. Hill, *J. Mech. Phys. Solids*, **1**: 265 (1953). The problem of oblique penetration has been treated by S. A. Meguid and I. F. Collins, *Int. J. Mech. Sci.*, **19**: 361 (1977).

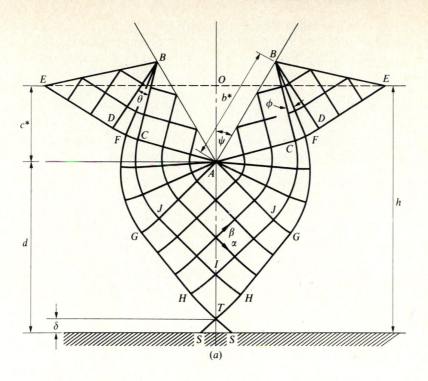

(a)

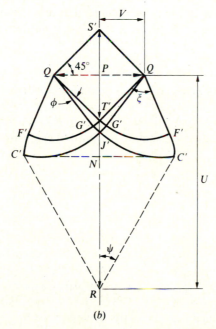

(b)

Figure 8.13 Critical penetration of a wedge into a block of finite depth ($\psi \geqslant 22.17°$). (a) Slipline field; (b) hodograph.

where θ is defined by (9). The above relation enables us to calculate ϕ for any given ψ. The longitudinal tensile force Q acting across AT is obtained from Eq. (147), Chap. 6, where it is necessary to write $\pi/2 - \psi$ for ψ, ϕ for α, b^* for $2a$, and $-p_0$ for p_0. Since the mean compressive stress at C is $p_0 = q - k$, the result is

$$Q = 2kb^*[-4L(\phi, \xi) + I_0(2\sqrt{\phi\xi})] \qquad \xi = \frac{\pi}{2} - \psi + \phi$$

in view of (20). The condition of zero resultant horizontal force across $STAB$ requires $Q + 2k\delta = qb^* \cos \psi$. Substituting in the above equation, we get

$$\delta = b^*[(1 + \theta) \cos \psi + 4L(\phi, \xi) - I_0(2\sqrt{\phi\xi})] \qquad (21)$$

By Eq. (146), Chap. 6, where $s = d - \delta$ and $2a = b^*$, the height of the vertex A above the foundation is

$$d = b^*[(1 + \theta) \cos \psi + 2F_1(\phi, \xi) + 4L(\phi, \xi)] \qquad (22)$$

For a given wedge angle, δ/b^* and d/b^* can be computed from (21) and (22), using interpolations based on Table A-3. The proposed slipline field holds for $\phi \geqslant 0$, which is equivalent to

$$\theta + \psi \geqslant \pi/2 - 1 \qquad \text{or} \qquad \psi \geqslant 22.17° \qquad \theta \geqslant 10.53°$$

The limiting case $\phi = 0$ corresponds to $d/b^* = (1 + \theta) \cos \psi \simeq 1.096$, giving $h/b^* \simeq 1.820$ and $c^*/h \simeq 0.398$, the point T being at a height 0.096h above the foundation.

The construction of the hodograph begins with the vector PQ of magnitude V, representing the unknown velocity of the rigid ends. There is a velocity discontinuity of amount $\sqrt{2}V$ across the sliplines $BFGTS$ on either side. The triangle STS moves vertically upward with a speed V, and the particle just above T moves downward with the same speed. The remainder of the hodograph is defined by the equal circular arcs $T'G'F'$ centered at Q and having an angular span ξ. The velocity of the wedge is represented by the vector PR, where R is located by drawing $C'R$ at an angle ψ to the vertical. If the speed of the wedge is denoted by U, then from geometry,

$$\frac{U}{V} = \frac{PN}{PQ} + \frac{NC'}{PQ} \cot \psi$$

where $C'N$ is perpendicular to PR. Since the ratios PN/PQ and NC'/PQ can be obtained from the appropriate table (Appendix), the ratio V/U can be computed for any given ψ. As ϕ tends to zero, V/U approaches the value $\tan 11.09° \simeq 0.196$.

The ratio δ/b^* given by (21) vanishes when $\psi \simeq 29.5°$ and $\psi \simeq 54.5°$, its value for $\psi = 60, 75,$ and $90°$ being 0.07, 0.37, and 0.91 respectively. For $29.5° < \psi < 54.5°$, the vertical principal stress at T becomes compressive and the triangle STS disappears. The height of the wedge tip is then given by

$$d = b^*[I_0(2\sqrt{\phi\xi}) + 2F_1(\phi, \xi)]$$

in view of (146), Chap. 6. The angle ϕ must be determined from the condition that the resultant horizontal thrust $qb^* \cos \psi$ on the wedge face is equal to the longitudinal tensile force Q across AT. Using (147), Chap. 6, it is easily shown that

$$\frac{I_0(2\sqrt{\phi\xi}) - 4L(\phi, \xi) - (1 + \theta) \cos \psi}{I_0(2\sqrt{\phi\xi}) + 2F_1(\phi, \xi)} = 1 + \theta - \phi - \xi \qquad (23)$$

where $\xi = \pi/2 - \psi + \phi$. When ϕ has been estimated from (23) by trial and error, d/b^* and h/b^* can be evaluated. For $54.5° < \psi \leqslant 90°$, the plastic triangle reappears at the base, and Eqs. (20) to (22) again hold for the geometry of the field.

For $\psi \leqslant 22.17°$, the plastic zone extending from the wedge tip to the foundation consists of an isosceles right triangle as shown in Fig. 8.14a. The sideways thrust on each contact face is then balanced by a uniform tension $2k$ across the height of the triangle. Equating $qb^* \cos \psi$ to $2k(h - c^*)$, and using (7) and (10), we obtain

$$h = b^*[(2 + \theta) \cos \psi - \sin (\psi - \theta)] \qquad \psi \leqslant 22.17° \qquad (24)$$

The material in the triangle FAF rises vertically with a speed $V = U \tan \psi$, which is also the speed of separation of each rigid half. The material in the corner CAF is not overstressed, since the pressure difference $2k(1 + \theta)$ across the faces AC and AF is less than $2k(\pi/2 - \psi)$ when $\psi < 22.17°$. The calculated values of h/b^*, c^*/h, V/U, and ϕ over the whole range of values of ψ are given in Table 8.2. A graphical plot of c^*/h against ψ is included in Fig. 8.8. The effect of friction on the wedge face is to decrease the critical penetration by an amount $\mu(h - c^*) \cos^2 \psi$ approximately.

If a block of thickness $2h$ is indented on opposite sides by a pair of identical wedges with their vertices approaching one another, the point T of Fig. 8.13 must always lie on the horizontal axis of symmetry, which replaces the smooth foundation. The principal stress normal to this axis has intensity $2k(\phi + \xi - \theta - 1)$, where ϕ is given by (23). This stress is tensile in the range $29.5° < \psi < 54.5°$, and compressive outside it. The limit of validity of the field corresponds to $\phi = 0$, which is equivalent to

$$\theta + \psi + (1 + \theta) \cos \psi = \frac{\pi}{2} \qquad \text{or} \qquad \psi \simeq 19.48°$$

The values of h/b^* and c^*/h for this limiting case are 1.752 and 0.429 respectively. For smaller values of ψ, an expression for h/b^* may be written down on the basis of a uniform tensile stress $t = 2k(\pi/2 - \psi - \theta)$ acting across the section between the wedge tips. The condition of horizontal equilibrium is $t(h - c^*) = qb^* \cos \psi$, which gives

$$\frac{h}{b^*} = \frac{\pi/2 + 1 - \psi}{\pi/2 - \theta - \psi} - \sin (\psi - \theta) \qquad \psi \leqslant 19.48° \qquad (25)$$

in view of the relations (7) and (10). The ratio h/b^* according to this formula

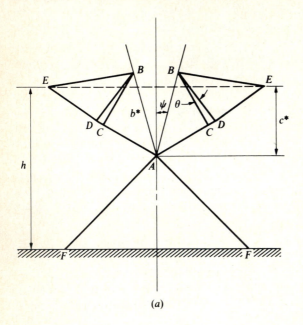

(a)

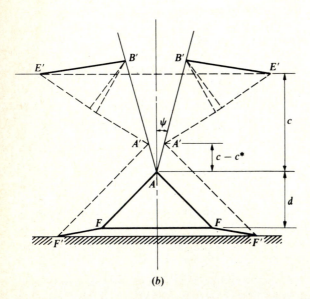

(b)

Figure 8.14 Critical penetration and subsequent separation due to indentation ($\psi \leqslant 22.17°$). (a) Onset of separation; (b) a later instant during the process.

steadily decreases with ψ, reaching the limiting value 1.637 when ψ is vanishingly small.†

(v) *Cutting of metal strips* The above analysis is directly applicable to the cutting operation in which a metal strip or sheet is split in two by a smooth knife-edged tool. The initial penetration of the tool gives rise to the piling-up mode of deformation in which the plastic zone does not extend to the base. When the penetration reaches the critical depth, the two halves begin to separate, and no further material is raised to the upper surface.‡ If the sheet is placed on a rigid base, and the semiangle of the wedge is less than 22.2° (which is the case of practical interest), the upper plastic zone $ACDEB$ begins to unload, while a depression appears along the base of a plastic triangle FAF having its vertex at the wedge tip. During the period of separation, consider an instant when the tool has penetrated by an additional distance $c - c^*$, as depicted in Fig. 8.14b, the speed of the tool being taken as unity. The material at each instant is sheared along the momentary plastic boundaries terminating at the tool tip. The triangle FAF is stressed to the yield point by a uniform tension $2k$, the material outside this region being nonplastic. Since the upward speed of the plastic material at each stage is $\tan \psi$, the total height of the depression is $(c - c^*) \tan \psi$, which is the distance between FF and $F'F'$. The perpendicular distance of F' from AF is $(c - c^*)(1 + \tan \psi)/\sqrt{2}$, while the displacement of F' parallel to AF is $\sqrt{2}(c - c^*) \tan \psi$. Hence the instantaneous simple shear at each stage is of amount

$$\gamma = \frac{2 \tan \psi}{1 + \tan \psi}$$

which increases from 0 to 0.534 as ψ is increased from 0 to 20°. The horizontal distance between F and F' is $(c - c^*)(1 + 2 \tan \psi)$, and the height of A above FF is

$$d = (h - c^*) - (c - c^*)(1 + \tan \psi) \tag{26}$$

The separation is complete when d vanishes. The corresponding value of c/h can be calculated from the above equation, using Table 8.2. In actual practice, failure by fracture will occur before this limiting penetration is attained. The horizontal thrust on each contact face must be equal to $2kd$ for equilibrium. Consequently, the vertical load per unit width during the process is $4kd \tan \psi$, which decreases linearly with the depth of penetration. Since the load increases linearly with penetration for $c \leqslant c^*$, the greatest value of the load is $4k(h - c^*) \tan \psi$, which occurs at the moment when the separation begins. The variation of the load P with the

---

† The effect of friction along the wedge face on the critical penetration for indentation by opposed wedges has been investigted by W. Johnson and H. Kudo, *Int. J. Mech. Sci.*, **2**: 294 (1961). The corresponding problem for a block resting on a smooth foundation has been considered by B. Dodd and K. Osakada, *Int. J. Mech. Sci.*, **16**: 931 (1974).

‡ R. Hill, op. cit. The separation process in the cutting of a strip by a pair of opposed tools has been treated by H. Kudo and K. Tamura, *JSME, Semi-Int. Symp.*, **168**: 139 (1967).

penetration c may be written from (26) as

$$\frac{P}{4kh} = \left(\frac{c}{c^*} - \frac{c}{h}\right) \tan \psi \qquad\qquad c \leqslant c^*$$

$$\frac{P}{4kh} = \left(1 - \frac{c}{h}\right) \tan \psi - \frac{c - c^*}{h} \tan^2 \psi \qquad c \geqslant c^* \tag{27}$$

The work done per unit width of the strip before the beginning of separation is $2kc^*(h - c^*) \tan \psi$, and that during the separation process is $k\gamma(h - c^*)^2$. The total work done during the entire cutting operation does not differ significantly from the value $kh^2 \tan \psi$. When there is friction on the tool face, the cutting load is multiplied by the factor $(1 + \mu \cot \psi)$ approximately.

(vi) *Estimation of the critical width* If the width of the block is small compared to its thickness, the plastic zone spreads out to the sides of the block as the indentation continues. For a given depth of penetration, the critical width $2w$ is such that a slipline $AGHJKL$ extends from the wedge tip to the lateral surface of the block (Fig. 8.15). At this moment, the deformation mode involves a lateral displacement of the rigid corner by sliding over the slipline JKL. Since two rigid bodies can slide

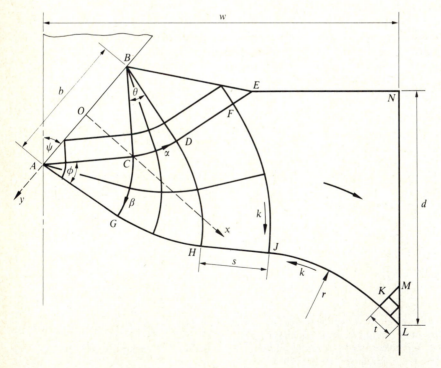

Figure 8.15 Slipline field for a block of critical width corresponding to given penetration by a rigid wedge.

over one another only if the surface of contact is cylindrical, JK must be a circular arc of some radius r. It is also necessary to include straight segments GH and KL of lengths s and t respectively, and a plastic triangle KLM in a state of uniaxial compression $2k$ parallel to the surface.† The field $CDFJHG$ is constructed in the usual manner, starting from the circular arcs CD and CG with angular spans θ and ϕ respectively.

By simple geometry, the net angle turned through in going from L to A along the slipline $LKJHGA$ is $\pi/2 - (\psi + \phi)$ in the counterclockwise sense. Since the mean compressive stress is equal to k at both E and L, the Hencky equations indicate that the stress will be continuous at any point, such as C, if

$$\phi = \frac{\pi}{4} - \frac{\psi - \theta}{2}$$

where θ is given by (9). It follows from Hencky's first theorem that the angle turned through by JK is also equal to ϕ. The rigid material in the corner A will be stressed below the yield limit so long as AG is inclined at an angle greater than $\pi/4$ to the downward vertical. The validity of the slipline field therefore requires

$$\phi + \psi \leqslant \pi/2 \qquad \text{or} \qquad \psi \leqslant 50.6°$$

The three unknown quantities r, s, and t can be determined from the conditions of overall equilibrium of the rigid shoulder. This is ensured by the vanishing of the resultant force components and the resultant couple due to the tractions distributed over $BAGHJKL$. When the unknowns have been found, the critical width of the block is obtained from simple geometry, using the fact that HJ makes a clockwise angle $(\psi - \theta)/2$ with the horizontal. The perpendicular distance of H from the axis of symmetry is equal to $x_0 \cos \psi + (b/2 - y_0) \sin \psi$, where (x_0, y_0) are the rectangular coordinates of H, their values being obtainable from the appropriate table (Appendix).

The distribution of normal and shear stresses over the circular arc JK is statically equivalent to the forces $kr(\cos \phi + \sin \phi - 1)$ and $kr(2\phi + \cos \phi - \sin \phi - 1)$ per unit width acting at its center of curvature T (not shown), along TJ and parallel to JH respectively, together with a counterclockwise couple equal to $kr^2\phi$ per unit width. The distribution of tractions along the slipline AGH is equivalent to the specific forces $-P - k(1 + 2\theta)(b/2 - y_0)$ and $Q - k(1 + 2\theta)x_0$ acting at H in the x and y directions respectively, together with the specific clockwise couple

$$M - Py_0 - Qx_0 + k\left(\frac{1}{2} + \theta\right)\left\{x_0^2 + \left(\frac{b}{2} - y_0\right)^2\right\}$$

where $-P$, Q, and M are the equivalent forces and couple at the origin O corresponding to zero hydrostatic stress in ABC. The ratios P/kb, Q/kb, and $4M/kb^2$ are identical to those of Table A-1 for the same angular coordinates (θ, ϕ). In setting up the equations of equilibrium, it is generally convenient to take

† R. Hill, *Phil. Mag.*, **41**: 745 (1950).

moment about H, and resolve the forces parallel to the coordinate axes, noting the fact that HJ makes a counterclockwise angle of $(\psi + \theta)/2$ with the x axis.

Using the equations of force equilibrium, s/b and t/b can be expressed as linear functions of r/b. In the limiting case of $\psi = 50.6°$, for which $\theta = \phi = 39.4°$, these relations are found to be

$$\frac{s}{b} = 1.3882 - 0.4565 \frac{r}{b} \qquad \frac{t}{b} = 0.9832 - 0.4113 \frac{r}{b}$$

Inserting them into the equation of couple equilibrium reduces it to the quadratic

$$0.0989 \frac{r^2}{b^2} + 0.3309 \frac{r}{b} - 1 = 0$$

and the solution for the particular geometry of the field finally becomes

$$\frac{r}{b} = 1.920 \qquad \frac{s}{b} = 0.512 \qquad \frac{t}{b} = 0.194$$

These values furnish $w = 3.201b$, which means that the critical width is 4.142 times the width of the impression. The point L is found to be at a distance $2.078b$ below the corner N. For a knife-edged indenter ($\psi \simeq 0$), the inner boundary of the rigid shoulder consists of a pair of circular arcs, and an explicit analytical solution is possible, the results being

$$\frac{r}{b} = \frac{4}{\pi} \qquad \frac{s}{b} = \frac{4}{\pi} - \frac{1}{\sqrt{2}} \qquad \frac{t}{b} = \sqrt{2} - \frac{4}{\pi} \qquad (28)$$

The critical semiwidth is $w = b(\pi + 4)/\pi$, which is 2.273 times the depth of penetration. The rigid shoulder in this case is symmetrical about the bisector of the angle at the corner N.

When the semiangle of the wedge is greater than $50.6°$, the plastic zone spreads around the vertex A to include a finite length of the axis of symmetry. The fan angle at A is then equal to $\pi/2 - \psi$, so that AG is inclined at $\pi/4$ to the axis, while the circular base of the rigid shoulder has an angular span equal to $\theta > \pi/2 - \psi$. The continuation of this slipline to the left meets the axis at a point where it is intersected by the slipline through B that makes an angle $\pi/4 - (\psi - \theta)/2$ with BD. No calculation has been made for the critical width based on the modified slipline field, but its value cannot differ appreciably from that given by the simple formula

$$w = b(2e^\theta - \sin \psi) \qquad \frac{\pi}{4} < \psi \leqslant \frac{\pi}{2} \qquad (29)$$

For $\psi = 50.6°$ and $\psi = 90°$, this formula coincides with the result obtained from the statically admissible extension of the basic slipline field involving a stress-free boundary generated through E.

8.3 Compression of a Wedge by a Flat Die

(i) *Yield point of a blunt wedge* An infinite wedge of angle 2ψ is truncated by a horizontal plane perpendicular to the axis of symmetry. The wedge is vertically compressed by a flat die which is at least as wide as the plane section $AA = 2a$. The deforming zone at the yield point is represented by the field $ACDBA$ (Fig. 8.16), in which the central triangle ACA moves vertically downward with the same speed as that of the die. The material within $CDBA$ is instantaneously displaced sideways, with the streamlines coinciding with the sliplines parallel to CDB. The pressure on

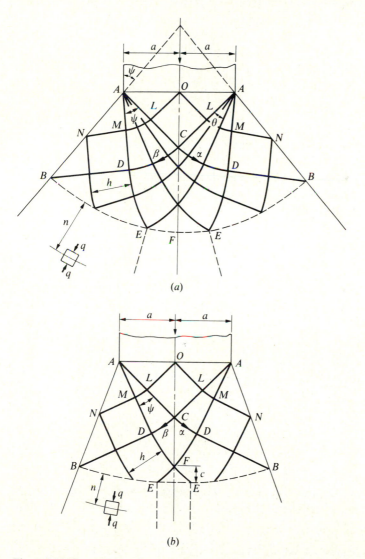

(a)

(b)

Figure 8.16 Compression of a trancated wedge by a flat die. (a) $\psi \geqslant 27.34°$; (b) $\psi \leqslant 27.34°$.

the die is uniformly distributed and is given by

$$q = 2k(1 + \psi)$$

which is independent of the frictional condition along the die. The result has been found to be in good agreement with experiment.[†] The die pressure is the same for all fields whose boundaries lie between the slipline $OLMN$ and $ACDB$ provided the die is perfectly smooth.

To show that the rigid region is able to support the stresses across the boundaries $ACDB$, the slipline field is extended[‡] as far as the principal stress trajectory $BEFEB$ passing through the terminals B. The extended field is uniquely defined by the equal circular arcs CD and their straight continuations DB. Let the absolute values of the radii of curvature of the α and β lines of the field defined by CD be denoted by R and S respectively. From (83), Chap. 6, we may express R and S in the form

$$R(\alpha, \beta) = \sqrt{2}\,a\left\{I_0(2\sqrt{\alpha\beta}) + \frac{\partial}{\partial\alpha}\left[I_0(2\sqrt{\alpha\beta})\right]\right\}$$

$$S(\alpha, \beta) = \sqrt{2}\,a\left\{I_0(2\sqrt{\alpha\beta}) + \frac{\partial}{\partial\beta}\left[I_0(2\sqrt{\alpha\beta})\right]\right\}$$

(30)

where (α, β) are the angles turned through from C along the base sliplines. Considering the field BDE to the left of the axis of symmetry, let h denote the length of the straight segment of the β line through any point of the boundary BE. The local radius of curvature of the α line through this point is numerically equal to $R(\alpha, \psi) + h(\alpha)$, so that

$$dh = -[R(\alpha, \psi) + h(\alpha)]\,d\alpha$$

along the boundary, Substitution for $R(\alpha, \psi)$ from (30) leads to the differential equation

$$\frac{d}{d\alpha}[\sqrt{2}\,a I_0(2\sqrt{\alpha\psi}) + h] = -[\sqrt{2}\,a I_0(2\sqrt{\alpha\psi}) + h]$$

Since $h = \sqrt{2}\,a$ at B, where $\alpha = 0$, the integration of the above equation results in

$$h = \sqrt{2}\,a[2e^{-\alpha} - I_0(2\sqrt{\alpha\psi})]$$

(31)

For sufficiently large values of ψ, h vanishes at E, which corresponds to $\alpha = \theta \leqslant \psi$ (Fig. 8.16a), the relationship between θ and ψ being

$$I_0(2\sqrt{\theta\psi})\exp(\theta) = 2$$

(32)

The angles θ and ψ are equal to one another when $\psi \simeq 27.34°$. For $\psi \geqslant 27.34°$, θ decreases as ψ increases, reaching the value $\theta \simeq 16.39°$ when $\psi = \pi/2$. For

† A. Nadai, Z. angew. Math. Mech., **1**: 20 (1921).
‡ M. Sayir, Z. angew. Math. Phys., **20**: 298 (1969).

$\psi \leqslant 27.34°$, the point E corresponds to $\alpha = \psi$ (Fig. 8.16b), and the slipline field involves an isosceles triangle EFE of height

$$c = a[2e^{-\psi} - I_0(2\psi)]$$

The triangle is in a uniform plastic state of stress, the principal stress normal to the axis of symmetry being tensile and of amount $2k\psi$.

When ψ exceeds $27.34°$, the central part EFE of the trajectory is defined by the relation $R \, d\alpha = -S \, d\beta$ along its length. Using (30) for R, the equation for the curve EFE may be written as

$$d[I_0(2\sqrt{\alpha\beta})] = -I_0(2\sqrt{\alpha\beta}) \, d(\alpha + \beta)$$

Since the trajectory passes through the point $(0, \psi)$ satisfying (32), the above equation integrates to

$$I_0(2\sqrt{\alpha\beta}) \exp(\alpha + \beta - \psi) = 2 \tag{33}$$

The point F, where the trajectory intersects the axis of symmetry, corresponds to $\alpha = \beta = \phi$ (say), where

$$I_0(2\phi) \exp(2\phi - \psi) = 2 \tag{34}$$

When $\psi = \pi/2$, representing the indentation of a flat surface, (34) furnishes $\phi \simeq 47.67°$. It follows from (34) that $2\phi \geqslant \psi$ for all values of $\psi \leqslant 103.6°$.

The normal pressure q_0 transmitted across the trajectory varies along its length. At any point of BE, the pressure is equal to $2k(1 - \alpha)$, where α is the angle turned through along this curve. For $\psi \geqslant 27.34°$, the normal pressure at any point of the curve EE is $2k(1 + \psi - \alpha - \beta)$, its greatest value being $2k(1 + \psi - 2\phi)$ occurring at F. The intensity of q_0 is therefore nowhere greater than $2k$ whenever ψ is less than $103.6°$. The stress distribution in the rigid region lying below the trajectory will be statically admissible if we assume a typical element in this region to be in a state of uniaxial compression q in a direction normal to the trajectory. The curve $BEEB$ then becomes a line of stress discontinuity. For equilibrium, q must vary along the normal so as to be inversely proportional to the radius of curvature of the transverse principal stress trajectory. Since the normal stress must be continuous across the discontinuity, we have

$$q\left(1 + \frac{n}{\rho}\right) = q_0$$

where n is the normal distance of the element from the discontinuity, and ρ the radius of curvature of the discontinuity where it is intersected by the normal. The value of ρ at any point of BE is $\sqrt{2}$ times the radius of curvature of the curved sliplines through that point. Along the curve EFE, it is easy to show that ρ is equal to $\sqrt{2} \, RS/(R + S)$. Since ρ is discontinuous at the points E, the stress is also discontinuous across the normals through these points.

(ii) *Finite compression for a pointed vertex* Consider the symmetrical compression

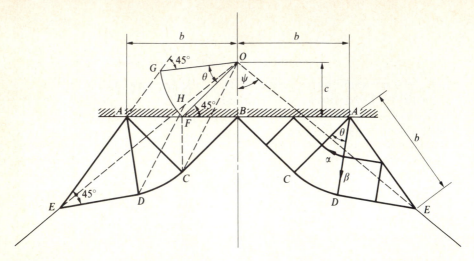

Figure 8.17 Slipline field and unit diagram for finite compression of a wedge with a smooth flat die.

of an infinite wedge of semiangle ψ by a smooth flat die whose plane face is normal to the bisector of the wedge angle.† The wedge initially has a pointed vertex O, so that the deformation at each stage is geometrically similar. At any instant during the compression, let c be the depth of the flattened surface AA below O, and b the semiwidth of the surface of contact (Fig. 8.17). The deformed part of the wedge therefore forms another wedge of semiangle $\theta < \psi$. The sliplines through E meet at B on the axis of symmetry, the portion CD of each of these sliplines being a circular arc of angular span θ. The material being incompressible, the triangles OBE and ABE must be equal in area. Since the perpendicular distances of E from BA and OB are $b \cos \theta$ and $b(1 + \sin \theta)$ respectively, we have

$$b^2 \cos \theta = cb(1 + \sin \theta)$$

or
$$\frac{b}{c} = \frac{1 + \sin \theta}{\cos \theta} = \tan\left(\frac{\pi}{4} + \frac{\theta}{2}\right) \tag{35}$$

The height of O above E is $c + b \cos \theta$. The condition that E lies on the original wedge face therefore gives

$$\tan \psi = \frac{b(1 + \sin \theta)}{c + b \cos \theta} = \frac{(1 + \sin \theta)^2}{\cos \theta(2 + \sin \theta)} \tag{36}$$

in view of (35). If the downward speed of the die is taken as unity, the particles in the deforming region move with uniform speed $\sqrt{2}$ along the sliplines parallel to $BCDE$. It follows from the Hencky equations that the uniform die pressure is of

† R. Hill, *Proc. 7th Int. Congr., Appl. Mech., London* (1948). For asymmetric compression, see K. L. Johnson, *J. Mech. Phys. Solids*, **16**: 395 (1968); I. F. Collins, *Int. J. Mech. Sci.*, **22**: 735 (1980).

amount

$$q = 2k(1 + \theta)$$

As ψ is decreased, θ also decreases, eventually vanishing for $\tan \psi = \frac{1}{2}$, or $\psi \simeq 26.6°$. The displaced surface AE is then vertical and of length equal to c. The die pressure in this limiting case is equal to the plane strain yield stress $2k$. The solution is not valid for smaller wedge angles. Values of θ, b/c, and $q/2k$ for various semiangles of the wedge are given in Table 8.5.

The material originally occupying the triangle OBC has been sheared parallel to BC to assume the final position BFC, where OF is inclined at $45°$ to the die face. Similarly, the material finally in the triangle ADE has been sheared parallel to DE from its initial position HDE, where H is the point of intersection of OE and the parallel to ED through A. The region $OCDH$ in the initial state has been displaced into the region $FCDA$ in the final state. The left-hand half of Fig. 8.17 may be regarded as the unit diagram in which the foci lie on the circular arc FG of angular span θ. Since the projection of AB perpendicular EA is equal to the sum of the projections of OB and OG in view of (35), G lies on EA produced.

There is another possible solution which is valid for any condition of friction over the die face. The displaced surface AE is still straight, but the slipline through E passes through the opposite end A of the surface of contact (Fig. 8.18). A wedge-shaped cap of dead metal ACA is attached to the die as it moves downward with increasing compression. From geometry, AE is equal to the width of the surface of contact, denoted by $2b$. Equating the areas of the triangles OBE and ABE as before, and using the geometry of the figure, it is easily shown that

$$\frac{b}{c} = \frac{1 + 2\sin\theta}{2\cos\theta} \qquad \tan\psi = \frac{(1 + 2\sin\theta)^2}{4\cos\theta(1 + \sin\theta)} \tag{37}$$

This solution is valid for $\theta \geqslant 0$, or $\psi \geqslant \tan^{-1}(1/4) \simeq 14°$. The normal pressure on the die is still equal to $2k(1 + \theta)$, but its value for a given wedge angle is slightly higher than that in the preceding solution.[†]

(iii) *Yielding of a tapered projection* A block of material having a straight-sided symmetrical projection is vertically compressed by a smooth flat die.[‡] The

Table 8.5 Results for the finite compression of a wedge

ψ	30°	40°	50°	60°	70°	80°	90°
θ	5.62°	21.05°	35.59°	49.57°	63.21°	76.66°	90°
b/c	1.103	1.456	1.945	2.716	4.199	8.551	∞
$q/2k$	1.098	1.367	1.621	1.865	2.103	2.338	2.571

[†] The situation where the die and the wedge are of comparable hardness has been discussed by W. Johnson, F. U. Mahtab, and J. B. Haddow, *Int. J. Mech. Sci.*, **6**: 329 (1964).

[‡] Slipline fields for both smooth and rough dies have been proposed by W. Johnson and P. B. Mellor, *Engineering Plasticity*, pp. 623–624, Van Nostrand, London (1973).

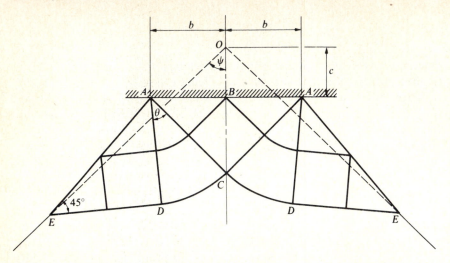

Figure 8.18 Slipline field for finite compression of a wedge with a rough flat die.

semiwidth b of the horizontal surface of contact exceeds the length a of the inclined faces of the projection. Figure 8.19 shows one half of the slipline field and the associated hodograph when the load has reached the yield point. The angle ψ of the centered fan ACD is equal to the angle of inclination of the side AB with the vertical. The usual field $ABCDE$ is extended by considering a second fan BCF of angular span θ. The circular arcs CD and CF of equal radii uniquely define the field $CDGF$ and its extension $DGHE$. The remaining field EHO is constructed from the slipline EH and the frictionless condition along EO. The angle θ must be such that the bounding slipline through B terminates at the die center O.

The die pressure has the constant value $2k(1 + \psi)$ along AE, but its intensity steadily increases from E to reach the value $2k(1 + \psi + 2\theta)$ at O. The mean pressure acting over OA can be written down from Eq. (147), Chap. 6, by setting $p_0 = k$, $s = b$, and replacing $2a$ by a. Setting $\psi + \theta = \xi$, the mean die pressure q may be written as

$$\frac{q}{2k} = (\theta + \xi) + \frac{a}{b}[I_0(2\sqrt{\theta\xi}) - 4L(\theta, \xi)] \tag{38}$$

where

$$\frac{b}{a} = I_0(2\sqrt{\theta\xi}) + 2F_1(\theta, \xi)$$

For a given angle of inclination ψ, the above expressions furnish the relationship between $q/2k$ and b/a parametrically through θ, the quantities I_0, L, and F_1 being obained from Table A-3 or A-4. The rigid material in the corner FBJ is found not to be overstressed if

$$\theta \leqslant \frac{\pi}{4} - \frac{\psi}{2}$$

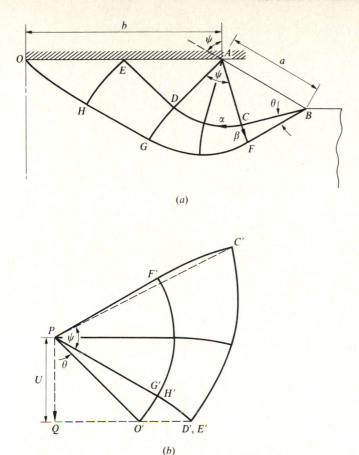

(a)

(b)

Figure 8.19 Plastic compression of a tapered projection in a large block. (a) Slipline field (b) hodograph.

The plastic state of stress at the corner for the limiting value of θ involves a singularity at B over an angular span θ, lying between a rectangular zone and a triangular zone of constant stress. The results based on (38) are shown graphically in Fig. 8.20 for various values of ψ, the limit of applicability of the solution being shown by the broken line.

The velocity of the rigid die is represented by the vertical vector PQ of magnitude U in the hodograph (Fig. 8.19). A velocity discontinuity of amount $\sqrt{2}\,U$ occurs across the slipline through O, which is mapped as the circular arc $O'F'$ of angular span $\psi + \theta$. The field $O'F'C'E'$ of the hodograph is defined by $O'F'$ and its reflection in $O'E'$, which is the image of the die face OA. The material triangle ABC is instantaneously displaced with a velocity that is represented by the vector PC'. The sense of rotation of the sliplines and their corresponding images indicates that the plastic work rate is nowhere negative.

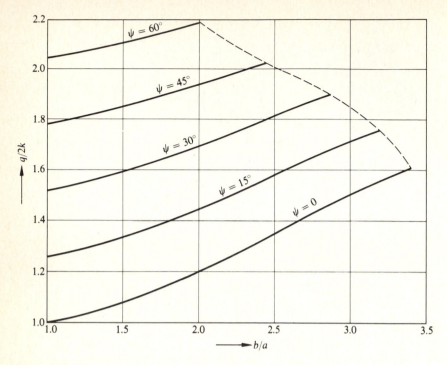

Figure 8.20 The mean die pressure at the yield point for a tapered projection.

8.4 Cylindrical Depression in a Large Block

(i) *Indentation by a cylindrical indenter* A semi-infinite block of metal is indented by the curved surface of a smooth circular cylinder of radius R. The depth of indentation is assumed small in comparison with the width of the area of contact between the cylinder and the block. At any instant during the process, let c denote the depth of the lowest point of the cylinder below the original surface, and a the corresponding semiwidth of the cylindrical depression (Fig. 8.21). The cylinder exerts normal pressure over the surface of contact CA, while the remainder of the surface is traction free. The problem is to determine the position of the point A, the shape of the raised surface AB, and the pressure distribution along CA. The downward speed of the cylinder is taken as unity with c as the time scale.

The associated slipline field $ACDEB$ may be considered as an appropriate modification of the field $A'OD'E'B$, which corresponds to the indentation by a flat punch of width equal to $2a$. If the velocity field for the flat punch is assumed as a first approximation for the cylindrical indenter, the deformed free surface at each stage moves with a velocity whose vertical component is unity.† Let c_1 be the value of c for which deformation first occurs at a given surface particle specified by

---

† A. J. M. Spencer, *J. Mech. Phys. Solids*, **10**: 17 (1962).

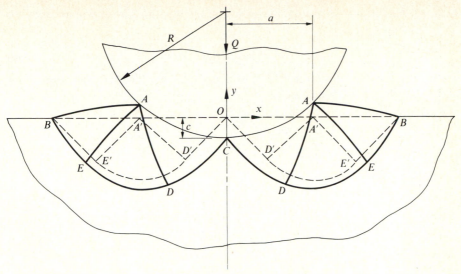

Figure 8.21 Indentation of a plane surface by a cylindrical punch. The perturbed slipline field is only diagrammatic.

its distance x from the vertical axis, and c_2 the instant when the same particle comes in contact with the indenter. If y denotes the vertical height of the particle above the original surface, then

$$\frac{dy}{dc} = 1 \quad (c_1 < c < c_2) \qquad\qquad \frac{dy}{dc} = -1 \quad (c > c_2)$$

During the interval $0 < c < c_1$, the particle remains at rest. Since geometry changes are neglected, we immediately get on integration,

$$y = c - c_1 \qquad\qquad c_1 \leqslant c \leqslant c_2$$
$$y = -c + 2c_2 - c_1 \qquad c \geqslant c_2 \tag{39}$$

If the configuration of Fig. 8.21 corresponds to $c > c_2$, the considered particle lies on AC, and the height of the particle above C is $2c_2 - c_1$ by (39). Then from geometry

$$x^2 = 2R(2c_2 - c_1)$$

The distance x is related to the value of a corresponding to the instants c_1 and c_2 by the expression

$$x = a(c_2) = 2a(c_1)$$

The above equations can be satisfied by assuming c_2/c_1 to have a constant value m (say). Then

$$a(c_1) = \sqrt{R\left(m - \frac{1}{2}\right)c_1} \qquad a(c_2) = \sqrt{2R\left(2 - \frac{1}{m}\right)c_2}$$

which will be mutually consistent if $m = 4$. The semiwidth of the region of contact is therefore given by

$$a(c) = \sqrt{7Rc/2} \qquad (40)$$

Substituting the values $c_1 = x^2/(14R)$ and $c_2 = 4c_1$ in (39), the equation of the deformed surface is obtained as

$$
\begin{aligned}
y &= -c + \frac{x^2}{2R} \qquad 0 \leqslant x \leqslant a \\[2mm]
y &= c - \frac{x^2}{14R} \qquad a \leqslant x \leqslant 2a
\end{aligned}
\qquad (41)
$$

The height of A above the original surface is therefore equal to $\frac{3}{4}c$. The slipline field $ACDEB$ can now be constructed from the known shapes of AB and AC, and the given conditions along them. It may be verified that the volume of the raised lips above the original surface is equal to the volume of the depression below this surface, as required by the incompressibility of the material.

Consider the α line through a typical point on the surface of contact, the position of the point being specified by the distance x. The point of intersection of this slipline with AB is situated at a distance $2a - x$ from the axis of symmetry, to the above order of approximation. By (41), the slopes of AC and AB at the extremities of this slipline are numerically equal to x/R and $(2a - x)/7R$ respectively. The angle turned through by this slipline is therefore equal to

$$\frac{\pi}{2} - \frac{2(a + 3x)}{7R}$$

Substituting from (40), and using Hencky's equation, the normal pressure q at a generic point on the surface of contact is obtained as

$$\frac{q}{2k} = \left(1 + \frac{\pi}{2}\right) - \left\{\sqrt{\frac{2c}{7R}} + \frac{6x}{7R}\right\} \qquad (42)$$

The vertical load applied to the block is $2k(2 + \pi)a$ approximately, where a is given by (40). In order to improve this approximation, and to derive the slipline field around the indenter, it is necessary to consider a second approximation for the velocity field.†

(ii) *Expansion of a semicylindrical cavity* Consider the related problem in which a semicylindrical cavity is expanded from a point O on the plane surface of an infinitely extended medium.‡ The shape of the cavity is maintained at each stage

---

† The plastic deformation of a plane surface produced by rolling contact with a rigid circular cylinder has been examined by E. A. Marshall, *J. Mech. Phys. Solids*, **16**: 243 (1968).

‡ R. Hill, *Proc. 7th Int. Congr., Appl. Mech.*, London (1948); *The Mathematical Theory of Plasticity*, pp. 223–226, Clarendon Press, Oxford (1950).

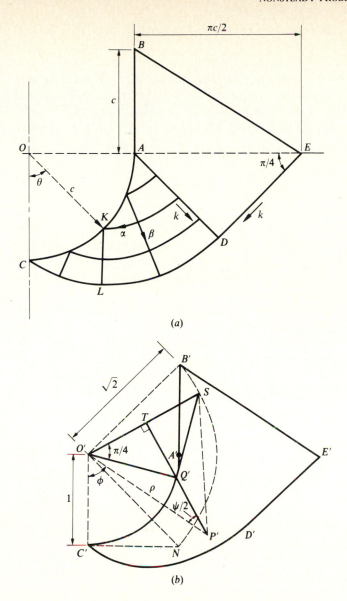

Figure 8.22 Expansion of a semicylindrical cavity. (*a*) Slipline field; (*b*) unit diagram.

by the application of a suitably distributed radial pressure over the cylindrical surface AC of radius c (Fig. 8.22). The surface of the coronet ABE is free from external forces, and the region $ADEB$ is carried outward as a rigid body, a part of it having been unloaded from a previous plastic state. The sliplines AD and DE are assumed straight, inclined at $45°$ to the original plane surface. The normal and shear stresses across these sliplines are each equal to k, with the direction of the shear stress as shown, so that conditions of force and moment equilibrium of $ADEB$ are identically satisfied.

All β lines in the field ACD are straight, meeting the cavity surface at $45°$. The α lines are the involutes of a circle with center O and radius $c/\sqrt{2}$. These lines may therefore be generated by unwinding a taught string from the circular evolute. Let θ denote the angular position of any point K on the cavity surface with respect to the point C. The angle between the β lines through C and K is also θ, and consequently the length KL is equal to $c\theta/\sqrt{2}$. The normal pressure at K must have the value

$$q = 2k\left(1 + \frac{\pi}{2} - \theta\right) \tag{43}$$

by Hencky's equations. The greatest pressure therefore occurs at the deepest point C of the cavity. Since the areas OAC and ABE must be equal to one another, while AD is of length $\pi c/2\sqrt{2}$, the length AB must be equal to the cavity radius c. The entire configuration remains geometrically similar during the formation of the cavity. It can be shown that the rigid material below CDE is capable of supporting the stresses acting across the boundary.

Since the normal component of velocity vanishes along the plastic boundary CDE, the velocity component along the straight sliplines is zero throughout the field ACD. The particles therefore move along the curved sliplines with a constant speed $\sqrt{2}$, the rate of radial expansion of the cavity being unity when c is the time scale. This is compatible with the assumed rigid body motion of the coronet, which slides over DE with a speed equal to $\sqrt{2}$. In the unit diagram, the foci lie on the circular arc $B'N$ with center O' and radius $\sqrt{2}$, where $O'B'$ and $O'N$ are parallel to the α directions at D and C respectively. The figure $O'C'NB'$ is in fact the hodograph of the process with O' representing the pole and $O'C'$ the radial velocity of the cavity surface at C.

The trajectories of the particles in the unit diagram are straight in the region $A'D'E'B'$ and curved in the region $A'C'D'$. The tangent at any point P' on a curved trajectory passes through the corresponding focus S. This is located by drawing the tangent $P'T$ to the circle with center O' and radius $1/\sqrt{2}$, and then producing the radius OT to meet the focal circle at the required point. Since $O'S$ is inclined at $45°$ to $O'Q'$, where Q' is the point of intersection of $P'T$ and $C'A'$, it follows that $Q'S$ is tangential to the boundary circle and is of unit length. Let (ρ, ϕ) be the polar coordinates of P' with respect to O' and the downward vertical, and let ψ denote the angle $O'P'S$. Since $SQ' = O'Q' = 1$, it follows from the geometry of the triangle $O'Q'P'$ that

$$\rho^2(1 - \cos\psi) = 1 \quad \text{or} \quad \cos\psi = \frac{\rho^2 - 1}{\rho^2}$$

The differential equation of the trajectory in the region $A'C'D'$ is

$$\frac{d\phi}{d\rho} = -\frac{\tan\psi}{\rho} = -\frac{\sqrt{2\rho^2 - 1}}{\rho(\rho^2 - 1)} \tag{44}$$

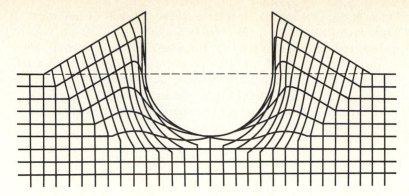

Figure 8.23 Distortion of a square grid around an expanded semicylindrical cavity (*after R. Hill*).

The arc element ds, measured in the direction of increasing ϕ, is $-d\rho \sec \psi$. Substitution in (12) gives

$$\frac{dc}{c} = \frac{ds}{\rho} = -\frac{d\rho}{\rho} \sec \psi = -\frac{d\rho}{\rho^2 - 1}$$

Let c_0 be the value of the cavity radius when a particle, whose image is P' in the final state, is overtaken by the plastic boundary. If (ρ_0, ϕ_0) are the coordinates of the point P'_0 where the trajectory meets the boundary $C'D'$, the integration of the above equation furnishes

$$c^2(\rho^2 - 1) = c_0^2(\rho_0^2 - 1) \tag{45}$$

To find the final position P of a corner P_0 in a square grid, c_0 is first obtained by drawing $O'P'_0$ parallel to OP_0, which is of length $c_0\rho_0$. Equation (45) then gives ρ, and the final angle ϕ is obtained by the integration of (44). The distortion of the square grid computed in this way is shown in Fig. 8.23.

It follows, from the unit diagram, that the material which finally occupies the triangle BDE has always moved parallel to DE after becoming plastic. Hence, this material was originally contained in the triangle ODE, and has suffered a simple shear parallel to DE on crossing the plastic boundary. The elements that form the cylindrical surface of the cavity were originally situated along OC, and have suffered heavy distortions of a complex character. Due to the nature of the loading assumed in this problem, the height of the coronet is somewhat greater than that expected when a semicylindrical indenter is used to form the cavity.

8.5 Compression Between Smooth Platens

A rectangular block of material is compressed symmetrically by a pair of rigid parallel platens under conditions of plane strain. Zones of plastic material are initiated at the corners of the platens, the width of which is assumed smaller than the width of the block. Since the material is rigid/plastic, the platens cannot approach one another until the plastic zones have spread through the thickness of the block. When the load attains the yield point value, the plastic compression of the block begins, and the rigid overhangs are thrust apart. In a real material, with a low rate of work-hardening, the load-compression curve has a pronounced bend, which marks the beginning of large plastic strains. The mean load corresponding to the bend is very closely equal to the rigid/plastic yield point load. The following solution applies to the initial yielding and the continued deformation of an ideally plastic material with a well-defined yield point.[†]

(i) *Width/height ratios between 1 and 2* Consider the slipline field in the upper left-hand quadrant of the block when the ratio of the width $2w$ of the platens to the current height $2h$ of the block lies between 1 and 2 (Fig. 8.24). Depending on the value of the ratio w/h, there are two types of field, both involving a stress singularity at the edge of the platen. All sliplines meet the axes of symmetry AF and DF, and the frictionless boundary OA, at an angle of $45°$. It follows from this boundary condition that the angle turned through by each curved slipline of the various domains is equal to the angle ψ of the centered fan OBC. The field shown in (a) holds for $w/h \leqslant \sqrt{2}$ and contains a region $BCGHED$ where one family of sliplines is straight having a constant length λ_1. The field shown in (b) holds for $w/h \geqslant \sqrt{2}$, and involves a region $CEGHIJ$ in which one family of sliplines is again straight and of length λ_2. For given values of h and ψ, the semiwidths of the platen corresponding to the two fields are denoted by w_1 and w_2 respectively. The straight part of the exit slipline OBD disappears when $w_1 = w_2 = \sqrt{2}\,h$, while the curved part of the exit slipline vanishes for $\psi = 0$, giving $w_1 = h$ and $w_2 = 2h$.

To analyze the field of Fig. 8.24a, the sliplines OC and FE are represented by vectors σ_1 and σ_1' respectively (Sec. 6.5(iii)). Then the curves GC and HE, generated on the concave and convex sides of OC and FE, are represented by $T_\psi^{-1}\sigma_1$ and $T_\psi\sigma_1'$ respectively. Since the radius of curvature at any point of GC numerically exceeds that at the corresponding point of HE by the amount λ_1, we have

$$T_\psi^{-1}\sigma_1 = T_\psi\sigma_1' + \lambda_1 c$$

where c is the vector representing a circular arc of unit radius. Similarly, the fact that BC and DE are given by $Q_{\psi\psi}\sigma_1$ and $T_\psi^{-1}\sigma_1'$ respectively, leads to the relation

$$Q_{\psi\psi}\sigma_1 = T_\psi^{-1}\sigma_1' - \lambda_1 c$$

where $Q_{\psi\psi} = (T_\psi - T_\psi^{-1})/2$. The elimination of c and σ_1' from the above equations readily furnishes

$$\sigma_1' = \tfrac{1}{2}\sigma_1 \qquad (T_\psi^{-1} - \tfrac{1}{2}T_\psi)\sigma_1 = \lambda_1 c \tag{46}$$

The first relation of (46) shows that the sliplines OC and FE are similar in the ratio $2:1$, and the associated fields OCG and FED are therefore similar in the same ratio.

† A. P. Green, *Phil. Mag.*, **42**: 900 (1951).

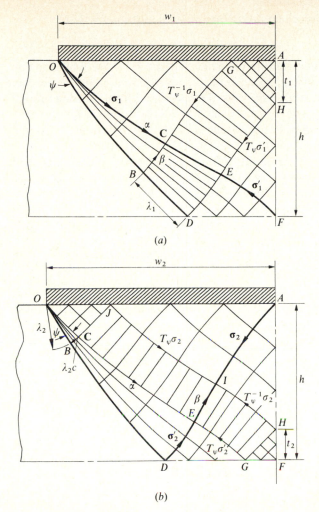

(a)

(b)

Figure 8.24 Slipline fields for compression of a block between smooth platens. (a) $1 \leqslant w/h \leqslant \sqrt{2}$; (b) $\sqrt{2} \leqslant w/h \leqslant 2$.

Turning now to the field of Fig. 8.24b, let σ_2 and σ_2' represent the curves AI and DE respectively. The curves HI and GE are then given by $\mathbf{T}_\psi^{-1}\sigma_2$ and $\mathbf{T}_\psi\sigma_2'$ respectively. Since the normal distance between the two curves is λ_2, we have

$$\mathbf{T}_\psi^{-1}\sigma_2 = \mathbf{T}_\psi\sigma_2' + \lambda_2\mathbf{c}$$

The vectors representing the curves CB and CE are $\lambda_2\mathbf{c}$ and $\mathbf{T}_\psi\sigma_2 + \lambda_2\mathbf{c}$ respectively. By (77), Chap. 6, σ_2' may be written as

$$\sigma_2' = \mathbf{P}_{\psi\psi}(\lambda_2\mathbf{c}) + \mathbf{Q}_{\psi\psi}(\mathbf{T}_\psi\sigma_2 + \lambda_2\mathbf{c})$$

$$= \mathbf{T}_\psi(\lambda_2\mathbf{c}) + \mathbf{Q}_{\psi\psi}(\mathbf{T}_\psi\sigma_2)$$

since $\mathbf{P}_{\psi\psi} + \mathbf{Q}_{\psi\psi} = \mathbf{T}_\psi$. Premultiplying this equation by the operator $\mathbf{T}_\psi^{-1}$ gives

$$\mathbf{T}_\psi^{-1}\sigma_2' = \mathbf{Q}_{\psi\psi}\sigma_2 + \lambda_2\mathbf{c}$$

Substituting for $\mathbf{Q}_{\psi\psi}$, and proceeding as before, it is readily shown that

$$\boldsymbol{\sigma}_2' = \tfrac{1}{2}\boldsymbol{\sigma}_2 \qquad (\mathbf{T}_\psi^{-1} - \tfrac{1}{2}\mathbf{T}_\psi)\boldsymbol{\sigma}_2 = \lambda_2\mathbf{c} \qquad (47)$$

The first of these relations indicates that the fields AIJ and DEG are similar to one another in the ratio $2:1$, which means that $AJ = 2DG$, and that the vertical distance of I from the platen is twice the vertical distance of E from the longitudinal axis of the block.

It follows from (46) and (47) that $\boldsymbol{\sigma}_1/\lambda_1$ and $\boldsymbol{\sigma}_2/\lambda_2$ satisfy the same operational equation, and are therefore equal to one another. In other words, the sliplines represented by $\boldsymbol{\sigma}_1$ and $\boldsymbol{\sigma}_2$ are similar in the ratio $\lambda_1:\lambda_2$, and so are the fields defined by them on either side. Thus, the straight face OG of (a) is proportional to the length AH of (b) in the ratio $\lambda_1:\lambda_2$. Since the field FEH of (a) is one half of the similar field generated by OC on its convex side (not shown), the length AJ of (b) and FH of (a) are in the ratio $2\lambda_2:\lambda_1$. Consequently,†

$$\frac{w_1}{h} = \frac{\lambda_1}{\lambda_2} \qquad \frac{w_2}{h} = \frac{2\lambda_2}{\lambda_1} \qquad w_1 w_2 = 2h^2 \qquad (48)$$

When $\lambda_1 = \lambda_2 = 0$, we obtain $\boldsymbol{\sigma}_1 = \boldsymbol{\sigma}_2 = \boldsymbol{\sigma}$, which is found to satisfy the matrix equation $\mathbf{T}_\psi^2\boldsymbol{\sigma} = 2\boldsymbol{\sigma}$. This indicates that $\boldsymbol{\sigma}$ is the eigenvector of the matrix $\mathbf{T}_\psi$, the corresponding eigenvalue being equal to $\sqrt{2}$. The value of ψ that corresponds to the eigenfield is found to be $19.67°$ approximately.

For a selected value of $\psi < 19.67°$, the vector $\boldsymbol{\sigma}_1/\lambda_1$ or $\boldsymbol{\sigma}_2/\lambda_2$ can be determined from (46) or (47) by a process of matrix inversion. The coefficients of the vector $\boldsymbol{\sigma}_1/\lambda_1$ having been found, it is a straightforward matter to compute the ratios w_1/λ_1 and h/λ_1 from the construction of the domains OCG and FEH. When w_1 and λ_1 are thus found in terms of h, the corresponding quantities w_2 and λ_2 are known from (48). The conditions of zero horizontal force across the slipline OBD gives the hydrostatic pressure at O, considered on this slipline. The stress distribution throughout the field than follows from Hencky's equations.

(ii) *Further properties of the field* Let p_1 and p_2 denote the hydrostatic pressures in the uniformly stressed regions AGH and OCJ of the fields (a) and (b) respectively. The mean normal pressures exerted on the plane boundary by the regions of variable stress in the two fields are denoted by s_1 and s_2 respectively, when the hydrostatic pressure origins are taken at G and J. If the mean normal pressures on OA corresponding to the fields (a) and (b) are $\bar{q}_1$ and $\bar{q}_2$ respectively, the vertical equilibrium requires

$$\bar{q}_1 w_1 = s_1(w_1 - t_1) + kt_1 + p_1 w_1$$

$$\bar{q}_2 w_2 = s_2(w_2 - t_2) + kt_2 + p_2 w_2$$

where $t_1 = \lambda_1/\sqrt{2}$ and $t_2 = \lambda_2/\sqrt{2}$. In view of the similarities established above, the

---

† A. P. Green, *Phil. Mag.*, **42**: 900 (1951); I. F. Collins, *Proc. R. Soc.*, A, **303**: 317 (1968).

mean normal pressures acting across HF in (a) and AH in (b) are $-s_2$ and $-s_1$ respectively. Since the resultant horizontal force across AF is zero, we have

$$-s_2(h - t_1) - kt_1 + p_1h = 0$$

$$-s_1(h - t_2) - kt_2 + p_2h = 0$$

Eliminating s_1 and s_2 between the two sets of equations, and using (48), we arrive at the result

$$\bar{q}_1 = \bar{q}_2 = p_1 + p_2 \qquad (49)$$

Thus, for a given value of ψ, the mean normal pressures in the two cases are identical. To calculate the value of this mean pressure, it is only necessary to find the hydrostatic pressures in the regions of constant stress. The results of the computation are presented in Table 8.6, and a graphical plot for the mean pressure is shown in Fig. 8.25.

For reasons of symmetry, velocity discontinuities can only occur along $OCEF$ in (a), and along $OBDEIA$ in (b). The outward speed of the rigid overhang is $(w_1/h)U$ in the first case and $(w_2/h)U$ in the second case, where U is the downward speed of the upper platen. In (a), the velocity is continuous across the exist slipline OBD which is mapped into a single point B' in the hodograph having its pole at P (Fig. 8.26). The particles immediately below and above the discontinuity OCF are mapped into the parallel curves $O'C'F'$ and $O''C''F''$ respectively. The velocity is constant along each straight slipline, the triangle AGH being carried down the

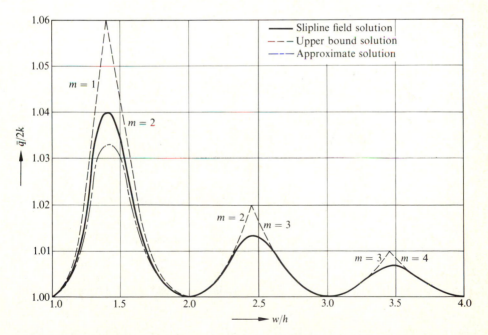

Figure 8.25 Variation of the mean die pressure with the width/height ratio for frictionless compression.

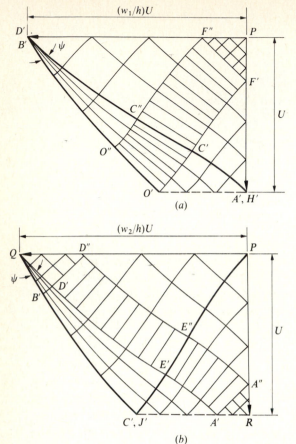

Figure 8.26 Hodographs for compression between smooth parallel platens. (a) $1 \leqslant w/h \leqslant \sqrt{2}$; (b) $\sqrt{2} \leqslant w/h \leqslant 2$.

platen with velocity U. In (b), the particles immediately to the right of the slipline OBD are mapped into the circular arc $B'D'$, whose center Q maps the rigid overhang. The velocity discontinuity propagating along the slipline $DEIA$ gives rise to the parallel curves $D'E'A'$ and $D''E''A''$ in the hodograph. The slipline field

Table 8.6 Data for plane strain compression of a block between smooth flat dies

ψ°	λ_1/h	λ_2/h	w_1/h	w_2/h	p_1/k	p_2/k	$\bar{q}/k$
0.00	1.4142	1.4142	1.0000	2.0000	1.0000	1.0000	2.0000
4.00	1.1959	1.0550	1.1335	1.7644	0.9787	1.0359	2.0146
8.00	0.9361	0.7491	1.2497	1.6004	0.9077	1.1345	2.0422
12.00	0.6388	0.4767	1.3401	1.4924	0.7781	1.2879	2.0660
16.00	0.3126	0.2238	1.3968	1.4318	0.5849	1.4928	2.0778
19.67	0	0	1.4142	1.4142	0.3533	1.7263	2.0796

and the hodograph in each case form identical networks, and the plastic work rate is found to be everywhere positive.

Consider now the artificial deformation mode in which the hodograph of Fig. 8.24a is represented by Fig. 8.26b by suitably modifying the velocity boundary conditions. The artificial mode differs from the actual one in that there is a velocity discontinuity along $OBDEH$, instead of one along $OCEF$, and in addition there is a uniform normal velocity on AH. This additional velocity is represented by the vector RA' in the hodograph, and its magnitude is equal to $(t_2/h)U$. Equating the rate at which material enters the deforming zone across OA and HA, and leaves it across OD, we get

$$w_1 + \frac{t_1 t_2}{h} = w_2$$

since the velocity of the rigid overhand is now of magnitude $(w_2/h)U$. The above relation may be combined with (48) to obtain[†]

$$w_1^2 + \tfrac{1}{2}\lambda_1^2 = w_2^2 - \lambda_2^2 = 2h^2 \tag{50}$$

The new mode of deformation introduced above has no physical significance. Such artificial modes are, however, occasionally useful for establishing certain relations in a straightforward manner.

(iii) *An explicit solution for* $1 \leqslant w/h \leqslant 2$ Since the fan angle ψ is fairly small, it is a good approximation to assume the curved part of the exit slipline to be a circular arc, its radius of curvature being denoted by $\sqrt{2}\rho_1$ for the first field and $\sqrt{2}\rho_2$ for the second field.[‡] Considering the field (a) of Fig. 8.27, let CL and EN be drawn perpendiculars from the vertices to the opposite sides of the similar domains OCG and FED. By Eqs. (149), Chap. 6, the position of C relative to O in field (a) is given by the distances

$$OL \simeq \tfrac{1}{2}(w_1 - t_1)(1 + \sin \psi) \qquad CL \simeq (h - t_1)(1 - \sin \psi)$$

Since $EN = \tfrac{1}{2}CL$, and the height of C above E is $t_1(\cos \psi - \sin \psi)$, it follows from simple geometry that

$$\tfrac{3}{2}(h - t_1)(1 - \sin \psi) + t_1(\cos \psi - \sin \psi) = h$$

The horizontal distance between C and E is $t_1(\cos \psi + \sin \psi)$. Since $FN = \tfrac{1}{2}OL$ in view of the similarity of OCG and FED, we have

$$\tfrac{3}{4}(w_1 - t_1)(1 + \sin \psi) + t_1(\cos \psi + \sin \psi) = w_1$$

The above equations immediately furnish the ratios t_1/h and w_1/h in terms of the

[†] I. F. Collins, *J. Mech. Phys. Solids*, **16**: 137 (1968).

[‡] J. Chakrabarty, unpublished work (1978). For a different approximation, see W. Johnson and I. E. McShane, *Appl. Sci. Res.*, **A9**: 169 (1960).

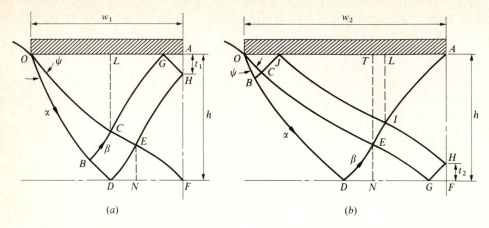

Figure 8.27 Compression between smooth platens ($1 \leqslant w/h \leqslant 2$). The curved part of the exit slipline is assumed as a circular arc of radius ρ_1 in (a) and ρ_2 in (b).

angle ψ. Thus

$$\frac{t_1}{h} = \frac{1 - 3\sin\psi}{3 - 2\cos\psi - \sin\psi} \qquad \frac{w_1}{h} = \frac{4\cos\psi + \sin\psi - 3}{3 - 2\cos\psi - \sin\psi} \tag{51}$$

As ψ increases from zero, t_1/h decreases from unity and w_1/h increases from unity. By (148), Chap. 6,

$$\frac{\rho_1}{h} = \frac{1 - t_1/h}{1 - \cos\psi + \sin\psi} = \frac{2}{3 - 2\cos\psi - \sin\psi} \tag{52}$$

The hydrostatic pressure at O on OB is $p_0 = p_1 + 2k\psi$ by Hencky's equation, where p_1 is the pressure in GAH. The hydrostatic pressure along the straight segment DB is equal to p_1. Using Eq. (150), Chap. 6, the condition of zero horizontal thrust across ABD may be written as

$$(h - t_1)(p_1 + 2k\psi) + k[\tfrac{1}{2}(w_1 - t_1) - 2\rho_1\psi] + t_1(p_1 - k) = 0$$

or

$$\frac{p_1}{k} = 2\psi\left(\frac{\rho_1 + t_1}{h} - 1\right) + \frac{3t_1 - w_1}{2h}$$

Substituting from (51) and (52), the expression for p_1/k in terms of ψ is obtained as

$$\frac{p_1}{k} = 1 - \frac{4(1 + \psi)\sin\psi - 4\psi\cos\psi}{3 - 2\cos\psi - \sin\psi} \tag{53}$$

The region $BCED$ in the field (b) may be constructed by the vectorial superposition of the centered fans defined by BC and BD. Considering the perpendicular drawn from E on OA, it can be shown to the above order of approximation that

$$OT \simeq \tfrac{1}{2}w_2(1 + \sin\psi) + t_2(1 - \cos\psi) \qquad ET \simeq h(1 - \sin\psi) + t_2(1 - \cos\psi)$$

Let L and N be the feet of the perpendiculars drawn from I and E on OA and DF respectively. Since the vertical and horizontal projections of EI are $t_2(\cos \psi + \sin \psi)$ and $t_2(\cos \psi - \sin \psi)$ respectively, we obtain from geometry,

$$IL = (h + t_2)(1 - \sin \psi) - 2t_2 \cos \psi \qquad EN = h \sin \psi - t_2(1 - \cos \psi)$$

$$JL = (\tfrac{1}{2}w_2 - t_2)(1 + \sin \psi) \qquad GN = \tfrac{1}{2}w_2(1 - \sin \psi) - t_2(2 - \cos \psi)$$

In view of the similarity of the regions JIA and GED, we have $IL = 2EN$ and $JL = 2GN$. The above relations therefore furnish

$$\frac{t_2}{h} = \frac{1 - 3 \sin \psi}{4 \cos \psi + \sin \psi - 3} \qquad \frac{w_2}{h} = \frac{2(3 - 2 \cos \psi - \sin \psi)}{4 \cos \psi + \sin \psi - 3} \tag{54}$$

It may be verified that (51) and (54) identically satisfy (48) and (50), where $\lambda_1 = \sqrt{2}\,t_1$ and $\lambda_2 = \sqrt{2}\,t_2$. Since the height of B above D is $h - t_2(\cos \psi + \sin \psi)$, it follows from (148), Chap. 6, that

$$\frac{\rho_2}{h} = \frac{1 - (t_2/h)(\cos \psi + \sin \psi)}{1 - \cos \psi + \sin \psi} = \frac{3 \cos \psi}{4 \cos \psi + \sin \psi - 3} \tag{55}$$

The hydrostatic pressure along OB is $p_0 = p_2 - 2k\psi$, where p_2 is the pressure in the triangle OCJ. The horizontal compressive force acting on the slipline BD is given by Eq. (150), Chap. 6, with a and b denoting the horizontal and vertical projections of BD. Equating the resultant horizontal force across OBD to zero, we get

$$h(p_2 - 2k\psi) + k\left[\tfrac{1}{2}w_2 - 2t_2(\cos \psi - \sin \psi)\right] - 2k\rho_2\psi = 0$$

Inserting from (54) and (55), and simplifying the resulting expression, we arrive at the formula

$$\frac{p_2}{k} = 1 + 2\left\{\psi + \frac{3 \cos \psi(\psi - \sin \psi) - \sin \psi(1 - 3 \sin \psi)}{4 \cos \psi + \sin \psi - 3}\right\} \tag{56}$$

According to this approximate analysis, $\lambda_1 = \lambda_2 = 0$ when the fan angle ψ is equal to $\sin^{-1}(1/3) \simeq 19.47°$, and this corresponds to $w_1 = w_2 = \sqrt{2}\,h$ and $\rho_1 = \rho_2 \simeq 2.56h$. The value of $\bar{q}_1 = \bar{q}_2$ in this case is $2.067k$, which may be compared with the exact value $2.080k$ obtained by the matrix analysis. The approximate solution is represented by the chain-dotted curve in Fig. 8.25.

(iv) General width/height ratios greater than unity For all integral width/height ratios greater than unity, the block instantaneously deforms as a number of rigid units sliding along a crisscross of straight sliplines inclined at $45°$ to the smooth platens.† There is a constant velocity discontinuity of amount $\sqrt{2}\,U$ initiated at the block center and propagated along the length of the block by successive reflections from the platens. This simple solution cannot be valid for nonintegral

---

† When the dies overlap the block, the die pressure is $2k$ for all width/height ratios, the sliplines being straight with an inclination of $45°$ to the die face.

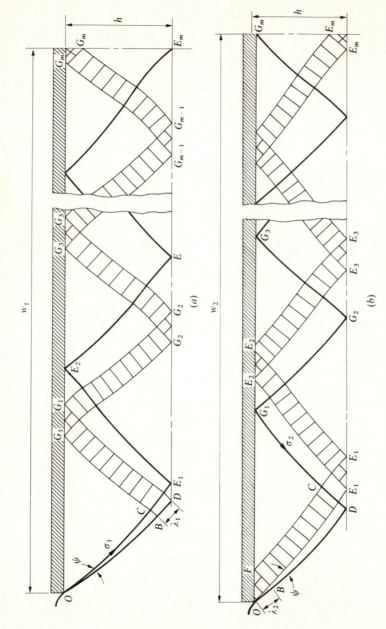

Figure 8.28 Two types of slipline field for frictionless compression of a block with arbitrary width/height ratios.

values of w/h, since the velocity discontinuity then terminates on the exit sliplines, a situation which is incompatible with the rigid body motion of the overhangs. The solution for intermediate width/height ratios must therefore be such that it degenerates into the crisscross pattern when w/h approaches an integral value.

The obvious extensions of the fields of Fig. 8.24 for w/h greater than 2 are shown in Fig. 8.28. Each field involves m number of curvilinear triangles bordering the platen.[†] The boundaries of the successive domains of the field meet the platen at G_1, E_2, etc., and the horizontal axis of symmetry at E_1, G_2. etc. The unknown initial slipline is determined from the conditon that G_m and E_m lie on the same vertical axis. If the initial slipline of the centered fan in (a), or the first regular domain in (b) is represented by $\boldsymbol{\sigma}$, it can be shown that

$$[P_{\psi\psi} - (1 + 2m)Q_{\psi\psi}]\boldsymbol{\sigma} = 2m\lambda\mathbf{c} \tag{57}$$

where λ is the length of the straight segment of the exit slipline. For a given value of m, the fields (a) and (b) coincide when $\lambda = 0$, the corresponding vector being an eigenvector satisfying the equation

$$mT_\psi^2\boldsymbol{\sigma} = (1 + m)\boldsymbol{\sigma}$$

The eigenvalue of the matrix T_ψ is evidently equal to $\sqrt{(1 + m)/m}$. The width/height ratio corresponding to this eigenfield is found to have the value $\sqrt{m(1 + m)}$. The generalizations of (48) and (50) are easily shown to be

$$\frac{w_1}{h} = \frac{m\lambda_1}{\lambda_2} \qquad \frac{w_2}{h} = \frac{(1 + m)\lambda_2}{\lambda_1} \qquad w_1w_2 = m(1 + m)h^2$$

$$w_1^2 + \tfrac{1}{2}m\lambda_1^2 = w_2^2 - \tfrac{1}{2}(1 + m)\lambda_2^2 = m(1 + m)h^2 \tag{58}$$

where the subscripts 1 and 2 refer to the fields (a) and (b) respectively. The mean pressure on the platen in both cases is readily shown to be equal to the sum of the hydrostatic pressures in the uniformly stressed regions as before. The variation of $\bar{q}/2k$ with w/h is shown in Fig. 8.25. The pressure oscillates between maxima and minima corresponding to eigenfields and integral width/height ratios respectively. The oscillatory nature of the solution has been confirmed by experiment.[‡]

An upper bound solution for the compression problem can be obtained by assuming the block to be divided into a number of independent rigid units formed by a symmetrical crisscross of lines inclined at a constant acute angle θ to the horizontal (Fig. 8.29a). The deformation of the block is produced by sliding of the rigid triangles along their boundaries, the velocity discontinuity being of amount $U \operatorname{cosec} \theta$, where U is the speed of compression. If there are m triangular units in contact with each platen, the rate of energy dissipation for a quarter of the block is $mkhU \operatorname{cosec}^2 \theta$, the dimension of the block in the direction of zero strain being taken as unity. According to the upper bound theorem, this energy must exceed

† A. P. Green, *Phil. Mag.*, **42**: 900 (1951); I. F. Collins, *Proc. R. Soc.*, A, **303**: 317 (1968).

‡ A. B. Watts and H. Ford, *Proc. Inst. Mech. Eng.*, **1B**: 448 (1952). See also B. B. Murdi and K. N. Tong, *J. Mech. Phys. Solids*, **4**: 121 (1956).

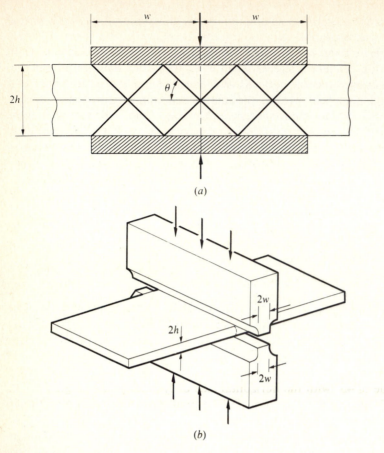

Figure 8.29 Compression between frictionless platens. (a) Velocity discontinuity pattern; (b) plane strain compression test.

the rate of external work which is equal to qwU. Since $\tan \theta = mh/w$, the upper bound formula becomes†

$$\frac{\bar{q}}{2k} = \frac{1}{2}\left(\frac{w}{mh} + \frac{mh}{w}\right) \tag{59}$$

For any given ratio w/h, the integer m in (59) should be such that the corresponding value of $\bar{q}/k$ is the least. The largest width/height ratio for a given m is equal to $\sqrt{m(1 + m)}$, which corresponds to a peak value of the pressure. The upper bound pressure given by (59) is shown by the broken curve in Fig. 8.25. A minimum value of $2k$ for q occurs for integral values of w/h, as in the exact solution. An obvious

† The upper bound solution, due to R. Hill, has been presented by A. P. Green, op. cit.

lower bound is $\bar{q} = 2k$, which is the minimum pressure required for plastic compression of the block.

The stress-strain curve of a strip metal is accurately obtained by the plane strain compression test, in which the strip is compressed between a pair of opposed dies completely spanning its width (Fig. 8.29b). The dies are sufficiently long and narrow, so that the constraint of the nonplastic material on either side inhibits lateral spread, and the deformation is essentially plane strain. The ratio of the die width to the strip thickness should be maintained between 2 and 4 by appropriately changing the dies during the test. The load is applied incrementally, the specimen being removed and measured at each stage with careful lubrication of the strip before each load increment. The current value of the yield stress is obtained by extrapolating to zero deformation for each particular test involving 1 to 2 percent reduction. This technique ensures that friction is not able to build up during the test. The reduction at each increment of load must be sufficient to surmount the knee of the stress-strain curve at each stage.†

8.6 Compression Between Rough Platens

(i) *Perfectly rough platens* ($w/h \leqslant 3.64$) Consider the plastic compression of a block between a pair of rough parallel platens, when there is an overhang of rigid material on either side of the compressed region. For sufficiently small values of the width/height ratio greater than unity,‡ the solution involves a wedge-shaped rigid material extending over the whole platen (Fig. 8.30). The exit slipline AC is straight, making an angle of 45° with the horizontal axis of symmetry. The angle θ of the centered fan ABC depends on the ratio w/h. The circular arc CB and its reflection in the axis define the remaining field CBD, where D is at the center of the block. The equilibrium of the overhang requires the mean compressive stress on AC to be equal to k. Since we do not know the stress distribution in the rigid region above the slipline ABD, only an average pressure over the platen can be calculated. By Eq. (99), Chap. 6, the average value of the normal pressure is given by

$$\frac{\bar{q}}{2k} = \frac{h}{w} \{I_0(2\theta) + 2\theta[I_0(2\theta) + I_1(2\theta)]\} \tag{60}$$

where w/h corresponding to any given θ is obtained directly from the table (Appendix). The above pressure is identical to that for the indentation of a block between a pair of flat dies with a height/width ratio equal to w/h (Sec. 8.1(ii)). The solution is valid for $0 \leqslant \theta \leqslant \pi/4$, which is equivalent to $1 \leqslant w/h \leqslant 3.64$. In the limiting case of $\theta = \pi/4$, the radial slipline AB is coincident with the surface of the platen, and the frictional stress is then equal to the yield stress in shear.

If the speed of each platen is denoted by U, there is a tangential velocity discontinuity of amount $\sqrt{2}\,U$ along the boundary of the rigid zone. The particle

---

† A. B. Watts and H. Ford, *Proc. Inst. Mech. Eng.*, **169**: 1141 (1955).
‡ The slipline field is essentially due to L. Prandtl, *Z. angew. Math. Mech.*, **3**: 401 (1923).

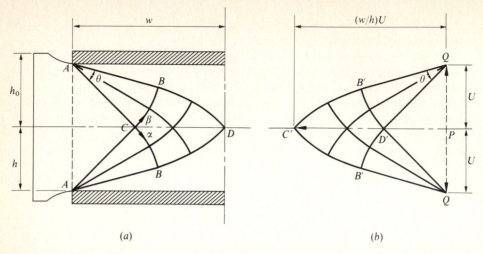

Figure 8.30 Compression of a block between perfectly rough parallel platens when $1 \leqslant w/h \leqslant 3.64$. (a) Slipline field; (b) hodograph.

situated immediately to the left of D moves horizontally with a velocity represented by the vector PD' in the hodograph. The slipline DBA is mapped into the circular arc $D'B'$ which defines the remaining field $D'B'C'$ of the hodograph. The velocity is continuous across the slipline AC, and is represented by the vector PC'. It follows from the similarity of the slipline field and the hodograph that the vector PC' is of magnitude $(w/h)U$, as required by the incompressibility of the material. The plastic work rate is found to be everywhere positive.

For a given value of w/h, the coefficient of friction μ on the platens must exceed a critical value μ^* in order that the above solution can be valid. The minimum coefficient of friction is the least possible ratio of tangential stress to the normal pressure for which the rigid corner at A reaches the yield limit. Using (33) and (35), Chap. 6, with $\lambda_1 = \pi/4$, $\lambda_2 = 0$, $\alpha = \pi/4 - \theta$, and $q_1 = k(1 + 2\theta)$, it is found that

$$\mu^* = \frac{\sin 2\theta}{1 + 2\theta + \cos 2\theta} \tag{61}$$

As θ increases from zero, μ^* also increases from zero, reaching the limiting value 0.39 when θ is $\pi/4$. Numerical values of w/h, $\bar{q}/2k$, and μ^* for different values of θ are given in Table 8.7.

Table 8.7 Results for compression between rough parallel platens

θ	0°	10°	20°	30°	40°	45°
w/h	1.0	1.383	1.853	2.440	3.190	3.644
$\bar{q}/2k$	1.0	1.049	1.172	1.342	1.551	1.667
μ^*	0	0.149	0.261	0.343	0.383	0.389

The solution is independent of the amount of compression, provided h is taken to be the current semiheight of the block. The width/height ratio therefore increases as the plastic compression continues. To find the equation of the contour of the material squeezed out during a finite compression, let $2t$ be the thickness of the overhang at a distance z from the exit plane AA. The considered section of the overhang coincided with AA when the block thickness was $2t$. During an incremental compression specified by the thickness change $2dh$, the same section moves out through a distance $dz = -(w/h)\, dh$. By the time the thickness is reduced to $2h$, this section has moved to the distance z, where

$$z = w \ln \left(\frac{t}{h} \right) \qquad \text{or} \qquad t = h \exp \left(\frac{z}{w} \right) \tag{62}$$

It follows from above that the tangent to the free surface at the edge of the platen passes through the block center. The total displacement of the overhang is equal to $w \ln (h_0/h)$, where $2h_0$ is the initial thickness of the block.

(ii) *Perfectly rough platens* $(w/h \geqslant 3.64)$ For $w/h \geqslant 3.64$ and $\mu \geqslant 0.39$, the α lines meet the platens tangentially† over a certain length d (Fig. 8.31), where the frictional stress has its greatest value k. The ratios d/h and w/h are defined by the angle θ between radial sliplines AC and AD. The rigid region above MN moves down with the platen, losing material to the plastic region as the compression proceeds. The region CBE is one half of the field defined by the equal circular arcs CB, while $BEFM$ is part of the field defined by BE and the limiting state of friction along the platen (Sec. 6.7(i)). The slipline EF and the condition of symmetry about the horizontal axis define the remaining field EFN. By an analytical treatment for this region, it can be shown that‡

$$\frac{w}{h} = I_0(2\xi) + 2F_1(\xi, \xi) - 2\left\{ \sqrt{\frac{\theta}{\eta}} I_1(2\sqrt{\theta\eta}) + 2F_2(\theta, \eta) \right\} \tag{63}$$

where F_1 and F_2 are mathematical functions introduced in Sect. 6.6, and

$$\xi = \frac{\pi}{4} + \theta \qquad \eta = \frac{\pi}{2} + \theta$$

The first two terms appearing on the right-hand side of (63) represent the value of x/h at the point (ξ, ξ) of the field defined by equal circular arcs through C, which is taken as the origin of angular coordinates. Since the normal pressure across the horizontal axis at N is equal to $2k(1 + 2\xi)$ by Hencky's equations, the resultant

---

† The slipline field was given by L. Prandtl, op. cit., in relation to the compression between a pair of overlapping platens. Computations based on Prandtl's field have been carried out by R. Hill, E. H. Lee, and S. J. Tupper, *J. Appl. Mech.*, **18**: 46 (1951). For experimental evidence, see G. T. Van Rooyen and W. A. Backofen, *J. Mech. Phys. Solids*, **7**: 163 (1959).

‡ The result follows from a direct analysis using the conditions of continuity and symmetry. The solution has been obtained by J. Chakrabarty in an unpublished work (1978).

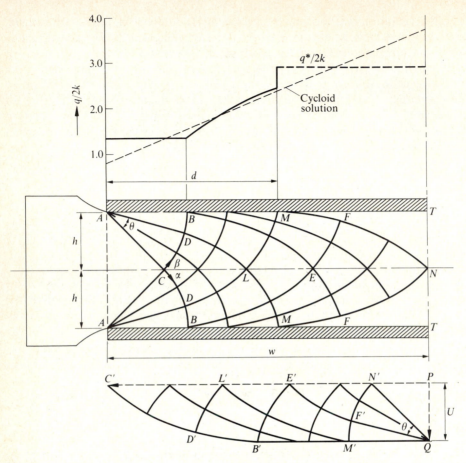

Figure 8.31 Slipline field, hodograph, and pressure distribution in compression between perfectly rough parallel platens ($w/h \geqslant 3.64$). The hodograph corresponds to the upper quadrant of the block.

vertical force P acting over ACN per unit width may be expressed as

$$\frac{P}{2kh} = I_0(2\xi) + 2\xi[I_0(2\xi) + I_1(2\xi)] - 2\left\{(1 + 2\xi)f(\theta) - 2\int_0^\theta f(\alpha)\,d\alpha\right\}$$

where $f(\theta)$ denotes the expression in the curly bracket of (63). The first two terms of the above equation have been written down by analogy with (60). Evaluating the integral, we have†

$$\frac{P}{2kh} = (1 + 2\xi)\left\{I_0(2\xi) - 2\sqrt{\frac{\theta}{\eta}}I_1(2\sqrt{\theta\eta})\right\} + 2\xi I_1(2\xi)$$

$$+ 4[F_1(\theta, \eta) - \theta I_0(2\sqrt{\theta\eta})] \quad (64)$$

---

† The integral is readily found on using the relations $dF_2/d\theta = \sqrt{\theta/\eta}\,I_1(2\sqrt{\theta\eta})$ and $dN/d\theta = F_2(\theta, \eta)$, in view of (62), Chap. 6, where N is given by (95), Chap. 6.

Let Q denote the vertical load per unit width over AM of length d. The expressions for d and Q in terms of the angle θ are obtained from Eqs. (127) and (128), Chap. 6, with $\chi = \pi/4$ and $p_0 = k(1 + \pi/2)$. The result is

$$\frac{d}{h} = \sqrt{2}\left\{I_0(2\sqrt{\theta\xi}) - \sqrt{\frac{\theta}{\xi}}\, I_1(2\sqrt{\theta\xi}) + 2F_1(\theta, \xi) - 2F_2(\theta, \xi)\right\} \tag{65}$$

$$\frac{Q}{2kh} = \frac{1}{\sqrt{2}}\left\{\left(1 + \frac{\pi}{2}\right)I_0(2\sqrt{\theta\xi}) + \left(\frac{\pi}{2} - 1\right)\sqrt{\frac{\theta}{\xi}}\, I_1(2\sqrt{\theta\xi}) + 2F_1(\theta, \xi) + 2F_2(\theta, \xi)\right\} \tag{66}$$

When θ is a multiple of $15°$, d/h is directly obtained from Table A-10. The normal pressure q on the platen at a generic point of BM is equal to $k(1 + \pi/2 + 4\alpha)$, and the distance of this point from A is given by the right-hand side of (65) with α written for θ, and ξ denoting the quantity $\pi/4 + \alpha$. The pressure distribution on the platen is represented by the upper solid curve in Fig. 8.31, only an average value of the pressure being considered over the rigid part MT. If the mean pressures over AT and MT are denoted by $\bar{q}$ and q^* respectively, then

$$\bar{q} = \frac{P}{w} \qquad q^* = \frac{P - Q}{w - d}$$

The calculated values of w/h, d/h, $\bar{q}/2k$, and $q^*/2k$ covering the whole range of values of θ are given in Table 8.8. The empirical formula

$$\frac{\bar{q}}{2k} = \frac{3}{4} + \frac{w}{4h} \qquad \frac{w}{h} \geqslant 1 \tag{67}$$

predicts the mean pressure on the platen to within 1 percent when $w/h \geqslant 2.5$. By an extension of the slipline field of Fig. 8.31, we can obtain the solution for $w/h > 6.72$, but no calculations have been made for this range of width/height ratios. The relation (67) agrees with experiments when the rate of work-hardening is small.

The velocity of the upper platen relative to the horizontal axis is represented by the vector PQ, of magnitude U, in the hodograph (Fig. 8.31). The rigid/plastic boundary NM is mapped into the circular arc $N'M'$ of radius $\sqrt{2}\, U$, which is the magnitude of the velocity discontinuity across this boundary. The shape of the hodograph is identical to that of the slipline field, except that certain boundaries

Table 8.8 Compression of a block between perfectly rough platens

θ	$0°$	$10°$	$20°$	$30°$	$40°$	$45°$
w/h	3.644	4.282	4.956	5.658	6.369	6.718
d/h	1.414	1.881	2.414	3.035	3.771	4.194
$\bar{q}/2k$	1.667	1.828	1.999	2.179	2.360	2.447
$q^*/2k$	1.908	2.218	2.534	2.867	3.221	3.412

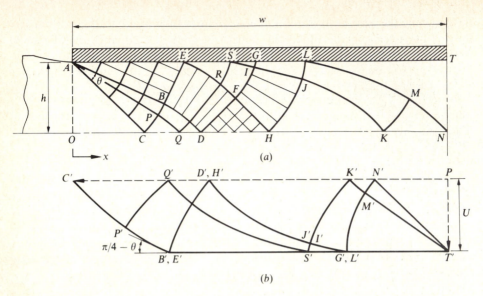

Figure 8.32 Compression of a block between partially rough platens. (*a*) Slipline field; (*b*) hodograph.

are interchanged. The vector PC', representing the velocity of the rigid overhang, is therefore of magnitude $(w/h)U$. Since the horizontal component of velocity over each vertical section increases from the platen to the axis of symmetry, a vertical line is deformed into a curve which is convex outward.†

The slipline fields of Figs. 8.30 and 8.31 also apply to the compression of a rectangular block of width $2w$ between a pair of overlapping platens under identical frictional conditions. The triangular region ACA at the end of the block is then uniformly stressed to the yield point, and is moved outward as a rigid whole. The free ends of the block therefore remain plane, and the solution continues to hold as the compression proceeds. The particles on the free surface gradually move round the corners to come in contact with the platens during the compression.

(iii) *Partially rough platens* Consider the situation where the frictional stress is a constant fraction of the normal pressure, whenever the magnitude of the shear stress is less than the yield stress (Fig. 8.32). The angular span θ of the fan ABC depends on the coefficient of friction μ which is given by the right-hand side of (61). The triangle ABE is uniformly stressed to the yield point with the α lines meeting the platen at an angle $\pi/4 - \theta$. The construction of the field CBD furnishes the slipline EF, which in conjunction with the given frictional condition along EG

---

† The compression of a block between two rough plates of unequal widths has been investigated by W. Johnson and H. Kudo, *Int. J. Mech. Sci.*, **1**: 336 (1960). See also I. F. Collins, *J. Mech. Phys. Solids*, **16**: 73 (1968), who treated the compression of a block resting on a foundation.

defines the field *EFG*. Since the normal pressure increases along *EG*, the frictional stress would reach the limiting value *k* at some point *S*, provided μ is sufficiently high. Beyond this point, the sliplines must meet the platen tangentially and normally as in the preceding solution. The field may be continued as far as the slipline *LN* passing through the assumed block center *N*. The construction of the slipline field is facilitated by the graphical method explained in Sec. 6.4(iv). In Mohr's stress plane, the normal and shear stresses transmitted to the platen at any point between *E* and *S* are represented by a point *X* which lies on a straight line passing through the origin and inclined at an angle $\tan^{-1} \mu$ with the negative σ axis,† the corresponding pole of the circle being on the horizontal through *X*.

When the semiwidth of the platen has a value between *OK* and *ON*, the velocity discontinuity initiated at the centre of the block disappears on the surface of the platen, and the shape of the hodograph remains unchanged. If, on the other hand, the block center is located somewhere between *H* and *K*, the velocity discontinuity is reflected at the platens and finally terminated at the edge *A*. The hodograph is then modified by the presence of subsidiary regions where the characteristics of one or both families are straight. The solution is not valid when the semiwidth of the platen is less than *OH*, since the velocity discontinuity then terminates on the exit slipline *AC*, which renders the velocity distribution incompatible with the motion of the rigid overhang.‡ For such values of the width/height ratio, the exit slipline must be assumed curved as in the case of smooth platens.

As the coefficient of friction decreases, the position of the point where the frictional stress first attains the value *k* moves farther from the edge of the platen. Unless the platen is sufficiently wide, the boundary of the central rigid zone would not meet the platen tangentially. The entire plastic material in contact with the platen would therefore consist of alternate regions of constant and variable stresses. The magnitude of the velocity discontinuity is gradually reduced as it propagates by successive reflections at the platens. When the rigid/plastic boundary meets the platen in a region of variable pressure, the velocity discontinuity terminates at the edge of the platen, thus giving an acceptable solution. The calculated pressure distributions for $w/h = 7$ and suitable values of μ are shown graphically in Fig. 8.33, the mean pressures over the central rigid zone being indicated by broken lines. The lowest die pressure in each case occurs near the edge of the platen, and is of magnitude $2k(\theta + \cos^2 \theta)$. It may be noted that the frictional stress is equal to *k* over a substantial part of the platen for the higher values of μ.

If we assume, for simplicity, that the frictional condition on the compression platen induces a constant shear stress mk ($m \leqslant 1$), the sliplines meet the platens at constant angles. In this case, the slipline field and the hodograph form identi-

---

† The graphical solution and associated numerical results have been given by J. M. Alexander, *J. Mech. Phys. Solids*, **3**: 233 (1955).

‡ When the platens overlap the block, the proposed slipline field holds for all geometrically possible width/height ratios, since the velocity discontinuity can terminate on the edge of the block, causing a change in shape of the plastic edge. See J. F. W. Bishop, *J. Mech. Phys. Solids*, **6**: 132 (1958).

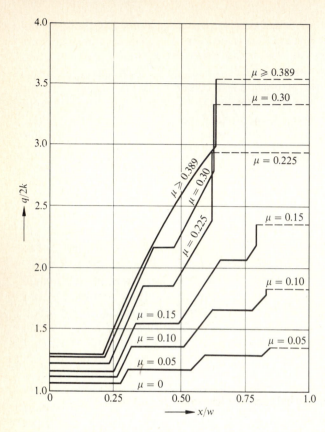

Figure 8.33 Pressure distribution on the platens for $w/h = 7$ (*after J. M. Alexander*). The broken lines indicate mean pressures over the central rigid zone.

cal networks.† The angle which the α lines make with the platens is λ, where $\cos 2\lambda = m$. The magnitude of the velocity discontinuity propagated from the center of the block is multiplied at each reflection from the platens by a constant factor $\tan \lambda$. The discontinuity is therefore progressively diminished toward the edge, and the straight line regions of the slipline field are correspondingly reduced in length toward the center. The solution is only valid for those width/height ratios for which the discontinuities terminate at the edges of the platens.

(iv) *Prandtl's cycloid solution* When the width of the block is large compared to its height $2h$, the slipline field at sufficiently large distances from either end of the block would approach a limiting configuration. The limiting field is most conveniently obtained by considering the rectangular components of the stress, where the x axis is taken along the horizontal axis of symmetry and the y axis through the left-hand edge of the block. Since the slipline directions are independent of x, it follows that τ_{xy} and $\sigma_x - \sigma_y$ are also independent of x. We

† A. P. Green, *J. Mech. Phys. Solids*, **2**: 73 (1954).

therefore write

$$\tau_{xy} = kf(y) \qquad \sigma_x - \sigma_y = 2k\sqrt{1 - f^2}$$

in view of the yield criterion. Substituting for σ_x and τ_{xy} into the equilibrium equations (3), Chap. 6, we have

$$\frac{\partial \sigma_y}{\partial x} + kf'(y) = 0 \qquad \frac{\partial \sigma_y}{\partial y} = 0$$

The second of these equations indicates that σ_y is independent of y. In view of the symmetry condition $f(0) = 0$, the first equation furnishes

$$f(y) = \frac{y}{b} \qquad \sigma_y = -k\left(\frac{x}{b} + c\right)$$

where b and c are constants. Considering the generalized frictional condition $\tau_{xy} = mk$ on $y = h$, where $0 \leqslant m \leqslant 1$, we get $b = h/m$. The stress distribution therefore becomes†

$$\sigma_x = -k\left(c + \frac{mx}{h}\right) + 2k\sqrt{1 - \frac{m^2 y^2}{h^2}}$$

$$\sigma_y = -k\left(c + \frac{mx}{h}\right) \qquad \tau_{xy} = k\left(\frac{my}{h}\right) \tag{68}$$

The constant c can be determined from the condition that the resultant horizontal thrust on a vertical section must balance the frictional resistance $2mkx$. Hence

$$c = \frac{1}{m}\sin^{-1} m + \sqrt{1 - m^2} \tag{69}$$

It may be noted that the derivative $\partial\sigma_x/\partial y$ along the interface $y = \pm h$ tends to infinity as m tends to unity. A perfectly rough platen corresponds to $m = 1$, for which $c = \pi/2$. Assuming the solution (68) to hold as far as the edge $x = 0$, the pressure distribution for $m = 1$ is shown by the broken curve in Fig. 8.31. It is evident that the above solution is a good approximation even up to a distance h from the edge. The actual pressure distribution approaches the limiting distribution in a quasi-oscillatory manner. A similar approach would result for $m < 1$, though probably more slowly the smaller the value of m. If (68) is assumed to hold for $0 \leqslant x \leqslant w$, the average pressure on the platen becomes

$$\bar{q} = k\left(c + \frac{mw}{2h}\right)$$

When the platens are perfectly rough, the above formula differs only slightly from the empirical equation (67).

† L. Prandtl, *Z. angew. Math. Mech.*, **3**: 401 (1923); A. P. Green, *J. Mech. Phys. Solids*, **2**: 73 (1954).

The slipline field corresponding to (68) may also be expressed analytically. It follows from Mohr's circle for the stress that the slope of the α line at any point of the field is

$$\frac{dy}{dx} = -\frac{2(k - \tau_{xy})}{\sigma_x - \sigma_y} = -\sqrt{\frac{1 - my/h}{1 + my/h}} \tag{70}$$

The slope of the β line is given by the reciprocal of above with a change in sign. Setting $my/h = \sin 2\psi$ on the right-hand side, we have

$$\frac{dy}{dx} = \mp\left(\frac{1 \mp \tan \psi}{1 \pm \tan \psi}\right) = \mp\tan\left(\frac{\pi}{4} \mp \psi\right)$$

where the upper sign corresponds to the α lines and the lower sign to the β lines. It follows from the above equation that ψ is the counterclockwise angle turned through along a slipline from its point of intersection with the axis of symmetry. Substituting for y on the left-hand side, and integrating, we obtain the equations of the sliplines in the parametric form

$$x = d \mp \frac{h}{m}[2\psi \pm (1 - \cos 2\psi)] \qquad y = \frac{h}{m}\sin 2\psi \tag{71}$$

The constant d is evidently equal to the distance between the origin and the point of intersection of a given slipline with the axis of symmetry. The sliplines are cycloids (Fig. 8.34), generated by a circle of radius h/m with its center on the axis. The radii

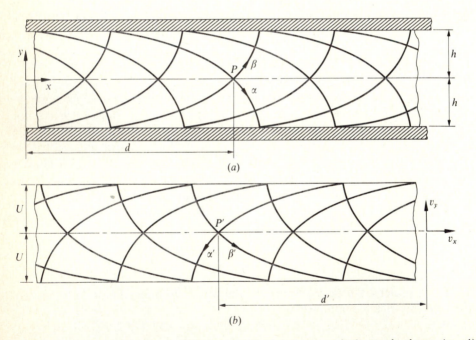

Figure 8.34 Compression of a sufficiently wide block between perfectly rough platens ($m = 1$). (a) Slipline field; (b) hodograph.

of curvature of the sliplines are numerically equal to

$$\frac{2h}{m}\sqrt{2\left(1 \pm \frac{my}{h}\right)} = \frac{4h}{m}\cos\left(\frac{\pi}{4} \mp \psi\right)$$

The sliplines meet the platens at constant angles λ and $\pi/2 + \lambda$, where $\cos 2\lambda = m$. When $m = 1$, the horizontal projection of each slipline between the two platens is πh, the radius of curvature of a slipline at the point of its tangency with a platen being $4h$. If the solution is assumed to hold in the central part of a sufficiently large block, the semiwidth of the rigid wedge would be $(1 + \pi/2)h$ when $m = 1$.

At sufficiently large distances from the center of the block, the velocity distribution should approach a limiting state, such that the strain rates are independent of x. In view of the boundary conditions $v_y = -U$, $y = h$, and $v_y = 0$, $y = 0$, the velocity components v_x and v_y satisfying the incompressibility condition may be written as

$$v_x = U\left[\frac{x}{h} + g(y)\right] \qquad v_y = -U\left(\frac{y}{h}\right) \tag{72}$$

where $g(y)$ must be determined from the stress-strain relation expressed by Eq. (5), Chap. 6. Using (68), we get

$$g'(y) = \frac{2my/h^2}{\sqrt{1 - m^2y^2/h^2}}$$

and the integration of this equation results in

$$g(y) = c' - \frac{w}{h} - \frac{2}{m}\sqrt{1 - \frac{m^2y^2}{h^2}} \tag{73}$$

where c' is a constant. The term w/h has been included for convenience. The rate of horizontal flow across a vertical section must be equal to the rate at which material to the right is displaced by the platens. Hence

$$-\int_0^h v_x \, dy = (w - x)U$$

The substitution for v_x results in $c' = c/m$, where c is given by (69). From (71), (72), and (73), the velocity components may be expressed as

$$v_x = d' \mp \frac{U}{m}[2\psi \mp (1 - \cos 2\psi)] \qquad v_y = -\frac{U}{m}\sin 2\psi \tag{74}$$

where

$$d' = -U\left(\frac{2-c}{m} + \frac{w-d}{h}\right)$$

Equations (74) define the hodograph net, which consists of two orthogonal families of cycloids. The upper sign applies to the α' curves and the lower sign to the β' curves shown in Fig. 8.34.

If the rate of change following the motion is denoted by a dot, so that $\dot{y} = v_y$ and $\dot{h} = -U$, the second equation of (72) gives

$$\frac{dy}{dh} = -\frac{\dot{y}}{U} = \frac{y}{h} \quad \text{or} \quad \frac{d}{dh}\left(\frac{y}{h}\right) = 0$$

Thus, a given particle remains at the same relative distances from the axis and the platen. Equally spaced horizontal lines therefore continue to be equally spaced. Consider, now, the horizontal displacement of a generic particle P during the compression. At any instant of time, when the block thickness is $2h$, let ξ be the distance by which P is in advance of the surface particle Q which was vertically above P in the initial state. The difference between the horizontal velocities of P and Q is $\dot{\xi}$, which gives

$$-\frac{d\xi}{dh} = \frac{\dot{\xi}}{U} = \frac{\xi}{h} + \frac{2}{m}\left\{\sqrt{1 - \frac{m^2 y^2}{h^2}} - \sqrt{1 - m^2}\right\}$$

in view of (72) and (73). Since y/h remains constant during the compression, the above equation is immediately integrated. The result may be expressed as

$$\left(\frac{mh\xi}{h_0^2 - h^2} + \sqrt{1 - m^2}\right)^2 + \frac{m^2 y^2}{h^2} = 1 \tag{75}$$

where $2h_0$ is the initial thickness of the block. It follows from (75) that an original vertical line is distorted into an ellipse with semimajor axis h/m and semiminor axis $(h_0^2 - h^2)/mh$. When $m = 1$, the ellipse is tangential to the platens at the extremities of its major axis. For $m < 1$, the block may be imagined as part of a thicker block of height $2h_0/m$. The calculated distortion for perfectly rough platens has been found to be in good agreement with experiment.[†]

(v) *Approximate solutions* An upper bound solution for plane strain compression between partially rough platens may be obtained by assuming an instantaneous sliding of rigid blocks over a crisscross of planes equally inclined to the die face.[‡] The number of discontinuities in each quadrant is denoted by m, the discontinuity pattern and the hodograph when m is odd being shown in Fig. 8.35. The velocity of each triangular block which slips over the die face is determined from the condition that the component of this velocity in the direction of the resultant traction is constant along the die face (see Sec. 2.6(iii)). It follows that the lower vertices of the velocity triangles must lie on a straight line inclined at an angle $\lambda = \tan^{-1} \mu$ to the horizontal. The magnitude of the velocity discontinuity across the lines through a typical point on the axis of symmetry is $\sqrt{1 + d^2/h^2}$ times the vertical component of velocity of the included material triangle. By the geometry of the hodograph,

† See, for example, A. Nadai, *Theory of Flow and Fracture of Solids*, p. 537, McGraw-Hill, New York (1950); J. F. Nye, *J. Appl. Mech.*, **18**: 337 (1951).
‡ I. F. Collins, *J. Mech. Phys. Solids*, **17**: 323 (1969).

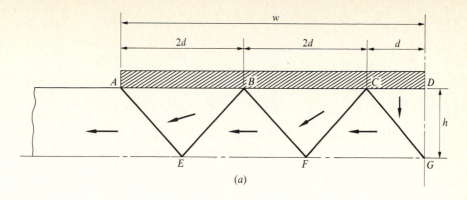

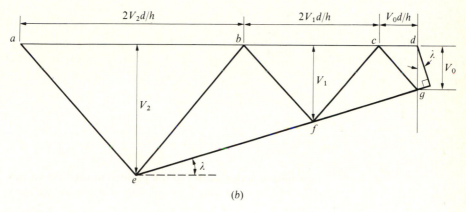

Figure 8.35 Upper bound modes for compression with Coulomb friction. (*a*) Discontinuity pattern; (*b*) hodograph.

these velocity components are

$$V_1 = \left(\frac{1+v}{1-v}\right)V_0, \; V_2 = \left(\frac{1+v}{1-v}\right)^2 V_0, \ldots, V_n = \left(\frac{1+v}{1-v}\right)^n V_0$$

where V_0 is the vertical speed of compression, and

$$v = \frac{\mu d}{h} = \frac{\mu w}{mh} \qquad n = \frac{m-1}{2}$$

Since the length of each discontinuity is equal to $\sqrt{d^2 + h^2}$, the upper bound on the mean normal pressure $\bar{q}$ is given by

$$\bar{q}wV_0 = k\left(h + \frac{d^2}{h}\right)(V_0 + 2V_1 + \cdots + 2V_n)$$

Substituting for the velocities, and summing up the geometric series, the die

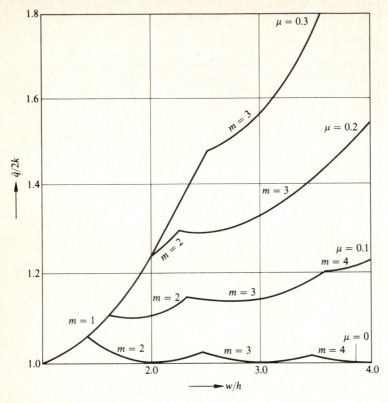

Figure 8.36 Upper bounds on the mean pressure for compression between partially rough platens.

pressure is obtained as

$$\frac{\bar{q}}{2k} = \frac{1}{2}\left(\frac{w}{m^2 h} + \frac{h}{w}\right)\left\{\frac{1+v}{v}\left(\frac{1+v}{1-v}\right)^n - \frac{1}{v}\right\} \tag{76}$$

When m is even, say equal to $2n$, the velocity discontinuity emanates from the center of the die face, while the material triangle containing the vertical centerline is at rest. The associated hodograph may be constructed as before, using the condition that the rigid blocks sliding over the die face have a constant component $V_0 \cos \lambda$ in the direction of the resultant traction. The application of the upper bound theorem then leads to

$$\bar{q} w V_0 = k\left(h + \frac{d^2}{h}\right)(2V_1 + 2V_2 + \cdots + 2V_n)$$

where $V_1, V_2, \ldots$ are the vertical velocity components given by

$$V_r = \frac{V_0}{1-v}\left(\frac{1+v}{1-v}\right)^{r-1} \qquad r = 1, 2, \ldots$$

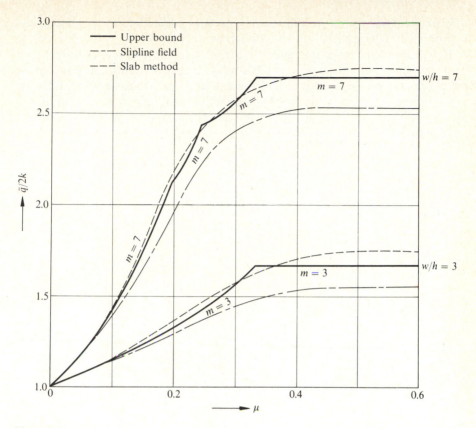

Figure 8.37 Comparison of approximate solutions with exact solution for plane strain compression with friction.

The parameter v is the same as before, and the dimensionless die pressure finally becomes

$$\frac{\bar{q}}{2k} = \frac{1}{2}\left(\frac{w}{m^2 h} + \frac{h}{w}\right)\left\{\frac{1}{v}\left(\frac{1+v}{1-v}\right)^n - \frac{1}{v}\right\} \tag{77}$$

If the friction is vanishingly small, v tends to zero, and expressions in the curly brackets of (76) and (77) reduce to m. Both these expressions then coincide with (59), derived for frictionless compression. Figure 8.36 shows the predicted variation of the upper bound pressure with the width/height ratio for various values of the coefficient of friction.

For μ exceeding a certain critical value, improved upper bounds would result from the assumption of sticking friction occurring along the die face on one or more of the sliding blocks. The corresponding hodograph must be modified by using the condition that the vertical component of velocity of each of these blocks is equal to V_0. The modified upper bounds are incorporated in the appropriate graphical plots of Fig. 8.37. The last two segments of the rising part of the upper solid curve correspond to partial sticking friction, while the horizontal segment corresponds to sticking friction over the whole die face.†

For practical purposes, it is useful to consider an approximate solution based on assumptions similar to those in von Karman's theory of rolling. Let q denote the normal pressure on the platens at a distance x from the left-hand edge, and p the mean horizontal compressive stress on the corresponding vertical section. If x is less than a critical distance x_0, the frictional stress is equal to $\mu q < k$, and the equilibrium of a vertical slice of thickness dx requires

$$\frac{dp}{dx} = \frac{\mu q}{h}$$

If the friction is sufficiently small, the yield criterion may be written approximately as

$$q - p = 2k$$

Eliminating p between the above equations, and using the boundary conditions $q = 2k$ at $x = 0$, we obtain the pressure distribution

$$q = 2k \exp\left(\frac{\mu x}{h}\right) \qquad x \leqslant x_0 \qquad (78a)$$

The derivation of (78a) actually involves the relation $dp/dx = dq/dx$, which may be considered as reasonably accurate for fairly large frictional stresses. When the platens are sufficiently wide, the frictional stress μq becomes equal to the shear yield stress k at $x = x_0$, and the above equation furnishes

$$x_0 = \frac{h}{\mu} \ln \frac{1}{2\mu}$$

For $x \geqslant x_0$, the frictional stress has a constant value k, and the equilibrium equation becomes

$$\frac{dq}{dx} \simeq \frac{dp}{dx} = \frac{k}{h}$$

Since the pressure must be continuous at $x = x_0$, the integration of the above equation gives the linear pressure distribution

$$q = k\left\{\frac{x}{h} + \frac{1}{\mu}\left(1 - \ln \frac{1}{2\mu}\right)\right\} \qquad x \geqslant x_0 \qquad (78b)$$

The analysis holds equally good for compression between overlaping platens provided x is measured from the edge of the block. The pressure steadily increases with x, attaining a peak value at the center of the platen. Substituting from (78) into

---

† The upper bound results for partial and full stictions will be found in Probs. 8.16 and 8.17.

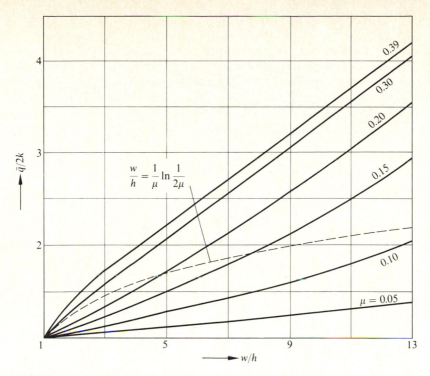

Figure 8.38 Variation of mean die pressure with width/height ratio for compression between rough platens (technological theory).

the relation $\bar{q} = (\int_0^w q\,dx)/w$, the mean pressure on the platens is obtained as[†]

$$\frac{\bar{q}}{2k} = \frac{h}{\mu w}\left\{\exp\left(\frac{\mu w}{h}\right) - 1\right\} \qquad \frac{\mu w}{h} \leqslant \ln\frac{1}{2\mu}$$

$$\frac{\bar{q}}{2k} = \frac{h}{\mu w}\left\{\left(\frac{1}{4\mu} - 1\right) + \frac{1}{4\mu}\left(1 + \frac{\mu w}{h} - \ln\frac{1}{2\mu}\right)^2\right\} \qquad \frac{\mu w}{h} \geqslant \ln\frac{1}{2\mu}$$

$$(79)$$

The dimensionless mean pressure (79) is plotted as a function of the width/height ratio in Fig. 8.38 for various values of μ. The technological solution is compared with the slipline field and upper bound solutions in Fig. 8.37. It is evident that the approximate theories are reasonably good for relatively small values of μ. When w/h and μ are both small, the mean pressure exceeds $2k$ by the factor $\mu w/2h$ to a close approximation.[‡]

(vi) *Compression between rough inclined platens* A mass of rigid/plastic material is

---

[†] J. M. Alexander, *J. Mech. Phys. Solids*, **3**: 233 (1955); J. F. W. Bishop, *ibid.*, **6**: 132 (1958).

[‡] The compression of a strip without edge constraints has been discussed by R. Hill, *Phil. Mag.*, **41**: 733 (1950).

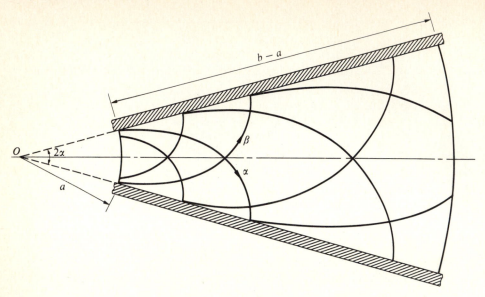

Figure 8.39 Slipline field for compression of a plastic mass between perfectly rough inclined platens.

symmetrically compressed between a pair of platens that are inclined to one another at a small angle 2α (Fig. 8.39). The platens are so rough that the material is caused to shear along them over the region of relative sliding. If the compressed material extends sufficiently far, the stress distribution at the yield point would attain a limiting state, as in the case of parallel platens. Using polar coordinates (r, θ) with respect to the axis of symmetry and the virtual apex O, and neglecting terms of order $1/r$, the equilibrium equations may be written approximately as

$$\frac{\partial \sigma_r}{\partial r} + \frac{1}{r}\frac{\partial \tau_{r\theta}}{\partial \theta} \simeq 0 \qquad \frac{\partial \tau_{r\theta}}{\partial r} + \frac{1}{r}\frac{\partial \sigma_\theta}{\partial \theta} \simeq 0$$

Since the slipline directions for the limiting field are independent of r, we assume

$$\tau_{r\theta} = \frac{k\theta}{\alpha} \qquad \sigma_r - \sigma_\theta = 2k\sqrt{1 - \frac{\theta^2}{\alpha^2}}$$

in the region of converging flow of the plastically compressed material. These expressions identically satisfy the yield criterion, and are consistent with the equations of equilibrium. A simple integration leads to the stress distribution[†]

$$\sigma_r = -k\left(A + \frac{1}{\alpha}\ln\frac{r}{a}\right) + 2k\sqrt{1 - \frac{\theta^2}{\alpha^2}}$$

$$\sigma_\theta = -k\left(A + \frac{1}{\alpha}\ln\frac{r}{a}\right) \qquad \tau_{r\theta} = \frac{k\theta}{\alpha}$$

(80)

† The solution for the stresses is essentially due to A. Nadai, *Z. Phys.*, **30**: 106 (1930).

where a is the radius to the left edge of the block, and A is a constant. Since σ_θ is independent of θ, it is numerically equal to the normal pressure on the platen at any radius r. If the solution is assumed to hold at $r = a$, the condition of zero resultant horizontal force across this cylindrical surface would furnish A. Since $\cos \theta \simeq 1$, we have

$$\int_0^\alpha \sigma_r \, d\theta = -k\alpha A + 2k \int_0^\alpha \sqrt{1 - \frac{\theta^2}{\alpha^2}} \, d\theta = k\alpha\left(-A + \frac{\pi}{2}\right) = 0$$

giving $A = \pi/2$. The normal pressure varies along the platens as the logarithm of the radial distance r, the pressure at $r = a$ being equal to $\pi k/2$. By analogy with (70), the differential equations for the two families of sliplines are

$$r\frac{d\theta}{dr} = \mp \sqrt{\frac{\alpha \mp \theta}{\alpha \pm \theta}}$$

where the upper sign holds for the α lines and the lower sign for the β lines. Performing the integration, the polar equations of the sliplines are obtained as

$$\ln\left(\frac{r}{r_0}\right) = -\alpha + \sqrt{\alpha^2 - \theta^2} \mp \alpha \sin^{-1}\left(\frac{\theta}{\alpha}\right) \tag{81}$$

where r_0 is the length of the radius vector to the point of intersection of a slipline with the axis of symmetry. The radial distances from O to the extremities of any slipline between the platens are in the ratio $\exp(\pi\alpha)$. When α is vanishingly small, (80) reduces to (68) with $m = 1$, and (81) becomes equivalent to (71).

Let the radial and circumferential velocities of a typical particle be denoted by u and v respectively. In view of the incompressibility equation

$$\frac{\partial u}{\partial r} + \frac{u}{r} + \frac{1}{r}\frac{\partial v}{\partial \theta} = 0$$

the velocity distribution for the compressed material must be of the form

$$u = \frac{U}{\alpha}\left[1 + \frac{f(\theta)}{r}\right] \qquad v = -\frac{U\theta}{\alpha}$$

where U is the inward speed of each compression platen normal to its plane. The components of the strain rate are readily found as

$$\dot{\varepsilon}_r = -\dot{\varepsilon}_\theta = -\frac{Uf(\theta)}{\alpha r^2} \qquad \dot{\gamma}_{r\theta} \simeq \frac{Uf'(\theta)}{2\alpha r^2}$$

The fact that the principal axes of stress and strain rate coincide is expressed by

$$\frac{2\dot{\gamma}_{r\theta}}{\dot{\varepsilon}_r - \dot{\varepsilon}_\theta} = \frac{2\tau_{r\theta}}{\sigma_r - \sigma_\theta} \qquad \text{or} \qquad \frac{f'(\theta)}{f(\theta)} = -\frac{2\theta}{\sqrt{\alpha^2 - \theta^2}}$$

This equation integrates to $f(\theta) = -c\exp\left(2\sqrt{\alpha^2 - \theta^2}\right)$, where c is a positive constant, and the velocity field becomes

$$u = \frac{U}{\alpha}\left\{1 - \frac{c}{r}\exp\left(2\sqrt{\alpha^2 - \theta^2}\right)\right\} \qquad v = -\frac{U\theta}{\alpha} \tag{82}$$

The relative velocity of sliding between the material and the platen is of magnitude $(U/\alpha)(1 - c/r)$, which changes sign at the radius $r = c$.

When the compressed material extends from $r = a$ to $r = b$, the stress distribution in the region $a \leqslant r \leqslant c$ is directly given by (80), while that in $c \leqslant r \leqslant b$ is obtained from (80) by changing the sign of $\tau_{r\theta}$ and replacing $\ln(r/a)$ by $\ln(b/r)$. The constant A is again equal to $\pi/2$ for the resultant longitudinal force to vanish across the end $r = b$. The continuity of the radial stress at $r = c$ requires $c = \sqrt{ab}$. The distribution of normal pressure on the platen represents a friction hill with a peak value occurring at $r = c$. The mean die pressure $\bar{q}$ is given by

$$\frac{\bar{q}}{2k} = \frac{\pi}{4} + \frac{1}{2\alpha}\left(\frac{\sqrt{\rho} - 1}{\sqrt{\rho} + 1}\right) \tag{83}$$

where ρ denotes the ratio b/a, which is equal to the ratio of the end thicknesses of the block. For a given value of the mean width/height ratio, the mean pressure on the die increases as the wedge angle is increased.†

8.7 Yielding of Notched Bars in Tension

A flat bar of ductile material is symmetrically notched on opposite sides and is pulled in tension along the longitudinal axis under conditions of plane strain. The bar is sufficiently long so that the stress distribution around the notch is independent of the end conditions. As the load is increased to a critical value, yielding occurs at the roots of the notches due to the existence of local stress concentration. With further increase in load, the plastic zones spread inward, while the overall extension of the bar remains of the elastic order of magnitude. At some stage during the loading, a second plastic zone begins to form on the longitudinal axis. The yield point state is eventually reached as a result of fusion of the primary and secondary plastic zones, and large deformations become imminent. The average logitudinal stress across the minimum section at the yield point is $2k$ multiplied by a certain factor, known as the constraint factor, which depends on the geometry of the notch.

(i) *Bars of supercritical thickness* When the notches are sufficiently deep, the initial deformation at the yield point is confined to a region in the neighborhood of the notch root. The constraint factor is then independent of the ratio of the thickness of the bar to that of the minimum section. We begin with a notch whose shape is a circular arc of radius c, the thickness of the neck being denoted by $2a$. The yield

---

† A slipline field solution for compression between perfectly rough inclined platens has been presented by W. Johnson and H. Kudo, *Appl. Sci. Res.* **A9**: 206 (1960).

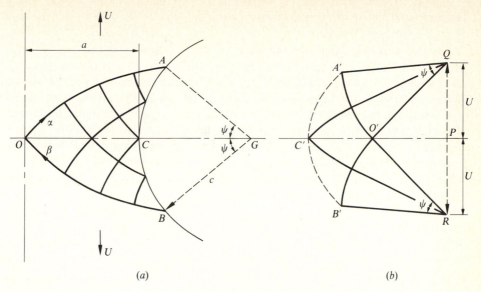

Figure 8.40 Tension of a bar with symmetrical circular notches. (*a*) Slipline field; (*b*) hodograph.

point corresponds to the moment when the sliplines from the points *A* and *B* on the notch surface meet the geometric center *O* of the bar (Fig. 8.40*a*). At this stage, the ends of the bar are free to move apart, and the extension can continue momentarily under constant load, provided the rate of work-hardening is small. The sliplines in the field *AOB* are logarithmic spirals (Sec. 5.8(ii)), the principal stress trajectories being radial lines through *G* and concentric circular arcs. If 2ψ is the angle subtended by the arc *AB* at the center, then

$$a = c(e^\psi - 1) \qquad \text{or} \qquad \psi = \ln\left(1 + \frac{a}{c}\right) \tag{84}$$

The hydrostatic tension has the value k along the notch surface. Hence the longitudinal tensile stress across the minimum section is $\sigma = 2k(1 + \phi)$ at a point whose distance from the notch root is $z = c(e^\phi - 1)$, where ϕ denotes the corresponding angular distance along the notch. The stress steadily increases as we go inward from the notch root. The yield point load per unit width is

$$T = 2\int_0^a \sigma\, dz = 4kc\int_0^\psi (1 + \phi)e^\phi\, d\phi = 4kc\psi e^\psi$$

The elastic stress concentration is dispersed by the plastic yielding which reverses the stress gradient near the root. Substituting from (84), we get

$$T = 4ka\left(1 + \frac{c}{a}\right)\ln\left(1 + \frac{a}{c}\right) \tag{85}$$

The ratio $T/4ka$ is equal to the constraint factor, which rises from unity as the value of a/c is increased from zero.

There is a velocity discontinuity of amount $\sqrt{2}\,U$ across the sliplines OA and OB, where U denotes the speed with which the ends of the bar are drawn outward. The velocities of the rigid ends are represented by the vectors PQ and PR in the hodograph (Fig. 8.40b). The image of the point O, considered to the right of the vertical axis, is defined by the vertex O' of the isosceles right triangle $QO'R$. The sliplines OA and OB are mapped into the circular arcs $O'A'$ and $O'B'$, which define the field $O'A'B'$. The broken curve $A'B'$ forms the image of the notch surface, and is given by the relation $\alpha + \beta = \psi$, where (α, β) are the angular coordinates referred to the circular base curves. Since the velocity vector is known in direction and magnitude at each point of the bar, the distortion of a square grid can be easily calculated.

Consider, now, a V-notched bar of included angle $\pi - 2\psi$, having a circular fillet of radius c (Fig. 8.41). The circular root of angle 2ψ joins smoothly to the straight sides of the notch at A and B. When $a/c \leqslant e^{\psi} - 1$ the plastic zone does not extend to the straight sides, and the constraint factor is the same as that given by (85). For $a/c > e^{\psi} - 1$, the slipline field involves triangular regions AGK and BHL adjacent to the boundary. The obvious continuation of the field across the neck

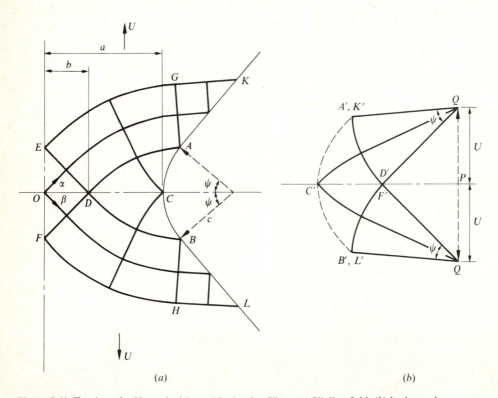

(a) (b)

Figure 8.41 Tension of a V-notched bar with circular fillets. (a) Slipline field; (b) hodograph.

leads to the uniformly stressed region *DEF*, having a depth

$$b = a - c(e^\psi - 1) \geqslant 0$$

The longitudinal stress across the minimum section varies from *C* to *D* as before, but along *DO* it has the constant value $2k(1 + \psi)$. The total longitudinal load per unit width at the yield point is[†]

$$T = 4k[c\psi e^\psi + b(1 + \psi)] = 4ka\left[(1 + \psi) - \frac{c}{a}(e^\psi - 1 - \psi)\right] \qquad (86)$$

which holds in the range $a/c \geqslant e^\psi - 1$. The expression in the square bracket of (86) is equal to the constraint factor, which steadily increases with the angle ψ. The dependence of the constraint factor on the notch angle and root radius is shown graphically in Fig. 8.42. When the notch is sharp ($c = 0$), the constraint factor reduces to the value $1 + \psi$, the slipline field in that case being identical to that for the plastic compression of a truncated wedge of semiangle ψ (Sec. 8.3(i)).

The material in *DEF* moves horizontally as a rigid whole with a speed *U*, while velocity discontinuities of amount $\sqrt{2}\,U$ propagate along the sliplines *EK* and *FL*. The resultant velocity is constant along each straight slipline in the regions *ADEG* and *BDFH*. An inspection of the slipline field and the hodograph indicates that the rate of plastic work is everywhere positive. There is an alternative mode of deformation, giving the same constraint factor, in which the bounding sliplines are considered through the point *O* as indicated in the figure. However, the actual potentially deformable region consistent with the yield point load must extend as far as the sliplines *EK* and *FL*.

The preceding solutions are based on the assumption that the bar is long enough to permit the deformation to be confined in the central region. If the bar is too short, the deformation spreads to the ends of the bar,[‡] and the yield point load then depends on the ratio l/a, where *l* is the semilength of the bar. Suppose that the longitudinal velocity at the ends is constrained to be uniform, while the tractions are kept shear free. For a sharply notched bar, the longitudinal stress across the minimum section is then numerically equal to the pressure required for indenting a block of height $2l$ by opposed punches of width $2a$ (Sec. 8.1(ii)). The stress attains the value $2k(1 + \psi)$ when *l* increases to a critical value l^*, beyond which the deformation becomes localized in the neck. The critical length ratios for various values of ψ are given in the following table:

ψ	15°	30°	45°	60°	75°	90°
l^*/a	2.16	3.09	4.14	5.38	6.90	8.72

† R. Hill, *Q. J. Mech. Appl. Math.*, **2**: 40 (1949). The solution for elliptic notches has been discussed by P. S. Symonds, *J. Appl. Phys.*, **20**: 107 (1949). The tension of rectangularly notched bars has been investigated by E. H. Lee, *J. Appl. Mech.*, **21**: 140 (1954).

‡ This problem has been examined by J. Salencon, *C. R. Acad. Sci., Paris*, **A264**: 613 (1967).

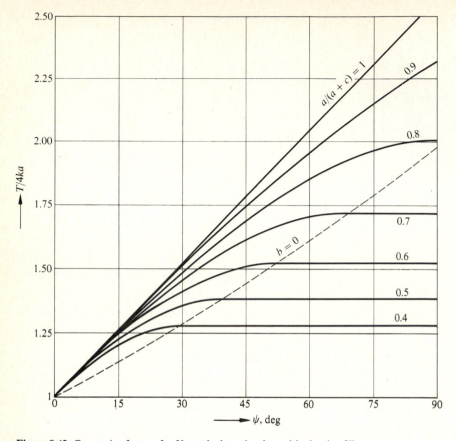

Figure 8.42 Constraint factors for V-notched tension bar with circular fillets.

(ii) *Critical and subcritical thicknesses* For a given depth of the minimum section, the deformation is localized in the neck so long as the thickness of the bar is greater than a critical value. The critical thickness can be estimated by suitably extending the plastic stress field giving the same constraint factor. For a bar with sharp notches, a statically admissible extension of the Prandtl field *ABCD* is shown in Fig. 8.43*a*. The field generated by the circular arc *BC* is continued as far as *FGHD*, where *F* is defined by the bisector of the fan *ABC*. The region *DHK* is associated with the traction-free boundary *DK* whose tangent at *K* is parallel to the longitudinal axis. The tangent may therefore be taken as the flat side of the bar. By Eq. (139), Chap. 6, the corresponding semithickness is

$$h^* = a(2e^\psi - 1) \tag{87}$$

All stresses are set to zero in the region to the right of *DK*, which is therefore an admissible discontinuity. The material beneath the slipline *FGHK* can be shown to be capable of supporting the boundary tractions without being overstressed. The

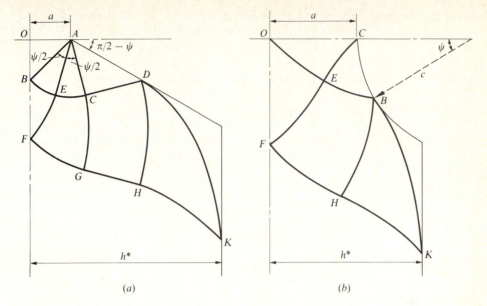

Figure 8.43 Quarter slipline fields giving upper bounds on the critical thickness. (*a*) Sharply notched bar; (*b*) round-notched bar.

possible deformation is of course restricted to the zone $ABCD$. Since the stress field is statically admissible, the yield point load is underestimated for a notched bar of thickness $2h^*$. Equation (87) therefore provides an upper bound to the critical thickness. A lower bound estimate, based on a field involving instantaneous rotation of rigid shoulders, is found to be remarkably close to the upper bound value.† It follows that (87) is a very good approximation to the actual critical thickness, the error being less than 0.2 percent over the whole range of values of ψ.

When the notch is circular, an upper bound to the critical thickness can be obtained by considering the field of Fig. 8.43*b*. The region $OBHF$ is defined by the spiral OB and its reflection in the longitudinal axis, the position of F being determined by the slipline through C. The angles subtended by the segments OE, CE, and EF are each equal to $\psi/2$. The slipline BH defines the traction-free boundary BK and the associated field BHK. The tangent to BK at K is parallel to OF, and is therefore taken as the side of the bar, the material to the right of BK being assumed stress free. It can be shown that the rigid part of the bar below FHK is able to support the tractions across this slipline. By (143), Chap. 6, with $b = ce^\psi$, we have

$$h^* = ce^\psi(e^\psi - 1) = ae^\psi = a\left(1 + \frac{a}{c}\right) \qquad (88)$$

---

† Both upper and lower bound estimates for the critical thickness of a V-notched bar have been obtained by D. J. F. Ewing and R. Hill, *J. Mech. Phys. Solids*, **15**: 115 (1967). See also J. E. Neimark, *J. Appl. Mech.*, **35**: 111 (1968).

on using (84). For a semicircularly notched bar,[†] the critical semithickness is therefore $2a$ (since $c = a$), and the corresponding value of the constraint factor is $2 \ln 2 \simeq 1.389$. The upper bound formula (88) predicts the critical thickness with an error not exceeding 0.1 percent. Both the fields of Fig. 8.43 have the interesting property that the distance of K from the horizontal axis of symmetry is twice the distance OF.

Consider now a V-notched bar with a circular root of radius c and angle 2ψ. When the deformation spreads to the straight sides of the notch (Fig. 8.41), the slipline field can be extended into the rigid region by generating a stress-free boundary beginning at L. The boundary is continued through an angle ψ to the point where the tangent to the boundary is parallel to the vertical axis. The resulting slipline field is the vector resultant of the field for a circular notch of radius c, and the field for a sharp notch with neck depth $2b$, corresponding to the same angular measure ψ (Sec. 6.8). The upper bound to the critical thickness is therefore obtained by an identical combination of the results for circular and sharp notches. Using (87) and (88), we get

$$h^* = b(2e^\psi - 1) + ce^\psi(e^\psi - 1)$$

$$= a(2e^\psi - 1) - c(e^\psi - 1)^2 \qquad (89)$$

This formula holds for $b \geqslant 0$, which is equivalent to $a/c \geqslant e^\psi - 1$, and provides a close approximation for the critical thickness of the bar.[‡] For $a/c \leqslant e^\psi - 1$, the critical thickness is identical to that for a circular notch, and is given by the last expression of (88).

When the thickness of the bar is less than critical, the deformation spreads out to the sides of the bar, the effect of which is to decrease the constraint factor for decreasing thickness. Indeed, any removal of material from the potentially deformable region must lower the actual yield point load. The construction of Fig. 8.43 can be used to generate a statically admissible stress field for any specimen geometrically contained in that having the critical thickness. For a sharp V-notched bar, the field involves a smaller fan angle θ, giving a lower bound value $1 + \theta$ for the constraint factor. It is only necessary to write h for h^* and θ for ψ in (87) to obtain the required geometrical relationship. The lower bound formula for a sharp notch therefore becomes

$$\frac{T}{4ka} = 1 + \ln\left\{\frac{1}{2}\left(1 + \frac{h}{a}\right)\right\} \qquad h \leqslant h^* \qquad (90)$$

[†] This particular geometry has been considered by D. N. de G. Allen and R. V. Southwell, *Phil. Trans. R. Soc.*, **A242**: 270 (1950), to obtain an elastic/plastic solution using the relaxation method. The solution shows a constraint factor of about 1.22, which is somewhat smaller than the rigid/plastic value. The discrepancy is apparently due to an inherent flaw in their method of solution. For a full discussion of this point, see R. Hill, *The Mathematical Theory of Plasticity*, pp. 245–247, Clarendon Press, Oxford (1950).

[‡] The closeness of the approximation is verified by a lower bound estimate based on a kinematically admissible field. See D. J. F. Ewing, *J. Mech. Phys. Solids*, **16**: 81 (1968).

When the V-notched bar has a circular root, the semiangle of the root covered by the statically admissible field may be denoted by θ. The relationship between θ, h/a, and c/a is given by (89), and the corresponding lower bound for the constraint factor by (86), with θ written for ψ and h written for h^*. Thus

$$\frac{T}{4ka} = (1 + \theta) - \frac{c}{a}(e^\theta - 1 - \theta)$$

$$h \leqslant h^* \qquad (91)$$

$$\frac{h}{a} = (2e^\theta - 1) - \frac{c}{a}(e^\theta - 1)^2$$

These relations express the constraint factor as a function of h/a and c/a parametrically through θ. In the subcritical range, θ must be less than the smaller of the two quantities ψ and $\ln(1 + a/c)$. Equations (91) also apply to a notch of circular shape, for which $h \leqslant h^*$ means $\theta \leqslant \ln(1 + a/c)$. For a given h/a less than the appropriate critical value, θ must be determined by trial and error before the constraint factor can be calculated. The lower bound value cannot differ from the actual constraint factor by more than 0.2 percent. It is evident that the constraint factor of the notched bar is independent of the notch angle in the subcritical range. The radius of the notch root has only a marginal effect on the constraint factor when the thickness is less than critical.†

(iii) *Necking of a V-notched bar* If a notched bar made of nonhardening material is stretched beyond the yield point, the thickness of the minimum section gradually decreases as the deformation proceeds. We shall analyze the development of the neck for a sharp V-notched bar on the assumption that the deforming part of the notch retains its shape while the included angle is progressively reduced.‡ At any stage during the necking, let the notch angle be equal to $\pi - 2\theta$ when the neck thickness is reduced to $2t$. The upper end of the bar is assumed fixed, the lower end being moved downward with a constant speed. The instantaneous position of the minimum section AE is denoted by its height c above the x axis as shown in Fig. 8.44a. The normal velocity on AE is unity, if the increment of time scale is $-dc$.

Since the normal velocity vanishes on BCD, the velocity component along the straight β lines in the region $ABCD$ must be zero. The magnitude of the velocity therefore has a constant value along each α line in this domain. The velocity field will be compatible with the assumed deformation mode if we write

$$u = -C\left(\sqrt{2} - \frac{r}{t}\right) \qquad v = 0 \qquad \text{in } ABC$$

---

† Upper bound solutions for subcritical thicknesses have been discussed by D. J. F. Ewing and R. Hill, *J. Mech. Phys. Solids*, **15**: 115 (1967) for the sharp notch, and by D. J. F. Ewing, op. cit., for the circular notch.

‡ This solution is due to O. Richmonds, *J. Mech. Phys. Solids*, **17**: 83 (1969). An alternative, but somewhat less realistic, solution has been presented by E. H. Lee, *J. Appl. Mech.*, **19**: 331 (1952), on the assumption that the plastic zone remains geometrically similar throughout the process.

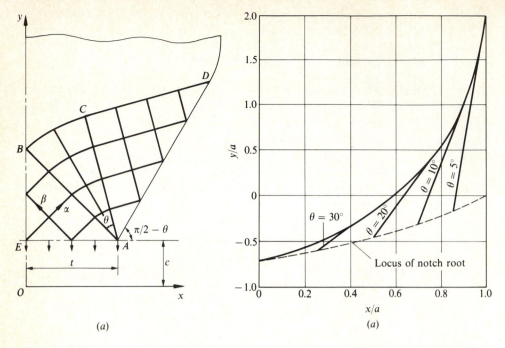

Figure 8.44 Finite extension of a V-notched bar. (*a*) Slipline field; (*b*) neck profiles for $c_0 = \psi = 0$.

where r is the radius to a circular arc, and C is a constant. The velocity is therefore continuous across the rigid/plastic boundary. In the region ABE, the components u and v are constant along the straight lines $x - y = \text{const}$ and $x + y = \text{const}$ respectively. The velocity field which is continuous across AB, and satisfies the symmetry condition $u = v$ along BE, may be written as

$$u = -\frac{C}{\sqrt{2}}\left(1 + \frac{x - y + c}{t}\right) \qquad v = -\frac{C}{\sqrt{2}}\left(1 - \frac{x + y - c}{t}\right) \qquad \text{in } ABE$$

The boundary condition on AE is $u + v = -\sqrt{2}$, which gives $C = 1$. The strain rate is constant in this region, the rate of extension at the minimum section being $1/t$. The nonzero component u within the triangle ACD is readily found from the condition of continuity across AC. In particular, the velocity of the free surface is given by

$$u = -\sqrt{2}\left\{1 - \frac{1}{2}\left(\frac{x}{t} - 1\right)\operatorname{cosec}\theta\right\} \qquad v = 0 \qquad \text{on } AD \qquad (92)$$

The magnitude of the velocity decreases from $\sqrt{2}$ at A to zero at D. The equation of the free surface is $y = c + (x - t)\cot\theta$. The rate of change of this equation, following the motion of the surface, becomes

$$-v_y + \left(v_x + \frac{dt}{dc}\right)\cot\theta + (x - t)\operatorname{cosec}^2\theta\,\frac{d\theta}{dc} = 1$$

where v_x and v_y are the components of the surface velocity referred to the rectangular axes. Evidently, $v_x = u\cos(\pi/4 - \theta)$ and $v_y = u\sin(\pi/4 - \theta)$, where u is given by (92). Substitution in the above equation furnishes

$$\frac{dt}{dc} = \tan\theta + \sec\theta \qquad \frac{d\theta}{dc} = -\frac{1}{2t} \tag{93}$$

Eliminating c between these two equations, and using the initial conditions $t = a$ and $\theta = \psi$, we obtain the integral

$$\frac{t}{a} = \frac{(1 - \sin\theta)^2}{(1 - \sin\psi)^2} \tag{94}$$

As the deformation proceeds, t/a steadily decreases from unity. Inserting (94) into the second equation of (93), and integrating, we get

$$\frac{c_0 - c}{a} = \frac{f(\theta) - f(\psi)}{(1 - \sin\psi)^2}$$

$$f(\theta) = 3\theta + \cos\theta(4 - \sin\theta) \tag{95}$$

where c_0 is an arbitrary initial value of c. The elongation of the bar at any stage is evidently equal to $2(c_0 - c)$. As the thickness of the neck decreases, the extent of the deforming zone is progressively reduced. A part of the material undergoing deformation at the initial yielding therefore becomes rigid, forming a curved profile given by the locus of the point D. The equation of the profile may be written in the parametric form

$$x = t(1 + 2\sin\theta) \qquad y = c + 2t\cos\theta$$

Typical neck profiles for an unnotched bar ($\psi = 0$) are shown in Fig. 8.44b. They are found to be in broad agreement with experimentally measured profiles.[†] If the necking could be continued down to a point ($t = 0$) without fracture, the displacement of the minimum section would finally reach the value $0.71a$ approximately. The longitudinal stress at the minimum section steadily increases, its value at any stage being $2k(1 + \theta)$. The total longitudinal force decreases, however, throughout the necking process.[‡]

Since the longitudinal strain rate at the minimum section is twice the rate

---

[†] O. Richmond, op. cit. Similar experimental results have also been reported by G. T. Hahn and A. R. Rosenfeld, *Acta Met.*, **13**: 293 (1965).

[‡] A solution in which the notch is assumed to become circular has been discussed by E. H. Lee and A. J. Wang, *Proc. 2d U.S. Nat. Congr. Appl. Mech.*, p. 489 (1954). The necking of a bar with symmetrical circular notches has been investigated by A. J. Wang, *Q. Appl. Math.*, **11**: 427 (1953). See also, M. Toulios and I. F. Collins, *Int. J. Mech. Sci.*, **24**: 61 (1982).

of change of θ in view of (93), the total strain at any stage is equal to $2(\theta - \psi)$. It is interesting to note that although the strain is uniform in the minimum section, its magnitude is not given by $\ln (a/t)$. This is due to the fact that a portion of material on the original section moves round the notch root to reach the free surface of the notch during the stretching. Available experimental results tend to suggest that work-hardening does not have a significant effect on the shape of the neck profile.

8.8 Bending of Single-Notched Bars

(i) *Deep wedge-shaped notches* A wide rectangular bar of metal, having a wedge-shaped notch on one side, is subjected to pure bending under conditions of plane strain. The bar is assumed long enough for the yield point couple to be independent of the precise distribution of tractions at the ends. It is supposed that the bar is bent to open the notch, so that the stress across the minimum section is tensile near the root and compressive near the opposite side. If the notch is sufficiently deep, the region of plastic deformation is confined in the neck, and the state of stress at the yield point is independent of the thickness of the bar.†

Depending on the notch angle $\pi - 2\psi$, there are two possible solutions given by the slipline fields of Fig. 8.45. In the neighborhood of the notch, the sliplines are straight, meeting the surface at $45°$. The field is continued round the singularity A to form the centered fan ADE on either side of the central axis. Near the plane surface FF, the sliplines are again straight, and the state of stress is a uniform compression $2k$ parallel to the surface. In the solution corresponding to field (a), the two slipline domains are extended across the minimum section to meet at a neutral point N, the region $AENE$ being in a state of uniform tension $2k(1 + \psi)$ acting across AN. The position of N is determined from the condition that the resultant horizontal tension across the minimum section is zero. Denoting the height of the triangle FNF by d, we get

$$d(2 + \psi) = a(1 + \psi)$$

where a is the depth of the minimum section. The yield point couple M per unit width of the bar is given by

$$\frac{M}{ka^2} = \frac{d}{a} = \frac{1 + \psi}{2 + \psi} \tag{96}$$

An unnotched bar of thickness a will yield under a moment equal to $\frac{1}{2}ka^2$. The ratio of the actual yield moment of the notched bar to that of the unnotched bar is known as the constraint factor, denoted by f. In the case of a sharp notch, the constraint factor is therefore twice the right-hand side of (96). The stress changes discontinuously at N, which is evidently a point of stress singularity, the jump in the hydrostatic pressure being $2k(1 + \psi)$. By (40), Chap. 6, the pressure jump must

---

† The slipline fields for the bending of bars with wedge-shaped and circular notches are due to A. P. Green, *Q. J. Mech. Appl. Math.*, **6**: 223 (1953), who produced some experimental evidence in support of the theory.

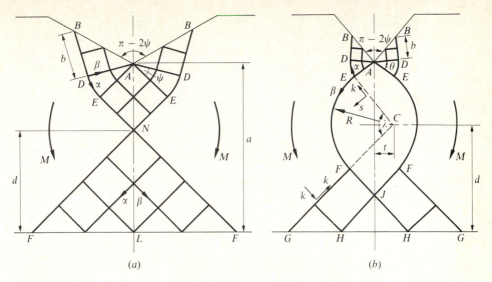

Figure 8.45 Pure bending of a sharply notched bar. (a) $\psi \leqslant 32.7°$; (b) $\psi \geqslant 32.7°$.

not exceed the value πk in order to avoid overstressing of the rigid corners at N. The field (a) is therefore valid for $\psi \leqslant \pi/2 - 1 \simeq 32.7°$. In other words, the semiangle of the notch must be less than or equal to 1 rad.

When $\psi \geqslant 32.7°$, the solution involves the field (b), where the slipline domains defined by the stress-free surfaces are connected by a pair of curved sliplines EF. The rigid ends of the bar rotate by sliding along the curves EF, which must therefore be circular arcs of some radius R. The fan angle θ at A cannot be greater than ψ, since the rigid material in the corner EAE must not be overstressed. If the angular span of EF is denoted by λ, it follows from geometry and Hencky's equations that

$$\lambda - (\psi - \theta) = \frac{\pi}{2} \qquad \lambda - \theta = 1$$

Since $\theta \leqslant \psi$, we have $\lambda \geqslant \pi/2$. The above equations can be solved for λ and θ to give

$$\lambda = \frac{\pi}{4} + \frac{\psi}{2} + \frac{1}{2} \qquad \theta = \frac{\pi}{4} + \frac{\psi}{2} - \frac{1}{2} \tag{97}$$

For a given notch angle, the slipline field is completely specified by the radii b and R of the circular arcs DE and EF, and the height d of the center C above GG. One of the three relations necessary for finding these parameters is provided by the fact that the sum of the vertical projections of GC, CE, and EA must be equal to a. Thus

$$d + R \sin\left(\lambda - \frac{\pi}{4}\right) + b \cos\left(\lambda - \frac{\pi}{4}\right) = a \tag{98}$$

The other two relations are furnished by the conditions of zero horizontal and vertical resultants of the tractions across the interface $AEFG$. Setting up the equations is simplified by continuing the plastic stress field into the rigid region to form the fictitious fan CEF. The normal stresses acting on CF and CE are then equal to $-k$ and s respectively, where $s = k(2\lambda - 1)$. Considering the tractions along the path $AECG$, we obtain

$$R(\lambda \sin \lambda - \theta \cos \lambda) + b(\lambda \sin \lambda + \theta \cos \lambda) = \sqrt{2}\,d$$
$$R(\theta \sin \lambda + \lambda \cos \lambda) - b(\theta \sin \lambda - \lambda \cos \lambda) = 0 \tag{99}$$

In view of the second equation, $(R - b) \sin \lambda$ is equal to $\sqrt{2}\,\lambda$ times the distance t of C from the vertical axis of symmetry. When the ratios d/a, R/a, and b/a have been computed from (98) and (99), the yield point couple and the constraint factor can be found from the formula

$$f = \frac{2M}{ka^2} = \frac{2d^2}{a^2} - \left(\frac{R - b}{a}\right)^2 + 2\lambda\left(\frac{R^2 + b^2}{a^2}\right) \tag{100}$$

which is obtained by taking moment of the tractions about C. When $\lambda = \pi/2$, we have $R = b$, and the yield moment becomes identical to that gives by (96). The constraint factor increases with increasing ψ until b vanishes. The condition $b \geqslant 0$ is equivalent to

$$\tan \lambda \geqslant -\frac{\lambda}{\lambda - 1} \qquad \text{or} \qquad \psi \leqslant 86.79°$$

in view of the second equation of (99). When $b = 0$, the constraint factor has a maximum value of 1.261. For all sharper notches ($\psi > 86.79°$) the slipline field and the constraint factor are independent of the notch angle. The geometrical parameters and the constraint factors calculated from above are given in Table 8.9.

Since the stress is discontinuous across N in the field (a), there cannot be a velocity discontinuity along the sliplines passing through this point. The rigid ends of the bar instantaneously rotate about N, inducing a velocity distribution along the boundaries of the deforming zones. The normal component of velocity is readily shown to be constant along the free surface FF, which therefore remains plane during the incipient distortion. The field (b), on the other hand, involves a velocity discontinuity of constant amount propagating along the sliplines $AEFJH$.

Table 8.9 Results for pure bending of sharply notched bars

$\psi°$	$\lambda°$	R/a	b/a	d/a	t/a	f	$h*/a$
15.00	....	....	0.313	0.558	....	1.116	1.132
32.70	90.00	0.275	0.275	0.611	0	1.222	1.299
45.00	96.15	0.330	0.191	0.623	0.058	1.245	1.353
60.00	103.65	0.368	0.109	0.628	0.099	1.257	1.395
75.00	111.15	0.385	0.043	0.629	0.116	1.260	1.417
86.79	117.04	0.389	0	0.630	0.130	1.261	1.423

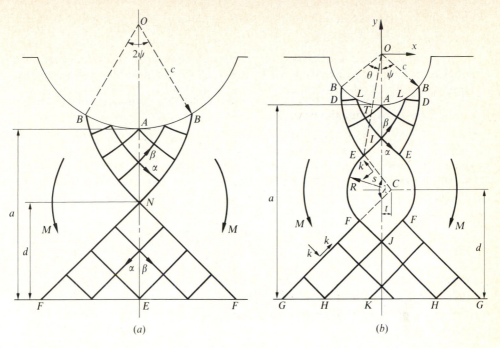

Figure 8.46 Slipline fields for pure bending of a bar with a circular notch. (a) $a/(a + c) \leqslant 0.64$; (b) $a/(a + c) \geqslant 0.64$.

The velocity normal to the bar is constant along HH but increases linearly along HG, while suffering an abrupt change across H. Depending on whether the bar is bent to open or close the notch, either a bulge or a hollow should develop on the surface at HH. The deformation mode and the constraint factor have been found to be in close agreement with experiment.†

(ii) *Notches of circular profile* Consider the pure bending of a bar having a deep notch in the form of a circular arc of radius c. The sliplines near the notch surface are logarithmic spirals, which make an angle of 45° with any radial line drawn through the center, extending over an angle 2ψ. For sufficiently small values of the ratio a/c, the sliplines through B meet at a point N on the minimum section, as shown in Fig. 8.46a. If the bar is bent to open the notch, the stress acting across the minimum section is tensile above N and compressive below N. The tensile stress increases from $2k$ at the notch root to $2k(1 + \psi)$ at the neutral point. The compressive stress has the constant magnitude $2k$, the sliplines in the triangle FNF being straight. The resultant tension across AN is equal to $2kc\psi e^{\psi}$, and this must be balanced by the resultant compression $2kd$ acting across NE. Hence

$$d = c\psi e^{\psi} = a - c(e^{\psi} - 1)$$

---

† The deformation pattern is revealed by etching, which darkens the regions of plastic flow. See A. P. Green and B. B. Hundy, *J. Mech. Phys. Solids*, **4**: 128 (1956).

where the last expression follows from simple geometry. The relationship between a/c and ψ therefore becomes

$$\frac{a}{c} = (1 + \psi)e^\psi - 1 \tag{101}$$

The yield point couple is obtained by taking moment about N of the stress distribution across the minimum section. Thus

$$M = kd^2 + 2kc^2 \int_0^\psi (1 + \phi)(e^\psi - e^\phi)e^\phi \, d\phi$$

where ϕ is the angle corresponding to any point on AN. Performing the integration and substituting for d, we obtain

$$f = \frac{2M}{ka^2} = \frac{c^2}{a^2} \{2\psi(1 + \psi)e^{2\psi} - (e^{2\psi} - 1)\} \tag{102}$$

Equations (101) and (102) give the constraint factor f as a function of a/c. It is easy to see that f tends to unity as a/c tends to zero. The range of application of the solution is determined by the condition that the hydrostatic pressure jump across N cannot exceed πk. This gives $\psi \leqslant \pi/2 - 1 \simeq 32.7°$ as before, or equivalently,

$$\frac{a}{c} \leqslant \frac{\pi}{2} \exp\left(\frac{\pi}{2} - 1\right) - 1 \simeq 1.78 \qquad \text{or} \qquad \frac{a}{a + c} \leqslant 0.64$$

The hodograph corresponding to the right-hand half of the field is shown in Fig. 8.47a, the instantaneous angular velocity of the rigid end being denoted by ω. The boundary of the deforming region is mapped as the geometrically similar curve $B'N'F'$ rotated through $90°$ in the clockwise sense. The region $N'A'B'$ of the hodograph is defined by the logarithmic spiral $N'B'$ and its reflection in the vertical through N'. The broken curve $A'B'$ intersects the characteristics at $45°$, and

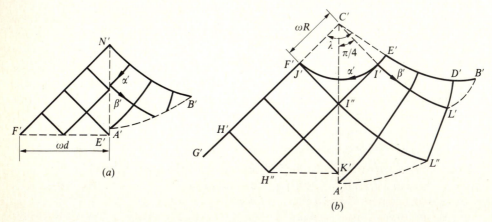

Figure 8.47 Hodographs for the bending of a bar with a circular notch. (a) $\psi \leqslant 32.7°$; (b) $\psi \geqslant 32.7°$.

represents the image of the free surface of the notch. The triangle $N'F'E'$ consists of straight characteristics, the base $E'F'$ forming the image of the free surface EF.

For higher values of a/c, the slipline field of Fig. 8.46b becomes applicable. The upper slipline domain is bounded by the spirals IE through a point I on the minimum section. These sliplines are continued as circular arcs EF to join with the straight-line field on the lower side. The material on the concave side of $IEFJ$ acts as a rigid pivot, permitting the ends of the bar to rotate as rigid bodies. If the angles turned through by BE and EF are denoted by θ and λ respectively, the compatibility of the hydrostatic pressures at B and F requires $\lambda - \theta = 1$ by Hencky's equations. From geometry, OE is inclined at an angle $\lambda - \pi/2$ to the vertical. The angles λ and θ are therefore given by (97) in terms of the angle subtended by the arc AB. Since the sum of the vertical projections of GC, CE, and OE is equal to $a + c$, the geometrical relationship becomes

$$d + R \sin \left(\lambda - \frac{\pi}{4} \right) + c(e^\theta \sin \lambda - 1) = a \tag{103}$$

where d is height of the center C above the base GG.

The normal tensile stress across the radial line TE is $2k(1 + \phi)$ at a point situated at a distance ce^ϕ from 0, where ϕ denotes the angular position of the corresponding point on the notch surface. The counterclockwise moment about O of the stress distribution along TE due to the material on its right is

$$G_0 = 2kc^2 \int_0^\theta (1 + \phi)e^{2\phi} \, d\phi = kc^2[\theta e^{2\theta} + \tfrac{1}{2}(e^{2\theta} - 1)] \tag{104}$$

per unit width. The resultant tension across TE per unit width is readily found to be $2kc\theta e^\theta$. The components of this force along Ox and Oy are

$$F_x = 2kc\theta e^\theta \sin \lambda \qquad F_y = -2kc\theta e^\theta \cos \lambda$$

Thus, the stress distribution across TE is statically equivalent to a force $2kc\theta e^\theta$ acting through O, together with a counterclockwise couple G_0. The moment of the stresses about C, acting in the clockwise sense, therefore has the value

$$G = kc^2[\theta e^{2\theta} - \tfrac{1}{2}(e^{2\theta} - 1)] + \sqrt{2} \, kRc\theta e^\theta \tag{105}$$

Assuming the plastic region to extend into the fan CEF as before, and considering the tractions along the path $TECG$, the equations of horizontal and vertical equilibrium may be written as

$$\sqrt{2}(d - c\theta e^\theta \sin \lambda) - R(\lambda \sin \lambda - \theta \cos \lambda) = 0$$
$$\sqrt{2} \, c\theta e^\theta \cos \lambda + R(\theta \sin \lambda + \lambda \cos \lambda) = 0 \tag{106}$$

Since λ and θ are obtained from (97) for any chosen value of $\psi \geqslant 32.7°$, Eqs. (103) and (106) can be easily solved for the ratios d/a, R/a, and c/a. The distance $2t$ between the centers C can be found from the expression

$$t = R \cos \left(\lambda - \frac{\pi}{4} \right) - ce^\theta \cos \lambda$$

The triangle HJH is of height $d + t - \sqrt{2}\,R$ by simple geometry. Considering the moment about C of the tractions acting along the entire path $TECG$, and using (105), it is easy to show that

$$f = \frac{2M}{ka^2} = \frac{2d^2 + c^2}{a^2} + (1 + 2\theta)\frac{R^2}{a^2} + 2\sqrt{2}\,\frac{Rc}{a^2}\,\theta e^{\theta} + (2\theta - 1)\frac{c^2}{a^2}\,e^{2\theta} \quad (107)$$

The above solution is applicable to any notch having a circular root, so long as the deformation remains confined to the circular contour. When c is vanishingly small, ψ attains the limiting value $86.79°$, corresponding to $\tan\lambda = -\lambda/(\lambda - 1)$, the limiting constraint factor being 1.261. The slipline field dimensions and the constraint factor are given in Table 8.10 as functions of the parameter $a/(a + c)$, which varies between zero and unity.

The material in the central pivot is mapped into the pole C' of the hodograph shown in Fig. 8.47b, which refers to the right-hand half of the slipline field. The boundary $B'E'F'G'$ of the hodograph is geometrically similar to the boundary of the deforming region, the radius of the circular arc $E'F'$ being equal to the velocity discontinuity across the slipline $LIEFJH$. The segment EI is mapped into the circular arc $E'I'$, of angular span $\lambda - \pi/2$, and the field $E'I'L'B'$ is constructed from the base curves $E'B'$ and $E'I'$. The particles immediately to the left of IL are mapped into $I''L''$, which defines the field $I''A'L''$, with the characteristics inclined at $45°$ to $A'I''$ and $A'L''$. The broken curves $B'L'$ and $L''A'$ form the image of the notch surface AB. The remainder of the hodograph is self-explanatory, the distance $J'H'$ being ω times the corresponding distance JH. The rate of plastic work is found to be everywhere positive.

(iii) V-notch with circular fillet

Suppose that a deep wedge-shaped notch is provided with a circular fillet of radius c and angular span 2ψ. If the ratio a/c is sufficiently large, for a given value of ψ, the slipline field extends to the straight sides of the notch covering a length $AB = \sqrt{2}\,b$ on either side of the central axis. For $\psi \leqslant 32.7°$, the regions of tension and compression meet at a neutral point N on the minimum section as shown in Fig. 8.48a. The longitudinal tensile stress in the region $EFEN$ has the constant value $2k(1 + \psi)$, the depth of this region being equal

Table 8.10 Pure bending of bars with circular notches on one side

$a/(a + c)$	$\psi°$	$\lambda°$	R/a	d/a	t/a	f	h^*/a
0.285	10.00	....	......	0.522	......	1.030	1.045
0.477	20.00	....	......	0.542	......	1.061	1.096
0.640	32.70	90.00	0	0.568	0	1.100	1.166
0.757	45.00	96.15	0.132	0.589	0.015	1.141	1.238
0.870	60.00	103.65	0.251	0.609	0.052	1.190	1.313
0.944	75.00	111.15	0.338	0.623	0.092	1.234	1.378
1.000	86.79	117.04	0.389	0.630	0.120	1.261	1.423

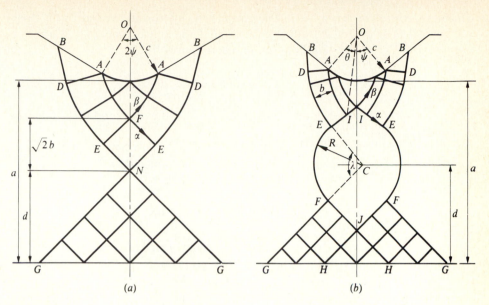

Figure 8.48 Pure bending of a V-notched bar with a circular fillet. (a) $\psi \leqslant 32.7°$; (b) $\psi \geqslant 32.7°$.

to $\sqrt{2}\,b$. It follows from simple geometry that

$$d + \sqrt{2}\,b + c(e^{\psi} - 1) = a$$

where d is the height of the triangle GNG, having a uniform horizontal compression $2k$. The condition of zero resultant tension across the minimum section gives

$$-d + \sqrt{2}\,b(1 + \psi) + c\psi e^{\psi} = 0$$

Solving the above equations, the deep notch parameters b and d can be expressed in terms of the notch geometry as

$$\sqrt{2}\,b = \frac{(a + c) - c(1 + \psi)e^{\psi}}{2 + \psi}$$

$$d = \frac{(a + c)(1 + \psi) - ce^{\psi}}{2 + \psi} \tag{108}$$

The stresses across the minimum section are statically equivalent to a pure couple, whose magnitude per unit width is

$$M = k(d^2 + 2b^2) + k\psi(\sqrt{2}\,b + ce^{\psi})^2 - \tfrac{1}{2}kc^2(e^{2\psi} - 1)$$

Substituting from (108), and simplifying the expression, the constraint factor is obtained as

$$f = \frac{2M}{ka^2} = \left\{1 - \frac{c}{a}(e^{\psi} - 1)\right\}^2 + \frac{\psi}{2 + \psi}\left\{1 + \frac{c}{a}(e^{\psi} + 1)\right\}^2 - \frac{c^2}{a^2}(e^{2\psi} - 1) \tag{109}$$

which reduces to (96) when $c = 0$. The jump in the hydrostatic pressure at N does not exceed πk if $\psi \leqslant 32.7°$, which ensures that the rigid corners at N are not overstressed. The validity of the solution also requires $b \geqslant 0$, or

$$\frac{a}{c} \geqslant (1 + \psi)e^\psi - 1$$

For values of a/c lower than the above critical value, the solution for the circular notch should hold.

The slipline field for $\psi \geqslant 32.7°$ is shown in Fig. 8.48b, where the tension and compression regions are connected by circular arcs of radius R and angular span λ. From geometrical considerations, it is easily shown that

$$d + R \sin\left(\lambda - \frac{\pi}{4}\right) + b \cos\left(\lambda - \frac{\pi}{4}\right) + c(e^\theta \sin \lambda - 1) = a \qquad (110)$$

where d is the height of the center C, and θ the angle turned through by the slipline AI. The angles λ and θ are defined by the given angle ψ through (97). The remaining equations involving d, R, and b are obtained by adding $\sqrt{2}$ times $c\theta e^\theta \sin \lambda$ and $c\theta e^\theta \cos \lambda$ to the left-hand side of the first and second equations respectively of (99), expressing the force equilibrium. The corresponding modification of (100), that includes the moment of the tractions along AI, leads to the constraint factor[†]

$$f = \frac{2d^2 + c^2}{a^2} - \left(\frac{R - b}{a}\right)^2 + 2\lambda\left(\frac{R^2 + b^2}{a^2}\right)$$

$$+ 2\sqrt{2}\frac{(R + b)c}{a^2}\theta e^\theta + (2\theta - 1)\frac{c^2}{a^2}e^{2\theta} \qquad (111)$$

which reduces to (107) when $b = 0$. For a given angle ψ, only those values of a/c are relevant for which $b \geqslant 0$. If a/c is less than that for which $b = 0$, the solution for the circular notch should apply. The variation of the constraint factor with the fillet angle for various values of $a/(a + c)$ is shown in Fig. 8.49. The region to the right of the broken curve ($b = 0$) corresponds to the circular notch, and the constraint factor is then independent of the notch angle.[‡]

(iv) **The critical bar thickness** The ratio of the notch depth to the core thickness has a critical value above which the deep notch solution is applicable.[§] As in the

---

[†] The constraint factors have been obtained numerically by G. Lianis and H. Ford, *J. Mech. Phys. Solids*, **7**: 1 (1958). Lower bound solutions based on discontinuous stress fields have also been given by these authors, who produced some experimental results supporting the theory.

[‡] The slipline field and lower bound solutions for a trapezoidal notch have been presented by G. Lianis and H. Ford, *J. Mech. Phys. Solids*, **7**: 1 (1958).

[§] For shallower notches, the deformation spreads to the surface of the bar, as has been shown experimentally by B. B. Hundy, *Metallurgia*, **49**: 109 (1954). Possible slipline fields have been suggested by A. P. Green, *J. Mech. Phys. Solids*, **6**: 259 (1956).

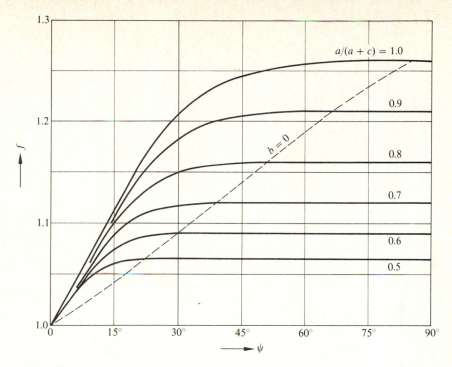

Figure 8.49 Constraint factors for single-notched bars having a deep V notch with circular fillet.

tension bar problem, an accurate estimate of the critical thickness can be obtained by continuing the plastic stress field into the rigid region. When the notch is either wedge-shaped or circular, a logical extension of the field containing a stress singularly N is illustrated in Fig. 8.50. The construction in each case involves a stress-free boundary beginning from the notch surface at B, and continued as far as K where the tangent is perpendicular to the axis of symmetry. The thickness of the bar is such that the tangent coincides with the surface of the bar, the material between BK and the notch surface being assumed stress free.

The angle turned through by each of the sliplines BJ and JK must be half the angle ψ turned through by BK. By Hencky's first theorem, the angle of the fan centered at N is therefore equal to $\psi/2$. The wedge-shaped notch involves a regular field $DEGH$ defined by a pair of circular arcs of radius b, and its natural extension $DHJB$. By (139), Chap. 6, the position of the point K is given by

$$\sqrt{2}\,w = b\,(2e^\psi - 1) \qquad d = \sqrt{2}\,b[I_0(\psi) + A_0(\psi)]$$

where $\sqrt{2}\,b = a/(2 + \psi)$ in view of (96). Hence, the corresponding thickness of the bar is

$$h^* = a + \sqrt{2}\,b(e^\psi - 1) = a\left(1 + \frac{e^\psi - 1}{2 + \psi}\right) \tag{112}$$

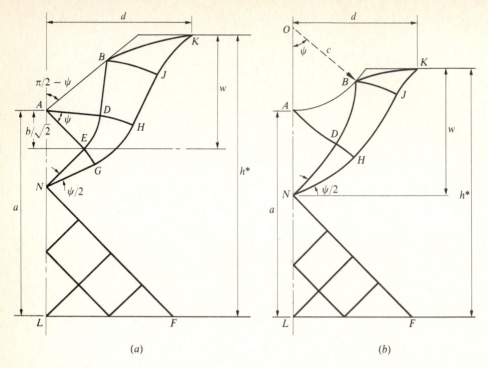

Figure 8.50 Slipline fields giving an approximation to the critical thickness. (*a*) Wedge-shaped notch; (*b*) circular notch.

It is reasonable to suppose that the rigid material outside the field is able to support the tractions acting across the boundary $FNGHJK$. The above formula then provides an upper bound on the critical thickness.† In any case, the actual critical value cannot differ from that obtained from (112) by more than 0.2 percent.

For a notch with a circular profile, the singular field NBJ of angular span $\psi/2$ at N can be determined in analytical terms following the method of Sec. 6.6(ii). The stress-free boundary BK is also found analytically by the method of Sec. 6.7(ii). The radius of curvature of the stress-free surface increases from $c(e^\psi - 1)$ at B to $ce^\psi \sinh \psi$ at K, where the center of curvature is on the horizontal axis through N. The position of K relative to N is given by

$$w = \frac{c}{2}(e^{2\psi} - 1) \qquad d = ce^\psi A_0(\psi)$$

where ψ depends on c/a according to (101). The approximation to the critical

---

† Slipline fields giving both upper and lower bound approximations to the critical thickness have been discussed by A. P. Green, op. cit., and D. J. F. Ewing, *J. Mech. Phys. Solids*, **16**: 205 (1968).

thickness is obtained as

$$h^* = a + w - c(e^\psi - 1) = a + \frac{c}{2}(e^\psi - 1)^2 \tag{113}$$

When a V-notched bar has a circular fillet of semiangle $\psi \leqslant 32.7°$, the expressions for h^* and d can be written down by using the principle of superposition. Thus, for $b \geqslant 0$, we write

$$h^* = a + \sqrt{2}\,b(e^\psi - 1) + \frac{c}{2}(e^\psi - 1)^2$$

$$d = \sqrt{2}\,b[I_0(\psi) + A_0(\psi)] + ce^\psi A_0(\psi) \tag{114}$$

where b is given by (108) in terms of a, c, and ψ. The slipline field in this case is obtained by combining the sharp-notch and round-notch fields, following the principle of vectorial superposition.†

For $\psi \geqslant 32.7°$, the singular field of N is replaced by a regular field in which one of the base sliplines is a circular arc of radius R and angular span $\psi/2$. No simple algebraic expressions are available for this range, since the angles θ and ψ are unequal, although an analytical treatment is still possible along the lines of Sec. 6.7(ii). The calculated values of h^*/a for sharp- and round-notched bars are included in Tables 8.8 and 8.9 respectively. The value of $h^*/a - 1$ for a V-notched bar with circular fillets, when $b \geqslant 0$, may be obtained by adding b/a times the sharp notch value of $(a/b)(h^*/a - 1)$ to c/a times the round notch value of $(a/c)(h^*/a - 1)$, using identical values of the angle ψ.

(v) *The Izod and Charpy tests* The analysis for the pure bending of single-notched bars is easily modified to deal with notch-bend tests, where the bar is subjected to transverse impact loading that tends to open the notch. The purpose of these tests is to assess the toughness of metals under severe conditions of service. In the Izod test, the specimen has a square cross section with a 45° V-notch on one face, and is loaded as a cantilever firmly clamped up to the minimum section. The Charpy test differs from the Izod test in that the specimen is loaded as a simply supported beam under symmetrical three-point loading, the specimen being struck centrally on the opposite flat surface. Since the breadth of the region of deformation is everywhere considerably smaller than the width of the bar, a plane strain condition is approximately achieved. The following analysis is based on the assumption that the material is rigid nonhardening, and that the effect of impact loading is separable from the plastic flow problem.‡

The slipline field corresponding to the Izod test is shown in Fig. 8.51a, neglecting the small fillet radius which is usually present at the root. It is assumed

† The combined bending and tension of double- and single-notched bars have been treated by B. Dodd and M. Shiratori, *Int. J. Mech. Sci.*, **20**: 451 and 465 (1978), **22**: 127 (1980).

‡ A. P. Green and B. B. Hundy, *J. Mech. Phys. Solids*, **4**: 128 (1956); J. M. Alexander and T. J. Komoly, ibid., **10**: 265 (1962).

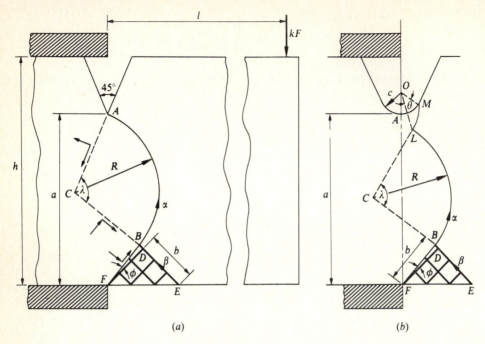

Figure 8.51 Slipline fields for the Izod test. (*a*) Sharp notch-root; (*b*) rounded notch-root.

that the notch is sufficiently deep and that the deformation does not spread to the sides of the notch. There is a stress singularity at the point F on the flat surface where the clamping block ends. The region DEF is uniformly stressed in compression by an amount $2k$ parallel to the surface. The free end of the bar rotates by sliding along the circular arc AB of radius R and angular span λ. The resultant velocity is constant in magnitude and direction along each radial line in BDF, and also along each slipline normal to DE. The hodograph consists of a curve geo-metrically similar to $ABDE$, rotated through 90° in the clockwise sense.

Let ϕ denote the angle of the fan BDF, and b the radius of the circular arc BD. Since A is directly above F, and the distance between the two points is a, we have the geometrical relations

$$b + R \sin \lambda = a \cos\left(\frac{\pi}{4} - \phi\right) \qquad R(1 - \cos \lambda) = a \sin\left(\frac{\pi}{4} - \phi\right)$$

which give

$$\frac{R}{a} = \frac{\sin(\pi/4 - \phi)}{1 - \cos \lambda} \qquad \frac{b}{a} = \frac{\sin(\phi + \lambda/2 - \pi/4)}{\sin(\lambda/2)} \tag{115}$$

If the fan CAB is assumed to be plastically stressed as before, the normal pressure on CB and CA are $k(1 + 2\phi)$ and $k(1 + 2\phi - 2\lambda)$ respectively. The resultant traction across $FBCA$ is most conveniently resolved along FB and CB, leading to the force relations

$$b - R\sin\lambda + (1 + 2\phi)R(1 - \cos\lambda) + 2\lambda R\cos\lambda = F\cos\left(\frac{\pi}{4} - \phi\right)$$

$$(116)$$

$$(1 + 2\phi)(b + R\sin\lambda) + R(1 - \cos\lambda) - 2\lambda R\sin\lambda = -F\sin\left(\frac{\pi}{4} - \phi\right)$$

where kF is the applied shearing force at a distance l from the minimum section. Since the resultant moment about C of the tractions acting along FBA must balance the moment of the applied force, we get

$$\frac{1}{2}(1 + 2\phi)b^2 + R(b + R) = \left[l + R\cos\left(\lambda + \phi - \frac{\pi}{4}\right)\right]F \qquad (117)$$

The set of five equations (115) to (117) may be solved for the five unknowns b, R, λ, ϕ, and F, using a trial-and-error procedure. The bending moment at the minimum section is kFl, and the constraint factor is $f = 2Fl/a^2$. In the standard Izod test, $a = 8$ mm and $l = 22$ mm, giving $f \simeq 1.224$. The thickness of the specimen is $h = 10$ mm, which is found to be slightly higher than the critical thickness (about 9.8 mm) for the deep notch solution.†

It is not difficult to allow for a circular fillet of radius c. The sliplines in the neighborhood of the fillet are logarithmic spirals meeting the notch surface at 45°, as shown in Fig. 8.51b. The tractions across the boundary LM, having an angular span θ, give rise to a resultant tension $2kc\theta e^\theta$ in the direction perpendicular to OL, the moment of the tractions about C being given by (105). The line of action of the resultant traction makes an acute angle of $\lambda - \pi/4$ with the direction CB. If kP and kQ denote the resolved components of this resultant along BF and BC respectively, due to the material on the left of OL, then

$$P = 2c\theta e^\theta \sin\left(\lambda - \frac{\pi}{4}\right) \qquad Q = 2c\theta e^\theta \cos\left(\lambda - \frac{\pi}{4}\right) \qquad (118)$$

The force relations (116) must now be modified by the addition of the quantities P and Q on the right-hand side. The geometrical relations (115) are modified to

$$b + R\sin\lambda + ce^\theta \cos\left(\lambda - \frac{\pi}{4}\right) = (a + c)\cos\left(\frac{\pi}{4} - \phi\right)$$

$$(119)$$

$$R(1 - \cos\lambda) + ce^\theta \sin\left(\lambda - \frac{\pi}{4}\right) = (a + c)\sin\left(\frac{\pi}{4} - \phi\right)$$

where $\theta = \lambda - \phi - 1$ by Hencky's pressure relations along $DBLM$. The equation of equilibrium of the moments about the point C becomes

$$\frac{1}{2}(1 + 2\phi)b^2 + R(b + \lambda R) + G = \left[l + R\cos\left(\eta - \frac{\pi}{4}\right) + ce^\theta \cos\eta\right]F \qquad (120)$$

---

† The critical thickness ratio is about 1.22, which is slightly lower than the actual ratio of 1.25. The critical thickness ratio for the same bar under pure bending is about 1.39 (D. J. F. Ewing, op. cit.). The standard Izod specimen is therefore not deep enough for the pure bending solution to apply.

where $\eta = \lambda + \phi$, and kG represents the right-hand side of (105). The angle subtended by the arc AM is equal to $2\lambda - \pi/2 - 1$, which must be less than the semiangle ψ of the fillet for the validity of the solution. This condition is satisfied for the standard Izod notch, where $\psi = 67.5°$ and $c = 0.25$ mm. The results of the numerical computation with and without the fillet† are shown in Table 8.11.

The state of stress in the minimum section near the notch root is of some practical interest. The tensile stress σ across the minimum section increases from $2k$ at the root to a maximum value which occurs at the elastic/plastic boundary. The position of this boundary is most likely to lie between the points S and T (not shown), where LS is the circular principal stress trajectory through L, and LT is the continuation of the slipline ML. The stress is constant along LS and has the magnitude $2k(\lambda - \phi)$. If the plastic region is assumed to extend as far as T, the stress at this point is equal to $2k(2\lambda - \pi/2)$. The maximum stress should lie between these two values of σ irrespective of the root radius. Experiments seem to indicate that fracture occurs in notched specimens when the maximum tensile stress reaches a critical value.‡

The slipline field for the idealized (quasi-static) problem of the Charpy test is symmetrical about the minimum section, each half of the field being identical to that for the Izod test. In practice, the central load would have to be supported over a finite length of the surface, which would separate the two singularities at the base by a distance equal to the indenter width. The effect of neglecting the indenter width is, however, unlikely to lead to significant errors in the overall mode of deformation and the stress distribution near the notch root.§ In the standard Charpy test, the end load kF acts at a distance $l = 20$ mm from the minimum section, the corresponding results being included in Table 8.11. The theoretical predictions have been verified by experiments on the yielding of bars in slow bend tests.¶

Table 8.11 Theoretical results for Izod and Charpy tests (all dimensions are in millimeters)

	l	R	b	$\lambda°$	$\phi°$	F	f
$c = 0$	22	3.963	2.478	103.55	7.31	1.780	1.224
	20	4.022	2.032	102.38	7.38	1.945	1.216
$c = 0.25$	22	3.844	2.533	101.44	7.69	1.771	1.218
	20	3.909	2.460	100.37	7.76	1.936	1.210

† These results are due to D. J. F. Ewing, *J. Mech. Phys. Solids*, **16**: 205 (1968). Similar calculations have been made earlier by J. M. Alexander and T. J. Komoly, ibid., **10**: 265 (1962).

‡ J. F. Knott, *J. Iron Steel Inst.*, **204**: 104 (1966); *J. Mech. Phys. Solids*, **15**: 97 (1967).

§ D. J. F. Ewing (op. cit.) calculated the field for finite widths of the indenter, assuming it to be flat-topped, but neglected its effect on the bending moment at the minimum section. His constraint factors for nonzero indenter widths are therefore somewhat approximate.

¶ A. P. Green and B. B. Hundy, *J. Mech. Phys. Solids*, **4**: 128 (1956). See also T. R. Wilshaw and P. I. Pratt, ibid., **14**: 7 (1966).

8.9 Bending of Double-notched Bars

Consider the yielding of a rectangular bar containing a pair of identical deep notches with their common axis perpendicular to the longitudinal sides. Before examining the slipline fields under pure bending, it would be instructive to obtain elementary bounds on the constraint factor without reference to the notch geometry. To determine an upper bound, let us assume an incipient rigid body rotation of the ends of the bar about a central pivot formed by a pair of equal circular arcs of angle 2λ (say) passing through the notch roots. The radius of curvature of each circular arc is $R = \frac{1}{2}a \operatorname{cosec} \lambda$, where a is the depth of the minimum section. Since the velocity discontinuity that occurs along the circular arcs is of amount ωR, where ω is the angular velocity of rotation, the rate of internal energy dissipation due to shearing along the circular arcs is equal to $4k\omega R^2\lambda$. Equating this to the rate of external work $2M\omega$ done by the applied couples M, we obtain the upper bound

$$M = 2kR^2\lambda = \tfrac{1}{2}ka^2\lambda \operatorname{cosec}^2 \lambda$$

The yield couple has a minimum value when $\tan \lambda = 2\lambda$, or $\lambda \simeq 66.8°$, and the upper bound then becomes $M = 0.69ka^2$. An obvious lower bound is $M = 0.5ka^2$, furnished by the stress distribution corresponding to pure bending of a uniform bar of thickness a. A pair of longitudinal stress discontinuities are considered through the notch roots, the material above and below these lines being assumed stress free. It follows that the constraint factor $2M/ka^2$ must lie between 1 and 1.38, irrespective of the shape of the notch. These bounds also apply to single-notched bars with arbitrary notch profiles.

(i) *Wedge-shaped and circular notches* In the pure bending of a bar which is symmetrically notched on opposite sides, the slipline field becomes symmetrical about both the longitudinal axis and the minimum section. For deep wedge-shaped notches with a sufficiently large included angle $\pi - 2\psi$, the stress across the minimum section is $2k(1 + \psi)$ above the neutral point and $-2k(1 + \psi)$ below the neutral point, giving a constraint factor of $1 + \psi$ as the tension bar problem. The jump in the hydrostatic pressure across the neutral point is $2k(1 + 2\psi)$, which must not exceed πk for the solution to be valid, leading to the restriction $\psi \leqslant 16.35°$. the minimum thickness of the bar for which the deep-notch solution is applicable is ae^ψ, where a is the thickness at the notch root.

When the notch profile is a circular arc of radius c, there are two identical fields of logarithmic spirals meeting at the center of the minimum section. The relationship between the yield point moment M and the geometrical parameters a and c may be expressed as

$$M = kc^2\{1 - (1 - 2\psi)e^{2\psi}\} \qquad a = 2c(e^\psi - 1)$$

where ψ is the semiangle of the notch covered by the slipline field. Eliminating ψ

from above, we get[†]

$$f = \frac{2M}{ka^2} = \left(1 + \frac{2c}{a}\right)^2 \ln\left(1 + \frac{a}{2c}\right) - \left(\frac{2c}{a} + \frac{1}{2}\right) \tag{121}$$

The solution is valid for $\psi \leqslant 16.35°$, which is equivalent to $a/c \leqslant 0.661$, or $a/(a + c) \leqslant 0.398$. The critical thickness of the bar is twice the dimension w shown in Fig. 8.50b, and consequently,

$$h^* = c(e^{2\psi} - 1) = a\left(1 + \frac{a}{4c}\right)$$

When a V-notched bar is provided with a circular fillet of radius c and semiangle $\psi \leqslant 16.35°$, the length of the straight sides of the notch covered by the slipline field for sufficiently large values of a/c is

$$\sqrt{2}\,b = \frac{a}{2} - c(e^\psi - 1)$$

The stress distribution across the minimum section is equivalent to a pure couple M, and the constraint factor becomes

$$f = \frac{2M}{ka^2} = \left\{1 - \frac{2c}{a}(e^\psi - 1)\right\}^2 + \psi\left(1 + \frac{2c}{a}\right)^2 - \frac{2c^2}{a^2}(e^{2\psi} - 1) \tag{122}$$

which holds for $b \geqslant 0$. The critical thickness is readily obtained by the super-position of the results for $c = 0$ and $b = 0$, and is expressed by the relation

$$h^* = 2\sqrt{2}\,be^\psi + c(e^{2\psi} - 1) = ae^\psi - c(e^\psi - 1)^2 \tag{123}$$

The effect of a circular root in a V-notched bar is therefore to reduce the critical thickness as well as the constraint factor.

Figure 8.52 shows the slipline fields for pure bending of a bar with symmetrically formed sharp and round notches corresponding to $\psi \geqslant 16.35°$. Each field involves a pair of identical circular arcs EF of radius R and angular span 2λ, with centers C on the longitudinal axis of the bar. The hydrostatic pressure continuously changes from $-k$ at D to k at G, vanishing at the center of the arc EF. If the angle turned through by DE is denoted by θ, geometrical considerations and Hencky's equations lead to

$$\lambda - (\psi - \theta) = \frac{\pi}{4} \qquad \lambda - \theta = \frac{1}{2}$$

Thus, $\lambda \geqslant \pi/4$ in the relevant range $\theta \leqslant \psi$, the expressions for λ and θ in terms of ψ being

$$2\lambda = \frac{\pi}{4} + \psi + \frac{1}{2} \qquad 2\theta = \frac{\pi}{4} + \psi - \frac{1}{2} \tag{124}$$

[†] A. P. Green, *Q. J. Mech. Appl. Math.*, **6**: 223 (1953); G. Lianis, *Ing.-Arch.*, **29**: 55 (1960).

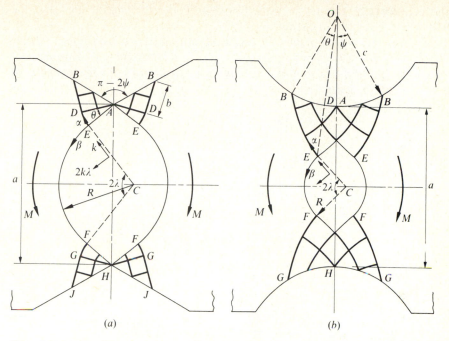

Figure 8.52 Pure bending of double-notched bars with $\psi \geqslant 16.35°$. (a) Wedge-shaped notch; (b) circular notch.

In view of the symmetry of the fields, the equations of horizontal equilibrium are identically satisfied. The sharp notch field is completely defined by the radii R and b, which are obtained from the geometry of the field and the condition of vertical equilibrium. Assuming the plastic region to extend into the fan CEF, with the normal pressure on CF equal to $2k\lambda$, it is easy to show that

$$R(2\lambda \cos \lambda - \sin \lambda) - b(2\lambda \sin \lambda - \cos \lambda) = 0$$

$$2(R \sin \lambda + b \cos \lambda) = a$$

These equations are readily solved for R and b to give

$$\frac{2R}{a} = \frac{2\lambda \sin \lambda - \cos \lambda}{2\lambda - \sin 2\lambda} \qquad \frac{2b}{a} = \frac{2\lambda \cos \lambda - \sin \lambda}{2\lambda - \sin 2\lambda} \tag{125}$$

The resultant moment about C of the tractions acting along CE and EA must be equal to half the yield moment M, giving the constraint factor

$$f = \frac{2M}{ka^2} = 4\left\{ \frac{bR}{a^2} + \lambda\left(\frac{b^2 + R^2}{a^2} \right) \right\} \tag{126}$$

As the notch angle decreases, the constraint factor increases, until ψ reaches the value 59.91°, which corresponds to $b = 0$, or $\tan \lambda = 2\lambda$. The constraint factor at this stage has a maximum value of 1.38, and the slipline field reduces to a pair of circular arcs through the notch roots. For $\psi \geqslant 59.91°$, the slipline field and the constraint factor remain unchanged.

In the case of circular notches, the resultant tension transmitted across DE per unit width is equal to $2kc\theta e^\theta$ acting at right angles to DE, the vertical component of the force being $2kc\theta e^\theta \sin(\lambda - \pi/4)$. The equilibrium equations and the geometrical relation therefore become

$$R(2\lambda \cos \lambda - \sin \lambda) - 2c\theta c^\theta \sin\left(\lambda - \frac{\pi}{4}\right) = 0$$

$$R \sin \lambda + ce^\theta \cos\left(\lambda - \frac{\pi}{4}\right) - c = \frac{1}{2}a$$

Using the fact that $2\theta = 2\lambda - 1$, the ratios R/c and a/c may be expressed in the form

$$\frac{R}{2c} = \frac{\theta e^\theta \sin(\lambda - \pi/4)}{2\lambda \cos \lambda - \sin \lambda} \qquad 1 + \frac{a}{2c} = \frac{\sqrt{2}\, e^\theta(\lambda - \sin^2 \lambda)}{2\lambda \cos \lambda - \sin \lambda} \tag{127}$$

The counterclockwise moment about C due to the tractions on CE is $k\lambda R^2$, while that due to the tractions on DE exerted by the material on its left is given by (105). The constraint factor is easily shown to be

$$f = \frac{2M}{ka^2} = 4\left(\lambda \frac{R^2}{a^2} + \sqrt{2}\frac{Rc}{a^2}\theta e^\theta\right) + \frac{2c^2}{a^2}\{1 - 2(1 - \lambda)e^{2\theta}\} \tag{128}$$

where λ and θ are given by (124). The results for both sharp and round notches are shown in Table 8.12. The constraint factor for the double-notched bar is compared with that for the single-notched bar in Fig. 8.53.

When a V-notched bar is provided with a circular fillet of radius c and angular span 2ψ, and the deformation spreads to the straight faces of the notch, the radii R and b defining the slipline field can be determined from the equations

$$R(2\lambda \cos \lambda - \sin \lambda) - b(2\lambda \sin \lambda - \cos \lambda) = 2c\theta e^\theta \sin\left(\lambda - \frac{\pi}{4}\right)$$

$$R \sin \lambda + b \cos \lambda + ce^\theta \cos\left(\lambda - \frac{\pi}{4}\right) - c = \frac{a}{2} \tag{129}$$

Table 8.12 Results for pure bending of double-notched bars

$\psi°$	$\lambda°$	Sharp notch			Circular notch		
		R/a	b/a	f	R/a	$a/(a+c)$	f
10.00	.....	.....	0.354	1.175	.....	0.276	1.061
16.35	45.00	0.354	0.354	1.285	0	0.398	1.102
20.00	46.82	0.399	0.306	1.314	0.084	0.459	1.129
30.00	51.82	0.480	0.198	1.356	0.262	0.624	1.203
45.00	59.32	0.533	0.082	1.379	0.437	0.841	1.306
59.91	66.78	0.544	0	1.380	0.544	1.000	1.380

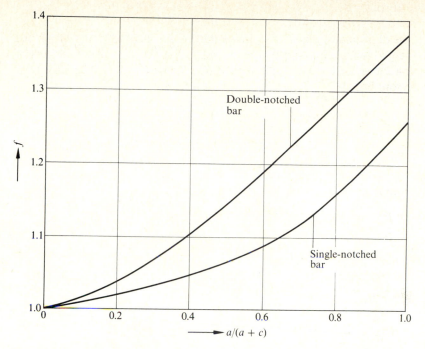

f

$a/(a + c)$

Figure 8.53 Constraint factors for the bending of bars containing circular notches.

where λ and θ are related to ψ through (124). When R/a and b/a have been calculated for given c/a and ψ, the constraint factor can be found from the formula

$$f = 4\left\{\frac{bR}{a^2} + \sqrt{2}\frac{(R + b)c}{a^2}\,\theta e^\theta + \lambda\left(\frac{b^2 + R^2}{a^2}\right)\right\} + \frac{2c^2}{a^2}\{1 - 2(1 - \lambda)e^{2\theta}\} \quad (130)$$

For a given ψ, the solution is valid for those values of c/a which correspond to $b \geqslant 0$. The results are shown graphically in Fig. 8.54, where the broken curve represents the limit of applicability of (130).

(ii) *Solution for trapezoidal notches* A bar subjected to pure bending is provided with a symmetrical pair of trapezoidal notches having their oblique faces inclined at an angle ψ to the longitudinal axis. The applied couples tend to open the upper notch so that the stresses are tensile in the upper half and compressive in the lower half of the bar. Let c denote the semiwidth of the horizontal free surface of the notch. If c/a is less than a critical value, that depends on the angle ψ, the deformation spreads to the inclined free surfaces involving a length $AB = \sqrt{2}\,b$ on either side. The existence of the plastic stress field around the notch requires that the fans ACD and AEF centered at A must have an angular span equal to $\psi/2$.

When ψ is sufficiently small, the sliplines through B meet in a single point N on the horizontal axis, Fig. 8.55a. The hydrostatic stress changes discontinuously

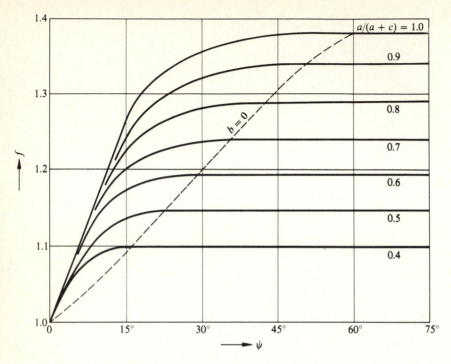

Figure 8.54 Variation of the constraint factor with notch angle for double-notched bars with circular fillets.

across the neutral point N, the magnitude of the stress jump being equal to $2k(1 + 2\psi)$. To avoid yielding of the rigid corners at N, the discontinuity in stress must be less than the amount πk, giving $\psi \leqslant 16.35°$ as the condition for validity of the field. The longitudinal stress across the minimum section increases from $2k$ to $2k(1 + \psi)$ along CL, where $CDLD$ is defined by equal circular arcs of radius $\sqrt{2}\,c$. If the depth OL is denoted by d, then

$$\sqrt{2}\,b = \frac{a}{2} - d \geqslant 0$$

The resultant horizontal tension per unit width across ADL is $Q + kd$, where Q denotes the force corresponding to zero hydrostatic stress at O. The clockwise moment about O of the tractions along ADL, due to the material on its left, is equal to $G + \frac{1}{2}k(d^2 - c^2)$ per unit width, where G is the clockwise moment when the hydrostatic stress vanishes at O. Since the resultant moment of the tractions along $ADLN$ about N is equal to half the applied moment M, we have

$$M = (kd + Q)a - k(d^2 - c^2) - 2G + 4k(1 + \psi)b^2$$

Values of Q/kc and G/kc^2, corresponding to the angular coordinates $(\psi/2, \psi/2)$, are directly found from Table A-1. The ratio d/c is also obtained from the same table,

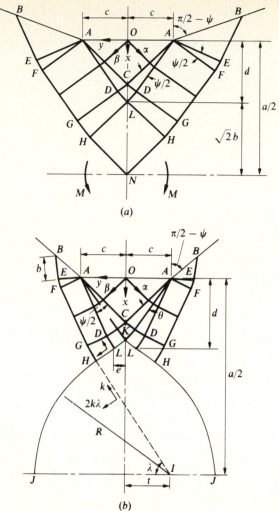

(a)

(b)

Figure 8.55 Pure bending of double-notched bars with symmetrical trapezoidal notches. (a) $\psi \leqslant 16.35°$; (b) $\psi \geqslant 16.35°$.

and the constraint factor is calculated from the formula

$$f = \frac{2M}{ka^2} = 1 + \psi\left(1 - \frac{2d}{a}\right)^2 + 2\left(\frac{Q}{ka} - \frac{d}{a}\right) + 2\left(\frac{c^2 + d^2}{a^2} - \frac{2G}{ka^2}\right) \qquad (131)$$

For a given ψ, the ratio c/a must be less than that corresponding to $2d = a$. The limiting value of $2c/a$ in this solution decreases from 1.0 to 0.77 as ψ increases from zero to 16.35°. When $c = 0$, the constraint factor reduces to the sharp-notch value $1 + \psi$ as expected.

The slipline field for $\psi \geqslant 16.35°$ is shown in Fig. 8.55b. The deforming zone around the notch is extended as far as $BEFGHLK$, the angle turned through by the segment GH being denoted by θ. The sliplines KLH are continued as circular arcs

HJ of radius R and angular span λ. The centered fans ACD and AEF have an angular span $\psi/2$, while b denotes the radius of the circular arc EF. In view of the opposing nature of the stresses above and below the horizontal axis of symmetry, the hydrostatic pressure must vanish at J. It is easy to show that

$$\lambda + \theta = \frac{\pi}{4} + \frac{\psi}{2} \qquad \lambda - \theta = \frac{1}{2}(1 + \psi)$$

The first of these relations follows from simple geometry, while the second relation is a consequence of Hencky's equations. Thus

$$2\lambda = \frac{\pi}{4} + \frac{1}{2} + \psi \qquad 2\theta = \frac{\pi}{4} - \frac{1}{2}$$

Let $(-P, Q)$ be the rectangular components of the resultant force per unit width across ADL, and G the clockwise moment of the tractions on ADL about O due to the material on the left of this line, when the hydrostatic stress at C is taken as zero. The hydrostatic tension k existing at C is responsible for additional forces $k(c - e)$ and kd, and an additional moment about O equal to $\frac{1}{2}k(d^2 - c^2 + e^2)$. If M' denotes the counterclockwise moment per unit width about the point I due to the tractions acting on ADL, then

$$\frac{M'}{kc^2} = \frac{t}{c}\left(1 - \frac{e}{c}\right) + \frac{1}{2}\left(1 - \frac{d^2 + e^2}{c^2}\right) + \frac{a}{2c}\left(\frac{d}{c} + \frac{Q}{kc}\right) - \frac{Pt + G}{kc^2} \qquad (132)$$

where (d, e) are the rectangular coordinates of L, and t is the distance of I from the vertical axis of symmetry. Evidently, $t = R \cos \lambda - b \sin \lambda - e$. The equilibrium condition is most conveniently established by considering the fictitious plastic fan HIJ, which gives a uniform normal tension equal to $2k\lambda$ along IH. The geometrical relationship and the equation of vertical equilibrium are easily shown to be

$$R \sin \lambda + b \cos \lambda + d = \frac{a}{2}$$

$$R(2\lambda \cos \lambda - \sin \lambda) - b(2\lambda \sin \lambda - \cos \lambda) = -(c - e) + \frac{P}{k} \qquad (133)$$

For a given notch geometry, the ratios R/c and b/c can be determined from (133) using the appropriate table. The constraint factor is obtained by taking moment of the tractions acting along the path ADLHI about the center I. In terms of the moment M' given by (132), the result may be expressed as

$$f = \frac{2M}{ka^2} = 4\left\{\frac{bR}{a^2} + \lambda\left(\frac{b^2 + R^2}{a^2}\right) + \frac{M'}{ka^2}\right\} \qquad (134)$$

The solution holds only for those values of c/a and for which $b \geqslant 0$. Outside this range, the deformation does not extend to the inclined faces of the notch, and the constraint factor is then independent of the notch angle. The variation of f with ψ for different values of $2c/a$ is shown in Fig. 8.56. The broken curve gives the relationship between c/a and ψ that corresponds to $b = 0$. When the semiangle of

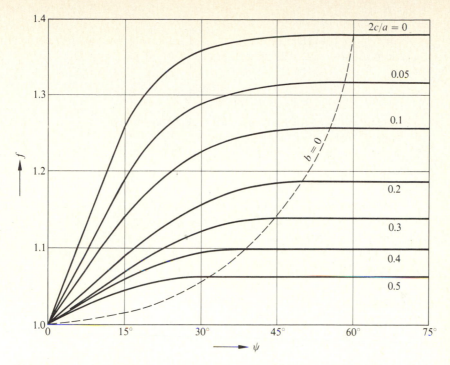

Figure 8.56 Constraint factors for double-notched bars with trapezoidal notches.

the notch is less than $30.09°$ (which includes the rectangular notch), the constraint factor is independent of ψ for all c/a ratios.

8.10 Bending of Beams and Curved Bars

(i) *The yield point of a cantilever* The influence of transverse shear on the yield point load of wide beams can be estimated by using the theory of slipline fields. We begin by considering a uniform cantilever of rectangular cross section under a terminal load. The width of the beam is assumed to be at least five times its thickness, so that the incipient deformation is essentially one of plane strain. The load which the cantilever can carry naturally depends on the manner in which it is supported at the built-in end.† When the support is perfectly strong, and the length l of the beam is not too large compared to its thickness h, the slipline field is symmetrical about the centroidal axis, as shown in Fig. 8.57a. The deformation mode consists of a rotation of the rigid end of the cantilever by sliding over the circular arc DE which connects the regions of tension and compression. The plastic

---

† This solution has been discussed by A. P. Green, *J. Mech. Phys. Solids*, **3**: 1 (1954), for the more general case of a symmetrically tapered cantilever. See also, E. T. Onat and R. T. Shield, *Proc. 2d. U.S. Nat. Congr. Appl. Mech.*, 535 (1954).

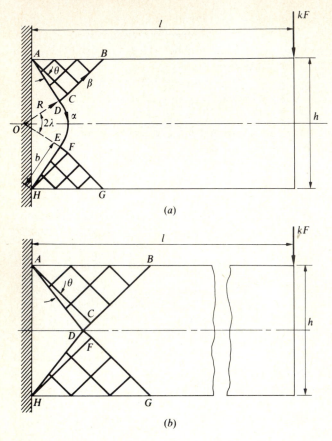

(a)

(b)

Figure 8.57 Slipline fields for an end-loaded cantilever with strong support. (a) $l/h \leqslant 13.74$; (b) $l/h \geqslant 13.74$.

triangles ABC and HGF are uniformly stressed, with the sliplines meeting the free surfaces at an angle of $45°$.

If the angle subtended by the arc DE at the centre O is denoted by 2λ, the angular span of each of the fans centered at A and H is $\theta = \pi/4 - \lambda$. Since the hydrostatic pressure varies from $-k$ to k along $CDEF$, the Hencky equations immediately furnish the second relation $2(\lambda - \theta) = 1$. Thus

$$\lambda = \frac{\pi}{8} + \frac{1}{4} \simeq 36.82° \qquad \theta = \frac{\pi}{8} - \frac{1}{4} \simeq 8.18°$$

The pressure naturally vanishes at the center of the arc DE. If the radii OD and AD are denoted by R and b respectively, then by geometry,

$$2(R \sin \lambda + b \cos \lambda) = h$$

The resultant horizontal force across the plastic boundary is identically zero. The condition for vertical equilibrium is most conveniently established by considering DOE as a fictitious plastic fan, the normal pressure on OE being then equal to $2k\lambda$. If the applied shearing force per unit width is denoted by kF, then the distribution of tractions along OEH furnishes

$$R(2\lambda \cos \lambda - \sin \lambda) + b(\cos \lambda - 2\lambda \sin \lambda) = \frac{F}{2}$$

From the last two equations, the ratios R/h and b/h can be expressed in terms of F/h. Using the value of λ, we get

$$\frac{R}{h} = -0.0461 + 1.2283 \frac{F}{h} \qquad \frac{b}{h} = 0.6592 - 0.9197 \frac{F}{h} \qquad (135)$$

The distance of O from the plane AH is $R \cos \lambda - b \sin \lambda$. Considering moment of the tractions and the applied shear force about the point O, it is easy to show that

$$2bR + 2\lambda(b^2 + R^2) = F(l + R \cos \lambda - b \sin \lambda)$$

The expression in the bracket is the distance of the load point from O. Substituting for R, b, and λ, we obtain the quadratic

$$0.5005 - \left(\frac{l}{h} - 0.4320\right)\frac{F}{h} - 0.7674 \frac{F^2}{h^2} = 0 \qquad (136)$$

which is easily solved for F/h corresponding to any given value of l/h, the results being shown in Table 8.13. As the length/height ratio increases, R/h decreases and b/h increases. The condition for the above solution to be valid is $R \geqslant 0$, which is equivalent to

$$\frac{F}{h} \geqslant 1 - 2\lambda \tan \lambda \simeq 0.0375 \qquad l/h \leqslant 13.74$$

The hodograph would be identical to the slipline field, rotated through $90°$ in the clockwise sense, the rigid rotating part of the beam being mapped into a

Table 8.13 Yield point state of a uniform cantilever under a terminal load

l/h	Strong support			Weak support			
	F/h	R/h	b/h	F/h	R/h	b/h	d/h
1.00	0.518	0.591	0.183	0.498	0.529	0.174	0.594
1.50	0.370	0.409	0.319	0.358	0.357	0.298	0.554
2.00	0.281	0.298	0.401	0.273	0.252	0.373	0.552
3.00	0.185	0.181	0.489	0.181	0.139	0.454	0.535
5.00	0.108	0.086	0.560	0.107	0.046	0.520	0.521
7.595	0.0694	0.039	0.596	0.0689	0	0.554	0.514
13.74	0.0375	0	0.625	0.0374	0	0.588	0.508

geometrically similar figure. The centroid of the loaded end has a vertically downward motion at the incipient collapse.

For higher values of l/h, the slipline field is modified to that of Fig. 8.57b, where the fan angles are less than 8.18°, and point D is a singularity of stress. The yield criterion is not violated in the rigid corners at D, since the hydrostatic pressure jump across this point is less than $2k$ times each included angle. The relationship between F/h and l/h is closely approximated by the empirical formula

$$\frac{h}{F} \simeq \frac{2l}{h} - \frac{7}{8} \qquad \frac{l}{h} \geqslant 13.8$$

The angle θ varies approximately as the square root of F/h, which tends to zero as l/h tends to infinity.

Suppose that the cantilever fits into a horizontal slot in a rigid support. The beam is supported by upward and downward reactions acting over narrow regions near the ends of the inserted length.† If the unsupported length l is not too large compared to the thickness h, the slipline field is of the form shown in Fig. 8.58a. The field is no longer symmetrical about the horizontal axis of the beam, the center of rotation O of the rigid end being at some distance $d > \frac{1}{2}h$ from the upper free surface. The plastic triangle near the upper surface is not anchored at the corner A, and the singularity at this point is absent. If λ is the angle made by OE with the horizontal, and θ the angle of the fan centered at H, it is readily shown that

$$\lambda = \frac{1}{2} \simeq 28.65° \qquad \theta = \frac{\pi}{4} - \frac{1}{2} \simeq 16.35°$$

The hydrostatic pressure increases along the circular arc CE from $-k$ at C to $\lambda k/2$ at E. Proceeding as before, the geometrical relation and the force equations may be written as

$$R \sin \lambda + b \cos \lambda = h - d$$

$$R\left(\frac{\pi}{2} \sin \lambda + \cos \lambda\right) + b\left(\frac{\pi}{2} \cos \lambda + \sin \lambda\right) = 2d \qquad (137)$$

$$R\left(\frac{\pi}{2} \cos \lambda - \sin \lambda\right) + b\left(\cos \lambda - \frac{\pi}{2} \sin \lambda\right) = F$$

Substituting the value of λ in the above equations, it is possible to express R/h, b/h, and d/h in terms of F/h. Thus

$$\frac{R}{h} = -0.0851 + 1.2348\frac{F}{h} \qquad \frac{b}{h} = 0.6145 - 0.8850\frac{F}{h}$$

$$\frac{d}{h} = 0.5015 + 0.1847\frac{F}{h} \qquad (138)$$

---

† A. P. Green, *J. Mech. Phys. Solids*, **3**: 1 (1954). The bending problem has also been treated by C. Anderson and R. T. Shield, *Int. J. Solid Struct.*, **3**: 935 (1967).

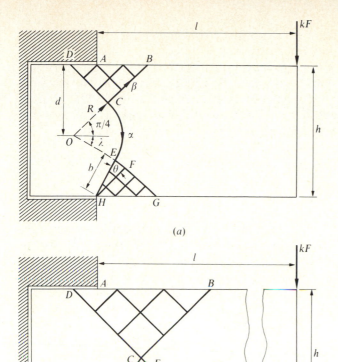

(a)

(b)

Figure 8.58 Slipline fields for an end-loaded cantilever with weak support. (a) $l/h \leqslant 7.60$; (b) $l/h \geqslant 7.60$.

Taking moment about O, the equation of couple equilibrium is written in the form

$$bR + \frac{\pi}{4}(b^2 + R^2) + d^2 = F(l + R\cos\lambda - b\sin\lambda)$$

The relationship between F/h and l/h now follows on substitution from (138). The result may be expressed as

$$0.5015 - \left(\frac{l}{h} - 0.3693\right)\frac{F}{h} - 0.7540\frac{F^2}{h^2} = 0 \tag{139}$$

When F/h has been found by solving (139) for any given l/h, the dimensions of the field can be calculated from (138). The results are given in Table 8.12. The limit of applicability of the field corresponds to $R = 0$, which gives $F/h \simeq 0.069$ and $l/h \simeq 7.60$. For longer beams, the slipline field reduces to that shown in Fig. 8.58b, where the fan angle at H is less than 16.35°. A sufficiently accurate empirical

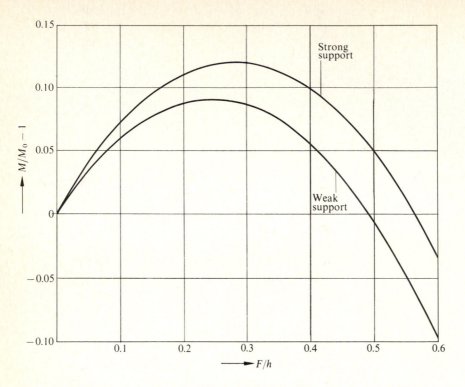

Figure 8.59 Influence of shear on the yield moment for the plane strain bending of beams.

formula in this range is

$$\frac{h}{F} = \frac{2l}{h} - \frac{3}{4} \qquad \frac{l}{h} \geqslant 7.6$$

The influence of the shearing force on the yield point state is most conveniently demonstrated by considering the relationship between M/M_0 and F/h, where M is the bending moment per unit width at the built-in end, and M_0 the fully plastic moment under pure bending. Referring to (136) and (139), and noting the fact that $2Fl/h^2 = M/M_0$, the required relationship may be written as

$$\frac{M}{M_0} \simeq 1 + 1.532 \frac{F}{h} \left(0.564 - \frac{F}{h} \right) \qquad \text{strong support}$$

$$\frac{M}{M_0} \simeq 1 + 1.508 \frac{F}{h} \left(0.492 - \frac{F}{h} \right) \qquad \text{weak support}$$

$$(140)$$

These expressions are correct to within 0.3 percent over the range $0 \leqslant F/h \leqslant 0.6$. The curve for M/M_0 against F/h shows a maximum for $F/h \simeq 0.28$ with strong support, and $F/h \simeq 0.25$ with weak support, the maximum values of M/M_0 being 1.12 and 1.09 respectively (Fig. 8.59). The simple beam theory assumes $M \simeq M_0$, whatever the magnitude of the shearing force.

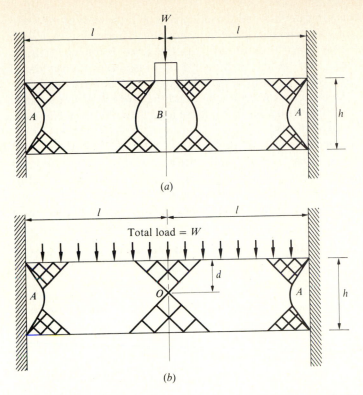

Figure 8.60 Yield point of fixed-ended beams with strong supports. (*a*) Central loading; (*b*) uniform loading.

The theory developed for a concentrated load is directly applicable to the case of uniform loading if the effect of the surface pressure on the slipline field is neglected. The error involved in this approximation is found to be insignificant.[†] Then, for a given total load kF per unit width, the greatest bending moment for an end-loaded cantilever of length l is identical to that for a uniformly loaded cantilever of length $2l$. Consequently, the solution for the distributed loading is obtained from Table 8.12, with l representing the semilength of the beam. Equations (140) evidently remain unchanged when the load is uniformly distributed.

(ii) *Fixed-ended and simply supported beams* Consider a beam of uniform rectangular cross section with thickness h and length $2l$, and having built-in support at both ends. Fig. 8.60a represents the situations where the beam is brought to the yield point by a central load W per unit width, the length over which its acts being negligible compared to $2l$. The collapse mode consists of rotation of the two halves

† See, for example, W. Johnson and R. Sowerby, *Int. J. Mech. Sci.*, **9**: 433 (1967).

of the beam about the central pivot B and the end pivots A. If the ends of the beam are prevented from moving laterally, there is a horizontal thrust at each end, which can be determined from the condition that the pivot B moves vertically downward. This requires that the line joining the two instantaneous centers of rotation of each half relative to its support and the pivot must be horizontal. The larger the end thrust, the higher is the center at A and the lower is the center at B. However, the effect of the end thrust on the collapse load is small, and is therefore neglected in the following analysis.†

The deformation on either side of pivot B will be identical to that of a cantilever of length l with weak support and under an end load $W/2$ per unit width. At A, there is either a strong or a weak support, and W is dependent on this support condition. If the magnitudes of the bending moments at A and B are M_A and M_B respectively, the equilibrium requirement for each half of the beam furnishes

$$\frac{M_A + M_B}{2M_0} = \frac{Wl}{4M_0} = \frac{W}{W_0} = \frac{W}{2kh}\left(\frac{l}{h}\right)$$

where W_0 is the load predicted by the elementary theory that assumes $M_A = M_B = M_0$. In the above expression, M_B/M_0 is given by the second relation of (140) with F replaced by $W/2k$, while M_A/M_0 is given by the first or the second relation according to whether there is a strong or weak support at A. The relationship between W/W_0 and l/h therefore becomes

$$\left(\frac{W_0}{W}\right)^2 - \left(1 - 0.803\,\frac{h}{l}\right)\frac{W_0}{W} - 1.520\,\frac{h^2}{l^2} = 0 \qquad \text{strong support}$$

$$\left(\frac{W_0}{W}\right)^2 - \left(1 - 0.742\,\frac{h}{l}\right)\frac{W_0}{W} - 1.508\,\frac{h^2}{l^2} = 0 \qquad \text{weak support}$$

(141)

For sufficiently long beams, W_0/W decreases almost linearly with h/l, the last term of each equation being negligible in that case.

Suppose, now, that the load W carried by the beam per unit width is uniformly distributed over its length $2l$ (Fig. 8.60b). The deforming zone near the central section is bounded by a pair of straight sliplines intersecting at a point O on this section. For l/h greater than about 3, the effect of the surface pressure on the regions of deformation may be neglected. Then for strong end support, O lies on the horizontal axis of the beam, and the end thrust is zero. In the case of weak end supports, there is a positive end thrust which we neglect. The bending moment at the central section of the beam is then equal to M_0. The overall equilibrium of each half of the beam requires

$$1 + \frac{M_A}{M_0} = \frac{Wl}{4M_0} = \frac{2W}{W_0} = \frac{W}{2kh}\left(\frac{l}{h}\right)$$

---

† A. P. Green, *J. Mech. Phys. Solids*, **3**: 143 (1954). The corresponding problem in plane stress has also been discussed by Green in the same paper.

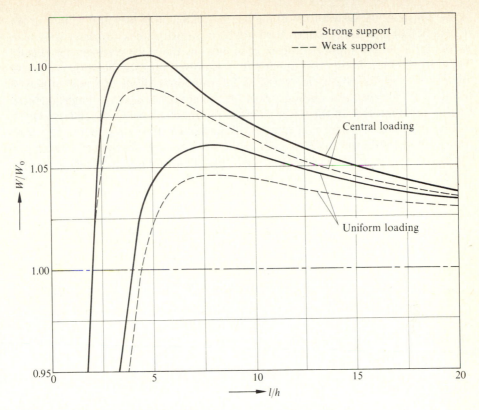

Figure 8.61 Graphical representation of the relationship between W/W_0 and l/h for fixed-ended beams.

where W_0 is the load corresponding to the simple theory, based on the assumption $M_A = M_0$. Substituting for M_A/M_0, which is given by the appropriate expression in (140) with F replaced by $W/2k$, the relationship between W/W_0 and l/h may be written as

$$\left(\frac{W_0}{W}\right)^2 - \left(1 - 0.864\frac{h}{l}\right)\frac{W_0}{W} - 3.064\frac{h^2}{l^2} = 0 \qquad \text{strong support}$$

$$\left(\frac{W_0}{W}\right)^2 - \left(1 - 0.742\frac{h}{l}\right)\frac{W_0}{W} - 3.016\frac{h^2}{l^2} = 0 \qquad \text{weak support}$$

(142)

The variation of W/W_0 with l/h for fixed-ended beams under both concentrated and distributed loads is shown graphically in Fig. 8.61. The ratio W/W_0 is greater than unity over the practical range of values of l/h. A maximum value of W/W_0 is attained for l/h equal to about 4 with concentrated loading, and about 8 with distributed loading.

The collapse load for a beam which is simply supported at both ends can be determined in a similar manner. When the beam is loaded by a concentrated load at the midspan, the distribution of the bending moment is identical to that in the central half of a centrally loaded fixed-ended beam with weak supports. The ratio W/W_0 for a simply supported beam of length l is therefore given by the second equation of (141). If the load applied to a simply supported beam is uniformly distributed over its length, the region of deformation consists of a pair of plastic triangles of the type shown in Fig. 8.60b. The height of the upper triangle, denoted by d, is less than that of the lower triangle, the ratio d/h being dependent on the surface pressure w. There is a uniform compressive stress $2k + w$ in the upper triangle, and a uniform tensile stress $2k$ in the lower triangle, both acting parallel to the axis of the beam. The condition of zero resultant horizontal thrust gives

$$d = \frac{h}{2 + w/2k}$$

Equating the moment of the stresses across the central section to that produced by the external forces, we get

$$(2k + w)d^2 + 2k(h - d)^2 = wl^2$$

where $2l$ denotes the length of the beam. If the effect of the surface pressure on the plastic region is ignored, the triangles are identical in size, and the corresponding value of the surface pressure is $w_0 = k(h/l)^2$, as predicted by the elementary theory. Substituting for d, the last equation may be expressed in the dimensionless form

$$\left(\frac{w_0}{w}\right)^2 - \left(1 - 0.5\frac{h^2}{l^2}\right)\frac{w_0}{w} - 0.25\frac{h^2}{l^2} = 0 \qquad (143)$$

which gives w/w_0 as a function of l/h. The effect of the surface pressure is to strengthen the beam when it is sufficiently short. For example, w exceeds w_0 by 6.2, 2.8, and 1.6 percent, when l/h is equal to 2, 3, and 4 respectively. For higher values of l/h, the elementary theory is sufficiently accurate.

(iii) *Cantilever under shear load and axial thrust* A cantilever of length l and uniform thickness h is rigidly supported at one end, and is subjected to a shear load kF and an axial compressive force kN per unit width at the other end.† For sufficiently small values of N/h, in relation to the length ratio l/h, the slipline field involves a pair of centered fans as shown in Fig. 8.62a. The radii of the circular arcs CD, EF, and DE are denoted by a, b, and R respectively. If the angles made by OD and OE with the horizontal are ψ and λ respectively, the angular spans of the fans at A and H are $\phi = \pi/4 - \psi$ and $\theta = \pi/4 - \lambda$. The Hencky relations therefore furnish

$$\lambda + \psi = \frac{\pi}{4} + \frac{1}{2} \qquad \lambda - \phi = \psi - \theta = \frac{1}{2}$$

---

† The slipline fields and the associated numerical results have been given by A. P. Green, *J. Mech. Phys. Solids*, **3**: 1 and 143 (1954). The bending of an I-beam under combined loading has been treated by A. S. Ranshi, N. R. Chitkara and W. Johnson, *Int. J. Mech. Sci.*, **18**: 375 (1976).

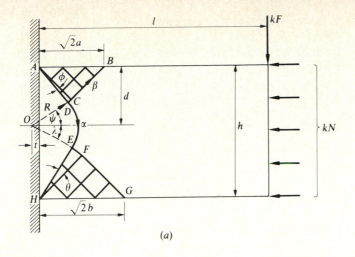

(a)

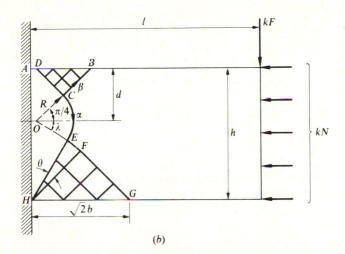

(b)

Figure 8.62 Yielding of a cantilever under combined shear and axial loading with two of four possible slipline fields.

The specification of any one of the angles fixes the others. Since A is directly above H, and AH is equal to h, we have

$$a \sin \psi - b \sin \lambda + R(\cos \lambda - \cos \psi) = 0$$

$$a \cos \psi + b \cos \lambda + R(\sin \lambda + \sin \psi) = h$$

(144)

To set up the equilibrium equations, we extend the plastic stress field into the fan ODE as before. Then the hydrostatic tension along OD is $2k\lambda$, and the hydrostatic compression along OE is $2k\psi$. Using (144), and considering the tractions along

$ADOEH$, the equations of force equilibrium may be expressed as

$$\left(1 + \frac{\pi}{2}\right)(b \sin \lambda - R \cos \lambda) + 2R(\sin \lambda + \sin \psi) = h - F$$

$$\left(1 + \frac{\pi}{2}\right)(b \cos \lambda + R \sin \lambda) + 2R(\cos \lambda - \cos \psi) = 2\lambda h + N \tag{145}$$

When $N = 0$, the field is symmetrical, with $\lambda = \psi = \pi/8 + 1/4$. As N increases from zero for a given l/h, ψ increases and λ decreases, until ϕ vanishes with $\psi = \pi/4$ and $\lambda = 1/2$. For any assumed value of λ or ψ lying between these limits, the ratios a/h, b/h, R/h, and N/h can be determined in terms of F/h by solving (144) and (145). The corresponding relationship between l/h and F/h is then obtained from the moment equation

$$\lambda(R^2 + a^2) + \psi(R^2 + b^2) + R(a + b) = F(l + t) + N\left(\frac{h}{2} - d\right) \tag{146a}$$

where t and d denote the horizontal and vertical distances between points O and A. By simple geometry,

$$t = R \cos \psi - a \sin \psi \qquad d = R \sin \psi + a \cos \psi \tag{146b}$$

The results of the computation are presented† in Table 8.14. It is found that R/h decreases with increasing l/h for a given N/h, or with increasing N/h for a given l/h. A limit of applicability of the solution therefore corresponds to $R = 0$, which gives

$$\frac{F}{h} = 1 - m\left(1 + \frac{\pi}{2}\right) \sin \lambda \sin \psi \qquad \frac{N}{h} = m\left(1 + \frac{\pi}{2}\right) \cos \lambda \sin \psi - 2\lambda$$

$$\frac{l}{h} = m^2(\lambda \sin^2 \lambda + \psi \sin^2 \psi)\frac{h}{F} + m\left(\sin \psi + \frac{N}{F} \cos \psi\right) \sin \lambda - \frac{N}{2F} \tag{147}$$

Table 8.14 Numerical results for cantilevers yielding under terminal load and axial thrust

l/h	F/h	N/h	θ	ϕ	R/h	b/h	d/h
8	0.0656	0.093	0.163	0.234	0.033	0.616	0.412
	0.0646	0.279	0.203	0.083	0.021	0.657	0.394
	0.0595	0.652	0.227	0	0	0.767	0.350
4	0.136	0.002	0.163	0.123	0.120	0.552	0.366
	0.134	0.275	0.203	0.083	0.107	0.593	0.349
	0.131	0.459	0.243	0.043	0.082	0.643	0.337
	0.124	0.658	0.285	0	0.041	0.706	0.332
2	0.280	0.089	0.163	0.123	0.297	0.419	0.272
	0.278	0.267	0.203	0.083	0.284	0.461	0.255
	0.273	0.445	0.243	0.043	0.257	0.512	0.244
	0.264	0.638	0.285	0	0.214	0.577	0.240

† The last row of results for $l/h = 8$ are based on relations that are given in Prob. 8.39.

where $m \simeq 1.0422$. These expressions relate F/h and N/h to the ratio l/h. As N increases from zero, l/h corresponding to $R = 0$ decreases from the value 13.74. For l/h equal to 12, 10, and 8, the limiting values of N/h are 0.22, 0.33, and 0.45 respectively.† Another limit of the solution is reached when $\phi = 0$, which furnishes the relations

$$\frac{R}{h} = -0.1139 + 1.2409\frac{F}{h} \qquad \frac{N}{h} = 0.6757 - 0.1437\frac{F}{h}$$

$$0.4456 - \left(\frac{l}{h} - 0.5180\right)\frac{F}{h} - 0.7831\frac{F^2}{h^2} = 0 \tag{148}$$

Both R and ϕ are zero when $F/h \simeq 0.092$, giving $N/h \simeq 0.663$ and $l/h \simeq 5.30$. As the length ratio decreases from this limiting value, F/h increases and N/h decreases when $\phi = 0$. The values of N/h for l/h equal to 3.0 and 2.0 are 0.651 and 0.638 respectively, so long as ϕ vanishes.

When N/h exceeds the value predicted by (148) for a given $l/h < 5.30$, the slipline field of Fig. 8.62b becomes applicable. The field is not anchored at the corner A, so that the distance AD progressively increases as the axial force is increased. The angles λ and θ have constant values, equal to $1/2$ and $\pi/4 - 1/2$ respectively. The geometrical parameters R, b, and d can be expressed in terms of F and N, using (137), with N added to the right-hand side of the second equation. The results are easily shown to be

$$\frac{R}{h} = -0.0851 + 1.2348\frac{F}{h} - 0.0425\frac{N}{h}$$

$$\frac{b}{h} = 0.6145 - 0.8850\frac{F}{h} + 0.3073\frac{N}{h} \tag{149}$$

$$\frac{d}{h} = 0.5015 + 0.1847\frac{F}{h} - 0.2493\frac{N}{h}$$

Since the effect of axial force is to expand the region of compression and contract the region of tension, b increases and d decreases with increasing N. The relative magnitudes of F and N at the yield point depend on the length of the beam according to the equation

$$bR + \frac{\pi}{4}(b^2 + R^2) + d^2 = F(l + R\cos\lambda - b\sin\lambda) + N\left(\frac{h}{2} - d\right)$$

for moment equilibrium. Substituting from (149), and using the value of λ, we

---

† The possibility of buckling is not considered here. In reality, this would impose a further limitation on the l/h ratio.

obtain the relation

$$0.5015 - \left(\frac{l}{h} - 0.3693\right)\frac{F}{h} - 0.7540\frac{F^2}{h^2} = \frac{N}{h}\left(0.1246\frac{N}{h} - 0.1846\frac{F}{h} - 0.0015\right)$$

$$(150)$$

which may be compared with (139), where $N = 0$. For a given value of l/h, both F/h and R/h decrease with increasing N/h. A limit is therefore reached when $R = 0$, and this corresponds to

$$\frac{F}{h} = 0.0689 + 0.0344\frac{N}{h} \qquad \frac{l}{h} \simeq 7.594 - 3.458\frac{N}{h}$$

The preceding solutions are not valid for l/h less than a certain critical value, that depends on N/h, for which yielding extends to the free end of the cantilever. This critical value is approximately given by the condition that the mean shear stress across the axis of the beam is equal to k. The critical length ratio is found to increase linearly with the axial pressure.

When the beam is longer than that for which R vanishes under a given axial force, the slipline field becomes similar to that of either Fig. 8.57b or Fig. 8.58b, depending on whether N is small or large. In the first case, the fans centered at A

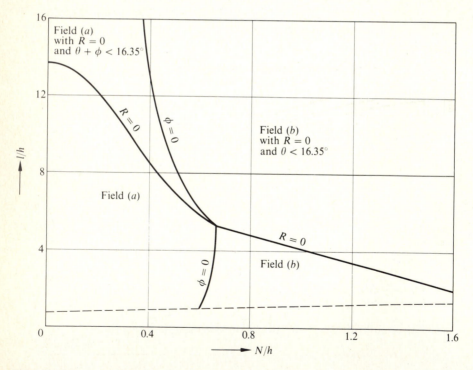

Figure 8.63 Range of validity of the various fields for a cantilever under combined end load and axial thrust.

and H are of unequal sizes, the sum of their angular spans being less than $\pi/4 - 1/2$ to avoid yielding of the rigid corners at D. In the second case, the upper fan is absent, and the angle of the lower fan is less than $\pi/4 - 1/2$. The ranges of applicability of the different types of solution are indicated in Fig. 8.63. The region below the broken line represents yielding extended to the loaded end of the cantilever.

Let M denote the bending moment kFl per unit width at the built-in cross section. A relationship between M, N, and F can be written down on the basis of (150), the range of applicability of which may be extended by a slight adjustment of some of the coefficients. For practical purposes it is sufficiently accurate to use the expression

$$\frac{M}{M_0} = 1 - \left(\frac{N}{2h}\right)^2 + 0.742 \frac{F}{h}\left(1 + \frac{N}{2h}\right) - 1.508 \frac{F^2}{h^2} \tag{151}$$

provided the axial force is large enough for the slipline field to be detached from the top corner of the beam. The above formula would be free from such a restriction if the beam were weakly supported. The last two terms on the right-hand side of (151) represent the influence of transverse shear on the bending of beams in the presence of axial forces.

The variations of F/h and $M/M_0(= 2Fl/h^2)$ with N/h are shown graphically in Fig. 8.64 for three different values of l/h. The former set of curves represent parts of the yield loci for the cantilever, giving possible yield point states under combined

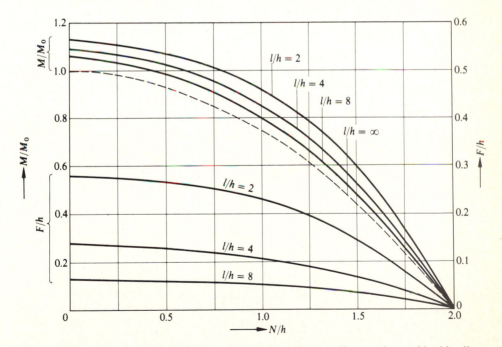

Figure 8.64 Variations of F/h and M/M_0 with N/h for a uniform cantilever under combined loading.

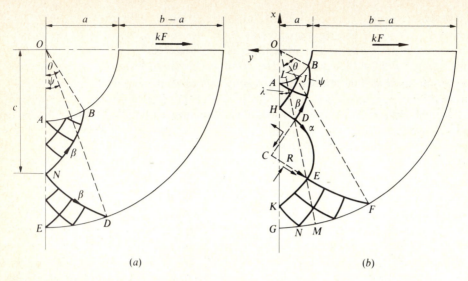

Figure 8.65 Yielding of a semicircular bar due to radial shear loading, only one-half of the slipline field being shown. (*a*) $b/a \leqslant 5.37$; (*b*) $b/a \geqslant 5.37$.

loading. The latter set of curves indicate the effect of transverse shear on the relationship between the bending moment and the axial force. The broken curve has the equation

$$\frac{M}{M_0} = 1 - \left(\frac{N}{2h}\right)^2$$

given by the elementary theory which neglects the effect of the shearing force. For N/h less than about 0.7, M/M_0 is within 12 percent of the value given by this equation, which strictly holds for an infinitely long beam.[†]

(iv) *Yielding of a semicircular bar* A curved bar, whose cylindrical surfaces are formed by concentric semicircles of radii a and b, is brought to the yield point by radially outward shearing forces kF per unit width applied at the ends.[‡] Depending on the ratio b/a, there are two types of solution, one-half of the slipline field in each case being shown in Fig. 8.65. The field (*a*) consists entirely of logarithmic spirals having their pole at the center O, while a stress discontinuity occurs across the neutral point N whose distance from O is denoted by c. If the angles subtended by AB and DE are denoted by θ and ψ respectively, the geometry

---

[†] The plastic collapse of knee-frames has been considered by W. Johnson, *Appl. Sci. Res.*, **A11**: 318 (1961), and by W. Johnson and R. Sowerby, *Int. J. Mech. Sci.*, **9**: 433 (1967). The bending of a cantilever containing holes of various shapes has been investigated by A. S. Ranshi, N. R. Chitkara, and W. Johnson, *Int. J. Mech. Sci.*, **15**: 15, 324 (1973).

[‡] This solution is due to W. Johnson and R. Sowerby, *Int. J. Mech. Sci.*, **9**: 433 (1967).

of the field requires

$$\theta = \ln \frac{c}{a} \qquad \psi = \ln \frac{b}{c} \qquad \theta + \psi = \ln \frac{b}{a}$$

The tensile stress across AN increases outward from $2k$ at A, and the compressive stress across EN decreases inward from $2k$ at E. The resultant tensile and compressive forces acting across AN and EN per unit width are $2kc\theta$ and $2kc\psi$ respectively. The condition of horizontal equilibrium therefore gives

$$F = 2c(\theta - \psi) = 2c \ln \frac{c^2}{ab} \tag{152}$$

The normal stress distribution along the line AN due to the material on its left produces a clockwise moment about O, and its magnitude per unit width is given by (104), with c replaced by a. The moment of the tractions along EN, considered in the same sense, is also obtained from (104) by writing b and $-\psi$ for c and θ respectively. Since the sum of the two moments must vanish for equilibrium, we get

$$a^2(1 + 2\theta)e^{2\theta} + b^2(1 - 2\psi)e^{-2\psi} = a^2 + b^2$$

or

$$\frac{2c^2}{ab}\left(1 + \ln \frac{c^2}{ab}\right) = \frac{b}{a} + \frac{a}{b} \tag{153}$$

from which c/a can be calculated for any given b/a. The collapse load then follows from (152). The limit of validity of the solution is defined by the condition that the hydrostatic pressure jump across N must not exceed πk. The pressures immediately below and above N are $k(1 - 2\psi)$ and $-k(1 + 2\theta)$ respectively, giving $\theta - \psi \leqslant \pi/2 - 1$. Substituting for θ and ψ, and using (153), this restriction may be written as

$$\frac{b}{a} + \frac{a}{b} \leqslant \pi \exp\left(\frac{\pi}{2} - 1\right) \qquad \text{or} \qquad \frac{b}{a} \leqslant 5.374$$

Higher values of b/a are associated with the field (b), in which the tension and compression regions are connected by a circular arc DE of radius R and angular span $\pi/2$. The sliplines in $AHDB$ and $GKEF$ are logarithmic spirals with pole at O. The points D and E lie on a radial line inclined at an angle λ to the vertical axis. If θ and ψ denote the angles turned through by BD and EF respectively, it follows from geometry and Hencky's equations that

$$\sqrt{2}\, R = be^{-\psi} - ae^{\theta} \qquad \theta - \psi = \frac{\pi}{2} - 1 \tag{154}$$

To set up the equations of equilibrium, it is convenient to suppose that the region CDE is a plastically stressed centered fan. Then the hydrostatic tension along CD is $k(1 + 2\theta)$ and the hydrostatic compression along CE is $k(1 - 2\psi)$. The stresses acting across CD and CE are statically equivalent to a force passing through D, having components kP and kQ along Ox and Oy, respectively, together with a couple kG in the counterclockwise sense. It is easily shown that

$$P = \sqrt{2}\,R(\theta - \psi) \qquad Q = \sqrt{2}\,R(\theta + \psi) \qquad G = kR^2(1 - \theta - \psi)$$

The resultant tensile and compressive forces per unit width transmitted across LD and ME are equal to $2ka\theta e^{\theta}$ and $2kb\psi e^{-\psi}$ respectively. The conditions of equilibrium of the forces, resolved along and perpendicular to OG, give the relations

$$F \sin \lambda = P \qquad F \cos \lambda = 2(a\theta e^{\theta} - b\psi e^{-\psi}) + Q$$

Substituting for P and Q, and using (154), the above relations may be expressed as

$$F \sin \lambda = \left(\frac{\pi}{2} - 1\right)(be^{-\psi} - ae^{\theta})$$

$$\tag{155}$$

$$F \cos \lambda = \left(\frac{\pi}{2} - 1\right)(be^{-\psi} + ae^{\theta})$$

The counterclockwise moment of the tractions on DCE about O is equal to $(G - Qae^{\theta})$ per unit width. The moments of the stresses acting across LD and ME are obtained as before. The condition of zero resultant moment of the tractions along $LDCEM$ about the point O furnishes

$$a^2(1 + 2\theta)e^{2\theta} + b^2(1 - 2\psi)e^{-2\psi} = a^2 + b^2 + 2(G - Qae^{\theta})$$

The substitution for G and Q, and the subsequent elimination of R, result in

$$\left(\frac{\pi}{2} - 1\right)e^{4\theta} - \left\{\frac{b^2}{a^2} - \frac{2b}{a}\exp\left(\frac{\pi}{2} - 1\right) + 1\right\}e^{2\theta} + \left(\frac{\pi}{2} - 1\right)e^{\pi - 2}\left(\frac{b}{a}\right)^2 = 0$$

$$\tag{156}$$

This is a quadratic in $e^{2\theta}$, and is readily solved for any given value of b/a. When θ has been found, the quantities ψ, R/a, F/a, and λ can be calculated from (154) and (155). The results are summarized in Table 8.15. The ratio M/M_0 appearing in the last column represents the quantity $F(b + a)/(b - a)^2$. It is the ratio of the bending moment about the center of the vertical section, to the fully plastic moment under pure bending.

Table 8.15 Yield point of a semicircular bar with end loads

b/a	θ	ψ	c/a	R/a	λ	$F/(b - a)$	M/M_0
2.0	0.404	0.289	1.498	.....	.....	0.344	1.032
3.0	0.685	0.414	1.983	.....	.....	0.537	1.074
4.0	0.898	0.488	2.455	.....	.....	0.671	1.119
5.374	1.126	0.555	3.084	0	0	0.805	1.173
6.0	0.803	0.232	.....	1.786	0.347	0.848	1.188
7.0	0.667	0.097	.....	3.207	0.488	0.898	1.193

The field (*a*) is associated with a continuous distribution of velocity vanishing at *N*. The material outside the field instantaneously rotates as a rigid body about the point *N*. The field (*b*) involves a velocity discontinuity of amount ωR propagating along *JHDEKN*, where ω is the angular velocity of the rigid end of the bar rotating about the point *C*. In both cases, a hollow develops at the upper free surface, and a bulge develops at the lower free surface, during the deformation at the plastic collapse. The hodograph may be constructed in the same way as that for the bending of notched bars, and the plastic work rate shown to be nowhere negative.

8.11 Large Bending of Wide Sheets

(i) *Bending without tension* Consider the pure bending of a metal sheet which is so wide that the strain in the width direction is negligible. The material is assumed to be nonhardening, and the strains in the plane of bending are supposed to be of unlimited magnitude. In the region well away from the loaded ends, the distributions of stress and strain must be identical for all transverse sections of the sheet. The original plane surfaces of the sheet are assumed to be cylindrical due to bending, the radii of curvature of the internal and external surfaces at any stage being *a* and *b* respectively (Fig. 8.66). The principal stresses in the plane of bending act in the radial and circumferential directions, and the equation of equilibrium is

$$\frac{\partial \sigma_r}{\partial r} = \frac{\sigma_\theta - \sigma_r}{r} \tag{157}$$

where (r, θ) are the polar coordinates with respect to the instantaneous center of curvature. If *c* denotes the radius of the instantaneous neutral surface, the yield criterion becomes

$$\sigma_r - \sigma_\theta = 2k \qquad a \leqslant r \leqslant c$$

$$\sigma_\theta - \sigma_r = 2k \qquad c \leqslant r \leqslant b$$

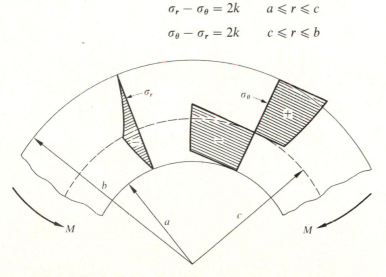

Figure 8.66 Stress distribution in the plane strain bending of a wide sheet.

All fibers inside the neutral surface are momentarily compressed, and those outside the neutral surface are momentarily extended. Substituting in (157) from above, and using the boundary conditions $\sigma_r = 0$ on $r = a$ and $r = b$, we obtain

$$\sigma_r = -2k \ln \frac{r}{a} \qquad \sigma_\theta = -2k\left(1 + \ln \frac{r}{a}\right) \qquad a \leqslant r \leqslant c$$

$$\sigma_r = -2k \ln \frac{b}{r} \qquad \sigma_\theta = 2k\left(1 - \ln \frac{b}{r}\right) \qquad c \leqslant r \leqslant b$$
(158)

The magnitude of the hoop stress steadily increases toward $r = c$ in the region of compression, and decreases toward $r = c$ in the region of tension. The condition of continuity of the radial stress across the neutral surface gives

$$\frac{c}{a} = \frac{b}{c} \qquad \text{or} \qquad c = \sqrt{ab}$$

The radial stress has its greatest numerical value $k \ln (b/a)$ on the neutral surface, where the jump in the circumferential stress is of the amount $4k$. It follows from (157) that

$$\int_a^b \sigma_\theta \, dr = \int_a^b \frac{\partial}{\partial r}(r\sigma_r) \, dr = [r\sigma_r]_a^b = 0$$

The condition of zero resultant force across any section is therefore automatically satisfied. The applied couple per unit width is

$$M = \int_a^b \sigma_\theta r \, dr = \frac{1}{2} k(a^2 + b^2 - 2c^2) = \frac{1}{2} k(b - a)^2 \tag{159}$$

The associated velocity field should correspond to a circumferential extension of fibers outside the neutral surface, and contraction of fibers inside the neutral surface. Since the material is incompressible and isotropic, the components of the strain rate must satisfy the relations $\dot{\varepsilon}_r + \dot{\varepsilon}_\theta = 0$ and $\dot{\gamma}_{r\theta} = 0$. If the radial and circumferential velocities are denoted by u and v respectively, the velocity equations become

$$\frac{\partial u}{\partial r} + \frac{u}{r} + \frac{1}{r}\frac{\partial v}{\partial \theta} = 0 \qquad \frac{\partial v}{\partial r} - \frac{v}{r} + \frac{1}{r}\frac{\partial u}{\partial \theta} = 0$$

The solution of these equations, giving an axially symmetrical distribution of strain rate, with the center of curvature assumed to be momentarily fixed in space, may be written as

$$u = -\frac{1}{2\alpha}\left(r + \frac{c^2}{r}\right) \qquad v = \frac{r\theta}{\alpha} \tag{160}$$

where θ is measured from the vertical axis of symmetry, and α denotes the angle of bending per unit length. The components of the corresponding strain rate are found to be

$$-\dot{\varepsilon}_r = \dot{\varepsilon}_\theta = \frac{1}{2\alpha}\left(1 - \frac{c^2}{r^2}\right) \qquad \dot{\gamma}_{r\theta} = 0$$

Since u depends only on r, the surfaces of the sheet remain cylindrical, in accordance with the initial assumption. In view of the linear variation of v with r, the radial planes remain plane during the bending. It follows from (160) that the instantaneous rates of change of the internal and external radii are

$$\dot{a} = -\frac{1}{2\alpha}\left(a + \frac{c^2}{a}\right) = -\frac{a+b}{2\alpha} \qquad \dot{b} = -\frac{1}{2\alpha}\left(b + \frac{c^2}{b}\right) = -\frac{a+b}{2\alpha} \qquad (161)$$

Thus $\dot{b} - \dot{a} = 0$, which indicates that the sheet thickness $h = b - a$ remains unchanged. The couple M is therefore independent of the amount of bending, the material being nonhardening. Denoting the original length of the sheet by l, and equating the initial and final volumes, we get

$$l(b - a) = \frac{1}{2}(b^2 - a^2)l\alpha \qquad \text{or} \qquad \alpha = \frac{2}{a+b}$$

It follows that the fiber which currently coincides with the central surface has undergone zero resultant change in length due to equal amounts of compression and extension during the bending.

Consider the movement of a typical fiber situated at a distance $\frac{1}{2}eh$ from the central plane in the unbent state $(-1 \leqslant e \leqslant 1)$. Let r denote the current radius of the fiber in the bent state. Since the fiber divides the section into areas that remain constant during the bending,

$$\frac{r^2 - a^2}{b^2 - r^2} = \frac{1+e}{1-e}$$

or

$$r^2 = \tfrac{1}{2}\{(b^2 + a^2) + e(b^2 - a^2)\} \qquad (162)$$

The original central fiber $(e = 0)$ moves to the convex side during bending, its radius of curvature in the final state being

$$r_0 = \sqrt{\frac{a^2 + b^2}{2}}$$

The initial position of the fiber finally coinciding with the neutral surface $(r = c)$ corresponds to $e = -(b - a)/(b + a)$. The neutral surface itself approaches the inner boundary from its initial position which coincides with the central plane. Hence, all fibers initially above the central plane are progressively extended, while all fibers finally inside the neutral surface are progressively compressed. The fibers which are contained in the region corresponding to

$$-\left(\frac{b-a}{b+a}\right) < e < 0 \qquad \text{or} \qquad c < r < \sqrt{\frac{a^2 + b^2}{2}}$$

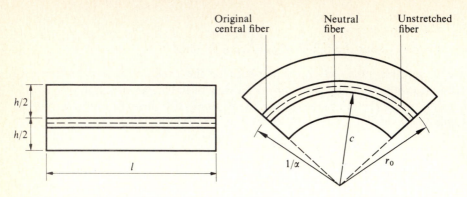

Figure 8.67 Relative movement of selected longitudinal fibers in sheet bending.

are first compressed and then extended, being overtaken by the neutral surface at some intermediate stage. In actual practice, therefore, these fibers would be subject to the Bauschinger effect. The initial position of the fiber that suffers equal amounts of extension and contraction is given by $2e = -(b - a)/(b + a)$, which is obtained by setting $r = \frac{1}{2}(a + b)$ in (162). Since the bending occurs under constant couples, the work done per unit width is $M\alpha l$, and its value per unit volume is equal to $k(b - a)/(b + a)$. Figure 8.67 shows the initial and final positions of the unstretched fiber as well as those which define the region of reversed loading.†

(ii) *Bending under tension* Consider, now, the situation where the sheet is bent by couples M and tensions T per unit width applied at the ends. The tensions are assumed to act in the circumferential directions at each stage, their inward resultant being balanced by a uniform normal pressure p applied over the inner boundary. The condition for equilibrium is evidently

$$T = ap$$

On the convex side of the neutral surface, the stresses are still given by (158), while on the concave side, the stresses are modified to

$$\sigma_r = -p - 2k \ln \frac{r}{a} \qquad \sigma_\theta = -p - 2k\left(1 + \ln \frac{r}{a}\right) \qquad a \leqslant r \leqslant c$$

The radius of the neutral surface is obtained from the condition that σ_r must be continuous across $r = c$. Thus

$$c^2 = ab \exp\left(-\frac{p}{2k}\right) = ab \exp\left(-\frac{T}{2ka}\right) \tag{163}$$

† R. Hill, *The Mathematical Theory of Plasticity*, pp. 287–294, Clarendon Press, Oxford (1950). The analysis for the stress distribution has also been given by J. D. Lubahn and G. Sachs, *Trans. ASME*, **72**: 201 (1950).

which shows that the effect of tension is to move the neutral surface toward the inner boundary. If T acts along the tangent to the central fiber, the equilibrium of the moments about the center of curvature requires

$$M = \int_a^b \sigma_\theta r \, dr - \tfrac{1}{2}T(a + b)$$

Substituting for σ_θ and integrating, we obtain

$$M = \frac{1}{2} k \left\{ a^2 + b^2 - 2ab \exp\left(-\frac{T}{2ka} \right) \right\} - \frac{1}{2} Tb \tag{164}$$

The velocity distribution is again given by (160), but the thickness of the sheet no longer remains constant. If the thickness at any stage is denoted by h, then by (161), its rate of change is

$$\dot{h} = \dot{b} - \dot{a} = -\frac{1}{2\alpha}(b - a)\left(1 - \frac{c^2}{ab} \right) = -\frac{h}{2\alpha}(1 - e^{-p/2k})$$

which shows that the thickness is decreased by the application of the tension. Dividing the above relations by a similar one for $\dot{a}$, we get

$$\frac{dh}{da} = \frac{h(e^{p/2k} - 1)}{h + a(e^{p/2k} + 1)} = \frac{h(e^{\lambda h/a} - 1)}{h + a(e^{\lambda h/a} + 1)}$$

where $2k\lambda$ denotes the applied tensile stress T/h. The last equation may be expressed in the more convenient form

$$\frac{dh}{h} = -\left(\frac{e^{\lambda h/a} - 1}{2 + h/a} \right) \frac{a}{h} d\left(\frac{h}{a} \right)$$

If the variation of λ is prescribed during the bending, the thickness variation can be found by numerical integration. Assuming λ to be maintained constant, and denoting the initial thickness by h_0, we obtain the solution

$$\ln \frac{h_0}{h} \simeq \lambda \ln\left(1 + \frac{h}{2a} \right) - \frac{1}{2}\left(\frac{\lambda h}{2a} \right)^2 \tag{165}$$

to a sufficient accuracy when $\lambda h/a$ is less than about 0.5. The thickness progressively decreases for all positive values of λ, which must be less than unity to avoid necking of the sheet. Figure 8.68 indicates how the thickness of the sheet decreases with continued bending under constant values of the applied tensile stress.

When h/a is less than about 0.2, the transverse stresses may be neglected as a first approximation. The neutral surface is then at a distance $\tfrac{1}{2}\lambda h$ from the central surface. The thickness of the sheet at any stage is less than the original thickness by a factor $\lambda h/2a$ approximately. The longitudinal strain in any fiber at a distance y from the neutral surface is of amount y/a to the same order. This strain may be regarded as produced by a continuous extension or contraction of the fiber. The work done per unit volume of the sheet is easily shown to be $(1 + \lambda^2)kh/2a$. If the

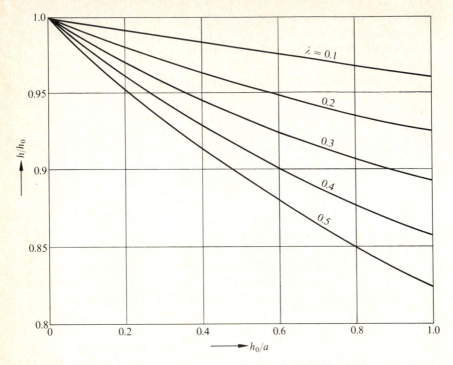

$$\frac{h}{h_0}$$

Figure 8.68 Variation of thickness ratio with progressive bending of a sheet under tension.

bent sheet is straightened under the same tension, there is a further thinning of an equal amount, involving the same amount of work. The total thinning during the process of bending and unbending is therefore a fraction $\lambda h/a$ of the original sheet thickness.

(iii) *Inclusion of strain-hardening* If the material work-hardens, the thickness of the sheet progressively decreases as the bending proceeds even when no tension is applied. The velocity field (160) continues to hold, but c is now less than $\sqrt{ab}$, while α is greater than $2/(a + b)$. The movement of the longitudinal fibers is still represented by Fig. 7.67, where $r_0^2 = (a^2 + b^2)/2$ as before. The condition of the constancy of volume of the material readily furnishes

$$\alpha = \frac{2h_0}{b^2 - a^2} = \frac{h_0}{h}\left(\frac{2}{a + b}\right) = \frac{h_0}{h\rho}$$

where ρ is the mean radius of the bent sheet, while h_0 and h are the initial and final thicknesses. Introducing dimensionless variables

$$\xi = \frac{b - a}{b + a} = \frac{h}{2\rho} \qquad \eta = \frac{h}{h_0}$$

the ratios of the final and initial lengths of fibers coinciding with the inner, outer, and neutral surfaces may be written as

$$a\alpha = \frac{1-\xi}{\eta} \qquad b\alpha = \frac{1+\xi}{\eta} \qquad c\alpha = \frac{c}{\rho\eta} \tag{166}$$

In view of (161), the thickness of the sheet changes at the rate

$$\frac{dh}{d\alpha} = \dot{b} - \dot{a} = -\frac{h}{2\alpha}\left(1 - \frac{c^2}{ab}\right)$$

Using (166), and the fact that $d\alpha/\alpha = d\xi/\xi - 2dh/h$, the above equation can be expressed in the dimensionless form

$$\frac{d\eta}{d\xi} = \left\{1 - (1-\xi^2)\frac{\rho^2}{c^2}\right\}\frac{\eta}{2\xi} \tag{167}$$

Initially, $\xi = 0$ and $\eta = c/\rho = 1$, while $d\eta/d\xi$ vanishes in the limit. In order to solve (167) numerically, it is necessary to establish a second relationship between c/ρ, ξ, and η, depending on the strain-hardening property of the material.

The material below the neutral surface $r = c$ has been hardened by progressive compression, while that above the radius $r = r_0$ has been hardened by progressive extension. The stresses in these regions may be calculated by expressing the uniaxial stress-strain law in the form

$$\sigma = \sqrt{3}k\left\{1 + m\left[1 - \exp\left(-\frac{\sqrt{3}}{2}n\varepsilon\right)\right]\right\}$$

where m and n are empirical constants, and k is the initial yield stress in shear. In view of the plane strain condition, ε is equivalent to $(2/\sqrt{3})\varepsilon_\theta$ for $r \geqslant r_0$, and to $-(2/\sqrt{3})\varepsilon_\theta$ for $r \leqslant c$. Since $\varepsilon_\theta = \ln{(r\alpha)}$, the yield criterion may be written as

$$\begin{aligned}
\sigma_\theta - \sigma_r &= 2k\{1 + m[1 - (r\alpha)^{-n}]\} & r_0 \leqslant r < b \\
\sigma_\theta - \sigma_r &= -2k\{1 + m[1 - (r\alpha)^n]\} & a \leqslant r \leqslant c
\end{aligned} \tag{168}$$

The hardening process is rather complex over the region $c \leqslant r \leqslant r_0$, where each fiber has suffered a reversal of stress from compression to tension at some stage during the bending. The Bauschinger effect that would result in a real material may be approximately allowed for by assuming the yield stress to vary linearly[†] with $\ln{(r/c)}$. Taking the initial yield stress in reversed loading as $2k$, we write

$$\sigma_\theta - \sigma_r = 2k\left\{1 + m\frac{\ln{(r/c)}}{\ln{(r_0/c)}}[1 - (r_0\alpha)^{-n}]\right\} \qquad c \leqslant r \leqslant r_0 \tag{169}$$

so that the yield stress becomes continuous at $r = r_0$. The distribution of σ_r follows

---

† P. Dadras and S. A. Majilessi, *J. Eng. Ind., Trans. ASME*, **104**: 224 (1982). For an analysis that neglects the Bauschinger effect, see H. Verguts and R. Sowerby, *Int. J. Mech. Sci.*, **17**: 31 (1975). An approximate analysis based on linear work-hardening has been given earlier by F. Proksa, *Der Stahlbau*, **2**: 29 (1959).

from a straightforward integration of (157), using (168) and (169). The position of the neutral surface in the absence of tension is given by

$$\int_a^b (\sigma_\theta - \sigma_r) \frac{dr}{r} = 0$$

carrying out the integration separately over the three regions, and using the fact that

$$r_0 \alpha = \alpha \sqrt{\frac{a^2 + b^2}{2}} = \frac{\sqrt{1 + \xi^2}}{\eta} \tag{170}$$

in view of (166), the result may be expressed in the dimensionless form

$$\left(\frac{c}{\rho\eta}\right)^n + 2n\left(1 + \frac{1}{m}\right)\ln\left(\frac{\rho}{c}\sqrt{1 - \xi^2}\right) - \frac{n}{2}\left\{1 + \left(\frac{\eta}{\sqrt{1 + \xi^2}}\right)^n\right\}\ln\left(\frac{\rho}{c}\sqrt{1 + \xi^2}\right)$$

$$= \left(\frac{1 - \xi}{\eta}\right)^n - \left(\frac{\eta}{1 + \xi}\right)^n + \left(\frac{\eta}{\sqrt{1 + \xi^2}}\right)^n \tag{171}$$

Equation (167) must be considered simultaneously with (171) to obtain c/ρ and η as functions of ξ. The differential equation (167) may be solved by the Runge-Kutta numerical method, the ratio c/ρ for approximate values of ξ and η being found from (171) using an iterative procedure.

The continued bending of a work-hardening sheet following the yield point requires the application of an increasing bending moment. The magnitude of the moment is most conveniently obtained from the equation (Sec. 5.5(i))

$$M = \frac{1}{2}\int_a^b (\sigma_\theta - \sigma_r)r \, dr$$

Substituting from (168) and (169), the bending moment per unit width (assuming $n \neq 2$) may be written as[†]

$$\frac{M}{k\rho^2} = \left(1 - \frac{c^2}{\rho^2} + \xi^2\right)\left\{(1 + m) - \frac{(m/4)[1 - (\eta/\sqrt{1 + \xi^2})^n]}{\ln\left[(\rho/c)\sqrt{1 + \xi^2}\right]}\right\}$$

$$+ \frac{m\eta^2}{n + 2}\left\{\left(\frac{c}{\rho\eta}\right)^{n+2} - \left(\frac{1 - \xi}{\eta}\right)^{n+2}\right\} + \frac{mc^2}{2\rho^2}$$

$$- \frac{m\eta^2}{n - 2}\left\{\frac{n}{2}\left(\frac{\eta}{\sqrt{1 + \xi^2}}\right)^{n-2} - \left(\frac{\eta}{1 + \xi}\right)^{n-2}\right\} \tag{172}$$

on using relations (166) and (170). For a nonhardening material, $m = 0$ and $c^2 = \rho^2(1 - \xi^2)$, reducing (172) to the result $M = \frac{1}{2}kh^2$ obtained previously.

[†] J. Chakrabarty, unpublished work (1983). A similar solution based on a different strain-hardening law has been presented by P. Dadras and S. A. Majilessi, op. cit.

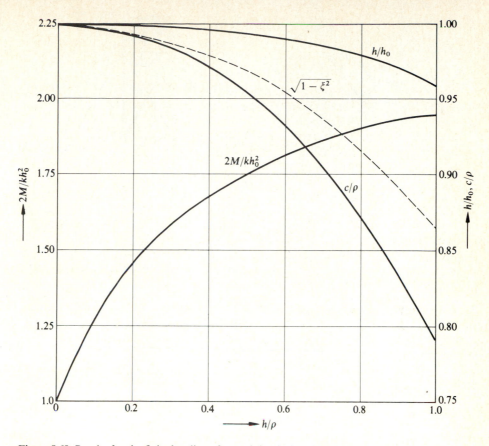

Figure 8.69 Results for the finite bending of a work-hardening sheet ($m = 1.5$, $n = 4.0$).

The results of the computation based on $m = 1.5$ and $n = 4.0$ are presented in Fig. 8.69. The broken curve represents the variation of c/ρ for a nonhardening material. In the absence of tension, the thickness change that occurs in bending is seen to be fairly small. The ratio $c/\sqrt{ab}$ steadily decreases from unity to reach a value of 0.91 approximately when ρ becomes equal to h. The depth of the zone of reversed loading, when $\rho = h$, is found to be about 17 percent of the total thickness of the sheet.

Problems

8.1 The velocity discontinuity pattern of Fig. A may be used to obtain an upper bound solution for the plane strain indentation of a semi-infinite block by a smooth flat punch. Find the angle θ that minimizes the mean punch pressure q, and evaluate the best upper bound in terms of the shear yield stress k.

    Answer: $\theta = \tan^{-1}\sqrt{2}$, $q = 4\sqrt{2}\,k$.

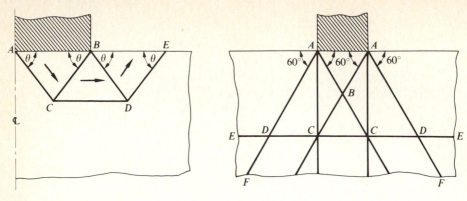

Figure A   **Figure B**

8.2 A lower bound solution for the indentation of a semi-infinite block by a smooth flat punch may be obtained by using the stress discontinuity pattern of Fig. B. Construct a statically admissible stress field based on the proposed discontinuities, and estimate the lower bound on the mean punch pressure.

Answer: $q = 5k$.

8.3 Figure C shows a possible upper bound mode for the indentation of a block of finite width by a smooth flat punch. The rate of deformation is constant in the region $ABDC$, outside which a simple block sliding occurs. Minimizing the punch pressure with respect to b/a, obtain the upper bound

$$\frac{q}{k} = 2 + \sqrt{\frac{2w}{a} - 1} \qquad \frac{w}{a} \leqslant 7.19$$

Show that for a sufficiently narrow block, an improved bound is $q = k(1 + w/a)$, obtained on the basis of a single discontinuity originating from a corner of the punch and terminating on the opposite vertical side of the block.

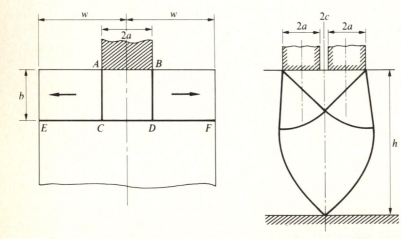

Figure C   **Figure D**

8.4 The slipline field of Fig. D furnishes an upper bound solution for the co-indentation of a block, resting on a smooth foundation, by a pair of identical punches separated by a distance $2c$. Show that the mean punch pressure q is given by

$$\frac{q}{2k} = \left(1 + \frac{c}{2a}\right)f\left(\frac{h}{2a+c}\right) \qquad \frac{h}{a} \geqslant 2 + \frac{c}{a}$$

where $2kf(h/a)$ denotes the pressure corresponding to indentation by a single punch. Assuming $c = a$, find the range of values of h/a for which this solution gives a lower punch pressure than that produced by the two punches acting independently, as well as that in which the deformation is localized around the punch.

Answer: $3.0 \leqslant h/a \leqslant 11.6$.

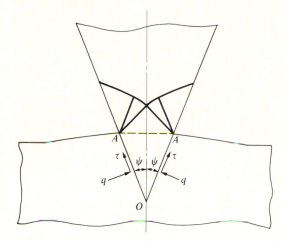

Figure E

8.5 In the indentation of a semi-infinite block by a rigid wedge, shown in Fig. E, let q denote the normal pressure and τ the frictional stress acting on the wedge faces. Assuming a uniform state of stress in the region AOA of the wedge, and a stress discontinuity across AA, show that the wedge will remain rigid if

$$q + \tau \cot \psi < 2k'(1 + \psi) \qquad \tau < k' \sin 2\psi$$

where k' is the shear yield stress of the wedge material. Plot the critical value of k'/k against the semiangle $\psi \geqslant 15°$ for the limiting cases of perfectly smooth and perfectly rough wedge faces.

8.6 In the orthogonal indentation of a semi-infinite block by a symmetrical wedge of semiangle ψ, a mean shear strain γ may be defined as that whose product with k is equal to the plastic energy per unit volume of the deformed material. Prove that

$$\gamma = \frac{c}{b}\left(\frac{2}{2+\theta}\right)\left(\frac{q}{k}\sin\psi + \frac{\tau}{k}\cos\psi\right)$$

where q is the normal pressure and τ the frictional stress on the wedge faces, the quantities c, b, and θ being the same as those introduced in Sec. 8.2. Show a graphical comparison of the dependence of γ on $\psi \leqslant \pi/4$ for a fictionless wedge with that for a perfectly rough wedge.

8.7 Let (x_0, y_0) be the initial coordinates of a typical particle in a large block indented by a smooth rigid wedge of semiangle ψ. If (x, y) are the final coordinates of the particle, when the depth of penetration is c, show that a linear relationship of the type

$$x = Ax_0 + By_0 + Dc \qquad y = Ex_0 + Fy_0 + Gc$$

holds for all particles finally occupying the regions ACG and BDE (Fig. 8.7), the axes of x and y being taken along OE and OA respectively. Find the coefficients of the above equations for both these regions when $\psi = 30°$.

Answer: $A = 0.817$, $B = -0.683$, $D = 0.683$, $E = 0.049$, $F = 1.183$, $G = -0.183$ in *ACG*, and $A = 0.766$, $B = -0.457$, $D = 0.610$, $E = 0.157$, $F = 1.268$, $G = -0.357$ in *BDE*.

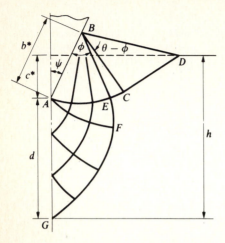

Figure F

8.8 Figure F shows one-quarter of the slipline field for the indentation of a block of thickness $2h$ by a pair of perfectly rough opposed wedges, with $\psi \leqslant \pi/4$, when the depth of penetration is critical. Using the results of Prob. 6.13, show that the angle ϕ defining the critical field is given by

$$\frac{B + 2[L(\phi, \xi) + N(\phi, \xi)]}{F_1(\phi, \xi) + F_2(\phi, \xi)} = \phi + \xi - \theta$$

where $2B = \sqrt{2}\,\theta \cos \psi + \sin (\pi/4 - \psi)$, and $\xi = \pi/4 - \psi + \phi$, the expression in the denominator being equal to $d/\sqrt{2}\,b^*$. Draw the associated hodograph, and compute the values of ϕ, h/b^*, and c^*/h when $\psi = 15°$.

Answer: $\phi = 46.24°$, $h/b^* = 3.180$, $c^*/h = 0.263$.

8.9 A block of metal of thickness $2h$ is indented by a pair of opposed wedge-shaped tools of semiangle $\psi \geqslant 45°$, the frictional condition being such that a $90°$ cap of dead metal is attached to the wedge face. Draw the slipline field corresponding to the critical depth of penetration c^*, and show that the fan angle ϕ defining the ratio c^*/h is given by

$$\{I_0(2\phi) + 2\phi[I_0(2\phi) + I_1(2\phi)]\}\frac{a}{H} = 1 + \theta$$

where a is the semiwidth of the impression, H the height of the raised lip above the horizontal axis of symmetry, and θ the fan angle corresponding to the piling-up mode. Compute the values of ϕ, h/b^*, and c^*/h for a perfectly rough $90°$ wedge.

Answer: $\phi = 73.3°$, $h/b^* = 5.339$, $c^*/h = 0.099$.

8.10 Consider the indentation of a block of critical semiwidth w by a rough obtuse-angled wedge of semiangle ψ, when a false cap of dead metal is attached to the wedge face. Construct a statically admissible extension (similar to Fig. 8.43) of the field corresponding to the piling-up mode, based on a stress-free boundary that meets the sides of the block at a depth d below the original free surface. Assuming $\psi = 50.6°$, calculate the ratios w/a and d/c, where a is the semi-width of the impression, and c the depth of penetration for which the width is critical.

Answer: $w/a = 7.68$, $d/c = 10.63$.

8.11 Figure G shows a statically admissible extension of the slipline field for the indentation of a block by a smooth rigid wedge of semiangle $\psi = 50.6°$, giving an estimate of the critical semiwidth w. The

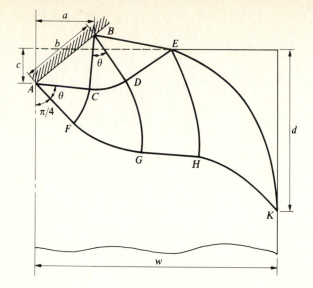

Figure G

extended field is symmetrical about the perpendicular bisector of the surface of contact AB between the wedge and the block, the material to the right of EK being assumed stress-free. Using Eqs. (138), Chap. 6, show that

$$\frac{w}{b} = 2e^{\theta} - \cos\theta \qquad \frac{d}{b} = 2\sin\theta(1 + \sin\theta)$$

where θ is the angle of the fan. Complete the numerical values of w/a and d/c, using the appropriate value of θ.

Answer: $w/a = 4.148$, $d/c = 4.710$.

8.12 Draw the slipline field and the hodograph for the indentation of a block whose width $2w$ is critical for the depth of penetration c of a knife-edged indenter of vanishingly small angle. Establish the equations of equilibrium of the rigid shoulder that undergoes an incipient rotation at the yield point, and evaluate the parameters that define the geometry of the field. Hence deduce the results

$$\frac{w}{c} = 1 + \frac{4}{\pi} \qquad \frac{d}{c} = 2 - \frac{4}{\pi}(\sqrt{2} - 1)$$

where d is the depth of the point where the line of velocity discontinuity meets the vertical side of the block, and c the depth of penetration.

8.13 Figure H represents the initial stages of the symmetrical extrusion of a billet through a lubricated tapering die of semiangle ψ. As the billet moves on, the square corner of the billet is deformed into the region $ACDEB$, whose shape remains geometrically similar. Derive an expression for the mean punch pressure p_e in terms of c/H, ψ, and θ, where c is the punch travel and H the semithickness of the undeformed billet. Assuming $\psi = 30°$, and an extrusion ratio of 2, find the range of values of c/H and $p_e/2k$ for which the solution is valid. Obtain also the range of values when the die is perfectly rough.

Answer: $c/H \leqslant 0.438$, $p_e/2k \leqslant 0.441$; $c/H \leqslant 0.360$, $p_e/2k \leqslant 0.669$.

8.14 A perfectly rough circular cylinder of radius R fits exactly into a preformed cavity on the surface of a semi-infinite medium (Fig. I). The proposed slipline field for the incipient indentation involves a 90° cap of dead metal attached to the cylinder. Show that the required vertical force on the cylinder per unit width corresponding to an angle of contact $2\psi \geqslant \pi/2$ is

$$P = 2kR\left\{\left(1 + \frac{3\pi}{2} - 2\psi\right)\sin\psi - 3\cos\psi + 2\sqrt{2}\right\}$$

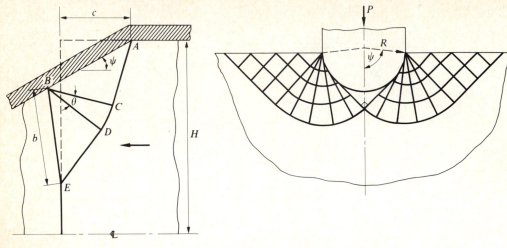

Figure H   **Figure I**

Is it possible to associate a consistent velocity field with the assumed slipline field? Comment on the nature of this solution.

8.15 A composite block of width $2w$ is made of an inner layer of material of thickness h_1 and yield stress $2k_1$, bonded between two equal outer layers of another material of total thickness h_2 and yield stress $2k_2$. The block is symmetrically compressed by a pair of rough platens under conditions of plane strain. Using the technological method, show that the variation of the die pressure q with the distance x from the central vertical axis is given by

$$\frac{q}{2k} = \exp\left\{2\mu\left(\frac{w - x}{h_1 + h_2}\right)\right\} \qquad \frac{2w}{h_1 + h_2} \leqslant \frac{1}{\mu}\ln\left(\frac{k_2}{2\mu k}\right)$$

where k denotes the quantity $(k_1h_1 + k_2h_2)/(h_1 + h_2)$. Assuming $\mu = 0.15$, $k_2/k_1 = 0.8$, $w/h_1 = w/h_2 = 7$, calculate the value of $\bar{q}/2k_1$, where $\bar{q}$ is the mean die pressure.
  Answer: 1.592.

8.16 An upper bound solution for the plane strain compression of a block between perfectly rough platens may be obtained by using the velocity discontinuity pattern of Fig. 8.35. Draw the associated hodograph and deduce the upper bound formula

$$\frac{\bar{q}}{2k} = \frac{1}{2}\left(\frac{mh}{w} + \frac{w}{mh}\right) + \left(1 - \frac{1}{m^2}\right)\frac{w}{4h}$$

where m is the number of discontinuities, assumed odd. Plot the upper bound mean pressure against the width/height ratio, and compare it with that predicted by the slipline field solution.

8.17 Referring to the velocity discontinuity pattern of Fig. 8.35, let the coefficient of friction μ be large enough for sticking friction to occur on n number of sliding blocks. Draw the modified hodograph for $n = 1$ and $n = 2$ using seven discontinuities. Assuming $w/h = 7$, show that the mean die pressure for $n = 1$ and $n = 2$ are given by

$$\frac{\bar{q}}{2k} = \frac{5}{7} + \frac{4(1 + 3\mu)}{7(1 - \mu)^2} \qquad \text{and} \qquad \frac{\bar{q}}{2k} = 1 + \frac{6}{7}\left(\frac{1 + \mu}{1 - \mu}\right)$$

respectively. For what value of μ would Coulomb's law of friction give way to sticking friction over the whole platen according to the upper bound solution?
  Answer: $\mu = 1/3$.

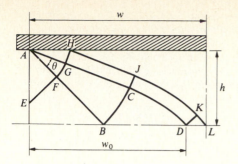

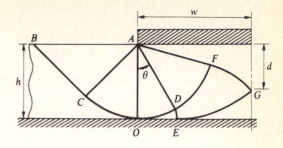

Figure J   **Figure K**

8.18 Figure J shows a quarter slipline field for the compression of a block between partially rough overlapping platens, the fan angle θ being dependent on the coefficient of friction μ. Show that the mean die pressure $\bar{q}$ may be written as

$$\frac{\bar{q}}{2k} = \frac{w_0}{w} f\left(\frac{w_0}{h}\right) + (1 + 2\theta)\left(1 - \frac{w_0}{w}\right)$$

where $f(w_0/h)$ is the dimensionless mean pressure when $w = w_0$. Assuming $\mu = 0.2$, find the range of values of w/h for which the field is applicable, and estimate the coresponding range of values of $\bar{q}/2k$. Construct the associated hodograph, and compute the horizontal component of velocity of the rigid triangle AEF per unit speed of compression.

Answer: $1.568 \leqslant w/h \leqslant 2.762$, $1.102 \leqslant \bar{q}/2k \leqslant 1.272$, $V/U = 2.163$.

8.19 Figure K shows one half of the slipline field for the compression of a block, resting on a perfectly rough foundation, using a sufficiently rough punch, the range of values of w/h being fairly small. Show that the mean punch pressure $\bar{q}$ is given by

$$\frac{\bar{q}}{2k} = \frac{1 + \pi}{2} + 2\theta + \frac{h}{w}\left\{\frac{d}{2h} - \int_0^\theta \left(\frac{x}{h}\right) d\alpha\right\}$$

where the integral is taken along the segment FG. Using Table A-9, compute the values of w/h and $\bar{q}/2k$ corresponding to $\theta = 15$, 30, and 45°, and determine in each case the least coefficient of friction μ required on the punch face. Discuss the hodograph associated with the slipline field.

Answer: $w/h = 1.069$, 1.517, 2.097; $\bar{q}/2k = 2.662$, 2.870, 3.153; $\mu = 0.126$, 0.210, 0.242.

8.20 The slipline field for the compression of a block resting on a rigid perfectly rough foundation is shown in Fig. L, when the punch is perfectly smooth and the width/height ratio is moderate. Explain how the domain HJL may be constructed from an extension of $EDFG$, using the principle of vectorial superposition. Show that the dimensionless mean punch pressure is

$$\frac{\bar{q}}{2k} = 1 + \frac{\pi}{2} + 2\theta - \frac{4h}{w}\int_0^\theta f\left(\alpha, \frac{\pi}{4} + \alpha\right) d\alpha$$

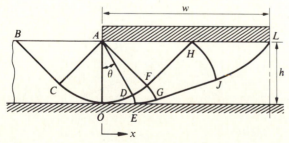

Figure L

where $f(\alpha, \beta)$ represents x/h for the field $OFGE$ with the base point at O. Use Table A-9 to calculate w/h and $\bar{q}/2k$ corresponding to $\theta = 15$, 30, and 45°, following a numerical method of integration.

Answer: $w/h = 2.138, 3.034, 4.194$; $\bar{q}/2k = 2.659, 2.993, 3.278$.

8.21 Figure M shows one half of a possible slipline field for the compression of a symmetrical tapered projection on a large block of material by a rough flat die. The frictional condition is such that a cap of rigid material is attached to the whole die face. If the mean normal pressure on the die is denoted by q, show that

$$\frac{q}{2k} = \left(1 + \frac{b}{a}\operatorname{cosec}\psi\right)\ln\left(1 + \frac{a}{b}\sin\psi\right)f\left\{\psi\left(1 + \frac{2b}{a}\operatorname{cosec}\psi\right)\right\};$$

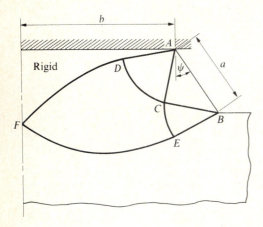

Figure M

the function f being the same as that defined in Table 7.5. Compute the value of $q/2k$ when $b/a = 2$ and $\psi = 30°$. Draw the associated hodograph, and estimate the angle η which the direction of motion of the free surface AB makes with the horizontal.

Answer: $q/2k = 2.124, \eta = 10.26°$.

8.22 A bar of thickness $2h$, containing symmetrical notches of arbitrary shape, is pulled longitudinally under conditions of plane strain. An upper bound on the yield point load is obtained by assuming a deformation mode that consists of block sliding along a pair of mutually perpendicular lines OA and OB as shown in Fig. N. Show that the constraint factor is

$$f = \frac{1}{2}\left(\cot\theta + \frac{h}{a}\tan\theta\right)$$

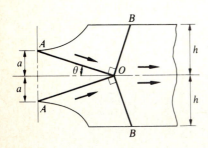

Figure N

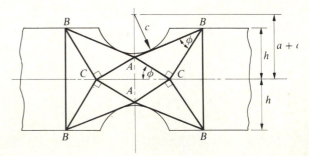

Figure O

Find the value of θ that minimizes the constraint factor, and obtain the least upper bound $\sqrt{h/a}$ for the constraint factor f.

8.23 Figure O shows a stress discontinuity pattern for the yielding of a tension bar of semithickness h containing a symmetrical pair of circular notches of radius c. Construct a statically admissible stress field based on the assumed discontinuities, and obtain the lower bound $f = (h/a)(1 - \cos 2\phi)$ for the constraint factor, the angle ϕ being given by

$$\left(1 + \frac{c}{a}\right) \cot \phi - \frac{h}{a} \tan \phi = \frac{c}{2a} \operatorname{cosec}^2 \phi$$

Assuming the notches to be semicircular with $c = a$, compute the value of ϕ and the constraint factor.
Answer: $\phi = 34.2°$, $f = 1.264$.

8.24 Draw the slipline field and the hodograph for the yielding of a tension bar of minimum thickness $2a$, containing a pair of deep trapezoidal notches of included angle $\pi - 2\psi$ and root width $2c$, assuming the plastic deformation to extend to the inclined faces of the notch, show that the yield point load T is given by

$$\frac{T}{4ka} = \left(1 + \frac{\psi}{2}\right)\left(1 - \frac{d}{a}\right) + \frac{Q}{2ka}$$

where Q is the longitudinal force per unit width across the minimum section between the notch root and the apex of a plastic triangle of height $a - d$, having its base on the longitudinal axis. Assuming $c/a = 0.41$, draw a graph showing the variation of the constraint factor with ψ over the range $0 \leqslant \psi \leqslant \pi/2$.

8.25 In the necking of a V-notched bar under longitudinal tension, a possible deformation mode is one in which the slipline field remains geometrically similar. The initial and final configurations are indicated by broken and full lines respectively in Fig. P, the lower end of the bar being assumed fixed during the extension. Show that the angle θ which the deformed free surface makes with the vertical axis is given by

$$\tan \theta = \frac{(\sec \psi + 2 \tan \psi)^2}{4(\sec \psi + \tan \psi)}$$

where $\pi/2 - \psi$ denotes the semiangle of the notch. Show also that the thickness of the neck decreases by an amount $c(\sec \psi + 2 \tan \psi)$ when the minimum section has moved vertically through a distance c.

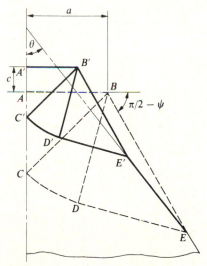

Figure P

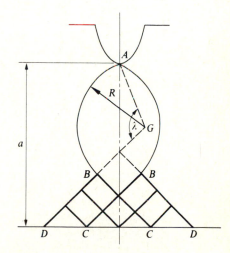

Figure Q

8.26 A bar containing a symmetrical pair of opposed V notches of semiangle $\pi/2 - \psi$ is brought to the yield point by an axial force T and a bending moment M. The notches are sufficiently deep, and the notch angle large enough for a stress discontinuity to occur on the notch axis. Draw the slipline field, and deduce the relationship

$$\left(\frac{T}{T_0}\right)^2 + (1 + \psi)\frac{M}{M_0} = (1 + \psi)^2$$

where $T_0 = 2ka$ and $M_0 = \frac{1}{2}ka^2$, the thickness of the minimum section being denoted by a. Draw a graph showing the variation of M/M_0 with T/T_0 for $\psi = 15°$, and compare it with that for $\psi = 0$. Construct the hodograph associated with the slipline field.

8.27 An upper bound solution for the pure bending of a single notched bar with an arbitrary notch profile may be obtained by assuming the deformation mode associated with the slipline field of Fig. Q. There is a velocity discontinuity initiated at A and terminated at C. Using kinematical arguments, derive the upper bound expression

$$M = k\left(\lambda - \frac{1}{2}\right)R^2 + k\left[a - R\sin\left(\lambda - \frac{\pi}{4}\right)\right]^2$$

Find the values of λ and R/a for which the yield moment M is a minimum, and calculate the corresponding constraint factor.

Answer: $\lambda \simeq 117°$, $R/a = 0.389$, $f = 1.261$.

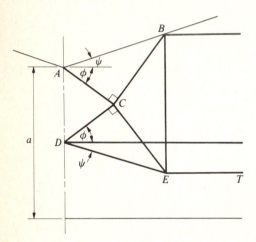

Figure R

8.28 In the pure bending of a bar containing a deep V notch, a lower bound solution is furnished by the stress discontinuity pattern of Fig. R. The triangle DCE is the mirror image of ACB with respect to the horizontal through C. Assuming the material below DET to be in a state of uniaxial compression $2k$, show that the lower bound on the constraint factor is

$$f = 2\left(\frac{1 + \sin\psi}{2 + \sin\psi}\right) \qquad \psi \leqslant \frac{\pi}{6}$$

What would be the lower bound value when $\psi \geqslant \pi/6$? Obtain a graphical plot for the lower bound f against ψ, and compare it with that given by the slipline field.

8.29 The discontinuity pattern of Fig. R may be used to obtain a lower bound solution for a curved notch by taking AB as tangential to the notch surface. Assuming a circular notch of radius c, and maximizing the constraint factor with respect to the angle ψ which AB makes with the horizontal, show that the relationship between f and c/a may be written parametrically as

$$f = \frac{2[\cos\psi - (c/a)(1 - \cos\psi)]^2}{(1 - \sin\psi)(2 + \sin\psi)} \qquad 1 + \frac{a}{c} = \frac{(1 + \sin\psi)(1 + 2\sin\psi)}{\cos\psi\,(1 - \sin\psi)}$$

Find the range of values of c/a for which this solution is valid, and modify the lower bound outside this range.

    Answer: $c/a \geqslant 0.169$, $f = 1.2(1 - 0.155c/a)^2$.

8.30 Draw the slipline field for pure bending of a bar containing a deep rectangular notch whose root is of width $2c$. The ratio c/a (where a is the minimum thickness) is such that the state of stress involves a discontinuity on the notch axis at some distance d below the notch root. Show that the constraint factor is given by

$$f = 2\left(1 - \frac{d}{a}\right)^2 + \frac{c^2 + d^2}{a^2} + 2\left(\frac{Qd - G}{ka^2}\right) \qquad \frac{a}{c} = \frac{Q}{2kc} + \frac{3d}{2c}$$

where Q and G have the same significance as that in Sec. 8.9(ii). Find the range of values of $2c/a$ for which the field is valid, and obtain the corresponding range of values of f.

    Answer: $1 \geqslant 2c/a \geqslant 0.565$, $1 \leqslant f \leqslant 1.029$.

8.31 Consider the plane strain yielding of a rigid/plastic bar with an Izod notch under pure bending. A solution to this problem may be obtained on the assumption that the notch is deep enough for the slipline field of Fig. 8.48 to apply. Using the geometry of the Izod notch, compute the dimensions R, b, and d of the slipline field for the standard value of a, and compute the constraint factor. Find also the critical thickness of the bar in pure bending.

    Answer: $R = 2.813$ mm, $b = 0.394$ mm, $d = 5.003$ mm, $f = 1.243$, $h^* = 11.10$ mm.

8.32 Figure S shows the slipline field for the bending of a symmetrically tapered cantilever under a concentrated end load kF per unit width, when the ratio l/h is increased to a critical value. Show that the depth of the cross section at C is equal to h, and that

$$\frac{2l}{h} = \cot\psi(\sin 2\psi + \cos 2\psi) \qquad \frac{F}{h} = 2\sin 2\psi$$

Discuss the solution for l/h ratios exceeding the critical value for a given angle ψ. Prove that the solution is valid for $\psi \leqslant 15°$, and construct a statically admissible stress field for $\psi \geqslant 15°$.

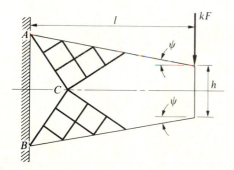
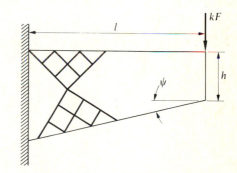

Figure S   **Figure T**

8.33 The limiting slipline field for the bending of a cantilever, whose bottom face is inclined at an angle ψ to the horizontal, is shown in Fig. T. The plastic triangles forming the field are of unequal sizes, and the field is not anchored to the bottom corner of the beam. Prove that

$$\frac{l}{h} = \frac{\tan\psi + 2(1 + \cot\psi)}{2 + \tan^2\psi} \qquad \frac{F}{h} = \frac{4\tan\psi}{2 + \tan^2\psi}$$

at the yield point, and find the range of values of ψ for which the solution is valid. Comment on the possible slipline field and the yield point load when l/h exceeds the limiting value for a given ψ.

Answer: $\psi \leqslant 30.36°$.

8.34 For sufficiently small values of the length/thickness ratio, the yield point state of a tapered cantilever, which is symmetrical about its longitudinal axis, is represented by the slipline field of Fig. U. Assuming $\psi = 10°$, show that the collapse load is given by

$$0.2725 - \left(\frac{l}{h_0} - 1.0336\right)\frac{F}{h_0} - 1.1641\frac{F^2}{h_0^2} = 0$$

where h_0 is the beam thickness at the built-in end. Find the greatest value of l/h for which the field is valid, and compute the corresponding values of F/h and b/h.

Answer: $l/h = 3.416$, $F/h = 0.685$, $b/h = 1.297$.

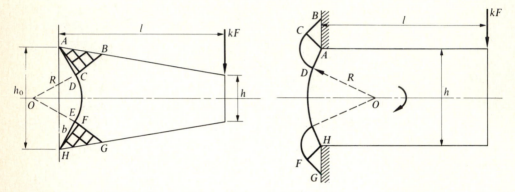

Figure U   **Figure V**

8.35 When the beam and its vertical support are made of one piece of material, the slipline field for a sufficiently short cantilever, carrying a concentrated load at the free end, is that shown in Fig. V. Assuming a uniform thickness h, obtain the relationship between F/h and l/h in the form

$$0.6950 - \left(\frac{l}{h} - 0.1863\right)\frac{F}{h} - 0.0971\frac{F^2}{h^2} = 0$$

Find the range of values of F/h and l/h for which the solution is geometrically possible and statically admissible.

Answer: $0.590 \leqslant F/h \leqslant 1$, $1.307 \geqslant l/h \geqslant 0.784$.

8.36 An upper bound solution to the preceding problem for any l/h ratio may be obtained by assuming an incipient rigid body rotation of the beam caused by sliding over a concave circular arc passing through the corners A and H. Show that the relationship between F/h and l/h may be expressed parametrically in the form

$$\frac{F}{h} = 1 - 2\lambda\cot\lambda \qquad \frac{2l}{n} = \cot\lambda + \frac{\lambda h}{F}\operatorname{cosec}^2\lambda$$

where 2λ is the angle subtended by the circular arc at its center of curvature, which is on the longitudinal axis of the beam. Compute the upper bound value of F/h when $l/h = 1$, and compare it with that given by the slipline field.

Answer: $F/h = 0.782$.

8.37 A uniform beam of length $2l$ and thickness h is strongly built-in at one end and simply supported at the other. The beam is loaded to the point of collapse by a load W per unit width at the midspan. Show that

$$\frac{W}{W_0} = 1 + 0.783\frac{F}{h} - 1.516\frac{F^2}{h^2}$$

where kF is the magnitude of the shearing force per unit width at the built-in end, given by

$$1.520\frac{F^2}{h^2} + \left(\frac{l}{h} - 0.803\right)\frac{F}{h} - 1 = 0$$

and W_0 is the collapse load according to the elementary theory. Compute the values of F/h and W/W_0 when $l/h = 4$.

Answer: $F/h = 0.276$, $W/W_0 = 1.101$.

8.38 Suppose that the beam of the preceding problem carries a load W per unit width, uniformly distributed over its length $2l$. Neglecting the effect of surface pressure on the regions of plastic deformation, show that

$$\frac{W}{W_0} = \left(\frac{\sqrt{\rho} + 1}{\sqrt{2} + 1}\right)^2 \qquad \frac{2l}{h} = (\sqrt{\rho} + \rho)\frac{h}{F}$$

at the incipient collapse, where $(\rho - 1)M_0$ and kF are the magnitudes of the bending moment and shearing force respectively per unit width at the built-in end. Assuming $l/h = 4$ find the numerical values of F/h and W/W_0.

Answer: $F/h = 0.441$, $W/W_0 = 1.024$.

8.39 Draw the slipline field for a strongly supported cantilever yielding under a terminal shear load kF and an axial thrust kN per unit width, when the stress state involves a discontinuity with the tensile plastic triangle detached from the supporting wall. Using the notation of Fig. 8.62, show that

$$\frac{F}{\sqrt{2}b} = \sin\theta - \theta(\cos\theta - \sin\theta)$$

$$\frac{N + 2h}{\sqrt{2}b} = \cos\theta + (1 + \theta)(\cos\theta + \sin\theta)$$

where b/h depends on l/h and θ according to the relationship

$$[2(1 + \theta) + \sin 2\theta]\frac{b}{h} = \left(\frac{N}{b} + \frac{2h}{b}\right) - \frac{2Fl}{bh}$$

Assuming $l/h = 8$, find the angle θ for which the upper plastic zone becomes attached to the corner of the beam, and calculate the corresponding ratios F/h and N/h.

Answer: $\theta = 13.3°$, $F/h = 0.059$, $N/h = 0.523$.

8.40 A symmetrical frame of channel section, shown in Fig. W, is subjected to shear loads kF per unit width across the end sections. For an optimum design of the frame, its horizontal and vertical members are intended to reach the yield point simultaneously when the load attains a critical value. Using the slipline fields as indicated, show that the collapse load is given by

$$\frac{F}{2h} = -\left(1 + \frac{2a}{h}\right) + \sqrt{1 + \left(1 + \frac{2a}{n}\right)^2}$$

Assuming $a/h = 2$, determine the thickness ratio h'/h that corresponds to the proposed optimum design.

Answer: $h'/h = 0.851$.

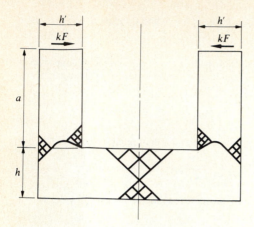

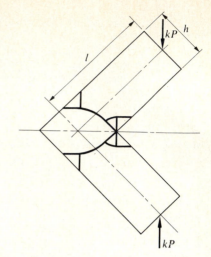

Figure W

Figure X

8.41 Figure X shows a possible slipline field for the plastic collapse of a 90° symmetrical knee-frame under equal and opposite compressive forces kP per unit width at the end sections. The field is made up of two isosceles triangles and two circular arcs on either side of the axis of symmetry. Show that the relationship between P/h and l/h is

$$0.694\frac{P^2}{h^2} + \left(1.414\frac{l}{h} - 0.524\right)\frac{P}{h} - 1.003 = 0$$

Draw a graph showing the variation of P/h with l/h over the range $1 \leqslant l/h \leqslant 5$. Find the greatest l/h ratio for which the solution is applicable.

Answer: $l/h \leqslant 7.35$.

8.42 The cross section of a cylinder of constant wall thickness h consists of two concentric squares, the length of each side of the inner square being $2a$. At the onset of plastic collapse, due to the application of an internal pressure p, sliplines emanating from the inner corners meet the external surface in a manner indicated in Fig. X. In addition, deformation occurs in each member through a pair of central triangular zones of the type shown in Fig. W. Show that

$$0.819\left(\frac{p}{k}\right)^2 + \left(0.5 - 0.371\frac{h}{a}\right)\frac{p}{k} - \left(\frac{h}{a}\right)^2 = 0$$

to a close approximation. Compute the dimensionless collapse pressure when $h/a = 0.25, 0.5,$ and 0.75.

Answer: $p/k = 0.123, 0.393, 0.705$.

TABLES ON SLIPLINE FIELDS

Table A-1 Slipline field (10° equiangular) defined by circular arcs of equal radii (Fig. 6.14)

(α^0, β^0)	x/a	y/a	$\int_0^\alpha \dfrac{x}{a}\,dt$	$\int_0^\alpha \dfrac{y}{a}\,dt$	$P/2ka$	$Q/2ka$	M/ka^2
(10, 10)	1.38332	0.00000	0.22220	0.01713	0.50000	0.76067	1.6298
(10, 20)	1.57730	0.23137	0.24971	0.05618	0.50455	0.96049	2.2247
(10, 30)	1.73236	0.50019	0.27033	0.10005	0.59173	1.18900	3.0534
(10, 40)	1.84131	0.79987	0.28323	0.14751	0.77302	1.42368	4.1338
(10, 50)	1.89806	1.12245	0.28777	0.19721	1.05457	1.63865	5.4850
(10, 60)	1.89793	1.45876	0.28357	0.24768	1.43630	1.80580	7.1114
(10, 70)	1.83785	1.79872	0.27051	0.29736	1.91132	1.89626	9.0036
(10, 80)	1.71656	2.13158	0.24872	0.34470	2.46558	1.88186	11.1382
(10, 90)	1.53472	2.44626	0.21864	0.38817	3.07791	1.73683	13.4778
(20, 20)	1.85262	0.00000	0.54940	0.07724	0.50000	1.24397	3.2065
(20, 30)	2.09558	0.28998	0.60453	0.17015	0.60710	1.57921	4.5695
(20, 40)	2.29410	0.63439	0.64393	0.27405	0.83920	1.94125	6.3825
(20, 50)	2.43655	1.02661	0.66542	0.38628	1.21358	2.29807	8.7025
(20, 60)	2.51224	1.45763	0.66731	0.50374	1.74042	2.61163	11.5740
(20, 70)	2.51193	1.91614	0.64841	0.62301	2.42109	2.83925	15.0264
(20, 80)	2.42827	2.38877	0.60814	0.74039	3.24668	2.93561	19.0680
(20, 90)	2.25626	2.86043	0.54660	0.85206	4.19690	2.85490	23.6943
(30, 30)	2.44045	0.00000	1.00076	0.19663	0.50000	2.05487	6.7079
(30, 40)	2.75042	0.37015	1.08423	0.36321	0.76609	2.59096	9.6027
(30, 50)	3.00816	0.81758	1.14022	0.54898	1.22398	3.15118	13.3881
(30, 60)	3.19622	1.33579	1.16458	0.74947	1.89971	3.68814	18.1924
(30, 70)	3.29774	1.91460	1.15386	0.95930	2.80944	4.14448	24.1329
(30, 80)	3.29717	2.54004	1.10551	1.17239	3.95646	4.45459	31.3103
(30, 90)	3.18117	3.19459	1.01807	1.38201	5.32851	4.54726	39.8045
(40, 40)	3.19044	0.00000	1.60306	0.39709	0.50000	3.35163	14.0424
(40, 50)	3.59178	0.47948	1.71618	0.66415	1.01224	4.18605	19.9700
(40, 60)	3.93022	1.06725	1.78593	0.96148	1.81912	5.03998	27.6681
(40, 70)	4.18047	1.75724	1.80519	1.28227	2.95989	5.84212	37.4273
(40, 80)	4.31724	2.53793	1.76789	1.61812	4.45967	6.50515	49.5398
(40, 90)	4.31632	3.39212	1.66929	1.95924	6.32460	6.92807	64.2927
(50, 50)	4.16156	0.00000	2.39319	0.70812	0.50000	5.37010	28.9370
(50, 60)	4.68742	0.62844	2.53785	1.11209	1.39095	6.64714	40.8364
(50, 70)	5.13587	1.40755	2.61736	1.56150	2.73267	7.93540	57.2701
(50, 80)	5.47104	2.33214	2.62031	2.04645	4.58426	9.12701	75.8899
(50, 90)	5.65607	3.38925	2.53668	2.55444	6.98430	10.08908	100.3936
(60, 60)	5.43401	0.00000	3.42154	1.16981	0.50000	8.45941	58.6747
(60, 70)	6.12961	0.83152	3.60020	1.76042	1.96931	10.39168	82.3086
(60, 80)	6.72822	1.87186	3.68369	2.41738	4.12428	12.32054	113.0190
(60, 90)	7.17956	3.11739	3.65436	3.12667	7.05361	14.08301	152.2482
(70, 70)	7.11722	0.00000	4.75666	1.83668	0.50000	13.13398	117.5106
(70, 80)	8.04448	1.10870	4.97238	2.68222	2.84634	16.03367	164.1977
(70, 90)	8.84843	2.50624	5.05131	3.62287	6.22928	18.90505	225.0838
(80, 80)	9.36106	0.00000	6.49179	2.78427	0.50000	20.14922	233.4019
(80, 90)	10.60500	1.48762	6.74819	3.97780	4.16776	24.47367	325.4187
(90, 90)	12.37126	0.00000	8.75376	4.11483	0.50000	30.61170	461.3014

(*continued*)

Table A-1 (*Continued*)

(α^0, β^0)	x/a	y/a	$\int_0^\alpha \frac{x}{a} dt$	$\int_0^\alpha \frac{y}{a} dt$	$P/2ka$	$Q/2ka$
(10, 100)	1.29502	2.73171	0.18094	0.42630	3.72038	1.43947
(10, 110)	1.00217	2.99726	0.13658	0.45777	4.35902	0.97372
(10, 120)	0.66203	3.17302	0.08674	0.48140	4.95496	0.33066
(10, 130)	0.28553	3.31019	0.03281	0.49625	5.46591	−0.49023
(20, 100)	1.99362	3.31467	0.46455	0.95414	5.23957	2.55253
(20, 110)	1.64111	3.73423	0.36347	1.04287	6.33053	1.99299
(20, 120)	1.20279	4.10157	0.24551	1.11470	7.41432	1.14268
(20, 130)	0.68609	4.39947	0.11351	1.16645	8.42545	−0.01708
(30, 100)	2.93927	3.85743	0.89123	1.58103	6.89550	4.34905
(30, 110)	2.56471	4.50493	0.72601	0.76210	8.60800	3.78830
(30, 120)	2.05507	5.11128	0.52477	1.91789	10.39655	2.79975
(30, 130)	1.41283	5.64930	0.29127	2.04135	12.17195	1.32945
(40, 100)	4.15581	4.29688	1.50633	2.29465	8.53719	6.99978
(40, 110)	3.81735	5.22382	1.27788	2.61240	11.05235	6.60401
(40, 120)	3.28742	6.13958	0.98499	2.90000	13.79425	5.62541
(40, 130)	2.55846	7.00658	0.63106	3.14474	16.65451	3.95676
(50, 100)	5.65465	4.55761	2.35834	3.07055	9.94323	10.66669
(50, 110)	5.43276	5.80729	2.07965	3.57771	13.43582	10.68783
(50, 120)	4.96062	7.09980	1.69797	4.05709	17.39435	9.97060
(50, 130)	4.21463	8.38852	1.21406	4.48860	21.70310	8.33232
(60, 100)	7.43079	4.55369	3.49635	3.87017	10.81620	15.47748
(60, 110)	7.42864	6.15439	3.19645	4.62581	15.42909	16.26758
(60, 120)	7.12198	7.88039	2.74501	5.36790	20.85566	16.18914
(60, 130)	6.46450	9.67964	2.13681	6.06766	26.99456	14.96047
(70, 100)	9.45897	4.19164	4.96688	4.63926	10.78175	21.50263
(70, 110)	9.80113	6.14905	4.69470	5.70552	16.59299	23.52194
(70, 120)	9.79797	8.34567	4.21400	6.78956	23.69032	24.60446
(70, 130)	9.37403	10.73037	3.50902	7.85351	32.02063	24.34732
(80, 100)	11.69017	3.37440	6.81001	5.30685	9.39390	28.72855
(80, 110)	12.51924	5.66370	6.63786	6.74432	16.37699	32.54568
(80, 120)	12.98655	8.33838	6.19504	8.25363	25.25168	35.45737
(80, 130)	12.98195	11.35760	5.44967	9.78869	36.06112	36.95285
(90, 100)	14.04879	2.00646	9.05499	5.78509	6.14764	37.02934
(90, 110)	15.51978	4.56449	9.08124	7.64681	14.12631	43.31076
(90, 120)	16.64932	7.68420	8.77483	9.66259	24.73173	48.90600
(90, 130)	17.28909	11.34762	8.08135	11.78104	38.16630	53.13019

Table A-2 Slipline field (15° equiangular) defined by circular arcs of equal radii (Fig. 6.14)

$(\alpha°, \beta°)$	x/a	y/a	$\int_0^\alpha \frac{x}{a} dt$	$\int_0^\alpha \frac{y}{a} dt$	$P/2ka$	$Q/2ka$	$-R/\sqrt{2}\,a$
(15, 15)	1.60541	0.00000	0.37164	0.04090	0.50000	0.96895	1.34060
(15, 30)	1.91528	0.40491	0.42950	0.13968	0.61237	1.37230	1.70216
(15, 45)	2.12146	0.90445	0.46096	0.25365	0.96882	1.80840	2.08539
(15, 60)	2.19542	1.47161	0.46203	0.37572	1.60112	2.17596	2.49098
(15, 75)	2.11580	2.07087	0.43064	0.49775	2.30406	2.35742	2.91967
(15, 90)	1.87091	2.66013	0.36688	0.61106	3.62855	2.23167	3.37218
(15, 105)	1.46073	3.19343	0.27322	0.70704	4.87903	1.68941	3.84928
(15, 120)	0.89812	3.62413	0.15438	0.77778	6.11605	0.64970	4.35174
(15, 135)	0.20913	3.90864	0.01720	0.81665	7.16450	0.92401	4.88033
(30, 30)	2.44045	0.00000	1.00076	0.19663	0.50000	2.05487	1.89226
(30, 45)	2.88691	0.58452	1.11593	0.45396	0.96924	2.87076	2.41384
(30, 60)	3.19622	1.33579	1.16458	0.74947	1.89971	3.68814	2.98997
(30, 75)	3.31111	2.22250	1.13451	1.06585	3.35366	4.32217	3.62402
(30, 90)	3.18117	3.19459	1.01807	1.38201	5.32851	4.54726	4.31949
(30, 105)	2.76883	4.18467	0.81330	1.67427	7.73744	4.11828	5.08003
(30, 120)	2.05507	5.11128	0.52477	1.91789	10.39655	2.79975	5.90946
(30, 135)	1.04421	5.88403	0.16397	2.08889	13.02181	0.40114	6.81175
(45, 45)	3.64393	0.00000	1.97186	0.53657	0.50000	4.25200	2.77261
(45, 60)	4.30787	0.87046	2.14535	1.04635	1.65558	5.81401	3.54894
(45, 75)	4.78130	2.02238	2.18847	1.63077	3.68655	7.31428	4.43210
(45, 90)	4.96154	3.41994	2.07375	2.25687	6.71115	8.40663	5.43136
(45, 105)	4.75043	4.99259	1.78192	2.88235	10.73082	8.65837	6.55657
(45, 120)	4.06556	6.63459	1.30500	3.45755	15.58708	7.58567	7.81819
(45, 135)	2.85214	8.20819	0.64880	3.92844	20.92823	4.70737	9.22731
(60, 60)	5.43401	0.00000	3.42154	1.16981	0.50000	8.45942	4.17299
(60, 75)	6.44410	1.32557	3.65497	2.08138	2.95498	11.36579	5.35480
(60, 90)	7.17956	3.11739	3.65436	3.12667	7.05362	14.08300	6.72829
(60, 105)	7.46466	5.33563	3.36483	4.24800	13.01725	15.96404	8.31425
(60, 120)	7.12198	7.88039	2.74501	5.36790	20.85566	16.18914	10.13510
(60, 135)	5.99181	10.58699	1.77370	6.39201	30.27937	13.82527	12.21503
(75, 75)	8.15773	0.00000	5.56721	2.26986	0.50000	16.28967	6.40387
(75, 90)	9.72015	2.05192	5.86139	3.82252	5.37585	21.60466	8.23274
(75, 105)	10.87560	4.86968	5.75159	5.60559	13.29815	26.48027	10.39236
(75, 120)	11.32964	8.41166	5.13494	7.52168	24.67692	29.71845	12.92463
(75, 135)	10.77297	12.53548	3.92972	9.43481	39.54042	29.78754	15.87547
(90, 90)	12.37126	0.00000	8.75376	4.11483	0.50000	30.61170	9.96930
(90, 105)	14.81817	3.21530	9.10614	6.69403	9.82539	40.21826	12.83452
(90, 120)	16.64932	7.68420	8.77483	9.66259	24.73173	48.90601	16.25925
(90, 135)	17.37652	13.36870	7.57409	12.85879	45.97204	54.47900	20.32313
(105, 105)	18.97064	0.00000	13.52919	7.15272	0.50000	56.50182	15.68904
(105, 120)	22.83928	5.08568	13.92919	11.38185	17.91480	73.71759	20.22026
(105, 135)	25.76136	12.22120	13.18442	16.26217	45.45436	89.11389	25.68809
(120, 120)	29.40511	0.00000	20.76709	12.10816	0.50000	102.90486	24.89870
(120, 135)	35.56721	8.10339	21.18790	18.99868	32.49360	133.55544	32.11773
(135, 135)	46.02632	0.00000	31.86576	20.15697	0.50000	185.52815	39.77868

Table A-3 Some mathematical functions associated with slipline fields (10° equiangular net)

$(\alpha°, \beta°)$	$I_0(2\sqrt{\alpha\beta})$	$I_1(2\sqrt{\alpha\beta})$	$F_1(\alpha, \beta)$	$F_2(\alpha, \beta)$	$L(\alpha, \beta)$	$N(\alpha, \beta)$
(10, 10)	1.030694	0.177205	0.176313	0.015347	0.015308	0.000890
(10, 20)	1.061858	0.254422	0.179005	0.015503	0.015464	0.000897
(10, 30)	1.093494	0.316325	0.181725	0.015661	0.015622	0.000904
(10, 40)	1.125609	0.370768	0.184472	0.015819	0.015780	0.000911
(10, 50)	1.158207	0.420752	0.187247	0.015979	0.015939	0.000918
(10, 60)	1.191293	0.467794	0.190050	0.016140	0.016100	0.000925
(10, 70)	1.224872	0.512784	0.192881	0.016302	0.016262	0.000932
(10, 80)	1.258949	0.556297	0.195741	0.016466	0.016425	0.000939
(10, 90)	1.293528	0.598728	0.198629	0.016631	0.016590	0.000946
(20, 20)	1.125609	0.370768	0.363505	0.062805	0.062176	0.007219
(20, 30)	1.191293	0.467794	0.374578	0.064082	0.063446	0.007330
(20, 40)	1.258949	0.556297	0.385875	0.065379	0.064735	0.007442
(20, 50)	1.328616	0.640362	0.397400	0.066695	0.066043	0.007555
(20, 60)	1.400334	0.722044	0.409157	0.068032	0.067372	0.007669
(20, 70)	1.474145	0.802549	0.421149	0.069386	0.068720	0.007785
(20, 80)	1.550091	0.882657	0.433380	0.070766	0.070089	0.007903
(20, 90)	1.628212	0.962911	0.445852	0.072163	0.071479	0.008021
(30, 30)	1.293528	0.598728	0.573459	0.146764	0.143516	0.024930
(30, 40)	1.400334	0.722044	0.599462	0.151225	0.147917	0.025502
(30, 50)	1.511843	0.842615	0.626252	0.155787	0.152418	0.026085
(30, 60)	1.628212	0.962911	0.653848	0.160453	0.157023	0.026678
(30, 70)	1.749569	1.084404	0.682267	0.165224	0.161731	0.027282
(30, 80)	1.876067	1.208073	0.711528	0.170102	0.166547	0.027897
(30, 90)	2.007856	1.334622	0.741650	0.175090	0.171470	0.028523
(40, 40)	1.550091	0.882657	0.820172	0.275045	0.264483	0.061021
(40, 50)	1.708553	1.043706	0.869138	0.286116	0.275295	0.062886
(40, 60)	1.876067	1.208073	0.920065	0.297521	0.286437	0.064798
(40, 70)	2.052987	1.377542	0.973012	0.309269	0.297916	0.066756
(40, 80)	2.239681	1.553356	1.028041	0.321369	0.309742	0.068761
(40, 90)	2.436525	1.736469	1.085212	0.333828	0.321921	0.070815
(50, 50)	1.919401	1.249907	1.121081	0.459701	0.432953	0.124209
(50, 60)	2.145089	1.464591	1.203301	0.482606	0.455043	0.128947
(50, 70)	2.386341	1.689962	1.289593	0.506373	0.477973	0.133831
(50, 80)	2.643907	1.927660	1.380110	0.531026	0.501768	0.138862
(50, 90)	2.918564	2.179034	1.475009	0.556591	0.526454	0.144045
(60, 60)	2.436525	1.736469	1.498741	0.718262	0.660265	0.225772
(60, 70)	2.751674	2.026500	1.627776	0.760679	0.700564	0.236078
(60, 80)	3.091896	2.336920	1.764379	0.804992	0.742694	0.246760
(60, 90)	3.458608	2.669688	1.908886	0.851269	0.786722	0.257828
(70, 70)	3.151136	2.390779	1.983040	1.075568	0.962297	0.380655
(70, 80)	3.586986	2.785893	2.177079	1.148565	1.030463	0.400842
(70, 90)	4.061601	3.214701	2.384184	1.225329	1.102267	0.421883
(80, 80)	4.132711	3.278872	2.614174	1.566356	1.360992	0.608956
(80, 90)	4.732810	3.819994	2.897875	1.685740	1.470452	0.645659
(90, 90)	5.477845	4.491456	3.446708	2.238922	1.886494	0.937955

(Continued)

Table A-3 (*Continued*)

$(\alpha°, \beta°)$	$I_0(2\sqrt{\alpha\beta})$	$I_1(2\sqrt{\alpha\beta})$	$F_1(\alpha, \beta)$	$F_2(\alpha, \beta)$	$L(\alpha, \beta)$	$N(\alpha, \beta)$
(10, 100)	1.328616	0.640362	0.201546	0.016796	0.016755	0.000953
(10, 110)	1.364216	0.681412	0.204492	0.016964	0.016922	0.000960
(10, 120)	1.400334	0.722044	0.207467	0.017132	0.017091	0.000967
(10, 130)	1.436976	0.762388	0.210472	0.017302	0.017260	0.000974
(20, 100)	1.708553	1.043706	0.458570	0.073582	0.072890	0.008141
(20, 110)	1.791156	1.125346	0.471538	0.075022	0.074321	0.008263
(20, 120)	1.876067	1.208073	0.484759	0.076483	0.075774	0.008386
(20, 130)	1.963329	1.292083	0.498237	0.077966	0.077248	0.008511
(30, 100)	2.145089	1.464591	0.772652	0.180189	0.176504	0.029161
(30, 110)	2.287926	1.598418	0.804554	0.185401	0.181650	0.029810
(30, 120)	2.436525	1.736469	0.837376	0.190728	0.186910	0.030470
(30, 130)	2.591052	1.879067	0.871138	0.196173	0.192287	0.031143
(40, 100)	2.643907	1.927660	1.144590	0.346656	0.334464	0.072918
(40, 110)	2.862227	2.127607	1.206240	0.359861	0.347377	0.075071
(40, 120)	3.091896	2.336920	1.270228	0.373452	0.360671	0.077275
(40, 130)	3.333337	2.556168	1.336622	0.387438	0.374355	0.079532
(50, 100)	3.211118	2.445268	1.574451	0.583097	0.552058	0.149384
(50, 110)	3.522405	2.727451	1.678604	0.610571	0.578608	0.154881
(50, 120)	3.853290	3.026622	1.787639	0.639042	0.606131	0.160542
(50, 130)	4.204671	3.343795	1.901732	0.668539	0.634657	0.169349
(60, 100)	3.853290	3.026622	2.061646	0.899580	0.832716	0.269295
(60, 110)	4.277487	3.409478	2.223021	0.949997	0.880747	0.281172
(60, 120)	4.732810	3.819994	2.393384	1.002595	0.930888	0.293472
(60, 130)	5.220936	4.259915	2.573124	1.057451	0.983124	0.306207
(70, 100)	4.577475	3.679975	2.605027	1.306017	1.177799	0.443809
(70, 110)	5.137230	4.184478	2.840308	1.390792	1.257238	0.466650
(70, 120)	5.743616	4.731017	3.090753	1.479822	1.340746	0.490439
(70, 130)	6.399521	5.322469	3.357121	1.573280	1.428491	0.515208
(80, 100)	5.391235	4.413396	3.203014	1.812083	1.586472	0.684128
(80, 110)	6.112165	5.063302	3.530835	1.945713	1.709365	0.724436
(80, 120)	6.900012	5.774074	3.882639	2.086971	1.839459	0.766654
(80, 130)	7.759435	6.550255	4.259789	2.236212	1.977094	0.810860
(90, 100)	6.302675	5.235114	3.853727	2.427148	2.055662	1.001117
(90, 110)	7.213649	6.057226	4.294650	2.627550	2.236148	1.067679
(90, 120)	8.217512	6.964348	4.771660	2.840770	2.428559	1.137794
(90, 130)	9.321431	7.963378	5.287051	3.067476	2.633529	1.211621

Table A-4 Some mathematical functions associated with slipline fields (15° equiangular net)

$(\alpha°, \beta°)$	$I_0(2\sqrt{\alpha\beta})$	$I_1(2\sqrt{\alpha\beta})$	$F_1(\alpha, \beta)$	$F_2(\alpha, \beta)$	$L(\alpha, \beta)$	$N(\alpha, \beta)$
(15, 15)	1.069722	0.270874	0.267842	0.034861	0.034664	0.003021
(15, 30)	1.141848	0.396203	0.277073	0.035662	0.035461	0.003074
(15, 45)	1.216431	0.501694	0.286516	0.036476	0.036273	0.003126
(15, 60)	1.293528	0.598728	0.296173	0.037304	0.037098	0.003180
(15, 75)	1.373197	0.691604	0.306050	0.038147	0.037938	0.003234
(15, 90)	1.455494	0.782485	0.316148	0.039003	0.038792	0.003289
(15, 105)	1.540479	0.872645	0.326473	0.039875	0.039660	0.003345
(15, 120)	1.628212	0.962911	0.377028	0.040761	0.040543	0.003401
(15, 135)	1.718754	1.053859	0.347816	0.041661	0.041441	0.003459
(30, 30)	1.293528	0.598728	0.573459	0.146764	0.143516	0.024930
(30, 45)	1.455494	0.782485	0.612758	0.153493	0.150155	0.025792
(30, 60)	1.628212	0.962911	0.653848	0.160453	0.157023	0.026678
(30, 75)	1.812166	1.145918	0.696791	0.167649	0.164126	0.027588
(30, 90)	2.007856	1.334622	0.741650	0.175090	0.171470	0.028523
(30, 105)	2.215797	1.530998	0.788490	0.182781	0.179063	0.029484
(30, 120)	2.436525	1.736469	0.837376	0.190728	0.186910	0.030470
(30, 135)	2.670591	1.952162	0.888378	0.198940	0.195019	0.031483
(45, 45)	1.718754	1.053859	0.962585	0.359377	0.342163	0.088595
(45, 60)	2.007856	1.334622	1.059851	0.383882	0.365955	0.093193
(45, 75)	2.324529	1.632581	1.163695	0.409641	0.390976	0.097982
(45, 90)	2.670591	1.952162	1.274453	0.436705	0.417276	0.102968
(45, 105)	3.047949	2.296934	1.392472	0.465126	0.444908	0.108158
(45, 120)	3.458608	2.669688	1.518115	0.494960	0.473925	0.113557
(45, 135)	3.904669	3.073026	1.651761	0.526264	0.504385	0.119174
(60, 60)	2.436525	1.736469	1.498741	0.718262	0.660265	0.225772
(60, 75)	2.918564	2.179034	1.695111	0.782595	0.721396	0.241371
(60, 90)	3.458608	2.669688	1.908886	0.851269	0.786722	0.257828
(60, 105)	4.061601	3.214701	2.141233	0.924520	0.856473	0.275182
(60, 120)	4.732810	3.819994	2.393384	1.002595	0.930888	0.293472
(60, 135)	5.477845	4.491456	2.666634	1.085750	1.010220	0.312742
(75, 75)	3.603255	2.800611	2.277236	1.301627	1.147899	0.484119
(75, 90)	4.388336	3.509446	2.636430	1.444326	1.280029	0.525736
(75, 105)	5.284345	4.317061	3.034519	1.598892	1.423423	0.570198
(75, 120)	6.302675	5.235114	3.474713	1.766121	1.578846	0.617667
(75, 135)	7.455634	6.275774	3.960436	1.946854	1.747109	0.668313
(90, 90)	5.477845	4.491456	3.446708	2.238922	1.886494	0.937955
(90, 105)	6.746982	5.635961	4.069817	2.525787	2.144453	1.033963
(90, 120)	8.217512	6.964348	4.771660	2.840770	2.428559	1.137794
(90, 135)	9.913270	8.499616	5.559888	3.186103	2.740929	1.249978
(105, 105)	8.483823	7.205212	5.243406	3.741912	3.006923	1.705405
(105, 120)	10.533010	9.061593	6.292744	4.284116	3.479366	1.906840
(105, 135)	12.936870	11.245519	7.491859	4.887019	4.006998	2.127252
(120, 120)	13.311905	11.586790	8.046604	6.155953	4.719110	2.976105
(120, 135)	16.624721	14.606950	9.784528	7.140300	5.549216	3.372169
(135, 135)	21.088695	18.689985	12.468811	10.044348	7.360324	5.056309

Table A-5 Field defined by identical logarithmic spirals (Fig. 6.15)

$(\alpha°, \beta°)$	x/b	y/b	$-R/\sqrt{2}\,b$	$-S/\sqrt{2}\,b$	$G(\alpha, \beta)$	$H(\alpha, \beta)$
(10, 10)	0.35263	0.00000	1.03069	1.03069	0.16267	0.19352
(10, 20)	0.50848	0.17798	1.20089	0.92283	0.16522	0.19639
(10, 30)	0.61576	0.37087	1.35388	0.83311	0.16780	0.19928
(10, 40)	0.68301	0.55487	1.49258	0.75864	0.17040	0.20220
(10, 50)	0.71435	0.73129	1.61944	0.69698	0.17303	0.20515
(10, 60)	0.71453	0.89663	1.73650	0.64609	0.17568	0.20813
(10, 70)	0.68794	1.04826	1.84550	0.60425	0.17837	0.21113
(10, 80)	0.63861	1.13494	1.94786	0.57003	0.18108	0.21417
(20, 20)	0.72701	0.00000	1.12561	1.12561	0.31405	0.44219
(20, 30)	0.90062	0.20656	1.31665	1.06594	0.32405	0.45478
(20, 40)	1.02915	0.42887	1.49846	1.01944	0.33426	0.46762
(20, 50)	1.12886	0.63435	1.64224	1.01499	0.34468	0.48070
(20, 60)	1.15529	0.89624	1.84251	0.95815	0.35532	0.49404
(20, 70)	1.15536	1.13008	2.00807	0.94022	0.36617	0.50764
(20, 80)	1.11541	1.35687	2.17111	0.92907	0.37724	0.52150
(30, 30)	1.14692	0.00000	1.29353	1.29353	0.47130	0.77806
(30, 40)	1.35089	0.24303	1.51817	1.28250	0.49389	0.80981
(30, 50)	1.50884	0.51659	1.74181	1.28189	0.51718	0.84247
(30, 60)	1.61718	0.81435	1.96615	1.29027	0.54120	0.87607
(30, 70)	1.67268	1.12942	2.19265	1.30648	0.56595	0.91061
(30, 80)	1.67259	1.45420	2.42260	1.32954	0.59145	0.94613
(40, 40)	1.64034	0.00000	1.55009	1.55009	0.65129	1.24500
(40, 50)	1.88910	0.29670	1.50729	1.59289	0.69250	1.30941
(40, 60)	2.08842	0.64230	2.10668	1.64545	0.73543	1.37623
(40, 70)	2.22947	1.03049	2.39910	1.70687	0.78012	1.44551
(40, 80)	2.30375	1.45190	2.70296	1.77640	0.82663	1.51734
(50, 50)	2.24216	0.00000	1.91940	1.91940	0.87241	1.90419
(50, 60)	2.55424	0.37328	2.26362	2.02660	0.93998	2.02096
(50, 70)	2.81097	0.81797	2.62748	2.14520	1.01103	2.14301
(50, 80)	2.99678	1.32986	3.01304	2.27478	1.08568	2.27052
(60, 60)	2.99748	0.00000	2.43653	2.43653	1.15669	2.84200
(60, 70)	3.39779	0.47813	2.87802	2.62533	1.26095	3.04001
(60, 80)	3.73320	1.06053	3.35386	2.82994	1.37154	3.24840
(70, 70)	3.96608	0.00000	3.15114	3.15114	1.53214	4.18084
(70, 80)	4.48774	0.62335	3.72662	3.44735	1.68720	4.50248
(80, 80)	5.22835	0.00000	4.13271	4.13271	2.03572	6.09475

Table A-6 Field defined by circular arcs of unequal radii (Fig. 6.16)

γ = 0

(α°, β°)	x/a	y/a	P/2ka	Q/2ka
(15, 0)	0.00000	0.00000	0.00000	1.00000
(15, 15)	0.53569	-0.06972	-0.29226	1.17669
(15, 30)	1.10787	0.00644	-0.00644	-0.52513
(15, 45)	1.67622	0.24277	-0.58917	2.22802
(15, 60)	2.19542	0.64228	-0.37461	3.03787
(15, 75)	2.61842	1.19504	0.21295	3.92072
(15, 90)	2.90010	1.87752	1.23596	4.73775
(30, 0)	0.00000	0.00000	0.00000	1.00000
(30, 15)	0.51680	-0.21232	-0.39503	1.23480
(30, 30)	1.14692	-0.29353	-0.78112	1.77375
(30, 45)	1.85772	-0.19809	-1.01977	2.66016
(30, 60)	2.60026	0.11161	-0.94010	3.88423
(30, 75)	3.31111	0.65971	-0.35750	5.36645
(30, 90)	3.91609	1.45181	0.90053	6.94728
(45, 0)	0.00000	0.00000	0.00000	1.00000
(45, 15)	0.45978	-0.34969	-0.54917	1.25036
(45, 30)	1.10430	-0.61371	-1.17984	1.88174
(45, 45)	1.92517	-0.71875	-1.73723	3.01478
(45, 60)	2.88811	-0.58866	-1.99136	4.72096
(45, 75)	3.93047	-0.15299	-1.64978	6.98632
(45, 90)	4.96154	0.64316	-0.38536	9.67740
(60, 0)	0.00000	0.00000	0.00000	1.00000
(60, 15)	0.36610	-0.47161	-0.73922	1.20022
(60, 30)	0.97205	-0.93176	-1.70361	1.84833
(60, 45)	1.84876	-1.29022	-2.74860	3.16706
(60, 60)	2.99748	-1.43653	-3.60430	5.35509
(60, 75)	4.38036	-1.24892	-3.86964	8.53059
(60, 90)	5.91362	-0.60716	-3.02902	12.66685
(75, 0)	0.00000	0.00000	0.00000	1.00000
(75, 15)	0.23998	-0.56824	-0.94076	1.06631
(75, 30)	0.74832	-1.22250	-2.30940	1.61106
(75, 45)	1.60592	-1.87321	-4.01449	2.97792
(75, 60)	2.86960	-2.38931	-5.79017	5.54115
(75, 75)	4.55447	-2.60326	-7.15606	9.63359
(75, 90)	6.61712	-2.32267	-7.40313	15.45114
(90, 0)	0.00000	0.00000	0.00000	1.00000
(90, 15)	0.08830	-0.63094	-1.12283	0.83392
(90, 30)	0.43438	-1.45967	-2.92853	1.11929
(90, 45)	1.18475	-2.41994	-5.44035	2.31010
(90, 60)	2.45502	-3.38333	-8.46974	5.00032
(90, 75)	4.34494	-4.15494	-11.52931	9.82565
(90, 90)	6.89342	-4.47785	-13.76240	17.34921

γ = 15°

(α°, β°)	x/a	y/a	P/2ka	Q/2ka
(15, 0)	0.51764	-0.13397	0.10566	1.00154
(15, 15)	1.07055	-0.06038	-0.00311	1.25583
(15, 30)	1.61975	0.16780	-0.12478	1.72892
(15, 45)	2.12146	0.55404	-0.01851	2.38013
(15, 60)	2.53021	1.08818	0.40941	3.12623
(15, 75)	2.80240	1.74766	1.22530	3.84447
(15, 90)	2.90010	2.49696	2.45301	4.38121
(30, 0)	0.50000	-0.26795	0.17985	1.19481
(30, 15)	1.10786	-0.34629	-0.17210	1.42162
(30, 30)	1.79368	-0.25420	-0.42647	2.09732
(30, 45)	2.51020	0.04466	-0.42782	3.09092
(60, 0)	0.36603	-0.50000	0.22116	1.14486
(60, 15)	0.94596	-0.94036	-0.87402	1.53497
(60, 30)	1.78615	-1.28387	-1.78549	2.57865
(60, 45)	2.88810	-1.42419	-2.56956	4.38915
(60, 60)	4.21573	-1.24406	-2.87142	7.09025
(60, 75)	5.68870	-0.62752	-2.22520	10.67829
(60, 90)	7.17955	0.52480	-0.09213	14.96097
(75, 0)	0.25882	-0.58226	0.25323	1.15793
(75, 15)	0.74260	-1.20484	-1.35561	1.37248
(75, 30)	1.56068	-1.82548	-2.81960	2.46235
(75, 45)	2.76818	-2.31861	-4.38714	4.59551

γ = 15°

(α°, β°)	x/a	y/a	P/2ka	Q/2ka
(30, 60)	3.19622	0.57361	-0.00403	4.34178
(30, 75)	3.78013	1.33811	1.01001	5.71461
(30, 90)	5.18510	2.32176	2.74336	6.99683
(45, 0)	0.44829	-0.39279	0.20810	1.11280
(45, 15)	1.06800	-0.64664	-0.46731	1.53345
(45, 30)	1.85772	-0.74768	-0.97710	2.41680
(45, 45)	2.78454	-0.62245	-1.25448	3.81639
(45, 60)	3.78819	-0.20295	-1.03537	5.73488
(45, 75)	4.78129	0.56387	-0.01825	8.06687
(45, 90)	5.65272	1.70637	2.09671	10.57332
(75, 60)	4.38034	-2.52328	-5.64044	8.08936
(75, 75)	6.35608	-2.25448	-5.94732	13.14459
(75, 90)	8.58433	-1.32087	-4.47853	19.74038
(90, 0)	0.13397	-0.63397	0.30211	1.15110
(90, 15)	0.46501	-1.41740	-1.85502	1.00451
(90, 30)	1.17332	-2.32856	-3.99966	1.94970
(90, 45)	2.38194	-3.24510	-6.64192	4.19304
(90, 60)	4.18430	-3.98066	-9.36609	8.32158
(90, 75)	6.61710	-4.28891	-11.41745	14.86854
(90, 90)	9.63232	-3.87792	-11.66858	24.16340

γ = 30°

(α°, β°)	x/a	y/a	P/2ka	Q/2ka
(15, 0)	0.96593	-0.24118	0.51858	0.87635
(15, 15)	1.46209	-0.03486	0.28688	1.17980
(15, 30)	1.91527	0.31385	0.29157	1.64937
(15, 45)	2.28442	0.79623	0.55175	2.22654
(15, 60)	2.53021	1.39173	1.13257	2.81075
(15, 75)	2.61842	2.06824	2.06390	3.27227
(15, 90)	2.52339	2.78342	3.32876	3.46393
(30, 0)	1.00000	-0.50000	0.36603	1.02360
(30, 15)	1.61974	-0.41678	0.11437	1.47195
(30, 30)	2.26714	-0.14676	0.04245	2.20022
(30, 45)	2.88691	0.33111	0.30021	3.16801
(30, 60)	3.41436	1.02169	1.03541	4.27001
(30, 75)	3.78012	1.91013	2.36838	5.33237
(30, 90)	3.91609	2.95966	4.36411	6.11838
(45, 0)	0.96593	-0.75882	0.24755	1.13517
(45, 15)	1.67621	-0.84969	-0.25488	1.73769
(45, 30)	2.51020	-0.73700	-0.53399	2.77881
(45, 45)	2.31364	-0.35938	-0.43561	4.26924
(45, 60)	4.30786	0.33111	0.29950	6.13496
(45, 75)	5.09271	1.36011	1.93497	8.18691
(45, 90)	5.65272	2.72274	4.69419	10.10818
(60, 0)	0.86603	-1.00000	0.17121	1.20345
(60, 15)	1.61554	-1.30639	-0.82347	1.89193
(60, 30)	2.60025	-1.43175	-1.49124	3.25207
(60, 45)	3.78819	-1.27055	-1.80346	5.35800
(60, 60)	5.10756	-0.71826	-1.36914	8.22650
(60, 75)	6.44408	0.31481	0.27278	11.71603
(60, 90)	7.64291	1.88807	3.59836	15.47931
(75, 0)	0.70711	-1.20711	0.14223	1.22379
(75, 15)	1.42920	-1.75484	-1.56180	1.84506
(75, 30)	2.49873	-2.19151	-2.83582	3.45470
(75, 45)	3.93045	-2.37320	-3.90108	6.18271
(75, 60)	5.68869	-2.13393	-4.23667	10.22130
(75, 75)	7.67496	-1.30163	-3.14502	15.58088
(75, 90)	9.72011	0.28034	0.19889	21.99270
(90, 0)	0.50000	-1.36603	0.16258	1.19481
(90, 15)	1.11749	-2.16043	-2.40670	1.51481
(90, 30)	2.17724	-2.96359	-4.51208	3.20442
(90, 45)	3.76433	-3.61113	-6.74104	6.42451
(90, 60)	5.91360	-3.88332	-8.48174	11.64602
(90, 75)	8.58433	-3.51915	-8.79912	19.17994
(90, 90)	11.63734	-2.23892	-6.44819	29.01780

Table A-7 Unsymmetrical field defined by circular arcs of equal radii (Fig. 6.17)

$(\alpha°, \beta°)$	x/a	y/a	$R/\sqrt{2}a$	$-S/\sqrt{2}a$	$F_1^*(\alpha, \beta)$	$F_2^*(\alpha, \beta)$
(0, 0)	1.00000	0.00000	1.00000	1.00000	0.00000	0.00000
(0, 10)	0.81116	0.15846	0.82547	1.00000	0.00000	0.00000
(0, 20)	0.59767	0.28171	0.65093	1.00000	0.00000	0.00000
(0, 30)	0.36603	0.36603	0.47640	1.00000	0.00000	0.00000
(0, 40)	0.12326	0.40883	0.30187	1.00000	0.00000	0.00000
(0, 50)	−0.12326	0.40883	0.12734	1.00000	0.00000	0.00000
(10, 0)	1.15846	0.18884	1.00000	1.17453	0.17365	0.01519
(10, 10)	0.91129	0.33168	0.79788	1.14166	0.17101	0.01504
(10, 20)	0.65062	0.42668	0.60146	1.10927	0.16840	0.01489
(10, 30)	0.38523	0.47359	0.41065	1.07736	0.16581	0.01473
(10, 40)	0.12353	0.47370	0.22536	1.04593	0.16326	0.01458
(10, 50)	−0.12664	0.42970	0.04552	1.01497	0.16072	0.01444
(20, 0)	1.28171	0.40233	1.00000	1.34907	0.34202	0.06031
(20, 10)	0.97747	0.51332	0.77073	1.27854	0.33160	0.05909
(20, 20)	0.67551	0.56682	0.55359	1.21004	0.32139	0.05789
(20, 30)	0.38546	0.56706	0.34823	1.14354	0.31139	0.05670
(20, 40)	0.11568	0.51972	0.15431	1.17898	0.30160	0.05554
(30, 0)	1.36603	0.63397	1.00000	1.52360	0.50000	0.13397
(30, 10)	1.01002	0.69705	0.74400	1.41072	0.47697	0.12990
(30, 20)	0.67567	0.69755	0.50729	1.30260	0.45465	0.12592
(30, 30)	0.37204	0.64439	0.28904	1.19910	0.43301	0.12204
(30, 40)	0.10570	0.54780	0.08846	1.10009	0.41205	0.11824
(40, 0)	1.40883	0.87674	1.00000	1.69813	0.64279	0.23396
(40, 10)	1.01011	0.87732	0.71770	1.53827	0.60279	0.22444
(40, 20)	0.65508	0.81525	0.46253	1.38721	0.46253	0.21522
(40, 30)	0.35024	0.70488	0.23298	1.24461	0.52765	0.20628
(40, 40)	0.09879	0.56024	0.02761	1.11015	0.49240	0.19762
(50, 0)	1.40883	1.12325	1.00000	1.87266	0.76604	0.35721
(50, 10)	0.98003	1.04842	0.69182	1.66127	0.70530	0.33896
(50, 20)	0.68396	0.89344	0.41928	1.46414	0.64769	0.32140
(50, 30)	0.32499	0.74898	0.17995	1.28061	0.59307	0.30452
(60, 0)	1.36603	1.36602	1.00000	2.04720	0.86603	0.50000
(60, 10)	0.92285	1.20575	0.66635	1.77978	0.78147	0.46910
(60, 20)	0.56915	1.01085	0.37750	1.53365	0.70220	0.43963
(60, 30)	0.26765	0.77826	0.12984	1.30760	0.62792	0.41154

Table A-8 Bessel functions of the first kind and of integral orders

$(\alpha°, \beta°)$	$J_0(2\sqrt{\alpha\beta})$	$J_1(2\sqrt{\alpha\beta})$	$J_2(2\sqrt{\alpha\beta})$	$J_3(2\sqrt{\alpha\beta})$	$J_4(2\sqrt{\alpha\beta})$	$J_5(2\sqrt{\alpha\beta})$
(10, 10)	0.969770	0.171888	0.015077	0.000312	0.000010	0.000000
(10, 20)	0.939998	0.239384	0.029848	0.002468	0.000153	0.000008
(10, 30)	0.910682	0.288696	0.044317	0.004500	0.000342	0.000021
(10, 40)	0.881815	0.328227	0.058486	0.006875	0.000604	0.000042
(10, 50)	0.853394	0.361292	0.072361	0.009535	0.000938	0.000074
(10, 60)	0.825413	0.389620	0.085943	0.012439	0.001342	0.000115
(20, 20)	0.881815	0.328227	0.058486	0.006875	0.000604	0.000042
(20, 30)	0.825413	0.389620	0.085943	0.012439	0.001342	0.000115
(20, 40)	0.770757	0.435898	0.112246	0.018858	0.002356	0.000235
(20, 50)	0.717810	0.472020	0.137421	0.025951	0.003637	0.000406
(20, 60)	0.666538	0.500628	0.161494	0.033589	0.005173	0.000633
(30, 30)	0.744072	0.455031	0.124973	0.022329	0.002964	0.000313
(30, 40)	0.666538	0.500628	0.161494	0.033589	0.005173	0.000633
(30, 50)	0.592694	0.532854	0.195594	0.045859	0.007934	0.001089
(30, 60)	0.522424	0.555201	0.227360	0.058887	0.011214	0.001692
(40, 40)	0.568879	0.541273	0.206438	0.050128	0.008971	0.001274
(40, 50)	0.477509	0.565724	0.247281	0.067895	0.013674	0.002180
(40, 60)	0.392166	0.578329	0.284217	0.086479	0.019205	0.003368
(50, 50)	0.371742	0.579977	0.292862	0.091214	0.020709	0.003710
(50, 60)	0.274897	0.580842	0.332705	0.115227	0.028901	0.005703
(60, 60)	0.169794	0.568870	0.373436	0.144341	0.040071	0.008720

Table A-9 Field defined by a circular arc and a straight limiting line (Fig. 6.18)

$(\alpha°, \beta°)$	x/a	$-y/a$	$\int_0^\beta \frac{x}{a}\,dt$	$-\int_0^\beta \frac{y}{a}\,dt$	$-R/a$	$-S/a$
(15, 15)	0.26481	0.00000	0.06893	0.00301	1.03466	0.00000
(30, 15)	0.54173	0.03687	0.07279	0.02132	1.10602	2.28016
(30, 30)	0.54819	0.00000	0.28054	0.02459	1.14349	0.00000
(45, 15)	0.81715	0.15141	0.20105	0.00332	1.17975	0.57931
(40, 30)	0.86389	0.04245	0.42229	0.08212	1.29863	0.31945
(45, 45)	0.87131	0.00000	0.64990	0.08592	1.34182	0.00000
(60, 15)	1.06908	0.34526	0.25486	0.11128	1.25601	0.89809
(60, 30)	1.19650	0.18125	0.55312	0.17995	1.46430	0.68088
(60, 45)	1.25200	0.05164	0.87501	0.20919	1.60586	0.38527
(60, 60)	1.26101	0.00000	1.20454	0.21382	1.65821	0.00000
(75, 15)	1.27455	0.61379	0.29438	0.17897	1.33484	1.23718
(75, 30)	1.51724	0.42864	0.66157	0.31629	1.64096	1.08711
(75, 45)	1.67307	0.22784	1.08117	0.40202	1.89585	0.84303
(75, 60)	1.74240	0.06576	1.52995	0.43895	2.07263	0.48725
(75, 75)	1.75386	0.00000	1.98834	0.44486	2.13951	0.00000
(90, 15)	1.41152	0.94567	0.31586	0.25705	1.41628	1.59724
(90, 30)	1.79197	0.78843	0.73635	0.48610	1.82912	1.54109
(90, 45)	2.09707	0.55547	1.24748	0.66320	2.21408	1.38039
(90, 60)	2.29657	0.29812	1.82508	0.77477	2.54057	1.08990
(90, 75)	2.38683	0.08696	2.44032	0.82332	2.77097	0.64073
(90, 90)	2.40196	0.00000	3.06813	0.83116	2.85935	0.00000

Table A-10 Field defined by a noncircular slipline and an inclined limiting line (Fig. 6.19)

ψ = 0

(α°, β°)	x/a	y/a	$\int_0^\beta \frac{x}{a}\,dt$	$\int_0^\beta \frac{y}{a}\,dt$
(15, 15)	2.13815	1.00000	0.55865	0.25325
(30, 15)	3.01369	0.88076	0.77552	0.19457
(45, 15)	4.00492	0.46540	1.00577	0.06132
(30, 30)	3.03448	1.00000	1.56855	0.44567
(45, 30)	4.16709	0.84475	2.07942	0.23621
(45, 45)	4.19412	1.00000	3.17562	0.48403

ψ = 15°

(α°, β°)	x/a	y/a	$\int_0^\beta \frac{x}{a}\,dt$	$\int_0^\beta \frac{y}{a}\,dt$
(15, 15)	2.18524	1.58553	0.57344	0.40498
(30, 15)	3.25127	1.72024	0.84673	0.40410
(45, 15)	4.58814	1.53776	1.16955	0.31696
(60, 15)	6.12284	0.89411	1.51922	0.11486
(30, 30)	3.23806	1.86763	1.69621	0.87992
(45, 30)	4.66080	2.04635	2.38396	0.79132
(60, 30)	6.50055	1.79407	3.17488	0.46879
(45, 45)	4.64303	2.24409	3.60186	1.36117
(60, 45)	6.59960	2.48871	4.89822	1.03633
(60, 60)	6.57500	2.76176	6.62281	1.73493

ψ = 30°

(α°, β°)	x/a	y/a	$\int_0^\beta \frac{x}{a}\,dt$	$\int_0^\beta \frac{y}{a}\,dt$
(15, 15)	2.07022	2.19524	0.54654	0.56378
(30, 15)	3.25620	2.67832	0.86159	0.64686
(45, 15)	4.89905	2.88566	1.27260	0.64584
(60, 15)	6.97534	2.60186	1.76479	0.51329
(75, 15)	9.37980	1.59287	2.30466	0.20144
(30, 30)	3.19518	2.84474	1.70420	1.37694
(45, 30)	4.82404	3.50705	2.54800	1.48974
(60, 30)	7.13822	3.79792	3.61977	1.35542
(75, 30)	10.12609	3.38809	4.86947	0.85197
(45, 45)	4.73984	3.73655	3.79734	2.44769
(60, 45)	7.03302	4.66768	5.47850	2.47332
(75, 45)	10.35735	5.08415	7.56144	1.96693
(60, 60)	6.91406	4.99183	7.30507	3.75150
(75, 60)	10.20677	6.32735	10.25864	3.47441
(75, 75)	10.03541	6.79394	12.90314	5.21172

ψ = 45°

(α°, β°)	x/a	y/a	$\int_0^\beta \frac{x}{a}\,dt$	$\int_0^\beta \frac{y}{a}\,dt$
(15, 15)	1.78384	2.78384	0.47529	0.71806
(30, 15)	2.98186	3.69073	0.80735	0.90773
(45, 15)	4.88133	4.44401	1.28711	1.03467
(60, 15)	7.39945	4.76792	1.91813	1.03349
(75, 15)	10.65984	4.32188	2.68051	0.82849
(90, 15)	14.45794	2.72746	3.52453	0.34123
(30, 30)	2.86158	3.86158	1.56798	1.90382
(45, 30)	4.55170	5.13968	2.51628	2.29805
(60, 30)	7.21420	6.22284	3.83851	2.47941
(75, 30)	10.97579	6.69592	5.52659	2.27227
(90, 30)	15.82512	6.03094	7.50702	1.47890
(45, 45)	4.38144	5.38143	3.67959	3.68583
(60, 45)	6.81353	7.21914	5.67615	4.25152
(75, 45)	10.70739	8.80177	8.37577	4.31510
(90, 45)	16.28666	9.50179	11.73085	3.51407
(60, 60)	6.56787	7.56787	7.41909	6.20233
(75, 60)	10.12359	10.25282	11.10462	6.82343
(90, 60)	15.89198	12.59552	15.95898	6.42124
(75, 75)	9.76368	10.76368	13.69518	9.59670
(90, 75)	15.02977	14.73803	20.00985	10.02563
(90, 90)	14.49573	15.49571	23.85592	14.01617

ψ = 60°

(α°, β°)	x/a	y/a	$\int_0^\beta \frac{x}{a}\,dt$	$\int_0^\beta \frac{y}{a}\,dt$
(15, 15)	1.32958	3.30289	0.36029	0.85537
(30, 15)	2.39941	4.67665	0.67539	1.16715
(45, 15)	4.29293	6.11033	1.18819	1.46068
(60, 15)	7.20903	7.29810	1.93160	1.65790
(75, 15)	11.26159	7.80906	2.91441	1.65649
(30, 30)	2.21268	4.83247	1.27207	2.41899
(45, 30)	3.76078	6.81883	2.24087	3.16336
(60, 30)	6.54324	8.92420	3.73830	3.79181
(75, 30)	10.88585	10.69162	5.82946	4.08402
(90, 30)	16.99348	11.46013	8.50774	3.76055
(45, 45)	3.48955	7.04475	3.17982	4.98838
(60, 45)	5.76422	9.96108	5.34718	6.27878
(75, 45)	9.90310	13.09111	8.56041	7.21232
(90, 45)	16.43424	15.74751	12.90717	7.33037
(105, 45)	25.71302	16.91304	18.32006	6.06649

ORTHOGONAL CURVILINEAR COORDINATES

In the solution of special problems, it is frequently necessary to express the basic equations in terms of a system of orthogonal curvilinear coordinates. The following method is convenient for developing equations in curvilinear coordinates from the corresponding equations in rectangular coordinates. Alternatively, recourse may be made to general tensor calculus, not treated in this book, which furnishes equations that are valid in any coordinate system.

We denote the curvilinear coordinates by α, β, γ, and associate them with an orthogonal triad of unit vectors $\mathbf{e}^\alpha$, $\mathbf{e}^\beta$, $\mathbf{e}^\gamma$ forming a right-handed system. These vectors represent the directions of the maximum rate of change of the respective coordinates at each point. The components of the unit base vectors with respect to a fixed set of rectangular axes x_j will be denoted by e_j^α, e_j^β, e_j^γ. They are indentical to the components $a_{\alpha j}$, $a_{\beta j}$, $a_{\gamma j}$ of the transformation tensor a_{ij} defined by Eqs. (20), Chap. 1. To avoid confusion, the summation convention will be used only for the italic subscripts, denoting the rectangular components of vectors and tensors. If v_α, v_β, v_γ denote the curvilinear components of the velocity vector v_j of a typical particle, then

$$v_j = e_j^\alpha v_\alpha + e_j^\beta v_\beta + e_j^\gamma v_\gamma \tag{a}$$

It is convenient to begin with the gradient of a scalar function $\phi = \phi(\alpha, \beta, \gamma)$. Denoting the line elements along the coordinate curves by $h_1 \, d\alpha$, $h_2 \, d\beta$, and $h_3 \, d\gamma$, where h_1, h_2, and h_3 are functions of (α, β, γ), we have

$$\operatorname{grad} \phi = \frac{\mathbf{e}^\alpha}{h_1} \frac{\partial \phi}{\partial \alpha} + \frac{\mathbf{e}^\beta}{h_2} \frac{\partial \phi}{\partial \beta} + \frac{\mathbf{e}^\gamma}{h_3} \frac{\partial \phi}{\partial \gamma} \tag{b}$$

Since the rectangular components of grad ϕ are $\partial \phi / \partial x_i$, the operator $\partial / \partial x_i$ is equal to the ith component of the vector operator on the right-hand side of (b). The application of this operator on Eq. (a) furnishes $\partial v_j / \partial x_i$, and hence the strain rate $\dot{\varepsilon}_{ij}$, provided due account is taken of the variation of the base vectors along the coordinate curves. The curvilinear components of the strain rate are then obtained by comparison with the expression

$$\dot{\varepsilon}_{ij} = e_i^\alpha e_j^\alpha \dot{\varepsilon}_{\alpha\alpha} + e_i^\beta e_j^\beta \dot{\varepsilon}_{\beta\beta} + e_i^\gamma e_j^\gamma \dot{\varepsilon}_{\gamma\gamma} + (e_i^\alpha e_j^\beta + e_i^\beta e_j^\alpha) \dot{\varepsilon}_{\alpha\beta}$$
$$+ (e_i^\beta e_j^\gamma + e_i^\gamma e_j^\beta) \dot{\varepsilon}_{\beta\gamma} + (e_i^\gamma e_j^\alpha + e_i^\alpha e_j^\gamma) \dot{\varepsilon}_{\gamma\alpha} \tag{c}$$

A similar expression can be written down for the stress tensor σ_{ij}. The condition $\partial \sigma_{ij} / \partial x_i = 0$ for equilibrium, in the absence of body forces, leads to three equations of equilibrium in the curvilinear coordinate system.

Consider, for example, cylindrical coordinates (r, θ, z), where r is the perpendicular distance from a fixed axis, coinciding with the z axis, and θ is the angle measured round this axis. In this case, $h_1 = h_3 = 1$ and $h_2 = r$, giving

$$\frac{\partial}{\partial x_i} = e_i^r \frac{\partial}{\partial r} + e_i^\theta \frac{1}{r} \frac{\partial}{\partial \theta} + e_i^z \frac{\partial}{\partial z} \tag{d}$$

Since the base vectors are unit vectors, their variations along any curve are orthogonal to these vectors. It follows from simple geometry that

$$\frac{\partial \mathbf{e}^r}{\partial \theta} = \mathbf{e}^\theta \qquad \frac{\partial \mathbf{e}^\theta}{\partial \theta} = - \mathbf{e}^r \qquad \frac{\partial \mathbf{e}^z}{\partial \theta} = 0 \tag{e}$$

The derivatives of the base vectors with respect to r and z are identically zero. Applying (d) on Eq. (a) with (α, β, γ) replaced by (r, θ, z), and using (e), we get

$$\frac{\partial v_j}{\partial x_i} = e_i^r e_j^r \frac{\partial v_r}{\partial r} + e_i^r e_j^\theta \frac{\partial v_\theta}{\partial r} + e_i^r e_j^z \frac{\partial v_z}{\partial r}$$
$$+ \frac{1}{r} \left[e_i^\theta e_j^r \left(\frac{\partial v_r}{\partial \theta} - v_\theta \right) + e_i^\theta e_j^\theta \left(v_r + \frac{\partial v_\theta}{\partial \theta} \right) + e_i^\theta e_j^z \frac{\partial v_z}{\partial \theta} \right]$$
$$+ e_i^z e_j^r \frac{\partial v_r}{\partial z} + e_i^z e_j^\theta \frac{\partial v_\theta}{\partial z} + e_i^z e_j^z \frac{\partial v_z}{\partial z} \tag{f}$$

The expression for $\partial v_i / \partial x_j$ is obtained by interchanging the subscripts i and j in (f). A similar operation on the stress tensor σ_{ij}, expressed in terms of the cylindrical components and the base vectors, furnishes

$$\frac{\partial \sigma_{ij}}{\partial x_i} = e_j^r \left[\frac{\partial \sigma_{rr}}{\partial r} + \frac{1}{r} \left(\frac{\partial \sigma_{r\theta}}{\partial \theta} + \sigma_{rr} - \sigma_{\theta\theta} \right) + \frac{\partial \sigma_{rz}}{\partial z} \right]$$
$$+ e_j^\theta \left[\frac{\partial \sigma_{r\theta}}{\partial r} + \frac{1}{r} \left(\frac{\partial \sigma_{\theta\theta}}{\partial \theta} + 2\sigma_{r\theta} \right) + \frac{\partial \sigma_{\theta z}}{\partial z} \right]$$
$$+ e_j^z \left[\frac{\partial \sigma_{rz}}{\partial r} + \frac{1}{r} \left(\frac{\partial \sigma_{\theta z}}{\partial \theta} + \sigma_{rz} \right) + \frac{\partial \sigma_{zz}}{\partial z} \right] \tag{g}$$

in view of the orthogonality of the base vectors. The curvilinear components of the strain rate, the associated equations of equilibrium, and the components of the spin tensor are finally obtained from the fact that

$$\dot{\varepsilon}_{ij} = \frac{1}{2}\left(\frac{\partial v_i}{\partial x_j} + \frac{\partial v_j}{\partial x_i}\right) \qquad \frac{\partial \sigma_{ij}}{\partial x_i} = 0 \qquad \omega_{ij} = \frac{1}{2}\left(\frac{\partial v_i}{\partial x_j} - \frac{\partial v_j}{\partial x_i}\right)$$

In the case of spherical coordinates (r, θ, ϕ), where r is the length of the radius vector, θ the meridional angle measured from a fixed polar axis, and ϕ the angle measured round this axis, we have $h_1 = 1$, $h_2 = r$, $h_3 = r \sin \theta$, leading to the expression

$$\frac{\partial}{\partial x_i} = \mathbf{e}_i^r \frac{\partial}{\partial r} + \mathbf{e}_i^\theta \frac{1}{r}\frac{\partial}{\partial \theta} + \mathbf{e}_i^\phi \frac{1}{r \sin \theta}\frac{\partial}{\partial \phi} \qquad (h)$$

From geometry, the nonzero derivatives of the base vectors are easily shown to be

$$\frac{\partial \mathbf{e}^r}{\partial \theta} = \mathbf{e}^\theta \qquad \frac{\partial \mathbf{e}^\theta}{\partial \theta} = -\mathbf{e}^r \qquad \frac{\partial \mathbf{e}^r}{\partial \phi} = \sin \theta \mathbf{e}^\phi$$

$$(k)$$

$$\frac{\partial \mathbf{e}^\theta}{\partial \phi} = \cos \theta \mathbf{e}^\phi \qquad \frac{\partial \mathbf{e}^\phi}{\partial \phi} = -(\sin \theta \mathbf{e}^r + \cos \theta \mathbf{e}^\theta)$$

Using the above relations, $\partial v_j/\partial x_i$ and $\partial \sigma_{ij}/\partial x_i$ can be expressed in terms of the curvilinear components of the field quantities and their derivatives with respect to r, θ, and ϕ. These expressions furnish the strain rates and the equations of equilibrium in spherical coordinates. The end results for cylindrical and spherical coordinates are summarized below, using σ_r, σ_θ, $\tau_{r\theta}$, etc., for the components of the stress, and $\dot{\varepsilon}_r$, $\dot{\varepsilon}_\theta$, $\dot{\gamma}_{r\theta}$, etc., for the components of the strain rate.

Cylindrical coordinates Setting $v_r = u$, $v_\theta = v$, and $v_z = w$, the components of the strain rate in cylindrical coordinates (r, θ, z) may be written as

$$\dot{\varepsilon}_r = \frac{\partial u}{\partial r} \qquad\qquad 2\dot{\gamma}_{r\theta} = \frac{1}{r}\frac{\partial u}{\partial \theta} + \frac{\partial v}{\partial r} - \frac{v}{r}$$

$$\dot{\varepsilon}_\theta = \frac{u}{r} + \frac{1}{r}\frac{\partial v}{\partial \theta} \qquad 2\dot{\gamma}_{\theta z} = \frac{\partial v}{\partial z} + \frac{1}{r}\frac{\partial w}{\partial \theta}$$

$$\dot{\varepsilon}_z = \frac{\partial w}{\partial z} \qquad\qquad 2\dot{\gamma}_{rz} = \frac{\partial w}{\partial r} + \frac{\partial u}{\partial z}$$

In the absence of body forces, the stress equations of equilibrium in cylindrical coordinates are

$$\frac{\partial \sigma_r}{\partial r} + \frac{1}{r}\frac{\partial \tau_{r\theta}}{\partial \theta} + \frac{\partial \tau_{rz}}{\partial z} + \frac{\sigma_r - \sigma_\theta}{r} = 0$$

$$\frac{\partial \tau_{r\theta}}{\partial r} + \frac{1}{r}\frac{\partial \sigma_\theta}{\partial \theta} + \frac{\partial \tau_{\theta z}}{\partial z} + \frac{2\tau_{r\theta}}{r} = 0$$

$$\frac{\partial \tau_{rz}}{\partial r} + \frac{1}{r}\frac{\partial \tau_{\theta z}}{\partial \theta} + \frac{\partial \sigma_z}{\partial z} + \frac{\tau_{rz}}{r} = 0$$

The cylindrical components of the vector $\boldsymbol{\omega} = \frac{1}{2}$ curl $\mathbf{v}$, given by the non-zero components of the associated spin tensor, are easily shown to be of the form

$$2\omega_r = \frac{1}{r}\frac{\partial w}{\partial \theta} - \frac{\partial v}{\partial z}, \qquad 2\omega_\theta = \frac{\partial u}{\partial z} - \frac{\partial w}{\partial r}, \qquad 2\omega_z = \frac{\partial v}{\partial r} + \frac{v}{r} - \frac{1}{r}\frac{\partial u}{\partial \theta}$$

Spherical coordinates Setting $v_r = u$, $v_\theta = v$, and $v_\phi = w$, the strain rates in spherical coordinates (r, θ, ϕ) may be written as

$$\dot{\varepsilon}_r = \frac{\partial u}{\partial r} \qquad\qquad 2\dot{\gamma}_{r\theta} = \frac{1}{r}\frac{\partial u}{\partial \theta} + \frac{\partial v}{\partial r} - \frac{v}{r}$$

$$\dot{\varepsilon}_\theta = \frac{u}{r} + \frac{1}{r}\frac{\partial v}{\partial \theta} \qquad\qquad 2\dot{\gamma}_{\theta\phi} = \frac{1}{r\sin\theta}\frac{\partial v}{\partial \phi} + \frac{1}{r}\frac{\partial w}{\partial \theta} - \frac{w}{r}\cot\theta$$

$$\dot{\varepsilon}_\phi = \frac{u}{r} + \frac{v}{r}\cot\theta + \frac{1}{r\sin\theta}\frac{\partial w}{\partial \phi} \qquad\qquad 2\dot{\gamma}_{r\phi} = \frac{\partial w}{\partial r} - \frac{w}{r} + \frac{1}{r\sin\theta}\frac{\partial u}{\partial \phi}$$

The corresponding equations of statical equilibrium, when body forces are absent, take the form

$$\frac{\partial \sigma_r}{\partial r} + \frac{1}{r}\frac{\partial \tau_{r\theta}}{\partial \theta} + \frac{1}{r\sin\theta}\frac{\partial \tau_{r\phi}}{\partial \phi} + \frac{1}{r}(2\sigma_r - \sigma_\theta - \sigma_\phi + \tau_{r\theta}\cot\theta) = 0$$

$$\frac{\partial \tau_{r\theta}}{\partial r} + \frac{1}{r}\frac{\partial \sigma_\theta}{\partial \theta} + \frac{1}{r\sin\theta}\frac{\partial \tau_{\theta\phi}}{\partial \phi} + \frac{1}{r}(\sigma_\theta\cot\theta - \sigma_\phi\cot\theta + 3\tau_{r\theta}) = 0$$

$$\frac{\partial \tau_{r\phi}}{\partial r} + \frac{1}{r}\frac{\partial \tau_{\theta\phi}}{\partial \theta} + \frac{1}{r\sin\theta}\frac{\partial \sigma_\phi}{\partial \phi} + \frac{1}{r}(3\tau_{r\phi} + 2\tau_{\theta\phi}\cot\theta) = 0$$

The non-zero components of the spin tensor in spherical coordinates furnish the corresponding components of the dual vector $\boldsymbol{\omega}$ as

$$2\omega_r = \frac{1}{r}\frac{\partial w}{\partial \theta} + \frac{w}{r}\cot\theta - \frac{1}{r\sin\theta}\frac{\partial v}{\partial \phi}$$

$$2\omega_\theta = \frac{1}{r\sin\theta}\frac{\partial u}{\partial \phi} - \frac{\partial w}{\partial r} - \frac{w}{r}$$

$$2\omega_\phi = \frac{\partial v}{\partial r} + \frac{v}{r} - \frac{1}{r}\frac{\partial u}{\partial \theta}$$

When the stress distribution is symmetrical about a given axis, which is taken as the z-axis, the equations in cylindrical coordinates are simplified by the fact that $\tau_{r\theta} = \tau_{\theta z} = 0$ at each point in the absence of torsion, while the remaining stresses are independent of θ. If spherical coordinates are used, the situation of torsionless symmetry about the polar axis $\theta = 0$ is characterized by $\tau_{r\phi} = \tau_{\theta\phi} = 0$, and the independence of the stresses with respect to ϕ.

SUBJECT INDEX

Absolute minimum weight, 276, 285
Adiabatic heating, 15
Alternating plasticity, 263
Anisotropy, 1, 68
Anisotropic flow rule, 88
Anisotropic hardening, 67
Annealing, 2
Anticlastic curvature, 150
Arches, limit analysis of, 290
Associated flow rule, 74, 80, 86
Autofrettage, 326

Bauschinger effect, 5, 68, 747
Beams:
   continuous, 233
   fixed-ended, 173, 230, 729
   simply supported, 169, 731
   propped cantilever, 176
Bending:
   beams, 126, 148, 160, 230, 723
   curved bars, 350, 738
   notched bars, 700, 715
   wide sheets, 741
   influence of work-hardening, 151, 172, 746
Bending under tension, 744
Boundary conditions, 41, 45
Buckling of columns, 294
Burgers vector, 2

Cam plastometer, 15, 583
Cantilever:
   end-loaded, 163, 723
   uniformly loaded, 167
   combined axial and shear loads, 732
Centered fan, 431
Characteristics:
   plane strain, 409
   plastic torsion, 186, 206, 210
Charpy test, 711
Circular hole:
   radial expansion, 373, 378
   stress concentration, 369
Collapse load, 229
Collapse mechanism, 229
Cold rolling:
   Bland and Ford theory, 561
   Von Karman theory, 557
   influence of tension, 567
   minimum thickness, 571
   roll flattening, 554
   roll force and torque, 552
Combination of mechanisms, 243
Combined bending and tension, 215, 287
Combined bending and torsion, 140, 212
Combined tension and torsion, 133, 138
Compatibility equation, 38, 335, 342, 351, 359, 369, 408

NAME INDEX